D1281135

Environmental
Impact
Analysis
Handbook

OTHER McGRAW-HILL HANDBOOKS OF INTEREST

Environmental Impact Analysis Handbook

Edited by
JOHN G. RAU and DAVID C. WOOTEN
University of California at Irvine

McGraw-Hill Publishing Company

New York   St. Louis   San Francisco   Auckland   Bogotá   Caracas
Hamburg   Lisbon   London   Madrid   Mexico   Milan
Montreal   New Delhi   Oklahoma City   Paris   San Juan
São Paulo   Singapore   Sydney   Tokyo   Toronto

Library of Congress Cataloging in Publication Data

Main entry under title:

Environmental impact analysis handbook.
    Includes bibliographies and index.
    1.   Environmental impact analysis—United States—
Handbooks, manuals, etc.   I.   Rau, John G.   II.   Wooten,
David C.
TD194.6E585   333.7   79-11125
ISBN 0-07-051217-5

5 6 7 8 9   VBVB   99876543210

*The editors for this book were Harold B. Crawford, Joseph Williams,
and Esther Gelatt, the designer was Naomi Auerbach, and the production
supervisors were Teresa F. Leaden and Paul A. Malchow. It was set in Caledonia
by University Graphics, Inc.*

*To the National Environmental Policy Act,
whose passage made this handbook necessary.*

Contents

Preface

In a move to institutionalize the concern for the protection of the environment, Congress passed the National Environmental Policy Act of 1969 (NEPA), establishing a national policy for the protection of the environment. In recognizing the effect of people's activities on the environment, NEPA laid down environmental impact statement requirements for federal agencies which propose to undertake activities that are likely to affect environmental quality. These requirements specify that under certain circumstances a detailed statement should be prepared which considers: (1) the environmental impact of the proposed action, (2) any adverse environmental effects which cannot be avoided should the proposal be implemented, (3) alternatives to the proposed action, (4) the relationship between local short-term uses of the environment and the maintenance and enhancement of long-term productivity, and (5) any irreversible and irretrievable commitments of resources which would be involved in the proposed action should it be implemented. In response to this federal legislation, many states have adopted similar legislation for projects which they undertake or for which local or state governmental agencies have discretionary approval.

In general, the types of environmental impact can be expected to include population growth, high-density urbanization, industrial expansion, resource exploitation, air, water, and noise pollution, undesirable land use patterns, damage to life systems, threats to health, and other consequences adverse to environmental goals. The assessment, both qualitative and quantitative, of these environmental impacts requires the use of many disciplines ranging from economics, urban planning, law, and social sciences to civil, mechanical, and soils engineering, in addition to geology, chemistry, biology, and archaeology. As a result, the conduct of the environmental impact assessment must be performed by a multidisciplined team of individuals who use the appropriate tools and techniques of their trade. However, those individuals who are responsible for reviewing and evaluating environmental impact statements (or reports) do not necessarily have the benefit of being multidisciplined in the required areas. Consequently, they need some general familiarity with the tools and techniques which are used in each area, the corresponding limitations, and the extent of their applicability. Furthermore, since the requirement for environmental impact statements is relatively new, the state of the art in performing impact analyses and the development of standardized tools and techniques are basically in the infancy period. There is thus a valid need to develop techniques and tools for performing environmental impact analyses which can be used by both the environmental analysts and the reviewers of environmental impact assessments.

This reference handbook is designed to provide environmental planners, analysts, and decision-makers with specific techniques and tools that can be used to assess and predict the environmental impact of projects such as residential, commercial, and industrial developments, new communities, urban renewal or redevelopment, park and recreational facility development, dams and flood control projects, wastewater and sewer treatment plants, new airport construction, and power generating stations. The contents of this handbook are an outgrowth of course notes used by the authors in training programs and courses presented over the last few years through the University of California at Irvine. Furthermore, the tools and techniques presented have all been and are currently being used by the authors in conducting environmental impact studies.

Each chapter is designed to be self-contained to provide the reader with a concise and complete summary of applicable tools and techniques in each of the chapter areas. In this regard, this handbook can also be a valuable reference source to those individuals whose interests lie only in a few of the specific environmental impact areas rather than in all areas presented.

Chapter 1 is designed to set the stage for the entire handbook by providing the legislative and legal background, a discussion of the general topics to be found in an environmental impact statement, and an identification of the typical kinds of environmental impacts for various types of projects and activities. Chapter 2 is concerned with tools and techniques for assessing the socioeconomic-related environmental impacts, such as the demand for public services, revenues, and costs to local governmental agencies, employment, and population growth. Chapter 3 presents a detailed discussion of air quality impacts and the procedures to be used in measuring and predicting these types of impacts. Chapter 4 provides information on the effects of noise on the environment and how these effects can be measured and mitigated, if necessary. Chapter 5 deals with a relatively new impact area of concern, namely, the effects on energy—both demand and consumption. In this chapter the basic techniques for measurement and prediction of energy requirements by project type are presented. Chapter 6 focuses on water quality considerations in environmental impact assessments and the general procedures that are used in measuring and predicting water quality impacts. Chapter 7 addresses the approach to assessing the impacts on vegetation and wildlife (i.e., flora and fauna) as the result of land use activities. Finally, Chapter 8 presents a discussion of the various types of techniques used to "put all the pieces together" in the sense of being able to simultaneously consider both the positive and negative environmental impacts of projects with the desire to obtain an overall conclusion as to the "total impact" of the project.

We are indebted to our colleagues and chapter contributors, Dr. Gary S. Samuelsen, Dr. David York, Dr. James Speakman, Dr. Blair Folsom, Dr. Ted Hanes, and Mr. Vince Mestre, for their many hours of effort in preparation of the handbook material. Collectively, we represent an example of the multidisciplined team necessary for the conduct of environmental studies. We are also indebted to the many students who, over the years, have provided comments and guidance on the material presented and have had a stimulating influence on all of our efforts. We make no claim that this handbook is a complete compendium of tools and techniques for environmental impact assessment—it was not intended to be. The state of the art in this area is changing continuously as we become more aware of our environment and the many factors which affect it. It is our intention that the material presented in these eight chapters will provide both an introduction and starting point for those interested in knowing more about the conduct of environmental impact assessments.

J. G. Rau and D. C. Wooten

Contributors

Dr. Blair Folsom *Member of Professional Staff, Energy and Environmental Research Corporation, Irvine, Calif.*

Dr. Ted Hanes *Professor of Biology, California State University, Fullerton, Calif.*

Mr. Vince Mestre *Consulting Professional Engineer, Irvine, Calif.*

Mr. John Rau *Co-editor,* Environmental Impact Analysis Handbook; *Lecturer in Civil and Environmental Engineering, University of California at Irvine; and Manager of Operations Research and Economic Analysis, Ultrasystems, Inc., Irvine, Calif.*

Dr. Gary Samuelsen *Assistant Professor of Mechanical Engineering, University of California at Irvine*

Dr. James S. Speakman *Director of Operations, Mississippi Valley Region, Moore, Gardner & Associates, Inc., Memphis, Tenn.*

Dr. David York *Clark, Dietz & Associates—Engineers, Inc., Urbana, Ill.*

Dr. David Wooten *Co-editor,* Environmental Impact Analysis Handbook; *Lecturer in Mechanical and Environmental Engineering, University of California at Irvine; and Vice President of Engineering, Environsonics, Inc., Los Angeles, Calif.*

Environmental Impact Analysis Handbook

Chapter 1

Concepts of Environmental Impact Analysis

JOHN G. RAU

On January 1, 1970, the National Environmental Policy Act (NEPA) became Public Law 91-190. Few, if any, other pieces of federal legislation or acts of the U.S. Congress have so significantly altered the workings of the federal bureaucracy or engendered such controversy as NEPA. Because many states responded with the passage of similar, or parallel, legislation, these ramifications have extended to the local governmental levels. Beginning with the actions and policies of federal agencies, NEPA seeks to insure that environmental values will receive proper placement among the socioeconomic and technical priorities considered in making decisions which affect the quality of the human environment. As a result, to insure that all federal agencies take into consideration these environmental values in their actions, the requirement for an environmental impact statement (EIS) was established. This necessitated the preparation of guidelines and procedures by federal agencies regarding the preparation, review, evaluation, and distribution of EIS's with the basic objective of evaluating the environmental impacts of these actions. This chapter then focuses on the concepts involved in this process of environmental impact analysis.

KEY FEATURES OF NEPA

The four key features of the National Environmental Policy Act of 1969 deal with:
1. The purpose of the act
2. The establishment of policy
3. The establishment of the requirement for environmental impact statements
4. The creation of a Council on Environmental Quality.

The act had the following purposes: "To declare a national policy which will encourage productive and enjoyable harmony between man and his environment; to promote efforts which will prevent or eliminate damage to the environment and biosphere and stimulate the health and welfare of man; to enrich the understanding of the ecological systems and natural resources important to the Nation; to establish a Council on Environmental Quality."

Section 101(b) of the act states that:

In order to carry out the policy set forth in this Act, it is the continuing responsibility of the Federal Government to use all practicable means, consistent with other essential considerations of national policy, to improve and coordinate Federal plans, functions, programs, and resources to the end that the Nation may—

(1) fulfill the responsibilities of each generation as trustee of the environment for succeeding generations;

(2) assure for all Americans safe, healthful, productive, and esthetically and culturally pleasing surroundings;

(3) attain the widest range of beneficial uses of the environment without degradation, risk to health or safety, or other undesirable and unintended consequences;

(4) preserve important historic, cultural, and natural aspects of our national heritage, and maintain, wherever possible, an environment which supports diversity, and variety of individual choice;

(5) achieve a balance between population and resource use which will permit high standards of living and a wide sharing of life's amenities; and

(6) enhance the quality of renewable resources and approach the maximum attainable recycling of depletable resources.

The most important feature of the act can be found in Section 102 in which it is stated that:

(2) all agencies of the Federal government shall . . .

(c) include in every recommendation or report on proposals for legislation and other major Federal actions significantly affecting the quality of the human environment, a detailed statement by the responsible official on—

(i) the environmental impact of the proposed action,

(ii) any adverse environmental effects which cannot be avoided should the proposal be implemented,

(iii) alternatives to the proposed action,

(iv) the relationship between local short-term uses of man's environment and the maintenance and enhancement of long-term productivity, and

(v) any irreversible and irretrievable commitments of resources which would be involved in the proposed action should it be implemented.

This section of the act, in essence, directs the preparation of the environmental impact statement and describes its contents in general terms.

The Council on Environmental Quality (CEQ) is created, and its functions, responsibilities, and funding are described in Title II of the act. Its prime purposes are to evaluate all federal programs in terms of the national policy, to advise and assist the President in environmental matters, and to develop national policies in this area.

In support of NEPA, the Environmental Quality Improvement Act of 1970 became Public Law 91-224 on April 3, 1970. This act authorizes an Office of Environmental Quality which provides the professional and administrative staff for the Council on Environmental Quality. In addition, on March 5, 1970, Executive Order 11514, Protection and Enhancement of Environmental Quality, was issued to establish the policy for federal agencies to initiate measures needed to direct their policies, plans, and programs so as to meet national environmental goals.

HISTORICAL PERSPECTIVE OF FEDERAL ENVIRONMENTAL PROTECTION LEGISLATION, LAWS, AND ACTS

The development of environmental protection and enhancement measures in the United States has been determined to a considerable extent by federal legislation. This has encompassed the whole range of environmental factors from air pollutants to solid waste, but has for the most part been formulated as an array of single-purpose legislative instruments, each directed toward some specific pollution problem. Typical problem areas are air quality, energy, environmental quality, fish and wildlife resources, historical preservation, land use, noise quality, solid waste, transportation, and water quality. A brief summary of typical environmental protection and quality legislation is provided in Table 1.1.

Air Quality Legislation

Federal legislation related to air pollution began in July 1965 when Congress authorized a federal program of research on air pollution and technical assistance to state and local

TABLE 1.1   Brief Summary of Environmental Legislation and Regulations

Subject	Legislation/regulation	Selected features
Air Quality	1. Clean Air Act of 1963 2. Motor Vehicle Act of 1965 3. Air Quality Act of 1967 4. National Emissions Standards Act 5. Clean Air Act Amendments of 1970, 1971, and 1977	*a.* Authorizes the regulation of both mobile and stationary sources of pollution. *b.* Established national air quality standards. *c.* Requires states to submit implementation plans to meet standards. *d.* Provides the Environmental Protection Agency (EPA) with authority to issue compliance orders or to bring civil suits.
Energy	1. Federal Energy Administration Act of 1974 (Public Law 93-275)	*a.* Created the Federal Energy Administration (FEA). *b.* Requires that if any proposed FEA regulation is likely to have a substantial impact on the nation's economy or on large numbers of individuals or businesses, an opportunity for oral presentation of views, data, and arguments shall be afforded. *c.* Requires any proposed FEA regulation to be submitted for review to the Administrator of the Environmental Protection Agency if the proposed regulation affects the quality of the environment.
	2. Energy Supply and Environmental Coordination Act of 1974 (Public Law 93-319)	*a.* Established a means for assistance in meeting the essential needs for fuels in a manner consistent with existing national commitments to protect and improve the environment. *b.* Granted the authority to the Administrator of the FEA to: (1) prohibit certain power plants and major fuel-burning installations from burning natural gas or petroleum products as their primary energy source; (2) require that certain power plants in the early planning process be designed and constructed to be capable of using coal as their primary energy source; and (3) allocate coal to any power plant or major fuel-burning installation that has been prohibited by an FEA order from burning natural gas

TABLE 1.1   Brief Summary of Environmental Legislation and Regulations (Continued)

Subject	Legislation/regulation	Selected features
		or petroleum products as its primary energy source. *c.* Stated that no action taken under the Clean Air Act shall be deemed a major federal action significantly affecting the quality of the human environment within the meaning of NEPA.
Environmental quality	1. National Environmental Policy Act of 1969 (Public Law 91-190)	*a.* Established a national policy for the environment. *b.* Created the Council on Environmental Quality (CEQ). *c.* Established environmental impact requirements for federal agencies which propose to undertake activities that are likely to affect environmental quality.
	2. Executive Order 11514, Protection and Enhancement of Environmental Quality, 5 March 1970	*a.* Established the policy for federal agencies to initiate measures needed to direct their policies, plans, and programs so as to meet national environmental goals. *b.* Stated the responsibilities of the CEQ.
	3. Environmental Quality Improvement Act of 1970 (Public Law 91-224)	*a.* Created the Office of Environmental Quality which provides the professional and administrative staff for the CEQ.
Fish and wildlife resources	1. Endangered Species Conservation Act of 1969 (Public Law 91-135)	*a.* Provides a program for the conservation, protection, restoration, and propagation of selected species of native fish and wildlife, including migratory birds that are threatened with extinction. *b.* Provides the Secretary of the Interior with the authority to implement such a program.
	2. Migratory Bird Treaty Act	*a.* Provides assistance in the restoration of game birds and other wild birds in those parts of the United States where they have become scarce or extinct. *b.* Regulates the introduction of American or foreign birds or animals in localities where they have not heretofore existed.

Subject	Legislation/regulation	Selected features
	3. Fish and Wildlife Coordination Act (Public Law 85-624)	*a.* Stated that wildlife conservation shall receive equal consideration and be coordinated with other features of water-resource development programs through the effectual and harmonious planning, development, maintenance, and coordination of wildlife conservation and rehabilitation. *b.* Provides assistance via the Secretary of the Interior to public or private agencies and organizations in the development, protection, rearing, and stocking of all species of wildlife and their habitat. *c.* Established the requirement that there shall be included in any report to Congress supporting a recommendation for authorization of any new project for the control or use of water an estimation of the wildlife benefits or losses to be derived therefrom. *d.* Established the requirement that whenever the waters of any stream or other body of water are impounded, or diverted, or the channel is deepened, or the stream or other body of water is otherwise controlled or modified for any purpose whatever, including navigation and drainage, by any department or agency of the United States, adequate provisions shall be made for the use and management of wildlife resources and habitat, including the development and improvement of these wildlife resources.
	4. Wild and Scenic Rivers Act (Public Law 90-542)	*a.* Established the policy that selected rivers which, with their immediate environments, possess outstanding and remarkable scenic, recreational, geologic, fish and wildlife, historic, cultural, or other similar values, shall be preserved in free-flowing condition.

TABLE 1.1   Brief Summary of Environmental Legislation and Regulations *(Continued)*

Subject	Legislation/regulation	Selected features
	5. Anadromous Fish Conservation Act (Public Law 91-249)	*a.* Provides for the conservation, development, and enhancement of the anadromous fishery resources of the nation that are subject to depletion from water resources developments and other causes.
	6. Migratory Bird Conservation Act	*a.* Created the Migratory Bird Conservation Commission with the authority to consider and pass upon any area of land, water, or land and water that may be recommended by the Secretary of the Interior for purchase or rental for use as inviolate sanctuaries for migratory birds.
	7. Federal Aid in Wildlife Restoration Act	*a.* Authorizes the Secretary of the Interior to cooperate with the states in the selection, restoration, rehabilitation, and improvement of areas of land or water adaptable as feeding, resting, or breeding places for wildlife.
	8. Recreational Use of Conservation Areas Act	*a.* Authorizes the Secretary of the Interior to acquire areas of land which are suitable for fish and wildlife oriented recreational development, or for the protection of natural resources.
Historical preservation	1. Executive Order 11593 (36 FR 8921)	*a.* Requires that federal plans and programs contribute to the preservation and enhancement of sites of historical, architectural, and archaeological significance.
	2. National Historic Preservation Act of 1966	*a.* Requires that, prior to approval of federal activities, departments shall take into account the effect of the undertaking on any district, site, building, structure, or object that is included in the National Register, and give the Advisory Council on Historic Preservation a reasonable opportunity to comment with regard to such undertaking.

Subject	Legislation/regulation	Selected features
Land use	1. Coastal Zone Management Act of 1972 (Public Law 92-583)	*a.* Established the policy to preserve, protect, develop, and, where possible, to restore or enhance the resources of the nation's coastal zone. *b.* Provides assistance to states in the development and implementation of management programs to achieve wise use of the land and water resources of the coastal zone giving full consideration to ecological, cultural, historic, and aesthetic values, as well as the need for economic development.
	2. Flood Disaster Protection Act of 1973 (Public Law 93-234)	*a.* Required that all communities identified by the Department of Housing and Urban Development (HUD) as having special flood hazards must enter HUD's insurance program by a specified date or lose all federal assistance for "acquisition or construction purposes" in the flood hazard area.
	3. Rivers and Harbors and Flood Control Act of 1970 (Public Law 91-611)	*a.* Established the requirement for the Army Corps of Engineers to consider economic, social, and environmental effects of civil works projects.
	4. National Flood Insurance Act of 1968	*a.* Authorized a flood insurance program by means of which flood insurance, over a period of time, can be made available on a nationwide basis. *b.* Encouraged state and local governments to make appropriate land use adjustments to constrict the development of land which is exposed to flood damage and minimize damage caused by flood losses. *c.* Authorized the Secretary of Housing and Urban Development to conduct studies and investigations involving: laws, regulations, or ordinances relating to encroachments and obstructions on stream chan-

TABLE 1.1   Brief Summary of Environmental Legislation and Regulations *(Continued)*

Subject	Legislation/regulation	Selected features
		nels and floodways; the orderly development and use of floodplains of rivers or streams, floodway encroachment lines, and floodplain zoning; building codes, building permits, and subdivision or other building restrictions.
	5. Executive Order 11296 (31 FR 10663)	*a.* Provides for agency evaluation of flood hazards in planning of facilities, construction of buildings and facilities, disposal of lands and properties, and land use planning.
	6. Flood Control Act of 1936	*a.* Directed the Corps of Engineers to participate in flood control and the improvement of navigable waterways if the benefits which accrue are in excess of the estimated costs, and if the lives and social security of people are otherwise adversely affected.
	7. Land and Water Conservation Fund Act of 1965 (Public Law 88-578)	*a.* Provides funds for and authorizes federal assistance to the states in planning, acquisition, and development of needed land and water areas and facilities.
	8. Wilderness Act (Public Law 88-577)	*a.* Established a National Wilderness Preservation System to be composed of federally owned areas designated as wilderness areas.
	9. Multiple-Use and Sustained Yield Act of 1960	*a.* Established the policy that national forests be established and administered for outdoor recreation, range, timber, watershed, and wildlife and fish purposes. *b.* Authorizes the Secretary of Agriculture to cooperate with interested state and local governmental agencies and others in the development and management of the national forests.
Noise quality	1. Noise Pollution and Abatement Act of 1970 (Public Law 91-604)	*a.* Established the Office of Noise Abatement and Control (ONAC) within EPA to investigate and study noise and its effect upon the public health and welfare.

Subject	Legislation/regulation	Selected features
	2. Noise Control Act of 1972 (Public Law 92-574)	*a.* Gives authority to ONAC to specify noise limits for products distributed in commerce. *b.* Gives responsibility to ONAC to conduct and evaluate noise control activities of other branches of the federal government.
Solid waste	1. Solid Waste Disposal Act (Public Law 89-272)	*a.* Authorizes a research and development program to promote the demonstration, construction, and application of solid waste management and resource recovery systems. *b.* Provides financial and technical assistance to states and local governments in the planning and development of resource recovery and solid waste disposal programs.
	2. Resource Recovery Act of 1970 (Public Law 91-512)	*a.* Authorizes funds for demonstration grants for recycling systems and for studies of methods to encourage resource recovery. *b.* Requires the EPA to publish guidelines for construction of solid waste disposal systems.
Transportation	1. Department of Transportation Act of 1966	*a.* Declared the national policy that special effort should be made to preserve the natural beauty of the countryside and public park and recreation lands, wildlife and waterfowl refuges, and historic sites. *b.* Placed restrictions on approval of programs or projects which required the use of publicly owned land currently used as a park, recreation area, or wildlife and waterfowl refuge.
	2. Airport and Airway Development Act of 1970	*a.* Requires airport development projects to be reasonably consistent with plans for development of the area in which the airport is located. *b.* Requires airport development projects to protect the natural resources and environmental quality of the nation. *c.* Requires public hearings to be held for consideration of

TABLE 1.1   Brief Summary of Environmental Legislation and Regulations *(Continued)*

Subject	Legislation/regulation	Selected features
		the economic, social, and environmental effects of airport development projects. *d.* Requires airport development projects to comply with applicable air and water quality standards.
	3. Federal-Aid Highway Act of 1970 (Public Law 91-605)	*a.* Requires the Secretary of Transportation to promulgate guidelines designed to assure that possible adverse economic, social, and environmental effects relating to any proposed project on any federal-aid system have been fully considered in developing such project. Further, it requires that the final decisions on the project are made in the best overall public interest.
	4. Urban Mass Transportation Act of 1964	*a.* Provides assistance to public and private transit agencies by making available capital grants and loans for acquisition, construction, or improvement of facilities and equipment in urban areas. *b.* Prohibits the Secretary of Transportation from approving any application for assistance unless he can make a finding in writing that either no adverse environmental effect is likely to result from such project, or that there exists no feasible and prudent alternative to such effect and all reasonable steps have been taken to minimize such effect.
Water quality	1. Water Pollution Control Act of 1948, as amended in 1956 2. Water Pollution Control Act of 1961	*a.* Provides an enforcement procedure for water pollution abatement.
	3. Water Quality Act of 1965	*a.* Requires states to establish and submit water quality standards for all interstate waters.
	4. Water Pollution Control Act of 1972	*a.* Requires the use of area-wide or regional approaches for handling pollution sources ranging from urban industrial

Subject	Legislation/regulation	Selected features
		sewage to wastes from agricultural and forestry operations. *b.* Requires the EPA to provide financial assistance and guidelines for both planning and management operations to ensure that local agencies meet these requirements. *c.* Gives EPA authority to assume responsibility itself in those states or areas that do not meet its guidelines.
	5. Water Bank Act of 1970 (Public Law 91-559)	*a.* Authorizes and directs the Secretary of Agriculture to formulate and carry out a continuous program to prevent the serious loss of wetlands, and to preserve, restore, and improve such lands.
	6. Clean Water Restoration Act of 1956	*a.* Provides for research and construction funds in wastewater treatment.
	7. Water Quality Improvement Act of 1970	*a.* Requires federally regulated activities to have state certification that they will not violate water quality standards.
	8. Water Resources Planning Act (Public Law 89-80)	*a.* Established the Water Resources Council with responsibility for assessing the adequacy of water supplies; studying the administration of water resources; and developing principles, standards, and procedures for federal participants in the preparation of comprehensive regional or river basin plans. *b.* Established the framework for state and federal cooperation through a series of river basin commissions.
	9. Rivers and Harbors Act of 1899	*a.* Gives the authority to the Army Corps of Engineers to grant permits for construction, dredging, and filling in the navigable waters.
	10. National Water Commission Act (Public Law 90-515)	*a.* Established the National Water Commission with the duties to: (1) review present anticipated national water resource problems, making

TABLE 1.1   Brief Summary of Environmental Legislation and Regulations *(Continued)*

Subject	Legislation/regulation	Selected features
		such projections of water requirements as may be necessary; (2) consider economic and social consequences of water resource development; and (3) advise on such specific water resource matters as may be referred to it by the President and the Water Resources Council.
	11. Federal Water Project Recreation Act (Public Law 89-72)	*a.* Established the requirement that full consideration be given to outdoor recreation and fish and wildlife enhancement opportunities afforded by federal navigation, flood control, reclamation, hydroelectric, and multiple-purpose water resource projects.
	12. Marine Protection, Research, and Sanctuaries Act of 1972 (Public Law 92-532)	*a.* Established the policy of the United States to regulate the dumping of all types of materials into ocean waters and to prevent or strictly limit the dumping into ocean waters of any material which would adversely affect human health, welfare, or amenities, or the marine environment, ecological systems, or economic potentialities.
	13. Clean Water Act of 1977 (Public Law 95-217)	*a.* Revises the Water Pollution Control Act of 1972 and includes several provisions related to development. *b.* If the plans are approved by the EPA, states can substitute their own water quality management plans developed under Section 208 of the act for federal controls. *c.* Gives states authority to determine priorities for construction of publicly owned treatment works projects. *d.* If EPA approves, states can administer their own programs for non-navigable waters and small wetlands.

governments. The Clean Air Act of 1963 and the Motor Vehicle Act of 1965, augmented by the Air Quality Act of 1967 and the Clean Air Act Amendments of 1970, 1971, and 1977 represent the most significant federal legislation regarding air quality. The 1970 amendments, as the strongest air pollution control legislation, authorize the regulation of both mobile and stationary sources of pollution. The most important sections of these programs deal with establishing national air quality standards, describing a framework for the states to meet these standards, and improving procedures for federal enforcement. The Environmental Protection Agency (EPA) has thus far set national air quality standards for particulate matter, sulfur oxides, carbon monoxide, photochemical oxidants, hydrocarbons, and nitrogen dioxide. In effect, the standards represent a federal ceiling on pollution. They are an effort to make pollution control uniform and to set adequate bases for abatement. They do not, however, prevent states from setting more stringent standards if their regions require them.

In addition, the Clean Air Act requires each state to formulate an air pollution abatement plan for each air quality control region in the state or, in the case of a region that crosses state boundaries, for the part that lies within the state boundaries. These implementation plans must describe in detail how the state intends to achieve and maintain the national ambient air quality standards, both primary (i.e., adequate to protect human health) and secondary (i.e., adequate to protect human "welfare" values), in the required time.

There are three other crucial elements to the state plans. First, there is the control strategy, which must include a survey of each region's existing air quality and a detailed inventory of the emissions from all pollution sources in the region to determine what kind of air pollution problem exists. The control strategy must set forth all the measures that will be taken to assure that the region's air quality meets the national standards. These measures must include emission limitations on particular sources and other reasonable control procedures, such as process changes, fuel controls, and land use and transportation controls, if they are necessary to meet the standards. The control strategy must also give timetables for compliance with its control measures. Second, the plan must include both a system for monitoring emissions from individual sources and a network for sampling ambient air quality. Finally, the state must be able to review new sources and their effects on ambient air quality and to revise its strategy as control techniques and air quality standards change in the future. In addition, states must include in their implementation plans effective measures to notify the public when air pollution levels exceed primary standards and to educate the public as to hazards involved and corrective measures available.

The Clean Air Act Amendments of 1970, 1971, and 1977 are an example of a recent shift in the burden of proof in pollution control. When the EPA now specifies that an air pollutant is a health hazard, industry must either comply with the emission standard or prove the health hazard does not exist.

Energy Legislation

Most federal statutes conferring authority over energy usage and development were enacted when environmental protection was not an important objective and, as a result, contain little or no reference to the environment. However, three major federal statutes enacted since 1968 make environmental protection a pervasive concern in the implementation of energy development activities. The National Environmental Policy Act requires that all federal agency activities be carried out with environmental protection in mind, and imposes procedural requirements to assure this. The Clean Air Act Amendments of 1970 prescribe a national plan to reduce air pollution. Hence, this affects many energy-related activities, including fuel extraction and processing, electricity production, and energy consumption. The Federal Water Pollution Control Act Amendments of 1972 similarly lay down a national plan for water pollution abatement, which also will affect energy-related activities.

It must be recognized that energy production, transportation, and consumption have a multiplicity of impacts on the environment. For example, stricter environmental protection standards for offshore oil production may decrease the supply from that source and cause a shift either to imported oil, with the associated hazards of marine pollution from tankers, or to coal, some of which will be obtained by environmentally destructive stripmining. Tighter environmental controls on nuclear power plants may delay construction and cause a shift to stripmined coal or offshore or imported oil; restrictions on coalfired plants may necessitate a switch to nuclear power or oil. Imports also have harmful environmental consequences, even without considering the impact of their production on the environment abroad. Since oil and gas imports will come from overseas, this will necessitate increased tanker traffic and construction of substantial new port facilities—thus creating the potential for oil pollution of the marine environment.

More recently, the Energy Supply and Environmental Coordination Act (ESECA) of 1974 provided the authority to prohibit certain power plants and major fuel-burning installations from burning natural gas or petroleum products as their primary energy source. This is an attempt to reduce the requirements for petroleum resources to generate energy, thus increasing the requirements for other resources such as coal. Because of the Clean Air Act standards on pollution levels, there is an increased demand for low-sulfur coal. In addition, this act states that no action taken under the Clean Air Act by the Environmental Protection Agency to implement ESECA will be regarded as a major federal action affecting the quality of the human environment within the meaning of NEPA; hence, no environmental impact statement would be required.

Fish and Wildlife Resources Legislation

The greatest disturbance to wildlife is alteration of habitat by humans. In some cases, human activities benefit certain types of wildlife, whereas, for other cases, loss or degradation of habitat poses a fundamental threat to continued existence. All human activities affect wildlife habitat in some way: directly, as in logging, farming, and channelization; or indirectly, as with livestock grazing, pesticide use, and introduction of exotic species.

It is of particular concern in the preparation and passage of legislation relative to fish and wildlife resources to prevent the elimination of fish or wildlife species due to human activities, to insure that fish and wildlife populations do not drop below self-perpetuating levels, and to preserve representations of all plants and animal communities for future generations. Possible impacts of human activities would include: noise effects on migratory patterns, breeding habits, animal behavior; air pollution effects on plant life; disruption of one or more of the food web chains; elimination of any rare or endangered species; interference with the movement of any resident or migratory fish or wildlife species; change in existing features of any lagoon, bay, or tideland.

The National Environmental Policy Act lists among its purposes " . . . to promote efforts which will prevent or eliminate damage to the environment and biosphere . . ." of which wildlife is surely an integral part, and " . . . to enrich the understanding of the ecological systems and natural resources important to the Nation . . ." In the declaration of national environmental policy, NEPA pledges " . . . to use all practicable means and measures . . . to create and maintain conditions under which man and nature can exist in productive harmony . . . to the end that the Nation may . . . fulfill the responsibilities of each generation as trustee of the environment for succeeding generations . . ." Ostensibly, these purposes and policies embrace wildlife protection.

Previously passed federal legislation deals with either the specific protection of wildlife types (e.g., the Endangered Species Conservation Act, the Migratory Bird Treaty Act, and the Fish and Wildlife Coordination Act) or the protection of wildlife habitat area (e.g., the Wild and Scenic Rivers Act, the Anadromous Fish Conservation Act, the Migratory Bird Conservation Act, the Federal Aid in Wildlife Restoration Act, and the Recreational Use of Conservation Areas Act).

In particular, the Fish and Wildlife Coordination Act requires that fish and wildlife receive "equal consideration with other project purposes," provides for the "enhancing of fish and wildlife values," and authorizes "compensatory wildlife features where some damage is inevitable." This act was responsible for the construction of certain mitigative fishery measures such as fish ladders and fish lifts in the Bonneville Dam across the Columbia River in Oregon, and facilities for the preservation of salmon and other fish below the Grand Coulee Dam in the state of Washington.

Also, the Wild and Scenic Rivers Act effectively controls those activities that in the past have been harmful to the waters and associated fish and wildlife resources. Those activities include, for example, dredging and filling, channelization, and alternatives to streams resulting from highway construction. It also provides for the preservation and maintenance of the aesthetic and recreational values of the identified area.

To support these laws, many agencies have established specific operational policies and procedures relative to the protection of wildlife resources. For example, the statutes providing for withdrawals, reservations, and restrictions of public lands for defense purposes provide that whenever the Department of Defense wishes to withdraw an area exceeding 5000 acres for a single project, it must file an application outlining certain specifications. The application must state whether, and if so to what extent, the proposed use will affect continuing full operation of the public land laws and federal regulations relating to conservation, utilization, and development of mineral resources, fish and wildlife resources, water resources, and scenic, wilderness, recreational, and other values. Such a withdrawal might also be considered a major federal action in the context of NEPA and thus require an environmental impact statement.

For the most part, however, the federal government leaves the responsibility for game conservation and management to the states. This is true even with regard to federally owned lands.

Historical Preservation Legislation

It is the basis of the National Environmental Policy Act that, prior to any major federal action, the concerned federal agency identify the nature of the resource and evaluate the impact of its decision on it. One element of that resource is the historic or cultural patterns that exist in an area. NEPA specifically states that the nation may "preserve important historic, cultural, and natural aspects of our national heritage, and maintain, wherever possible, an environment which supports diversity and variety of individual choice."

Under NEPA all federal agencies must identify the historic and cultural patterns that exist in an area that will be impacted by their programs. The National Register of Historic Places is a central tool that permits agencies to accomplish this identification process. This document identifies districts, sites, buildings, structures, and objects significant to our historic and cultural heritage. The identification of our environment—both cultural and natural—does not constitute a taking of property interests.

The National Register of Historic Places concept was initially authorized by the Congress in Section 2 of the Historic Sites Act of 1935 and was begun through a series of surveys conducted by the National Park Service throughout the 1960s to locate sites of national significance. This aspect of the 1935 act is now implemented through the National Historic Landmark program also maintained by the National Park Service. All national historic landmarks are listed in the National Register of Historic Places. National Register of Historic Places listing is a Department of Interior decision; nomination must be accomplished via a proposal to the National Park Service.

The National Historical Preservation Act of 1966 authorized the Secretary of the Interior to "expand and maintain" such a national register to include sites significant in American history, architecture, archaeology, and culture. In addition, it was found and declared by Congress in this act:

(a) That the spirit and direction of the Nation are founded upon and reflected in its historic past.
(b) That the historic and cultural foundations of the Nation should be preserved as a living part of our Community life and development in order to give a sense of orientation to the American people.
(c) That, in the face of ever-increasing extensions of urban centers, highways, and residential, commercial, and industrial developments, the present governmental and nongovernmental historic preservation programs and activities are inadequate to insure future generations a genuine opportunity to appreciate and enjoy the rich heritage of our Nation; and
(d) That, although the major burdens of historic preservation have been borne and major effects initiated by private agencies and individuals, and both should continue to play a vital role, it is nevertheless necessary and appropriate for the Federal Government to accelerate its historic preservation programs and activities, to give maximum encouragement to agencies and individuals undertaking preservation by private means, and to assist State and local governments and the National Trust for Historic Preservation in the United States to expand and accelerate their historic preservation programs and activities.

General criteria for inclusion in the National Register require that:

The quality of significance in American history, architecture, archaeology, and culture is present in districts, sites, buildings, structures, and objects of state and local importance that possess integrity of location, design, setting, materials, workmanship, feeling, and association, and

1. Are associated with events that have made a significant contribution to the broad patterns of our history or

2. Are associated with the lives of persons significant in our past or

3. Embody the distinctive characteristics of a type, period, or method of construction, or represent the work of a master, or possess high artistic values, or represent a significant and distinguishable entity whose components may lack individual distinction or

4. Have yielded, or may be likely to yield, information important in prehistory or history.

Ordinarily, cemeteries, birthplaces, and graves of historical figures, properties owned by religious institutions or used for religious purposes, structures that have been moved from their original locations, reconstructed historic buildings, properties primarily commemorative in nature, and properties that have achieved significance within the past 50 years are not considered eligible for the National Register. However, such properties will qualify if they are integral parts of districts that do meet the criteria or if they fall within the following categories:

1. A religious property deriving primary significance from architectural or artistic distinction or historical importance

2. A building or structure removed from its original location but which is significant primarily for architectural value, or which is the surviving structure most importantly associated with a historic person or event

3. The birthplace or grave of a historical figure of outstanding importance if there is no appropriate site or building directly associated with his or her productive life

4. A cemetery which derives its primary significance from graves of persons of transcendent importance, from age, from distinctive design features, or from association with historic events

5. A reconstructed building when accurately executed in a suitable environment and presented in a dignified manner as part of a restoration master plan, and when no other building or structure with the same association has survived

6. A property primarily commemorative in intent if design, age, tradition, or symbolic value has invested it with its own historical significance

7. A property achieving significance within the past 50 years if it is of exceptional importance.

Since the 1966 act, the National Register has grown from a listing of several hundred properties that possessed national significance to a listing of about 10,000 properties of state, local, and national significance.

Land Use Legislation

Federal influence on the regulation and development of land is found in two general areas, namely: (1) the use of land directly under federal control, and (2) federal requirements accepted as part of programs that are largely initiated and managed by the states or their subdivisions but receive a considerable part of their financing from the federal government.

A great variety of federal laws and programs affect the pace, form, and distribution of urban development. Directly contracted federal public works projects and defense installations, to the extent that they create wealth or jobs, are the most obvious examples of federal activities with land use impacts. These impacts are manifest first on the land developed and second on the surroundings urbanized by people benefited or attracted as a result of the development. Federally assisted highways probably affect the appearance of the country more than any other federal program by enhancing access via automobile to center cities and remote wilderness areas. The use of land is incidental to so many endeavors that it is the exceptional federal law or program that does not affect land use in some way.

Although federal legislation affecting solid waste disposal, pesticides, and noise may affect land use, the most influential pollution control laws are the Federal Water Pollution Control Act Amendments of 1972 and the Clean Air Act.

Federal water pollution control legislation contains explicit and detailed requirements for land use planning and regulation at the state or regional level. Section 208 of the Federal Water Pollution Control Act Amendments recognizes that attention only to what comes out of the pipe will not clean up the waters, and that a focus on the land is also necessary. This section prescribes a strong regulatory framework in which states or areawide agencies must exercise control over the "location, modification, and construction of any facilities . . . which may result in any discharge in such area . . ." Methods of implementing this authority could include a permit system, or a review of local zoning, subdivision, and building approvals by a state or areawide planning body to assure the proper siting of "point" sources of water pollution. Moreover, the law requires the regulation of the more difficult to control "nonpoint" water pollution sources such as agricultural, forest, mine-related, and construction activities.

Under the Clean Air Act, the EPA administrator is required to set two uniform national standards for each pollutant which may lawfully be found anywhere in the nation. These ambient air quality standards consist of primary standards based on "the latest scientific knowledge useful in indicating the kind and extent of all identifiable effects on public health or welfare" and secondary standards which are intended to protect the public welfare from "any known or anticipated adverse effects associated with the presence of such air pollutant in the ambient air." The public welfare is broadly defined to include "effects on soils, water, crops, vegetation, man-made materials, animals, wildlife, weather, visibility, and climate, damage to and deterioration of property, and hazards to transportation, as well as effects on economic values and on personal comfort and well-being."

Referring to the section on Air Quality Legislation earlier in this chapter, the states are required to submit to the EPA administrator detailed implementation plans for meeting primary and secondary standards. In addition, the EPA administrator has established regulations requiring states to include in their plans provisions which will permit the review, and provide the authority to prevent, the construction, modification, or operation of complex sources at a location where emissions associated with such source would result in violation of a national standard or the state's control strategy. By a "complex source" is meant a facility that has or leads to secondary or adjunctive activity which emits or may emit a pollutant for which there is a national standard. These sources include, but are not limited to: shopping centers; sports complexes; drive-in theaters; parking lots and garages; residential, commercial, industrial, or institutional developments; amusement parks and recreational areas; highways; sewer, water, power, and gas lines; and other such facilities which will result in increased emissions from motor vehicles or other stationary sources. Thus, there is established the requirement for state control of large-scale developments and key facilities.

The nation's coastal zones contain some of the most critical ecological areas, many of which are extremely vulnerable to destruction. For example, the coastal wetlands provide a critical link between the terrestrial and aquatic ecosystems and provide a vital service for the marine ecosystem. It is estimated that over 70 percent of all commercially valuable marine fishes rely on the estuarine areas during at least part of their lives. Half of the biological productivity of the world's oceans occurs along the coasts, and the estuaries are the most productive areas known on earth. The Coastal Zone Management Act established the policy of protecting these resources and provided assistance to coastal states in the development of coastal management plans. In some states, this has led to the establishment of permit requirements for development within specified coastal areas.

With regard to noncoastal areas, the U.S. Congress authorized under the Wilderness Act the establishment of a Wilderness Preservation System to protect and administer specifically identified wilderness areas. In this act, a wilderness area, in contrast with areas where people and their own works dominate the landscape, is defined as an area where the earth and its community life are untrammeled by people, where people are visitors who do not remain. An area of wilderness is further defined to mean in this act one of undeveloped federal land retaining its primeval character and influence, without permanent improvements or human habitation, which is protected and managed so as to preserve its natural conditions and (1) generally appears to have been affected primarily by the forces of nature, with the imprint of people substantially unnoticeable; (2) has outstanding opportunities for solitude or a primitive and unconfined type of recreation; (3) has at least 5000 acres of land or is of sufficient size as to make practicable its preservation and use in an unimpaired condition; and (4) may also contain ecological, geological, or other features of scientific, educational, scenic, or historical value.

Legislation similar in its area of applicability is the Multiple-Use and Sustained-Yield Act of 1960 which declared that the national forests "shall be administered for outdoor recreation, range, timber, watershed, and wildlife and fish purposes." The act's multiple-use directive was defined to mean:

The management of all the various renewable surface resources of the national forests so that they are utilized in the combination that will best meet the needs of the American people; making the most judicious use of the land for some or all of these resources or related services over areas large enough to provide sufficient latitude for periodic adjustments in use to conform to changing needs and conditions; that some land will be used for less than all of the resources; and harmonious and coordinated management of the various resources, each with the other, without impairment of the productivity of the land, with consideration being given to the relative values of the various resources, and not necessarily the combination of uses that will give the greatest dollar return or the greatest unit output.

The Secretary of Agriculture was authorized in this act to cooperate with the states in the implementation of the act to manage the national forests.

Other land-related legislation deals with the protection of flood-prone areas and developments constructed in these areas. For example, the Flood Disaster Protection Act of 1973 established requirements for communities to receive the benefits of federal flood insurance programs. To enter such a program, communities must agree to develop land use control measures which are consistent with floodplain management criteria issued by the Department of Housing and Urban Development (HUD). These criteria require communities to enact floodplain ordinances, including zoning and building code provisions, which will reduce the likelihood of flood damage in hazardous areas. As HUD develops more specific information about the flood hazards within an area, more restrictive controls must be implemented by a community.

Furthermore, Section 122 of the Rivers and Harbors and Flood Control Act of 1970 requires that the Secretary of the Army

. . . promulgate guidelines designed to assure that possible adverse economic, social and environmental effects relating to any proposed project have been fully considered in developing such project, and that the final decisions on the project are made in the best overall public interest, taking into consideration the need for flood control, navigation and associated purposes, and the cost of eliminating or minimizing such adverse effects as the following:
  (1) air, noise and water pollution;
  (2) destruction or disruption of man-made and natural resources, esthetic values, community cohesion and the availability of public facilities and services;
  (3) adverse employment effects and tax and property value losses;
  (4) injurious displacement of people, businesses, and farms; and
  (5) disruption of desirable community and regional growth.

In conjunction with legislation of the type just described, there is a current drive for land use reform motivated primarily by a concern that many of the ecological and aesthetic treasures of the American landscape have been irreversibly destroyed by urban development and that more such despoliation will occur if strong measures are not taken to control urbanization. This concern has grown along with the general environmental awareness that has led to important pollution control reforms.

Finally, there is the growing conviction that land is more than a commodity to be subdivided and traded, but is also a resource like air and water to be carefully managed and conserved in the public interest. These convictions have received expression in a mode of land use planning that allocates and designs urban development in such a way that hydrological, topographical, geological, and other natural characteristics are analyzed and classified as constraints on the placement, kind, and amount of new development.

Noise Legislation

Of the commonly recognized forms of pollution, noise has been the last to receive much attention from governmental agencies. Congress addressed the problems of air and water pollution long before turning to noise. Water legislation passed in the late nineteenth century is still important today; however, the earliest federal legislation dealing with noise occurred in 1970. This was the Noise Pollution and Abatement Act of 1970 which established the Office of Noise Abatement and Control within the Environmental Protection Agency (EPA) to investigate and study noise and its effect upon the public health and welfare.

The Noise Control Act of 1972 was the next congressional action directed toward noise pollution. This act greatly expanded the responsibility and authority of the Office of Noise Abatement and Control. In particular, it authorized the EPA to coordinate all federal noise control programs and also instituted federal noise emission controls on all noisy products traded in commerce as well as federal controls on noise generated by interstate rail and motor carriers. The Noise Control Act does not assert for the federal government the comprehensive role that the government has assumed in controlling air and water pollution. Whereas the federal air and water quality laws regulate ambient conditions as well as emissions from the major sources of pollutants, the noise law, for the most part, controls only noise emissions at the source.

Noise pollution is like air and water pollution in that it can be attacked from a number of angles, including curbing emissions at the source, controlling the use of pollution sources, and controlling the use of land around them to reduce the impact of emissions. However, unlike air and water pollution, noise does not accumulate in the environment as a residual that flows from locality to locality or from state to state, but dissipates within a short time and distance from its point of generation. For this reason the legislative efforts have been directed toward control of noise at the source.

Regulations have been established in the past by various branches of the government to control noise and, indirectly, its effects. For example, the Federal Aviation Administration established the "Noise Standards: Aircraft Type Certification" (FAR Part 36), which places noise reduction requirements upon new jet aircraft. As a result, the new wide-bodied jets which meet these requirements are significantly quieter than the previously developed narrow-bodied aircraft. Further, the Federal Highway Administration has published noise standards that will limit the noise that future freeways may impose upon surrounding communities. There are a number of other federal noise regulations, such as the allowable noise exposure of industrial workers that is regulated by the Occupational Safety and Health Administration (OSHA) and the noise standards for residential housing published by the Department of Housing and Urban Development (HUD).

In particular, the environmental impact statement process under NEPA furnishes a means for HUD to identify and take into account all the environmental impacts, including noise, of its major actions. HUD has adopted procedures to implement Section 102(2)(C) of NEPA, which acknowledges the need to consider noise in environmental impact statements. As part of this procedure, HUD adopted uniform policies on noise exposure applicable to all of its programs. For example, HUD assistance for new construction will not be approved for new construction on sites having unacceptable noise exposures.

Other federal legislation has dealt indirectly with the problem of noise pollution. The Airport and Airway Development Act of 1970 and the Federal-Aid Highway Act identify noise as one factor among others to be considered in the planning, development, and construction of airports and highways. The Environmental Protection Agency is required to evaluate environmental factors involved in such projects and to report its findings to the Secretary of Transportation. He, in turn, must take them into consideration before making a final decision on the feasibility of a given project. In addition, the Clean Air Act of 1970 directed that substantial research be carried out to study a wide range of problems concerning the damage caused by noise.

Solid Waste Legislation

The term "solid waste" means garbage, refuse, and other discarded solid materials, including solid waste materials resulting from industrial, commercial, and agricultural operations, and from community activities. This classification does not include solids or dissolved material in domestic sewage or other significant pollutants in water resources, such as silt, dissolved or suspended solids in industrial wastewater effluents, dissolved materials in irrigation return flows, or other common water pollutants. The largest part of the total solid waste generated, by volume, are the "animal" and "mineral" wastes which each make up about 40 percent. The former represent wastes resulting from agriculture, food, and timber processing, whereas the latter are derived from the extraction and processing of ores and include the unused overburden of mining operations. While these wastes present severe environmental problems, it is municipal and industrial waste which affects the most people and generally receives the most attention.

The rate of solid waste generation has grown most rapidly in urban areas where disposal presents special problems due to population density and lack of space. This is especially significant since urban populations generate approximately 20 percent more waste per capita than nonurban populations. Recently enacted federal limits on air and water pollution place a new strain on solid waste disposal sites by increasing the quantity of waste which must be disposed of on land. The Federal Water Pollution Control Act, as amended in 1972, provides for controls on the amount of toxic and nontoxic substances that may be discharged into navigable waters. The Marine Protection, Research, and Sanctuaries Act of 1972 prohibits the dumping into oceans of high-level radioactive waste and radiological chemical warfare agents, and requires permits for dumping other materials which could adversely affect human health or the marine environment. The Clean Air Act establishes national ambient air quality standards prescribing the maximum emission levels of nonhazardous and hazardous substances. Most states in the implementation of the act have prohibited open burning at waste disposal sites. Thus, in many cases, waste which in the past might have been burned or liquefied must now be maintained in solid form.

The Solid Waste Disposal Act of 1965 was the first legislation aimed at solid waste management and is directed primarily at the loss of natural resources which solid waste represents. In this act, Congress found that technological progress leads to an ever-mounting increase in waste products and changes in their characteristics, that urban concentration aggravates the problem, and that inefficient and improper methods of disposal result in scenic blights and create serious hazards to the public health. These include pollution of air and water resources, accident hazards, and an increase in rodent and insect vectors of disease. Also, the act authorizes a research and development program with respect to solid waste to promote the demonstration, construction, and application of solid waste management and resource recovery systems. In addition, the act provides financial and technical assistance to state and local governments and interstate agencies in the planning and development of resource recovery and solid waste disposal programs, and promotes a national research and development program for improved solid waste management programs.

Congress amended the 1965 act with the Resource Recovery Act of 1970, which officially recognized the potential economic benefits of recovering a portion of the "trash" which was being discarded. This Act put a new emphasis on recycling and reusing waste materials by authorizing the funding of demonstration grants for recycling systems and for studies of methods to encourage resource recovery. Although the primary responsibility for the management of solid waste materials resides with state and local officials, federal activity was directed by Congress via the Resource Recovery Act into several areas, namely:

1. Construction, demonstration, and application of waste management and resource recovery systems for the preservation of air, water, and land resources

2. Technical and financial assistance to agencies in planning and developing resource recovery and waste disposal programs

3. National research and development programs to develop and test methods of dealing with collection, separation, recovery, recycling, and safe disposal of nonrecoverable waste

4. Guidelines for the collection, transportation, separation, recovery, and disposal of solid waste

5. Training grants in occupations involving design, operation, and maintenance of solid waste disposal systems.

In summary, municipal governments are confronted with an increasing rate of solid waste generation, increased disposal costs, decreased availability of landfill sites, particularly in urban areas, and new air and water pollution control restrictions which limit incineration and restrict disposal into oceans and waterways. With energy and materials shortages, the recovery of valuable resources from solid waste is becoming increasingly important.

Transportation Legislation

Over the past few years, there has been a growing concern over the impact of transportation-related activities on the environment—in particular, those impacts dealing with air pollution, noise, displacement of people and businesses, disruption of wildlife habitats, and overall growth-inducing effects. The construction of urban highways has not only affected those who were displaced or who lived adjacent to the route, but it has been a powerful force in shaping the growth and development of metropolitan areas. The urban highways extending from the inner city outward to the suburbs have played an important role in dispersing the population from the older in-town neighborhoods to the outskirts of the city and the burgeoning suburbs. Thus, those highways have impacts far removed from the mere transportation provided by automobiles and trucks. They affect an area's housing, its economic base, the availability of parks and open space, the integrity of neighborhoods, school districts, and other aspects of urban life.

While there have been substantial benefits flowing from the nation's highway system, it has been responsible for some of the most widespread and permanent damage to the natural environment and urban areas. The most immediate effect of highway construction in an urban area is the displacement of those who live or work in the path of the highway. Many people whose homes were not taken were left to live adjacent to unsightly, noisy freeways which divided neighborhoods and were traveled by cars emitting dangerous pollutants. In some instances, the highways threatened valuable urban parkland and open space, historic sites, and other recreation areas.

One of the earliest and most significant pieces of transportation legislation relative to environmental protection was the Department of Transportation act of 1966. This act gives specific recognition to the protection of water resources, recreation areas, and wildlife habitat in highway planning, design, and construction. Under this act, Congress declared it was in the best interests of the nation to develop a national transportation policy and a program that would provide for fast, safe, efficient, and convenient transportation. Furthermore, it stated that the transportation system should be consistent with other national objectives "including the efficient utilization and conservation of the Nation's resources." Under Section 4(f) of the act, as amended by the Federal-Aid Highway Act of 1968, it is stated that:

The Secretary shall not approve any program or project which requires the use of any publicly owned land from a public park, recreation area, or wildlife and waterfowl refuge of national, state, or local significance as determined by the Federal, State, or local officials having jurisdiction thereof, or any land from an historic site of national, state, or local significance as so determined by such officials unless (1) there is no feasible and prudent alternative to the use of such land, and (2) such program includes all possible planning to minimize harm to such park, recreation area, wildlife and waterfowl refuge, or historic site resulting from such use.

It is interesting to note that this act, as amended, formed the basis for the protection of the natural enviroment prior to the passage of the National Environmental Policy Act of 1969. The Department of Transportation Act of 1966 announced a national policy of preserving the environment and conservation of the nation's resources. However, it was not until the enactment of NEPA in 1970 that a conscientious effort was made to develop a policy of minimizing the adverse environmental effects of federally funded transportation programs.

The Federal-Aid Highway Act of 1970 further amended the Department of Transportation Act by establishing the requirement that the Secretary of Transportation promulgate guidelines designed to assure that possible adverse economic, social, and environmental effects deriving from any proposed project in any federal-aid system have been fully considered in developing such project. Further guidelines assure that the final decisions on the project are made in the best overall public interest, taking into consideration the need for fast, safe, and efficient transportation and public services, as well as the cost of eliminating or minimizing such adverse effects as the following: (1) air, noise, and water pollution; (2) destruction or disruption of synthetic and natural resources, aesthetic values, community cohesion, and availability of public facilities and services; (3) adverse employment effects, and tax and property value losses; (4) injurious displacement of people, businesses, farms; and (5) disruption of desirable community and regional growth.

The Federal-Aid Highway Act of 1970 is an amended version of the Federal-Aid Highway Act of 1956 which stated that any state highway department that submitted plans for a federal-aid highway project must have held public hearings, or at least provided the opportunity for hearings, and that these hearings should have been designed to consider the economic effects of the proposed highway. The latter Act was also amended in 1968 so as to provide that other than economic effects be considered in the highway plan. At this time, the law indicated that the social effects must be considered as well as the economic effects. Furthermore, consideration was to be given to the impact upon the environment and the consistency of the project with the goals and objectives of the community through which the highway passed. In the 1970 amendments, it was stated that in the certification of highway projects "such certification shall be accompanied by a report which indicates the consideration given to the economic, social, environmental, and other effects of the plan or highway location or design and various alternatives which were raised during the hearing or which were otherwise considered."

With regard to mass transit, the Urban Mass Transportation Act of 1964 established the requirement that project applications include a detailed statement on: (1) the environmental impact of the proposed project; (2) any adverse environmental effects which cannot be avoided should the proposal be implemented; (3) alternatives to the proposed project; (4) any irreversible and irretrievable impact on the environment which may be involved in the proposed project should it be implemented. It further established the requirement that the Secretary of Transportation make a finding after full and complete review of any hearing transcript that (a) adequate opportunity was afforded for the presentation of views by all parties with a significant economic, social, or environmental interest, and fair consideration has been given to the preservation and enhancement of the environment and to the interest of the community in which the project is located, and (b) either no adverse environmental effect is likely to result from such project, or there exists no feasible and prudent alternative to such effect and all reasonable steps have been taken to minimize such effect.

In the case of airports, the environmental consequences of constructing a new airport or a new runway make such projects major planning problems for any nearby community. An airport not only creates a noise problem in the area, but also has a wide effect on land use. Since a large area is required for a new airport, undeveloped areas or parks and recreation sites are likely targets. If a developed area is chosen for a new airport site, massive relocation is required, communities are divided, and the ground transportation network must be adjusted to accommodate the new facility. Finally, the construction of a new airport serves as a catalyst for commercial development in the area.

The Airport and Airway Development Act, passed by Congress in 1970, greatly expanded the federal role in airport planning and construction. Under this act, the Secretary of Transportation is directed to prepare a national airport system plan for the development of public airports. All airport projects are also required to be coordinated with areawide planning. Airport development projects must be "reasonably consistent" with the plans for development of the area in which the airport is located. In particular, Section 16 of the Airport and Airway Development Act of 1970 requires all airport development projects involving the location of an airport, an airport runway, or an extension of a runway, to provide an opportunity for a public hearing "for the purpose of considering the economic, social and environmental effects of the airport location and its consistency with the goals and objectives of such urban planning as has been carried out by the community."

Furthermore, the Clean Air Act Amendments of 1970 and the state implementation plans required by those amendments will have a significant impact upon future transportation planning. In the 1970 amendments, Congress required the states to adopt, and submit to EPA for approval, implementation plans setting forth means by which the states will meet the national ambient air quality standards. It is specifically stated that one measure to be considered in the implementation plans is land use and transportation controls. The transportation control plans which have been approved by EPA are basically designed to reduce vehicle miles traveled in an area. The plans contain a variety of features including parking restrictions and surcharges, improvements in existing transit systems and development of new systems, exclusive bus lanes, exclusive bus use on certain streets, mass transit incentive plans to be adopted by large employers, encouragement of car pooling, vehicle inspection and maintenance programs, and retrofitting of air pollution control devices on autos.

Water Quality Legislation

Water quality legislation at the federal level dates back to the nineteenth century when Congress enacted the River and Harbor Act of 1886, recodified in the Rivers and Harbors Act of 1899 (also known as "the Refuse Act"). The latter act prohibits obstructions to navigation and the dumping of any refuse (except municipal sewage) into navigable waters. It provides for abatement of actual and potential pollution. A person or corporation may lawfully discharge only after obtaining a permit from the Army Corps of Engineers.

The Environmental Protection Agency (EPA) reviews the applications for permits filed with the corps and makes recommendations. The permit program does not give a person the right to continue pouring wastes into water indefinitely. The EPA determines the environmental impact of an individual's activities and requires the polluter to stop the offending discharge within a reasonable time. A polluter who discharges without a permit or who violates a condition upon which a permit is granted may be prosecuted.

The Army Corps of Engineers has established criteria for the issuance of permits to build structures in the navigable waters. These are based on an evaluation of the probable impact on the public interest of the proposed structure, or work, and its intended use. The evaluation of the probable impact requires a careful weighing of all those factors which become relevant in each particular case. Among these are conservation, economics, aesthetics, general environmental concern, historic values, fish and wildlife values, flood damage prevention, land use classification, navigation, recreation, water supply, water quality, and, in general, the needs and welfare of the people. The decision, therefore, as to whether or not to authorize the proposal and the conditions under which it may be allowed to occur are determined by the outcome of the general balancing process.

Recognizing the threat that dirty water posed to the public health and welfare, Congress enacted the Federal Water Pollution Control Act in order to "enhance the quality and value of our water resources and to establish a national policy for the prevention, control and abatement of water pollution." This act and its amendments set out the basic legal authority for federal regulation of water quality. The original act was passed in 1948 and was subsequently amended in 1956, 1961, and 1970. The 1970 amendment imposed a new requirement on all applicants for a federal license or permit. If a proposed activity may result in a discharge into navigable waters, a certificate must be obtained from the affected state assuring that the activity would not violate state water quality standards.

The Federal Water Pollution Control Act accomplishes three basic tasks: (1) regulation of pollutant discharges from point sources, primarily industrial plants, municipal sewage-treatment plants, and agricultural feedlots; (2) regulation of spills of oil and hazardous substances; and (3) financial assistance for sewage treatment plant construction. Other provisions of the act regulate vessel sewage and disposal of dredged material. In addition, it provides federal support for various research and demonstration projects, and for state water pollution control programs. Basically, the Federal Water Pollution Control Act provides for two types of standards: effluent standards and water quality standards. An "effluent standard" is a measure of the allowable amount of a pollutant to be discharged in a time period, or it may specify a maximum permissible concentration in the effluent, or it may specify a maximum amount that may be discharged per unit of production (such as a specified number of suspended solids per ton of product produced). Effluent standards are generally based on the availability of the technology to control the effluent, although in some situations, as in the case of toxic pollutants, the effluent standard must be based on an assessment of the pollutant's effect in the receiving water. On the other hand, "water quality standards" describe the quality that will be required for a particular body of water, generally based on its designated use.

Reflecting basic state responsibility for water pollution control, the Federal Water Pollution Control Act requires the states to submit to EPA water quality standards for all interstate navigable waters. These state standards spell out water use classifications such as recreation, fish and wildlife propagation, public water supplies, or industrial and agricultural uses. States are then required to specify the water quality required to achieve these uses and outline detailed plans for maintaining the desired levels of quality. Consequently, the critical limits of any pollutant which may be discharged are determined at the state level.

Toxic pollutants and thermal discharges are specially treated in the 1970 amendments. The latter are especially important in the case of nuclear power plants since they present various environmental problems related to water. Two such problems are the effect of intake facilities on fish, and the diminished streamflows caused by the plant's water diversion and consumption demands. However, the problem which probably has received the most attention is the effect of discharges of heated water used in plant cooling, i.e., "thermal pollution." Whenever a party discharging heated water can show that a balanced, indigenous population of fish, shellfish, and wildlife may be maintained in the water body despite a greater thermal discharge than allowed by the technological control standards, the EPA may allow an appropriately greater thermal discharge. In addition, in reviewing thermal discharges, the EPA must assure that the location, design, construction, and capacity of cooling-water intake structures reflect the best technology available for minimizing adverse environmental impacts.

With regard to ocean dumping, only an estimated 10 percent or less of the total volume of pollutants entering the world's oceans enter through direct ocean dumping of wastes carried to sea for that purpose by vessels and aircraft. A far greater share of the entering pollutants come from land-based atmospheric sources or freshwater discharges. Still more are generated by vessels engaged in normal cargo-carrying operations. On the other hand, while many instances of ocean dumping are not particularly harmful and may, in fact, sometimes be beneficial to the marine environment, several aspects of the ocean dumping of wastes are particularly troublesome. First, some wastes dumped in the ocean are toxic or otherwise hazardous; they are dumped only because land-based disposal is either harmful to the environment or is quite expensive. Second, many crowded metropolitan coastal regions have limited land areas suitable for waste disposal and cannot rely on incineration or similar disposal methods because of their already pressing air pollution problems.

The basic federal legislation directed toward ocean dumping is the Marine Protection, Research and Sanctuaries Act of 1972, commonly referred to as the Ocean Dumping Act, which requires federal permits for: (1) the transportation from the United States of almost all materials for the purpose of dumping in ocean waters; (2) the actual dumping of such materials transported from outside the United States into the United States territorial sea or into the contiguous zone; and (3) the transportation of such materials by a United States agency or official from outside the United States for the purpose of dumping it into any ocean waters. No permit can be granted for the transportation for the purpose of dumping or the dumping itself of any radiological, chemical, or biological warfare agent, or any high-level radioactive waste.

In summary, the basic federal law of water pollution control is contained in the Federal Water Pollution Control Act. Other sources of federal law in this area are the Rivers and Harbors Act of 1899, the Marine Protection, Research and Santuaries Act of 1972, and, of course, the National Environmental Policy Act of 1969.

CONCEPT OF ENVIRONMENT

The term "environment" is not defined in the National Environmental Policy Act. However, it is clear from Section 102 of the act that the term is meant to be interpreted broadly to include physical, social, cultural, economic, and aesthetic dimensions. In this context we can define "environment" to mean the whole complex of physical, social, cultural, economic, and aesthetic factors which affect individuals and communities and ultimately determine their form, character, relationship, and survival. Examples of such factors are: air and water quality, erosion control, natural hazards, land use planning, site selection and design, subdivision development, conservation of plant and animal life, urban congestion, overcrowding, displacement and relocation resulting from public or private action or natural disaster, noise pollution, urban blight, code violations and building abandonment, urban sprawl, urban growth policy, preservation of cultural resources (including properties on the National Register of Historic Places), urban design and the quality of the constructed environment, and the impact of the environment on people and their activities.

The dimensions of the environment as defined are rather broad and can be further elaborated upon and categorized as follows:

1. Physical environment (natural and constructed)
   a. Land and climate
      (1) Soil—general characteristics, load-bearing capacity, existing and potential erosion, permeability, mineral content, shrink-swell potential
      (2) Topography—general characteristics, slope, grade of site, location and size of watershed
      (3) Subsurface conditions—geologic characteristics, geologic faults, aquifer recharge
      (4) Special conditions—flood plain; unique landscape; potential for mudslide, landslide, subsidence, or earthquake; aerial or underground transmission lines and right-of-way; gullied areas; irrigation
      (5) Climatic conditions—annual rainfall and seasonal distribution; average

annual temperature and temperature ranges; growing season; potential for flash floods, hurricanes, or tornadoes; wind conditions
  b. Vegetation, wildlife, and natural areas
    (1) Extent and type of vegetation and wildlife
    (2) Existence of on-site, or proximity to, unique natural systems such as stream systems, wildlife breeding areas, forests and wilderness areas
  c. Surrounding land uses and physical character of area
    (1) Type of development—single family or high-rise residential, industrial, commercial, open space, mixed, public and quasi-public
    (2) Densities—people per acre, dwelling units per acre, industrial and commercial square footage per acre
    (3) Building height, design, intensity, and lot sizes
  d. Infrastructure/public services
    (1) Water supply sources, quality, and distribution
    (2) Sanitary sewage and solid waste disposal facilities
    (3) Storm sewers and drainage
    (4) Energy resources—electricity, natural gas, oil
    (5) Transportation facilities servicing site—roads, public transit, parking, airports, heliports
  e. Air pollution levels
    (1) Major sources of air pollution in area
    (2) Extent of pollution (smog, dust, odors, smoke, hazardous emissions) in relation to local and state standards of health and safety
    (3) Frequency of inversions and air pollution alerts or emergencies
    (4) Conditions peculiar to the site and immediate area
  f. Noise levels
    (1) Noise sources—nearby airport, railway, highway
    (2) Ambient noise levels
    (3) Vibrations
  g. Water pollution levels
    (1) Ground and surface water relevant to site and area—drainage basin, source of water supply, water bodies with implications for health and recreational uses, existing water quality
    (2) Use and transportation of fertilizers and insecticides and their effect on euthrophication
    (3) Discharges from feed lots
    (4) Sewage disposal systems
    (5) Mine waste areas
2. Social environment
  a. Community facilities and services
    (1) Location and capacity of schools
    (2) Neighborhood, community, and regional parks servicing area
    (3) Recreational and cultural facilities
    (4) Police, fire, health, and social service facilities servicing the area
    (5) Local public transportation
  b. Employment centers and commercial facilities servicing area
  c. Character of community
    (1) Socioeconomic and racial characteristics
    (2) Community life—places to meet, management, organized activities
    (3) Population size and distribution
    (4) Housing conditions
3. Aesthetic environment
  a. Existence of on-site, or proximity, to significant historic, archaeological, or architectural sites or property
  b. Scenic areas, views, vistas, and natural landscape
  c. Architectural character of existing buildings
4. Economic environment
  a. Employment and unemployment levels
  b. Level and sources of income
  c. Economic base of area

    d. Land ownership including private, local, public, state, and federal
    e. Land values

CONCEPT OF ENVIRONMENTAL IMPACT

By definition (Refs. 28, 47, and 52) an "environmental impact" is any alteration of environmental conditions or creation of a new set of environmental conditions, adverse or beneficial, caused or induced by the action or set of actions under consideration. The attention given to environmental conditions, as referred to here, will vary according to the nature, scale, and location of the proposed action (or actions). Primary attention would be given to those factors most evidently affected, such as the effects on the resource base, including land, water quality and quantity, air quality, public services and energy supply, as well as other environmentally critical areas. For example, impact on the nesting grounds of an endangered species would be significant, while a similar impact on the nesting grounds of a species which is abundant may not be significant. Likewise, the significance of a high noise level is much different in a residential area than in an industrial area.

Generally, impacts can be categorized as either primary or secondary. This distinction is important for consideration of alternatives and ways to minimize adverse impacts in performing impact analysis. One way to describe the distinction is that project "inputs" generally cause primary impacts and project "outputs" generally cause secondary impacts. Primary impacts are generally easier to analyze and measure, while secondary impacts are usually more difficult to measure. Secondary impacts may, in fact, be more significant than primary impacts. For example, the primary impact may be a change in vegetative species composition, but the secondary consequence may be a significant reduction in a rare or endangered wildlife species.

"Primary impacts" are those that can be attributed directly to the proposed action. If the action is a field experiment, materials introduced into the environment which might damage certain plant communities or wildlife species would have a primary impact. If the action involves construction of a facility, such as a sewage treatment works, an office building, or a laboratory, the primary impacts of the action would include the environmental impacts related to construction and operation of the facility and land use changes at the facility site.

"Secondary impacts" are indirect or induced changes, and typically include the associated investments and changed patterns of social and economic activities likely to be stimulated or induced by the proposed action. If the action involves construction of a facility, the secondary impacts would include the environmental impacts related to induced changes in the pattern of land use, population density, and related effects on air and water quality or other natural resources. Also included would be any unplanned increase in growth rate or level experienced by the existing community as a result of the new construction. Such secondary effects, through their impacts on existing community facilities and activities, through inducing new facilities and activities, or through changes in natural conditions, may often be more substantial than the primary effects of the proposed action.

In the biophysical environment, the secondary impacts can be especially important. For example, removal of vegetation may cause excessive soil erosion which may cause excessive sediments in the receiving stream. This in turn will reduce the amount of sunlight that can penetrate the water, thus reducing the dissolved oxygen in the water. As a result, this will have an adverse effect on aquatic life and the water quality of the stream.

ENVIRONMENTAL IMPACT FACTORS AND
AREAS OF CONSIDERATION

The choice of impacts to be considered in performing an environmental impact analysis generally varies according to the type of project, development, or action under evaluation. However, the effects to be considered should include the following (Ref. 14) as applicable: air quality and air pollution control; weather modification; energy development, conservation, generation, and transmission; toxic materials; pesticides and herbicides; transportation and handling of hazardous materials; aesthetics; coastal area; historic and archaeological sites; flood plains and watersheds; mineral land reclamation; parks, forests and outdoor recreation; soil and plant life, sedimentation, erosion, and hydrologic conditions; noise control and abatement; chemical contamination of food products; food additives and food sanitation; microbiological contamination; radiation and radiological health; sanitation and waste systems; shellfish sanitation; urban planning and congestion; rodent control; water quality and water pollution control; marine pollution; river and canal regulation and stream channelization; and wildlife preservation.

For example, in the case of coal mining activities, typical impacts could be expected to include:

1. Physical environment impacts—alteration of atmospheric properties, atmospheric constituents, other atmospheric phenomena, landforms, soils, erosion patterns, solid waste loads, terrestrial ecosystems, aquatic ecosystems, and energy utilization efficiency; changes in surface water quantity, ground water quantity, and water quality; reduction of coal resources and noncoal mineral resources; encroachment into critical areas

2. Social environment impacts—alteration of demographic and population characteristics, health and safety characteristics, activity patterns, institutions and services, and community attitudes

3. Aesthetic environment impacts—alteration of aesthetic characteristics of the area

4. Economic environment impacts—alteration of land values, employment, tax base, regional income, and energy prices.

The general approach to identifying potential impacts and environmental factors which should be considered is to utilize a questionnaire checklist relative to the major impact areas. For example, such a checklist could be structured as follows:

1. Pollution effects
   a. Air quality
      (1) Will the action result in emissions into the atmosphere of toxic or hazardous substances or significant amounts of other pollutants?
      (2) How and to what extent will the action affect the air quality?
      (3) Will it contribute to a degradation of air quality?
      (4) Will it cause changes in chemical and physical composition?
   b. Water quality
      (1) How and to what extent will the action affect the availability, supply, use, and quality of water?
      (2) Will the action cause marine pollution or affect commercial fishery and shellfish sanitation?
      (3) Will it affect waterway regulation and stream modification activities?
      (4) Will the action divert water from one basin to another and have a significant effect on the quality or quantity of water in either basin?
      (5) Will the action contribute to a significant depletion or degradation of ground or surface water?
      (6) Will the action introduce toxic or hazardous substances or solid wastes into bodies of water?
      (7) Will the action significantly increase sedimentation in a body of water?
      (8) Will the action significantly alter the temperature of a body of water?
   c. Noise quality
      (1) Will the action result in the creation of excessive noise, considering the proximity of the likely effects of the noise on humans or wildlife?
      (2) Will the action result in kinds of noises and noise levels that will be disturbing or a nuisance in the immediate and overlying areas?
   d. Solid waste
      (1) How will the proposed action affect activities related to the creation, management, and disposal of solid waste materials?
      (2) What type of solid waste will be generated as a result of the action?
   e. Radiation
      (1) Will the proposed action create heat, noise, energy waves, electrical or radioactive effects, physical vibrations, or other thermal, electrical, or microwave activity that will be disturbing or a nuisance or create interference in the immediate and outlying areas?
   f. Hazardous substances
      (1) Will the proposed action create or generate any substances, materials, or activities that are dangerous because of toxicity, flammability, or explosive tendencies or characteristics?

(2) Will it create or generate substances that might result in contamination or deterioration of food, food sources, clothing, or other materials?

2. Vegetation and wildlife effects
   a. Will the action result in significant destruction of vegetation, wildlife, or marine life?
   b. Will the action substantially alter the breeding, nesting, or feeding grounds for birds?
   c. Will the action substantially alter the patterns of behavior of fish, mammals, amphibians, reptiles, and insects?
   d. Will the action significantly affect, beneficially or adversely, other forms of life or ecosystems of which they are a part?
   e. Will the action cause changes in biological productivity, including fish and wildlife habitat and population losses, impacts on rare and endangered species, and changes in species diversity?

3. Energy supply and natural resource effects
   a. Will the action require the use of nonrenewable energy sources in apparently excessive or disproportionate amounts?
   b. Will the action affect electric energy development, generation, transmission, and use?
   c. Will the action affect petroleum development, extraction, refining, transport, and use?
   d. Will the action affect natural gas development, production, transmission, and use?
   e. Will the action affect coal and minerals development, mining, conversion, processing, transport, and use?
   f. Will the action affect renewable resource development, production, management, harvest, transport, and use?
   g. Will the action affect energy and natural resources conservation?
   h. What are the patterns of allocation, utilization, and demand for energy consumption as the result of the action?

4. Natural hazards and geologic effects
   a. Will the action significantly affect soil quality?
   b. Will the action increase (or decrease) the stability or instability of the soils and/or geology of the site?
   c. Are the geologic or soils conditions of the site hazardous to building construction and human occupancy?
   d. Will the action increase the erosion or runoff potential of the site?
   e. Will the action increase the potential fire hazards of the site?
   f. Is the site subject to unusual terrain features such as steep slopes, abutting rock formations, or other conditions affecting construction, drainage, or livability?
   g. Are there unusual risks from natural hazards such as geologic fault, flash floods, volcanic activity, mudslides, or from the presence of ponds or other hazardous terrain features?

5. Land use and land management effects
   a. Recreation
      (1) Will the action have a significant effect on public parks or other areas of recognized scenic or recreational value?
   b. Historic, architectural, and archaeological preservation
      (1) Will the action have a significant effect on areas of recognized archaeological value or properties listed on, or being considered for nomination to, the National Register of Historical Places?
   c. Aesthetics
      (1) Will the action affect areas of unique interest or beauty?
      (2) Will the action alter the aesthetic qualities of the area?
   d. Socioeconomics
      (1) Will the action divide or disrupt existing land uses?
      (2) Will the action alter the economic base of the area?
      (3) Will the action increase traffic flow and congestion?
      (4) Will the action affect population density and congestion?
      (5) Will the action affect neighborhood character and cohesion?

(6) Will the action create employment opportunities?

(7) Will the action cause displacement and relocation of homes, families, and businesses?

(8) Will the action present new demands and requirements for public services?

(9) Will the action affect the quality of life of the residents of the area?

(10) Will the action have a significant effect on revenues and costs to local governmental agencies?

(11) Will the action induce population, commercial, industrial, or general economic growth of the area?

Factors for Consideration in Assessing Airport Impacts

Airport projects have potential impacts in five major areas, namely: noise, air quality, water quality, social impacts, and induced socioeconomic impacts.

According to Reference 54, noise impacts must be examined when project actions involve airport location, runway location, runway extension, first-time operation of jet aircraft, or runway strengthening to permit operation by a larger class of jet aircraft. The level of detail necessary for noise impact assessment will vary with the situation. However, consideration should be given, as appropriate, to the unique aspects of the land use situations such as the needs and desires of the community served by the airport, local life styles, construction and insulation characteristics specific to the area, and adjacent land use zoning. One very important purpose of the noise impact assessment will be to provide information to assure that appropriate restrictive action, including the adoption of zoning laws, has been or will be taken. To a reasonable extent, this action should restrict the use of land adjacent to or in the immediate vicinity of the airport to activities and purposes compatible with normal aircraft operations, including landing and takeoff of aircraft.

Air pollution should be examined by estimating the effect of the proposed action on pollutants associated with existing and future forecast operations. The effect on air pollutant concentrations, as well as on total amounts of pollutants, should be estimated and evaluated for consistency with state implementation plans for air quality under the Clean Air Act or other applicable standards. Air pollution effects of induced surface traffic resulting from increased air traffic should be estimated and considered when an action involves an airport location, a new runway, or a major runway extension. In addition, methods should be proposed for controlling and minimizing air pollution resulting from construction of the project.

Water-quality impacts are caused by surface runoff from areas of extensive grading and pavement due to new runways and runway extensions being constructed. In addition, airport projects generate requirements for water and wastewater disposal.

Social and community impacts include possible displacement of people and businesses and/or disruption of established communities. When these types of impacts occur, it is necessary to: (1) estimate the number and family characteristics of households to be displaced; (2) identify the effects of surface-traffic disruption, including effects on access to community facilities, recreation areas, and places of residence and business; (3) identify the impact on the neighborhood and housing to which relocation is likely to take place; and (4) describe the businesses to be displaced and the general effects of business dislocation on the economy of the community.

Induced socioeconomic impacts due to the location of an airport, a runway, or a major runway extension are typically secondary in nature. Some of these impacts include shifts in patterns of population movement and growth, public service demands, and changes in business and economic activity.

In addition, according to Section 4(f) of the Department of Transportation Act, it is necessary to avoid the taking or detrimental use, as from flyovers, of such land as public parks, recreation areas, wildlife and waterfowl refuges, historic sites, and areas of natural scenic beauty of local, state, or national significance.

Factors for Consideration in Assessing Highway Impacts

Highway construction projects have impacts in a number of areas, the most noteworthy of which are aesthetics, air quality, circulation and traffic patterns, noise, socioeconomics, water quality, and wildlife. Highways may stimulate or induce other actions (secondary impacts), such as more rapid land development or changed patterns of social and economic activities. Impacts associated with secondary actions, through their impacts on existing community facilities and activities, through inducing new facilities and activities, or through changes in natural conditions, may often be even more substantial than the primary impacts associated with construction of the highway. For example, the effect on population and area growth associated with the construction of new highways may be among the more significant impacts.

Of general concern relative to aesthetics are such impacts as: (1) blocking viewlines along visual corridors (such as valleys, stream courses, and streets); (2) blocking viewlines to landmarks in the community from residential areas, recreation areas, and commercial operations that benefit from view; (3) elevated or above-grade highway out of scale with adjacent urban development; (4) visual distraction and displeasing glare visible in recreational and residential areas; and (5) unattractive contrast between existing vegetation and revegetated or landscaped areas, between natural landforms and engineering features of the highway, and between urban or existing development patterns and the highway features.

Air quality impacts include: (1) dust and/or particulate matter on vegetation and structures surrounding the construction site or along hauling roads; (2) tire and exhaust particles coating roadside vegetation and structures; (3) increase in severity of existing smog conditions due to an increase in automobiles traveling through the area; and (4) generation of vehicle fumes and odors (such as from exhaust emissions, or tire and brake rubber).

Circulation impacts include: (1) blocking or impairing access along existing street patterns crossed by the highway, such as access to public and private services of residents and patrons within the service area, reinforcing or creating physical barriers between social groups, congesting through-street traffic by diverting traffic from dead-end or rerouted streets, and disrupting public transit routes; (2) dividing single land uses or resource areas such as agricultural operations, recreation areas, wildlife ranges or habitats; (3) increasing truck and construction equipment traffic on public roads during construction; (4) providing new or improved access to previously inaccessible or relatively inaccessible public and private lands; (5) providing or improving access to relatively undeveloped areas outside urban centers, thus inducing commercial and industrial operations to locate outside urban centers; and (6) increasing traffic traveling through the area and thus causing an increased demand for travel related services.

Noise impacts generally involve the area within sound of the traffic such as: (1) disturbance of surrounding passive recreational activities requiring quiet and serene conditions for their enjoyment; (2) disturbance of educational, health care, and cultural activities or institutions particularly sensitive to noise, such as schools, churches, hospitals, sanitariums, auditoriums, and theaters; (3) disturbance to operation or patronage of commercial activities requiring or benefiting from quiet surroundings; and (4) disturbance to surrounding residential development.

Socioeconomic impacts include: (1) removal of residential, commercial, and industrial land uses and displacement of both residents and jobs; (2) removal of structures or sites of scenic, architectural, archaeological, or historic significance; (3) loss of sites having unique potential or suitability for commercial or industrial activities; (4) loss of taxable private land revenues; (5) relocation costs to displaced residents greater than compensation paid; and (6) severance of interpersonal ties of displaced residents to former neighborhood/community (family ties, ethnic bonds, or neighborhood friendships).

Water quality impacts involve one or more of the following: (1) turbidity and silting of adjacent streams and reservoirs caused, for the most part, by the erosion of the raw soils exposed during construction and maintenance operations (the primary impact of these effects generally involve increased operating costs or shortened life of affected reservoirs and channels; damage or elimination of fish and other aquatic life; and possible damage to buildings, roads, and bridge foundations); (2) watershed modification caused by the impingement of the road system and its construction on estuarine areas, marshes, wooded swamps, and streams—in particular, in estuarine areas disturbance of natural flows can affect ecological determinants such as sedimentation patterns, mixing of fresh and salt waters, nutrient flows, shellfish beds, fish and wildlife, and local vegetative patterns; (3) highway runoff contamination caused by runoff containing oil, fuel, tar, pesticides, fertilizer, deicing salts, animal and human wastes, and the products of combustion which can affect water quality, wildlife, and roadside vegetation; (4) sanitary wastes from temporary and permanent waste disposal facilities (Note: Waste disposal is accomplished through portable toilets during construction and permanent rest areas after construction); in either case, raw or inadequately treated discharges can have an impact on local water systems; and (5) contamination of surface and ground water supplies and recharge areas by polluted fill material, where the use of polluted fill material can affect the concentrations of biological, physical, chemical, and radiological contaminants in water supplies.

Wildlife impacts would generally include: (1) loss or degradation of unique or highly productive wildlife, fish, or shellfish habitats; (2) division of wildlife ranges and migratory patterns; (3) displacement of wildlife to other ranges; (4) impairing or blocking migration and/or movement of aquatic biota; and (5) visual disturbance of wildlife on adjoining lands.

All the aforementioned factors would generally be considered in the environmental impact assessment of highway projects. However, according to Reference 55, a state highway department request for location and design approval shall be accompanied by reports and other documents showing that the development of the project has taken many factors into consideration: the need for fast, safe, and efficient transportation; highway costs; traffic benefits; and public services (including provisions of national defense). Further, these materials must discuss the anticipated economic, social, and environmental effects of the proposal and alternatives under consideration, to the extent applicable, on the following:

1. "Regional and community growth" including general plans and proposed land use, total transportation requirements, and status of the planning process

2. "Conservation and preservation" including soil erosion and sedimentation, the general ecology of the area as well as manufactured and natural resources such as park and recreational facilities, wildlife and waterfowl areas, historic and natural landmarks

3. "Public facilities and services" including religious, health and educational facilities, public utilities, fire protection, and other emergency services

4. "Community cohesion" including residential and neighborhood character and stability, highway impacts on minority and other specific groups and interests, and effects on local tax base and property values

5. "Displacement of people, businesses, and farms" including relocation assistance, availability of adequate replacement housing, and economic activity (employment gains and losses, etc.)

6. "Air, noise and water pollution" including consistency with approved air quality implementation plans, FHWA noise level standards, and any relevant federal or state water quality standards

7. "Aesthetic and other values" including visual quality, such as "view of the road" and "view from the road," and the joint development and multiple use of space.

Factors for Consideration in Assessing Power Project Impacts

With regard to major power projects and the construction of natural gas pipeline facilities, the variety of potential environmental impacts (Refs. 22, 23) which could occur depends upon the phasing in the project's life cycle. Specifically, the impacts may be different during the construction, operation (including maintenance, breakdown, and malfunction), and termination, or abandonment, phases of the project.

During construction, the types of impacts which can be expected to occur, or at least considered, are:

1. Land features and uses
   a. Impacts on present or future land use, including commercial use, mineral resources, recreational areas, public health and safety, and the aesthetic value of the land and its features
   b. Temporary restrictions on land use due to construction
   c. Effects of construction related activities upon local traffic patterns, including roads, highways, ship channels, and aviation patterns
2. Species and ecosystems
   a. Impact of construction on the terrestrial and aquatic species and habitats in the area, including clearing, excavation, and impoundment

  b. Possible major alteration to the ecosystem
  c. Potential loss of an endangered species
 3. Socioeconomics
  a. Effects on labor, housing, local industry, and public services during construction
  b. Relocation and displacement of families and businesses in the area
  c. Effects on local economic base from new business, housing development, and payrolls
  d. Impacts on the human elements, including the need for increased public services such as schools, health facilities, police and fire protection, housing, waste disposal, markets, transportation, communication, energy supplies, and recreational facilities
  e. Impacts on present and future recreational use and potential of the local area or region
 4. Air quality
  a. Qualitative and quantitative effects of air pollution emissions generated during construction
 5. Water quality
  a. Effects on local water quality in the area, including sedimentation, erosion, and water runoff
 6. Noise quality
  a. Levels and types of construction noise created
 7. Waste disposal
  a. Impacts of disposal of all waste material, such as spoils, vegetation, and construction materials

During the operation and maintenance phase, the types of impacts which should be addressed are:
 1. Land features and uses
  a. Restrictions on existing and potential land use in the vicinity of the proposed action, including mineral and water resources
  b. Effects of operation-related activities upon local traffic patterns including roads, highways, ship channels, and aviation patterns, and the possible need for new facilities
 2. Species and ecosystems
  a. Impact of operation upon terrestrial and aquatic species and habitats, including plant and animal species having economic or aesthetic value
  b. Impact on animal migrations, foods, and reproduction
  c. Ecosystem imbalances caused by the action
  d. Possible major alteration to an ecosystem or the loss of an endangered species
  e. Cumulative effects of this action in conjunction with other similar existing projects or proposed actions
 3. Socioeconomics
  a. Effects on local socioeconomic development in relation to labor, housing and population growth trends, relocation, local industry and industrial growth, and public services
  b. Economic benefits resulting from the services and products, energy, and other results of the action (such as tax benefits to local and state governments, growth in local tax base from new business and housing developments, and payrolls)
  c. Impacts on human elements, including any need for increased public services (such as schools, police and fire protection, housing, waste disposal, markets, transportation, communication, and recreational facilities)
  d. Extent to which maintenance of the area is dependent upon new sources of energy or the use of such vital resources as water
 4. Air quality
  a. Impact of air pollution emissions generated on the present air quality levels
 5. Water quality
  a. Impact on present water quality, including sedimentation due to reservoir operations, downstream water releases, power-peaking operations, location of outlet works, and sanitary, waste, and process effluents
 6. Noise quality

  a. Impact on present noise levels due to project-related noise
7. Solid wastes
  a. Impacts from accumulation of solid wastes and by-products that will be produced
8. Use of resources
  a. Impact of obtaining and using resources such as water, energy, and raw products for project operations
9. Maintenance
  a. Impact of maintenance programs such as subsequent clearing or treatment of rights-of-way
  b. Potential impact of major breakdowns and shutdowns of the facilities
10. Accidents and catastrophes
  a. Potential impacts from accidents and natural catastrophes which might occur.

Finally, in the termination and abandonment phase the major considerations would be given to the impact on land use and the aesthetics of the area as the result of such an action.

Prior to the issuance of a construction permit or an operating license for a nuclear power plant, each applicant is required (Ref. 3) to submit a report on the potential environmental impacts of the proposed plant and associated facilities. Of particular concern are the environmental effects of plant operation on surface water bodies, ground water, air, and land. Typical impact areas would include the following:

1. Effects of operation of heat dissipation system
  a. Effect that the heated effluent will have on the temperature of the receiving body of water
  b. Effects of released heat on marine and freshwater life
  c. Potential hazards of the cooling-water intake and discharge structures to fish species and food base organisms
  d. Effects of passage through the condensor on zooplankton, phytoplankton, mero-plankton, and small nektonic forms such as immature fish and the resultant implications for the important species
  e. Potential biological effects of modifying the natural circulation of the water body, especially where water is withdrawn from one region or zone and discharged into another
  f. Plant-induced changes in the temperature of the discharged water subsequent to environmental stabilization, such as would occur as the result of reactor shutdown
  g. Effects of heat dissipation facilities such as cooling towers, lakes, spray ponds, or diffusers on the local environment and on agriculture, housing, highway safety, airports, or other installations with respect to noise and meteorological phenomena, including fog or icing, and cooling-tower blowdown and drift.
  h. Impacts on ground water such as the alteration of water table levels, recharge rates, and soil permeability
2. Radiological impact on biota other than man
  a. Extent and nature of possible effects of radiation exposure pathways on local vegetation and local and migratory animals
  b. Effects on land areas and on vegetation in the environs of all waters that receive any liquid radioactive effluent
  c. Estimation of maximum radionuclide concentrations that may be on the ground surface and present in important local vegetation and local migratory animals and the dose rates that may result from those concentrations
3. Radiological impact on man
  a. Identification of possible pathways for radiation exposure of man such as drinking (milk and water), swimming, fishing, eating of fish, game, invertebrates, and plants
  b. Expected annual average concentrations of radioactive nuclides in receiving water at locations where water is consumed or otherwise used by human beings or where it is inhabited by biota of significance to human food chains
  c. Expected radionuclide concentrations in aquatic and terrestrial organisms significant to human food chains.

    d. Estimated total body and significant organ doses to individuals at the point of maximum ground level concentrations off-site from radioactive gases and particulates

    e. Deposition of radioactive halogens and particulates on food crops and pasture grass

    f. Estimated total external dose and total population external dose received by individuals outside the facility from direct radiation (namely, gamma radiation emitted by turbines and radioactive waste vessels)

4. Effects of chemical and biocide discharges

    a. Comparison of specific concentrations of chemical and biocide discharges at the points of discharge with natural ambient concentrations without the discharge

    b. Projected effects of the effluents for both acute and chronic exposure of the biota, including any long-term buildup in sediments and in the biota

    c. Effects on terrestrial and aquatic environments from oil or chemical wastes which contaminate surface and/or ground water

    d. Effects of chemicals in cooling-tower blowdown and drift on the environment

    e. Anticipated chemical or biocide contamination of domestic water supplies (from surface water bodies or ground water)

5. Effects of operation and maintenance of the transmission system

    a. Effects on plant life, wildlife habitat, land resources, and scenic values

    b. Changes in land and water use at the plant site

    c. Interaction effects of the plant with other existing or projected neighboring plants

    d. Effect of ground water withdrawal on ground water resources in the vicinity of the plant

    e. Effects of solid and liquid waste disposal.

Other types of power-related projects include electric transmission lines, terminals, and substations. For these types of facilities, the environmental factors and impact areas to be considered generally consist of the following (Ref. 38):

    1. Impact on soils—clearing for rights-of-way, access roads; substation and tower sites; excavation for pole holes; slope alterations

    2. Impact on vegetation—the traversal and removal of vegetation from the rights-of-way; effects and limitations on use of area chemical spraying; possibility of fire hazards during construction; effects on sensitive, rare, or endangered species; disposal of construction material and land clearance debris

    3. Impact on animals—effects on the local and migratory wildlife, including birds and fish; effects on feeding, grazing, mating, nesting, migration, and habitation due to construction activity and noise; altered food supply, increased access to hunters, and loss of cover protection and roosting sites; effects on rare or endangered species

    4. Impact on aesthetics—visual impact as viewed from highways, parks, developed communities, etc.

    5. Impact on water resources—effects of construction-caused erosion and drainage on pollution of water resources; effects of traversing lakes, rivers, streams, etc.

    6. Impact on formally classified areas which the proposed facility will traverse, abut, or affect environmentally—wilderness areas, primitive areas, wild and scenic rivers, national recreation areas, natural areas, scenic areas, historical and archaeological areas, geological areas, national trails, national parks and monuments, and wildlife refuges

    7. Impact on aviation—effects on airports and paths of low-flying aircraft

    8. Impact on human activity—effects of closing or opening of affected areas to farming, recreation, hunting, or other activities; resulting multiple-use opportunities of proposed rights-of-way for various activities involving people; potential hazards to human safety

    9. Impact on economy of the area—economic impacts on the affected areas and industries which the proposed facilities will serve; economic impacts during construction, including increased local employment opportunities, payments for rights-of-way, and income from sale of goods and services to contractors; continuing economic impacts, including local employment opportunities for operation and maintenance, increased tax revenues, and increased availability of electric power; anticipated impact on development of local industries

    10. Impact from noise and electromagnetic radiation—effects of noise pollution from transformer hum, operation of circuit breakers, and corona (wet and dry weather); ozone

and electrical interference with radio and television reception or communication circuits; induced voltages in metal fences, gates, underground and surface piping, etc.

Factors for Consideration in Assessing Wastewater Treatment Facilities Impacts

Environmental considerations relative to water quality management plans and municipal wastewater treatment facilities cover a broad range of potential impact areas. Typically, however, these types of impacts can be categorized as those due to construction, those which are long-term, and those which are secondary in nature.

For example, potential construction impacts would include:

1. Erosion and earthwork impacts such as significant alterations to landforms, streams, or natural drainage
2. Clearing impacts such as removal of groundcover, vegetation and trees, and the use of herbicides, defoliants, blasting, cutting, bulldozing, or burning
3. Methods used for and resulting impacts of the disposal of soil
4. Impacts resulting from site acquisition such as displacement of people, condemnation of property, and effects on neighboring property values
5. Water quality impacts such as the effects of dredging, tunneling, or trenching on turbidity and suspended solids, and the cleanup and disposal of fuels, grease, oil, equipment washwater, and excess herbicides and insecticides
6. Air quality impacts such as dust and smoke from burning
7. Noise and vibration impacts such as those due to noisy construction methods such as pile driving, jack hammers or blasting, and effects of construction noise on residences, businesses, and wildlife
8. Effects of blasting an aquatic and animal life
9. Disruption of vehicular and pedestrian traffic in the area
10. Construction hazards to the public safety.

Long-term impacts are those which generally occur after the construction phase and would include such factors as:

1. Land use impacts such as the elimination of beneficial uses, restrictions on future adjacent land uses, and changes in land form of the area
2. Changes in the aesthetics of the area such as alteration in the natural or present character and obscuration of natural views
3. Air quality impacts such as the prevailing wind patterns and the effects of odors and emissions on parks, residences, businesses, highways, or other public access areas
4. Water quality impacts such as the effect of percolation of treated wastewater on ground water supplies, subsurface drainage into surface waters, and effects on stream biota and aquatic habitats
5. Solid waste impacts due to the methods used for disposal of grit, ash, and sludges
6. Impacts on special areas such as disturbance or encroachment on historic sites, recreational areas, or natural preserves
7. Noise levels from the project in terms of decibels, time of noise, duration and types of noise, and vibration
8. Effects of the use of insecticides as they affect insect proliferation, and land and water quality
9. Alteration of wildlife, birdlife, and aquatic life habitats
10. Effects on land-based ecosystems near the facility site such as stream bank cover, and vegetal and wooded growth on rights-of-way
11. Effects of lighting and light levels on nearby residents
12. Potential flooding hazards due to sewer overflows or the location of the facility in a floodplain area
13. Potential adverse health effects or the creation of public health problems
14. Social disruption due to relocation of people, disruption of employment opportunities, or impairment of public services.

Secondary impacts of a treatment works or water quality management plan would involve induced changes, either absolute changes or increases in the rate of change, in industrial, commercial, agricultural, or residential land use concentrations or distributions. Assessing changes of this type would include the consideration of a variety of factors such as: the vacant land subject to increased development pressure as a result of the treatment works; the increases in population which may be induced; the compatibility of the project with the type of growth desired by the residents of the area; the faster rate of change of population; changes in population density; the potential for overloading sewage treatment works; the extent to which landowners may benefit from the areas subject to increased development; the nature of land use regulations in the affected area and their potential effects on development; and deleterious changes in the availability or demand for energy.

Factors for Consideration in Assessing the Impacts of Water-Related Projects

Water-related projects include such actions as channelization, dredging, irrigation, levees, dikes and bank stabilization projects, water storage, and small boat basins. Each of these actions generally has its own peculiar set of impacts which needs to be considered.

The general term "channelization" is often used to refer to a broad spectrum of manmade alterations in and around stream and river beds. These alterations may include such activities as channel relocation, clearing and snagging, diking, and dredging. Depending on the responsible agency, these alteration activities may take on varying definitions. For instance, to some agencies "clearing and snagging" means simply removal of trees and debris from the channel, including trees on the immediate bank which are weak, dead, or undercut and probably will fall into the channel in the near future. To other agencies "clearing and snagging" includes a much broader spectrum of activities, such as removal of gravel bars, pools, and riffles from the stream channel as well as deforestation of the immediate banks for some distance. The common end product of such channelization activities is a straight flume-like channel denuded of vegetation with enlargement in width and/or depth to approximate a trapezoidal cross section.

Alterations to the natural stream ecosystem that are severe enough to disturb its functioning or structure will create changes within the system. Most stream perturbations, depending on the type and extent, affect the kinds and numbers of species, and the relative sizes of populations. Channelization and associated alteration such as dredging can affect the overall productivity of the watershed by disturbing and removing solid substrates and by creating eroding sediments and unstable river beds. This can collectively lead to a decrease in light penetration into the water. Channelization may cause a shift in relative sizes of populations of aquatic communities with the more tolerant becoming very common. Continued perturbations over an extended period may bring about the elimination of those species with narrow ranges of tolerance and perhaps an increase in species that thrive in the perturbed conditions. Clearing of bankside vegetation during channelization operations can cause an increase in stream temperatures which brings about a change in the biota of the stream. Furthermore, downstream habitats and aquatic life are often adversely affected by upstream channelization projects because of increased sediment deposits, increased assimilation of nutrients, increased potential of flash flooding in these areas, increased flow velocities, and decreased retention times within channelized areas.

"Irrigation" is the practice of applying water to land by controlled artificial means to promote growth of selected crops in areas in which the natural hydrologic cycle may preclude such growth. For this type of water project, one is generally concerned with: (1) the amount and locations of runoff to be expected; (2) the water quality of irrigation return flows with regard to nutrient loading, dissolved oxygen, temperature, pesticides, total solids, salinity, sediment loading, and turbidity; (3) the impacts of irrigation flows on the quality of the receiving water; and (4) low flow problems associated with irrigation diversions.

Levees, dikes, and bank stabilization projects are closely related to channelization projects and have similar types of impacts. Levees and dikes are generally single-purpose flood protection structures built adjacent to the banks of streams susceptible to flooding. Bank stabilization projects include such features as reduction of erosion from riverbanks and removal of natural growth and debris from channels to allow passage of greater flows. Generally, impacts from these activities would include the effects on water quality due to construction and operation, the effects on wildlife due to reduction of riparian habitat, increased turbidity and velocity due to channelization of flood flows, and effects on the natural drainage system.

Water storage projects can be considered as either being single-purpose (e.g., water supply only) or multiple-purpose (e.g., projects designed to have more than one function (such as storage for power, navigation, flood control, recreation, fish and wildlife enhancement, and water supply). Regardless of the function of the water storage project, there is a rather broad range of potential impacts including: alterations in habitats and species diversity of the area; alterations in visual appearance and overall aesthetics of the area; removal of land from present use; changes in unique natural or artificial resources; alteration in adjacent property values; increase in availability of local water resources; changes in downstream recreational activities; potential induced growth in residential, commercial, and industrial land uses in the surrounding area; and changes in water quality due to impoundment such as dissolved oxygen depletion in bottom layers, seasonal temperature stratifications, effects on sediment transport, and potential eutrophication.

Finally, "small boat basins" are facilities which provide boat launching, storage, supplies, and services for small pleasure craft. The potential impacts of this type of project would include: impacts on the facilities used for the collection, treatment, and disposal of domestic sewage and other liquid wastes generated by users of the small boat basin; possibility of spillage due to the handling movement of petroleum products and other hazardous materials; potential for water quality deterioration due to venting of bilge water; or impacts on existing patterns of circulation and water movement due to construction of breakwaters and impedance of flushing in the basin.

MEASUREMENT OF ENVIRONMENTAL IMPACTS

A fundamental step in the conduct of environmental impact analyses is to identify quantitative measures for evaluating the extent of the impacts and to develop formulas for computing these measures. These formulations are based upon the type of data that is generally available, or can easily be collected, for various types of land uses. This data includes such items as: people per dwelling unit (DU), number of dwelling units, water consumption per capita for residential projects, gross leasable area (GLA), gross floor area (GFA), number of employees per 1000 square feet GLA, electricity consumption per 1000 square feet GLA, etc. for commercial and industrial projects. In the following sections are presented illustrative, typical quantitative measures and their corresponding formulations for impacts in the natural/physical environment, social and economic areas. *These measures are intended to be illustrative and are not to be interpreted as being the only ones which can be or should be used.* A further and more detailed discussion of impact measures is presented in subsequent chapters of this handbook.

Measures of Natural/Physical Environment Impacts

Air Pollution The two basic sources of air pollution are: (1) mobile emissions and (2) stationary emissions. Mobile emissions are generated as the result of vehicle travel in the project area and are a function of such factors as vehicle trips per day, average vehicle speed, and vehicle trip lengths. Stationary emissions are generated as the result of factors such as domestic and commercial heating, by-products of industrial processes that are vented to the air, solid waste incineration, powerplant emissions, and smoke and odors from industrial activities.

A measure of the air pollution impact is given by the total emissions per day. This measure could be formulated as follows:

CASE: Mobile emissions (expressed in grams per day)

For residential projects:

$$\sum_{\text{all pollutants}} \left(\begin{array}{c} \text{Pollutant} \\ \text{emission} \\ \text{factor in} \\ \text{g/mi} \end{array} \right) \left(\begin{array}{c} \text{Usage rate} \\ \text{in vehicle} \\ \text{miles} \\ \text{traveled} \\ \text{per day} \end{array} \right) \left(\begin{array}{c} \text{Number of} \\ \text{vehicles} \\ \text{per} \\ \text{dwelling} \\ \text{unit} \end{array} \right) \left(\begin{array}{c} \text{Number of} \\ \text{dwelling} \\ \text{units} \end{array} \right)$$

For commercial and industrial projects:

$$\sum_{\text{all pollutants}} \left(\begin{array}{c} \text{Pollutant} \\ \text{emission} \\ \text{factor in} \\ \text{g/mi} \end{array} \right) \left(\begin{array}{c} \text{Gross} \\ \text{leasable} \\ \text{area (GLA)} \\ \text{in 1000 ft}^2 \end{array} \right) \left(\begin{array}{c} \text{Average} \\ \text{daily} \\ \text{trips per} \\ \text{1000 ft}^2 \\ \text{GLA} \end{array} \right) \left(\begin{array}{c} \text{Usage rate} \\ \text{in vehicle} \\ \text{miles} \\ \text{traveled} \\ \text{per day} \end{array} \right)$$

CASE: Stationary emissions (expressed in pounds per time period)
For residential projects:

$$\sum_{\text{all pollutants}} \left(\begin{array}{c} \text{Pollutant} \\ \text{emission} \\ \text{factor in} \\ \text{lb/1000 ft}^3 \end{array} \right) \left(\begin{array}{c} \text{Natural gas} \\ \text{consumption} \\ \text{in 1000 ft}^3 \text{ per} \\ \text{dwelling unit} \end{array} \right) \left(\begin{array}{c} \text{Number of} \\ \text{dwelling} \\ \text{units} \end{array} \right)$$

For commercial and industrial projects:

$$\sum_{\text{all pollutants}} \left(\begin{array}{c} \text{Pollutant} \\ \text{emission} \\ \text{factor in} \\ \text{lb/1000 ft}^3 \end{array} \right) \left(\begin{array}{c} \text{Gross floor} \\ \text{area (GFA)} \\ \text{in 1000 ft}^2 \end{array} \right) \left(\begin{array}{c} \text{Natural gas} \\ \text{consumption} \\ \text{in 1000} \\ \text{ft}^3 \text{ per 1000} \\ \text{ft}^2 \text{ GLA} \end{array} \right)$$

(NOTE: These formulas can be modified accordingly in the case of the use of heating oil.)

Upon multiplying by the appropriate conversion factor, these quantities could be converted to the units of tons per day. Then they could be added to obtain the total emissions in tons per day.

Water Pollution   Land developments potentially affect water quality in many ways. For example, the amounts and nature of wastes may overwhelm local sewage treatment facilities. Where septic tanks are used, the wastes generated may exceed the capability of the soil to remove or degrade wastes, and may thus affect underground and surface waters. Changes in land contours, vegetation, and permeable land cover during and after construction may increase the amount and content of storm runoff. Thus, the assessment of the impact on water quality must consider all such possibilities.

A measure of the water pollution impact is given by the percent change in water quality of the area as the result of the project, which could be formulated by

$$\frac{\left(\begin{array}{c} \text{Water quality} \\ \text{in area before} \\ \text{project} \end{array} \right) - \left(\begin{array}{c} \text{Water quality} \\ \text{in area after} \\ \text{project} \end{array} \right)}{\left(\begin{array}{c} \text{Water quality} \\ \text{in area before} \\ \text{project} \end{array} \right)} \times 100$$

This measure could be computed for each of the basic water quality components, namely: suspended and dissolved solids, dissolved oxygen, and toxic materials.

Noise Pollution   Land developments may affect noise levels in the short term by construction, and in the long term by changes in vehicular and pedestrian traffic, industrial processes, and other activities. Since noise increases will be most pronounced in the immediate vicinity of the project, impact measurement efforts should be concentrated there. Usually, the areawide effects will be diffuse. However, in some situations the effects may be far-reaching, such as in the case of airport development where the impacts extend to those areas within the take-off and landing patterns of aircraft, or in the case of the development of a new shopping center which draws so much traffic that noise levels rise along all corridors leading to it.

A measure of the noise pollution impact is given by the percent change in the number of people bothered by excessive noise and vibration as the result of the project, which could be formulated by

$$\frac{\left(\begin{array}{c} \text{Number of people} \\ \text{bothered by excessive} \\ \text{noise and vibration} \\ \text{before project} \end{array} \right) - \left(\begin{array}{c} \text{Number of people} \\ \text{bothered by excessive} \\ \text{noise and vibration} \\ \text{after project} \end{array} \right)}{\left(\begin{array}{c} \text{Number of people} \\ \text{bothered by excessive} \\ \text{noise and vibration} \\ \text{before project} \end{array} \right)} \times 100$$

The term "excessive noise" could be based on a comparison of noise levels observed relative to acceptable noise level standards.

Solid Waste The generation of solid waste implies the requirement for solid waste disposal. Thus, the use of land sites for dumping occurs with the resulting air pollution emissions from burning. Depending on the type of project being considered, the solid waste generated can vary significantly. For example, residential developments generate such items as newspapers, cans, bottles, and other rubbish, whereas manufacturing facilities generate such items as scrap materials, wrapping and shipping materials, and various manufacturing by-products.

A measure of the solid waste impact is given by the total solid waste generated, which could be formulated by

$$\left(\begin{matrix}\text{Solid waste}\\ \text{in pounds}\\ \text{per capita}\end{matrix}\right)\left(\begin{matrix}\text{Number of}\\ \text{people per}\\ \text{dwelling unit}\end{matrix}\right)\left(\begin{matrix}\text{Number of}\\ \text{dwelling}\\ \text{units}\end{matrix}\right)$$

for residential projects, and by

$$\left(\begin{matrix}\text{Solid waste}\\ \text{in pounds per}\\ \text{employee}\end{matrix}\right)\left(\begin{matrix}\text{Number of}\\ \text{employees}\\ \text{per business}\end{matrix}\right)\left(\begin{matrix}\text{Number of}\\ \text{businesses}\end{matrix}\right)$$

or

$$\left(\begin{matrix}\text{Solid waste}\\ \text{in lb per}\\ \text{employee}\end{matrix}\right)\left(\begin{matrix}\text{Number of}\\ \text{employees per}\\ \text{1000 ft}^2\text{ GLA}\end{matrix}\right)\left(\begin{matrix}\text{Total GLA}\\ \text{in 1000 ft}^2\end{matrix}\right)$$

for commercial and industrial projects.

Sewage As in the case of solid waste, the generation of sanitary sewage implies the requirement for special facilities—specifically, those for wastewater treatment. The amount of sewage generated varies considerably depending upon the type of project. For example, chemical processing, animal rendering, and food products plants are usually sources of relatively larger quantities of wastewater than are residential projects.

A measure of the sewage impact is given by the total sewage generated, which could be formulated by

$$\left(\begin{matrix}\text{Gal of}\\ \text{sewage per}\\ \text{capita}\end{matrix}\right)\left(\begin{matrix}\text{Number of}\\ \text{people per}\\ \text{dwelling unit}\end{matrix}\right)\left(\begin{matrix}\text{Number of}\\ \text{dwelling}\\ \text{units}\end{matrix}\right)$$

for residential projects, and by

$$\left(\begin{matrix}\text{Gal of}\\ \text{sewage per}\\ \text{employee}\end{matrix}\right)\left(\begin{matrix}\text{Number of}\\ \text{employees}\\ \text{per business}\end{matrix}\right)\left(\begin{matrix}\text{Number of}\\ \text{businesses}\end{matrix}\right)$$

or

$$\left(\begin{matrix}\text{Gal of}\\ \text{sewage per}\\ \text{employee}\end{matrix}\right)\left(\begin{matrix}\text{Number of}\\ \text{employees per}\\ \text{1000 ft}^2\text{ GLA}\end{matrix}\right)\left(\begin{matrix}\text{Total GLA}\\ \text{in 1000 ft}^2\end{matrix}\right)$$

for commercial and industrial projects.

Another possible measure would be the peak rate of sewage discharge, in which case the first term in each of these formulas would be replaced by the corresponding peak (or maximum) discharge rate.

Water Demand A very important concern relative to new development is the demand for water and its resulting impact on available supplies. If development affects the water quality of the area by polluting surface and ground water sources, then an increasing demand for water would imply that supplies be drawn from alternate or perhaps even inferior sources. Added demand from development may also occasionally cause shortages, necessitating rationing of water for bathing and watering lawns, or even for drinking.

Projecting possible water shortages can be based on known usage rates for various business and household characteristics applied to development plans. Projected usage by the development would be compared to available supplies, expected rainfall, and general use trends. This suggests that a measure of the impact on water demand would be given by the number of gallons of water consumed over a specified time period (day, month, or year), which could be formulated by

$$\left(\begin{matrix}\text{Gal of}\\\text{water per}\\\text{capita}\end{matrix}\right)\left(\begin{matrix}\text{Number of}\\\text{people per}\\\text{dwelling unit}\end{matrix}\right)\left(\begin{matrix}\text{Number of}\\\text{dwelling}\\\text{units}\end{matrix}\right)$$

for residential projects, and by

$$\left(\begin{matrix}\text{Gal of}\\\text{water per}\\\text{employee}\end{matrix}\right)\left(\begin{matrix}\text{Number of}\\\text{employees}\\\text{per business}\end{matrix}\right)\left(\begin{matrix}\text{Number of}\\\text{businesses}\end{matrix}\right)$$

or

$$\left(\begin{matrix}\text{Gal of}\\\text{water per}\\\text{employee}\end{matrix}\right)\left(\begin{matrix}\text{Number of}\\\text{employees per}\\1000\text{ ft}^2\text{ GLA}\end{matrix}\right)\left(\begin{matrix}\text{Total GLA}\\\text{in }1000\text{ ft}^2\end{matrix}\right)$$

for commercial and industrial projects.

Greenery and Open Space   The amount of greenery and open space in a community is often directly changed and affected by the development of an area. Greenery has significant economic, social, psychological, and aesthetic benefits. It can reduce visual pollution, save energy (via shade and wind screening), increase privacy, improve the climate in its immediate vicinity, and make life more pleasant. Open space affects aesthetics, recreation opportunities, the microclimate, and human perceptions of crowdedness.

Since many urban areas are trying to increase their amount of greenery and open space, or at least minimize losses, a measure of the impact of development on greenery and open space is given by the percent change in greenery and open space acreage, which could be formulated by

$$\frac{\left(\begin{matrix}\text{Greenery and}\\\text{open space}\\\text{acreage before}\\\text{project}\end{matrix}\right)-\left(\begin{matrix}\text{Greenery and}\\\text{open space}\\\text{acreage after}\\\text{project}\end{matrix}\right)}{\left(\begin{matrix}\text{Greenery and}\\\text{open space}\\\text{acreage before}\\\text{project}\end{matrix}\right)} \times 100$$

Vegetation and Wildlife   Development may physically destroy vegetation and wildlife by altering or destroying habitats. The effects may be both on the site itself as a result of construction and in the surroundings as a result of pollution from the development and secondary effects in the community. The localized destruction of species (common, rare, or endangered) may significantly affect the quality of life in and near a development and may contribute to the alteration of species diversity. The diversity of species is also thought to be an indicator of the stability of the local ecosystem.

The impact of development on wildlife and vegetation may be evaluated by site surveys before and after development to determine the diversity and abundance of major species at various times of the year. To inventory plant diversity, the populations within species, numbers of different species, and their interspersion and spread over space should be noted. Of particular interest would be any changes in the abundance of nuisance species of vegetation and wildlife.

The preceding discussion suggests that there are several measures which could be used to assess the impact on vegetation and wildlife, such as the number of endangered or threatened wildlife (vegetation) species, which can be formulated by[1]

---

[1]An alternative formulation could be based on the concept of "value" where the biologist establishes a numerical value scale for each type of species and then computes the number of each type threatened times its value and adds these products to obtain a measure of the "total" impact.

$$\sum_{\text{all types}} \begin{pmatrix} \text{Number of endangered} \\ \text{and threatened wildlife} \\ \text{(vegetation) species of} \\ \text{each type} \end{pmatrix}$$

and the percent of the local wildlife (vegetation) habitat that would be destroyed or adversely affected by the development, which can be formulated by

$$\frac{\begin{pmatrix} \text{Local wildlife} \\ \text{(vegetation)} \\ \text{habitat acreage} \\ \text{before project} \end{pmatrix} - \begin{pmatrix} \text{Local wildlife} \\ \text{(vegetation)} \\ \text{habitat acreage} \\ \text{after project} \end{pmatrix}}{\begin{pmatrix} \text{Local wildlife} \\ \text{(vegetation)} \\ \text{habitat acreage} \\ \text{before project} \end{pmatrix}} \times 100$$

Views A new development may either interfere with people's views of the scenery or remove obstructions. For example, a new tall building may block the view of green space from an existing building, or, on the other hand, razing buildings to create a plaza may open up previously blocked views. This suggests that a measure of the impact on views would be provided by the number of people whose views or sightlines are blocked, degraded, or improved. The sightlines blocked by the development and the nature of the old and new vistas can be determined geometrically from plans, maps and photos, and by site visits. Distinctions might be made between sightline changes for residential and nonresidential buildings, when it is assumed the occupants will place different values on them. Certain businesses and tourist attractions are dependent on their views, and harm to them should be specifically noted.

For private residences other than highrises, all occupants might be counted. For highrises and business establishments, the approximate number of the occupants or daily users whose views will be affected might be noted. People affected can be roughly estimated using the average number of occupants per type of housing unit, times the number of housing units, or by using census or other population sources.

In the case of commercial areas, one could use the technique of drawing diagrams to show the areas from which views are obscured, including public areas such as streets. The number of people affected could then be roughly estimated.

Historical, Archaeological, and Paleontological Sites Land developments may destroy, impair access to, or crowd landmarks of cultural significance such as those with historical, archaeological, or paleontological value. The importance of these landmarks is determined in terms of rarity, distance to closest similar example, interest to tourists and the public, and interest to scholars. In addition, such a landmark might also appear on the National Register of Historic Preservation.

Considerations of this type suggest the establishment of a value or importance scale for such sites based on the determination of the significance of each of these factors. For example, rarity may be expressed as the number of existing examples essentially equivalent to the threatened landmark. Distance to closest similar example may be expressed in miles or travel time to indicate the degree to which removing the landmark would curtail opportunity for enjoyment or learning. Importance or interest to the public as to whether or not a landmark is worth saving can be obtained by a survey of citizens. Tourist usage and enjoyment may be expressed in two ways, namely: (1) in annual attendance figures, and (2) in subjective ratings based on attention in domestic and foreign guidebooks, queries to travel agents, and surveys of tourists visiting the landmark. Interest to scholars, such as how critical a landmark is for research or teaching, may be determined by seeking opinions from historians, artists, scientists, and members of architectural and historical review boards.

The preceding discussion suggests that a measure of the impact on historical, archaeological, and paleontological sites is given by the total value of the sites disturbed, which can be formulated by

$$\sum_{\text{all sites}} \left[\begin{pmatrix} \text{Value of each} \\ \text{historical site} \\ \text{disturbed} \end{pmatrix} + \begin{pmatrix} \text{Value of each} \\ \text{archaeological} \\ \text{site disturbed} \end{pmatrix} + \begin{pmatrix} \text{Value of each} \\ \text{paleontological} \\ \text{site disturbed} \end{pmatrix} \right]$$

Measures of Social Impacts

Housing In order to realistically assess the impact on housing, it is necessary to recognize that housing objectives differ in many communities. Some that have undergone rapid recent growth are attempting to slow or stop new housing. Others, attempting to cash in on growth, are promoting primary or vacation residences. Still others are attempting to improve housing for low-to-moderate income families.

Because of this diversity, communities must devise their own sets of housing impact measures. These measures should consider the effects of new development on substandard housing, changes in the mix of housing, and the impact on housing needs of existing residents. A community could then convert these considerations into measures more closely attuned to their desired housing mix or desired direction of change. Preferably, proposed changes in housing would be related quantitatively to community housing mix needs and objectives. If this is too complex in some circumstances, an alternative approach is to make ad hoc judgments on whether the new mix is desirable.

Residential development obviously affects the housing stock in the community by providing new housing units. Commercial and industrial developments affect housing indirectly by their effect on jobs and the local economy, which in turn may alter housing demand, prices, rates of abandonment, upkeep, and crowding. It is also possible that, as the result of development of any kind, there may be a change in the existing community housing stock by destroying existing housing on the site.

The preceding discussion suggests that several plausible measures of housing impact could be defined and formulated. Typical measures would be: (1) the percent change in substandard housing units, which can be formulated by

$$\frac{\begin{pmatrix} \text{Number of} \\ \text{substandard} \\ \text{housing units} \\ \text{before project} \end{pmatrix} - \begin{pmatrix} \text{Number of} \\ \text{substandard} \\ \text{housing units} \\ \text{after project} \end{pmatrix}}{\begin{pmatrix} \text{Number of} \\ \text{substandard} \\ \text{housing units} \\ \text{before project} \end{pmatrix}} \times 100$$

(2) the percent change in the number of housing units available for low-to-moderate income families, which can be formulated by

$$\frac{\begin{pmatrix} \text{Number of available} \\ \text{low-to-moderate} \\ \text{income housing} \\ \text{units before project} \end{pmatrix} - \begin{pmatrix} \text{Number of available} \\ \text{low-to-moderate} \\ \text{income housing} \\ \text{units after project} \end{pmatrix}}{\begin{pmatrix} \text{Number of available} \\ \text{low-to-moderate} \\ \text{income housing} \\ \text{units before project} \end{pmatrix}} \times 100$$

(3) the percent change in the number of vacant housing units in the community, which can be formulated by

$$\frac{\begin{pmatrix}\text{Number of}\\\text{vacant housing}\\\text{units before}\\\text{project}\end{pmatrix} - \begin{pmatrix}\text{Number of}\\\text{vacant housing}\\\text{units after}\\\text{project}\end{pmatrix}}{\begin{pmatrix}\text{Number of}\\\text{vacant housing}\\\text{units before}\\\text{project}\end{pmatrix}} \times 100$$

Parks   Many communities have established park standards for local and neighborhood parks expressed in terms of acres per capita. Usually, these standards serve only as goals, and multiplying such a standard by existing population does not necessarily yield the current inventory of park acreage. If there is presently a deficiency in park acreage relative to the adopted community standards, then an increase in population as the result of new residential development can only increase this deficiency. Generally speaking, new commercial and industrial developments have little impact on park requirements except insofar as they induce new people to move to the community and thus become potential users of existing facilities. Consequently, a measure of the impact on parks is given by the additional number of park acres required, which can be formulated by

$$\begin{pmatrix}\text{Community}\\\text{standard in}\\\text{number of}\\\text{park acres}\\\text{per capita}\end{pmatrix}\begin{pmatrix}\text{Number of}\\\text{people per}\\\text{dwelling}\\\text{unit}\end{pmatrix}\begin{pmatrix}\text{Number of}\\\text{dwelling}\\\text{units}\end{pmatrix}$$

and applies only to residential projects.

Schools   New development can change the number of students, age distribution of students, and school enrollments in a community. These effects can be estimated and are useful in determining how many pupils will have to switch schools, how many will walk and how many will ride buses, how much crowding will occur, how will student-teacher ratios change, and whether or not present classroom capacity can serve greater demands.

For evaluating proposed developments, most communities maintain statistics on the number of students expected for elementary, junior high, and high schools per housing unit by type of housing. This suggests that a measure of the impact on schools due to residential developments is given by the number of new students generated, which can be formulated by

$$\left[\begin{pmatrix}\text{Number of}\\\text{elementary}\\\text{students per}\\\text{dwelling}\\\text{unit}\end{pmatrix} + \begin{pmatrix}\text{Number of}\\\text{junior high}\\\text{students per}\\\text{dwelling}\\\text{unit}\end{pmatrix} + \begin{pmatrix}\text{Number of}\\\text{high school}\\\text{students per}\\\text{dwelling}\\\text{unit}\end{pmatrix}\right]\begin{pmatrix}\text{Number of}\\\text{dwelling}\\\text{units}\end{pmatrix}$$

In the case of commercial and industrial projects the number of new students generated can be formulated by

$$\left[\begin{pmatrix}\text{Number of}\\\text{elementary}\\\text{students}\\\text{per}\\\text{employee}\end{pmatrix} + \begin{pmatrix}\text{Number of}\\\text{junior high}\\\text{students}\\\text{per}\\\text{employee}\end{pmatrix} + \begin{pmatrix}\text{Number of}\\\text{high school}\\\text{students}\\\text{per}\\\text{employee}\end{pmatrix}\right]\begin{pmatrix}\text{Number of}\\\text{employees}\\\text{per 1000}\\\text{ft}^2\text{ GLA}\end{pmatrix}\begin{pmatrix}\text{Total}\\\text{GLA in}\\1000\text{ ft}^2\end{pmatrix}$$

Strictly speaking, in order to estimate new students, one should count only new employees who reside in the community, as well as any current employees who move to the community.

Public Transportation   Land development may affect the accessibility, use, and convenience of public transportaion by altering demand patterns, and thus routing, scheduling, and crowdedness. It may generate enough demand to allow additional service on existing public transit lines, the creation or rerouting of lines, or even the start of a new public transit system.

Two key areas of concern in assessing the impact on public transportation are: (1) the accessibility of public transit to existing residents, and (2) the accessibility of public transit to commercial and industrial employees. This suggests two impact measures (Ref. 34) given by (1) the percent change in the number of residents with access to public transit within x feet of their residences, which can be formulated by

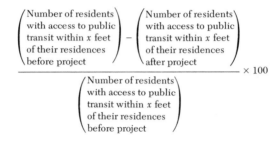

and (2) the percent change in the number of employees who can get within y distance of work location by public transit, which can be formulated by

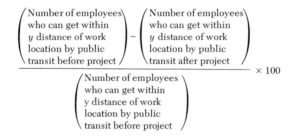

Traffic Flow and Congestion   Land development can also affect vehicular travel by changing the number and length of car trips needed, by changing local street and road patterns, by creating the need for additional traffic controls, and by changing the supply and demand for parking space. The result could be an effect on travel times and the duration and severity of congestion in the immediate vicinity of the project or on the nearby portions of roads radiating from the area of the development, which receive the full impact of the increased traffic load.

For proposed developments, the existing street capacities and traffic volumes are usually known or measurable. To this base the expected change in car trips caused by the development could be added. The number of car trips generated by each household unit can be estimated based on the type and price of unit, expected socioeconomic characteristics of the occupants, and the expected modal split between car and other means of transportation. Similar estimates can be made by type of business for industrial and commercial developments.

The preceding discussion suggests that one measure of the impact on traffic flow is given by the average daily trips generated, which could be formulated by

$$\left(\begin{array}{c}\text{Average}\\\text{daily trips}\\\text{per dwelling}\\\text{unit}\end{array}\right)\left(\begin{array}{c}\text{Number of}\\\text{dwelling}\\\text{units}\end{array}\right)$$

for residential projects and by

$$\begin{pmatrix} \text{Average} \\ \text{daily trips} \\ \text{per 1000} \\ \text{ft}^2 \text{ GLA} \end{pmatrix} \begin{pmatrix} \text{Total GLA} \\ \text{in 1000 ft}^2 \end{pmatrix}$$

for commercial and industrial projects.

For proposed developments, estimating precise changes in travel times between specific points is difficult. The most common general approach is to estimate the number of vehicular trips that will be generated by the development, based on past trips per household or business and choices between public and private modes. The expected traffic volumes on various roads can be computed by making assumptions as to the distribution of the added trips throughout the road network during rush hours and non-rush hours, and adding them to existing volumes, taking account of expected changes by current users. The average travel speeds can then be estimated from projected traffic volumes and known street characteristics.

The two aspects of congestion which are important are its severity and its duration. The severity of congestion can be defined as the ratio of the maximum time to travel between two points relative to the "no traffic" or off-peak, law-abiding travel time between those points. The duration of congestion can be defined as the length of time during which travel times between two points is some percentage above the off-peak travel time. Consequently, two measures (Ref. 34) of the impact on traffic congestion relative to a fixed pair of points would be given by the percent change in severity of congestion and the percent change in duration of congestion. These measures can be respectively computed as follows:

$$\frac{\begin{pmatrix} \text{Severity of} \\ \text{congestion} \\ \text{before project} \end{pmatrix} - \begin{pmatrix} \text{Severity of} \\ \text{congestion} \\ \text{after project} \end{pmatrix}}{\begin{pmatrix} \text{Severity of} \\ \text{congestion} \\ \text{before project} \end{pmatrix}} \times 100$$

and

$$\frac{\begin{pmatrix} \text{Duration of} \\ \text{congestion} \\ \text{before project} \end{pmatrix} - \begin{pmatrix} \text{Duration of} \\ \text{congestion} \\ \text{after project} \end{pmatrix}}{\begin{pmatrix} \text{Duration of} \\ \text{congestion} \\ \text{before project} \end{pmatrix}} \times 100$$

Displacement of People   In some cases, such as highway development through an established area or in redevelopment projects, the project may uproot current residents by physically displacing them from their homes and may also displace workers by removing existing stores and other enterprises. Secondary effects would include causing people to move because of the project's effect on taxes or on the physical or social environment. Similarly, when certain jobs are eliminated and not moved to a convenient new location or are substituted for by new jobs in the development, the net loss of employment could cause some people to leave the community entirely.

This discussion suggests that a measure of the impact on displacement of people is given by the number of people (residents and employees) displaced, which could be formulated by

$$\begin{pmatrix} \text{Number of} \\ \text{businesses} \\ \text{displaced} \end{pmatrix} \begin{pmatrix} \text{Number of} \\ \text{employees} \\ \text{per business} \end{pmatrix} + \begin{pmatrix} \text{Number of} \\ \text{dwelling units} \\ \text{displaced} \end{pmatrix} \begin{pmatrix} \text{Number of} \\ \text{people per} \\ \text{dwelling unit} \end{pmatrix}$$

Unfortunately, this measure, as formulated, only provides an estimate of the individuals who are directly displaced. It is considerably more difficult to estimate the amount of displacement due to secondary effects.

Crowdedness   Crowdedness is based on the perceptions of the residents and workers in an area relative to the specific elements of crowding that may be especially annoying, such as too many people on the streets, traffic jams, crowding at local informal or formal recreational centers, long lines at stores, excessive waiting times at restaurants, etc. The likelihood that people will feel crowded may vary a great deal depending on the arrangement of space, the adequacy of services, and personal living styles.

Population density (usually in terms of people per acre) is often used as a measure of crowdedness. However, this measure does not accurately reflect the considerations mentioned in the preceding paragraph. A more direct measure of individual overall perception of community crowdedness is thus desirable in order to capture the impressions that are not adequately reflected in such a simple measure as population density. A measure of the impact on crowdedness is given by (Ref. 34) the percent change in people who perceive their neighborhood as too crowded, which could be formulated by

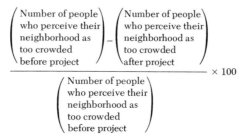

The computation of this measure would require surveys of people in the project area both before and after development.

Police Protection   Potential changes in crime rates and perceived crime hazards in an area are generally of concern in evaluating the impact of land development projects. Developments may affect crime rates in a community by adding or removing targets (people, businesses, and residences) and by changing physical or social conditions that may breed crime or make crimes easier to commit. The potential targets for economic crime may change in number, density, vulnerability, or lucrativeness. For example, a new office building may increase the opportunities for larceny. Crime attracted by the development may spill over into the existing community or the development might attract crime away from the existing community. In either case, there is the potential for a net change in the crime activity in the area.

The ability of police to deter crime might be affected by development. For example, increases in noncrime duties such as traffic control may reduce the time available for patrol duties. The size of the police force may not increase fast enough to keep up with the new growth. People in the new community may not cooperate as willingly with police on crime prevention or detection. And response time to crime calls may be lengthened due to increased traffic congestion.

This discussion suggests that a measure of the impact on police protection is given by the percent change in crime rate in the project area, which can be formulated by

$$\frac{\left(\begin{array}{c}\text{Crime rate}\\ \text{in project area}\\ \text{before project}\end{array}\right) - \left(\begin{array}{c}\text{Crime rate}\\ \text{in project area}\\ \text{after project}\end{array}\right)}{\left(\begin{array}{c}\text{Crime rate}\\ \text{in project area}\\ \text{before project}\end{array}\right)} \times 100$$

Fire Protection   Fire hazard from new developments may be considered in two parts, namely: (1) the change in the likelihood of a fire getting started in the first place, as measured by fire rates, and (2) the change in the likely spread and risk to life of fire once started. The issue goes beyond risks to occupants of the new development. It concerns other members of the community who may visit, work, or shop there, and it involves the potential spread of fire to the rest of the community. It may mean added risks for firemen, and it may increase fire protection costs generally.

Fire incidence may be affected by many aspects of new development such as: types of construction and materials; equipment, processes, and activities in the development; education level, age distribution, and attitudes of new entrants into the community (which have a bearing on arson, false alarms, and accidental fires); and character of previous development (removal of blighted buildings would probably reduce fire hazards.)

Fire spread may be affected by such factors as: overall design, street layout, and proximity of buildings to each other; hydrant locations and built-in private fire defenses; distance to nearest fire station; and adequacy of amount and type of fire fighting equipment and personnel in light of the needs of the new development (such as special fire towers for highrise buildings.)

Fire incidence rates and the likelihood of fire spread and its extent both contribute to the fire insurance rating of the area. Thus, this suggests that a measure (Ref. 34) of the impact on fire protection is given by the percent change in the fire rating of the project area, which can be formulated by

$$\frac{\left(\begin{array}{c}\text{Fire rating of}\\\text{project area}\\\text{before project}\end{array}\right) - \left(\begin{array}{c}\text{Fire rating of}\\\text{project area}\\\text{after project}\end{array}\right)}{\left(\begin{array}{c}\text{Fire rating of}\\\text{project area}\\\text{before project}\end{array}\right)} \times 100$$

Measures of Economic Impacts

Community Fiscal Posture   A new development's fiscal impact on local government is determined by comparing the public revenues with the operating and capital expenditures. The determination of these monetary amounts depends to a considerable extent on whether the government will maintain or change its level and quality of services to the new development and to the rest of the community after the development is completed. Concurrently, the level of service to be provided is likely to depend to some extent on the estimated fiscal impacts, that is, the community chooses a level of service based in part on its perception of what it can afford. To further complicate matters, maintaining the same expenditures per capita is not necessarily synonymous with maintaining the same quality of service, since the demands for services and the costs of supplying them may change faster or slower than the rate of residential or business population growth.

Local revenues can be grouped into four categories, namely: (1) revenues associated with real property wealth such as property taxes; (2) revenues associated with income and level of consumption such as local income, sales, and utility taxes; (3) per capita, per pupil, or other per "population unit" revenues, which are derived from either a per capita tax, or redistribution from higher levels of government; and (4) miscellaneous revenues, which include fees, user charges, fines, licenses, and minor items.

On the other hand, local costs can be grouped into the two basic categories of (1) operating expenditures, and (2) capital expenditures. Local operating expenditures can be grouped into those incurred in supplying services used (1) primarily by households, such as education, libraries, health and welfare, and recreation; and those used (2) by both business enterprises and households, such as fire and police, utilities, general government, and transportation. Capital expenditures associated with new development can be grouped into two categories, namely: (1) facilities linked directly with the development, such as new schools, sewer lines, water lines, fire stations, and other new facilities to be utilized primarily by the new development; and (2) facilities constructed or expanded as part of a capital improvement program which will be shared by existing and new residents or enterprises in the community.

The biggest difficulty in determining the costs and revenues associated with new development is the allocation of capital costs between the existing community and the new development. If a new facility is part of a capital improvement program and is initially underutilized in expectation of future growth, only the share of the total cost needed to meet the demands created by the new development should be allocated to it. If the new development uses available space in existing facilities, some analysts would allocate only the short-term incremental cost, some the long-term incremental cost, and some the average cost. Which method to use depends on the viewpoint and purpose of the analysis. The short-term incremental cost reflects the out-of-pocket additional expense for the facility. The long-term incremental cost reflects the costs attributed to new development over the long run. The average cost concept assumes each user bears an equal burden.

In some cases, a new development triggers a new capital investment that will be used by the entire community and will raise the per capita cost to the community for a service. In this case, it would be reasonable to distribute the cost of such facilities equally among all users.

Facilities fully utilized prior to new development, such as public schools, should not be considered as part of the capital cost attributable to the new development.

A measure of the impact on the community fiscal posture as the result of new development is given by the revenue-to-cost ratio. This could be computed on a one-time basis, such as in the first year after completion of the project, or on an annual basis projected over some future time period of interest. The latter method is particularly attractive since it enables one to amortize the capital costs on an annual basis. In this ratio, the revenue amount would represent that which is derived directly from the project, and the cost amount would represent the operating and capital cost allocations directly attributed to the project. Another positive feature of the revenue-to-cost ratio concept is that it enables the analyst to estimate how many dollars of revenue will be derived for every dollar spent by the community. Hence, a measure of whether or not the project pays its own way is provided.

Employment   The ability to measure the impact of new development on employment depends to some extent on the type of development being considered. For example, on a short-term basis, construction-related jobs are created by residential, commercial, and industrial projects. However, on a long-term basis, it is not clear how to estimate jobs created—especially, in the case of residential projects.

Commercial and industrial projects generate long-term direct employment as the result of new or expanded business enterprises and create secondary impacts due to increased economic activity stimulated by the development. Employment impacts of industrial and commercial developments in the long run depend on the number of new jobs created, the availability in the community of the skills needed, and the proportion of these jobs likely to be filled by community residents, commuters, or immigrants. It further depends on whether local persons who fill the new jobs would otherwise be unemployed, would just be entering or reentering the labor force, or would be changing employment within the community. In the last case, the impact further depends on whether the jobs left behind are filled, and if so, whether from within the community or outside.

Employment may also result from the so-called "multiplier effect." Each new primary job created by industrial and commercial development may stimulate added service jobs. There also may be increased demand for labor in industries supplying materials for construction or for use by the new industry, though these jobs may fall largely outside the immediate community.

The traditional measure of the impact of a new development on employment is the estimated total number of jobs created. This measure is of interest and is relatively easy to estimate, but it does not directly reflect the impact of the development on employment opportunities for the present citizens of the community. It does not indicate whether the new jobs will be taken by persons from outside the community or by persons within the community. Furthermore, as previously discussed, consideration must also be given to the impact of the development on unemployment and underemployment, as well as on the new long-term and short-term jobs added to the community. Also, it is important to note that a net addition of jobs to a community reduces the percentage unemployed even if no one currently unemployed gets a job. For this reason, both absolute and percentage changes should be identified.

Natural Disaster Risk   Each year, natural disasters take a large toll in life and property and cost large sums of money. Yet, much construction goes on in danger zones in the path of these disasters. Whenever this type of development occurs, one must recognize and estimate the risks involved in order to assess the potential impacts of the development.

Of particular concern in the development of areas known to lie in floodplains, or subject to natural hazards such as stream overflow and mudslides, is the potential exposure of property values and lives to such risks. A new development might also increase the hazards to the rest of the community by removing natural barriers, changing land contours, and contributing to changes in vegetation and soil permeability.

By reviewing local floodplain maps and other related information, coupled with the plans for building locations, heights, terrain configuration, construction materials, and building design, one can estimate the expected flood damage. This measure could be formulated by

$$
\sum_{\text{all flood intensities}} \begin{pmatrix} \text{Probability of} \\ \text{a flood of} \\ \text{intensity } i \\ \text{occurring} \end{pmatrix} \begin{pmatrix} \text{Value of project} \\ \text{damage should a} \\ \text{flood of intensity} \\ i \text{ occur} \end{pmatrix}
$$

Also of concern is whether or not a community is in a high-risk area for damage from tremors or earthquakes and, if so, what is the corresponding risk exposure. This can be determined by examining geological maps of the area to identify where landslides have occurred in the past and where geologic fault lines lie. Unfortunately, many areas of the country do not have such detailed risk maps, in which case special surveys and studies would have to be performed. In any case, a measure of the seismic risk exposure is given by the expected earthquake damage, which could be formulated by

$$
\sum_{\text{all earthquake intensities}} \begin{pmatrix} \text{Probability of} \\ \text{an earthquake} \\ \text{of intensity } i \\ \text{occurring} \end{pmatrix} \begin{pmatrix} \text{Value of property} \\ \text{damage should an} \\ \text{earthquake of} \\ \text{intensity } i \text{ occur} \end{pmatrix}
$$

Knowledge of construction materials and techniques for the new development and local meteorological conditions can be used to assess risk from wind, hurricanes, and other meteorological conditions.

Energy Consumption The effect of new development on the consumption of scarce resources, especially energy and fuels, is of increasing concern. Individual developments are unlikely to make a significant difference in local energy consumption, except where the development is very large or if shortages are very critical. However, new developments can collectively influence local energy and fuel consumption as the result of cumulative effects. Of special interest are the requirements for electricity and natural gas.

Electrical energy requirements for proposed residential and commercial developments can usually be estimated from established local averages for analogous recent developments in the community, or from design details of the development and assumptions about the behavior of the development's users. Adjustments can be made for anticipated energy-saving features such as extra-heavy insulation, or for energy-wasting features such as glass walls. Estimates for industrial development depend on the nature and level of planned production, and this data will probably be available from the industry itself. In any case, a measure of the impact on electricity consumption would be given by the electricity consumption in kilowatt-hours (kWh), over a specified time period (day, month, or year), which could be formulated by

$$
\begin{pmatrix} \text{Number of} \\ \text{kWh per} \\ \text{dwelling unit} \end{pmatrix} \begin{pmatrix} \text{Number of} \\ \text{dwelling} \\ \text{units} \end{pmatrix}
$$

for residential projects, and by

$$
\begin{pmatrix} \text{Number of} \\ \text{kWh per} \\ \text{1000 ft}^2 \text{ GLA} \end{pmatrix} \begin{pmatrix} \text{Total GLA} \\ \text{in 1000 ft}^2 \end{pmatrix}
$$

for commercial and industrial projects.

Nature gas consumption impacts can be measured in a similar way, using the measure given by the natural gas consumed in 1000 ft^3 (MCF) over a specified time period, which could be formulated by

$$
\begin{pmatrix} \text{Number of} \\ \text{MCF per} \\ \text{dwelling unit} \end{pmatrix} \begin{pmatrix} \text{Number of} \\ \text{dwelling} \\ \text{units} \end{pmatrix}
$$

for residential projects, and by

$$\left(\begin{array}{c}\text{Number of}\\ \text{MCF per 1000}\\ \text{ft}^2\ \text{GLA}\end{array}\right)\left(\begin{array}{c}\text{Total GLA}\\ \text{in 1000 ft}^2\end{array}\right)$$

for commercial and industrial projects.

Upon multiplying the above quantities by the appropriate conversion factor and then adding the consumption estimates, one can express the total energy consumption in British thermal units (BTUs), a standard measure used in energy analyses.

CONCEPT OF SIGNIFICANT EFFECT

Section 102(2)(C) of the National Environmental Policy Act established the requirement for an environmental impact statement for "major Federal actions significantly affecting the quality of the human environment." No definitions are provided in the act as to what is meant by the various terms "quality of the human environment," "major Federal action," and "significant affect." However, numerous federal agencies have responded in their guidelines for the preparation of environmental impact statements with interpretations of these terms.

The "human environment" is defined (Ref. 58) as the aggregate of all external conditions and influences (aesthetic, ecological, cultural, social, economic, historical, etc.) that affect human life.

"Major Federal action" means (Refs. 28 and 52) any federal action which requires the substantial commitment of resources or triggers such a substantial commitment by another (or others.) According to Ref. 58, major actions are those of superior, large, and considerable importance involving substantial planning, time, resources, or expenditures. Any action that is likely to precipitate significant forseeable alterations in land use, planned growth, development patterns, traffic volumes, travel patterns, transportation services (including public transportation), and natural and artificial resources would be considered a major action.

For example, Federal Highway Administration actions which are ordinarily considered to be major actions are (Ref. 58):

1. A new freeway or expressway
2. A highway which provides new access to an area and is likely to precipitate significant changes in land use or development patterns
3. A new or reconstructed arterial highway which provides substantially improved access to an area and is likely to precipitate significant changes in land use or development patterns
4. A new circumferential or belt highway which bypasses a community
5. A highway which provides new access to areas containing significant amounts of exploitable natural resources
6. Added interchanges to a completed freeway or expressway which provide new or substantially improved access to an area and are likely to precipitate significant changes in land use or development patterns
7. A project that warrants a "major action" classification because it has been given national recognition by Congress, even though it is not included in previous entries in this list.

In addition, the Army Corps of Engineers' operation, maintenance, and management activities which could be major actions having an impact on the environment are (Ref. 41):

1. Disposal of dredged material in wetlands or marshlands
2. Disposal of polluted dredged material in unconfined or open water areas
3. Debris collection and disposal activities
4. Resource management programs involving the cutting, sale, and/or disposal of forest resources; extensive plant disease eradication; predator or vector control; and aquatic plant control
5. Reservoir regulation in which some environmental benefits must be sacrificed for other environmental benefits or economic considerations, such as drawdown to provide water for power and for downstream water quality control
6. Leases, licenses, rights-of-way, administrative permits, and other actions involving use by others of project resources, if impact is significant and not otherwise covered in another environmental statement

7. Redesignation of project land under management by the corps from scenic buffer or "green belt," undeveloped natural area, or wildlife management area to more intensive type of public use or some other type of use.

According to Refs. 14 and 58, an action "significantly affecting the quality of the human environment" is an action in which the overall cumulative primary and secondary consequences significantly alter the quality of the human environment, curtail the choices of beneficial uses of the human environment, or interfere with the attainment of long-range human environmental goals. Significant effects can include actions which may have both beneficial and adverse effects, even if, on balance, the agency believes that the effect will be beneficial. The significance of a proposed action may also vary with the setting, with the result that an action that would have little impact in an urban area may be significant in a rural setting or vice versa. The words "major" and "significantly" are intended to imply thresholds of importance and impact that must be met before an environmental impact statement is required.

According to Ref. 28, in determining whether a proposed action will significantly affect the quality of the human environment, consideration must be given to:

1. The extent to which the action will cause environmental effects in excess of those created by existing uses in the area affected by it

2. The absolute quantitative environmental effects of the action itself, including the cumulative harm that results from its contribution to existing adverse conditions or uses in the affected area

3. The extent to which the proposed action is consistent with local land use plans.

The following are examples of the general types of actions which ordinarily would have a significant effect on the quality of the human environment (Refs. 33, 48, 52, 54, 58, 61, and 67):

1. Any action that is likely to be highly controversial on environmental grounds, because of relocation housing or any other reason

2. Any action that is likely to have a significantly adverse impact on natural, ecological, cultural, or scenic resources of national, state, or local significance

3. Any matter falling under section 4(f) of the Department of Transportation Act, requiring protection of publicly owned land from a public park, recreation area, or wildlife and waterfowl refuge of national, state, or local significance

4. Any matter falling under section 106 of the National Historic Preservation Act of 1966, requiring consideration of the effect of the proposed action on any building included in the National Register of Historic Preservation and allowing a reasonable opportunity for the Advisory Council on Historic Preservation to comment on such action

5. Any action that is likely to affect the preservation and enhancement of sites of historical, architectural, or archaeological significance

6. Any action that (a) divides or disrupts an established community, by division of an existing use (such as cutting off residential areas from recreation areas or shopping centers), or disrupts orderly, planned development, or is inconsistent with plans or goals that have been adopted by the community in which the project is located; or (b) causes increased traffic and congestion levels on streets and highways

7. Any action that (a) involves inconsistency with any federal, state, or local law or administrative determination relating to the environment; (b) has a significantly detrimental impact on air or water quality or on ambient noise levels for adjoining areas; (c) involves a possibility of contamination of public resources such as a water supply system; or (d) affects ground water, flooding, erosion, or sedimentation

8. Any action involving significant taking of land, change in the use of land (particularly if it requires a change in zoning), or major construction

9. Any action with a substantial aesthetic or visual effect or which inflicts permanent damage on the visual landscape, especially on areas of unique interest or scenic beauty

10. Any action which causes substantial displacement of people and businesses

11. Any action which causes disturbance to the ecological balance of animal and vegetation habitats or the elimination of rare or endangered species, and substantially alters the behavior pattern of wildlife or interferes with important breeding, nesting, or feeding grounds

12. Any action which has a detrimental effect on the safety and health of the community or which causes development to occur in areas of known natural hazards

13. Minor actions which may set a precedent for future major actions with significant adverse impacts, or a number of actions with individually insignificant but cumulatively significant adverse impacts

14. Any other action that causes significant environmental impact by directly or indirectly affecting human beings through adverse impacts on the environment.

For example, the Environmental Protection Agency has established (Ref. 62) specific criteria for preparing impact statements on wastewater treatment facilities and water quality management plans. For the following specific situations, an impact statement shall always be prepared and processed:

1. When the project or plan is highly controversial for any reason, including the degree of treatment, method of final waste disposal, or the location of a plant or facility. In the case of location, where controversy centers only on the disruption incident to construction and serious effort is made to restore the environs as much as possible, an impact statement is not required.

2. When a project or plan defaces (either by physical presence or odor) a residential development or a recognized scenic area on either public or private lands. An impact statement shall not be required if the physical alteration is unobtrusive, or if a permanent industrial facility already exists in the area and the proposed project will not constitute a commitment of substantial additional land to industrial use providing all other effects are unobtrusive.

3. When a facility is to be sited on public land (or on private land in an urban area, which because of its natural beauty and wilderness state has potential or is being considered for public park development), and the loss of aesthetics incurred through construction and maintenance of the facility substantially reduces its value as a public park.

4. When the treated effluent being discharged will meet water quality standards applicable to the receiving waters but the public is using these waters for activities that require a classification higher than their present classification.

5. When a project will result in a substantial displacement of population.

6. When the environmental impact is the result of a number of projects impacting upon the same resource, as when a number of projects individually divert water from one river basin into another or discharge effluent into the ocean instead of using the effluent to recharge the groundwater aquifer.

7. When an existing plant is being refurbished to improve its treatment capabilities and the additions involve controversy or substantial additional impacts. However, where there is no substantial new land use, noise, or odor, an impact statement shall not normally be required.

8. When the project or plan will result in the installation of a major interceptor line that will provide service to undeveloped areas or permit expansion of already developed areas, and the effects that this will have on residential and commercial growth have not been adequately considered in the interim or final plan encompassing the project, or in the grant application and associated documents.

Many other federal agencies have identified those specific types of actions which they feel meet the criteria of significantly affecting the quality of the human environment.

For example, the Department of Housing and Urban Development (Ref. 45) has cited the following types of projects which clearly would require the preparation and dissemination of an environmental impact statement:

1. Projects involving maximum sound, signal, or substance emitting levels which exceed those established by law or competent regulation

2. Projects having an adverse impact (which cannot be satisfactorily removed or mitigated) upon any historic property

3. Projects involving any violation, which cannot be satisfactorily removed or mitigated, of criteria or standards under applicable laws or regulations governing environmental considerations, such as: the Clean Air Act, the Federal Water Pollution Control Act, HUD Noise Regulation (No. 1390.2), the Coastal Zone Management Act, the Fish and Wildlife Coordination Act, the National Flood Insurance Act, and similar federal and state laws and regulations

4. Projects involving the removal, demolition, conversion, or emplacement of a total of 500 or more dwelling units

5. Projects wherein or whereby water or sewer facilities will pass through, be adjacent to, or serve underdeveloped areas of 100 acres or more.

Furthermore, HUD has stated that all projects meeting or exceeding the following criteria of size or environmental impact should receive special consideration and review relative to the possible preparation of an environmental impact statement:

1. Projects which will result in a 50-percent change in the density, vehicular traffic, demand for energy, or demand for other public services in the area environmentally affected by the project

2. Neighborhood facilities projects having site acreage of 50,000 ft², or gross floor area of 30,000 ft²

3. All open space land projects involving (a) sanitary landfill; (b) impoundment of 2 surface acres or 25 acre/feet of water; (c) 50 acres, or more; and (d) conversions of open space land to non-open space land uses

4. All neighborhood development projects (NDP) and conversions from conventional urban renewal to neighborhood developments; all changes in NDP area plans

5. All above ground reservoirs and standpipes; all source development projects, including major river impoundments, raw-water reservoirs, well fields, treatment plants, treated water transmission or sewage collection lines which pass through, are adjacent to, or serve undeveloped areas of 50 acres or more.

The Department of Defense has cited (Ref. 42) types of actions which would require close environmental scrutiny because they may either affect the quality of the environment or create environmental controversy and thus require the preparation of an environmental impact statement. These are as follows:

1. Development or purchase of a new type of aircraft, ship, or vehicle, or of a substantially modified propulsion system for any aircraft, ship, or vehicle

2. Development or purchase of a new weapon system

3. Real estate acquisition, disposal, and outgrants

4. Construction projects

5. New installations (bases, posts, etc.)

6. Production, storage, transportation, testing, or disposal of lethal chemical munitions, pesticides, herbicides, and containers

7. Mission changes and troop developments which precipitate long-term population increases or decreases in any area, with special attention to the secondary impacts which may cause indirect environmental impact

8. Large quarrying or earth-moving operations

9. Constructing or installing fences or other barriers that might prevent migration or free movement of wildlife

10. Proposed construction of new sanitary landfills, incinerators, and sewage treatment plants

11. Existing, or changes to, master plans

12. Proposed construction or acquisition of new family housing

13. Dredging and other similar activities in the water

14. Exercises on or off federal property where significant environmental damage might occur, regardless of unit sizes

15. Opening areas that were previously closed to the public or closing or limiting of areas that previously were open to public use, such as roads or recreational areas

16. Proposed construction on floodplains, construction that may cause increased flooding, or activities on wetlands

17. Channelization of streams

18. Disposal of significant quantities of POL waste products

19. Proposed construction of roads, transmission lines, or pipelines

20. New, revised, or established regulations, directives, or policy guidance concerning activities that could have an environmental effect; regulations, directives, or policy guidance which limit any of the alternative means of performing the actions on this list

21. Any action which, because of real, potential, or purported adverse environmental consequences, is a subject of controversy among people who will be affected by the action, or which, although not the subject of controversy, is likely to create controversy when the proposed action becomes known by the public.

According to Ref. 28, the Interstate Commerce Commission has identified the following classes of actions as having the potential for significant environmental impact. These normally require an environmental impact statement; (1) rail line constructions, (2) commuter fare increases, (3) discontinuance of passenger trains, and (4) general rate increases. The following classes of actions would not normally require an environmental impact statement: (1) abandonment, acquisition, or operation of a line of railroads; (2) common use of rail terminals; (3) merger, purchase, control, or trackage rights proceedings; (4) water carrier certification; (5) investigation and suspension, rate complaint, or formal docket cases involving recyclable commodities; and (6) rulemaking.

The Federal Aviation Administration (Ref. 54) has identified the following projects as being typical of those that usually do not have consequences implying the environmental impact statement requirements:

1. Taxiway location or extension, runway or taxiway resurfacing, or reconstruction (not including major pavement strengthening)
2. Runway, taxiway, or apron lighting
3. Access roads and obstruction removal on airport property
4. Location, expansion, and reconstruction of aprons
5. Fire/crash buildings, maintenance equipment, and service buildings
6. Runway instrumentation under the Airport Aid Development Program (AADP).

OTHER CONCEPTS, TERMS, AND PHRASES IN THE IMPACT ASSESSMENT PROCESS

Consideration of Alternatives

Section 102(2)(C)(iii) of NEPA established the requirement for the consideration of "alternatives to the proposed action" in the preparation of an environmental impact statement. Also, Section 102(2)(D) refers to the need to "study, develop, and describe appropriate alternatives to recommend courses of action in any proposal which involves unresolved conflicts concerning alternative uses of available resources."

These requirements of NEPA suggest that in the environmental impact assessment process, it is particularly important to explore and evaluate the environmental impacts of all reasonable alternative actions, particularly those that might enhance environmental quality or avoid some or all of the adverse environmental effects. All reasonable alternatives would include those that have been or are likely to be suggested by agencies, groups, or individuals even though they would not meet project objectives. Alternatives which are deemed remote or speculative are not considered to be reasonable. The courts have ruled that agencies "need not indulge in crystal ball inquiry" in assessing the effects of alternatives.[2]

Typical alternatives that are reasonable for consideration are the following:

1. The alternative of taking no action
2. The alternative of postponing action pending further study
3. Alternatives requiring actions of a significantly different nature which would provide similar benefits with different environmental impacts (such as nonstructural alternatives to flood control programs, or mass transit alternatives to highway construction)
4. Alternatives related to different designs or details of the proposed action which would present different environmental impacts (such as cooling ponds versus cooling towers for a power plant, or alternatives that will significantly conserve energy)
5. Alternative measures to provide for compensation of fish and wildlife losses, including the acquisition of land, waters, and interests therein.

In the evaluation of alternatives relative to wastewater treatment facilities, the Environmental Protection Agency (Ref. 67) requires that the alternatives be screened with respect to: capital and operating costs; signficant primary and secondary environmental effects; physical, legal, or institutional constraints; and whether or not they meet regulatory requirements. In the formulation of reasonable alternatives, consideration should be given to:

1. Flow and waste reduction measures, including infiltration/inflow reduction
2. Alternative locations, capacities, and construction phasing of facilities
3. Alternative waste management techniques, including treatment and discharge, wastewater re-use, and land application
4. Alternative methods for disposal of sludge and other residual waste, including process options and final disposal options
5. Improving effluent quality through more efficient operation and maintenance.

In addition, if an EPA action may directly cause or induce the construction of buildings

---

[2]*National Resources Defense Council v. Morton*, 458F.2d827 (CADC, 1972)

or facilities in a floodplain, the responsible official is required to evaluate the flood hazards in connection with these facilities and, as far as practicable, to consider alternatives. These alternatives should preclude the uneconomic, hazardous, or unnecessary use of floodplains to minimize the exposure of facilities to potential flood damage, lessen the need for future federal expenditures for flood protection and flood disaster relief, and preserve the unique and significant public value of the floodplain as an environmental resource.

With regard to highway projects there are a variety of considerations relative to the formulation of reasonable alternatives such as: changes in scope or size of the project; alternative corridors, roadway configurations, or highway locations; alternative methods of construction if the methods could result in fewer environmental impacts (such as comparing bridges against culverts, and major cut-and-fill operations with alternative locations which would minimize the land disturbance).

Short-Term Versus Long-Term Effects

Section 102(2)(C)(iv) of NEPA established the requirement for the consideration of "the relationship between local short-term uses of man's environment and the maintenance and enhancement of long-term productivity." This expression has been interpreted (Ref. 67) to mean the extent to which the proposed action involves tradeoffs between short-term environmental gains and long-term gains, or vice-versa, and the extent to which the proposed action forecloses future options. Short-term impacts could be construed to mean those that are immediate impacts of short duration, such as those during construction and initial use. Conversely, long-term impacts could be construed to mean those lasting beyond construction and initial-use period or with implications for secondary impacts in the future.

For example, highway projects through estuaries, marshes, etc. may foreclose future choices of use and may permanently impair the natural activity of the area. The elimination of recreation and parklands can precipitate drastic changes in the social and economic character of the project area. Application of herbicides or pesticides may remove undesirable species, but the long lasting or cumulative effects of these agents may permanently damage other vegetative growth or result in disruption of ecological balance. Construction of a sewage-treatment plant may result in activities which create noise, dust, or erosion, but long-term aspects of the project include enhanced water quality in the receiving stream.

Typical short-term effects would include: construction noise, temporary soil erosion and siltation, dust, traffic disruption, unsightly areas during construction, and displacement of persons and businesses when such disruption is short-term and the displaced persons are relocated quickly.

Typical long-term effects would include: pollution resulting from the use of septic systems, poor drainage, incinerators, etc.; infrastructure overloads due to the project such as those which affect sewer, solid waste, traffic, schools, or social services; destruction or prolonged disruption of a cohesive community through displacement or other effects on the viability of a community environment; projects constituting an alteration in the community which could stimulate a trend, growth-inducing effect, or could substantially alter the character of a community (such as locating a high-rise building in a low-rise neighborhood, an increase in building intensity, or an increase in population density); loss of open space which could have been valuable as a part of a recreation area; significant alteration of a unique natural area; destruction of a historic building; and fundamental changes in the ecological structure of an area.

The importance of considering both the short-term and long-term effects is to enable the assessment of the cumulative impacts of the project which either significantly reduce or enhance the state of the environment for future generations.

Irreversible and Irretrievable Commitments of Resources

Section 102(2)(C)(v) of NEPA established the requirement for the consideration of "any irreversible and irretrievable commitments of resources which would be involved in the proposed action should it be implemented." According to Refs. 14, 43, and 58, the term "resources" does not mean only the labor and materials devoted to the action, but also includes the full range of natural and cultural resources committed to loss or destruction by the action. The terms "irreversible" and "irretrievable" apply primarily to nonrenewable resources. Endangered species, fossil fuels, minerals, or wilderness situations often involve irreversible effects. Irretrievable effects are the adverse effects on some value that will be lost and cannot be restored, such as an endangered or threatened animal that may become extinct as the result of the action.

Typical examples of such commitments would include: destruction of unique habitats for wildlife; increase of freshwater flow into an estuary, changing the balance of fresh and salt water; use of land previously reserved for special crops or unique vegetation; change of flow of water in a stream of particular scenic beauty; land committed to dams, spillways, lakes, pool areas, channels, and recreational facilities for water-oriented projects; and labor and capital investments.

It may also happen that a project will induce actions or cause other projects to occur which in themselves will lead to irreversible and irretrievable commitments of resources. For example, a transportation facility may precipitate other related actions, such as land development, exploitation of resources, travel, etc., that could induce a significant irreversible commitment which would curtail other use of the area.

SUMMARY OVERVIEW OF THE ENVIRONMENTAL IMPACT ASSESSMENT PROCESS

The major purpose of the National Environmental Policy Act of 1969 is to ensure that environmental quality is fully considered in the decision-making process. The vehicle used to accomplish this is the environmental impact statement whose specific objective is to provide a means for giving environmental quality careful and appropriate consideration in the planning and decision-making process. This document must be of sufficient detail to allow the responsible official(s) to make a decision with full knowledge and consideration of the environmental impacts expected. It is important that the impacts of a proposed action (or project) on the quality of the physical environment be objectively weighed with the impacts on the social, aesthetic, and economic environments, over both the short and the long run. Overall, the environmental impact statement process provides a formalized procedure for informing and taking account of comments from other individuals, agencies, and groups having expertise or interest in the subject area under consideration.

The key to providing this information in support of the decision-making process is the "impact assessment" which is an objective analysis conducted to identify and measure the likely economic, social, aesthetic, and environmental effects of the proposed action (activity or project) and the various reasonable alternatives. This requires the identification, measurement, and aggregation of the impacts to provide a "total" assessment.

The identification of impacts requires one to first describe and understand the conditions of the environment prior to the activity. There may be significant differences in impacts for a given activity in different areas. Geographical location is, therefore, one of the factors that affects the relative importance of an impact. For example, the impact a specific project on water quality in an area with abundant water supplies would differ significantly from the impact of the same project on an area with scarce water resources. Furthermore, the timing and duration of each significant impact should be determined. Impacts should be described to establish their effect on the immediate project area, within the adjacent area and the community (or region) as a whole. The timing of impacts should be identified to establish whether they are likely to occur during construction, shortly after the project is completed, or at some later time. The duration of impacts should be identified to establish whether they are reversible or irreversible, and whether they are short-term or long-term.

The measurement of impacts is the next logical step in the impact assessment process. Ideally, all impacts should be translatable into common units. However, this is not possible because of the difficulty of defining impacts (such as on income, on noise quality, and on rare or endangered species) in common units. Another difficulty is that in some cases the quantification of impacts may be beyond the state of the art. Therefore, one is generally faced with use of both quantitative and qualitative measurement techniques. In the latter case, it may be necessary to rely on expert judgment to answer the question of how the various dimensions of the environment are affected.

The final step involves the aggregation of project impacts into an overall assessment of the project's effects. One problem at this point is how to aggregate among the different measured impacts (quantitative and qualitative) to arrive at a single measure, or score, for project impact. This would involve expressing the various impact measures in common units or establishing a weighting-of-importance scheme to generate this single measure. This desire is perhaps more ideal than realistic or practical. However, even though subjective in nature, schemes do exist for accomplishing this objective. In many cases, however, it may suffice to present the measurement of the impacts in simple terms relative to each impact area such as air quality, community economy, energy, environmental quality, fish and wildlife resources, historical preservation, land use, noise quality, solid waste, transportation, and water quality.

REFERENCES

1. Allardice, David R., George E. Radosevich, Kenneth R. Koebel, and Gustav A. Swanson, *Water Law in Relation to Environmental Quality*, Completion Report No. 55, Colorado State University, 31 March 1974, PB 234 144.
2. Argonne National Laboratory, *Environmental Impact Handbook for Highway Systems*, Report No. 74–27, June 1974, PB 239 615.
3. Atomic Energy Commission, *Preparation of Environmental Reports for Nuclear Power Plants*, Regulatory Guide 4.2, March 1973.
4. ———, "Environmental Protection: Licensing and Regulatory Policy and Procedures," *Federal Register*, Vol. 38 No. 210, 1 Nov. 1973.
5. ———, "Environmental Statements–Operations (Revision)," *Federal Register*, Vol. 39, 14 Feb. 1974.
6. Barbaro, Ronald and Frank L. Cross, Jr., *Primer on Environmental Impact Statements*, Technomic Publishing Co., Inc., Westport, Conn., 1973.
7. Battelle Columbus Laboratories, *An Assessment Methodology for the Environmental Impact of Water Resource Projects*, Report No. EPA-600/5-74-016, July 1974, PB 240 002.
8. Brecher, Joseph J. and Manuel E. Nestle, *Environmental Law Handbook*, California Continuing Education of the Bar, 1970.
9. Burchell, Robert W. and David Listokin, *The Environmental Impact Handbook*, Center for Urban Policy Research, Rutgers–The State University, New Brunswick, N.J., 1975.
10. Bureau of Sport Fisheries and Wildlife, "Procedures for Preparation of 102(2)(c) Environmental Impact Statements," *Federal Register*, Vol. 37 No. 204, 20 Oct. 1972.
11. Christensen, K., D. Keyes, P. Schaenman, and T. Muller, *State-Required Impact Evaluation of Land Development, An Initial Look at Current Practices and Key Issues*, Land Use Working Paper No. 214-02, The Urban Institute, July 1974, PB 239 877.
12. CLM/Systems, Inc., *Airports and Their Environment: A Guide to Environmental Planning*, Report No. DOT P 5600.1, Sept. 1972, PB 219 957.
13. Council on Environmental Quality, *102 Monitor*, Vol. 1 No. 10, Nov. 1971.
14. ———, "Preparation of Environmental Impact Statements: Guidelines," *Federal Register*, Vol. 38 No. 147, 1 Aug. 1973.
15. ———, *Fifth Annual Report*, 1974.
16. Department of City Planning, City of Los Angeles, *EIR Manual for Private Projects*, Aug. 1975.
17. Dickerson, William D., *Guidelines for Review of Environmental Impact Statements*, "Vol. I: Highway Projects," Environmental Protection Agency, Sept. 1973, PB 229 726.
18. Dickert, Thomas G. and Katherine R. Domeny, *Environmental Impact Assessment: Guidelines and Commentary*, University Extension, University of California, Berkeley, 1974.
19. Dolgin, Eric L. and Thomas G. P. Guilbert, *Federal Environmental Law*, Environmental Law Institute, West Publishing Co., St. Paul, Minn., 1974.
20. Ellis, Robert H., Phillip B. Cheney, and David R. Zoellner, *A Method for Applying Research to Environmental Planning and Management: A Case Study of Issues Important to the Chesapeake Bay Region*, CEM Report No. 491, The Center for the Environment and Man, Inc., May 1973, PB 227 719
21. Federal Power Commission, "Environmental Impact Statements: Procedures for Preparation and Submission," *Federal Register*, Vol. 37 No. 212, 2 Nov. 1972.
22. ———, *Order Amending Part 2 of the General Rules to Provide Guidelines for the Preparation of Applicants' Environmental Reports Pursuant to Order No. 415-C*, Order No. 485, 7 June 1973.
23. ———, "Part 2–General Policy and Interpretations: Guidelines for Preparation of Applicants' Environmental Reports," *Federal Register*, Vol. 38 No. 117, 19 June 1973.
24. General Services Administration, "Environmental Impact Statements: Preparation Procedures," GSA Order Adm. 1095.1, *Federal Register*, Vol. 40, 4 April 1975.
25. Green, Harold P., *The National Environmental Policy Act in the Courts (January 1, 1970–April 1, 1972)*, The Conservation Foundation, Wash., D.C., May 1972.
26. Hemenway, Gail D., *Developer's Handbook–Environmental Impact Statements*, Associated Home Builders of the Greater Eastbay, Inc., Berkeley, Calif., 1973.
27. International Boundary and Water Commission, U.S. Section (El Paso, Tex.), "Proposed Opera-

1-58   Concepts of Environmental Impact Analysis

tional Procedures for Implementing Section 102(2)(c) of the National Environmental Policy Act of 1969," *Federal Register*, Vol. 37 No. 204, 20 Oct. 1972.
28. Interstate Commerce Commission, "National Environmental Policy Act of 1969: Guidelines for Implementation," *Federal Register*, Vol. 40 No. 166, 26 Aug. 1975.
29. Jain, R. K. and L. V. Urban, *A Review and Analysis of Environmental Impact Assessment Methodologies*, Technical Report No. E-69, Construction Engineering Research Lab., Champaign, Ill., June 1975, AD/A-013-359.
30. Jain, R. K., L. V. Urban, and G. S. Stacey, *Handbook for Environmental Impact Analysis*, Report No. CERL-TR-E-59, Construction Engineering Research Lab., Champaign, Ill., Sept. 1974, AD/A-006 241.
31. Johanning, James and Antti Talvitie, *State-of-the-Art of Environmental Impact Statements in Transportation*, Report No. UMTA-OK-11-0016-74-2, Dept. of Civil Eng. and Env. Eng., University of Oklahoma, Nov. 1974, PB 244-539.
32. Muller, Tom, *Fiscal Impacts of Land Development: A Critique of Methods and Review of Issues*, The Urban Institute, Wash., May 1975, PB 242 200.
33. National Highway Traffic Safety Administration, "Preparation of Environmental Impact Statements," *Federal Register*, Vol. 38 No. 245, 21 Dec. 1973.
34. Schaenman, Philip S. and Thomas Muller, *Measuring Impacts of Land Development, An Initial Approach*, Publication No. URI 86000, The Urban Institute, Nov. 1974, PB 242 211.
35. Tennessee Valley Authority, "Environmental Quality Management: Policy and Procedures," *Federal Register*, Vol. 39, 14 Feb. 1974.
36. U.S. Department of Agriculture, Farmer's Home Administration, "Program-Related Instructions: Environmental Impact Statements, Proposed Redesignation-Revision," *Federal Register*, Vol. 40 No. 246, 22 Dec. 1975.
37. ——, Forest Service, "Environmental Statements: Guidelines for Preparation," *Federal Register*, Vol. 39, 30 Oct. 1974.
38. ——, Rural Electrification Administration, "Rural Electric and Telephone Programs: Environmental Protection," REA Bulletin 20–21, *Federal Register*, Vol. 39, 27 June 1974.
39. ——, Soil Conservation Service, "Preparation of Environmental Impact Statements–Guidelines," *Federal Register*, Vol. 39, 3 June 1974.
40. U.S. Department of Army, Corps of Engineers, "Water and Related Land Resources; Feasibility Studies, Policies and Procedures, Part II," *Federal Register*, Vol. 40 No. 217, 10 Nov. 1975.
41. U.S. Department of Defense, Corps of Engineers, U.S. Army, "Administrative Procedure: Environmental Impact Statements," *Federal Register*, Vol. 39, 8 April 1974.
42. U.S. Department of Defense, Office of the Secretary, "Environmental Considerations in Department of Defense Actions," *Federal Register*, Vol. 39, 26 April 1974.
43. U.S. Department of Health, Education, and Welfare, Food and Drug Administration, "Environmental Impact Statements: Proposed Amendments to Procedures for Preparation," *Federal Register*, Vol. 39 No. 74, 16 April 1974.
44. U.S. Department of Housing and Urban Development, Office of the Secretary, "Protection and Enforcement of Environmental Quality: Policies, Procedures and Responsibilities," *Federal Register*, Vol. 39 No. 37, 22 Feb. 1974.
45. ——, Office of the Secretary, "Environmental Review Procedures: Proposed Policies and Procedures," *Federal Register*, Vol. 39 No. 198, 10 Oct. 1974.
46. ——, *Environmental Assessments for Project Level Actions, A Guidance Document*, HUD-CPD-66, Sept. 1974.
47. U.S. Department of Housing and Urban Development, "Procedures for Protection and Enhancement of Environmental Quality," Department Handbook 1390.1, *Federal Register*, Vol. 38, 18 July 1973 and Amendments, *Federal Register*, Vol. 39, 4 Nov. 1974.
48. U.S. Department of Interior, Bureau of Indian Affairs, "Environmental Impact Statements," *Federal Register*, Vol. 37 No. 204, 20 Oct. 1972.
49. ——, Bureau of Land Management, *Environmental Protection and Enhancement*, BLM Manual, 13 June 1974.
50. ——, Bureau of Reclamation, *Environmental Guidebook for Construction*, 1973.
51. ——, National Park Service," Part 60–National Register of Historic Places, Nominations by States and Federal Agencies," *Federal Register*, Vol. 41 No. 6, 9 Jan. 1976.
52. U.S. Department of Justice, Law Enforcement Assistance Administration, "Regulations Relating to the LEAA Implementation of the National Environmental Policy Act," *Federal Register*, Vol. 39, 6 Feb. 1974.
53. U.S. Department of Transportation, Federal Aviation Administration, *Instructions for Processing Airport Development Actions Affecting the Environment*, Order No. 5050.2A, 24 Feb. 1974.
54. ——, "Processing Airport Development Actions Affecting the Environment, Part II: References to Applicable Procedures," *Federal Register*, Vol. 40 No. 162, 20 Aug. 1975.
55. U.S. Department of Transportation, Federal Highway Administration, "Guidelines for Consideration of Economic, Social, and Environmental Effects," *Federal Register*, Vol. 37 No. 197, 11 Oct. 1972.
56. ——, "Part 771–Environmental Impact and Related Statements," *Federal Register*, Vol. 38 No. 210, 1 Nov. 1973.

57. ——, "Part 490–Special Programs, Economic Growth Center Development Highways," *Federal Register*, Vol. 39 No. 190, 30 Sept. 1974.
58. ——, "Design Approval and Environmental Impact," *Federal Register*, Vol. 39, 2 Dec. 1974.
59. ——, Office of the Secretary, "Procedures for Considering Environmental Impacts," *Federal Register*, Vol. 38 No. 210, 1 Nov. 1973.
60. ——, Saint Lawrence Seaway Development Corporation, "Procedures for Considering Environmental Impacts," *Federal Register*, Vol. 40, 26 April 1975.
61. ——, Urban Mass Transit Administration, "Implementation of Section 102(2)(c) of the National Environmental Policy Act of 1969, Section 4(f) of the Department of Transportation Act, Section 106 of the Historic Preservation Act, and Sections 3(d) and 14 of the Urban Mass Transportation Act of 1964," *Federal Register*, Vol. 37 No. 204, 20 Oct. 1972.
62. U.S. Environmental Protection Agency, "Environmental Impact Statements: Procedures for Preparation," *Federal Register*, Vol. 37 No. 13, 20 Jan. 1972.
63. ——, *The Challenge of the Environment: A Primer on EPA's Statutory Authority*, July 1972, PB 228 025.
64. ——, Region X, *Guidelines for Preparation of Environmental Statements for Reviewing and Commenting on Environmental Statements Prepared by Other Federal Agencies*, April 1973, PB 226 998.
65. ——, Region I, *Guides to Environmental Planning, Assessments and Impact Statements for Water Quality Management Plans and Municipal Wastewater Treatment Projects*, Feb. 1973 (revised Oct. 1974).
66. ——, "Environmental Impact Statements: Procedures for Voluntary Preparation," *Federal Register*, Vol. 39, 21 Oct. 1974.
67. ——, "Preparation of Environmental Impact Statements," *Federal Register*, Vol. 40, 14 April 1975.
68. Veterans Administration, "Environmental Impact Program," VA Manual MP-1, Part 1, Chap. 9, *Federal Register*, Vol. 37, 28 April 1972 and Amendments, *Federal Register*, Vol. 39, 17 June 1974.
69. Warner, Maurice L. et al, *Final Report on Energy from Coal: Guidelines for the Preparation of Environmental Impact Statements*, Battelle Columbus Laboratories, 20 April 1975, PB 242 960.
70. Wharton, James C., *Judicially Enforceable Rights Under NEPA*, Report No. UCRL-51828, Lawrence Livermore Laboratory, University of California, 21 May 1975.
71. Yarrington, Hugh J., *The National Environmental Policy Act*, Environment Reporter, Monograph No. 17, The Bureau of National Affairs, Inc., 4 Jan. 1974.

Chapter 2

Socioeconomic Impact Analysis

JOHN G. RAU

"Socioeconomic" is defined as being "of, relating to, or involving a combination of social and economic factors."[1] Within the National Environmental Policy Act of 1969, there are numerous references to social and economic factors such as:

1. Sec. 101 (a): "the profound influences of population growth, high-density urbanization, industrial expansion, resource exploitation" and "create and maintain conditions under which man and nature can exist in productive harmony, and fulfill the social, economic, and other requirements of present and future generations of Americans"

2. Sec. 101 (b)(2): "assure . . . esthetically and culturally pleasing surroundings"

3. Sec. 101 (b)(4): "preserve important historic, cultural, and natural aspects of our national heritage"

4. Sec. 101 (b)(5): "achieve a balance between population and resource use"

5. Sec. 102 (B): "insure that presently unquantified environmental amenities and values may be given appropriate consideration in decision making along with economic . . . considerations"

6. Sec. 201 (1): "the status and condition of . . . urban, suburban, and rural environment"

In general, socioeconomic factors that can be considered in the assessment of environmental impact range from social impact (population growth, density, aesthetics, standards of living, congestion, incompatibility with surrounding community, increase in recreational requirements, and conflict in lifestyles) to additional requirements for public services (water, sanitation, telephone, natural gas, electric facilities, police and fire protection, solid waste disposal, and overloading of schools). For example, every 1000 new people in a community will require:[2]

1. An additional supply of 100,000 to 200,000 gal of water daily, or 35 to 70 million gal/yr, or 300 individual water well systems, many equipped with water conditioners

[1]*Webster's Third New International Dictionary of the English Language Unabridged*, G & C Merriam Company, Publishers, Springfield, Mass., 1966

[2]Adapted from *United States Municipal News*, vol. 22, no. 16 (August 15, 1955); and "Environmental Health in Community Growth," *Am. J. Public Health*, vol. 53, no. 5 (May 1963). This is intended to be illustrative and not necessarily typical.

2. The collection and disposal of 4000 to 6000 lb of solid waste daily, or 730 to 1100 tons/yr

3. Recreational facilities to serve more people with more leisure time

4. Sewage treatment works to handle 100,000 to 150,000 gal/day, or 35 to 53 million gal of sewage/yr containing 170 lb of organic matter (biochemical oxygen demand) per day, or 62,000 lb/yr, and 70,000 lb of dry sewage solids/yr, or 300 additional septic tanks and appurtenant subsurface disposal facilities

5. A minimum of 4.8 new elementary schoolrooms, 3.6 new high school rooms, and additional teachers

6. At least 10.0 acres of land for schools, parks, and play areas

7. Approximately 1.8 policemen and 1.5 firemen; also new public service employees in public works, welfare, recreation, health, and administration

8. More than a mile of new streets

9. More streets to clean and free of snow and ice, and to drain

10. Two to four additional hospital beds, three nursing home beds, and appurtenant facilities

11. At least 1000 new library books

12. More automobiles, retail stores, services, commercial and industrial areas, county and state parks, and other private enterprises

Whether or not all of these factors need to be addressed in evaluating the environmental impact of a proposed project depends on which factors are "significant." Tools and techniques for measuring (or predicting) these types of impacts and the subsequent assessment of their significance comprise the basic subject of this chapter.

TYPES OF SOCIOECONOMIC IMPACTS

As discussed in the section Concept of Environment in Chapter 1, the dimensions of the environment are rather broad and can be categorized into various subcategories such as those concerned with the physical environment, the social environment, the aesthetic environment, and the economic environment. Dimensions which are socioeconomic-related would thus include:

Physical Environment

- Surrounding land uses and physical character of the area—type of development; densities; building height; design, intensity; lot sizes
- Infrastructure/public services—water supply sources, quality, and distribution; sanitary sewage and solid waste disposal facilities; storm sewers and drainage; energy resources; transportation facilities servicing site

Social Environment

- Community facilities and services—location and capacity of schools; neighborhood, community, and regional parks servicing area; recreational and cultural facilities; police, fire, health, and social service facilities servicing area; local public transportation
- Employment centers and commercial facilities servicing area
- Character of community—socioeconomic and racial characteristics; community life; population size; housing conditions

Aesthetic Environment

- On-site existence of or proximity to significant historic, archaeological, or architectural sites or property
- Scenic areas, views, vistas, and natural landscape
- Architectural character of existing buildings

Economic Environment

- Employment and unemployment levels
- Level and sources of income
- Economic base of area
- Land ownership including private, local, public, state, and federal
- Land values

For example (Ref. 8), the types of socioeconomic impacts which generally could be expected to occur from implementing a particular project are shown in Table 2.1.

TABLE 2.1   Illustrative Types of Socioeconomic Impacts

Impact area	Potential changes
General characteristics and trends in population for state, substate region, county, and city	Increase or decrease in population
Migrational trends in study area (study area is a function of alternatives and available data base)	Increase or decrease in migrational trends
Population characteristics in study area, including distributions by age, sex, ethnic groups, educational level, and family size	Increase or decrease in various population distributions, people relocations
Distinct settlements of ethnic groups or deprived economic/minority groups	Disruption of settlement patterns, people relocations
Economic history for state, substate region, county, and city	Increase or decrease in economic patterns
Employment and unemployment patterns in study area, including occupational distribution and location and availability of work force	Increase or decrease in overall employment or unemployment levels and change in occupational distribution
Income levels and trends for study area	Increase or decrease in income levels
Land use patterns and controls for study area	Change in land usage, may or may not be in compliance with existing land use plans
Land values in study area	Increase or decrease in land values
Tax levels and patterns in study area, including land, sales, and income taxes	Changes in tax levels and patterns resulting from changes in land usage and income levels
Housing characteristics in study area, including types of housing and occupancy levels and age and condition of housing	Changes in types of housing and occupancy levels
Health and social services in study area, including health, workforce, law enforcement, fire protection, water supply, wastewater treatment facilities, solid waste collection and disposal, and utilities	Changes in demand on health and social services
Public and private educational resources in study area, including K–12, junior colleges, and universities	Changes in demand on educational resources
Transportation systems in study area, including highway, rail, air, and waterway	Changes in demand on transportation systems, relocations of highways and railroads
Community attitudes and lifestyles, including history of area voting patterns	Changes in attitudes and lifestyles
Community cohesion, including organized community groups	Disruption of cohesion
Tourism and recreational opportunities in study area	Increase or decrease in tourism and recreational potential
Religious patterns and characteristics in study area	Disruption of religious patterns and characteristics
Areas of unique significance such as cemeteries or religious camps	Disruption of unique areas

In the recognition of these particular dimensions of the environment, and the assessment and identification of the corresponding environmental impacts (if any) of a project or activity, it is useful to have a checklist of potential impact areas. A typical checklist for socioeconomic impacts is presented in Table 2.2. This list is *not necessarily all-inclusive*, but includes areas of potential concern in performing a socioeconomic impact assessment.

Examples of Types of Socioeconomic Impacts

Airport Projects With regard to airport-related projects for the expansion of services (including new terminals, construction of new runways, extension of existing runways to accommodate wide-bodied aircraft), social impacts would generally include: housing required for new employees, increased risk of accidents to adjacent neighborhoods because of expanded aircraft activity, increase in traffic flow and congestion at and around airport, disruption in area due to construction activities, increased demand for local public transportation services, increased crowdedness in airport vicinity, increase in transient population in the area, health and lifestyle impairment because of noise effects, increase in population in the area, potential displacement of people, potential removal of businesses and/or homes, increase in air transportation service to area, and a change in character of the surrounding community. Similarly, economic impacts would generally include: new jobs created both from construction and long-term activity; general growth in commercial and industrial activity in the area; potential loss of taxable property due to acquisition of private lands; increased cost for public services such as police and fire protection, traffic control, and street expansion; change in adjacent property values; increased energy consumption at airport facilities; and an increase in local sales tax revenues and other tourist-oriented revenues.

As mentioned in the section on Factors for Consideration in Assessing Airport Impacts in Chapter 1, the Department of Transportation Act of 1966 as amended by the Federal-Aid Highway Act of 1968 Section 4(f) states that it is necessary to avoid the taking or detrimental use, as from flyovers, of such land as public parks, recreation areas, wildlife and waterfowl refuges, historic sites, and areas of natural scenic beauty of local, state, or national significance. The Secretary of Transportation is prohibited from approving any program or project requiring the use of such land unless there is no prudent and feasible alternative to its use and the program includes all possible planning to minimize harm. Thus, in making an assessment of the impact on these types of lands, it would be necessary to describe the size, activities, patronage, access changes, unique or irreplaceable qualities, and the relationship to other similarly used lands in the vicinity. From the point of view of socioeconomic impacts, the checklist in Table 2.2 would serve as a useful guide.

Coal Refinery Projects In assessing the socioeconomic impacts associated with the siting, construction, and operation of large-scale coal refining complexes (whose typical components would include the coal refinery, a power plant, the coal extraction and rehabilitation operation, water supply, transportation and transmission facilities, and contiguous communities), the major area of concern would generally include (Ref. 31):
1. creation of new industrial areas
2. influx of population
3. large land area commitment
4. interference with other routes of communication
5. destruction of scenic, historical, and archaeological sites
6. postdevelopment responsibility.

It is reasonable to expect that in the planning and siting of a large-scale coal refinery the area is being committed to a long-term industrial land use. This is the result of a number of events which can be expected to occur. The employment base created by the construction and operation of the coal refinery and power plant and the associated mining operations will be a nucleus for further industrialization of the area. Other primary manufacturing industries, primarily in the chemical field, will be attracted to the area by the availability of raw materials and power. Service industries will be created to support the refinery, power plant, mining, and other primary industries. Secondary employment opportunities will be created in the commercial, housing, construction, and related fields.

The introduction of a coal refinery complex and the subsequent industrialization will cause the population of the area to increase many times over the number of people directly engaged in the refinery, power plant, and mining operations. Residential devel-

TABLE 2.2 Socioeconomic Factor Checklist

COMMUNITY RESOURCES:

General and local community plans

Circulation	Safety
Conservation	Scenic highway
Housing	Seismic safety
Land use	Energy management/conservation
Noise	Air pollution control
Open space	Growth management

Human habitat

Aesthetics and visual effects	Growth patterns
Barriers and privacy	Health and safety
Community design	Population density and dispersion
Corridors	Urban congestion
Displacement and relocation	Community cohesion

Public services

Fire protection	Recreation and parks
Flood control and storm drains	Schools
Medical service	Public transportation
Police protection	Traffic flow and circulation
Postal service	

Utilities

Communications	Sewage waste disposal
Electrical power	Solid waste disposal
Natural gas	Water

CULTURAL AND SCIENTIFIC RESOURCES:

Community facilities

Cultural	Scientific
Religious	Special entertainment

Regional

Aesthetics	Historical sites
Archaeological sites	Paleontological sites

ECONOMY:

Community expenses

Public services	Others
Utilities	

Community revenue sources

Building permits	Sales tax
Fees (motor vehicles, etc.)	Utility connections
Fines	Utility service charges
Gas tax	Others
Property tax	

Regional expenses

Loans and subsidies	Utilities
Public assistance	Others
Public services	

Regional revenue sources

Fees	Taxes
Fines	Others

opment will occur commensurate with the expanded employment base. Public, social, and community services such as water supply, wastewater treatment, solid waste management, schools, police and fire protection, churches, recreation, etc., must be created to serve the population. Additional employment opportunities will become available through provision of public and community services which will further increase the population.

The commitments of land for large-scale coal refining complexes are varied and include those for water storage, mining operations, industrial development, commercial establishments, housing, public uses, and transportation corridors. Many of these commitments must be regarded as long-term, if not permanent, commitments of land. Transportation corridors would include: roads, rail lines, conveyors, and pipelines to supply coal to the refinery complex; roads, rail lines, conveyors, and slurry pipelines for the management and disposal of solid wastes; gas and liquid pipelines to supply products to export markets; transmission lines for the export of electrical energy; and roads, rail lines, and slurry pipelines for the export of solid products. Typically, right-of-ways are 150 to 200 ft wide and the required land area for transmission corridors will range from 3 to 4 mi^2 per 100 mi of transport. The commitment of land to transportation corridors generally represents a permanent and nonretrievable commitment of land resources.

With regard to other routes of communication, the imposition of transportation corridors may cause societal and economic conflicts. A network of transportation systems and corridors required to bring materials to and from the coal refinery complex may extend for many miles and can involve the commitment of significant land resources. Rail hauling of refinery products may have to compete for usage of the same rail system which currently handles the export of other products. The quantities of material to be exported from the refinery when combined with existing commerce could cause severe interference with intersecting transportation systems. In addition, the mining operations may also interfere with existing operations and require the relocation of highways, rail lines, gas pipelines, and other transportation systems.

Certain natural and aesthetic values in an area will be irretrievably lost by the creation of an industrial area resulting from the installation of a coal refinery. These values would potentially include scenic, historical, and archaeological sites, as well as other sites of cultural and scientific importance. A different environment will be created which may not be as desirable as the original environment. On the other hand, environmental costs and commitments can be minimized by the selection and preservation of selected areas. The benefits of these areas can also be enhanced by developing them and making them more accessible to a greater number of people. Furthermore, the temporal losses of aesthetic and human interest values due to mining operations can be minimized by reducing the time from initial exposure to acceptable rehabilitation and by rehabilitation to higher land uses.

One major environmental issue in the siting of coal refinery complexes deals with the future of the area beyond the design life of the coal refinery complex and the exhausting of recoverable coal in adjoining mines. The commitment of land to mining and the environmental effects of mining are temporal. With properly planned restoration, the time land is committed to this purpose can be minimized, the adverse environmental impacts can be alleviated, and the use of land can be retrieved. In most surface mining operations, the backfilling of the overburden materials has not been compacted. As a result, the restored areas have a diminution of postmining land uses. In most cases, the land must be committed to interim uses such as agriculture, silvaculture, or recreation until sufficient compaction is attained to permit higher land uses. If compaction is not used, diminution of land use must be accepted as an environmental effect and cost.

All of this suggests that long-range planning is required to assure the future vitality of the area. Such planning might include plans for rebuilding of the complex, the importation of coal from other areas for conversion, creation of new industries, and other options.

The development of a commercial coal refinery complex can also be expected to change the existing ownership pattern in surrounding areas from the small property owner to ownership by large companies. This would result from the need to acquire large acreages for mining operations, the siting of refinery facilities, and the development of service industries and urban areas.

Another significant impact would be the increase or decrease in property values in surrounding areas as a result of industrial, commercial, and urban development and mining operations. Although the latter activity can be expected to decrease property values in closely adjoining areas during the period of active mining and rehabilitation, those areas most suitable for industrial, commercial, and urban use should experience a notable increase in value.

The introduction of a large-scale coal refinery complex into any area will have far-reaching impacts on the surrounding environment. Some of the impacts on society will be positive, others will be negative, while still others will be either positive or negative depending upon each individual's social circumstances or set of values. An adverse socioeconomic impact to one segment of the population may be a positive impact to another. For example, the unemployed or under-employed worker would receive favorably the prospect of obtaining more stable and better paying work as offered by the refinery, but the small, local employer (the farmer or rancher) would consider the prospect of losing his source of labor a negative feature.

No matter what the pros and cons may be on various socioeconomic issues, coal refinery development will produce changes in existing lifestyle, land use, land ownership, economic base, population distribution, and level of urbanization.

Energy-Related Projects Energy- or power-related projects would generally be expected to include such activities as: the construction of natural gas pipeline facilities; electric transmission lines, terminals, and substations; new power plants; coastal energy activities such as Outer Continental Shelf (OCS) energy exploration and development; transportation, conversion, treatment, transfer, or storage of liquefied natural gas; and transportation, transfer, or storage of coal, oil, or natural gas. The socioeconomic impacts of these types of activities are varied and depend upon such factors as type of project, scope, size, duration of activity, and geographical location.

For example, as previously discussed in the section on Factors for Consideration in Assessing Power Project Impacts in Chapter 1, during construction the types of socioeconomic impacts which could be expected to occur would include: effects on labor, housing, local industry, and public services; relocation and displacement of families and businesses in the area; effects on a local basis from new business and housing development and payrolls; need for increased public services such as schools, health facilities, police and fire protection, housing, waste disposal, markets, transportation, communication, energy supplies, and recreational facilities; and impacts on present and future recreational use and potential of the local area or region. During the operation and maintenance phase, the types of socioeconomic impacts which could be expected to occur would include: restrictions on existing and potential land uses in the immediate vicinity; effects upon local traffic patterns including roads, highways, ship channels, and aviation patterns, and the possible need for new facilities; effects on labor, housing and population growth trends, relocation, local industry and industrial growth, and public services; economic benefits resulting from the services and products, energy, and other results of the action such as tax benefits to local and state governments, growth in local tax base from new business and housing developments, and payrolls; need for increased public services such as schools, police and fire protection, housing, waste disposal, markets, transportation, communication and recreational facilities; extent to which maintenance of the area is dependent upon new sources of energy; and potential impacts from accidents and natural catastrophes which might occur.

In the case of coastal-related energy activities, the U.S. Department of Commerce (Ref. 55) specifies that

the coastal zone of a coastal State is 'significantly affected' by the siting, construction, expansion, or operation of an energy facility if such siting, construction, expansion, or operation:
  (a) Causes population influxes in any area under the jurisdiction of a unit of general purpose local government;
  (b) Changes employment patterns in the coastal zone, including those in fishing and tourism;
  (c) Makes necessary new or improved public facilities or services in the coastal zone; or
  (d) Damages or threatens to damage any valuable environmental or recreational resource in the coastal zone, including degradation of air or water quality.

By "public facilities" in (c) is meant the following types of facilities:

1. Education—day care centers; primary, secondary, and general vocational schools, and school equipment; libraries, including books and equipment

2. Environmental protection—facilities and equipment used for air or water quality monitoring or to ensure continued viability of existing environmental resources

3. Government administration—facilities and equipment essential for local government administration

4. Health care—ambulances and associated equipment; clinic and hospital buildings and equipment

5. Public safety—detention centers; police equipment and stations; fire stations and fire-fighting equipment

6. Recreation—facilities and equipment for amateur sports; community recreational centers; local parks and playgrounds

7. Transportation—streets, roads, bridges, and road maintenance equipment; parking associated with public facilities; docks, navigation aids; air terminals in remote areas; mass transit limited to local bus systems

8. Public utilities—local electric generating plant and distribution systems; local natural gas distribution systems; solid waste systems; waste collection and treatment systems (including drainage); local water supply systems.

By "public services" in (c) is meant the salaries of personnel essential to the operation and maintenance of these public facilities.

Socioeconomic impacts of such projects as electric transmission lines, terminals, and substations would include: economic impacts on the affected areas and industries which the proposed facilities will serve, economic impacts during construction, local employment opportunities, payments for rights-of-way, income from sale of goods and services to contractors, increased tax revenues, increased availability of electric power, and development opportunities for local industries.

Prior to the issuance of a construction permit or an operating license for a nuclear power plant, it is necessary (Ref. 23) to conduct a cost-benefit analysis. This procedure considers and balances the environmental effects of the facility and the alternatives available for reducing or avoiding adverse environmental effects, as well as the environmental, economic, technical, and other benefits of the facility.

The primary benefits of a nuclear facility are those inherent in the value of the generated electricity which is delivered to consumers. Other primary benefits would be in the form of sales of steam or other products or services. There are other social and economic benefits which affect various political jurisdictions or interests to a greater or lesser degree. Some of these reflect transfer payments or other values which may partially, if not fully, compensate for certain services as well as external or environmental costs. Typical of these types of benefits would be the following: tax revenues to be received by local, state and federal governments; temporary and permanent new jobs created and payroll; incremental increase in regional product (value-added concept); enhancement of recreational values through making available for public use any parks, artificially created cooling lakes, marinas; enhancement of aesthetic values through any special design measures as applied to structures, artificial lakes or canals, parks; creation and improvement of local roads, waterways, or other transportation facilities; increased knowledge of the environment as a consequence of ecological research and environmental monitoring activities associated with plant operation, and technological improvements from the applicant's research program; creation of a source of heated discharge which may be used for beneficial purposes (e.g., in aquaculture, in improving commercial and sport fishing, and other water sports); and provision of public education facilities (e.g., a visitor's center).

There are also economic and social costs resulting from construction of a nuclear facility and its operation. Temporary or short-term costs would include: shortages of housing; inflationary rentals or prices; congestion of local streets and highways; noise and temporary aesthetic disturbances; overloading of water supply and sewage treatment facilities; crowding of local schools, hospitals, or other public facilities; overtaxing of community services; and the disruption of people's lives or the local community caused by acquisition of land for the proposed site. In the long-term, costs would include: impairment of recreational values (e.g., reduced availability of desired species of wildlife and sport fish, restrictions of access to land or water areas preferred for recreational use); deterioration of aesthetic and scenic values; restrictions on access to areas of scenic, historic, or cultural interest; degradation of areas having historic, cultural, natural, or archaeological value; removal of land from present or contemplated alternative uses; reduction of regional product due to displacement of persons from the land proposed for the site; lost income from recreation or tourism that may be impaired by environmental disturbances; lost income of commercial fishermen attributable to environmental degradation; decrease in real estate values in areas adjacent to the proposed facility; and increased costs to local governments for the services required by the permanently employed workers and their families.

Highway and Transportation-Related Projects   The general Factors for Consideration in Assessing Highway Impacts were discussed in the section so titled in Chapter 1. From the point of view of socioeconomic impacts, the general areas of concern would include the following: removal of residential, commercial, and industrial land uses; displacement of both residents and jobs; removal of structures or sites of scenic, architectural, archaeological, or historic significance; loss of sites having unique potential or suitability for commercial or industrial activities; loss of taxable private land revenues; relocation costs; severance of interpersonal ties of relocatees to former neighborhood/community (family ties, ethnic bonds, neighborhood friendships); circulation impacts such as blocking or impairing access along existing street patterns crossed by the highway, division of land uses, increased traffic flow, and congestion; disruption of neighborhood cohesion and identity; change in accessibility to schools, recreational facilities, and community services; alteration of the boundaries of a school district and the resultant tax base it is dependent upon as the result of right-of-way acquisition; barrier effects on access sensitive services such as emergency vehicle response time; ease of access to shopping and other commercial facilities; effects on future zoning or zone changes on adjacent lands; public safety effects such as changes in accident rates; aesthetic appearance of the area as regards viewlines and architectural contrasts; and financial costs to local community for relocation of utilities such as water, sewer, gas, and electric lines.

With regard to aesthetic impacts, they are probably the most elusive and subjective aspects of an environmental impact assessment. They refer to the artistic quality or natural beauty of the area as well as the appearance and architectural quality of the facility. The appearance of a transportation-related facility to the user and to the nonuser is not necessarily identical. An elevated section of highway or rail can offer panoramic views but may itself form a visual barrier. High fills impede horizontal views, while overhead spans cast ominous shadows and may or may not be an aesthetic liability. The existing landscape, including both natural and constructed features, surrounding proposed projects can also vary in aesthetic quality and importance. The impacts on the natural features will depend upon how highly an area is developed. For example, undeveloped areas have a greater potential for scenic views and, thus, large impacts are possible, whereas urbanized regions which have fewer natural sights, open spaces, and greenery can also be significantly impacted because of the rarity of untouched land. The most important impacts in urban areas are likely to concern architecturally, historically, or culturally valuable buildings or areas. A historical site can form a significant portion of the aesthetic appeal of an area and, thus, any change in access to a site or the displacement of a site are potentially significant adverse impacts.

According to Reference 56, a State Highway Department request for location and design approval must be accompanied by documentation which discusses the anticipated economic, social, and environmental effects on such items as: regional and community growth including general plans and proposed land use, total transportation requirements, and status of the planning process; public facilities and services including religious, health and educational facilities, public utilities, fire protection, and other emergency services; community cohesion including residential and neighborhood character and stability, highway impacts on minority and other specific groups and interests, and effects on local tax base and property values; displacement of people, businesses, and farms including relocation assistance, availability of adequate replacement housing, and economic activity (employment gains and losses, etc.); aesthetic and other values including visual quality such as "view of the road" and "view from the road," and the joint development and multiple use of space.

In summary, the effects on the community of transportation facility location and use must be measured according to how they relate to local transportation, community planning and environment, neighborhood and social structure, and community economic and fiscal structure.

OUTLINE OF THE BASIC STEPS IN PERFORMING THE SOCIOECONOMIC IMPACT ASSESSMENT

The basic steps (Ref. 8) to be followed in the prediction of changes in the socioeconomic environment and the assessment of the impact of these changes can be described as follows:

Step 1: Description of the Socioeconomic Environmental Setting

The first step is to collect pertinent data and information that will enable description of the environmental setting in terms of various selected socioeconomic impact areas such as those described in Table 2.2. Accomplishing this step will generally require the use of various sources of information such as the U.S. Bureau of Census, statistical abstracts of various governmental agencies, planning agencies, chambers of commerce, research departments of local banks and savings and loan companies, university libraries, and others.

Step 2: Identification of Critical Socioeconomic Factors

The second step involves identifying those socioeconomic impact areas that represent critical items relative to the human environment. Examples would include development in an area where public services are either near capacity or overburdened such as sewer and water systems, schools, streets, police and fire protection, recreational amenities, and so forth. To identify these critical impact areas generally requires a comparison of known design service levels with existing service levels, as well as the use of local planning factors and community design standards to assess the needs of the new development and its occupants.

Step 3: Prediction of Changes in Socioeconomic Factors

The third step involves the quantitative prediction, or at least qualitative description, of the changes in the socioeconomic environmental setting as the result of the new development. This requires that one have various tools, techniques, and methodologies for making these predictions. This means that it may be necessary to conduct special polls or surveys to collect data regarding the potential impacts and then infer what the consequences might be if development occurs.

Step 4: Discussion of Implications of Changes

The final step in the prediction and assessment of the impacts on the socioeconomic environment involves a discussion of the implications of these changes for each area impacted. These implications could be examined by comparing the values of the impact measures used, such as students generated prior to development versus students generated after development, or number of park acres required prior to development versus number of park acres required after development. Differences in these values would mean either an increase or decrease in some type of service, demographic characteristic, or other socioeconomic factor. Such changes must then be examined in terms of their impact, whether adverse (if so, then why and what are the consequences) or beneficial, on the socioeconomic environment. In the case of adverse changes, one should attempt to mitigate their impacts as much as possible.

The following sections of this chapter present a discussion of selected tools, techniques, and methodologies which can be utilized to assess impacts in particular socioeconomic areas. The discussion is not all-inclusive in the sense that each possible type of socioeconomic impact is discussed and a method is given for assessing the impact in this area. In fact, this is not possible because of the subjectivity and conjectural nature of many socioeconomic factors. Instead, the discussion presented is intended to be illustrative as to how a socioeconomic impact assessment might be approached in terms of data and informational needs, formulation of quantitative measures to assess the impact, and interpretation of the implications of the values of the quantitative measures. In essence, the following discussion addresses each of the pertinent elements of Steps 1–4.

ANALYSIS OF PUBLIC SERVICES AND FACILITIES IMPACTS

This section deals with measuring the impacts on the traditional public services such as health care, libraries, parks and recreation, police and fire protection, schools, and traffic circulation. Welfare, family planning, and other types of social services are potential impact areas which could be important for various types of projects such as elderly and/or low-to-moderate income housing. However, with the exception of those types of services addressed in this section, these social service impacts are difficult to measure in terms of the increased demand and overall change in quality of the existing level of service.

Land use developments affect the demand for these services, the environment in which they are provided, and thus the overall quality of services. The degree to which these services are affected by development is closely tied to the predevelopment quality of service, the resulting changes in public spending, and the remaining capacity of existing facilities. This suggests that one should not only consider the change in quality and level of service provided but also the fiscal impact (such as increased annual expenditures and the costs of new facilities needed) on the public agencies providing these services. In this section we will address the impacts on quality and level of service, and later in this chapter we will address the corresponding fiscal impacts.

Health Care

The main effect of most developments on the quality of local health care will be on the crowdedness of existing facilities and their accessibility to both the newcomers and the existing residents. In addition, large development projects, or the cumulative effect of many land use projects, may stimulate the construction of new hospitals, clinics, medical laboratories, etc. This may improve the quality of health services available to the existing community, as well as ease the problems of crowdedness and accessibility. The location of a development with respect to nearby hospitals and local traffic conditions will affect the travel time between an emergency room and the development. If, as a result of the development, there is a change in local traffic conditions, the travel time to health care facilities for persons in the surrounding community may be affected as well.

The preceding discussion suggests that measures, or indices, which could be used to assess the impact on health care facilities/service are provided by (Ref. 49):

    1. Average travel time for a person in the development to reach an emergency room

    2. Change in average travel time for a person in the surrounding community to reach an emergency room

    3. Change in area hospital bed need

    4. Hospital bed demand versus remaining hospital bed capacity of the area.

Items 3 and 4 require the ability to estimate hospital bed needs. In general (Ref. 4), a minimum standard of 4.5 beds per 1000 population can be used as an order-of-magnitude estimate of hospital bed needs.

Another indicator of crowdedness of local hospitals is reflected in the average waiting time for admission to a hospital for elective surgery. This particular measure is potentially not as accurate an indicator as those provided in the preceding list since it is generally influenced by other nondevelopment-related factors such as local admission policies, health insurance, personal income, and doctors' policies about hospitalization.

In terms of data sources and methods of computing these measures of impact on health care facilities and services, travel time estimates can be obtained from local surveys which give consideration to the locations of new facilities built and the expected changes in traffic flows and congestion. Changes in waiting times can be computed by assuming some representative frequency and duration of hospital visits both before and after the development occurs, and then computing the corresponding average waiting times. Remaining hospital bed capacities can be readily determined by surveying the hospitals serving the area and their average daily demand for bed space. An important consideration in estimating facility crowdedness is the fact that people tend to go to those facilities associated with their doctors, or which are reported to be especially good for the service they seek.

Libraries

Residential land use developments create the need for library services through the demand for books, reading materials, and educational films. This demand is generally met through two levels of service, namely: (1) the use of bookmobiles to meet rural/suburban needs not sufficient to warrant the construction of a library, and (2) the construction of community or branch libraries. The latter generally serves a population of 40,000–60,000 persons and has a service area of 2–5 miles, whereas bookmobiles generally provide services to areas containing 10,000–15,000 persons.

Generally speaking, the best one can do in estimating the impact on library facilities is to assess whether or not existing facilities are being overutilized or underutilized in terms of demand for reading materials and citizen utilization of space. An indicator of the former would be the number of books (or library items) checked out per capita, whereas an indicator of the latter would be the average percent occupancy of seats within the library.

TABLE 2.3   Typical Library Experience Factors

Population size	Book stock volumes per capita	No. of seats per 1000 population	Circulation volumes per capita	Total sq. ft. per capita
Under 10,000	3½–5	10	10	0.7–0.8
10,000–35,000	2¾–3	5	9.5	0.6–0.65
35,000–100,000	2½–2¾	3	9	0.5–0.6
100,000–200,000	1¾–2	2	8	0.4–0.5
200,000–500,000	1¼–1½	1¼	7	0.35–0.4
500,000 and up	1–1¼	1	6.5	0.3

SOURCE: Joseph L. Wheeler and Herbert Goldhor, *Practical Administration of Public Libraries*, Harper, New York, 1962, p. 554.

Development in a library service area where facilities are currently underutilized can be expected to have a relatively insignificant impact. However, when facilities are overutilized, a potentially significant adverse impact would occur.

Another way of approaching the assessment of the impact on library facilities would be to utilize existing "standards" or "experience formulas" to estimate such quantities as the number of new book stock volumes required, the number of new seats required, the number of new circulation volumes required, and the total new square footage of library space required. For example, Table 2.3 provides some illustrative factors based on a sample of libraries. Utilizing data of this type, one could estimate that each new 1000 people in a community of size 100,000–200,000 would generate the need for

1. 1750–2000 new book stock volumes
2. 2 new seats
3. 8000 new circulation volumes
4. 400–500 ft^2 of additional library space.

Parks and Recreation Facilities

Land development can affect the variety, accessibility, crowdedness, safety, and overall enjoyability of recreation and recreational facilities in the community by adding or eliminating facilities, by changing the numbers and types of potential users, and by changing the environment around these facilities. The facilities affected may include open space areas such as empty fields, easements and conservation areas, tennis courts, swimming pools, picnic areas, lawn bowling greens, playing fields, riding stables and horse trails, bicycle paths, boating areas, wetlands, woodland trails. It is not always easy to predict the direct impact on these types of facilities as the result of development. However, an indirect assessment of the impact of land developments on recreation is conducted through the use of planning standards. For example, in the case of parks the typical standard used is a specified number of acres of parks per 1000 population. Tables 2.4–2.6 provide illustrative examples of standards of this type.

Use of these standards in a formula of the form

$$\left(\begin{array}{c}\text{Number}\\ \text{of new}\\ \text{homes}\end{array}\right)\left(\begin{array}{c}\text{Number of}\\ \text{people per}\\ \text{home}\end{array}\right)\left(\begin{array}{c}\text{Number of}\\ \text{park acres}\\ \text{per capita}\end{array}\right)$$

would provide an estimate of the number of park acres required to support this new development. Comparing this required park acreage with that currently available in the

TABLE 2.4   Existing and Proposed Local and Regional Open Space Standards—Selected Metropolitan Areas
(in Acres per 1000 Population)

	Local	Regional
Association of San Francisco Bay Area Governments	15	15
P. H. Lewis: Recreation and Open Space in Illinois	20	40
National Recreation & Park Association	10	15
U.S. Bureau of Outdoor Recreation	5	12
Twin Cities Metropolitan Planning Commission (Minneapolis/St. Paul)	10	10
Arizona Outdoor Recreation Coordinating Commission	7	18
Wisconsin Outdoor Recreation Plan	25	80
Connecticut Department of Agriculture and Natural Resources	10	20
Baltimore, Maryland Regional Planning Council	14	15
Capital Region Planning Agency, Hartford, Conn.	15	50
New Jersey Division of State and Regional Planning	8	12
Central Massachusetts Regional Planning Commission	10	65

TABLE 2.5   Park Standards by Type and Population Served

Classification	Acres/1000 people	Size range	Population served	Service area
Playlots	N.A.	2500 ft² to 1 acre	500–2,500	Subneighborhood
Vest pocket parks	N.A.	2500 ft² to 1 acre	500–2,500	Subneighborhood
Neighborhood parks	2.5	Min. 5 acres up to 20 acres	2,000–10,000	¼–½ mile
District parks	2.5	20–100 acres	10,000–50,000	½–3 miles
Large urban parks	5.0	100+ acres	One for each 50,000	Within ½ hr. driving time
Regional parks	20.0	250+ acres	Serves entire pop. in smaller communities; should be distributed throughout larger metro areas	Within 1 hr. driving time
Special areas and facilities	N.A.	Includes parkways, beaches, plazas, historical sites, flood plains, downtown malls, and small parks, tree lawns, etc. No standard is applicable.		

SOURCE: National Recreation and Park Association

community would provide an indication of whether or not the existing level of park service is adequate to meet this demand. For example, using a park standard of 4.0 acres per 1000 population, this would imply the need for 6.4 acres to serve a new development of 500 homes with an average of 3.2 people per home:

$$(500 \text{ homes})(3.2 \text{ people per home})(0.004) = 6.4 \text{ acres}$$

TABLE 2.6   Illustrative Recreational Planning Standards

Activity	Participants	Total population
Beach activities	1 person per ½ linear ft of shoreline 1 person per 100 ft²	1 linear ft. of shoreline per 20 persons
Bicycling	40 persons per mile	
Boating and fishing	50 boats per ramp	
Boat access	150 persons per ramp	
Camping (improved)	4 persons per site 24 persons per acre	
Camping (highly developed)	4 persons per site 48 persons per acre	
Fishing	1 person per 10 linear ft of shoreline, pier, etc.	
Hiking	40 persons per mile	
Horseback riding	20 persons per mile	
Nature study and appreciation	1 person per 5 acres 40 persons per mile 100 persons per mile (guided)	
Picnicking	5 persons per site 100 persons per acre	1 acre per 6,000 persons 20 sites per 6,000 persons
Pool swimming	1 person per 15 ft²	1 pool per 10,000 persons
Hunting	1 person per 160 acres	
Baseball-regulation	20 persons per field	1 field per 6,000 persons
Baseball-junior	20 persons per field	1 field per 3,000 persons
Baseball-softball	20 persons per field	1 field per 3,000 persons
Tennis	4 persons per court	1 court per 2,000 persons
Basketball	10 persons per court	1 court per 500 persons
Recreation building		1 building per 15,000 persons
Outdoor theatre		1 theatre per 20,000 persons
Shooting range	2 persons per position	1 range per 50,000 persons
Golf course	8 persons per hole (regulation) 4 persons per hole (par 3)	1 course per 25,000 persons
Equipped play area	1 person per 10 ft²	1 acre per 3,000 persons
Football	25 persons per field	1 field per 10,000 persons
Multi-use court	20 persons per court	1 court per 500 persons
Shuffleboard	4 persons per court 8 courts per center	1 court per 6,000 persons

SOURCE: National Recreation and Parks Association

If there are currently 50 acres of parks serving the community and the total estimated demand based on applying this standard to the total residential population of the area is 40 acres, then the new total demand of 46.4 acres could theoretically be met. However, if the total estimated demand were 75 acres, then the new total demand of 81.4 acres could theoretically lead to increased crowdedness and lack of overall enjoyability at local park facilities.

One word of caution with regard to standards is that a standard for urban parks based on population is a most misleading planning tool because it considers all demand and need to be homogenous. In essence, it implies that 1000 residents living in housing on one-acre sites will have the same need for parks as 1000 residents living in high-density areas without any usable active or passive open space. On the face of it, this implied assumption is unrealistic.

When a park standard is derived, it is more likely that the basis for the standard is some estimate of demand for local recreational facilities, rather than some arbitrary considera-tion as would be used for planning purposes. A customary tool for estimating demand is a calculation of the number of visitor days occurring annually at present park sites. While current usage can be a helpful guideline in developing an estimate of need, the estimation of demand by using current-use data can lead to an erroneous view of the actual demand. First, an underestimation will result if present park lands are not reasonably convenient to the population they intend to serve. Furthermore, existing park lands may not have sufficient facilities to support normal recreational use, or the land and/or facilities pro-vided are currently reaching their support capacity. These factors will suppress an expression of demand through current use. Thus, a reservoir of latent demand will exist that cannot be estimated by the application of current-use data. Conversely, there can be factors present that lead to an overestimation of demand. This will be the case if recreational alternatives are limited or not available.

When other sources of recreational activities suffer the same characteristics just men-tioned for local parks, there will be a tendency to substitute the facilities provided by local parks when they offer the next best alternative. An example of these factors will illustrate the effect of skewing observable demand (current usage) from the actual demand. If insufficient or inadequate bicycle trails are provided when substantial demand exists to support them, current usage will be an underestimation of the actual demand for this facility. Additionally, the lack of sufficient trails for bicycle riding may generate an increase in the use of other facilities, such as local park playgrounds, due to the substitu-tion of an expressed demand for a latent one. Here, current-use information may lead to an overestimation of demand for local park facilities. Hence, undue reliance upon observed usage may further intensify an imbalance in the level or type of facilities provided.

Police and Fire Protection

The impact on police and fire protection services due to new development can generally be divided into two categories, namely: either a change in the level of "activity" requiring such protection, or a requirement for new facilities and/or personnel to provide the necessary level of protection.

For example, developments may affect crime rates in a community by adding or removing targets (people, businesses, residences, etc.) and by changing physical or social conditions that may breed crime or make crimes easier to commit. In this case, crime rate, as expressed in such terms as number of major crimes against residents per 100,000 people or number of major crimes against businesses per 1000 businesses, provides an indicator of the level of activity requiring police protection. Crime attracted by develop-ment may spill over into the existing community or the new development might attract crime away from the existing community. As a result, the ability of police to deter crime could be affected by development in the sense that increases in noncrime duties, such as traffic control, may reduce the time available for patrol duties. The extent to which this ability to provide crime protection is influenced by new development is a complex issue determined by many factors not always easily measured. This discussion suggests that a measure of the change in the level of activity is given by a comparison of the crime rates before and after development. This would require the conduct of surveys and the examination of police reports to ascertain whether or not there was a significant change in the crime rate for the development area as well as for the entire community. A major difficulty here, given a change in crime rate, is to show that this change was indeed caused by the development. Furthermore, without understanding all the factors in a community that affect the local crime rate, it is difficult to predict the after-development crime rate before development occurs.

Police department personnel needs are usually based on citywide needs, given deployment requirements on a geographical basis. This suggests the use of personnel allocation formulas requiring factors such as the number of sworn personnel per capita and the number of civilian personnel per capita. For example (Ref. 16), the City of Los Angeles Police Department projects a need for 3.0 police officers per 1000 population to provide the proper citizen/officer ratio and adequate reserves. Table 2.7 provides an illustration of these types of ratios according to city size.

TABLE 2.7   Median Number of Full-time Police Department Personnel (Uniformed and Civilian) Per 1000 Population

Classification	No. of reporting cities	Median (uniformed & civilian)	Median (uniformed only)
Total, all cities	1,447	1.70	1.50
Population group:			
Over 500,000	25	2.89	2.30
250,000–500,000	23	2.02	1.76
100,000–250,000	85	1.89	1.56
50,000–100,000	195	1.73	1.51
25,000–50,000	358	1.62	1.43
10,000–25,000	761	1.68	1.50

SOURCE: *Municipal Police Administration*, ICMA, 1971

Utilizing data such as that presented in Table 2.7 one can estimate the number of new police personnel required as the result of residential development as follows:

$$\begin{pmatrix}\text{Number} \\ \text{of new} \\ \text{homes}\end{pmatrix}\begin{pmatrix}\text{Number} \\ \text{of people} \\ \text{per home}\end{pmatrix}\begin{pmatrix}\text{Number of} \\ \text{police department} \\ \text{personnel per} \\ \text{capita}\end{pmatrix}$$

For example, given a 500-home development with an average of 3.2 persons per home in a city of 300,000 population, this would imply the need for approximately

$$(500 \text{ homes})(3.2 \text{ people per home})(0.00202 \text{ policemen}) = 3.23$$

new full-time police department personnel. This estimate, of course, is only an approximation since it uses the 50th percentile or median from a sample of cities rather than the actual ratio for the city in which this new development will occur. Furthermore, it is only valid if the 1600 residents are *new* residents in the community.

With regard to the impact on the level of fire protection activity due to development, there are two basic considerations: the change in the likelihood of a fire getting started (i.e., fire incidence) in the first place, as measured by fire rates; and the change in the likely spread and risk to life of fire once started. Fire incidence may be affected by many aspects of new development such as: types of construction and materials; equipment, processes, and activities in the development; socioeconomic characteristics of the occupants of the development; and character of previous development. These factors can also potentially affect the likelihood of fire spread which is also influenced by such factors as: overall design, street layout, and proximity of buildings to each other; hydrant locations and interior sprinkler systems; distance to nearest fire station; and adequacy of amount and type of fire fighting equipment and personnel available. This discussion suggests that measures of the impact of development on the level of fire protection activity would be given by the change in the likelihood of fire incidence and the change in the likelihood of fire spread in the development area. As in the case of crime rates, this would require the before- and after-development measurements (or estimations) of these likelihoods of occurrence. Since fire insurance ratings are generally a function of the likelihood of fire incidence and fire spread, another measure of the impact on the level of the fire protection activity would be given by the change in the fire insurance rating of the area.

Another consideration is the fact that fire station placement and overall fire protection for a given area is based upon required fire flows and response distances established by the local fire department. The required fire flow is the rate of flow (in gallons of water per minute) needed for fire-fighting purposes to confine a major fire to the buildings within a block or other group complex. The determination of this flow depends upon the building size, type of construction, actual occupancy use, and exposure to buildings within and surrounding the block or group complex. Therefore, a major impact consideration is whether or not there exist adequate fire flows and, if not, what facility improvements are necessary to provide them. In order to meet response time requirements, it may also be necessary to acquire new equipment and/or construct new facilities.

TABLE 2.8   Median Number of Full-time Fire Department Personnel (Uniformed and Civilian) Per 1000 Population

Classification	No. of reporting cities	Median
Total, all cities	1,211	1.43
Population Group:		
Over 500,000	26	1.72
250,000–500,000	23	1.69
100,000–250,000	82	1.67
50,000–100,000	184	1.55
25,000–50,000	321	1.50
10,000–25,000	575	1.29

SOURCE: *Municipal Fire Administration*, ICMA, 1967

Finally, as in the case of policemen, fire department personnel needs are usually based on citywide needs, given deployment requirements on a geographical basis. In this case, one could also use personnel allocation formulas requiring factors such as the number of sworn personnel per capita and the number of civilian personnel per capita. Table 2.8 provides an illustration of these types of ratios according to city size.

Utilizing data such as that presented in Table 2.8, one can estimate the number of new fire department personnel required as the result of residential development as follows:

$$\begin{pmatrix} \text{Number} \\ \text{of new} \\ \text{homes} \end{pmatrix} \begin{pmatrix} \text{Number} \\ \text{of people} \\ \text{per home} \end{pmatrix} \begin{pmatrix} \text{Number of} \\ \text{fire department} \\ \text{personnel per} \\ \text{capita} \end{pmatrix}$$

For example, given a 500-home development with an average of 3.2 persons per home in a city of 300,000 population, this would imply the need for approximately

$$(500 \text{ homes})(3.2 \text{ people per home}) (0.00169 \text{ firemen}) = 2.70$$

new full-time fire department personnel. As was the case in determining new police department personnel, this is only an approximation based on the use of a median ratio of personnel per capita from a statistical sample and is only valid if these 1600 residents are *new* residents in the community.

Schools

New residential development in a community has a direct impact on educational services by virtue of generating new students for existing schools, whereas new industrial and/or commercial development has an indirect impact on educational services because of employees relocating within the community. To estimate the impact of residential development on schools requires the use of student yield factors (sometimes referred to as pupil generation factors or bedroom multipliers) to predict the number of students expected at each grade level by type of dwelling unit. Additional factors which could influence this choice of student yields would include the number of bedrooms, family size, family income, ethnic mix, religious denomination, etc. Tables 2.9 and 2.10 provide illustrative examples of these types of factors.

TABLE 2.9   Students Per Dwelling Unit by Housing Type: Summary of 16 Community Studies

Locality	Students per housing unit (Grades K–12)					
	Single-family detached	Duplex	Town house	Mobile home	Garden apartments	High-rise apartments
Lombard, Illinois	1.20	....	...	....	0.30	....
Park Ridge, Illinois	1.58	1.45	1.48	....	1.21	....
Niles-Morton Grove, Illinois	1.14	1.99	1.65	0.25	0.28	0.03
Skokie, Illinois	1.67	....	1.61	....	1.25	0.90
Hinsdale, Illinois	1.05	....	...	....	0.20	....
Fairfax County, Virginia	1.08	1.08	0.65	0.37	0.21	0.09
Falls Church, Virginia	0.71	....	...	....	0.47	0.06
Prince George's County, Maryland	1.44	....	0.92	....	0.50	0.09
Montgomery County, Maryland	1.30	1.30	1.00	....	0.47	0.09
Philadelphia, Pennsylvania (suburban)	0.50	....	...	....	0.22	0.02
Windsor, Connecticut	1.50	....	...	....	0.13	....
Stamford, Connecticut	....	....	...	....	....	0.13
New Jersey (20 communities)	....	....	...	....	0.27	....
Bloomingdale, New Jersey	....	....	...	....	0.21	....
Framingham, Massachusetts	....	....	...	....	0.11	....
Nassau County, New York	....	....	...	....	0.24	....
North York Township, Ontario	....	....	...	....	....	0.16
Average:	1.17	1.46	1.28	0.31	0.40	0.18

SOURCE: American Society of Planning Officials, *School Enrollment by Housing Type*, Planning Advisory Service Report No. 210, 1966; Lee A. Syracuse, *Arguments for Apartment Zoning*, National Association of Home Builders, 1968; and individual studies from the five Illinois communities.

TABLE 2.10   Illustration of Estimated Children Per Dwelling Unit

Type of unit	Children per unit				
	Pre-school	Elementary grades	Junior high grades	Total grades	High school grades
	(0–4 years)	K–5 (5–10 years)	6–8 (11–13 years)	K–8 (5–13 years)	9–12 (14–17 years)
Detached single-family					
Three bedrooms	0.435	0.69	0.33	1.02	0.38
Four bedrooms	0.470	0.76	0.48	1.24	0.54
Five bedrooms	0.510	1.04	0.86	1.90	0.73
Attached single-family (townhouse, row house, quadriplex, etc.)					
One bedroom	...	...	...	...	...
Two bedrooms	0.680	0.18	0.03	0.210	0.050
Three bedrooms	0.716	0.56	0.17	0.730	0.210
Four bedrooms	1.000	1.11	0.45	1.560	0.540
Low-density apartments (to 15/acre)					
Efficiency	...	...	...	...	...
One bedroom	0.070	0.052	0.028	0.090	0.013
Two bedrooms	0.343	0.116	0.073	0.189	0.043
Three bedrooms	0.457	0.390	0.210	0.600	0.240
Four bedrooms	0.500	0.670	0.250	0.920	0.330
High-density apartments (16 +/acre)					
Efficiency	...	...	...	...	...
One bedroom	0.050	0.026	0.014	0.040	0.007
Two bedrooms	0.210	0.065	0.035	0.100	0.029
Three bedrooms	0.430	0.150	0.080	0.230	0.092

SOURCE: Illinois School Consulting Service

To illustrate the use of student yield factors, consider the case of constructing 2000 single-family dwelling units in a previously undeveloped area. It is estimated that the student yield will be as follows:

Grade level	Yield factor	×	No. DU's	=	Students
Elementary (K–6)	0.6		2,000		1,200
Junior High (7–9)	0.3		2,000		600
Senior High (10–12)	0.2		2,000		400
					2,200

Local school district data in terms of school size, remaining capacity on a district-wide basis, and new construction thresholds (i.e., once the enrollment reaches this level a new school is constructed) are as follows:

School type	Design capacity (students)	Remaining capacity (students)	New construction threshold
Elementary	600	200	900
Junior high	900	−50	1,350
Senior high	1,600	−2,000	2,400

Comparing the student yields from this residential development with the preceding school district data shows the following impacts:

$$\begin{pmatrix} 1200\ \text{new} \\ \text{elementary} \\ \text{students} \end{pmatrix} - \begin{pmatrix} 200 \\ \text{remaining} \\ \text{capacity} \end{pmatrix} = \begin{pmatrix} 1000 \\ \text{requiring} \\ \text{space} \end{pmatrix} \blacktriangleright \begin{array}{l} \text{One new school} \\ \text{based on 1 per} \\ \text{900 students} \end{array}$$

$$\begin{pmatrix} 600\ \text{new} \\ \text{junior high} \\ \text{students} \end{pmatrix} - \begin{pmatrix} -50 \\ \text{remaining} \\ \text{capacity} \end{pmatrix} = \begin{pmatrix} 650 \\ \text{requiring} \\ \text{space} \end{pmatrix} \blacktriangleright \begin{array}{l} \text{No new school} \\ \text{based on 1 per} \\ \text{1350 students} \end{array}$$

$$\begin{pmatrix} 400\ \text{new} \\ \text{senior high} \\ \text{students} \end{pmatrix} - \begin{pmatrix} -2000 \\ \text{remaining} \\ \text{capacity} \end{pmatrix} = \begin{pmatrix} 2400 \\ \text{requiring} \\ \text{space} \end{pmatrix} \blacktriangleright \begin{array}{l} \text{One new school} \\ \text{based on 1 per} \\ \text{2400 students} \end{array}$$

that is, one new elementary school and one new senior high school will be required to meet the educational needs of the area. It is important to recognize that this development is in itself *not solely responsible* for the requirement that two new schools be constructed, but is responsible in conjunction with other development within the educational service area for this new construction.

There are other impact areas which may not be as readily measurable, but which are equally important such as: the number of pupils who may have to switch schools, the number of pupils who formerly walked to school but will now have to use buses and vice versa, the amount of overcrowding which will occur and whether or not double sessions and/or year-round schooling will be required, changes in pupil-to-teacher ratios and how this might affect the overall quality of education, and any new types of educational opportunities that might occur which were not previously available in the community. Information regarding these areas can generally be obtained through discussions with representatives of the local school districts.

Traffic Circulation

An important area of consideration in assessing the environmental impact of development projects deals with the impact on existing circulation systems and traffic flows as the result of traffic generated by the project. This necessitates estimating the average daily trips (ADT) generated by the project and then determining the distribution of these trips along the local circulation routes. Factors which must be taken into consideration in conducting analyses of this type would include: time (season of year, day of week, hour of day, and periods of time within an hour); composition of traffic; directional distribution; lane distribution; classification of street or highway such as rural (interstate, primary state highway, secondary state highway, county, local roads) or urban (freeway, expressway, major arterial, collector street); type of service (commuter travel, recreational, farm-to-market, and other seasonal travel); type of geometric design (control of access, separation of opposing traffic, number of lanes, merging and intersecting areas); type and frequency of traffic controls (traffic signals, stop or yield signs, one-way movement); and capacity of street or highway in relation to volume.

Generally, the volume of traffic attracted to or produced by the various land uses is related to various parameters which describe the use itself. The traffic characteristics of

TABLE 2.11   Typical Average Daily Trip (ADT) Generation Factors

Type of facility	ADT
Apartments	6–10 per DU
Auto dealer	90–125 per acre
Department store	25–50 per 1000 ft² GLA
Drug store	20–45 per 1000 ft² GLA
Grocery store	30–100 per 1000 ft² GLA
Hospital	5–10 per bed
Hotel	3–5 per 1000 ft² GLA 10–15 per room
Industrial	2–4 per parking stall 4–9 per 1000 ft²GLA
Medical office	30–60 per 1000 ft² GLA
Neighborhood retail	60–80 per 1000 ft² GLA
Office	10–40 per 1000 ft² GLA
Park (community)	5–10 per acre
Restaurant	30–50 per 1000 ft² GLA for sit-down 200–500 per 1000 ft² GLA for fast foods
School	10–15 per acre 1–3 per student
Single-family, attached	7–10 per DU
Single-family, detached	10–15 per DU
Theater	0.7–1.2 per seat
Variety store	10–20 per 1000 ft² GLA

NOTE: GLA = gross leasable area
      DU = dwelling unit
SOURCE: Survey of traffic engineering consultants

residential uses tend to vary depending upon the number, size, and value of the units; the density of development; car ownership; proximity to commercial areas and schools; potential uses of mass transportation; and so forth. It has been found that normally the trip generation is higher in areas of high valuation. Typical single-family areas which do not have commercial schools or significant amounts of employment within the boundaries of the neighborhood itself tend to have generation rates of 10 to 14 vehicles per day per dwelling unit (total in both directions). This is less than it would be if there were employment centers within the area.

As an illustration, Table 2.11 presents *illustrative* planning factors for estimating daily trips generated by project type. However, for actual factors to be used for a specific type of project, it is suggested that consultation be made with a local traffic engineer or the city/county engineering department.

In order to demonstrate the use of these types of factors, consider the redevelopment of an area currently comprised of a medical building, a parking lot, commercial activities, and multifamily residential land uses into a totally integrated parking structure, offices, and shops. The medical building and apartments will be removed to make way for these improvements. Table 2.12 provides a comparison of the estimated trips generated before and after redevelopment, showing that the project increases the number of vehicular trips in and out of the area by 7920 − 4985 = 2935 trips for an approximate 59% increase. A further consideration in this example deals with the traffic flow eastbound and westbound on Waterfront Avenue, which is the main thoroughfare to and from the project area. A

TABLE 2.12   Comparison of Two-way Daily Trip Generation Before and After Redevelopment

Development	2-way ADT generation rate	Total area of existing development	Existing trip generation	Total area of proposed development	Proposed trip generation
Offices & bank	25/1000 ft²	28,400 ft²	710	55,836 ft²	1,400
Retail	80/1000 ft²	33,000 ft²	2,630	74,375 ft²	5,950
Restaurants	50/1000 ft²	11,400 ft²	570	11,400 ft²	570
Medical offices	40/1000 ft²	20,875 ft²	835	. . .	. . .
Apartments	10/unit	24 units	240	. . .	. . .
			4,985		7,920

TABLE 2.13   24-hour Traffic Count on Waterfront Avenue in Project Area

Time of day	Eastbound	Westbound	Total (2-way)
12–1 AM	65	70	135
1–2	15	30	45
2–3	15	30	45
3–4	5	25	30
4–5	5	10	15
5–6	10	20	30
6–7	30	55	85
7–8	180	205	385
8–9	390	345	735
9–10	400	365	765
10–11	525	465	990
11–12 NOON	590	470	1,060
12–1 PM	600	470	1,070
1–2	600	470	1,070
2–3	570	490	1,060
3–4	490	435	925
4–5	530	515	1,045
5–6	560	465	1,025
6–7	435	360	795
7–8	400	330	730
8–9	260	245	505
9–10	230	235	465
10–11	140	170	310
11–12	135	115	250
Total	7,180	6,390	13,570

recent traffic count survey (Table 2.13) shows that the noon till 2:00 P.M. period represents the peak hours of traffic on Waterfront Avenue. This table reveals that during the peak hours, Waterfront Avenue is carrying a maximum of 600 vehicles in two lanes in each direction. Since this particular street is designed to carry 600–800 vehicles per lane (or 1200–1600 in two lanes), there is surplus capacity available to carry the newly generated trips. During the noon till 2:00 P.M. peak period, nearly 16% (2140/13,570 × 100) of the daily trips occurs. Hence, approximately (0.16) (2,935) = 470 additional trips as the result of redevelopment of the area could be expected to occur, thus raising the peak period total to 1070 which is still less than design capacity.

Other traffic circulation-related types of impacts include such areas as travel times, degree of congestion, parking availability, public transit accessibility, and traffic safety. Travel time effects in residential areas can be assessed in terms of the change in travel time between major work or shopping destinations and the vicinity of the development. For developments in commercial and industrial areas, the changes in travel time between existing business and major residential locations or other business locations can be used.

TABLE 2.14   Typical Off-street Parking Standards

Residential	Spaces per DU
Single-family homes	2
Apartments/condominiums/townhouses (includes guest parking):	
Efficiency	1–½
1 bedroom	2
2 bedrooms	2–½
3 or more bedrooms	3
Mobile home park	2 tandem and 1 guest

Commercial	Spaces per 1,000 ft² GFA
Offices and banks	3.3
Business and professional services (including insurance) ..	3.3
Shopping center	4–5/1000 ft² GLA
Shopping foods (retail)	5
Convenience goods (retail)	5
Supermarkets (food stores)	4
Liquor stores	3
Personal services and repairs	5

*If townhouses are built with separate driveways (like single family homes), 2 stalls/DU can be used. If they are not, the apartments/condominiums/townhouses category should be used.

Note: GFA = gross floor area
       GLA = gross leasable area
       DU = dwelling unit
SOURCE: References 3 and 16.

Congestion can generally be assessed in terms of (1) its "severity," which is defined as the ratio of the maximum time to travel between two points relative to the "no traffic" or off-peak, law-abiding travel time between those points, and (2) its "duration," which is defined as the length of time during which travel times between two points is some percentage above the off-peak travel time. The effect of development on neighborhood or area parking will depend upon the size of the development, the number of new parking spaces provided by the development, the nature of the activities in the development, the availability of public transportation, existing parking demand in the area, and the prices charged for parking in the area. If the spaces provided are too few or too expensive, persons in the development may park in the surrounding neighborhood, thus increasing parking congestion. Table 2.14 provides some *illustrative* planning factors to be used in estimating parking requirements. New development may also generate enough demand to allow: additional service on existing public transit lines, the creation or rerouting of lines, or even the start of a new transit system. To assess this type of impact requires measuring (or estimating) such quantities as the number of residents within reasonable walking distance of public transit and the number of employees who can reach reasonable walking distance of their place of employment via public transit. Comparison of these quantities both before and after development will enable one to assess the impact on public transit facilities. Finally, the impact on traffic safety is generally measured by estimating the change in accident rates caused by increased traffic and any special hazards created or eliminated by the new development. This may require a before-and-after-development assessment in order to accurately measure the impact of the new development on traffic safety.

FISCAL IMPACT ANALYSIS

Introduction

The rapid development of suburbs in the 1960s and early 1970s near most larger American cities has caused substantial recent opposition to further growth. Municipalities' citizens have expressed profound frustration over heavier traffic, overcrowded schools, inadequate police protection, and soaring taxes. All of these evils have been attributed to the new residential subdivisions.

As a result, local governmental agencies have reacted in a variety of different ways. For example, in Loudoun County, Virginia, a zoning ordinance was passed requiring developers to assist in paying for new public services required by their projects. In Ramapo, New York, a suburb of New York City, a zoning ordinance was enacted in 1969 which requires a developer to finance the construction of new public services if they are not available before construction. Using a point system, the town assigns numerical weights to those existing or planned sewers, parks, school sites, roads, firehouses, and other services which the development will need. If fewer than 15 points are tallied, the developer must either finance construction of services sufficient to reach that score or face the town's refusal to issue a building permit. Other examples can be found across the country dealing with sewer moratoria and various types of no-growth or limited-growth policies.

The Ramapo method of control incorporates a quantitative approach to assessing the impact of development on the fiscal base. It also suggests the importance of analytical methods that yield more systematic evaluations than do traditional zoning practices.

Because of growth, many communities have become intensely interested in the potential impacts of specific projects upon the provision of basic local governmental services such as schools, utilities, streets and highways, parks, and open space. This is because the development of land for any purpose creates both an immediate demand for these services and a flow of revenues to the community from a variety of sources. Community officials need to know the costs of providing such services for new development projects, as compared with property taxes (and other public revenues) which those projects will generate. Also, they need to know the implications of cost-revenue surpluses or deficits of particular projects for the planning of other types of urban land use development.

The fiscal or cost-revenue impact of a new development is only one of a number of important criteria that should be considered by communities. For example, it may be determined that a low-cost housing project would generate revenue deficits. However, it might be beneficial to the city to allow such development if it achieved certain other community goals, such as providing decent housing for low-income families.

A city must not legislate against poverty, but rather alleviate the social costs of this condition by insuring that decent housing is made available within the financial reach of all income levels. The cost-revenue issue is thus inextricably tied to the philosophy of taxation and the issue of whether services should be provided based on the ability to pay or on the basis of need.

There are several factors that need to be considered when examining the fiscal impact of new developments. These are as follows:

1. *Inflation*—It is apparent that many cities throughout the nation are in dire financial condition because the costs of services are rising faster than revenues. If these cities are registering substantial growth, it is often concluded that such expansion is a generator of high costs. However, numerous studies have shown that inflation is a far greater culprit. Typically, municipal costs increase by 10 percent per year in the absence of any growth at all. Increases in salaries and in the costs of purchased goods and services are the primary reasons. Some of the worst fiscal problems are being experienced by municipalities which have shown no growth for several years. It is apparent that if the inflation among the existing stock of real estate does not match the inflation in costs, financial problems will result. Often, these increased costs have been paid for out of the cash surpluses resulting from growth in cities with a balanced real estate development program.

2. *Cost Averaging*—There is a general tendency to think of costs on a per capita or per household basis. This approach can be very reasonable, and it is one that is utilized in many studies where more refined data is not available. However, there are two basic problems. First, it is generally the older sections of any city—particularly the downtown area—that require far greater service than do the newer, more suburban sectors. Secondly, the additional cost of providing service to new areas is normally less than the average cost of providing services to existing units. The basic reason for this is that it is less expensive on a per household basis to expand an existing police department, for example, because a number of the more fixed, overhead costs do not increase proportionately.

3. *Price Averaging*—As with costs, there is often a tendency to think of revenues on an average per household basis. Actually, particularly during the past few years, a considerable portion of the new housing has been higher priced with correspondingly higher tax revenues than the existing housing stock.

4. *Planning for Growth*—A city's stage in the cycle of urbanization and the capacity of existing public facility systems are important factors in determining whether or not growth will be a deficit. The ability of city officials to anticipate future growth and the sizing of public facilities influences both the capability and the cost of providing public services to new development.

5. *Rising Quality of Services*—It might appear that growth does not pay if the governing agency has significantly raised the quality of services without a corresponding increase in the tax rate. Most often, the issue concerns parks and open space. If a general plan is adopted with increased park requirements per 1000 persons without an increase in either the tax rate or a park fee of some type, the average price of the new residential units must rise to offset the added costs. In the absence of this price increase, it would appear that the new units could not support themselves. While the goal of raising such standards is highly commendable, new developments are often burdened with requirements that could not possibly be met by the existing stock of residential real estate in the absence of a tax increase.

6. *Appraised Values*—The appraised value for tax purposes of units in new developments is generally very close to market value. However, this is often not the case in older subdivisions where appraisals and corresponding investments are out of date. Therefore, in such a situation, new developments may carry a disproportionate share of the tax load.

It must be recognized that a comparison of project costs and revenues provides only partial information concerning the impact of the project on the local economy. The analysis and discussion presented in the subsequent sections covers, in addition to costs, the property tax revenues and all other directly associated revenues for new development. There are also secondary elements such as increased employment in trade, service, and government sectors which in turn increases payrolls, income, and municipal revenues. Conversely, the restriction of growth may result in employment declines in construction and related industries, thus leading to reduced income and expenditures throughout a community's economy.

In the following sections, the term "cost-revenue analysis" will refer to only the fiscal impact of land development, as distinguished from cost-benefit analysis which is broader in scope and attempts to assign dollar values to other economic, social, or environmental impacts.

Local Government Revenue Categories

Local government revenues generally fall into three categories, namely:

1. Receipts from property tax which includes all revenue derived from the taxation of real estate and business personal property such as machinery, equipment and inventory.

2. Subventions from the state which includes receipts from sales tax, cigarette tax, liquor license tax, motor vehicle taxes, gas taxes, etc.

3. Miscellaneous revenues which includes receipts from charges and fees, bank, building and loan tax receipts, pension fund contributions, earnings on investments, grants, etc.

Table 2.15 provides a typical listing of public revenue sources.

Generally, in conducting environmental studies or performing environmental impact assessments, it is only necessary to determine project revenues for those revenue sources which are significant in lieu of a detailed cost-revenue analysis in which all revenue and

TABLE 2.15   Typical Community Revenue Categories

1. Property taxes
2. Taxes other than property: Redemptions, penalties, and interest
   Cigarette tax
   Transient occupancy tax
   Franchises
   City sales and use tax
   Property transfer tax
3. Licenses and permits: Business licenses
   Business license delinquencies
   Investigation fees
   Vending machines
   Bicycle licenses
   Dog licenses
   Building permits
   Lathing and plastering permits
   Plumbing and heating permits
   Trailer park permits
4. Fines, forfeitures, and penalties: Municipal court fines
5. Revenue from use of money and property: Interest on investments
   Rent, sale of equipment, property
6. Revenue from other agencies: State highway maintenance
   State motor vehicle license fees
   Trailer coach fees
   State liquor license fees
   Highway carrier's tax
   Federal grants
7. Charges for current service: Zoning and subdivision fees
   Plan checking fees
   Electrical maintenance fees
   Reinspection fees
   Sewer connection fees
   Engineering services fees
   Police service fees
   Fire service fees
   Taxi permits
   Equipment maintenance overhead
   Weed abatement fees
   Refuse collection and disposal charges
8. Other revenues: Sale of maps and publications
   Administrative services
   Refunds and rebates
   Donations and contributions
   Damage to city property
9. Parks and recreation revenues: Parking lots
   State highway maintenance
   Federal grants
   Recreation service fees
   Parkway tree inspection fee
   Sale of maps and publications
   Refunds and rebates
   Donations and contributions
   Damage to city property
10. Library revenues: Interest income
    Federal grants
    Library fines
    Library rental fees
    Sale of maps and publications
    Refunds and rebates

TABLE 2.15   Typical Community Revenue Categories (Continued)

Donations and contributions
Damage to city property
11. Retirement funds: Interest income
12. State gas tax funds: Interest income
State gas tax apportionment
State gas tax engineering aid
13. Arterial highway financing fund: Donations and contributions
14. Parking meter receipts: Parking meters
Off-street parking
Damage to city property
15. Contributions and donations
16. Water receipts: Sale of water
Meter turn-on charges
Connection charges
Interest income
Sale of property
Rental of property
Refunds and rebates
Donations and contributions
Damage to city property
17. Other

---

cost sources are specifically evaluated. In the case of residential projects, this would usually include property tax, sales tax, and certain subventions from other governmental agencies such as the state. For commercial and industrial projects, this would usually include property tax, sales tax, business license fees, and/or gross receipts taxes. Additional special revenues such as a utility tax on all electrical, telephone, gas, and water charges or even a city income tax can be considered as appropriate.

In the calculation of revenues to be allocated to private land uses, it is sometimes convenient to classify local (city or county) revenues as being differentially allocatable or nondifferentially allocatable. "Differentially allocatable" revenues (or costs) are those which result from, or are related to, land use and which are generated in a greater or lesser amount by one category of land use over another, i.e., revenues that are related specifically to type of land use. Examples would include property and sales taxes, as well as certain subventions. "Nondifferentially allocatable" revenues are comprised of those local revenues which cannot be directly related to one land use in a greater or lesser amount than to another land use, i.e., revenues that are not related specifically to type of land use. Miscellaneous revenues would generally be of this type. This means that we can regard local revenues as the sum of these two types of revenues.

The question then is how do we allocate the nondifferentially allocatable revenues to land use? Two common methods in use are: (1) to allocate on an acreage basis for each type of developed land in the city, and (2) to prorate on the basis of assessed valuation. In the first case, $x\%$ of these nondifferentially allocatable revenues would be allocated to the particular type of land use which contributes $x\%$ to the total city (of county) acreage. In the second case, it would be allocated to the particular type of land use which contributes $x\%$ to the total city (or county) assessed valuation. Illustrations of the use of these methods will appear in the discussions of the following sections.

Residential Revenue Determination   General formulations for the typically significant residential revenues are as follows:

1. Property tax

$$\begin{pmatrix} \text{Number} \\ \text{of dwelling} \\ \text{units} \end{pmatrix} \begin{pmatrix} \text{Average dwelling} \\ \text{unit market} \\ \text{value} \end{pmatrix} \begin{pmatrix} \text{Tax assessment} \\ \text{factor} \end{pmatrix} \begin{pmatrix} \text{Property} \\ \text{tax rate} \end{pmatrix}$$

where the tax assessment factor corresponds to the portion of the dwelling unit market value (as determined by the local assessor) which represents the assessed value.

2. Sales tax

$$\left(\begin{array}{l}\text{Number}\\ \text{of dwelling}\\ \text{units}\end{array}\right) \left(\begin{array}{l}\text{Number of}\\ \text{people per}\\ \text{dwelling unit}\end{array}\right) \left(\begin{array}{l}\text{Sales tax}\\ \text{revenue per}\\ \text{capita}\end{array}\right)$$

or, alternatively,

$$\left(\begin{array}{l}\text{Number of}\\ \text{dwelling}\\ \text{units}\end{array}\right) \left(\begin{array}{l}\text{Average family}\\ \text{income per}\\ \text{dwelling unit}\end{array}\right) \left(\begin{array}{l}\text{Fraction of}\\ \text{family income}\\ \text{for retail}\\ \text{expenditures}\end{array}\right) \left(\begin{array}{l}\text{Fraction of}\\ \text{retail sales}\\ \text{expenditures}\\ \text{received by}\\ \text{local jurisdic-}\\ \text{tion as tax}\end{array}\right)$$

3. Government subventions

$$\left(\begin{array}{l}\text{Number of}\\ \text{dwelling}\\ \text{units}\end{array}\right) \left(\begin{array}{l}\text{Number of}\\ \text{people per}\\ \text{dwelling unit}\end{array}\right) \left(\begin{array}{l}\text{Subvention}\\ \text{revenue per}\\ \text{capita}\end{array}\right)$$

Several comments are in order regarding these formulations.

First, if the project involves a widespread mix of dwelling unit values or prices, then the use of an average value would be misleading. In this case, if there are N different dwelling unit values, say, $V_1, \ldots, V_N$ and D_i dwelling units of value V_i ($i = 1, 2, \ldots, N$), then we would replace the quantity

$$\left(\begin{array}{l}\text{Number of}\\ \text{dwelling}\\ \text{units}\end{array}\right) \left(\begin{array}{l}\text{Average dwelling}\\ \text{unit market value}\end{array}\right)$$

by

$$\sum_{i=1}^{N} V_i D_i$$

Second, the first formulation given for sales tax revenue determination requires dividing the total sales tax revenue to the community in a specified period of time by the (average) population during this period to obtain an estimate of the sales tax revenue per capita. This is a rather simplistic formulation since it implies that sales tax revenue in the community is directly related to the population of the community, which may not be true at all. In fact, the commercial development in the community may be more of an "attraction" to residents of adjacent communities, thus generating retail expenditures which are in no way related to the residential population of the community. The alternative formulation given is an attempt to estimate those taxable retail expenditures that are generated directly by the project's population. The shortcoming of this formulation, however, is that these expenditures need not be made in the community but are spent outside in other commercial attraction areas. In either case, one should be aware of the inherent limitations in the use of either of these two formulas.

A further consideration with regard to the use of the second formula for estimating residential land use sales tax revenue is that this formula is based upon the consideration of how much a family spends in the community for taxable retail expenditures. This percentage of gross income spent on taxable items generally varies with a person's (or family's) income—specifically, as one's income increases, then a relatively smaller amount will be spent for taxable retail items (i.e., more will be spent on housing, savings, and investments). An *illustrative* variation of this percentage would be:

Income level	Annual income range	Percent of income for taxable retail expenditures
Low	<$8,000	47
Moderate	$ 8,000–$15,000	42
Middle-high	$15,000–$25,000	37
High	>$25,000	32

Third, the average family income per dwelling unit is typically assumed to be the qualifying income in order to purchase the dwelling unit. If one uses the "rule of thumb" which says that the house price should be 2½ times the income (which is a reasonable approximation to the requirement that monthly income should be five times the monthly payment for principal, interest, and taxes), then we would replace the term given by the average family income per dwelling unit by 0.4 times the average dwelling unit real value. If there are N different dwelling unit values, say, $V_1, V_2, \ldots, V_N$ and D_i dwelling units of value V_i ($i = 1, 2, \ldots, N$), then we would replace the quantity

$$\begin{pmatrix} \text{Number} \\ \text{of dwelling} \\ \text{units} \end{pmatrix} \begin{pmatrix} \text{Average family} \\ \text{income per} \\ \text{dwelling unit} \end{pmatrix}$$

by

$$\sum_{i=1}^{N} 0.4\, V_i D_i$$

A further comment about the above formulations of retail sales tax revenue is that it has been implicitly assumed that residents/consumers are new generators of retail sales and represent new sources of additional retail tax revenues. In reality, this would not be true.

TABLE 2.16   Illustrative Example of Property Tax Revenue for Residential Projects

Type of residential project	Density (units/acre)	Average value per unit	×	Tax assessment factor	×	Property tax rate	=	Total property tax revenue
Single-family, detached	5.0	$50,000		0.25		$1.60/100		$200
Single-family, attached	7.0	43,500		0.25		$1.60/100		174
	12.0	35,000		0.25		$1.60/100		140
	15.0	30,000		0.25		$1.60/100		120
Multifamily	15.0	15,000		0.25		$1.60/100		60
	20.0	12,000		0.25		$1.60/100		48
	30.0	9,000		0.25		$1.60/100		36

If all residents were new to the area, did not reside there previously, and had never spent any portions of their income in the community, then this would be valid. What actually happens is that, in many cases, families which already resided in the area are "stepping up." Thus, they would not be contributing any new retail sales. Furthermore, if there are adjacent communities offering "retail attractions," then these families may spend some of their income outside the community. Therefore, it would make sense to multiply the previously computed retail sales tax revenue per acre by a scaling factor to reflect this potential inequity. There are no hard and fast rules as to what the value of this factor should be and, as a result, a study of consumer buying behavior in the community versus place of residence would be necessary in order to determine the value of the scaling factor.

Finally, the average family size per dwelling unit typically ranges from a low of 1.75–2.00 people in retirement and mobile home communities to 3–5 people in other residential communities. Furthermore, it is reasonable to expect that there would be a high correlation between the number of people per dwelling unit and the number of bedrooms, or even the size of dwelling unit. Thus, depending upon the variation in house characteristics in a housing subdivision, the average family size per dwelling unit could vary.

As an example of the revenue computation for residential-type projects, consider three basic types, namely: (1) single-family detached; (2) single-family attached such as townhouses, condominiums, and planned unit developments; and (3) multifamily units or apartments. Tables 2.16 to 2.18 provide an illustration of the procedure for computing the revenues due to property taxes (at an assumed rate of $1.60 per $100 of assessed valuation), retail sales taxes (under the assumption that the city receives 1% of all retail sales), and state subvention revenues (under the assumption of a $22.50 per capita subvention for gas tax, in lieu of motor vehicle fees, cigarette tax, etc.), respectively, for different density configurations. The total revenue summary is shown in Table 2.19. This data indicates that even though the standard "separate house and lot" type of project generally yields more revenue per dwelling unit than "clustered" or multifamily types of projects, it yields less revenue on a per acre basis.

Commercial and Industrial Revenue Determination   General formulations for the typically significant commercial and industrial revenues are as follows:

1. Property tax

$$\left(\begin{array}{c}\text{Number} \\ \text{of acres}\end{array}\right) \left(\begin{array}{c}\text{Average} \\ \text{market value} \\ \text{per acre}\end{array}\right) \left(\begin{array}{c}\text{Tax assessment} \\ \text{factor}\end{array}\right) \left(\begin{array}{c}\text{Property} \\ \text{tax rate}\end{array}\right)$$

where the tax assessment factor corresponds to the portion of the commercial or industrial market value (as determined by the local assessor) which represents the assessed value. This need not be the same fraction, or percentage, as that for residential land uses, but generally depends upon state law.

TABLE 2.17   Illustrative Example of Retail Sales Tax Revenue for Residential Projects

	Density (units/acre)	Family income per unit ×	Fraction spent on taxable items	Total taxable = sales ×	Fraction returned to city	Sales tax = revenue
Single-family, detached	5.0	$20,000	0.37	7,400	0.01	$74.00
Single-family, attached	7.0	17,400	0.37	6,438	0.01	64.38
	12.0	14,000	0.42	5,880	0.01	58.80
	15.0	12,000	0.42	5,040	0.01	50.40
Multifamily	15.0	9,000	0.42	3,780	0.01	37.80
	20.0	8,000	0.42	3,360	0.01	33.60
	30.0	6,500	0.47	3,055	0.01	30.55

TABLE 2.18   Illustrative Example of State Subvention Revenue for Residential Projects

Type of residential project	Density (units/acre)	Family size per unit ×	Subvention revenue per capita =	Total state subvention revenue
Single-family, detached	5.0	3.5	$22.50	$78.75
Single-family, attached	7.0	3.2	22.50	72.00
	12.0	3.1	22.50	69.75
	15.0	3.0	22.50	67.50
Multifamily	15.0	2.5	22.50	56.25
	20.0	2.25	22.50	50.63
	30.0	2.0	22.50	45.00

TABLE 2.19   Illustrative Example of Total Residential Project Revenue

Type of residential project	Density (units/acre)	Property tax revenue +	Retail sales tax revenue +	State subvention revenue =	Total revenue Per DU	Total revenue Per acre
Single-family, detached	5.0	$200	$74.00	$78.75	$352.75	$1,764
Single-family, attached	7.0	174	64.38	72.00	310.38	2,173
	12.0	140	58.80	69.75	268.55	3,223
	15.0	120	50.40	67.50	237.90	3,569
Multifamily	15.0	60	37.80	56.25	154.05	2,311
	20.0	48	33.60	50.63	132.23	2,645
	30.0	36	30.55	45.00	111.55	3,347

2. Sales tax

$$\left(\begin{array}{l}\text{Number of}\\\text{square feet of gross}\\\text{leasable area}\end{array}\right)\left(\begin{array}{l}\text{Retail sales}\\\text{per square foot of}\\\text{gross leasable}\\\text{area}\end{array}\right)\left(\begin{array}{l}\text{Fraction of retail}\\\text{sales expenditures}\\\text{received by local}\\\text{jurisdiction as tax}\end{array}\right)$$

where gross leasable area (GLA) is generally on the order of 80–85% of the gross floor area.

3. Business license fees

$$\left(\begin{array}{l}\text{Business license}\\\text{fee per establishment}\end{array}\right)\left(\begin{array}{l}\text{Number of}\\\text{establishments}\end{array}\right)$$

4. Gross receipts tax

$$\left(\begin{array}{l}\text{Number of}\\\text{square feet of gross}\\\text{leasable area}\end{array}\right)\left(\begin{array}{l}\text{Retail sales}\\\text{per square foot of}\\\text{gross leasable}\\\text{area}\end{array}\right)\left(\begin{array}{l}\text{Gross receipts}\\\text{tax per unit}\\\text{of retail sales}\end{array}\right)$$

With regard to the sales tax formulation, one needs to estimate the retail sales per square foot of gross leasable area. For many industrial projects there may be no taxable retail sales, in which case this quantity would be zero. For commercial projects such as shopping centers, this estimate must be based on factors such as types of tenants and size of the shopping center. Table 2.20 presents typical data for determining the retail sales volume per square foot of gross leasable area for four types of shopping centers, namely: super regional centers, regional centers, community centers, and neighborhood centers. These centers can be described as follows:

1. Super regional centers—This type provides for extensive variety in general merchandise, apparel, furniture, and home furnishings as well as a variety of services and recreational facilities. It is built around at least three major department stores of generally not less than 100,000 ft² each. In theory, the typical size of a super regional center is about 750,000 ft² GLA, but in practice the size ranges to well over 1,000,000 ft².

2. Regional centers—This type provides for general merchandise, apparel, furniture, and home furnishings in depth and variety as well as a range of services and recreational facilities. It is built around one or two full-time department stores of generally not less than 100,000 ft². A typical size is on the order of 400,000 ft² GLA. The regional center is the second largest type of shopping center and provides services typical of a business district not yet as extensive as the super regional center.

3. Community centers—This type of center provides a wider range of facilities for the sale of soft lines (wearing apparel for men, women, and children) and hard lines (hardware and appliances) than the convenience goods and personal services normally found in a neighborhood shopping center. It is built around a junior department store or a variety store as the major tenant, and usually includes a supermarket. It does not normally have the full-line department store, though it may have a strong specialty store. In theory, the typical size is 150,000 ft² GLA.

4. Neighborhood centers—This type provides for the sale of convenience goods (foods, drugs, and sundries) and personal services (laundry and dry cleaning, haircutting, shoe repairing, etc.) for day-to-day living needs of the immediate neighborhood. It is built around a supermarket as the principal tenant. In theory, the neighborhood center has a typical gross leasable area close to 50,000 ft². It is the smallest type shopping center.

The gross leasable area is less than the gross floor area and is generally of the order of 80–85% of the gross floor area. Furthermore, retail sales tax revenue is usually allocated to commercial (as well as industrial) land uses to the extent that they capture net additional revenue from consumers. This includes consumers located within the community who were previously making those retail purchases elsewhere and consumers outside the city who are now making those purchases inside. This can be reflected in the assumed choice of the sales volume per square foot of gross leasable area.

At this point, a further comment is necessary regarding both business license fees and gross receipts tax revenue. Many local jurisdictions will have one or the other, but not both. In some cases, rather than base business-related fees on a flat fixed fee or annual gross receipts, such as in the previously described formulas for business license fees and gross receipts tax, a community may impose fees based on number of employees, square footage occupied, or other parameters.

For example, consider a regional shopping center with total value of $352,000 per acre with 2900 ft² per acre of department stores, 2640 ft² per acre of junior department stores, 2500 ft² per acre of variety stores, 4320 ft² per acre of supermarkets, and 2000 ft² per acre of

TABLE 2.20   Median Sales Volume ($) per Square Foot of Gross Leasable Area (GLA) for Shopping Centers by Type of Tenants

Tenant classification	Super regional	Regional	Community	Neighborhood
Food and food service:				
Candy, nuts	$102.83	$ 86.83	$ 42.99	. . .
Restaurant without liquor	91.77	80.59	64.09	$ 50.79
Restaurant with liquor	75.44	65.96	60.82	55.03
Fast food/carry-out	175.09	113.49	78.89	50.06
Supermarket	112.53	113.55	135.22	133.19
Ice cream parlor	114.56	100.56	62.08	53.70
General merchandise:				
Department store	75.03	61.41	. . .	. . .
Variety store	43.66	36.74	33.67	41.22
Junior department store	55.48	59.87	51.97	31.65
Clothing and Shoes:				
Ladies' specialty	87.85	81.00	62.08	69.65
Ladies' ready-to-wear	87.43	85.76	61.83	63.05
Menswear	100.88	96.66	65.55	67.96
Family wear	80.41	78.29	62.09	66.37
Family shoes	73.10	67.75	52.37	35.63
Ladies' shoes	68.21	67.46	35.44	. . .
Mens' and boys' shoe	120.50	107.87	72.89	. . .
Unisex/jean shop	157.71	110.11	66.12	32.91
Dry Goods:				
Yard goods	60.19	58.52	43.19	35.73
Other Retail:				
Books and stationery	101.10	79.23	58.45	52.16
Jewelry	185.11	141.21	77.93	76.91
Cosmetics	78.16	76.70	41.58	32.40
Drugs	85.65	79.83	78.95	64.82
Cards and gifts	78.16	70.97	48.90	37.29
Hardware	. . .	66.29	44.09	43.29
Liquor and wine	111.83	99.71	121.86	79.50
Services:				
Beauty shop	70.25	61.38	50.54	49.85
Optometrist	117.40	106.42	67.05	71.07
Barber shop	66.53	61.36	43.70	38.88
Cleaners and dyers	29.68	30.39	31.31	33.80
Coin laundries	. . .	54.11	16.56	15.03

SOURCE: *Dollars & Cents of Shopping Centers: 1975*, The Urban Land Institute.

drug stores. Assuming that the city property tax rate is $1.85 per $100 assessed valuation (25% of market value), the city receives 1% of all retail sales in property tax, there is a gross receipts tax of $1.25 per $1000 of retail sales, and using the median sales volume quantities of Table 2.20, the total revenue per acre would be:

$$\text{Property tax revenue per acre} = (\$1.85/100)(0.25)(\$352,000) = \$1628/\text{acre}$$

Retail sales
tax revenue
per acre

$$= (.01)[(2900)(\$61.41) + (2640)(\$59.87) + (2500)(\$36.74) + (4320)(\$113.55)(.30)* + (2000)(\$79.83)]$$

$$= \$9202$$

Gross receipts
tax revenue $= \$919$
per acre

$$\frac{\text{Total revenue}}{\text{per acre}} = \$8,267$$

Local Government Cost Categories

Local government costs generally fall into five categories, namely:

1. General government—administration, personnel, data processing, management and budget, public affairs, legal services, city/county clerk, zoning and planning, etc.

2. Public safety—police and fire protection, disaster services, civil defense, building regulation, etc.

3. Public works—roads and streets, parkway maintenance, electrical, engineering and plumbing services, refuse and sewage disposal, traffic engineering and parking, mechanical and hydraulic facilities, water service, surveying, etc.

4. Library services—bookmobiles, facilities, book stock and periodicals, films, etc.

5. Parks and recreation—recreational facilities, parks, recreation programs and activities, etc.

In most cases public safety and public works are the largest cost categories, followed by general government, parks and recreation, and libraries in this order. Table 2.21 provides a typical listing of public expenditure sources excluding capital improvements.

To assess the impact of new development on these categories also requires the recognition of the type or nature of the service costs incurred, such as whether they are direct or indirect. "Direct costs" are those which are allocated for services which the community will provide within the boundaries of the new development. This type of service is typified by such activities as police and fire protection, street sweeping, sewer system maintenance, water main maintenance, park development, and street construction. The nature of these direct services is that they will generate increased workloads as soon as the development is completed. "Indirect costs" are those which are allocated for services provided outside the new development which benefit all community residents to the same extent. Although new development will generate some increases in indirect costs, such as administration and overhead, this type of cost is generally allocated as a proportionate share of the community-wide total.

It is also important to recognize that differences in quality, quantity, range, and scale of public services offered affect public expenditures. One cannot draw conclusions about operation efficiency from expenditure data without considering all four factors. Efficiency implies lower expenditures, but expenditures can also be decreased by reducing the quality, quantity, range, or scale of public services. The most efficient small government unit will have higher per capita expenditures than an equally efficient larger unit because the larger one can better exploit equipment, specialized labor, and quantity purchasing. Beyond some size, such gains are offset by loss of management efficiency.

An analysis of public service expenditures must also include the effect on the welfare of the community as well as economic efficiency. Different social conditions among communities may require different levels of public services to maintain equal community welfare. For example, the breakdown of interpersonal relations associated with concentrated populations allegedly contributes to crime. Consequently, a large community may have to provide more public services, police and social programs, to maintain the same personal safety as a small community does with fewer public service programs.

A further consideration is that projects differ in their specific location, which may indicate higher city costs for outlying portions of the city. The socioeconomic characteristics of residents of a new development have a very real effect both on the demand for services such as police protection and on revenues generated such as through retail sales tax. The size of families as well as the age of children also affects the need for services,

*Assumes only 30% of sales are taxable.

particularly schools. The characteristics of a particular development also have an influence on the "effective" need for services. A planned residential development with private common areas and recreational facilities obviously reduces the effective need for public parks, but the developer is rarely given credit for these facilities to reduce an in-lieu park fee.

TABLE 2.21   Typical Community Operating Cost Categories

1. General government: Governing body
                      Clerk
                      Administrative
                      Personnel
                      Finance
                      Purchasing and warehousing
                      Attorney
                      General services
                      Elections
                      Nondepartmental
                      Retirement
                      Building maintenance
                      Health and welfare
                      Planning and zoning
                      Miscellaneous (insurance, public affairs, etc.)
2. Public safety: Police
                  Fire
                  Building regulation
                  Disaster services
                  Civil defense
3. Public works: Administration
                 Engineering
                 Electrical
                 General services (administration, field maintenance, traffic signs and
                          markings, refuse, equipment, maintenance)
                 Roads and streets
                 Sewers
                 Traffic and parking
                 Water service
                 Refuse collection
4. Libraries: Bookmobiles
              Reading materials
              Facilities
5. Parks and recreation: Administration
                         Recreation programs
                         Parks, parkways, etc.
                         Special facilities

Allocation of Public Service Costs to Land Uses   One of the most important issues—and one that has created the greatest controversy—is the question of cost allocation among the various land uses. On the revenue side, it is generally possible to make this allocation with a reasonable degree of precision. For expenditures, however, it is difficult in that very few communities have a program budget or a unit cost accounting system that allows a determination of the real cost of specific services.

If, for example, expenditures were allocated in the same proportion as revenues, then not only the total budget but also the "budget" for the industrial, commercial, and residential sectors would be in perfect balance. The general tendency among cost-revenue analysts—particularly those on the staff of cities—is to assign a dominant share of the costs to the residential sector. Many cities, for example, will simply divide the entire operating budget by the population to determine a per capita cost of providing services. This factor is then used to measure the impact of new projects.

A method which can be used to overcome the inadequacies of the per capita approach to allocating costs is the development value method. Where the per capita method relates increases in municipal expenditures to population growth, the development value method is based upon the premise that increases in expenditures are due to development in general. The critical step in this approach is the allocation of the expenditures (or budget) to residential and nonresidential development on the basis of value, usually assessed value. Typically, 70–90% of the value of communities is based on residential land use, and the remaining 10–30% of value is nonresidential land use. These proportions are then used to allocate the expenditures for municipal service. The residential portion can be calculated using the per capita technique, whereas the nonresidential portion can be calculated using the expenditure per dollar of development value ratio.

For example, consider a community with the following characteristics:

Total assessed value of community $200,000,000
Population 16,000
Existing nonresidential assessed value $40,000,000 (20%)
Existing residential assessed value $160,000,000 (80%)
Total annual expenditures for municipal services .. $1,750,000

Using the development value method, one would allocate (0.80) ($1,750,000) = $1,400,000, or $1,4000,000/16,000 = $88 per capita, for residential land use services and (0.20) ($1,750,000) = $350,000, or $350,000/$40,000,000 = $0.00875 per dollar of nonresidential assed value for nonresidential land use services. This means that for a new development in the community consisting of 200 single family homes generating 800 new residents and 15 acres of commercial development with assessed value of $10,000,000, the total estimated annual cost for public services would be determined as follows:

$$(800 \text{ residents}) \times (\$88 \text{ per capita}) \qquad = \$ \ 70,400$$

$$\left(\begin{matrix} \$0.00875 \text{ per } \$1 \\ \text{of assessed value} \end{matrix} \right) \times \left(\begin{matrix} \$10,000,000 \\ \text{assessed value} \end{matrix} \right) = \$ \ 87,500$$

$$\text{Total} = \$157,900$$

The per capita and development value approaches are indeed rather simplistic. The real issue in the allocation of costs to development is whether costs are apportioned to all beneficiaries (direct and indirect.) An admitted shortcoming in allocating costs only to direct beneficiaries is that many services are of indirect benefit to the community as a whole. It can be reasonably argued that the quality of such services as education, libraries, and parks constitute a real benefit to the industrial users. They do this by attracting the caliber of employees instrumental to success and by enhancing the quality of life and thereby increasing the efficiency and productivity of workers. While these benefits are real, it is extremely difficult to assign a dollar value to the benefit received. For this reason, one generally assigns costs to the direct beneficiary while acknowledging that this is an oversimplification.

In establishing the direct beneficiary of public services, it is important to recognize that local public services may be broadly grouped under two categories; namely:

1. "Services to property"—functions and facilities specifically identifiable with land use or development. Among these are facilities and activities substantially tied to land, such as street cleaning and maintenance, part of fire and police protection, waste collection and disposal, water supply and sewage disposal

2. "Services to people"—activities of community-wide or general benefit nature. This includes public health, welfare, education, and recreation, among other activities. These are relatively independent of location within a given community and, as a package, these services are essential for a viable community of a given population size regardless of where development takes place.

Within the broad categories of local government costs as stated in the earlier section of this chapter, Local Government Cost Categories, general government and public safety costs consist of both services to property and services to people. Public works costs are primarily services to property, whereas library service and parks and recreation costs are services to people.

As an illustration of how this information could be used to allocate public service costs, consider first the case of general government costs. Recognizing first that these costs are services to property, one could initially prorate these costs according to acreage of each type of land use. Then, further allocation of the residential portion could be handled on a per capita basis in recognition of the fact that these are also services to people. For example, consider a community with the following characteristics:

Number of developed acres:
Commercial   200
Industrial   100
Residential ....   <u>500</u>
                                       800 acres
Population   20,000
Total Annual Expenditures for General Government $10,000,000

On the basis of services to property allocation, we would allocate these costs to land use as

1. $\dfrac{200}{800} \times \$10,000,000 = \$2,500,000$

for commercial land use (or, equivalently, $12,500 per acre)

2. $\dfrac{100}{800} \times \$10,000,000 = \$1,250,000$

for industrial land use (or, equivalently, $12,500 per acre)

3. $\dfrac{500}{800} \times \$10,000,000 = \$6,250,000$

for residential land use.

Dividing $6,250,000 by 20,000 people, we obtain $312.50 per capita as the cost of services to people. Therefore, one could estimate the general government costs associated with new development by a formula of the form:

$$\left(\begin{array}{c}\$12,500 \\ \text{per acre}\end{array}\right) \left(\begin{array}{c}\text{Number of commercial} \\ \text{and industrial acres}\end{array}\right) + \left(\begin{array}{c}\$312.50 \\ \text{per capita}\end{array}\right) \left(\begin{array}{c}\text{Number of} \\ \text{new residents}\end{array}\right)$$

Public safety costs also relate to services to property as well as to services to people simply because protection is provided to property and human life. This suggests that a reasonable allocation of these costs would be based on the "relative value" of both property and human life. Even though reasonable as an approach, it has the shortcoming that it requires one to estimate the "value" of human life, which is a highly subjective determination. For this reason, assessed valuation is typically used as an indicator of value. For example, suppose community annual expenditures for public safety are $35,000,000 with assessed valuation distributed as follows:

Type of property	Percent	Assessed valuation
Commercial	56.0	$ 16 million
Industrial	17.5	5 million
Residential	26.5	7.5 million
		$28.5 million

On the basis of a services to property allocation, we would allocate these costs to land use as

1. $(0.56)(\$35,000,000) = \$19,600,000$

for commercial land use (or, equivalently, $1.228 per dollar of valuation)

2. $(0.175)(\$35,000,000) = \$6,125,000$

for industrial land use (or, equivalently, $1.228 per dollar of valuation)

3. $(0.265)(\$35,000,000) = \$9,275,000$

for residential land use.

Assuming that the 20,000 population in the community consists of 8000 dwelling units (DUs) with an average of 2.5 persons per DU and dividing $9,275,000 by 8000 dwelling units, we obtain $1,159.38 per dwelling unit as a measure of the cost of services to people. However, another measure of this type of service could be obtained upon dividing $9,275,000 by 20,000 people to obtain $463.75 per capita. Therefore, one could estimate the public safety costs associated with new development by a formula of the form:

$$\begin{pmatrix} \$1.228 \text{ per} \\ \text{dollar of} \\ \text{valuation} \end{pmatrix} \begin{pmatrix} \text{Total valuation} \\ \text{of commercial} \\ \text{and industrial} \\ \text{development} \end{pmatrix} + \begin{pmatrix} \$463.75 \text{ per} \\ \text{capita} \end{pmatrix} \begin{pmatrix} \text{Number of} \\ \text{new residents} \end{pmatrix}$$

An equivalent formula, assuming that the number of people per dwelling unit remained constant, would be to replace the right-hand term by $1159.38 per DU times the number of new dwelling units. The reader should recognize that in the case of public safety costs we have used the previously described development value method.

Public works costs are primarily services to property. Thus, these costs are typically allocated on a "per unit of property" basis such as per mile of street, per foot of sewer line, per foot of parkway, per street light, per tree, etc. For example, dividing the total annual cost for street and gutter maintenance by the total mileage of streets maintained would yield a cost factor in the form of street maintenance cost per mile of street. One could then estimate the street maintenance cost associated with new development as

$$\begin{pmatrix} \text{Street maintenance} \\ \text{cost per street mile} \end{pmatrix} \times \begin{pmatrix} \text{Total mileage of} \\ \text{streets in new} \\ \text{development} \end{pmatrix}$$

Finally, the cost of library service, and parks and recreation involve services to people and, as a result, are typically allocated on a per capita basis. This requires dividing the total annual cost for library service by the community population to obtain the library cost per capita and dividing the total annual cost for parks and recreation service by the community population to obtain the parks and recreation cost per capita. One could then estimate the library service and parks and recreation costs associated with new development as

$$\begin{pmatrix} \text{Library service} \\ \text{cost per capita} \end{pmatrix} \times \begin{pmatrix} \text{Number of} \\ \text{new residents} \end{pmatrix}$$

and

$$\begin{pmatrix} \text{Parks and recreation} \\ \text{cost per capita} \end{pmatrix} \times \begin{pmatrix} \text{Number of} \\ \text{new residents} \end{pmatrix}$$

respectively.

Allocation of Capital Expenditures   Capital expenditures present a special problem from the viewpoint of analysis. The items involved represent a one-time investment rather than a continuing expense and it can be extremely difficult to distinguish those capital expenditures attributable to urban expansion from those assignable to the refurbishment and upgrading of the existing buildings, streets, parks, sewers, etc. in the community. In addition, even those improvements which can be related to physical growth often represent the expansion of a facility or street which should have occurred years earlier. Operating costs from one year to the next cannot be postponed. However, capital expenditures can be and often are postponed. Therefore, one must examine closely not only the category of the expenditure but also the circumstances surrounding the timing of the fund expenditure.

Most capital expenditures are made from individual funds which have financing from sources independent of the general fund. As with the general fund, these individual funds do "balance" each year with costs equalling revenues. In general, the revenue sources relate to gas taxes, utility fees, or a separate property tax rate. The amount of this revenue which has been generated by "growth" can roughly be measured by the percentage of expansion which has occurred during the time period under consideration. For example, if an individual city grew by 20 percent over a four-year span, then 20 percent of the individual fund revenues could well be attributed to growth. If, in any given year, the capital expenditures which could be directly related to growth were in excess of the 20 percent figure, then the difference would, in effect, have to be covered by operating budget surpluses or other factors. To the extent that the capital expenditures related to growth were less than 20 percent, the growth element could be said to have generated a surplus.

Capital expenditures associated with new development include: (1) facilities linked directly with the proposed development, such as new schools, sewer lines, fire stations, and other new facilities; and (2) facilities to be constructed as part of a capital improvement plan which will be shared by existing residents or enterprises in the jurisdiction as well as those from the new development, such as junior colleges, new sewage and/or water treatment plants, and health care centers. Facilities in the second category are not usually triggered by a single new development project, but may be designed to incorporate increases in demand from projected development.

In order to determine whether a capital expenditure should be allocated to the existing residents exclusively or jointly to existing and new residents, the following questions should be answered:

- Would the facilities be built regardless of the approval of a proposed development?
- Would the proposed size of the facilities remain constant regardless of population increases resulting from new development?
- Would new residents be users, and thus beneficiaries, of the proposed facilities?

If the response to the first two questions is "yes," and to the third "no," then the project is unrelated to an increase in population. It is more likely, however, that the size of projects, such as schools, is based on projected growth in population. In this situation, the incremental (rather than average) cost should be allocated to the new development, if the data are available. It may be argued that since growth is already projected and thus a built-in factor, the approval of a specific development results in the addition of only part of the projected population. Thus, unless the population increment from new development does not exceed the projected level, the decision to provide the facility is independent of the approval of a request for new development.

Capital expenditure requirements are a crucial element in determining total costs associated with new development since, unless existing facilities are underutilized, increases in population require new public and private sector capital investments. An alternative to such investments given full utilization of facilities, would be a reduction in the level and quality of existing services, such as increased number of students in classrooms or double sessions, increased traffic congestion, or overcrowded recreational facilities.

In areas of rapid growth, public infrastructure investments frequently lag behind population increases because of public sector fiscal limitations, because of the initial diseconomies of scale associated with providing new facilities, or because of inadequate planning. As a result, there is usually at least a short-term degradation in the provision of public services.

The fiscal impact of increased capital expenditures on local government is often greater than the comparable impact of an increase in operating expenditures. This is because local government generally has to pay from its own resources for many major capital expenditures such as schools, and police and fire stations, while outlays to meet operating expenditures are frequently matched by the state and, in some cases, by the federal government. In estimating the capital needs of a community which is considering new development, capital expenditures include:

- School facilities directly linked with proposed new development
- Other locally financed facilities directly linked with new development
- Facilities designed for use by the total population, including residents or business enterprises in proposed developments

In the case of past capital expenditures, those linked to facilities fully utilized prior to new development, such as public schools, should not be considered as part of the capital cost attributable to the new development. Whether capital expenditures, which provide current as well as future benefits, should be paid from current revenues or over an extended time period involves issues of equity, since the composition of the population using the capital improvements undergoes change during the useful life of the investment.

Three typical methods of paying for major capital projects through long-term debt financing are: (1) use of general obligation bonds, (2) use of special tax bonds, and (3) use of revenue bonds. General obligation bonds are backed by the "full faith and credit" of the issuing municipality and are secured by an unconditional pledge of the jurisdiction's credit, including its taxing power. Special tax bonds are retired with the proceeds of special tax levies such as motor fuel, cigarette, or liquor taxes. Revenue bonds are those bonds secured with income received by a jurisdiction from the earnings of a revenue-producing enterprise, such as water works or user charges, tolls, or fees paid for use of the facility to be constructed with the proceeds of the issue.

In general, capital outlays financed from general revenue bonds are part of the local budget and are repaid from general tax revenues. That is, most jurisdictions allocate the cost of a new facility to the total community rather than to just the residents of the area which has caused the need for new capital expenditures. Certain jurisdictions have special taxing districts, where the repayment of capital outlays is allocated *only* to the area which utilizes the facilities. The likely users of most new facilities such as sewers or schools can usually be identified, although in some instances, such as transportation facilities, the user group is likely to extend beyond those who necessitated the capital outlay.

The method of financing chosen by a jurisdiction influences the short- and long-term costs of capital investment. In a slow-growing jurisdiction, a substantial portion of capital needs are frequently paid from current revenues on a "pay as you go" basis. Capital expenditures, such as those made for many utilities, are usually self-financing through revenue bonds, and thus impose no direct burden on the public sector fiscal structure. However, major capital costs, particularly in areas of rapid growth, usually cannot be financed from current funds. Therefore, general obligation bonds are issued for a selected payback period.

To illustrate the allocation of capital expenditures, suppose that a city is planning an initial capital investment of $500,000 to provide 75 acres of parks for a projected population of 18,000 new residents. If this investment is amortized over a 30-year period at 8 percent, then the equivalent annual cost would be

$$A = \$500,000 \left[\frac{0.08(1.08)^{30}}{(1.08)^{30}-1} \right]^* = (\$500,000)(0.08883)$$
$$= \$44,415$$

which amounts to $44,415/18,000 = \$2.47$ per capita, or $7.90 per dwelling unit if there is an average of 3.2 persons per dwelling unit.

If a city were to build a government services substation which is to utilize 6 acres and will include health, library, and public safety facilities at a total initial investment cost of $1,600,000, then the potential users would be both households and business enterprises in the area. A way of allocating this cost to the two categories of users would be a proration based on the percentage of the total assessed valuation of the city which each category represents. For example, if 60% of the community's assessed valuation can be attributed to residential land use and 30% to business (commercial or industrial), then we would allocate ($1,600,000)(0.60) = $960,000 to households and (1,600,000)(0.30) = $480,000 to businesses. The household allocation could be further allocated on a per capita or per dwelling unit basis. The business allocation could be spread on a per acre basis, with perhaps a further proration based on how much land use is commercial versus how much is industrial. These allocations can then be amortized on an annual basis depending upon the period of time involved and the assumed interest rate.

Examples of Fiscal Impact Analyses

Use of Tax Cost Index   One way of evaluating the fiscal impact of new development is to compare the public service costs associated with land use with the property tax rate necessary to generate revenues to cover these costs. This requires the use of the "tax cost index ratio" which is defined to be the net costs of governmental services divided by the assessed value of the specific land use. Net costs are the costs of governmental services for a particular land use which must be met from local property tax revenues, that is, the total cost for services less the nonproperty tax revenues. The assessed value represents the ability of a particular land use to generate property tax revenue. Specifically, the tax cost index ratio is defined as follows:

$$\left(\begin{array}{c} \text{Tax cost} \\ \text{index} \end{array} \right) = \frac{\left(\begin{array}{c} \text{Total cost of} \\ \text{government services} \end{array} \right) - \left(\begin{array}{c} \text{Total nonproperty} \\ \text{tax revenue} \end{array} \right)}{\text{Total assessed value}}$$

*Called the capital recovery factor over 30 years at 8 percent interest.

For example, if the net costs are $1000 and the assessed value is $50,000, then the tax cost index is $1000/$50,000 = 0.02. This means that a property tax rate of $2 per $100 of assessed value is necessary to raise $1000 in revenues to balance costs.

To illustrate this particular method, consider a project with assessed value equal to $48,000 and with costs and nonproperty tax revenues as shown in Table 2.22. The net cost is thus given by $3815 − $215 = $3600. Hence, the tax cost index is $3600/$48,000 = 0.075, which means that a property tax rate of $7.50 per $100 of assessed value is necessary to raise $3600 in revenues to cover the net costs. If the current community tax rate is less than $7.50 per $100, then the project does not pay its own way. However, if the current community tax rate is greater than $7.50 per $100, then the project generates revenues in excess of costs.

TABLE 2.22   Example Costs and Nonproperty Tax Revenues

Costs		Nonproperty tax revenues	
Police	$ 300	Sales tax	$ 25
Fire	325	Motor fuel	40
General government	410	Income tax	20
Schools	2200	Sewer/water	75
Sewer/water	105	Other	55
Other	475	Total	$215
Total:	$3815		

Annexation Versus No Annexation   Consider the situation (Ref. 44) where a city with population equal to 202,359 plans to annex 23 mi^2 of a neighboring unincorporated area with population equal to 47,262. The city council has expressed a desire to determine what the one-year fiscal impact of this annexation will be. The main concern is that if the annexation costs more than it returns in revenues, then it would not be a practical project for the city to undertake. For convenience, refer to the city without annexation as the "present city" and the county unincorporated area under consideration for annexation as the "annex area."

The revenue impacts can be analyzed as follows:

1. The assessed value of property in the present city is $1184 million and the assessed value in the annex area is $361 million. Based on a property tax rate of $1.97 per $100 of assessed value, this would imply property tax revenues of

$$(0.0197)(\$1,184,000,000) = \$23,324,800$$

for present city and

$$(0.0197)(\$361,000,000) = \$7,111,700$$

for annex area.

2. The total annual sales tax collection in the present city is estimated at $6,382,000. The total sales tax revenue generated via expenditures by residents from the annex area is $1,010,000. However, 20% of this is estimated to be already included in the $6,382,000 since some annex area residents currently shop in the present city. Therefore, annexation would only generate an additional $808,000 per year.

3. An additional source of local revenue is state sales tax redistribution which is revenue collected by the state and redistributed to local jurisdictions on the basis of their number of school-age children. The present city currently receives $3,009,900, and the annex area currently receives $800,100.

4. Personal property taxes are levied on the basis of number of automobiles and is currently levied at the rate of $111.34 per automobile. A recent survey of the county residents showed that residents in the present city owned 55,036 automobiles and residents in the annex area owned 22,475 automobiles. This implies personal property tax revenues of

$$(\$111.34)(55,036) = \$6,127,708$$

for present city and

$$(\$111.34)(22,475) = \$2,502,367$$

for annex area.

5. Machinery and tool tax receipts are based on industrial assessed valuation. In the present city this would amount to $1,665,200, and in the annex area this would amount to $144,800.

6. Utility taxes are imposed on both residential and nonresidential users. In the present city with 71,908 dwelling units (DUs), this would average $65 per DU, whereas in the annex area with 15,175 dwelling units, this would average $83 per DU. A survey of commercial and industrial enterprises in the two areas showed that $3,124,739 in utility taxes is generated in the present city and $425,260 would be generated in the annex area. This implies total utility tax revenues equal to

$$(71,908)(\$65) + \$3,124,739 = \$7,798,759$$

for present city and

$$(15,175)(\$83) + \$425,260 = \$1,684,785$$

for annex area.

7. Additional revenues would accrue from city taxes on prepared foods, transient lodging, entertainment admissions, etc. Of the total amount, estimated annually at $1.25 million, approximately 90% or $1.125 million would be from receipts in the present city, and the remaining $0.125 million would be from the annex area.

8. Fines and forfeitures estimated to be collected from both areas would amount to $1.9 million per year. Since this revenue source is basically population-oriented, this means that fines and forfeitures revenues would be

$$\left(\frac{202,359}{249,621}\right)(\$1,900,000) = \$1,540,263$$

for present city and

$$\left(\frac{47,262}{249,621}\right)(\$1,900,000) = \$359,737$$

for annex area.

9. Delinquent property tax payments for both areas are estimated to average $1.4 million per year. Since the present city contains 78% of the combined assessed valuation, it is estimated that $(0.78)(\$1,400,000) = \$1,092,000$ can be attributed to the present city and $(0.22)(\$1,400,000) = \$308,000$ can be attributed to the annex area.

10. It is estimated that the total annual revenue from business fees and professional licenses would be $8.8 million. Of this amount, half can be attributed to business enterprises and the remainder to professional occupations. A survey has shown that the annex area includes only 14% of the combined commercial activity and 5% of the professional businesses. Thus, the annual business-oriented revenues would be

$$(0.86)(\$4,400,000) + (0.95)(\$4,400,000) = \$7,964,000$$

for present city and

$$(0.14)(\$4,400,000) + (0.05)(\$4,400,000) = \$836,000$$

for annex area.

11. Vehicle license revenue, based on license fees for all motor vehicles, to local jurisdictions amounts to an average of $16.13 per automobile. Thus, this would provide revenue equal to

$$(\$16.13)(55,036) = \$887,731$$

for present city and

$$(\$16.13)(22,475) = \$362,522$$

for annex area.

12. Miscellaneous fees collected annually from such items as dog licenses, special permits, marriage licenses, etc. average about $2.31 per capita in the present city, thus yielding a total of ($2.31)(202,359) = $467,449. Assuming this factor also applies to the annex area, annual miscellaneous fees revenue would equal ($2.31)(47,262) = $109,175 for the annex area.

13. Utility sales tax payments are redistributed annually by the state to the local jurisdictions. For the present city, this amounts to $3.69 million annually, and for the annex area, this amounts to $810,000 annually.

14. Other miscellaneous revenues from such sources as alcoholic beverage control, inventory tax, and rolling stock tax are redistributed annually by the state to local jurisdictions on the basis of $8.25 per capita. This means that miscellaneous revenues will be equal to

$$(\$8.25)(202,359) = \$1,669,462$$

for present city and

$$(\$8.25)(47,262) = \$389,912$$

for annex area.

In summary, Table 2.23 provides a comparison of the expected annual revenues to be generated from each of the areas. The present city generates $66.744 million, and upon annexation, the combined areas would generate $83.097 million in annual revenues.

The cost impacts can be analyzed as follows:

1. A survey was conducted to determine how many additional employees would have to be hired if the annex area were annexed into the city. The results obtained are presented in Table 2.24 which shows that without annexation the annual operating cost would be $28.451 million and with annexation this cost would be $32.528 million.

2. Retirement plan contributions by the city to the employees' retirement fund average $1169 per employee on an annual basis. This means that the current cost to the present city is (3104)($1169) = $3,628,576, whereas the additional cost upon annexation would be (446)($1169) = $521,374, using the employee totals of Table 2.24.

3. In addition to the annual operating expenditures of items 1 and 2, capital expenditures are planned for construction and/or modification of existing facilities during the next year. Without annexation, these expenditures are estimated to total $38,000,000 plus an additional $8,000,000 to meet the increased demand of the new residents upon annexation.

TABLE 2.23   Comparison of Annual Revenues for Present City Versus Annex Area

Revenue source	Annual revenue (in $ million)	
	Present city	Annex area
Property tax	23.325	7.112
Sales tax	6.382	0.808
State sales tax redistribution	3.010	0.800
Personal property tax	6.128	2.502
Machinery and tool tax	1.665	0.145
Utility tax	7.799	1.685
Additional revenues	1.125	0.125
Fines and forfeitures	1.540	0.360
Delinquent taxes	1.092	0.308
Business fees and licenses	7.964	0.836
Vehicle license revenue	0.888	0.363
Miscellaneous fees	0.467	0.109
Utility sales tax	3.690	0.810
Miscellaneous revenues	1.669	0.390
Total	66.744	16.353

As a result, the impact on public service costs (in million dollars) can be summarized as follows:

	Present city	Annex area
Annual cost	32.080	4.598
Capital cost	38.000	8.000
	70.080	12.598

Comparing the above results with the revenue totals of Table 2.24 shows that the fiscal impact, under the assumption that capital costs are paid by cash, during the next year would be

	Without annexation	With annexation
Total revenue	66.744	83.097
Total cost	70.080	82.678
Revenue-cost	−3.336	0.419
Revenue-to-cost ratio	0.95	1.01

That is, without consideration of capital costs, both areas provide revenues in excess of capital costs. However, with consideration of capital costs, the present city would not generate revenues in excess of costs one year later unless the county area were annexed.

TABLE 2.24   Public Service Costs for Present City Versus Annex Area

Type of service	Existing number of employees	Additional number of employees	Operating cost per employee	Service cost (in $ million) Present city	Annex area
General government	146	41	$13,220	1.930	0.542
Judiciary	17	2	6,500	0.111	0.013
Library	85	15	7,400	0.629	0.111
Police	511	71	10,972	5.607	0.779
Fire	495	76	10,408	5.152	0.791
Other public safety	85	16	7,700	0.655	0.123
Public welfare	694	12	8,600	5.968	0.103
Public health	133	10	11,400	1.516	0.114
Public works	530	133	7,600	4.028	1.011
Recreation and parks	408	7	7,000	2.856	0.490
	3,104	446		28.451	4.077

Time-phased Fiscal Impact Analysis   To illustrate the approach and techniques used for fiscal impact analysis of time-phased projects (those whose development occurs over a period of time) consider a commercial project known as Mountain View Center, which is in the San Leandro Valley area of Los Angelo. This project consists of 10,000 ft² of retail shops to be completed in 4 years, 20,000 ft² of restaurant space to be completed in 3 years, 62,200 ft² of bank and savings and loan space to be completed in 3 years, and 44,000 ft² of office space to be completed in 4 years. The Mountain View Center economic analysis will focus on a comparison of revenues versus costs from the start of development, assumed now, through the first fiscal year in which 90% occupancy occurs in all buildings, assumed to be in 4 years. As a result, a time-phased cost-revenue analysis is developed.

For the purpose of this economic analysis, a number of assumptions must be employed. Specifically, these are as follows:

1. *Assessed value of land and improvements increases at the rate of 3% annually.* According to the *City of Los Angelo Budget for this fiscal year,* the city's assessed valuation for last year increased by 3.3% over the prior year, as compared to a 5-year average of 4.6% per annum. The county assessor's office indicates that the increase this year will be approximately 3%. Consequently, even though this is less than the 5-year average of 4.6%, this number will be used and can be regarded as a conservative estimate.

2. *City of Los Angelo property tax rate increases annually at 3%.* The city tax rate has increased from $2.031 per $100 to $2.7296 per $100 assessed valuation over the last 11 years. This amounts to an approximate increase of 3% per year over 11 years.

3. *City public service costs increase annually at 8%.* According to the *Los Angelo City Controller*, last year's expenditures were $564,719,540, whereas the expenditures were $248,568,466 for 9 years prior. This amounts to an approximate increase of 9% per year for 9 years. However, relative to the last two years, the increase is only 7.6%. Therefore, an 8% increase per annum is felt to be a conservative estimate. As a reference, Table 2.25 provides the currently adopted budget.

4. *Public service costs are distributed within the city proportionally to population and land use.* Since public service costs reflect basically the cost for services to people, it is plausible that the larger the population in a specific geographical area, the higher the costs should be for public services. Furthermore, for a given type of public service cost (with the exception of library costs which are normally distributed on a per capita basis), it also is plausible to prorate the cost according to land use. In other words, if $x\%$ of the land use is for residential, then $x\%$ of the cost should be allocated to residential land use. A similar statement can be made for commercial and industrial land uses.

TABLE 2.25   City of Los Angelo Budget for Current Year

Category	Cost
Library	$ 12,690,761
Police protection	143,451,209
Fire protection	63,532,366
Recreation and parks	35,358,741
Street maintenance, parkway trees, and maintenance, and street sweeping	31,791,389
Street lighting	7,452,505
Sewers (nonconnection expenses) and trash collection	37,094,113

5. *Percent city population residing in the San Leandro Valley area is expected to increase by 0.675% per year.* According to the Land Use Element of the city of Los Angelo General Plan, the city population residing in the *San Leandro Valley* area is expected to increase from 35.4 to 40.5% over the next 21 years, i.e., an approximate 0.675% increase per year. This also means that 35.4% of the presently estimated public service costs in Table 2.25 can be allocated to the San Leandro Valley area.

6. *Percent of developed land used for commercial projects in the San Leandro Valley area decreases by 2.2% per annum.* Currently, the total developed land (not including open space) in the San Leandro Valley is approximately 86,090 acres, of which 4492 acres are for commercial land use. The specific land use plan for this area shows that 18 years from now this total is expected to be 127,300 acres, of which 4500 acres will be for commercial land use. This represents a decrease from 5.2% to 3.5% (or a 2.2% decrease per year) in commercial land use in the San Leandro Valley area.

7. *Commercial square footage in the San Leandro Valley area increases at the approximate rate of 2.07% per year.* It was estimated that 2 years ago there were 59,972,800 ft^2 of commercial (office and retail) space in the San Leandro Valley area, and 18 years from now there will be approximately 90,453,400 ft^2 of commercial space. This means that the approximate increase per year over this 20-year period will be 2.073%.

Commercial Revenues.   There are three basic sources of revenue from the proposed commercial land use of the Mountain View Center, namely:

1. property tax
2. sales tax
3. business tax (both tax on sales and rental of commercial space).

The currently estimated market value for Mountain View Center land is $612,000 and, according to assumption 1, this is expected to increase by 3% per annum for valuation purposes. For valuation of improvements, a construction cost of $35.92/ft^2 today is assumed to increase at 3% per annum. This means that commercial space completed in 3 years will be valued at $(1.0927)(\$35.92) = \$39.25$ and in 4 years will be valued at $(1.1255)(\$35.92) = \$40.43$. Using the current tax rate of $2.7296 per $100 assessed valuation and assumption 2 (i.e., the tax rate increases annually at 3.0%), the total Mountain View Center annual property tax is derived as shown in Table 2.26.

TABLE 2.26   Annual Property Tax Revenue from Mountain View Center

	Current year	In 1 year	In 2 years	In 3 years	In 4 years
Market value of land	$612,000	$630,360	$649,271	$ 668,749	$ 688,811
Market value of improvements	0	0	0	3,226,350	5,506,566
Total Market Value	612,000	630,360	649,271	3,895,099	6,195,377
Assessed value (@ 25%)	153,000	157,590	162,318	973,775	1,548,844
Property tax	4,176	4,431	4,700	29,045	47,584

The sales tax computation is based on the assumption that 0.95% of all retail sales is returned to the city as sales tax. The derivation of retail sales is based on the use of estimated factors for sales/ft^2 gross leasable area (GLA), equal to 85% of gross floor area (GFA), by type of establishment. Thus, the annual sales tax revenue is as follows:

Retail shop(s)

$$(0.85)(10,000 \text{ ft}^2 \text{ GFA})(\$74.92 \text{ sales/ft}^2 \text{ GLA}) = \$686,820 \text{ sales}$$
$$\text{Sales tax} = (\$636,820)(0.0095) = \$6,050 \text{ in 4 years}$$

Restaurant

$$(0.85)(20,000 \text{ ft}^2 \text{ GFA})(\$72.85 \text{ sales/ft}^2 \text{ GLA}) = \$1,238,450 \text{ sales}$$
$$\text{Sales tax} = (\$1,238,450)(0.0095) = \$11,765 \text{ in 3 years and thereafter}$$

$$\text{Total Sales Tax} = \begin{cases} \$11,765 \text{ in 3 years} \\ \$17,815 \text{ in 4 years} \end{cases}$$

The business tax is derived from two sources, namely, tax on gross receipts and tax on rentals of commercial premises. The latter requires the use of annual per square foot rental factors by type of establishment. Furthermore, the derivation of gross receipts from commercial/office use will be based on the assumption of $30,000 sales per employee and 250 ft^2 GFA per employee. The applicable business tax rate in the city of Los Angelo is $5 per $1000 gross receipts and, in the case of commercial rental space, is $1.25 per $1000 rental receipts. Therefore, the annual gross receipts business tax revenue is as follows:

Commercial/office (including bank, and savings and loan)

(a) In 3 years,

$$\frac{(62,200 \text{ ft}^2 \text{ GFA})(\$30,000 \text{ sales per employee})}{(250 \text{ ft}^2 \text{ per employee})} = \$7,464,000 \text{ sales}$$
$$\text{Business tax} = (\$5/1000)(\$7,464,000) = \$37,320$$

(b) In 4 years,

$$\frac{(106,200 \text{ ft}^2 \text{ GFA})(\$30,000 \text{ sales per employee})}{250 \text{ ft}^2} = \$12,744,000 \text{ sales}$$
$$\text{Business tax} = (\$5/1000)(\$12,744,000) = \$63,720$$

Retail shop(s)

$$(0.85)(10,000 \text{ ft}^2 \text{ GFA})(\$74.92 \text{ sales/ft}^2 \text{ GLA}) = \$636,820 \text{ sales}$$
$$\text{Business tax} = (\$5/1000)(\$636,820) = \$3,184$$
$$\text{in 4 years}$$

Restaurant

$$(0.85)(20,000 \text{ ft}^2 \text{ GFA})(\$72.85 \text{ sales/ft}^2 \text{ GLA}) = \$1,238,450 \text{ sales}$$
$$\text{Business tax} = (\$5/1000)(\$1,238,450) = \$6,192$$
$$\text{in 3 years}$$
$$\text{and thereafter}$$

$$\text{Total Business Tax on Gross Receipts} = \begin{cases} \$43,512 \text{ in 3 years} \\ \$73,096 \text{ in 4 years} \end{cases}$$

The annual rental receipts business tax revenue is as follows:

Commercial/office

$$(0.85)(44{,}000 \text{ ft}^2 \text{ GFA})(\$7.20/\text{ft}^2 \text{ GLA}) = \$269{,}280 \text{ rent}$$
$$\text{Business tax} = (\$1.25/1000)(\$269{,}280) = \$337$$
$$\text{in 4 years}$$

Bank, and savings and loan

$$(0.85)(62{,}200 \text{ ft}^2 \text{ GFA})(\$3.05/\text{ft}^2 \text{ GLA}) = \$161{,}254 \text{ rent}$$
$$\text{Business tax} = (\$1.25/1000)(\$161{,}254) = \$202$$
$$\text{in 3 years}$$
$$\text{and thereafter}$$

Retail Shop(s)

$$(0.85)(10{,}000 \text{ ft}^2 \text{ GFA})(\$4.60/\text{ft}^2 \text{ GLA}) = \$39{,}100 \text{ rent}$$
$$\text{Business tax} = (\$1.25/1000)(\$39{,}100) = \$49$$
$$\text{in 4 years}$$

Restaurant

$$(0.85)(20{,}000 \text{ ft}^2 \text{ GFA})(\$4.28/\text{ft}^2 \text{ GLA}) = \$72{,}760 \text{ rent}$$
$$\text{Business tax} = (\$1.25/1000)(\$72{,}760) = \$91$$
$$\text{in 3 years}$$
$$\text{and thereafter}$$

$$\frac{\text{Total Business Tax}}{\text{on Rental Receipts}} = \begin{cases} \$293 \text{ in 3 years} \\ \$679 \text{ in 4 years} \end{cases}$$

In summary, we have

$$\text{Total Business Tax} = \begin{cases} \$43{,}805 \text{ in 3 years} \\ \$73{,}775 \text{ in 4 years} \end{cases}$$

Commercial Costs. The determination of commercial costs is based upon the use of the aforementioned assumptions 3–7 and the basic assumptions that police, fire, and sewer costs are allocated on a per 1000 ft² GFA basis, whereas street maintenance (including parkway trees and maintenance and street sweeping) and street lighting are allocated on a per acre basis. For example, using the public service costs in Table 2.25, the estimated commercial costs in the San Leandro Valley area for the current year and the annual growth factors are as follows:

1. Police protection

$$\frac{\left(\begin{array}{c}\text{Total city} \\ \text{police cost}\end{array}\right) \left(\begin{array}{c}\text{Fraction of city} \\ \text{population in San} \\ \text{Leandro Valley Area}\end{array}\right) \left(\begin{array}{c}\text{Fraction of developed land} \\ \text{for commercial use}\end{array}\right)}{(\text{Commercial space in 1000 ft}^2 \text{ GFA})}$$

$$= \frac{(\$143{,}451{,}209)(0.354)(0.052)}{(59{,}972.8)(1.02073)^2} = \$42.26/1000 \text{ ft}^2 \text{ GFA}$$

$$\text{Annual growth factor} = \frac{(1.08)(1.00675)(0.978)}{1.02073} = 1.04177$$

2. Fire protection

$$\frac{\left(\begin{array}{c}\text{Total city} \\ \text{fire cost}\end{array}\right) \left(\begin{array}{c}\text{Fraction of city} \\ \text{population in San} \\ \text{Leandro Valley Area}\end{array}\right) \left(\begin{array}{c}\text{Fraction of developed land} \\ \text{for commercial use}\end{array}\right)}{(\text{Commercial space in 1000 ft}^2 \text{ GFA})}$$

$$= \frac{(\$63{,}532{,}366)(0.354)(0.052)}{(59{,}972.8)(1.02073)^2} = \$18.72/1000 \text{ ft}^2 \text{ GFA}$$

$$\text{Annual growth factor} = \frac{(1.08)(1.00675)(0.978)}{1.02073} = 1.04177$$

3. Street maintenance, parkway trees and maintenance and street sweeping

$$\frac{\left(\substack{\text{Total city street} \\ \text{maintenance cost}}\right)\left(\substack{\text{Fraction of city} \\ \text{population in San} \\ \text{Leandro Valley Area}}\right)\left(\substack{\text{Fraction of} \\ \text{developed land} \\ \text{for commercial use}}\right)}{(\text{Number of commercially developed acres})}$$

$$= \frac{(\$31,791,389)(0.354)(0.052)}{4492} = \$130.31/\text{acre}$$

Annual growth factor $= (1.08)(1.00675)(0.978) = 1.06337$

4. Street Lighting

$$\frac{\left(\substack{\text{Total city} \\ \text{lighting cost}}\right)\left(\substack{\text{Fraction of city} \\ \text{population in San} \\ \text{Leandro Valley Area}}\right)\left(\substack{\text{Fraction of developed land} \\ \text{for commercial use}}\right)}{(\text{Number of commercially developed acres})}$$

$$= \frac{(\$7,452,505)(0.354)(0.052)}{4492} = \$30.55/\text{acre}$$

Annual growth factor $= (1.08)(1.00675)(0.978) = 1.06337$

5. Sewers (nonconnection expenses) and trash collection

$$\frac{\left(\substack{\text{Total city} \\ \text{sewers cost}}\right)\left(\substack{\text{Fraction of city} \\ \text{population in San} \\ \text{Leandro Valley Area}}\right)\left(\substack{\text{Fraction of developed land} \\ \text{for commercial use}}\right)}{(\text{Commercial space in 1000 ft}^2\ \text{GFA})}$$

$$= \frac{(\$37,094,113)(0.354)(0.052)}{(59,972.8)(1.02073)^2} = \$10.93/1000\ \text{ft}^2\ \text{GFA}$$

Annual growth factor $= \dfrac{(1.08)(1.00675)(0.978)}{1.02073} = 1.04177$

Table 2.27 provides a summary of the annual commercial cost factors for the current year over the next 4 years used to evaluate the public service costs as the result of the Mountain View Center planned land use.

Economic Summary.   Table 2.28 provides a summary of the Mountain View Center economic impact on the city of Los Angelo for the current year through 4 years. The revenue-to-cost ratio prior to 90% occupancy of any of the buildings ranges from 3.69 to 3.71. In three years when the bank, savings and loan, and restaurant space is occupied (at least 90%), this ratio jumps threefold to 10.53. It then increases to 10.73 in the next year when the office space is at least 90% occupied.

In terms of the present worth of revenues and costs, the present worth of the revenues at 8% per year (assumed to be the city's average rate of return on investments for the last 5 years) is given by

$$\substack{\text{Present worth} \\ \text{of revenues}} = \sum_{j=0}^{4} R_j (1.08)^{-j},$$

where R_j is the revenue in year j and year 0 is the current year. Using the yearly revenues of Table 2.28, we obtain

$$\substack{\text{Present worth} \\ \text{of revenues}} = \$181,775$$

Similarly,

$$\substack{\text{Present worth} \\ \text{of costs}} = \sum_{j=0}^{4} C_j (1.08)^{-j},$$

where C_j is the cost in year j and year 0 is the current year. Using the yearly costs of Table 2.28, we obtain

$$\frac{\text{Present worth}}{\text{of costs}} = \$19{,}244$$

The corresponding ratio of revenues to costs is then given by 9.45.

TABLE 2.27 Annual Cost Factors for Commercial Land Use in the San Leandro Valley Area

	Current year	In 1 year	In 2 years	In 3 years	In 4 years
Police protection (per 1000 ft² GFA)	$ 42.26	$ 44.03	$ 45.86	$ 47.78	$ 49.78
Fire protection (per 1000 ft² GFA)	18.72	19.50	20.32	21.17	22.05
Street maintenance (per acre)	130.31	138.57	147.35	156.69	166.62
Street lighting (per acre)	30.55	32.49	34.54	36.73	39.06
Sewers (per 1000 ft² GFA)	10.93	11.39	11.86	12.36	12.87

TABLE 2.28 Economic Impact of Mountain View Center

	Current year	In 1 year	In 2 years	In 3 years	In 4 years
Revenue sources:					
Property tax	$4,176	$4,431	$4,700	$29,045	$ 47,584
Sales tax	. . .	. . .	. . .	11,765	17,815
Business tax	. . .	. . .	. . .	43,805	73,775
Total	$4,176	$4,431	$4,700	$84,615	$139,174
Cost sources:					
Police protection*	. . .	. . .	. . .	$ 3,928	$ 6,780
Fire protection*	. . .	. . .	. . .	1,740	3,003
Street maintenance	$ 912	$ 170	$1,031	1,097	1,166
Street lighting	214	227	242	257	273
Sewers*	. . .	. . .	. . .	1,016	1,753
Total	$1,126	$1,197	$1,273	$ 8,039	$ 12,975
Revenue-to-cost ratio	3.71	3.70	3.69	10.53	10.73

*Not allocated to a fiscal year until 90% completion occurs.

School District Fiscal Impact Analysis

The most important factor affecting the fiscal impact of new development on local schools is the number of school-age children residing in a new development. Tables 2.9 and 2.10 provided illustrations of these pupil generation rates. However, it must be recognized that such rates can vary significantly with different family income levels, development characteristics (such as adult-oriented communities, retirement communities, low-income government housing projects), community characteristics (such as prestige or reputation of the local school district), and other variables.

Per capita operating expenditures can be calculated based upon the current operating budget of a district divided by the average daily attendance. This per capita figure is then applied to the projected number of students residing in the new development.

Capital costs are somewhat more difficult to project because they depend upon two factors, namely: (1) the specific characteristics of any building bond issues (interest rate and term) which will affect the annual payments, and (2) the growth characteristics of the entire school district which will affect both the quantity and timing of capital improvements.

An example illustrating the calculation of operating and capital costs is presented in Table 2.29.

School district revenues can be calculated on a fairly accurate basis, because they involve essentially three sources.

- Property tax revenues
- State (or federal) aid (operating and/or capital)
- Miscellaneous revenues

The single greatest source of revenue for the school district is the local property tax, which is based upon market value of the proposed development, the tax assessment rate, and the tax rate of the school district. An important factor related to the calculation of this revenue is the time lag associated with the distribution of these revenues to the school district. For example (Ref. 37), in Illinois, there is approximately an 18-month lag between the time a development is built and occupied (thus requiring educational services) and when the revenue is actually distributed to various taxing districts.

The type and amount of state aid depends upon the practices of a specific state. In some areas, the state pays for a portion of all capital costs. In general, most states contribute to

---

TABLE 2.29   Example of the Calculation of School District Operating and Capital Costs

Operating costs:
    Current operating budget ...... $923,000 per year
    Current enrollment .... 1,000 students
    Operating cost per pupil .... $923,000/1,000 = $923 per pupil
Capital costs:
    Elementary school   600 student capacity
    Costs:
        Land .... 11 acres at $12,000/acre = $132,000
        Building $1,750,000

Total annual (amortized) cost assuming financing by 20-year bonds at 6% interest

$$= (0.06)(\$1,882,000) + \frac{\$1,882,000}{20} = \$207,020$$

Amortized capital cost per elementary student:
    $207,020/600 = $345

Equivalent total annual cost:

$$\left(\begin{array}{c}\$923 \text{ per pupil}\\ \text{for operations}\end{array}\right) + \left(\begin{array}{c}\$345 \text{ per pupil}\\ \text{for capital}\\ \text{improvements}\end{array}\right) = \begin{array}{c}\$1,268 \text{ per}\\ \text{pupil}\end{array}$$

---

the operating budget of the local school system. The amount of aid is generally expressed on a per pupil basis and can be based on a number of factors (relative wealth of the district, local tax base, growth, etc.). For small developments, the state aid factor will not change; thus, the current per pupil figure for the district can be used to project revenues for new students. In the case of a large development bringing in substantial wealth (as expressed by the assessed value per pupil figure), the state aid for the district may actually decrease. A time lag factor of approximately 1 year is also frequently associated with state aid since revenues are generally based upon the previous year's enrollment.

The schools also receive revenues from a variety of sources including user charges, fines, and fees. These can generally be projected using the per capita technique, based upon the current experience of the district, and do not have a significant time lag associated with them.

An example of a fiscal impact analysis for a school district is presented in Table 2.30. In general, studies indicate that single-family development causes deficits for the school district, whereas high-density, multiple-family development usually generates surpluses over the long term.

As in the case of the impact on the cost of public services provided by cities and counties, a natural question to address is whether or not new development pays its own way relative to educational services. To illustrate an approach to answering this question let

$$C = \text{actual district cost per student,}$$

where C can be computed in one of two ways, namely:

$$1. \quad C = \left(\begin{array}{c}\text{Total per pupil}\\ \text{expenditure}\end{array}\right) - \left(\begin{array}{c}\text{Total per pupil assistance from}\\ \text{state and federal sources}\end{array}\right)$$

or

2. $C = \dfrac{\left(\begin{array}{l}\text{Total school year}\\\text{costs}\end{array}\right) - \left(\begin{array}{l}\text{Total annual federal and state}\\\text{assistance}\end{array}\right)}{(\text{Average annual school enrollment})}$

Theoretically, the cost per student should be prorated according to the type of land use and, specifically, according to the percent of the total assessed valuation (or property tax

TABLE 2.30   Example of Annual Fiscal Impact on Elementary School District

	Year 1	Year 2	Year 3	Year 4	Year 5
Annual students	72	72	54	40	. . .
Cumulative students	72	144	198	238	238
Total expenditures ($923 per pupil)	$66,456	$132,912	$182,754	$219,674	$219,674
State and federal aid ($415 per pupil)	0	29,880	59,760	82,170	98,770
Property tax revenues	0	57,918	112,943	158,753	181,733
Miscellaneous fees ($46 per pupil)	3,312	6,624	9,108	10,948	10,948
Total Revenues:	$ 3,312	$ 94,422	$181,811	$251,871	$291,451
Surplus (deficit) of new revenues over new operating expenditures	($63,144)	($ 38,490)	($ 943)	($ 32,197)	($ 71,777)
Capital costs (at $288 per pupil)	20,736	41,472	57,024	68,544	68,544
Surplus (Deficit):	($83,880)	($ 79,962)	($ 57,967)	($ 36,347)	$ 3,233

base) which is attributed to this type of land use. Typically, 70–80% of the property tax base in a school district is contributed by residential use. In general,

$$\begin{array}{l}\text{Cost per student}\\\text{for given land use}\end{array} = C \times P$$

where P = the fraction of the property tax base attributed to the given land use. If D is the density of students for the given land use, such as students per dwelling unit, then

$$\text{Total student cost per DU} = C \times P \times D$$

The property tax revenue is determined from the equation

$$\text{Property tax revenue per DU} = R \times A \times V$$

where R = school district property tax rate
   A = tax assessment factor
   V = market value of the dwelling unit
Therefore,
   1. If $R \times A \times V > C \times P \times D$, the project more than pays for the cost of education (to the district)
   2. If $R \times A \times V = C \times P \times D$, the project breaks even
   3. If $R \times A \times V < C \times P \times D$, the project does not pay its own way as regards educational expenses.
   To illustrate the preceding analysis, consider a school district in which the average cost per student is $1000, 70% of the cost is paid for by property taxes (i.e., 30% represents federal and state aid), and residential valuation in the school district is approximately 75%. Then C = ($1000)(0.70) = $700, and P = 0.75. This implies that for a tax rate of $6.50 per $100 assessed valuation and a tax assessment factor of 20%, the

$$\text{Property tax revenue per DU} = \left(\dfrac{6.50}{100}\right)(0.20)\, V = 0.013\, V,$$

and the

$$\text{Cost per DU} = (700)(0.75)\, D = 525\, D.$$

At breakeven, we must have

$$\begin{array}{l}\text{Property tax}\\\text{revenue per DU}\end{array} = \text{Cost per DU}$$

or, equivalently,

$$0.013\, V = 525\, D$$

which means that the breakeven market value for housing for selected student generation factors is as shown in Table 2.31.

TABLE 2.31   Illustrative House Value for Breakeven Versus Student Yield

Students per DU	House value
0.25	$ 10,096
0.50	20,192
0.75	30,288
1.00	40,384
1.50	60,576
2.00	80,768
3.00	121,152

Final Comments on Fiscal Impact Analysis

A fiscal impact analysis tells a local government agency how much additional money it must spend on services for a new development (over a period of years) and how much in additional revenue (taxes, fees, transfer payments, etc.) it will derive from the development. A comparison of these anticipated expenditures with the anticipated revenues enables the agency to assess whether or not the development is economically practical and, if not, what other courses of action might be taken.

A general approach which could be followed in evaluating the results of a fiscal impact analysis is:

Step 1

Revenues are first compared with the annual costs for operations and maintenance. If revenues do not cover these costs, then development is, in effect, premature. It would not make sense in this situation to require contributions from the developer, since an operations and maintenance deficit would most likely be expected to continue in the future.

Step 2

If revenues are greater than the annual costs for operations and maintenance, then the excess revenue should be compared with the expected capital costs. If this surplus revenue is greater than these costs, then the development is economically practical and pays its own way.

Step 3

If, in Step 2, the surplus revenue is less than the expected capital costs, then the development is presumed to be deficient only in the initial capital requirements. As a result, one could recommend that the net capital cost be paid by the developer as a condition of approval.

ANALYSIS OF SOCIAL IMPACTS

A basic approach to the assessment of social impacts as the result of new development consists of preparing a checklist for specific projects to be used as a guide in identifying typical impacts which might occur. Such a checklist would be a refinement of that presented in Table 2.2 and would be "project-peculiar." For example, Table 2.32 presents some illustrative, typical kinds of social impacts, according to type of project, with which analysts should be familiar. Table 2.33 presents a questionnaire type of checklist used by the state of Maryland Department of Transportation to identify social impacts resulting from new highway construction.

In addition to using a checklist as an aid to identifying potential impacts, one should also be aware of so-called "spillover" considerations. These considerations are sometimes referred to as "secondary effects" and, in many cases, are the real "significant" impacts of

TABLE 2.32   Illustrative Social Impacts by Type of Project

Project	Types of social impacts
Expansion of a major airport in a developed urban area, where "expansion" means the addition of new runways, increased terminal space and parking, and a general increase in landing and take-off activity.	1. Housing required for new employees 2. Increased risk of accidents to adjacent neighborhoods because of expanded aircraft activity 3. Increase in traffic flow and congestion at and around airport 4. Disruption in area due to construction activities 5. Increased demand for local public transportation services 6. Increased crowdedness in airport vicinity 7. Increase in transient population in the area 8. Health and lifestyle impairment because of noise effects 9. Increase in population of the area 10. Potential displacement of people 11. Potential removal of businesses and/or homes 12. Increase in air transportation service to area 13. Change in overall character of community
Construction of a nuclear power plant on the coastline in an undeveloped area which is known for its scenic beauty and is open to the general public for sunbathing, surfing, swimming, and fishing activities.	1. Reduction in land available for recreational use and thus reduced scope of recreational activities 2. Increase in traffic flow through the area 3. Risk to life and health due to nuclear accidents 4. New road construction
Redevelopment of a run-down residential area in the form of replacement of existing residential structures with a senior citizens' housing project	1. Provision of new housing in the area—in particular for senior citizens 2. Displacement of existing residents 3. Increased demand for local public transportation 4. Reduced school requirements 5. Change in neighborhood character and lifestyle 6. Demand for police and fire protection services 7. Increased demand for park and recreational facilities 8. Change in demand for social, health, and welfare services
Extension of an existing major freeway through an area whose existing use is for agriculture and preservation of natural wilderness areas	1. Increase in traffic flow and congestion in the area 2. Induced requirement for new local/arterial street construction 3. Change in land use in the area 4. Increased public access to natural areas
Closing down of a large military base in an established urban area	1. Reduction in school requirements 2. Reduction in local traffic flow and congestion 3. Change in overall character of the community 4. Change in population of the area 5. Reduction in off-base housing demand 6. Reduced need for public transportation

TABLE 2.33 Social Impact Questionnaire for Highway Projects

I. Community impact:
  A. Describe the community affected including type of neighborhood, income levels, land usage, etc.
  B. Does the alternate divide or disrupt an established community?
  C. What is the effect upon adjacent communities?
  D. What is the general effect of business, farm, and non-profit dislocation on the economy of the existing community including employment?
  E. Is there any adverse impact on particular groups such as the elderly and handicapped?
  F. How will the alternate affect the use of various community facilities and services such as hospitals, libraries, shopping areas, fire stations, police stations, schools, churches, and recreational facilities?
  G. To what extent will the alternate produce adverse effect on residential, commercial, and industrial development that is existing or planned?
  H. Will there be a significant change in population density or distribution?
  I. Will the adjacent property values be altered? Discuss (i.e., increased, decreased, zoning, development).

II. Estimated displacement:
  A. Give an estimate of the number of persons, families, and individuals to be displaced. Discuss their characteristics such as occupancy status, minorities, economic level, age, large families, handicapped, etc.
  B. How many and what type of businesses will have to be relocated? How many of these firms may be expected to discontinue?
  C. How many and what type of farm operations will be relocated? How many of these may be expected to discontinue operations?
  D. How many and what type of non-profit organizations will be affected?
  E. Will functional replacement be necessary? If so, discuss any additional displacement that may result.

III. Minority displacement:
  A. What is the racial character of the area affected, including the appropriate number by race of persons and families (affected means all persons directly displaced or located in areas directly adjoining the road)?
  B. What is the social and economic character of the area affected, including levels of income, whether the area is commercial or residential, and the approximate number of minority and non-minority owners of businesses and residences in the area?
  C. What is the racial character of the people employed in the area affected by the alternate?
  D. Are there any foreseeable problem areas or adverse impacts, such as rehousing difficulties, changes in income capabilities, mobility, or community cohesion?
  E. Will a minority area be by-passed or separated from contiguous areas by the alternate and, if so, what effect will this have on the minority community? To what extent will it perpetuate patterns of segregation, if at all?
  F. How will the alternate affect the use of various community facilities and services such as hospitals, libraries, shopping areas, fire stations, police installations, schools, churches, parks, and recreation centers by minority groups in the area?
  G. To what extent will the alternate produce an adverse effect on residential, commercial, and industrial development that is existing or planned within minority communities?

IV. Relocation plan:
  A. State the availability of DS&S housing which is within the financial means of those to be displaced that is normally available in the area. Will the housing be sufficient to meet the needs of those being displaced at the time displacement occurs? If not, describe the actions proposed to remedy the situation including housing of last resort. State the sources of this information.
  B. What will be the impact on the neighborhood or communities into which the displaced persons are likely to move?
  C. Give a statement of availability of replacement sites for businesses, farms, and non-profit organizations. State sources of this information.

IV. Relocation plan:
    D. Give an analysis of federal, state, and municipal programs that may affect the supply and demand for housing at the time displacement occurs.
    E. State the lead time required to complete relocation on the project. (i.e. from the Initiation of negotiations to the last person moved).
    F. Give a factual analysis showing that relocation can/cannot be resolved satisfactorily, and a statement that relocation can/cannot be accomplished in accordance with the requirements of the Uniform Relocation Assistance and Land Acquisition Policies Act of 1970 (P.L. 91-646).

SOURCE: Reference 24

the proposed project and thus are worthy of more attention. Typical "spillover" type considerations in assessing the community impact are as follows:
1. Changes in population patterns
2. Growth-inducing impact and new employment opportunities
3. Effect of residential overcrowding in terms of health, safety, delinquency, and family life
4. Creation of community disorganization by dividing or reducing access to educational, recreational, employment, or social areas
5. Dimunition of lifestyle quality by abuse of visual, cultural, and/or residential amenities
6. Effect on individuals on the basis of their employment, income, transport, or other living costs
7. Effect on property owners or commercial/industrial enterprises of increased cost of operation caused by limiting access, establishment of a monopoly location, or other restrictions on competitive effort
8. Effect on community tax structure as a result of increased costs of streets, utilities, and other municipal services.

For example, in the construction of a high-rise development, some of these types of effects can be delineated as follows:
1. an increase in commuter traffic and thus increased traffic flow, congestion and air pollution
2. potential increase in housing density of surrounding area because of new/more employment opportunities
3. destruction of neighborhood character
4. an increase in land values due to building intensification
5. an increase in transportation service facilities cost
6. attracts tenants from older existing office buildings, thus putting an economic squeeze on the latter
7. potential contribution to increased rents and home prices throughout the city
8. produces changes in the composition of the populace
9. potential for serious fire hazards with attendant increase in safety costs
10. provides construction employment activity
11. an increase in parking demand
12. an increase in car insurance due to increased auto density.

To illustrate these so-called secondary effects, consider the following two examples:

Example 1: Sewer Collection and Water Distribution Line Construction

Consider the construction on land now used for the grazing of cattle of a sewer collection line and a water distribution line to serve a 200-acre industrial site contiguous to and north of the city of Santa Rosa. The diameters of the sewer lines range from 12 to 36 in. and the diameter of the water line is 12 in. This particular project is under the sponsorship of the Economic Development Administration (EDA) since the Santa Rosa area is designated an EDA redevelopment area qualifying under Title IV of the Public Works and Economic Act of 1965, as amended. The criterion for designation is based on unemployment being 6 percent or more of the work force, and having averaged 6 percent or more for a specified period. In the case of the total county, the annual average unemployment rate has been at least 75 percent above the national average for 2 of the preceding 3 calendar years.

With the installation of the sewer and water service, the direct effect of the project will enable the city of Santa Rosa to provide water and sewer service to the Aerospace Manufacturing Company which will build a branch plant of its microwave division on the 200-acre site. The plant will serve as a training and manufacturing operation, and the Economic Development Administration estimates that approximately 690 people will be employed by the end of the first year's operations. This will mean approximately an 8 percent increase in manufacturing job opportunities. Total eventual employment at this plant is projected at 3000 workers. Consequently, the employment at the plant will assist in alleviating some of the persistent unemployment found in the city and the county. A secondary effect of the sewer and water project will be to enable the city to move ahead with plans to allow development of the residential and commercial areas known as Golden Hills and Meadowbrook Ranch. These areas comprise, respectively, 2000 and 900 acres of land north of the city. The 200-acre industrial site is part of the Golden Hills area. Only 120 acres will be initially occupied by Aerospace Manufacturing Company. Currently, the Meadowbrook Ranch area is developed as rural residential with single-family dwellings on large lots, serviced by individual septic systems, interspersed with orchards, vineyards, and pasture land. General plans for the Meadowbrook Ranch area call for development of residential units on the order of 1 to 5 units per acre, which will occur sooner now that the water and sewer service capability has been expanded.

Example 2: Industrial Park Development

Consider the renewal and development of a 74-acre industrial site known as the Fruit Industrial Park which is located in the southwestern portion of the city of Pablo. This site will be improved by the removal of structures, old automobile bodies, and other wastes. An industrial transportation network and utilities will be provided, and two open irrigation canals will be piped underground.

The immediate impact of the proposed Fruit Industrial Park will be on employment and income levels. An impact will also be felt on the community tax base both through increases in improved property values and through increases in consumer expenditure patterns resulting from increased income levels. To analyze each of these impact areas, we shall consider each of them separately as follows:

Employment Opportunity   Causal factors that will result in changes in employment opportunities consist of clearing, utilities, grading, surfacing, street widening, drainage, buildings, and commercial products. Consequently, the initial construction of park improvements, as well as the development of new industrial facilities and the related long-term maintenance requirements, will increase the demand for local construction workers.

The project area's existing industrial base presently employs a total of 119 workers. After proposed improvements, the present tenants expect to expand their total work force by 85 new employees.

Approximately 45 acres of the Fruit Industrial Park will be available for new industries. Assuming a conservative ratio of 10 employees per acre this would indicate that 450 new jobs would be generated by future long-term industrial development. This estimated employee-per-acre ratio is comparatively low in terms of normal industrial standards, but it is reasonable due to the less labor-intensive nature of the types of industries which may locate in the proposed project.

Assuming skill levels and educational requirements of existing industries and applying them to anticipated industries, it is concluded that a high proportion of jobs might be filled by local minority residents. Present employers have a 1/3 minority work force representation and have proposed to recruit over 40 percent for new jobs stemming from their expansion plans. These companies have also made the commitment to continue to provide apprentice and on-the-job training programs. It seems probable that the future expansion of 45 acres of industrial operations will also result in similar specialized programs for training unskilled local residents.

Income Levels   Commercial products will be the basic causal factor that impacts income levels in this area. Extrapolating from present wage scales of existing employers in the project area, it is estimated that average income levels of workers in the proposed Fruit Industrial Park will be approximately $10,000 annually, or about 75 percent higher than the prevailing average income of southwest Pablo.

The Fruit Industrial Park's projected total employment of 690 workers would indicate an annual payroll of $6.9 million, an increase of approximately $5.3 million over the existing payroll. The effect of this potential payroll would be to increase the existing aggregate of family incomes in southwest Pablo by 20 percent, or to raise the mean family income from $5775 to $7000.

Social impacts in a community as the result of development are not always easily measured. They are generally far reaching in scope and include such items as: perceived changes in "quality of life" or lifestyles; changes in recreational opportunities such as location and types of facilities, changes in desirability, and changes in use capacity, intensity, or duration of use; aesthetics such as changes in views and vistas, visual character of project, noise, odor, or other nuisance aspects; demand and availability of public services; protection of natural and historic resources; possible restrictions on private "rights," i.e., use of private property; housing demand, opportunities, and costs, such as changes in type and location of housing; changes in land use and neighborhood character; changes in neighborhood cohesion and stability; change in land development pattern such as induced growth, change in type and mixture of land uses, changes in density and general site design, and changes in extent and nature of suburban sprawl and urban redevelopment; potential relocation of families and business; effects on ethnic, religious, and other groups; and changes in population and general demographic patterns. Measurement of these types of impacts suggests the computation (or estimation) of the types of quantities shown in Table 2.34 (based on Ref. 49).

Measurements of the type just described generally require the use of demographic data such as that provided by census surveys, or the conduct of special polls and surveys. In the next section we discuss some of the typical social impact predictors which can be obtained from demographic data, and in a later section of this chapter we discuss the structuring and conduct of special polls and surveys.

Predictors of Social Impacts

There are a number of characteristics of area residents which can be used as predictors of social impacts. Included among these are:
1. Stability of households and housing tenure
2. Ethnic composition of area
3. Socioeconomic status of residents
4. Age distribution of residents
5. Transportation characteristics of residents.

The stability of a neighborhood can be regarded as an indicator of its quality or ability to function as a neighborhood and can generally be measured in terms of typical census data such as: (1) percent of residents living in their dwellings for 5 years or more, (2) median year residents moved to dwelling unit, and (3) percent of dwellings that are owned. The stability of residence is thus associated with neighborhood satisfaction and housing satisfaction, and also relates to home ownership. In addition, dependency on local facilities and services tends to increase with length of residence, thus implying that altering of the area through some form of development might lead to changes in everyday activity patterns.

The ethnic composition of the project area is important for projects which may have an impact on the ethnic concentration and mixing in an area, such as a major freeway which divides the area or redevelopment of a portion of the residential area. Here the term "ethnic" is used loosely to encompass racial, nationality, cultural, or linguistic groups. Based on a common interest to maintain ethnically mixed neighborhoods, a measure of ethnic mixing is provided[3] by the

$$\text{Index of qualitative variation for ethnicity} = \frac{\displaystyle\sum_{i=1}^{K}\sum_{j=i+1}^{K} n_i n_j}{\dfrac{K(K-1)}{2}\left(\dfrac{N}{K}\right)^2}$$

where n_i = number in ethnic category i

K = number of ethnic categories

N = total population in area (or census tract).

This index will always range between 0 and 1, where 1 indicates totally balanced mixing

[3]John H. Mueller, et al., *Statistical Reasoning in Sociology*, Houghton Mifflin, Boston, Mass., 1970.

TABLE 2.34   Illustrative Social Impact Measures for Selected Impact Areas

Impact area	Candidate social impact measure
Attractiveness	Rating of the physical and visual attractiveness of the development as determined by local citizens
	Percent of people who think the development improves or lessens overall neighborhood attractiveness, pleasantness, or uniqueness
Crowdedness	Change in percent of people perceiving their neighborhood as too crowded
Cultural, historical, and scientific resources	Rarity and perceived importance of cultural, historical, or scientific landmarks to be lost or made inaccessible
Housing	Change in number and percent of housing units that are substandard, and change in number and percent of people living in such units
	Change in number and percent of housing units by type relative to demand estimates or to number of families in various income classes in the community
People displacement	Number of people (residents or workers) displaced by development
Perceived satisfaction of residents	Change in percent of people finding community a good place to live
Population mix	Change in the population distribution by age, income, religion, race, or ethnic group and/or occupation
Privacy	Number and percent of people perceiving a loss in privacy
	Number and percent of people with change in visual or auditory privacy
Property values	Change in land values
Views	Number of people whose views are blocked, degraded, or improved
Recreation	Number of people x minutes or y miles closer to (or farther from) recreational facilities by type of facility
	Change in percent of capacity, waiting times, number of people turned away, persons per facility, or citizen perceptions of crowdedness at recreational facilities
	Change in perceived pleasantness of recreational experiences
Shopping	Change in the number of stores/services available within x minutes of y people by type of store/service
	Change in the percent of people generally satisfied with local shopping conditions (access, variety, crowdedness)
Sociability/friendliness	Change in frequency of visits to friends in existing neighborhood and frequency of visits between people in existing neighborhood and new development

of the ethnic groups. Conceptually, this index is based on the number of differences between categories divided by the maximum number of possible differences. For example, among 1500 residents of a census tract, suppose that 1000 are Anglo, 400 Mexican-American, and 100 black according to published census data. In this case we define

$$n_1 = 1000$$
$$n_2 = 400$$

$$n_3 = 100$$
$$K = 3$$
$$N = 1500$$

thus the

$$\text{Index of qualitative variation for ethnicity} = \frac{(1000)(400) + (1000)(100) + (400)(100)}{(3)(250{,}000)}$$
$$= 0.72$$

A change in the neighborhood ethnic composition as the result of new development activity such that this index value is lowered will result in a trend toward a less balanced mixing of the remaining ethnic groups.

The socioeconomic status of residents can be measured through the use of census tract descriptors such as: (1) median home value, (2) median gross rent, (3) median family income, (4) median years of school completed (for those 25 years of age and over), and (5) percent of white collar workers (i.e., those in professional, managerial, clerical, and sales occupational categories). Each of these descriptors is more meaningful when compared to a city or county as a whole, so that the socioeconomic status of any residential area is a relative characteristic. Residential areas containing residents with low income or educational levels tend to be more affected by any large-scale changes or development activities.

The family status and age distribution of an area are predictors of the degree of identification and ties to a residential area. These factors can be measured through the use of census data on (1) the percent of residents under 16 years of age, and (2) the percent of residents 60 and over to compute an index given by[4] the

$$\text{Dependency ratio} = \frac{\left(\begin{array}{c}\text{Percent of children}\\ \text{(Residents} < 16 \text{ yr of age)}\end{array}\right) + \left(\begin{array}{c}\text{Percent aged}\\ \text{(Residents} \geq 60 \text{ yr of age)}\end{array}\right)}{\left(\begin{array}{c}\text{Total active population}\\ \text{(16 yr of age} \leq \text{residents} \leq 59 \text{ yr of age)}\end{array}\right)} \times 100$$

This index provides a measure of the number of "dependent" residents. Therefore, the larger the index value, the more adverse would be the impact of any project activity which would disrupt or divide the community. When an area or census tracts under consideration consist primarily of multiple-family dwellings, another term given by the percentage of unmarried adults can be added to the numerator, since this group is particularly amenable to relocation and to large-scale neighborhood alterations.

The transportation characteristics of area residents are especially important for consideration of projects which may impact transportation routes and/or available modes of local transportation such as freeway construction projects. Households which are dependent on easy accessibility and which possess one or more automobiles are more receptive to freeway construction. However, because a greater proportion of lower class individuals, ethnic groups, and older persons have no automobile, as compared to the population as a whole, these groups often perceive the freeway as of little benefit to them. Some of the latter may profit by highway construction in terms of increased property values, but those of lower socioeconomic status and/or of ethnic minority backgrounds are primarily renters. Census data on neighborhood-level transportation and related characteristics can be used to determine project impacts on accessibility to transportation routes and the usage of various transportation modes. For example, such data would include transportation-to-work characteristics such as: (1) percent of workers per tract who drive a private vehicle to work, (2) percent of workers per tract who are passengers in a private vehicle, (3) percent of workers per tract who use buses, (4) percent of workers per tract who walk to work, and (5) location of work relative to census tract under consideration. These data can then be compared to the city or county as a whole to determine the extent of private vehicle dependency.

In summary, demographic data can be used as indicators of specific attitudinal and behavioral characteristics of residents of the project area. Hence, they can be regarded as measures of the lifestyle of a residential area. Stability of residence is a measure of an

---

[4]William Peterson, *Population*, Macmillan Co., New York, N.Y., 1969.

area's vitality and of the satisfaction of residents with their place of residence. Ethnic composition aids in predicting the degree of dependency or ties to a residential area. Socioeconomic status of residents provides an indication of the types of residential linkages and attitudinal characteristics of the area. The age distribution of residential areas provides an indication of the dependency of the residents upon special types of public services. Finally, the transportation characteristics of the residents of an area provide an indication of the reliance on personal vehicles for transportation and those households which are not tied to the immediate neighborhood for daily activities.

Use of Census Data

The major sources of census data of the type described in the section on Predictors of Social Impacts are the final reports of the various decennial censuses of the population prepared by the U.S. Bureau of the Census. These reports consist of standard published volumes, standard computer tapes, and custom reports prepared by the Bureau of the Census and generally include information of the following types:

- Demographic characteristics: total population, number of households, percentage of sizable minority groups, median age, persons per family
- Social characteristics: median years of school, percentage of single family units, percentage of owner-occupied units, median year houses built, major employment occupations, median income, median home value, median contract rent value

Table 2.35 provides a complete listing of the publications and computer tapes generally available from the Bureau of the Census.

Some words of caution in the use of census data are as follows. Changes in the definition of variables or ways of classifying data, changes in enumeration procedures, and changes in the boundaries of enumeration areas such as census tracts make it difficult to analyze trends in population characteristics. There are many instances where the Bureau of the Census has changed the definition of variables in order to make them more meaningful or because the old definitions have become obsolete. In some cases, an implicit redefinition of a variable occurs because of a change in the way the data were collected. This happened in the 1970 census, which was largely conducted by means of mailed, self-administered questionnaires, where it is estimated that a greater proportion of persons reported themselves as being black than were so reported in 1960. Finally, due to population shifts and changes in Bureau of the Census administrative procedures, the boundaries of census tracts and other enumeration areas are changed from one census period to the next, thus making it difficult to compare a census tract over different time periods.

In general, however, census data and other forms of related demographic information are widely used, are generally highly reliable, and are relatively inexpensive. These are the features which make the use of such information so attractive.

Conduct of a Social Survey

Designing, organizing, and conducting a sample survey to obtain social data needed for project impact assessment requires careful, detailed planning. Only by carefully planning a survey from initiation to completion can reliance be placed in the findings. Members of the Texas Transportation Institute at Texas A&M University (Ref. 27) have identified the basic steps involved in the planning and execution of a social survey. We repeat them here because of their value in the overall environmental impact assessment process.[5]

Step 1: Objectives of the Survey

Before undertaking a survey or choosing the methods by which it is to be conducted, it is important to formulate a clear statement of the objectives of the study. Without this step, an analyst might forget the objectives when involved in the details of planning and make decisions contrary to those objectives.

Step 2: Population to be Sampled

The word population is used to denote the aggregate from which the sample is chosen. The population to be sampled (the sampled population) should coincide with the population about which information is wanted (the target population). Sometimes, for reasons of practicality or convenience, the sampled population is more restricted than the target

[5]The material which follows is taken almost verbatim from Appendix D in Reference 27.

TABLE 2.35   **Publications and Computer Tapes Available for the 1970 Census of the Population**

Population Census Reports

VOLUME 1: CHARACTERISTICS OF THE POPULATION
This volume will consist of 58 "parts"—number 1 for the United States, numbers 2 through 52 for the 50 states and the District of Columbia in alphabetical order, and numbers 53 through 58 for Puerto Rico, Guam, Virgin Islands, American Samoa, Canal Zone, and Trust Territory of the Pacific Islands, respectively. Each part, which will be a separate clothbound book, will contain four chapters designated as A, B, C, and D. Each chapter (for each of the 58 areas) will first be issued as an individual paperbound report in four series designated as PC(1)-A, B, C, and D, respectively. The 58 PC(1)-A reports will be specially assembled and issued in a clothbound book, designated as Part A.

SERIES PC(1)-A: NUMBER OF INHABITANTS
Final official population counts are presented for states, counties by urban and rural residence, standard metropolitan statistical areas (SMSA's), urbanized areas, county subdivisions, all incorporated places, and unincorporated places of 1000 inhabitants or more.

SERIES PC(1)-B: GENERAL POPULATION CHARACTERISTICS
Statistics on age, sex, race, marital status, and relationship to head of household are presented for states, counties by urban and rural residence, SMSA's, urbanized areas, county subdivisions, and places of 1000 inhabitants or more.

SERIES PC(1)-C: GENERAL SOCIAL AND ECONOMIC CHARACTERISTICS
Statistics are presented on nativity and parentage, state or country of birth, Spanish origin, mother tongue, residence 5 years ago, year moved into present house, school enrollment (public or private), years of school completed, vocational training, number of children ever born, family composition, disability, veteran status, employment status, place of work, means of transportation to work, occupation group, industry group, class of worker, and income (by type) in 1969 of families and individuals. Each subject is shown for some or all of the following areas: states, counties (by urban, rural nonfarm, and rural farm residence), SMSA's, urbanized areas, and places of 2500 inhabitants or more.

SERIES PC(1)-D: DETAILED CHARACTERISTICS
These reports will cover most of the subjects shown in Series PC(1)-C, above, presenting the data in considerable detail and cross-classified by age, race, and other characteristics. Each subject will be shown for some or all of the following areas: states (by urban, rural-nonfarm, and rural-farm residence), SMSA's, and large cities.

VOLUME II: SUBJECT REPORTS
Each report in this volume, also designated as Series PC(2), will concentrate on a particular subject. Detailed information and cross-relationships will generally be provided on a national and regional level; in some reports, data for states or SMSA's will also be shown. Among the characteristics to be covered are national origin and race, fertility, families, marital status, migration, education, unemployment, occupation, industry, and income.

Housing Census Reports

VOLUME I: HOUSING CHARACTERISTICS FOR STATES, CITIES, AND COUNTIES
This volume will consist of 58 "parts"—number 1 for the United States, numbers 2 through 52 for the 50 states and the District of Columbia in alphabetical order, and numbers 53 through 58 for Puerto Rico, Guam, Virgin Islands, American Samoa, Canal Zone, and Trust Territory of the Pacific Islands, respectively. Each part which will be a separate clothbound book, will contain two chapters designated as A and B. Each chapter (for each of the 58 areas) will first be issued as an individual paperbound report in two series designated as HC(1)-A and B, respectively.

TABLE 2.35   *(Continued)*

Housing Census Reports

SERIES HC(1)-A: GENERAL HOUSING CHARACTERISTICS
Statistics on tenure, kitchen facilities, plumbing facilities, number of rooms, persons per room, units in structure, mobile home, telephone, value, contract rent, and vacancy status are presented for some or all of the following areas: states (by urban and rural residence), SMSA's, urbanized areas, places of 1000 inhabitants or more, and counties.

SERIES HC(1)-B: DETAILED HOUSING CHARACTERISTICS
Statistics are presented on a more detailed basis for the subjects included in the Series HC(1)-A reports, as well as on such additional subjects as year moved into unit, year structure built, basement, heating equipment, fuels, air conditioning, water and sewage, appliances, gross rent, and ownership of second home. Each subject is shown for some or all of the following areas: states (by urban, rural-nonfarm, and rural-farm residence), SMSA's, urbanized areas, places of 2500 inhabitants or more, and counties (by rural-farm residence).

VOLUME II: METROPOLITAN HOUSING CHARACTERISTICS
These reports, also designated as Series HC(2), will cover most of the 1970 census housing subjects in considerable detail and cross-classification. There will be one report for each SMSA, presenting data for the SMSA and its central cities and places of 50,000 inhabitants or more, as well as a national summary report.

VOLUME III: BLOCK STATISTICS
One report, under the designation Series HC(3), is issued for each urbanized area showing data for individual blocks on selected housing and population subjects. The series also includes reports for the communities outside urbanized areas which have contracted with the Census Bureau to provide block statistics from the 1970 census.

VOLUME IV: COMPONENTS OF INVENTORY CHANGE
This volume will contain data on the disposition of the 1960 inventory and the source of the 1970 inventory, such as new construction, conversions, mergers, demolitions, and other additions and losses. Cross-tabulations of 1970 and 1960 characteristics for units that have not changed and characteristics of the present and previous residence of recent movers will also be provided. Statistics will be shown for 15 selected SMSA's and for the United States and regions.

VOLUME V: RESIDENTIAL FINANCE
This volume will present data regarding the financing of privately owned nonfarm residential properties. Statistics will be shown on amount of oustanding mortgage debt, manner of acquisition of property, homeowner expenses, and other owner, property, and mortgage characteristics for the United States and regions.

VOLUME VI: ESTIMATES OF "SUBSTANDARD" HOUSING
This volume will present counts of "substandard" housing units for counties and cities, based on the number of units lacking plumbing facilities combined with estimates of units with all plumbing facilities but in "dilapidated" condition.

VOLUME VII: SUBJECT REPORTS
Each report in this volume will concentrate on a particular subject. Detailed information and cross-classifications will generally be provided on a national and regional level; in some reports, data for states or SMSA's may also be shown. Among the subjects to be covered are housing characteristics by household composition, housing of minority groups and senior citizens, and households in mobile homes.

Joint Population-Housing Reports

SERIES PHC(1): CENSUS TRACT REPORTS
This series contains one report for each SMSA, showing data for most of the population and housing subjects included in the 1970 census.

Joint Population-Housing Reports

SERIES PHC(2): GENERAL DEMOGRAPHIC TRENDS FOR METROPOLITAN AREAS, 1960 to 1970
This series consists of one report for each state and the District of Columbia, as well as a national summary report, presenting statistics for the state and for SMSA's and their central cities and constituent counties. Comparative 1960 and 1970 data are shown on population counts by age and race and on such housing subjects as tenure, plumbing facilities, value, and contract rent.

SERIES PHC(3): EMPLOYMENT PROFILES OF SELECTED LOW-INCOME AREAS
This series will consist of approximately 70 reports, each presenting statistics on the social and economic characteristics of the residents of a particular low-income area. The data relate to low-income neighborhoods in 54 cities and 7 rural poverty areas. Each report will provide statistics on employment and unemployment, education, vocational training, availability for work, job history, and income, as well as on value or rent and number of rooms in the housing unit.

Additional Reports

SERIES PHC(E): EVALUATION REPORTS
This open series will present the results of the extensive evaluation program conducted as an integral part of the 1970 census program, and relating to such matters as completeness of enumeration and quality of the data on characteristics.

SERIES PHC(R): PROCEDURAL REPORTS
This open series presents information on various administrative and methodological aspects of the 1970 census, and will include a comprehensive procedural history of the 1970 census. The first report issued focuses on the forms and procedures used in the data collection phase of the census.

Computer Summary Tapes

The major portion of the results of the 1970 census will be produced in a set of six tabulation counts. To help meet the needs of census users, these counts are being designed to provide data with much greater subject and geographic detail than it is feasible or desirable to publish in printed reports. The data so tabulated will generally be available—subject to suppression of certain detail where necessary to protect confidentiality—on magnetic computer tape, printouts, and microfilm, at the cost of preparing the copy.

First Count—source of the PC(1)-A reports: contains about 400 cells of data on the subjects covered in the PC(1)-B and HC(1)-A reports and tabulated for each of the approximately 250,000 enumeration districts in the United States.

Second Count—source of the PC(1)-B, HC(1)-A, and part of the PHC(1) reports: contains about 3500 cells of data covering the subjects in these reports and tabulated for the approximately 35,000 tracts and 35,000 county subdivisions in the United States.

Third Count—source of the HC(3) reports: contains about 250 cells of data on the subjects covered in the PC(1)-B and HC(1)-A reports and tabulated for approximately 1,500,000 blocks in the United States.

Fourth Count—source of the PC(1)-C, HC(1)-B, and part of the PHC(1) reports: contains about 13,000 cells of data covering the subjects in these reports and tabulated for the approximately 35,000 tracts and 35,000 county subdivisions in the United States; also contains about 30,000 cells of data for each county.

Fifth Count—will contain approximately 800 cells of population and housing data for 5-digit ZIP code areas in SMSA's and 3-digit ZIP code areas outside SMSA's; the ZIP code data will be available only on tape.

Sixth Count—source of the PC(1)-D and HC(2) reports: will contain about 260,000

TABLE 2.35 *(Continued)*

Computer Summary Tapes

---

cells of data covering the subjects in these reports and tabulated for states, SMSA's, and large cities.

The tapes will generally be organized on a state basis. To use the First Count and Third Count tapes, it will be necessary to purchase the appropriate enumeration district and block maps.

The term "cells" used herein to indicate the scope of subject content of the several counts refers to each figure or statistic in the tabulation for a specific geographic area. For example, in the Third Count, there are six cells for a cross-classification of race by sex: three categories of race (white, Negro, other race) by two categories of sex (male, female).

In addition to the above-mentioned summary tapes, the Census Bureau will make available for purchase certain sample tape files containing population and housing characteristics as shown on individual census records. These files will contain no names or addresses, and the geographic identification will be sufficiently broad to protect confidentiality. There will be six files, each containing a 1 percent national sample of persons and housing units. Three of the files will be drawn from the population covered by the census 15-percent sample and three from the population in the census 5-percent sample. Each of these three files will provide a different type of geographic information: One will identify individual large SMSA's and, for the rest of the country, groups of counties; the second will identify individual states and, where they are sufficiently large, will provide urban-rural and metropolitan-nonmetropolitan detail; and the third will identify state groups and size of place, with each individual record showing selected characteristics of the person's neighborhood.

SOURCE: U.S. Bureau of the Census, *Characteristics of the Population*, 1970.

population. If this occurs, it should be remembered that conclusions drawn from the sample apply only to the sampled population. Judgment about the extent to which the conclusion will also apply to the target population must depend on other sources of information. Any supplementary information that can be gathered about the nature of the difference between sampled and target population may be helpful. The population must be clearly defined by a set of rules. In sampling a neighborhood as a population, for example, rules must be set up to define what comprises that neighborhood. These rules must be usable in practice. The enumerator must be able to decide in the field, with little difficulty, whether or not a doubtful case belongs in the population.

Step 3: Data to be Collected

The person conducting the survey should be careful to verify that all data collected are relevant to the purposes of the survey, and that no essential data are omitted. There is a tendency to ask too many questions, some of which are never analyzed. A questionnaire that is too long lowers the quality of all questions, the important as well as the unimportant.

The survey designer should decide at this point what specific questions must be answered by the survey. Related questions which are "interesting" but not essential to the investigation should be considered on the basis of their importance, the additional cost of their inclusion, and whether or not they may take the place of more pertinent questions. The decisions made should be recorded so that the surveyor can have them when he drafts the detailed plans. If it is later proposed that still other questions be added, the surveyor can evaluate such proposals in light of the objectives and decisions previously made.

Step 4: Degree of Precision Required

The results of sample surveys are always subject to some uncertainty because only part of the population has been measured and because there are errors in measurement. This uncertainty can be reduced by: (1) taking larger samples, and (2) using superior instruments and extensive quality control procedures. Both of these usually involve some increase in expenditures in time and money. For this reason, it is important to specify the degree of precision that is desired so that costs can be held to a minimum. It is strongly advised that the services of a professional statistician be obtained at this point to determine the amount of sampling error that can be tolerated consistent with good decision making.

There are two conditions a sample must fulfill in a social survey:

1. It must be possible to make substantive statements, i.e., statements about variables in the population with a tolerable amount of certainty.

2. It must be possible to generalize from the sample to the target population with a tolerable amount of uncertainty.

Fortunately, these two requirements can be met simultaneously through proper sampling techniques. However, the procedures involved require a high degree of technical knowledge, and should be performed by an expert statistician.

Step 5: Methods of Measurement

There are three major alternative measurement instruments that may be utilized to obtain the information needed—specifically, these alternatives are: (1) personal (face-to-face) interview, (2) telephone interview, and (3) mailed questionnaire. The major advantages and disadvantages of each of these techniques are as follows:

The Personal Interview   The advantages of the personal interview are:

1. The personal interview usually yields a high percentage of returns, for most people are willing to cooperate.

2. It can be made to yield an almost perfect sample of the general population because practically everyone can be reached by and respond to this approach.

3. The information secured is likely to be more current than that secured by other techniques since the interviewer can clear up seemingly inaccurate answers by explaining the questions to the informant. If the latter deliberately falsifies replies, the interviewer may be trained to spot such cases and use special devices to get the truth.

4. The interviewer can collect supplementary information about the informant's personal characteristics and environment which is valuable in interpreting results and evaluating the representativeness of the persons surveyed.

5. Scoring and test devices can be used, with the interviewer acting as experimenter.

6. Visual material to which the informant is to react can be presented.

7. Return visits to complete items on the schedule or to correct mistakes can usually be made without annoying the informant. Thus, greater numbers of usable returns are assured than when other methods are employed.

8. The interviewer may catch the informant off guard and thus secure more spontaneous reactions than would be the case if a written form were mailed out for the informant to consider.

9. The interviewer can usually control which person or persons answer the questions, whereas in mail surveys several members of the household may confer before the questions are answered. On the other hand, group discussions can be held with the personal interview method if desired.

10. The personal interview may take long enough to allow the informant to become oriented to the topic under investigation. Thus, recall of relevant material is facilitated.

11. Questions about which the informant is likely to be sensitive can be carefully interspersed by the interviewer. By observing the informant's reactions, the investigator can change the subject if necessary or explain the survey problem further if it appears that the interviewee is about to rebel. In other words, a delicate situation can usually be handled more effectively by a personal interview than by other survey techniques.

12. More of the informant's time can be taken for the survey than would be the case if the interviewer were not present to elicit and record the information.

13. The language of the survey can be adapted to the ability or educational level of the person interviewed. Therefore, it is comparatively easy to avoid misinterpretations or misleading questions.

Some limitations of the personal interview are:

1. The transportation costs and the time required to cover addresses in a large area may make the personal interview method infeasible.

2. The human equation may distort the returns. If the interviewer has a certain economic bias, for example, the interview questions may unconsciously be posed in such a way as to secure confirmation of previously held views. In opinion studies especially, such biases may operate. To prevent such coloring of questions, most opinion surveyors instruct their interviewers to ask questions exactly as printed on the schedule.

3. Unless the interviewers are properly trained and supervised, the data recorded may be inaccurate and incomplete. A few poor enumerators may make a much higher percentage of returns unusable than if the informants filled out and mailed the interview form to survey headquarters.

4. The organization required for selecting, training, and supervising a field staff is more complex than that needed for surveys conducted by other methods.

5. It is usually claimed that costs per interview are higher when field investigators are employed than when telephone or mail surveys are used. This may not be true if the area to be covered is not too extensive. If the general public in a community is to be surveyed, the costs of securing a representative sample by telephone or mail inquiries will probably equal or exceed the cost by the personal interview method. This is because a personal follow-up will be necessary in the end to round out the sample.

6. The personal interview usually takes more time than the telephone interview, providing the persons who can be reached by telephone are a representative sample of the type of population to be covered by the survey. However, for a sample of the general public, a telephone inquiry is not a substitute for a personal interview. The lowest income groups may not have telephones.

7. If the interview is conducted in the home during the day, the majority of the informants will be women. If a response is to be obtained from a male member of the household, most of the field work will have to be done in the evening or on weekends. Since only an hour or two can be used for evening interviewing, the personal interview requires a larger staff for studies requiring contacts with an employed population.

The Telephone Interview   A summary of the merits of the telephone interview method follows:

1. The telephone interview is the quickest of the survey techniques. Interviewers can complete about 30 calls per hour if the calls are brief.

2. The refusal rate is usually low among people who are reached by phone.

3. It is easy to train and supervise interviewers since they can work in one room directly beside the supervisor.

4. The approach and questions are easy to standardize from one interviewer to another.

5. The cost per completed interview is low for the sample covered.

6. The geographic distribution of the sample can be easily controlled. An address listing of numbers is usually available and can be used for drawing the sample.

7. For studies of middle- and high-income groups, the telephone interview may be satisfactory because most people in those categories have telephones.

8. Interviews may be scattered over a wide area within a city without adding to the cost.

9. As compared with a mail questionnaire, the telephone survey is preferable because it usually costs less per return. Returns are higher on first solicitation, and they can be more effectively controlled from the point of neighborhood distribution.

The disadvantages of this method may be summarized as follows:

1. As a sample of the general population, telephone subscribers are not representative. Thus, unless the telephone interview is supplemented by a method that covers nonsubscribers, it should not be used.

2. Detailed data cannot be gathered by this method because the informants may soon become annoyed or impatient. If the schedule is too lengthy, the informant may either hang up or give unreliable answers.

3. When observation of the situation is an important element, the telephone interview is not useful. If the interviewer is supposed to evaluate the answers as to trustworthiness, there is very little in a short telephone conversation which will be useful.

4. Information about the respondent must be limited to one or two facts. Such items as age, nationality, income, etc., are difficult to secure by telephone.

5. Attitude scales must be used with caution. Also, opinions are less likely to be given freely since the informant cannot be certain of the credentials of the person calling.

6. The brevity of the introduction and questions does not give the informant much time to adjust to the subject matter of the survey. Reactions requiring careful thought—such as criticisms of existing transportation facilities, detailed opinions of proposed transportation facilities, etc.—should not be obtained by this technique.

7. The telephone situation neither encourages the respondent to amplify replies nor gives the interviewer much time to jot down the comments. A face-to-face interview is more conducive to a considered response.

8. The task of checking the no-answers, wrong numbers, busy signals, etc., is time-consuming but must be done if the sample is to be representative of telephone subscribers.

9. It is difficult to secure privacy on party lines.

10. There may be instances when the telephone technique is used by so many groups that informants develop an antagonism to all telephone inquiries.

11. The surveyor must be careful not to antagonize informants by phoning too early or too late in the day. One well-known survey agency makes it a policy never to call before 8:30 A.M. or after 10:30 P.M.

12. Misinformation is hard to detect and check in short inquiries.

The Mailed Questionnaire The advantages of the mailed questionnaire may be summed up as follows:

1. If mailed questionnaires are used, it is possible to cover a wider geographical area and to reach a much larger population with given funds than could be accomplished by personal interviews with each informant. This lower cost applies primarily if personal follow-ups are not made.

2. Mailing costs are relatively low compared with the transportation and time costs for a field staff.

3. The expensive and time-consuming task of training a staff of investigators is eliminated. This assumes, of course, that a large staff will not be needed to collect the schedule data from the people who do not answer the questionnaire.

4. The informant may answer questions more frankly by mail since anonymity is assured. On the other hand, some respondents may hesitate to put their ideas in writing for fear that their schedules may be identified even though unsigned. Actually, it is questionable whether anonymity is an advantage or a disadvantage.

5. The questionnaire may reach groups who are more or less protected from solicitors and investigators. In high-rent apartment houses or private homes where servants protect the occupants from solicitors and other doorbell ringers, for example, it is often difficult for investigators to gain admittance. Of course, the mere fact that the mail is received does not guarantee that it will not be thrown away by a secretary or even by the addressee as soon as the heading is read.

6. Personal antagonism to investigators which may lead to a refusal to give the desired information is avoided.

7. If time is not an important consideration and if the sample extends over a wide area, the cost of securing practically complete returns is probably lower than in the personal interview method.

8. The questions are standardized, whereas in the personal interview the investigator may alter them or suggest answers.

9. The questionnaire can be answered at the convenience of the respondent. This provides for the time to deliberate on each point, and if necessary to look up information needed to fill in the items. However, other members of the household may be consulted so the replies may be more representative of the family's point of view than of the respondent's alone.

10. It is claimed that the mail questionnaire brings many more returns from the male respondent than do the telephone or personal interview methods.

11. Where the persons to be reached are located in widely scattered areas of cities and are a mobile element of the population, it may be easier to locate them by mail (registered or special delivery) than by other methods.

Most of the advantages of the mailed schedule over the personal interview are offset by the following serious drawbacks:

1. The people who return questionnaires are not necessarily representative of the groups to whom the schedules are sent. This limitation is sufficiently great to outweigh almost all the advantages previously listed. Unless every effort is exerted to adjust for

nonresponse or to obtain practically complete returns from everyone solicited by mail, the technique should not be used.

2. The returns from mailed questionnaires sent to the general public may be very low, sometimes ranging from about 10 to 20 percent. The percentage of returns varies greatly, however, with different schedules and informants. One survey of medical doctors in New York State received about 50 percent returns without follow-ups. By continued effort the returns may be increased to 70 to 80 percent of the sample.

3. Since informants fill in the data on the questionnaire without the assistance of an investigator, they may misinterpret questions, omit essential items, or send in material which cannot be put in form for tabulation, thus making it necessary to discard many of the questionnaires.

4. The questions used must be simple and practically self-explanatory, since no training can be given the informant on their meaning and on how to fill out the schedule.

5. In most studies, the questionnaire must be relatively brief if high returns are to be obtained.

6. If the sample is to be unbiased, it is necessary to supplement the mailed returns with information obtained by personal interviews with the nonrespondents.

7. Checks on the honesty and reliability of returns are difficult to devise when the personal interviewer does not see and evaluate the informant.

8. It is practically impossible to return unsatisfactory or incomplete schedules to the informant for correction.

9. Because most people would rather talk than write, questionnaires must be made very interesting to induce responses.

10. An up-to-date address list of potential survey informants is difficult to find.

11. Mail returns from the last third of the respondents come in slowly. Hence, the mail survey must be spread over a relatively long period, if a high percentage of returns is to be secured.

12. Many questions which might antagonize the respondent cannot be included on the mail questionnaires but can be asked in personal interviews when the informant gradually can be led around to the subject.

These three measurement procedures are not mutually exclusive. For example, it may be desirable to combine telephone interviews with mailed questionnaires, or with personal interviews as a follow-up method when no response or incomplete responses are obtained from particular parts of the sample. Also, routine, nonsensitive data may be collected from one sample of the population by mail or phone, while sensitive information is obtained from another sample by personal interview.

Step 6: Construction of Record Forms

A major part of the preliminary work is the construction of the record forms on which the questions and answers are to be entered. Whenever possible, answers should be pre-coded, that is, entered in a manner in which they can be routinely transferred to mechanical equipment. In order to develop good record forms, it is necessary to know the precise structure of the final summary tables that will be used for drawing conclusions. For this reason, "dummy tables" or table "shells" for all summary tables should be constructed as a preliminary step in the construction of the record forms. These dummy tables also facilitate the evaluation of the questions included in the survey.

Step 7: Development of the Sampling Frame

Before the sample is selected, the population must be divided into parts which are called "sampling units." These units must: (1) cover the whole of the population, and (2) they must not overlap, in the sense that every element in the population belongs to one and only one unit. For the social survey used here, an appropriate sampling unit is the household.

The complete list of sampling units is called a "frame." The construction of the sampling frame is one of the major practical problems in conducting a survey. The person conducting the survey should have a critical attitude toward any list that has been routinely compiled for some purpose. Many times such lists are found to be incomplete or to contain an unknown amount of duplication. For example, if a city directory is used as the frame for the survey, households in housing units that have been built since the

directory was compiled will be omitted. Similarly, housing units that have been subdi- ·
vided into two or more apartments may only be listed as one household. The drawing of a
sample which is truly representative of the population requires an accurate sampling
frame. For this reason, considerable time and effort should be devoted to this task.

Step 8: Selection of the Sample

A variety of plans can be developed by which the sample may be selected. For each
sampling plan that is considered, rough estimates of the sample size can be made from the
requirements of the degree of precision determined earlier. From this, the relative costs
and time requirements for each sampling plan can be calculated and compared.

Step 9: The Pretest

It is advisable to try out the questionnaire and the field methods on a small scale. This
almost always results in improvement in the questionnaire and in the techniques of
administration. Very often, the pretest will reveal problems that will be serious on a large
scale. For example, a pretest may reveal that residents of a low socioeconomic neighborhood will not answer the door if the interviewer wears a tie or a dress because this is the
customary attire of bill collectors and door-to-door salespeople. The pretest also helps to
revise estimates of costs and time requirements. If the pretest sample is properly selected
from the population, it can be included as part of the overall survey, thus lowering costs
and improving the statistical accuracy of the survey.

Step 10: Organization of the Field Work

The size of the survey, the techniques used in collecting data, and the amount of
tabulation and analysis to be done will largely determine the extent and complexity of the
organization required. Useful rules which are applicable to every survey, regardless of
size and complexity, are:

   1. Whenever possible, put instructions in writing. This insures against making
hasty and inconsistent judgments and precludes errors arising from forgotten instructions.

   2. Guard against too much division of authority and overlapping of responsibility.
All persons should know (*a*) to whom they are responsible, (*b*) with whom they should
consult if their instructions do not cover a specific situation, (*c*) for what operations they
are responsible, and (*d*) what types of cases should be turned over to the person
specializing in "problem cases."

   3. Delegate some authority, but also keep in touch with the details of the operation.
The survey director is in the position of an executive of a business undertaking and should
always follow good rules of business administration.

   4. Keep account of the time and cost of the survey. Actual performance should be
closely compared with the plans. If the survey is running ahead of or behind the schedule
or allocated budget, the plans should be adjusted accordingly.

   5. Set up a routine check of the quality of every operation. It should be clearly
understood by all survey personnel that all work will be checked (including the survey
director's) and that incorrect work will have to be corrected. Standards of what constitutes
"allowable" error should be clearly stated before the survey begins and should be strictly
enforced. (Most sources of error should have been located in the pretest, and adjustments
should have been made before the main survey begins.)

   6. After errors have been detected, see that someone is responsible for making
corrections at every point affected. Often a staff worker corrects a mistake where it is
found without going back to other tabulations or data in which the same error may have
been made. The result is that when analysis is begun, figures that should agree do not.

   7. Require production reports periodically on the quantity and type of operations
performed. Frequent comparisons should be made between the amount of work planned
and the amount accomplished by each individual in the survey. Such comparisons allow
the survey director to spot problem areas and bottlenecks and to take appropriate action
before they become major problems.

   8. When possible, divide each job into several definite operations which can be
done by one worker and checked by another within a specified time.

   9. Avoid too many transcriptions of the data. The more the data are transcribed, the
greater the probability of mistakes in copying and verification.

10. Require the worker who does the operation to initial the record of it. This makes it possible to determine who makes an error or whose record requires further clarification.

Step 11: Summary and Analysis of the Data

If the previous steps have been carried out properly, this step should be almost routine. A final editing should be conducted at this stage in order to amend editing errors and to delete data that are obviously erroneous. Decisions about tabulating procedures are needed in cases where answers to questions were omitted by respondents and where erroneous data were deleted in the editing process. Then, the tabulations on the data are performed.

Step 12: Information Gained for Future Surveys

The more information that is initially known about a population, the easier it is to devise a sample that will give accurate estimates. Any completed sample is potentially a guide to improved future sampling. The information it provides about the means, the standard deviation, and the costs involved in getting the data should be quite helpful.

Also, things never go exactly as planned in a complex survey. The survey director, therefore, should keep a detailed record of the mistakes in execution of the survey and see that they do not occur in future surveys.

IMPACTS ON ECONOMIC PROFILE OF THE COMMUNITY

The economic profile of a community can be characterized by a number of variables (or descriptors), including employment and payrolls, square footage of commercial and industrial development, number of dwelling units, retail sales, valuation of land and

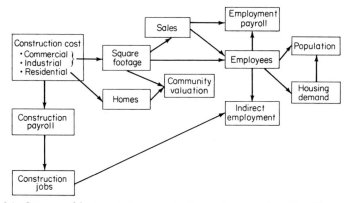

Figure 2.1   Overview of the impact of new construction on the economic profile of the community.

improvements, population, and rates of growth in development activities (as measured by housing starts, building permits issued, etc.). New construction activities have a variety of impacts on these variables as illustrated in Figure 2.1. For example, a portion of the construction cost is for labor thus implying construction payrolls and corresponding worker-years of employment. Construction provides new commercial and industrial square footage and/or new homes, thus increasing the community valuation. The construction of business square footage provides for sales, a portion of which represents labor and therefore jobs. With every worker-year of direct employment provided there are corresponding worker-years of secondary employment generated. Furthermore, the creation of new jobs means more employment opportunities; thus there is the increased likelihood of families moving to the area, creating an increase in population and a demand for housing.

Construction jobs are generally regarded as short-term jobs vis-à-vis long-term direct employment provided by new or expanded business enterprises. The number of construction jobs resulting from development activities can be estimated by several methods, namely: (1) comparison with similar projects previously completed, (2) estimating the labor portion of the total construction cost and then converting this into worker-years based on average construction labor costs, or (3) estimating the number of worker-hours for each type of labor skill (such as electricians, plumbers, carpenters, roofers, etc.) necessary. An illustration of the second method is as follows:

Total project construction cost $10,000,000
Estimated labor share of project construction cost 32%
Construction payroll ....(0.32)($10,000,000) = $3,200,000
Average annual construction salary $15,500
Construction employment (in worker-years) .... $\dfrac{\$3,200,000}{\$15,500} = 206$
Duration of construction activities .... 2 yr
Equivalent full-time annual jobs .... $\dfrac{206}{2} = 103$

The generation of construction jobs leads to the generation of indirect or nonconstruction jobs. For example, the materials used in construction must be purchased from various suppliers such as lumber companies, metal fabricators, machinery firms, etc. Similarly, these firms which supply the raw materials for construction also require goods and services from other firms in order to supply that which is demanded of them—a lumber company purchases timber from mills, a metal fabricator requires various iron products from steel companies, and the machinery producers must have parts with which to construct their machines. As a specific example, according to Reference 42 using data from the U.S. Department of Labor, a typical single family house with an average of 1622 ft² in 1972 required an average on-site labor of 1337 hours, an additional 1925 industry worker-hours off-site to produce construction material, and an additional 254 off-site hours in contractor's offices. Hence, the ratio of indirect to direct construction hours is approximately 1.63.

Residential development only creates jobs indirectly, with the exception of short-term construction jobs. These indirect jobs are created both in the public and private sector. For example, a high-rise building with elderly tenants is unlikely to increase the demand for teachers, but it will increase the demand for medical services. However, new detached housing units will mean more school-age children, and thus there will be an increase in the demand for teachers as well as recreational facilities and related skilled workers. The typical types of jobs which one could reasonably expect to be generated from residential development would include construction, retail services, and government-related jobs. To illustrate the computation of these estimates, consider a project consisting of 2000 homes with a market value of $90,000 each ($63,000 is actual construction cost), an average of 3.2 people per dwelling unit, and typical family income of $36,000. Employment estimates would be as follows:[6]

1. Construction jobs

$$\left(\begin{array}{c}2000 \\ \text{homes}\end{array}\right)\left(\begin{array}{c}\$63,000 \\ \text{construction} \\ \text{cost per} \\ \text{home}\end{array}\right)\left(\begin{array}{c}\text{Fraction of} \\ \text{construction} \\ \text{cost} \\ \text{representing} \\ \text{labor} = 0.2\end{array}\right) \Big/ \left(\begin{array}{c}\text{Average} \\ \text{Annual} \\ \text{construction} \\ \text{salary} = \$15,500\end{array}\right) = \begin{array}{c}1626 \\ \text{worker-years}\end{array}$$

2. Retail jobs

$$\left(\begin{array}{c}2000 \\ \text{homes}\end{array}\right)\left(\begin{array}{c}\$36,000 \\ \text{income} \\ \text{per home}\end{array}\right)\left(\begin{array}{c}\text{Fraction of} \\ \text{family income} \\ \text{for retail} \\ \text{expenditures} \\ = 0.32\end{array}\right)\left(\begin{array}{c}\text{Number of} \\ \text{retail employees} \\ \text{per dollar of} \\ \text{retail sales} \\ = 0.00005\end{array}\right) = \begin{array}{c}1152 \\ \text{jobs}\end{array}$$

[6]Factor values are for illustration only.

3. Government jobs

$$\left(\frac{2000}{\text{homes}}\right) \left(\frac{3.2 \text{ people}}{\text{per DU}}\right) \left(\begin{array}{c} \text{Number of} \\ \text{government} \\ \text{employees} \\ \text{per capita} \\ = 0.048 \end{array}\right) = 307 \text{ jobs}$$

Commercial and industrial developments create jobs directly as well as indirectly. Generally, nonresidential development brings three types of employment to the community, namely: (1) short-term jobs related to construction of the development and its associated public and private infrastructure such as utility lines, roads, sewers, water mains, etc.; (2) long-term direct employment provided by new or expanded business enterprises; and (3) secondary impacts due to increased economic activity stimulated by the development. Employment, particularly in retail business and services, may result from new development via the so-called multiplier effect.

Each new primary (or direct) job created by industrial and commercial developments may stimulate added service jobs. There also may be increased demand for labor in industries supplying materials for construction or for use by the new industry, even though these jobs may fall largely outside the community. The estimation of these induced or secondary employment impacts is not always easy to do and generally requires the use of employment multipliers. One type of multiplier, referred to as the "basic/ nonbasic ratio," is defined to be the number of primary (or direct) jobs in a firm divided by the number of secondary (or indirect) jobs created by virtue of the firm's sales. According to Reference 8, small cities usually have basic/nonbasic employment multipliers of about 1.0, medium-sized cities (10,000–30,000 in population) have multipliers of about 0.67, and large cities have multipliers of about 0.5.

Another multiplier used, referred to as the "regional employment multiplier," is defined to be the ratio between total employment (basic + nonbasic) and basic employment. Multiplying this ratio by the number of basic jobs gives the total jobs due to development. According to Reference 8, small cities have regional employment multipliers of about 2.0, medium-sized cities have multipliers of about 2.5, and large cities have multipliers of about 3.0.

One word of *caution* in using these multipliers is that they can be misleading. They cannot be used, necessarily, as valid indicators of local secondary employment or even regional secondary employment because of "leakage" effects—that is, those variables depend upon such factors as where the secondary spending is occurring, and the commuting patterns of workers.

Commercial employment includes two major categories, namely: employment in retail stores, and employment in offices. One method for estimating these types of employees is to compute employment on the basis of square footage in the business. For example, retail employees typically can be determined on the basis of 1 employee per 400–600 ft^2, whereas office employees can be determined on the basis of 1 employee per 100–250 ft^2. It is important to recognize that these ratios will differ by type of business establishment and location. For example, a self-service drug or department store will employ fewer persons per square foot than will a full-service department store. Similarly, large stores such as supermarkets will generally employ fewer persons per square foot than do small neighborhood convenience-goods stores.

Another method for estimating employees is on the basis of sales. For example, consider a furniture store with 15,000 ft^2 of space. The number of employees and estimated annual wages can be determined as follows:

Number of square feet 15,000
Average annual sales per square foot $90
Projected annual sales .... (15,000)($90) = $1,350,000
Average number of retail employees per dollar of retail sales 0.000016
Number of employees .... (0.000016)($1,350,000) = 22
Wages as percent of sales 15%
Projected annual payroll ....($1,350,000)(0.15) = $202,500

Average annual wage per employee .... $\dfrac{\$202,500}{22}$ = $9,250

In the case of industrial employment, employment levels cannot usually be reliably estimated by facility size since employment levels vary greatly within the industrial sector. For example, capital-intensive industries, such as chemical processing firms and utilities, employ relatively few people, whereas labor-intensive industries, such as the textile industry, tend to have a high employee-per-square-foot ratio. One method used to estimate industrial employment is to estimate the number of workers, based on capital investment per employee by type of industry. If a particular type of industry requires $100,000 in capital outlay per worker, then a $10 million facility will employ approximately 100 workers. Generally speaking, it is best to estimate industrial employees by comparison with known employment levels of similar industrial firms.

Airport Expansion Example

The expansion of a major airport generally has a variety of socioeconomic impacts which affect the overall economic profile of the community as well as the entire region. Such impacts could be expected to include passenger traffic, employment in terms of jobs and payrolls, tax revenues to local agencies, expenditures for goods and services, employee

TABLE 2.36   Projected Passenger Traffic Activity

Year	Passenger traffic			Increase relative to prior time
	Arrivals	Departures	Total	
Present	428,500	428,500	857,000	
Present +10	3,500,000	3,500,000	7,000,000	6,143,000
Present +20	7,150,000	7,150,000	14,300,000	7,300,000

TABLE 2.37   Projected Airport-Related Employment

Year	Number of employees	Increase relative to prior time
Present	335	
Present +10	$7,000,000/1,500 = 4,667$	4,332
Present +20	$14,300,000/1,500 = 9,533$	4,866

housing demand, and population growth. To illustrate these types of impacts, consider a project at International Airport, a major regional airport, consisting of (1) the construction of a new runway, 10,200 ft long by 150 ft wide, (2) a 2200-ft extension of the existing runway, which is 10,000 ft long and 150 ft wide, (3) the construction of new related taxiways, and (4) the expansion of passenger terminals and related facilities. The primary purpose of the project is to provide adequate runways and taxiways to accommodate the planned air travel demand of the residents living in the regional area. Another important purpose of the project is to reduce noise by permitting usage of wide-body, quiet aircraft which are presently prohibited due to strength limitations of the existing runway.

Passenger Traffic   At the present time, the Department of Airports estimates that total passenger traffic at International Airport is 857,000. As the result of this project, it is expected to increase to 7.0 million in 10 years and to 14.3 million in 20 years. Furthermore, it is generally assumed that 50 percent are departures and 50 percent are arrivals. Thus, the impact on passenger traffic is estimated as in Table 2.36. The estimated increase per year over this 20-year period is thus approximately $(14,300,000 - 857,000)/20 = 672,150$ passengers.

Employment   At the present time, there are 335 employees directly associated with airline operations. This implies a productivity of $857,000/335 = 2558$ passengers per employee per year. Because of the increase in airport size and type of facilities, it is expected that the productivity will be 1500 passengers per employee per year. Thus, the estimated number of direct airport-related employees is given in Table 2.37. The estimated increase in airport-related employment per year over the 20-year period is thus approximately $(9533 - 335)/20 = 460$ employees.

The airport construction activity over the next 10 years is estimated in Table 2.38. The average construction employment per year is approximately 2214.7/10 = 221 employees over the 10-year construction period.

Because of the increase in passenger traffic at International Airport, there will be an increased demand for hotel and motel rooms. It is assumed at the present time that 15 percent of the arriving passengers stay at the local area hotels and motels, whereas this is expected to increase to 20 percent as the result of the airport expansion. The average stay per person is estimated to be 1.5 days. As a result, the total demand in room-days can be computed as shown in Table 2.39. The number of hotel/motel employees is based on 0.5

TABLE 2.38 Projected Airport Construction Activity

Year	Construction cost	Percent labor	Total labor cost	Average annual labor rate	Equivalent jobs*
Present +1	$ 168,000	25	$ 42,000	$14,100	3.0
Present +2	373,000	25	93,250	14,805	6.3
Present +3	20,281,000	37	7,503,970	15,545	482.7
Present +4	15,828,000	27	4,273,560	16,323	261.8
Present +5	10,000,000	45	4,500,000	17,139	262.6
Present +6	10,000,000	45	4,500,000	17,996	250.1
Present +7	10,625,000	45	4,781,250	18,895	253.0
Present +8	10,625,000	45	4,781,250	19,840	241.0
Present +9	12,000,000	45	5,400,000	20,832	259.2
Present +10	9,480,000	45	4,266,000	21,874	195.0
Total	$99,380,000		$40,141,280		2,214.7

*By "Equivalent jobs" is meant "Total labor cost" divided by "Average annual labor rate."

TABLE 2.39 Projected Hotel/Motel Room Demand

Year	Arriving passengers	×	Fraction staying in area	=	Annual room demand	×	Average stay	=	Annual room-days sold
Present	428,500		0.15		64,275		1.5		96,413
Present +10	3,500,000		0.20		700,000		1.5		1,050,000
Present +20	7,150,000		0.20		1,430,000		1.5		2,145,000

employees per supportable room, where supportable rooms are the total available for rental. In estimating the number of rooms available for rental, one typically assumes an average annual occupancy factor from which one can compute this number of rooms as:

$$\frac{\text{Annual room-days sold}}{(\text{Occupancy factor})(365 \text{ days})} = \text{Number of supportable rooms}$$

Assuming a 70 percent occupancy factor, this means that the number of hotel/motel employees needed is given by

$$\begin{array}{c}\text{Number of} \\ \text{hotel/motel} \\ \text{employees}\end{array} = \frac{(0.5)(\text{Annual room-days sold})}{(\text{Occupancy factor})(365 \text{ days})}$$

Therefore, we can summarize the hotel/motel employment in Table 2.40. The average new hotel/motel employment per year is approximately 4011/20 = 201 employees over the 20-year period.

In addition, each employee in direct airport-related activities (airlines, government agencies, etc.) and in indirect airport-related activities (hotels, motels, travel agencies, construction, etc.) supports secondary or induced employment in trade, services, finance, and other supporting activities. This is the so-called "multiplier effect." In the regional area surrounding International Airport, this is approximately 3.0 in the sense that each direct or indirect airport employee generates three additional or secondary jobs in the area. The estimated impact on employment as the result of expansion can be summarized as shown in Table 2.41. Over the 20-year period, the estimated new employment per year is approximately 52,836/20 = 2642 jobs.

Employment Payroll for Primary Jobs The primary jobs discussed above are in the areas of airport-related, construction, and hotel/motel employment. At the present time, the average annual income per airport-related employee is $10,500, per construction employee is $13,429, and per hotel/motel employee is $4800. These incomes are expected

TABLE 2.40   Projected Hotel/Motel Employment

Year	Annual room-days sold	Supportable rooms	Number of employees	Increase relative to prior time
Present	96,413	377	187	
Present +10	1,050,000	4,110	2,055	1,868
Present +20	2,145,000	8,395	4,198	2,143
				4,011

TABLE 2.41   Projected Total Employment

Year	Airport-related	+	Construc-tion	+	Hotel/motel	=	Total	+	Indirect	=	Combined total	Increase relative to prior time
Present	335		0		187		522		1,566		2,088	
Present +10	4,667		195		2,055		6,917		20,751		27,668	25,580
Present +20	9,533		0		4,198		13,731		41,193		54,924	27,256
												52,836

TABLE 2.42   Projected Employment Payrolls

Employment data	Year		
	Present	Present +10	Present +20
Airport-related employees:			
Number	335	4,667	9,533
Salary per employee	$10,500	$17,103	$27,860
Total payroll ($10⁶)	3.518	79.820	265.589
Construction employees:			
Number	0	195	0
Salary per employee	$13,429	$21,874	$35,631
Total payroll ($10⁶)	0	4.265	0
Hotel/motel employees:			
Number	187	2,055	4,198
Salary per employee	$4,800	$7,819	$12,736
Total payroll ($10⁶)	0.898	16.068	53.466
Total Payroll ($10⁶)	4.416	100.153	319.055
Payroll Increase Relative to Prior Time ($10⁶)	. . .	95.737	218.902

to increase at the rate of 5 percent per annum. The estimated annual payrolls in these labor areas are shown in Table 2.42. Over the 20-year period, the estimated new payroll per year from primary jobs is approximately ($95,737,000 + $218,902,000)/20 = $15,731,950.

Local Taxes At the present time, airline and airline-related activities at International Airport produce about $0.48 per arriving passenger in local taxes, including state sales tax, motel/hotel bed tax, property tax, and other sales/use taxes. Because of the expansion in airport activities, this is expected to increase to $1.60 per arriving passenger. The estimated total annual tax revenue will be as shown in Table 2.43. Over the 20-year period, the estimated new local tax revenue per year is approximately $11,234,320/20 = $561,716.

Expenditures Local purchases of materials and services by airlines, concessionaires, and airline service organizations presently amount to $1.86 per passenger (enplaned plus deplaned). However, because of the expansion in airport activities, this is expected to increase to $4.00 per passenger. The estimated annual local expenditures for materials and services are shown in Table 2.44. Over the 20-year period, the estimated expenditures for materials and services per year are approximately $55,605,980/20 = $2,780,299.

TABLE 2.43   Projected Total Tax Revenue

Year	Number of arriving passengers	×	Tax revenue per passenger	=	Total tax revenue	Increase relative to prior time
Present	428,500		$0.48		$ 205,680	
Present +10	3,500,000		1.60		5,600,000	$ 5,394,320
Present +20	7,150,000		1.60		11,440,000	5,840,000
						$11,234,320

TABLE 2.44   Projected Expenditures for Materials and Services

Year	Number of passengers	×	Expenditures per passenger	=	Total expenditures	Increase relative to prior time
Present	857,000		$1.86		$ 1,594,020	
Present +10	7,000,000		4.00		28,000,000	$26,405,980
Present +20	14,300,000		4.00		57,200,000	29,200,000
						$55,605,980

TABLE 2.45   Projected Air-Visitor Local Expenditures

Year	Number of arriving passengers	Number of nonresidents	×	Expenditure per visitor	= Total expenditure	Increase relative to prior time
Present	428,500	214,250		$135	$ 28,923,750	
Present +10	3,500,000	1,750,000		242	423,500,000	$ 394,576,250
Present +20	7,150,000	3,575,000		433	1,547,975,000	1,124,475,000
						$1,519,051,250

In addition, there are expenditures made by air visitors to the area for goods and services—particularly by those visitors who are nonresidents in the area (estimated to be 50 percent of all arriving passengers). At the present time this is estimated to be $135 per visitor and is expected to increase by 6 percent per year. The estimated annual air-visitor expenditures are shown in Table 2.45. Over the 20-year period, the estimated local expenditures by air-visitors per year are approximately $1,519,051,250/20 = $75,952,563.

In total, local expenditures will average approximately $2,780,299 + $75,952,563 = $78,732,862 per year for purchase of materials, goods, and services.

Housing Demand It is assumed that new employees will reside in the immediate area adjacent to International Airport in the same proportion as the existing employees—specifically, 77 percent of these reside in the area at the present time. Furthermore, 76 percent of these reside in homes and the other 24 percent in apartments. This implies that the demand for housing will be as shown in Table 2.46. Over the 20-year period, the estimated average demand for housing will be 30,919/20 = 1546 homes and 9,764/20 = 488 apartments.

Population The average family size per employee is 3.5 persons. Thus, in the immediate area adjacent to International Airport, the population attributed to airport activities is described in Table 2.47. Over the 20-year period, the estimated average increase per year in population of the adjacent area is 142,391/20 = 7120 people.

Summary of Airport Expansion Impacts   In summary, the estimated impact on the economic profile of the area as the result of the expansion of International Airport is shown in Table 2.48.

TABLE 2.46   Projected Housing and Apartment Demand

Year	Total employment	Number of employees in area	Housing demand		Apartment demand	
			Total	Incremental	Total	Incremental
Present	2,088	1,608	1,222	. . .	386	
Present +10	27,668	21,304	16,191	14,969	5,113	4,727
Present +20	54,924	42,291	32,141	15,950	10,150	5,037
				30,919		9,764

TABLE 2.47   Projected Area Population

Year	Total employment	Number of employees in area	Total population in area	Incremental population
Present	2,088	1,608	5,628	
Present +10	27,668	21,304	74,564	68,936
Present +20	54,924	42,291	148,019	73,455
				142,391

TABLE 2.48   Projected Impact on Airport Area Economic Profile

Impact area	Type of impact
Passenger traffic	672,150 passengers per year increase
Employment	2,642 new jobs per year
Employment payroll	$15,731,950 new employment payroll per year
Local tax revenues	$561,716 increase per year
Expenditures for materials, goods and services	$78,732,862 increase per year
Housing	Demand for 1,546 new homes and 488 apartments per year
Population	7,120 new residents to area per year

SUMMARY

As mentioned at the beginning of this chapter, those dimensions of the environment which are socioeconomic-related include: (1) those aspects of the physical environment such as surrounding land uses, the physical character of the area, and the corresponding infrastructure/public services available; (2) those aspects of the social environment such as community facilities and services, employment centers and commercial facilities servicing the area, and the overall character of the community; (3) those aspects of the aesthetic environment such as historical and cultural sites, scenic views, and the architectural character and appearance of the area; and (4) those aspects of the economic base of the area, land ownership patterns, and land values.

In the preceding sections, tools, techniques, and methods of analysis have been presented for assessing impacts of land development projects in these various dimensions. Particular emphasis was placed on: public service areas such as health care, libraries, parks and recreation facilities, police and fire protection, schools, and traffic circulation; fiscal impact analysis for governmental agencies; uses of demographic data to predict social impacts and measure social conditions such as neighborhood stability, ethnic composition, residential dependency, transportation needs, and resident economic status; and measures of the economic profile of the community as characterized by employment levels, payrolls, population, housing, and sales of commercial and industrial businesses.

Socioeconomic assessment is a multifaceted type of analysis, covering many dimensions as described here. The ideas and methods presented in this chapter are intended to be illustrative as to how one could approach socioeconomic impact assessment in the specific areas considered. These ideas and methods are by no means all-inclusive; they were not intended to be. The individual practitioner must decide whether or not they are applicable to his/her situation. If not, then hopefully they are suggestive enough or can stimulate ideas for other approaches.

REFERENCES

1. Allen, Gary R., "Incorporating Economic Considerations in the Preparation of Environmental Impact Statements," Trans. Res. Record No. 583, *Social, Economic, and Environmental Implications in Transportation Planning,* pp. 55–70, 1976.
2. Arthur D. Little, Inc., *Economic Impact Report on the Pacific Terrace Hotel Project, City of Long Beach, California,* Jan. 1977.
3. Barton-Aschman Associates, Inc., "Parking Demand at the Regionals," *Urban Land,* May 1977, pp. 3–11.
4. ———, *St. Charles Communities Impact Evaluation Study,* "Report 1," *State of the Art in New Community Impact Research* and *Appendix.*
5. Bergmann, P. A., "Assessing the Consequences of Development: Clearwater, Florida's Community Impact Statement Requirement," *Environmental Comment,* Oct. 1976, pp. 10–13.
6. Bishop, A. B., C. H. Oglesby, and G. E. Willeke, *Socio-Economic and Community Factors In Planning Urban Freeways,* Project on Engineering-Economic Planning, Report EEP-33, Stanford University, Oct. 1969.
7. Buckley, Nora C., *An Overview of Studies of the Impact of Military Installations and Their Closings on Nearby Communities,* Report No. T-338, The George Washington University, 20 July 1976.
8. Canter, Larry W., *Environmental Impact Assessment,* McGraw-Hill, New York, 1977.
9. Castle III, G. H., "Evaluating the Impact of a New Community: Saint Charles Communities, Maryland," *Environmental Comment,* Oct. 1976, pp. 13–16.
10. City of Fresno, California, *Urban Growth Mangement Process,* 5 Dec. 1975.
11. Corwin, Ruthann, et al., *Environmental Impact Assessment,* Freeman, Cooper & Co., San Francisco, Calif., 1975.
12. Council on Environmental Quality, "Preparation of Environmental Impact Statements: Guidelines," *Federal Register,* Vol. 38 No. 147, 1 Aug. 1973.
13. Creighton, R. L. and C. W. Manning, "Coordinating Land Use and Transportation, How It Was Done at the Community Scale in Coventry, Connecticut," *Planners Notebook,* Vol. 5 Nos. 3–4, June–August 1975, published by the American Institute of Planners.
14. DeFerranti, D., C. Chew, I. Kobashi, E. Rolph, and J. Webb, *Municipal Services Pricing: Impact on the Spatial Location of Residential Development,* Report No. R 1878/4 NSF, The Rand Corp., Nov. 1975, PB-256 712.
15. deLeeuw, Frank, "The Demand for Housing: A Review of Cross-Section Evidence," *The Review of Economics and Statistics,* Vol. LIII No. 1, Feb. 1971, pp. 1–10.
16. Department of City Planning, *EIR Manual for Private Projects,* Los Angeles, Calif., Aug. 1975.
17. Department of the Army, Department of Defense, "Evaluation of Economic Benefits for Flood Control and Water Resource Planning," *Federal Register,* Vol. 39 No. 159, 15 Aug. 1974, pp. 29539–29550.
18. Dougharty, L. et al., *Municipal Service Pricing: Impact on Urban Development and Finance-Summary and Overview,* The Rand Corp., Report No. R-1878/1-NSF, Nov. 1975, PB-256 710.
19. Dougharty, L., S. Tapella, and G. Sumner, *Municipal Service Pricing: Impact on Fiscal Position,* Report No. R-1878/2-NSF, The Rand Corp., Nov. 1975, PB-256 711.
20. Environmental Protection Agency, *Environmental Impact Assessment Guidelines for Selected New Source Industries,* Oct. 1975, PB-258 527.
21. Federal Aviation Administration, Department of Transportation, *Instructions for Processing Airport Development Actions Affecting the Environment,* Instruction No. 5050.2A, 24 Feb. 1975.
22. Federal Highway Administration, Department of Transportation, *Social and Economic Effects of Highways,* 1974.
23. Federal Power Commission, *Order Amending Part 2 of the General Rules to Provide Guidelines for the Preparation of Applicants' Environmental Reports Pursuant to Order No. 415-C,* Order No. 485, 7 June 1973.
24. Finsterbusch, Kurt, *A Methodology for Social Impact Assessments of Highway Locations,* Report No. FHWA-MD-R-76-20, Maryland Department of Transportation, July 1976.
25. Garrison, C. B., "The Impact of New Industry: An Application of the Economic Base Multiplier to Small Rural Areas," *Land Economics,* Vol. XLVIII, No. 4, pp. 329–337, Nov. 1972.

26. Goldman, George and David Strong, *Government Costs and Revenues Associated wtih Implementing Coastal Plan Policies in the Half Moon Bay Subregion,* Spec. Pub. 3208, Division of Agricultural Science, University of California, Berkeley, Oct. 1976.

27. Guseman, P. K., J. M. Hall, T. K. Fuller, and D. Burke, *Social Impacts: Evaluation of Highway Project Development in Urban Residential Areas,* Report No. TTI-2-8-75-190-1, Texas Transportation Institute, Texas A&M University, April 1976, PB-255 684.

28. Harvard University, *Environmental Models for Planning and Policy Making,* Vol. III, *The Costs of Growth: Revenue and Expenditure Implications of Suburban Town Development,* Jan. 1975, PB-252 353.

29. Hellstrom, David I., *A Methodology for Preparing Environmental Statements,* Arthur D. Little, Inc., Cambridge, Mass., Aug. 1975.

30. Hemenway, G. D., *Developer's Handbook–Environmental Impact Statements,* Associated Home Builders of the Greater Eastbay, Inc., Berkeley, Calif., 1973.

31. Hittman Associates, Inc., *Environmental Effects, Impacts and Issues Related to Large Scale Coal Refining Complexes,* Report No. FE-1508-T2, May 1975.

32. Hyciak, Thomas J. and Thomas A. DeCoster, *An Information Base for Fiscal Decision-Making,* Report No. UO-LCCM-SOB-76-002, South Bend Urban Observatory, South Bend, Ind., June 1976.

33. Illinois Institute for Environmental Quality, *Environmental Impact Statements: A Handbook for Writers and Reviewers,* Aug. 1973, PB-226 276.

34. Isserman, Andrew M., "The Location Quotient Approach to Estimating Regional Economic Impacts," *AIP Journal,* Jan. 1977, pp. 33–41.

35. Johanning, J. et al., *State-of-the-Art of Environmental Impact Statements in Transportation,* Oklahoma Univ., Nov. 1974, PB-244 539.

36. Leigh, R., "The Use of Location Quotients in Urban Economic Base Studies," *Land Economics,* Vol. XLVI No. 2, pp. 202–205, May 1970.

37. Levin, M. S., "Cost-Revenue Impact Analysis: State-of-the-Art," *Urban Land,* June 1975, pp. 8–15.

38. Lustig, M. and J. R. Pack, "A Standard for Residential Zoning Based Upon the Location of Jobs," *AIP Journal,* Sept. 1974, pp. 333–345.

39. Mace, R. L. and W. J. Wicker, *Do Single-Family Homes Pay Their Way?, A Comparative Analysis of Costs and Revenues for Public Services,* Res. Mon. #15, Urban Land Institute, 1968.

40. Masser, Ian, *Analytical Models for Urban and Regional Planning,* David & Charles: Newton Abbot.

41. Muller, Thomas, "Assessing the Impact of Development," Environmental Comment, Oct. 1976, pp. 2–5.

42. ———, *Economic Impacts of Land Development: Employment, Housing and Property Values,* Report No. URI 25800, The Urban Institute, Wash., D.C., Sept. 1976.

43. Muller, T. and G. Dawson, *The Fiscal Impact of Residential and Commercial Development: A Case Study,* Paper No. 712-7-1, The Urban Institute, Dec. 1972.

44. ——— and ———, *The Impact of Annexation on City Finances: A Case Study in Richmond, Virginia,* Paper No. 712-11-1, The Urban Institute, May 1973.

45. New Jersey State, County and Municipal Government Study Commission, *Housing and Suburbs: Fiscal and Social Impact of Multifamily Development,* Oct. 1974, PB-244 785.

46. Real Estate Research Corp., *The Costs of Sprawl: Case Studies and Further Research,* Oct. 1975, PB-257 738.

47. Roberts, P., "Making Dollars and Sense Out of Fiscal Impact Analysis," *Environmental Comment,* Oct. 1976, pp. 5–7.

48. Rosen, S. J., *Manual For Environmental Impact Evaluation,* Prentice-Hall, Inc., Englewood Cliffs, N.J., 1976.

49. Schaenman, P. S. and T. Muller, *Measuring Impacts of Land Development, An Initial Approach,* Publication No. URI 86000, The Urban Institute, Nov. 1974.

50. Stenehjem, E. J., *Forecasting the Local Economic Impacts of Energy Resource Development: A Methodological Approach,* Argonne Nat. Lab., Report No. ANL/AA-3, Dec. 1975.

51. Stephens, Jr., G. M., "Fiscal Impact Model for Land Development: A Case Study," *Urban Land,* June 1975, pp. 16–23.

52. Stuart, D. G. and R. B. Teska, "Who Pays for What: A Cost-Revenue Analysis of Suburban Land Use Alternatives," *Urban Land,* March 1971, pp. 3–16.

53. The Urban Institute, *State-Required Impact Evaluation of Land Development: An Initial Look at Current Practices and Key Issues,* July 1974, PB-239 877.

54. Transportation Research Board, *Transportation Decision-Making, A Guide to Social and Environmental Considerations,* National Cooperative Highway Research Program Report No. 156, National Research Council, Wash., D.C., 1975, PB-244 947.

55. U.S. Department of Commerce, National Oceanic and Atmospheric Administration, "Coastal Energy Impact Program, Proposed Regulations for Financial Assistance to Coastal States," *Federal Register,* 22 Oct. 1976, pp. 46724–46740.

56. U.S. Department of Transportation, Federal Highway Administration, "Guidelines for Consideration of Economic, Social, and Environmental Effects," *Federal Register,* Vol. 37 No. 197, 11 Oct. 1972.

57. U.S. Department of Transportation, Federal Aviation Administration, "Processing Airport Devel-

opment Actions Affecting the Environment, Part II: References to Applicable Procedures," *Federal Register*, Vol. 40 No. 162, 20 Aug. 1975.

58. Weiss, S. J. and E. C. Gooding, "Estimation of Differential Employment Multipliers in a Small Regional Economy" *Land Economics*, Vol. XLIV, No. 2, pp. 235–244, May 1968.

59. Zycher, B., L. Dougharty, C. Chew, and I. Kobashi, *Municipal Service Pricing: Impact on the Growth of Residential Development*, Rand Corp., Report. No. R-1878/3-NSF, Nov. 1975, PB-256 200.

Chapter 3

Air Quality Impact Analysis

GARY S. SAMUELSEN

The analysis of air quality impact is a relatively new and developing art. Two reasons exist for its current status. First, a requirement for air quality impact analysis has been only recently established through new dimensions in environmental law. Second, an air quality impact analysis addresses the interaction of emissions, chemical changes, and transport of pollutant species in the atmosphere. The interaction of these factors is a complex subject of which we have at present only a rudimentary understanding.

Advances in air quality impact analysis techniques were first stimulated by the impact analyses required for siting nuclear power generation facilities.[93] The passage of the 1970 Federal Clean Air Amendments[18] and other federal acts—most notably, the National Environmental Policy Act,[70] and the Federal Highway Administration Act Amendments[89]—have stimulated additional advances in the development of methodologies. Finally, the passage of the 1977 Federal Clean Air Amendments[9] has provided further impetus for the development of impact analysis methodologies.

The goal of this chapter is to provide the background and direction necessary to construct and review an air quality impact analysis. Section 1 provides the background and general overview, Section 2 describes the structure of an air quality impact analysis and details the methodologies available for constructing an air quality impact analysis, and case studies are presented in Section 3. You will find that air quality impact analyses are not constructed from methods or information that are absolute. Rather, intuitive sense must be employed in both the construction and review.

The treatment of a practical situation usually involves the use of specific quantitative techniques along with a broad variety of assumptions engendered by the imperfect knowledge of the atmosphere and pollutant producing device. Although these assumptions may be justified on scientific grounds, their selection can be rationally made only by someone with a broader range of experience and knowledge than can be garnered from careful reading. . . . *

---

*From Reference 93, "Preface."

Section 1: Introduction

BACKGROUND

Technical Overview

The elements and interplay of an air quality impact analysis are presented in Fig. 3.1. The pollutant emissions from sources and atmospheric interactions determine the quality of air, and the effects on receptors establish the extent to which air quality is degraded. Were it not for the effects, air quality would be of passing interest.

Sources Anthropogenic sources of air pollutants are divided into two categories: mobile sources and stationary sources. "Mobile sources" include automobiles, trucks, buses, aircraft, ships, and railroads. "Stationary sources" include electric power utility boilers, commercial and domestic boilers and furnaces, refineries, and industrial processing. Sources which include a combination of mobile and stationary sources (airports, urban communities) are called "compound sources." Finally, sources which generate mobile source activity (shopping centers, sporting facilities) are called "indirect sources."

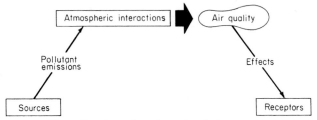

Figure 3.1 Air quality—the technical elements.

Atmospheric Interactions Sources emit a variety of particulate and gaseous pollutants into the atmosphere. Once the pollutants enter the atmosphere, many interactions occur. First, the pollutants are diluted upon injection into the atmosphere with a resultant decrease in concentration. Winds then act to transport the pollutants and promote additional mixing. Some pollutants may participate in chemical reactions while in transport. As a result, the quality of air at a receptor is determined by the type and amount of pollutants emitted upwind and the extent of atmospheric interaction (dispersion and chemical change) that has occurred during transport from the source to the receptor.

Receptors A receptor is any object affected adversely by the quality of air to which it is exposed. Receptors include human beings, vegetation, and materials.

The public typically judges air quality by visibility. Unfortunately, the quality of air is determined by more than visibility alone. Many of the pollutants in the ambient air are particles and gases that cannot be seen. Special instruments are required to measure the concentration of these particles and gases. The air quality is formally described by comparing the measured ambient concentration to established standards. Air quality standards are set for those pollutant species that affect the public health. Standards are also set to protect resources of interest to the public welfare such as agricultural crops, ornamental crops, and materials.

Sociopolitical Overview

The preparation of an air quality impact analysis is based on a technical assessment of the elements and processes shown in Fig. 3.1. However, in addition to the technical considerations of air quality, sociopolitical considerations are also a major factor. The sociopolitical process determines the social and political acceptability and potential success of local, state, and federal programs directed toward the improvement of air quality. The technical and sociopolitical interactions involved are shown graphically in Fig. 3.2. As shown, air quality is multidimensional in character with two distinctive arms: technical and sociopolitical. The development of an air quality impact analysis is primarily technical. The evaluation, review, and use of the analytical results in decision making occurs largely in the sociopolitical domain.

TYPICAL CONSIDERATIONS AND FACTORS

The air quality impact model of Fig. 3.1 identifies the general elements and processes involved in air quality impact analyses. An expanded pictorial representation of the air quality impact model is presented in Fig. 3.3. The purpose of this section is to outline the typical considerations and factors of interest in air quality impact analyses.

Air Pollutants

The conventionally accepted composition of "clean" air is presented in Table 3.1. A chemical species foreign to the composition of the "clean" atmosphere is called a "contaminant." A contaminant that can cause an adverse effect to a receptor and which

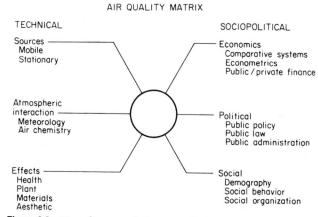

AIR QUALITY MATRIX

Figure 3.2 Air quality—a multidimensional subject. (*From Reference 86.*)

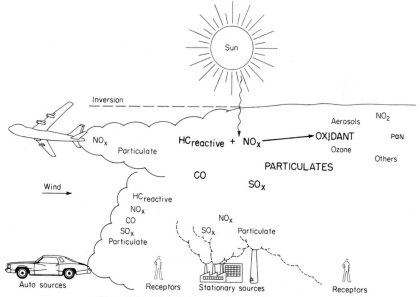

Figure 3.3 Air quality—a pictorial representation.

occurs in the atmosphere in concentrations sufficiently high to cause the adverse effect is called a "pollutant."

Air pollutants are classified into two categories: primary and secondary pollutants (primary and secondary pollutants are not the same nor are they related to primary and secondary air quality standards which will be discussed later). The distinction between the two pollutant categories is made according to the manner by which the pollutants are introduced into the atmosphere. "Primary pollutants" are those emitted directly from sources. "Secondary pollutants" are those formed by chemical processes in the atmosphere. Examples of primary pollutants are carbon monoxide (CO), nitric oxide (NO), nitrogen dioxide (NO_2), sulfur dioxide (SO_2), particulates, and various hydrocarbons (HC). Examples of secondary pollutants are photochemical oxidant and sulfates.

Another distinction is made between stable (conservative) and unstable (nonconservative) pollutants. "Stable pollutants" are those which do not participate in chemical

TABLE 3.1   Composition of Clean, Dry Air

Elements	ppm (vol)	$\mu g/m^3$
Nitrogen	780,900	8.95×10^8
Oxygen	209,400	2.74×10^8
Water	. . .	. . .
Argon	9,300	1.52×10^7
Carbon dioxide	315	5.67×10^5
Neon	18	1.49×10^4
Helium	5.2	8.50×10^2
Methane	1.0–1.2	6.56–7.87×10^2
Krypton	1.0	3.43×10^3
Nitrous oxide	0.5	9.00×10^2
Hydrogen	0.5	4.13×10^1
Xenon	0.08	4.29×10^2
Organic vapors	0.02	. . .

SOURCE: Reference 101.

processes in the atmosphere. "Unstable pollutants" are those which actively participate in the formation of secondary pollutants. Stable pollutants include carbon monoxide, sulfur dioxide, and particulates. Nitric oxide, nitrogen dioxide, ozone, and certain hydrocarbons are examples of unstable pollutants.

Examples of pollutant species that may be encountered in conducting an air quality impact analysis are listed in Table 3.2. The pollutant species considered for an impact analysis will depend upon the source. The pollutants most commonly encountered (CO, NO, NO_2, SO_2, HC, particulates, and photochemical oxidant) are those associated with sources that involve combustion (automobiles, trucks, buses, aircraft, fossil-fueled electric generation power plants, and boilers).

Sources

An air pollutant source is defined as any vehicle, facility, physical plant, installation, or activity that emits primary air pollutants into the atmosphere. This definition limits the consideration of sources to anthropogenic emissions. On a global basis, emissions from natural sources are with few exceptions (e.g., CO and SO_2) significantly greater than anthropogenic emissions. However, anthropogenic sources are concentrated in urban areas. In these areas, the contribution of anthropogenic sources to the total atmospheric burden is dominant. As a result, air quality impact analyses of projects located in urban areas will usually not consider the contribution from natural sources. In rural or pristine settings, however, an air quality impact analysis must often account for natural pollutant emissions as well.

The two categories of anthropogenic sources have already been introduced: mobile and stationary. You will recall that automobiles, trucks, buses, and aircraft are examples of mobile sources, and fossil-fueled electric generation plants, domestic and commercial boilers and furnaces, asphalt batching plants, dry cleaning operations, and auto painting establishments are examples of stationary sources. A more exhaustive listing of anthropogenic sources is presented in Table 3.3.

Mobile sources are presently a major contributor to air pollutant emissions in urban areas. As a result, projects that increase vehicle use are specially acknowledged by the label "indirect source."[35] An indirect source is any facility, plant, installation, or activity that has a significant amount of mobile source activity associated with its operation or use. Parking facilities, roadways, and airports are examples of indirect sources. Included in the parking facilities category are shopping centers, sports complexes, and large amusement facilities.

TABLE 3.2   Air Pollutants (not exclusive)

Common to urban environments:	
Sulfur dioxide (SO_2)	Nitrates
Particulate matter	Sulfates
Carbon monoxide (CO)	Lead
Hydrocarbons (HC)	Polynuclear organic matter
Nitrogen oxides (NO_x)	Fluorides
Photochemical oxidant*	Odors
Especially hazardous:	
Lead	Cadmium
Mercury	Asbestos
Confined to a specific source:	
Arsenic	Zinc
Chlorine gas	Barium
Hydrogen chloride	Boron
Copper	Chromium
Manganese	Selenium
Nickel	Pesticides
Vanadium	Radioactive substances
Others (example):	
Aeroallergens (pollens)	Biological aerosols
Aldehydes	Ethylene
Ammonia	Hydrochloric acid
Phosphorus	Hydrogen sulfide
Beryllium	Iron

*See Table 3.6 for constituents.

Some indirect sources may also include pollutant emissions from stationary sources. Large airports, for example, have stationary source emissions associated with refueling operations, as well as space heating and cooling of the terminal. Aircraft operations also contribute to the total emissions. Airports are examples of compound indirect sources. Not only must automobile activity be considered, but the contribution of other mobile source emissions and stationary source emissions must be considered as well.

For modeling purposes, sources are classified according to the following geometric configurations: point, line, and area. Examples of point sources are fossil-fueled electric power generating plants and large municipal incinerators. Roadways and airport flight patterns are classified and modeled as line sources. Oil refineries and residential tracts are typical area sources.

A listing of the common source types and the principal pollutants associated with each is presented in Table 3.4.

Atmospheric Interaction

In combination with the amount and type of primary pollutants emitted from the source, atmospheric interaction determines the quality of air at the receptor. Included are meteorology (dilution and mixing, and pollutant transport), and chemical processes (transformation of unstable primary pollutants to secondary pollutants). The extent of dilution, transport, mixing, and chemical change between the source and the receptor determines the quality of air. As the separation of source and receptor increases, the mixing and dilution of pollutants normally increases. Balancing the decrease in concentration of

TABLE 3.3   Anthropogenic Sources (not exclusive)

1. **External combustion sources**
   Bituminous coal combustion
   Anthracite coal combustion
   Fuel oil combustion
   Natural gas combustion
   Liquefied petroleum gas consumption
   Wood-waste combustion in boilers
   Lignite combustion
2. **Solid waste disposal**
   Refuse incineration
   Automobile body incineration
   Conical burners
   Open burning
   Sewage sludge incineration
3. **Internal combustion engine sources**
   Highway vehicles
     Light-duty, gasoline-powered vehicles
       (automobiles)
     Light-duty, diesel-powered vehicles
     Light-duty, gasoline-powered trucks and
       heavy-duty, gasoline-powered vehicles
     Heavy-duty, diesel-powered vehicles
     Gaseous-fueled vehicles
     Motorcycles
   Off-highway, mobile sources
     Aircraft
     Locomotives
     Inboard-powered vessels
     Outboard-powered vessels
     Small, general utility engines
     Agricultural equipment
     Heavy-duty construction equipment
     Snowmobiles
   Off-highway stationary sources
     Stationary gas turbines for electric
       utility power plants
     Heavy-duty, natural gas–fired pipeline
       compressor engines
     Gasoline and diesel industrial engines
4. **Evaporation loss sources**
   Dry cleaning
   Surface coating
   Petroleum storage
   Gasoline marketing
5. **Chemical process industry**
   Adipic acid
   Ammonia
   Carbon black
   Charcoal
   Chlor-alkali
   Explosives
   Hydrochloric acid
   Hydrofluoric acid
   Nitric acid
   Paint and varnish
   Phosphoric acid
   Phthalic anhydride
   Plastics

Printing ink
Soap and detergents
Sodium carbonate
Sulfuric acid
  Elemental sulfur-burning plants
  Spent-acid and hydrogen sulfide
    burning plants
  Sulfide ores and smelter gas plants
Sulfur
Synthetic fibers
Synthetic rubber
Terephthalic acid

6. **Food and agricultural industry**
   Alfalfa dehydrating
   Coffee roasting
   Cotton ginning
   Feed and grain mills and elevators
   Fermentation
   Fish processing
   Meat smokehouses
   Nitrate fertilizers
   Orchard heaters
   Phosphate fertilizers
   Starch manufacturing
   Sugar cane processing
7. **Metallurgical industry**
   Primary aluminum production
   Metallurgical coke manufacturing
   Copper smelters
   Ferroalloy production
   Iron and steel mills
   Lead smelting
   Zinc smelting
   Secondary aluminum operations
   Brass and bronze ingots
   Gray iron foundry
   Secondary lead smelting
   Secondary magnesium smelting
   Steel foundries
   Secondary zinc processing
8. **Mineral products industry**
   Asphaltic concrete plants
   Asphalt roofing
   Bricks and related clay products
   Calcium carbide manufacturing
   Castable refractories
   Portland cement manufacturing
   Ceramic clay manufacturing
   Clay and fly-ash sintering
   Coal cleaning
   Concrete batching
   Fiberglass manufacturing
     emissions and controls
   Frit manufacturing
   Glass manufacturing
   Gypsum manufacturing
   Lime manufacturing
   Mineral wool manufacturing

Perlite manufacturing
Phosphate rock processing
Sand and gravel processing
Stone quarrying and processing
9. **Petroleum industry**
    Petroleum refining
    Crude oil distillation
    Catalytic cracking
    Hydrocracking
    Catalytic reforming
    Polymerization, alkylation, and
    isomerization
    Hydrogen treating
    Chemical treating
    Physical treating

Blending
Natural gas processing
10. **Wood processing**
    Chemical wood pulping
    Pulpboard
    Plywood veneer and layout operations
    Woodworking operations
11. **Miscellaneous sources**
    Forest wildfires
    Fugitive dust sources
    Unpaved roads (dirt and gravel)
    Agricultural tilling
    Aggregate storage piles
    Heavy construction operations

SOURCE: Reference 24.

primary pollutants, however, is the additional time available for generation of secondary pollutants.

Air Basin   The concept of an air basin pertains to a large region that shares a common geographical area of sources and atmospheric interaction. The boundaries of an air basin are usually determined by mountains, large hills, and bodies of water. As a result of an EPA directive, the United States is divided into air basins.[32] For example, California has fourteen air basins, as shown in Fig. 3.4. The city of Los Angeles is located in the South Coast Air Basin.

Historically, local agencies charged with air pollution control have been established by county jurisdiction. The air basin concept recognizes that air quality transcends city and county boundaries and is a resource shared by a region. Since air quality programs at the regional level involve sociopolitical factors that are inconvenient, difficult, and often intractable in the absence of a regional agency, county agencies responsible for air pollution control are regularly encouraged to reorganize and form regional districts.

Regional, Subregional, and Local Air Quality   "Regional air quality" pertains to the air quality prevailing in the air basin. Regional air quality is influenced by the total emission of primary pollutants and generation of secondary pollutants throughout the air basin.

TABLE 3.4   Common Source Types

	HC	NO$_x$	CO	SO$_x$	Part
Point sources:					
Fossil-fueled electric power generating plants	x	o	x	o	o
Industrial boilers	o	o	o	o	o
Processing plants	o	x	x	x	o
Line sources:					
Highways, roadways	o	o	o	x	x
Aircraft	x	o	o	x	x
Railroads	x	o	o	o	o
Area sources:					
Indirect sources	o	o	o	x	x
Refineries	o	x	x	o	x
Residential tracts	x	o	x	o	x
Surface streets (aggregated)	o	o	o	x	x

o—primary emphasis
x—secondary emphasis

A spatial and temporal variation in regional air quality will always occur as a result of the spatial distribution of sources, meteorology, and topography in a given air basin. For example, while a prevailing wind from the ocean may maintain a high quality of air in communities adjoining the water, it will simultaneously transport primary pollutants from the ocean communities to the inland communities in the air basin and provide time for

Figure 3.4   Air basins. *(From Reference 14, p. 3.)*

secondary pollutants to form. For this reason, it is convenient to define the concept of a subregional air basin.

"Subregional air quality" pertains to the air quality prevailing in the subregional air basin. A subregional air basin is a subdivision of the air basin which encompasses a number of communities that share a uniform quality of air. The uniform quality of air will usually be based on topographical features of the area (such as enclosing hills), distribution of sources, and meteorology. In the California South Coast Air Basin, candidate areas for subregional air basins are shown in Fig. 3.5. The boundaries must reflect not only areas that share a uniform quality of air but also the availability of air monitoring data. Each subregional area should have at least one air monitoring station.

"Local air quality" pertains to that area in which a source increases the downwind concentrations above the upwind ambient concentrations (i.e., above the prevailing subregional air quality). An example of local air quality impact would be the increased pollutant concentrations immediately downwind of a freeway or large industrial source.

Meteorology   The importance of meteorology can be easily recognized by observing that the same amount and types of pollutants are emitted into the atmosphere of an air basin on a "good" air quality day as on a "bad" air quality day. The significant variable for these two days is the meteorology.

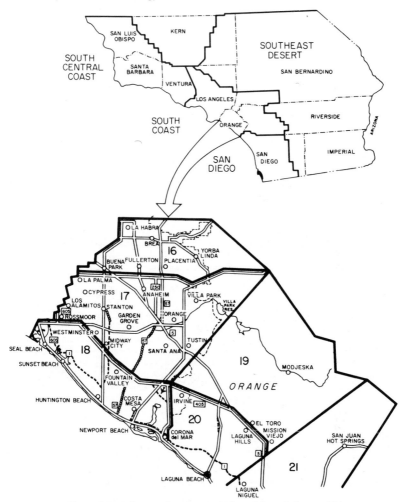

Figure 3.5   Subregional air basins. *(Adapted from Reference 96.)*

The principal meteorological variables on the regional level are horizontal convective transport (average wind speed and direction), and vertical convective transport (atmospheric stability, and the frequency, height, and strength of elevated inversions). On the local level, the principal variables are horizontal convective transport (wind speed and direction), vertical convective transport (atmospheric surface stability), and topography. Note that the difference between the regional and local level is the type of vertical convective transport.

Meteorology has introduced two terms to the air quality impact analysis terminology that are used interchangeably with "local" and "regional": microscale and mesoscale. Microscale is analogous to local and mesoscale to regional.

Horizontal Convective Transport.   "Horizontal convective transport" pertains to the pollutant dispersion resulting from microscale and mesoscale wind patterns. The wind is important to the dispersion of air pollutants horizontally and to determining the area of impact. An increase in the mean wind speed will lower the downwind ambient concentration of the pollutants by further diluting the pollutant emissions.

Certain frequently occurring wind patterns develop as a result of the topography of a region and the reoccurrence of certain meteorological conditions. Meteorological conditions control regional winds, sea and lake breezes, and land breezes. Large topographical features such as hills, canyons, and valleys influence wind behavior locally and promote channeling of winds, slope winds, and valley winds.

"Regional winds" are wind-flow systems produced by large-scale pressure patterns. A typical example is the prevailing westerly winds along the southern California coast which are traceable to the large anticyclone over the Pacific Ocean (see Fig. 3.13a). During certain periods of the year, another large pressure force can dominate, producing the famed easterly Santa Ana winds. In such cases, a high-pressure area exists within the highland plateaus with an associated low-pressure area in the lowlands. This creates a strong pressure gradient force. These large-scale pressure patterns can develop very strong wind speeds with gusts up to 50 mi/h.

"Sea" or "lake breezes" are produced when solar heating of a land surface adjacent to a ocean or large lake is much more rapid than the heating of the body of water. This results

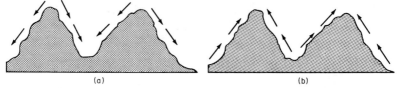

(a)   (b)

Figure 3.6   Valley airflows. (*a*) Nighttime airflow into valleys; (*b*) daytime airflow out of valleys.

in a temperature difference and an associated pressure difference between the air above the land surface and the air over the water. Because of the pressure gradient forces, the winds blow from the water towards the land. Sea or lake breezes are generally strongest during the summer months and are generally weaker during the winter months. The reason rests with the temperature of land surfaces which are highest during the summer months in contrast to the relatively constant temperatures of oceans and large lakes.

At night, the reverse will occur with the rapid radiative cooling of the land causing lower temperatures above the land surface than over the water. A "land breeze" may result, though the wind will generally be weaker than the lake or sea breeze. Land breezes are strongest during the winter months when nights are long and radiation cooling is at maximum.

"Slope winds" are associated with differential heating and cooling of hill and mountain slopes and valley floors. During the evening hours, radiation of heat from the earth's surface and consequent cooling of the ground and adjacent air cause density changes. The cooler, heavier air adjacent to the ground flows downhill under the influence of gravity. The downslope wind that results, called a drainage wind, is illustrated in Fig. 3.6a. The steeper the slopes of valleys, the stronger the downslope winds become. Vegetation will tend to reduce the flow due to frictional effects. Drainage winds are generally strongest during the winter months when nights are long and radiation cooling is at maximum. During daytime hours, the air adjacent to the ground becomes lighter and flows uphill. The upslope wind that results is illustrated in Fig. 3.6b.

"Valley winds" occur in well-defined valleys. At night, drainage winds flow down the valley sides, converge at the center, and flow toward the lower end of the valley. During daytime hours, weak flows up the center of the valley can also occur. When the general (regional) wind flow is directed approximately parallel to the valley, the surrounding hills can "channel" the wind through the valley. Channeling will often increase the local wind speed. Channeling can also result in a bidirectional wind frequency distribution.

The wind types just described act in a given area to establish wind patterns that are generally predictable but vary by season and time of day. Frequently occurring wind patterns for a given area reflect the cumulative effect of the wind types acting in the area. An example of late summer night and daytime regional wind patterns is presented in Fig. 3.7 for the California South Coast Air Basin.

A "wind rose" is defined as any one of a class of diagrams designed to show the distribution of wind direction experienced at a given location over an extended period of time. It thus shows the prevailing wind direction. The most common form consists of a circle from which sixteen lines emanate, one for each compass point. The length of each line is proportional to the frequency of wind from that direction, and the frequency of calm conditions is entered in the center. Many variations exist. Some indicate the range of wind speeds from each direction; some relate wind direction with other weather occurrences. Wind roses are constructed for a given period of time such as a particular day, month, season, or grouping of years. In constructing or interpreting wind roses, it is

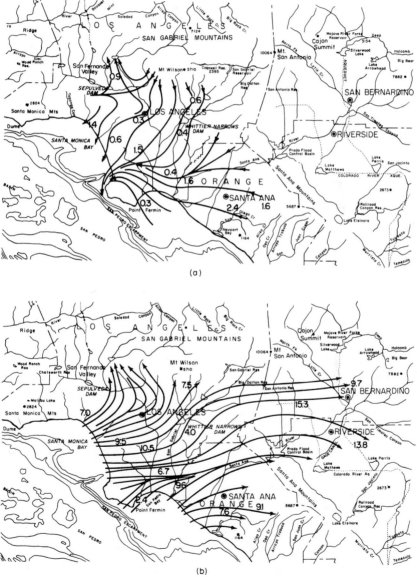

(a)

(b)

Figure 3.7   Wind patterns. (a) Typical nighttime (August, 0130 hours); (b) typical daytime (August, 1530 hours). The figures represent meters per second.

necessary to keep in mind the meteorological convention that *wind direction refers to the direction from which the wind is blowing.* A line or bar extending to the north on a wind rose indicates the frequency of winds blowing *from* the north, *not* the frequency of winds blowing toward the north.

An example of seasonal wind rose data is shown in Fig. 3.8. Wind rose data for other locations will be different and depend on local geographical and topographical features. A comparison of wind rose data with the local topography will often provide the evidence necessary to identify which of the wind types previously discussed are the dominant winds.

Vertical Convective Transport. The major determinant of vertical convective transport is the temperature gradient in the atmosphere which is called the lapse rate and given the symbol gamma, Γ, defined by

$$\Gamma = -\frac{dT}{dz} \tag{3.1}$$

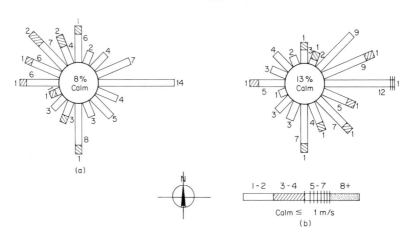

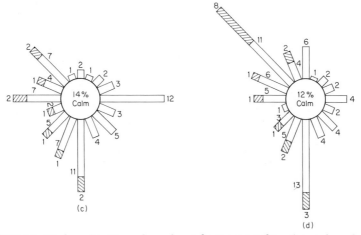

Figure 3.8   Wind rose data. Numerals near bars indicate percent of time that wind speed and direction prevail. (*a*) Autumn (September–November); (*b*) Winter (December–February); (*c*) Spring (March–May); (*d*) Summer (June–August).

In this section, the roles played by the lapse rate and factors associated with the lapse rate such as atmospheric stability, inversions, and measurement methods in the determination of vertical transport of pollutants are described.

LAPSE RATES. A discussion of lapse rates begins by a definition of three different types of lapse rates:

- *Environmental lapse rate,* Γ_{env}—The actual lapse rate in the atmosphere
- *Dry adiabatic lapse rate,* Γ_{dry}—The temperature history followed by a dry parcel of air as it rises or descends in the atmosphere due to buoyant or momentum forces
- *Saturated lapse rate,* Γ_{sat}—The temperature history followed by a parcel of air saturated with water vapor as it rises or descends in the atmosphere due to buoyant or momentum forces

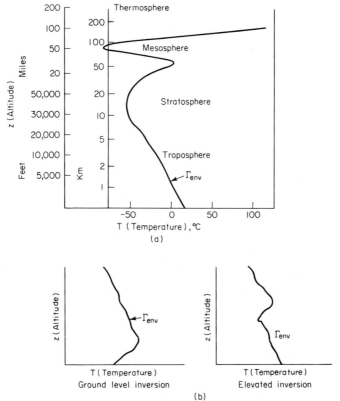

Figure 3.9 Atmospheric temperature profiles. (*a*) Prevailing temperature profile; (*b*) common temperature profiles near the ground. (*From Reference 21.*)

As shown in Fig. 3.9*a*, the prevailing environmental lapse rate is positive. That is, the prevailing temperature decreases with altitude in the lower atmosphere (0 to 5 km). Close to ground level, the environmental lapse rate often deviates from the prevailing. For example, the temperature may increase for a few hundred feet before joining the prevailing rate. In Fig. 3.9*b*, two examples of temperature increasing with altitude are shown. A temperature increasing with elevation (negative environmental lapse rate) is referred to as an "inversion" because the temperature profile is inverted from the prevailing environmental lapse rate.

To see how the environmental lapse rate promotes or inhibits vertical mixing, consider a parcel of combustion effluent emitted from an automobile. Our automobile is shown in Fig. 3.10 along with the environmental lapse rate. The parcel is exhausted at 35°C. The parcel buoyancy is positive because the parcel temperature is higher than the surrounding temperature. That is, the parcel will rise like a hot-air balloon.

As the parcel rises, the temperature of the parcel will change as a result of:

- *Heat transfer*—The temperature difference between the parcel and the surroundings will promote the transfer of energy from the parcel to the surroundings. All other factors being equal, the temperature of the parcel will decrease.

- *Expansion*—The pressure of the atmosphere decreases with altitude. As a result, the parcel will expand as it rises in the atmosphere. All other factors being equal, the expansion will cause the temperature of the parcel to decrease.

- *Condensation*—As the temperature of the parcel decreases, moisture in the parcel will condense to maintain saturation conditions. The condensation of the moisture is accompanied by a release of energy within the parcel equivalent to the enthalpy of

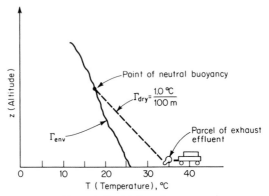

Figure 3.10 Adiabatic lapse rate.

condensation. All other factors being equal, the energy release will reduce the rate of temperature decrease.

Consider for the moment that the hot parcel is dry (i.e., no condensation), rises adiabatically (no heat transfer), and has the properties of air. (Such assumptions are not valid for combustion effluents but are made here for the sake of simplified discussion.) Our goal is to predict the temperature change due to expansion alone as the parcel rises. Thermodynamics provides the route for the analysis and examples of the approach are available (e.g., Reference 117). The result of the analysis is that a dry parcel of air rises or descends in the atmosphere with the following lapse rate:

$$\Gamma_{dry} = \frac{1°C}{100\ m} = \frac{5.4°F}{1000\ ft} \tag{3.2}$$

The dry adiabatic lapse rate is shown in Figure 3.10 as a dashed line. The line represents the temperature history that the parcel will experience as it rises and expands given the two constraints: (1) the parcel is dry, and (2) adiabatic conditions prevail. At some point, the parcel temperature becomes equivalent to the temperature of the surroundings. At the point of intersection, the parcel becomes neutrally buoyant relative to the surroundings, and the upward rise of the parcel terminates.

The dry adiabatic lapse rate is a convenient "yardstick" for evaluating the intensity of vertical convective mixing. However, the actual lapse rate of the parcel will depend upon the moisture content and the heat loss. Of the two, only the condensation of the moisture acts to increase the temperature. When condensation is taken into account, the lapse rate remains positive but the rate of temperature decrease is lowered. The so-called "saturated lapse rate" is approximately:

$$\Gamma_{sat} = \frac{0.6°C}{100\ m} = \frac{3.4°F}{1000\ ft} \tag{3.3}$$

In the following illustrations, the dry adiabatic lapse rate is used to describe the temperature history of the parcel. This follows convention in that it is assumed that the dry adiabatic lapse rate is the more representative of an actual parcel because heat transfer will partially offset the effects of condensation.

ATMOSPHERIC STABILITY. The effectiveness of the atmosphere in mixing or dispersing pollutants in the vertical direction is dependent on the "atmospheric stability." A stable atmosphere is one that suppresses vertical mixing. An unstable atmosphere is one which promotes vertical mixing. The difference between the two is governed by the environmental lapse rate near ground level.

Examples of the variation of the environmental lapse rate near ground level are presented in Fig. 3.9b. The relationship between atmospheric stability and environmental lapse rate is shown in Fig. 3.11.

The unstable case is characterized by an environmental lapse rate that is highly positive and diverges from the dry adiabatic lapse rate. As the parcel rises, the buoyancy of the parcel becomes more positive and vertical mixing is enhanced.

The neutral case is characterized by an environmental lapse rate that parallels the dry adiabatic lapse rate. As the parcel rises, the buoyancy of the parcel remains invariant.

The stable case is characterized by an environmental lapse rate with a slope less than the dry adiabatic lapse rate. As the parcel rises, the buoyancy of the parcel becomes less positive and vertical mixing is suppressed. At some point, the temperature of the parcel becomes equivalent to the environmental temperature and the vertical rise is terminated.

The strongly stable case is characterized by an environmental lapse rate that is negative relative to the adiabatic lapse rate. Not only does the buoyancy of the parcel decrease at a rapid rate as the parcel rises, but the extent of vertical mixing is sharply limited as well. This condition, as noted before, is popularly referred to as an "inversion" and represents the case for which vertical mixing is most restricted.

INVERSIONS. Inversions are typically of two types: ground level and elevated. The lapse

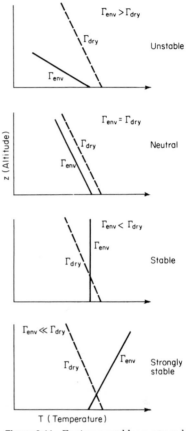

Figure 3.11 Environmental lapse rate and atmospheric stability.

rates for these two inversions are shown in Fig. 3.9b. The formation of the ground level inversion is illustrated in Fig. 3.12. The inversion is formed during nighttime hours as the ground level temperature decreases due to radiative heat loss to the universe. The air adjacent to the ground cools by conductive heat transfer. As the sun rises in the morning, the ground is warmed. The inversion is gradually weakened and is eventually destroyed. The ground level inversion is called a radiation inversion to acknowledge the mechanism of formation.

Elevated inversions are caused by the subsidence of air masses in high-pressure cells and the subsequent passage over warm land or water masses which heat the lower portions of these air masses. This process, shown in Fig. 3.13, results in a warm air mass above a cooler air mass. Figure 3.13a traces the history of a given air parcel from $t = t_0$ to $t = t_3$. As air is forced from the cell, the stream lines diverge at ground level and the velocity of the parcel accelerates horizontally. As a result, the height of the parcel of air subsides (descends) to fill the void left by the lateral spreading of air at ground level. As the parcel subsides, the air within is compressed and is thereby heated. The elevation change, hence the compression and temperature increase, is greater for the higher portions of the air mass. Figure 3.13a illustrates the temperature increase within the parcel assuming dry adiabatic lapse rate behavior. As the parcel moves further from the center of the cell, the inversion becomes more intense. This type of an inversion is termed a "subsidence inversion" and is formed over a distance approaching thousands of kilometers.

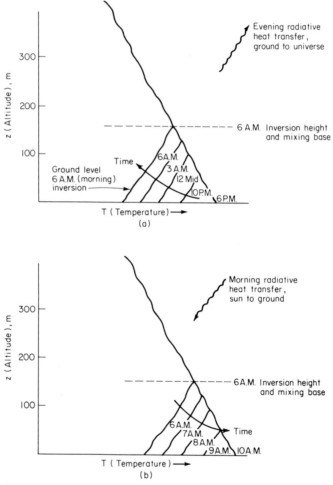

Figure 3.12   Ground-level radiation inversion. (a) Formation; (b) destruction.

The high-pressure cell responsible for the formation of the intense subsidence inversion in southern California is located over the Pacific Ocean. The cell is also responsible for the intense subsidence inversions experienced in the Orient.

A common occurrence as the parcel moves away from the cell is for the ground temperature to change. In the case of the semipermanent Pacific cell, the ocean temperature increases. The effect is shown in Fig. 3.13b. The profile developed is typical of the lapse rate observed in urban areas impacted by a subsidence-generated inversion.

As shown in Fig. 3.13c, the base of an elevated inversion is called the "mixing height" and the difference between the mixing height and ground level is called the "mixing depth." The atmospheric stability under an elevated inversion can vary and the dispersion of pollutants *within* the mixing volume can be restricted (for a stable lapse rate) or be relatively unrestricted (for an unstable lapse rate). In either event, the presence of the elevated inversion will suppress mixing above the mixing height. In contrast to the radiation inversion, subsidence inversions may persist for long periods (several days), whereas radiation inversions break up within hours after daybreak.

MEASUREMENT METHODS. Lapse rates are measured daily (sometimes twice daily) in most urban areas using "radiosondes." A radiosonde is a vessel about the size of a quart jar

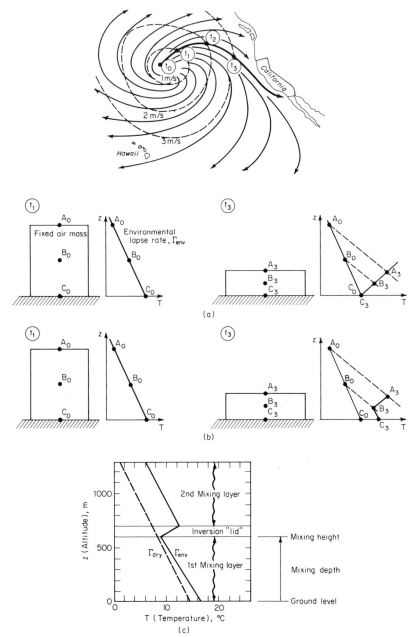

Figure 3.13 Elevated inversion. (*a*) Constant ocean temperature; (*b*) increasing ocean temperature; (*c*) elevated inversion. (*Adapted from Reference 3.*)

that contains sensors for temperature, pressure, and humidity, and a radio transmitter. The radiosonde is released from ground level attached to a helium-filled balloon and is allowed to rise through the atmosphere. The pressure, temperature, and humidity are measured simultaneously as the radiosonde rises, and the data are transmitted back to a

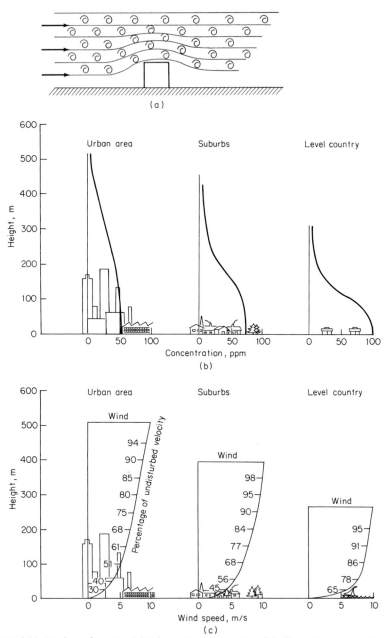

Figure 3.14 Mechanical mixing. (a) Schematic representation; (b) effect of terrain roughness on concentration profile; (c) effect of terrain roughness on wind speed profile. (*Adapted from Reference* 3.)

ground station. The pressure data are translated into altitude information. This altitude information and the temperature data are used to calculate the lapse rate.

New techniques for measuring the inversion height are gaining popularity. The most promising is the acoustic sounder. Located at ground level, the sounder transmits a pulsed, acoustic signal vertically. A discontinuity in atmospheric density (e.g., an inversion base) will produce a reflected pulse. The elapsed time between the transmission of the pulse and the return signal identifies the inversion height. The acoustic sounder has three distinct advantages over radiosondes. The capital investment and cost of operation is relatively inexpensive. The sounder can be run continuously. And sounders may be spatially distributed throughout a basin to measure the spatial and temporal variation in inversion height. A disadvantage to the sounder is the lack of detailed information on the lapse rate provided. One solution is to use sounders in combination with radiosondes.

Mechanical Mixing. In addition to large topographical features, the roughness of the terrain may promote vertical and lateral mixing as well. Such mixing, illustrated in Fig. 3.14a, is called mechanical mixing. Mechanical mixing depends on (1) the terrain roughness, and (2) the wind speed at groundlevel. The effect of terrain roughness on the pollutant-concentration gradient is shown in Fig. 3.14b. As the roughness increases, the effect of local turbulent dispersion increases and vertical mixing is enhanced. Offsetting this improved vertical mixing is a decrease in wind dilution caused by the decrease in ground level wind speed that accompanies an increase in surface roughness. The effect of terrain roughness on wind speed gradient is shown in Fig. 3.14c. As the roughness increases, the wind gradient becomes less steep and wind velocities close to the ground decrease. As a result, strong winds are required to both promote mechanical mixing and maintain reasonable wind dilution.

By varying building heights and by proper building spacing, the effective surface roughness can be increased. The subsequent increase in turbulence will enhance vertical and horizontal mixing. Tall buildings increase the surface roughness of the terrain and result in increased surface turbulence and increased surface wind friction. However, if proper account is not made of building spacing, pollutant dispersion in and about the pockets formed between buildings can be suppressed and result in the buildup of adverse levels of pollutant species known as the "street canyon effect."

In addition to the proper spacing and height of buildings, additional means are available to promote pollutant dispersion. For example, a building on stilts will induce convection at ground level and disperse locally produced pollutants (e.g., from automobile traffic). The effect of wind gusts upon pedestrians is a tradeoff. Pedestrian comfort, penalized as dispersion is improved, limits the extent to which building design can be used to promote dispersion.

Chemical Change The primary chemical change occurring in the atmosphere pertains to the formation of photochemical oxidant.* Oxidant formation involves a multistep chemical reaction mechanism. A simplistic model is shown in Table 3.5. The basic steps underlying oxidant formation are the interaction of oxides of nitrogen, reactive hydrocar-

TABLE 3.5   Oxidant Formation—Simplistic Model

NO_2	+	Light	→	NO	+	O
O	+	O_2	→	O_3		
O_3	+	NO	→	NO_2	+	O_2
O	+	HC	→	HCO^-		
HCO^-	+	O_2	→	HCO_3^-		
HCO_3^-	+	HC	→	Aldehydes, ketones, etc.		
HCO_3^-	+	NO	→	HCO_2^-	+	NO_2
HCO_3^-	+	O_2	→	O_3	+	HCO_2^-
HCO_2^-	+	NO_2	→	Peroxyacyl nitrates		

SOURCE: Reference 2.

*The terms "smog" or "photochemical smog" are labels often used to describe photochemical oxidant. Smog is an acronym coined in London, England in 1881 from a coupling of smoke and fog. The term smog has been popularly applied to the air pollution that besets Los Angeles though the air quality problem is generally created from mechanisms other than smoke and fog. As a result, the term smog lacks a clear definition.

bons, and ultraviolet light to produce ozone and other oxidants. The formation processes are shown schematically in Fig. 3.15. The role of time is illustrated in Fig. 3.15a for the major constituents. It takes time for oxidant to form once the reactants are present. In Fig. 3.15b, the role of some of the intermediates is shown as well as the role sulfur dioxide can play in the process to form visibility degrading aerosol. The pollutant species that constitute oxidant are shown in Fig. 3.3 and are listed in Table 3.6.

Despite the many species included in the oxidant family, ozone is often accepted (incorrectly, but understandably) as the index of photochemical oxidant. The reason? Acceptable measurement techniques are not yet available to readily measure species other than ozone. Ozone is the oxidant constituent commonly measured.

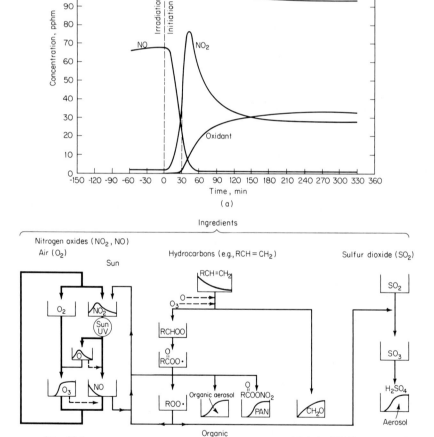

Figure 3.15 Oxidant formation. (a) Formation of oxidant in a controlled chamber; (b) schematic of oxidant formation. (*Adapted from Reference 99.*)

Our present understanding of the chemical kinetic mechanisms and photochemistry is a result of extensive smog chamber research efforts.[73] These research efforts are continuing in an effort to fill in information gaps in the oxidant formation process. Although these research results have provided a rudimentary understanding of the chemical mechanism occurring in the atmosphere, the actual atmospheric process is complicated by several additional variables. These variables include: (1) the role of trace pollutants not accounted for in the smog chamber; (2) the nonuniform addition of reacting pollutants into the mixture as oxidant is undergoing formation; (3) the residual oxidant present from the previous days; and (4) the role of meteorology in transporting the air parcel in which oxidant is being formed and in dispersing the air parcel once oxidant has been formed.

TABLE 3.6   Constituents of Oxidant

O_3 (ozone): Primary constituent. Colorless gas with strong odor. Irritates mucous membranes at common levels, can cause coughing, fatigue and interference with lung function at higher levels. Causes severe plant damage, cracking in rubber, and metal corrosion.

NO_2 (nitrogen dioxide): Light-brown gas with odor of bleach. Primary cause of brownish haze. Irritates eyes and nose and may cause increased susceptibility to infectious diseases. Restricts plant growth.

Aerosols: Micron-sized solid and liquid particles suspended in the air. Composition varies and is not well characterized. Primary cause of severe reduction of visibility on "smoggy" days. Carrier of other irritants (e.g., absorbed gases) into respiratory system.

PaN (peroxyacyl nitrates): A family of five compounds found in relatively low concentrations in the atmosphere (parts per billion compared to 100 parts per billion for ozone). Causes severe plant damage. Eye irritation attributed to some species. Effects on the public health not well established but likely to be significant and detrimental.

Other species: Acrolein, formaldehyde, and a multitude of additional species that await to be separated and characterized.

Air Pollutant Effects

Without effects, air pollution would not exist and assessing the impact on air quality would be unnecessary. Unfortunately, pollutants do produce adverse effects at levels frequently exceeded in the ambient air.

Health-related air pollution effects data come from animal exposure, epidemiological, and to a limited extent, human exposure studies. Active research is being pursued to (1) better establish the levels at which adverse effects are observed, (2) develop the correlation between animal and human response, (3) obtain more epidemiological information, and (4) bridge the information gaps and reduce the uncertainty associated with the current standards. In addition to evaluating the more common pollutants, research is directed to assess the synergistic effects of pollutants acting in combination and to assess the health-related effects of pollutants yet to be discovered or characterized (most notably, organic aerosols).

The current federal ambient air quality standards are presented in Table 3.7. Primary and secondary ambient air quality standards have been adopted for six pollutants: ozone, carbon monoxide, nitrogen dioxide, sulfur dioxide, nonmethane hydrocarbons, and particulate matter. The primary standards define the level of air quality to protect the public health and, in general, are set to protect that portion of the population (approximately 15 to 20 percent) especially susceptible to air pollutants. Secondary standards define the level of air quality which will protect the public welfare (e.g., materials, vegetation) from any known or anticipated adverse effects of air pollution.

The information upon which the federal ambient air quality standards are based is available in a series of documents.[23,74–78] The National Ambient Air Quality Standards (NAAQS) have been reviewed by the National Academy of Sciences[20,68] and the EPA.[43] The latter review resulted in the first change to the NAAQS since promulgation in 1971. The ambient air quality standard for oxidant was changed in 1979 to a standard specific to ozone, and the standard was relaxed from 0.08 ppm/1-hour to 0.12 ppm/1-hour.[44] The change from oxidant to ozone was primarily a result of the fact that, as mentioned earlier, ozone has been and continues to be the oxidant constituent commonly measured.

As shown in Table 3.7, the standards are expressed as concentrations averaged over fixed-time periods. Examples are 1-hour, 3-hour, 8-hour, and 24-hour averages. The annual arithmetic mean (AAM) and annual geometric mean (AGM) are also used for certain pollutants. In addition to the federal ambient air quality standards, an individual state may have in force an independent set of ambient air quality standards.

The ambient air quality standards serve a vital role in air quality impact analyses. The

TABLE 3.7 National Ambient Air Quality Standards

Pollutant	Averaging time	Primary* standard	Secondary† standard	General objectives
Ozone	1 hr	240 μg/m^3 (0.12 ppm)	240 μg/m^3 (0.12 ppm)	To prevent eye irritation and possible impairment of lung functions in persons with chronic pulmonary disease, and to prevent damage to vegetation.
Carbon monoxide	8 hr	10 mg/m^3 (9 ppm)	10 mg/m^3 (9 ppm)	To prevent interference with the capacity to transport oxygen to the blood.
	1 hr	40 mg/m^3 (35 ppm)	40 mg/m^3 (35 ppm)	
Nitrogen dioxide	Annual average	100 μg/m^3 (0.05 ppm)	100 μg/m^3 (0.05 ppm)	To prevent possible risk to public health and atmospheric discoloration
Sulfur dioxide	Annual average	80 μg/m^3 (0.03 ppm)	. . .	To prevent pulmonary irritation.
	24 hr	365 μg/m^3 (0.14 ppm)	. . .	
	3 hr	. . .	1300 μg/m^3 (0.5 ppm)	To prevent odor.
Suspended particulate matter	Annual geometric mean	75 μg/m^3	60 μg/m^3	To prevent health effects attributable to long continued exposures.
	24 hr	260 μg/m^3	150 μg/m^3	
Hydrocarbons (corrected for methane)	3 hr	160 μg/m^3 (0.24 ppm)	160 μg/m^3 (0.24 ppm)	To reduce oxidant formation.

*National Primary Standards: The levels of air quality necessary, with an adequate margin of safety, to protect the public health.

†National Secondary Standards: The levels of air quality necessary to protect the public welfare from any known or anticipated adverse effects of a pollutant.

SOURCE: Reference 29, p. 8186, and Reference 44.

standards provide the benchmark against which the impact is assessed. What is required now is the means for quantifying the impact.

ENVIRONMENTAL IMPACTS

Air Quality Modeling

The air quality impact of a source or collection of sources is evaluated by the use of models. The elements of a model are shown in Fig. 3.16. Simply stated, models simulate

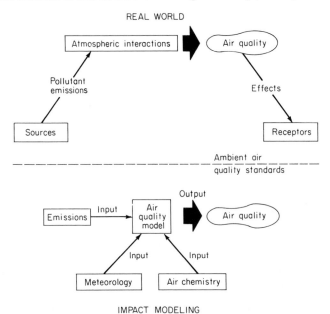

IMPACT MODELING

Figure 3.16 Relation between impact analysis and the real world.

the relationships between air pollutant emissions and the resulting impact on air quality. The inputs to the model include emissions, meteorology, and air chemistry, all of which are determined by formulating impact scenarios. The development of impact scenarios is addressed at the conclusion of this section. The discussion of modeling begins with an introduction to air quality models.

Air Quality Models When pollutants are emitted into the atmosphere, they are immediately diluted, transported, and mixed with the surrounding air. The role of air quality modeling is to represent these processes mathematically. Begin by considering a simple example.

Box Models. Consider that pollutants are being emitted into a volume of air that is bounded by imaginary walls as shown in Fig. 3.17. The box has a depth D, a width W, and an infinite length. Air is blowing through the box at a velocity $\bar{u}$. If we assume steady-state conditions and the pollutants mix very rapidly with the air, then the concentration throughout the box will be uniform and invariant with time, and the concentration is given by the following expression:

$$C_j = \frac{Q_j}{\bar{u}WD} \tag{3.4}$$

where $\bar{u}$ = wind velocity assumed constant, m/s
W = width of box normal to the wind direction, m
D = depth of box normal to the wind direction, m
Q_j = emission rate of species j, g/s

The box model makes a few useful points. First, it reminds us that a limited resource of air is available for diluting pollutants. Second, it tells us that concentration will increase as the volume of air is reduced or as the emission rate increases. Although the box model is not usually acceptable for formal air quality analysis, it is useful for qualitative estimates of a source impact.

One reason the box model is not acceptable for most analyses rests with its failure to account for dispersion of the pollutants laterally and vertically. The assumption of instantaneous mixing to bounded walls is not realistic. Simulation of the dispersion processes that actually occur in the atmosphere requires the use of dispersion models.

Dispersion Models.   Dispersion models are formulated from the fundamental differential equations governing the conservation of species. Dispersion models are more appropriate for the prediction of air quality because the models consider the point-by-point transport, dispersion, generation, and removal of pollutant species, and provide for spatial and temporal variation of these processes.

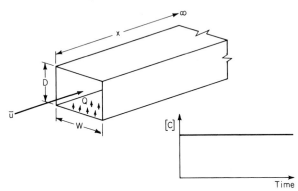

Figure 3.17   Box model.

The first step is to construct the basic mathematical equations upon which the models are founded. Consider the small control volume shown in Fig. 3.18. If we assume that a change in species concentration within the control volume can occur by transport and dispersion of the pollutant species within the volume, then the conservation of species (mass balance for species j) may be expressed in the form:

$$\frac{\delta}{\delta t}(VC_j) + \frac{\delta}{\delta x}(C_j\overline{u}\;dydz)\,dx + \frac{\delta}{\delta y}(C_j\overline{v}\;dzdx)\,dy + \frac{\delta}{\delta z}(C_j\overline{w}\;dxdy)\,dz$$
$$-\frac{\delta}{\delta x}\left(D_x\,dzdy\,\frac{\delta C_j}{\delta x}\right) - \frac{\delta}{\delta y}\left(D_y\,dzdx\,\frac{\delta C_j}{\delta y}\right) - \frac{\delta}{\delta z}\left(D_z\,dxdy\,\frac{\delta C_j}{\delta z}\right) - \emptyset_j\,dxdydz = 0$$

(3.5)

where V = the volume of the control volume, m³
   C_j = the concentration of pollutant species, kg/m³

   $\overline{u}$ = the average convective velocity in the x direction, m/s

   $\overline{v}$ = the average convective velocity in the y direction, m/s

   $\overline{w}$ = the average convective velocity in the z direction, m/s
   D_x = the turbulent mixing coefficient associated with the x direction, m²/s
   D_y = the turbulent mixing coefficient associated with the y direction, m²/s
   D_z = the turbulent mixing coefficient associated with the z direction, m²/s
   $\emptyset_j$ = a local term for the creation or removal of pollutant species j, kg/s
Now, if we recognize that $V = dxdydz$, Eq. 3.5 can be simplified:

$$\frac{\delta C_j}{\delta t} + \frac{\delta}{\delta x}(\overline{u}\,C_j) + \frac{\delta}{\delta y}(\overline{v}\,C_j) + \frac{\delta}{\delta z}(\overline{w}\,C_j) - \frac{\delta}{\delta x}\left(D_x\frac{\delta C_j}{\delta x}\right)$$
$$-\frac{\delta}{\delta y}\left(D_y\frac{\delta C_j}{\delta y}\right) - \frac{\delta}{\delta z}\left(D_z\frac{\delta C_j}{\delta z}\right) - \emptyset_j = 0 \qquad (3.6)$$

Further, if we consider the mixing coefficients, D_x, D_y, and D_z, to be approximately constant at a given time and location, Eq. 3.6 may be written as:

$$\frac{\delta C_j}{\delta t} + \bar{u}\frac{\delta C_j}{\delta x} + C_j\frac{\delta\bar{u}}{\delta x} + \bar{v}\frac{\delta C_j}{\delta y} + C_j\frac{\delta\bar{v}}{\delta y} + \bar{w}\frac{\delta C_j}{\delta z} + C_j\frac{\delta\bar{w}}{\delta z}$$

$$- D_x\frac{\delta^2 C_j}{\delta x^2} - D_y\frac{\delta^2 C_j}{\delta y^2} - D_z\frac{\delta^2 C_j}{\delta z^2} - \varnothing_j = 0 \qquad (3.7)$$

Application of the continuity equation for an incompressible fluid, $\dfrac{\delta\bar{u}}{\delta x} + \dfrac{\delta\bar{y}}{\delta y} + \dfrac{\delta\bar{w}}{\delta z} =$ 0, reduces Eq. 3.7 to the "basic dispersion equation" as follows:

$$\frac{\delta C_j}{\delta t} + \bar{u}\frac{\delta C_j}{\delta x} + \bar{v}\frac{\delta C_j}{\delta y} + \bar{w}\frac{\delta C_j}{\delta z} - D_x\frac{\delta^2 C_j}{\delta x^2} - D_y\frac{\delta^2 C_j}{\delta y^2} - D_z\frac{\delta^2 C_j}{\delta z^2} - \varnothing_j = 0 \qquad (3.8)$$

Equation 3.8 is the basic expression used to construct dispersion models for pollutant species interactions in the atmosphere. Solutions to Eq. 3.8 are not readily obtained

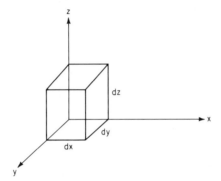

Figure 3.18   Control volume for mass balance.

without major assumptions. In fact, the models currently used to describe pollutant species transport and dispersion in the atmosphere differ significantly in the assumptions used to obtain a solution to Eq. 3.8.

The primary dispersion models constructed from Eq. 3.8 include Gaussian Models and direct numerical solutions.

GAUSSIAN MODELS. Gaussian models are constructed by application of the following assumptions to Eq. 3.8:

(1) $\dfrac{\delta C_j}{\delta t} = 0$ steady state condition; i.e., the pollutant concentration at a given point in space (x, y, z) is constant.

(2) $\bar{v}$, $\bar{w} = 0$ wind speed, $\bar{u}$, taken as uniform in space and invariant in time.

(3) $\bar{u}\dfrac{\delta C_j}{\delta x} \gg D_x\dfrac{\delta^2 C_j}{\delta x^2}$ transport in windward or x-direction controlled by convection.

Application of the above assumptions reduces Eq. 3.8 to the following form:

$$\underbrace{\bar{u}\frac{\delta C_j}{\delta x}}_{\text{Convection term}} = \underbrace{D_z\frac{\delta^2 C_j}{\delta z^2} + D_y\frac{\delta^2 C_j}{\delta y^2}}_{\text{Dispersion terms}} + \underbrace{\varnothing}_{\text{Source/removal term}} \qquad (3.9)$$

Equation 3.9 consists of a convection term, dispersion terms, and a source/removal term. The concentration (C_j) at a point (x, y, z) downwind of the source depends upon the convection of the species by wind transport, the lateral and vertical spread of species by turbulent dispersion, and the generation or removal of the species by chemical reaction.

Equation 3.9 may be solved if the following additional assumptions are applied:

- *Key Assumption:* The concentration in the x, z planes follow a Gaussian (normal) distribution in each of the two dimensions (Fig. 3.19).
- The pollutant species j is emitted from the source at a uniform rate, Q_j.
- The generation or removal of species within the flowfield is zero (i.e., nonreacting flowfield).

The application of the above assumptions to Eq. 3.9 results in the so-called Gaussian dispersion model depicted here by the following expression for a point source:

$$C_j(x,y,z) = \frac{Q_j}{2\pi \bar{u}\sigma_y\sigma_z} \exp\left[-\frac{1}{2}\left(\frac{y^2}{\sigma_y^2} + \frac{z^2}{\sigma_z^2}\right)\right] \tag{3.10}$$

where Q_j = emission rate of species j emitted by the source, g/s
σ_y = Gaussian coefficient for lateral dispersion, m
σ_z = Gaussian coefficient for vertical dispersion, m

The dispersion model represented by Eq. 3.10 is illustrated in Fig. 3.19a. The plume has a conical geometry. The assumption of a Gaussian distribution not only allows a solution to Eq. 3.8, but also allows the dispersion effects due to turbulent mixing to be conveniently incorporated into the model. Field observations are used to determine values of the Gaussian coefficients in the turbulent atmosphere.

Special versions of Eq. 3.10 are required for most air quality impact analyses. For example, point sources are typically ground level or elevated above ground level. For both cases, the ground serves as a barrier to dispersion and must be taken into account. For a source at ground level with perfect reflection, the Gaussian plume equation becomes:

$$C_j(x,y,z) = \frac{Q_j}{\pi \bar{u}\sigma_y\sigma_z} \exp\left[-\frac{1}{2}\left(\frac{y^2}{\sigma_y^2} + \frac{z^2}{\sigma_z^2}\right)\right] \tag{3.11a}$$

and, for a source elevated a distance H above ground level, the Gaussian plume equation becomes:

$$C_j(x,y,z) = \frac{Q_j}{2\pi \bar{u}\sigma_y\sigma_z} \exp\left(-\frac{y^2}{2\sigma_y^2}\right) \left[\exp\left(-\frac{(z-H)^2}{2\sigma_z^2}\right) + \exp\left(-\frac{(z+H)^2}{2\,\sigma_z^2}\right)\right] \tag{3.11b}$$

Both these expressions are derived on the basis that the ground is a perfectly reflecting surface.

A second example of a specialized version of Eq. 3.10 is used to model roadways. In Gaussian modeling, roadways are represented as a line source (an infinite series of point sources positioned side-by-side in a line). For a line source, the Gaussian plume equation becomes:

$$C_j(x,z) = \frac{2Q_j/L}{(2\pi)^{1/2}\bar{u}\sigma_z} \exp\left(-z^2/2\sigma_z^2\right) \tag{3.11c}$$

where Q_j/L is the emission per unit length of roadway.

The plumes produced by Eqs. 3.11a, 3.11b, and 3.11c are illustrated in Figs. 3.19b, 3.19c, and 3.19d, respectively, for a ground level point source, an elevated point source, and a line source. Notice that the Gaussian coefficients σ_y and σ_z increase in the downwind or x direction.

Application of these Gaussian plume models is the subject of the section on Air Quality Modeling presented later in this chapter.

DIRECT NUMERICAL SOLUTIONS. The limitations of the Gaussian plume and the box-model approaches, coupled with the increasing need for more precise representations of air quality (including chemical reactions), have prompted a move toward the numerical solution of the conservation-of-mass equation on a fixed three-dimensional grid.

Dispersion models that utilize a "direct numerical solution" provide the most detailed information, but they require the most extensive data base (geocoded emission inventories and detailed meteorological information). Since more complex meteorological conditions can be considered, the model can thus be applied to a greater variety of adverse situations occurring in the atmosphere over urban centers.

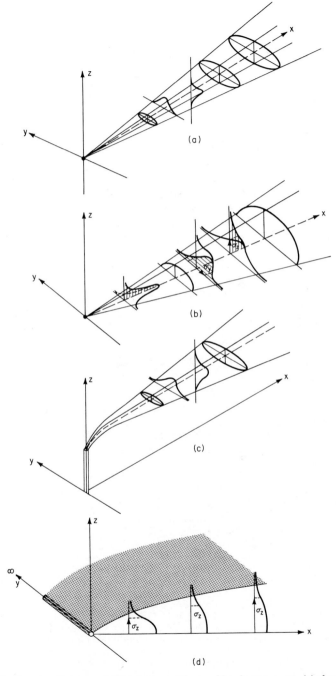

Figure 3.19 Gaussian dispersion. (*a*) Point source; (*b*) ground level point source; (*c*) elevated point source; (*d*) line source.

Above-average technical expertise is required to operate numerical models, and the necessary computer time is costly. The use of the models is most appropriate for large-area analyses, though applications to individual line and point sources are possible for complex geometries that lay beyond the reach of the less sophisticated models.

From a trend analysis point of view, the grid models are preferable since they output the region-wide air quality. Air pollution patterns may then be recognized, and trends may be more precisely analyzed. Further, comparison of the National Ambient Air Quality Standards to predicted values may be most appropriately addressed by a regional display of expected air quality.

Air Quality Impact Scenarios   As shown in Fig. 3.16, air quality is determined by a combination of pollutant emissions and atmospheric interaction. Combinations of emissions and atmospheric interactions that produce air quality impacts are referred to as impact scenarios. Before modeling can be conducted, impact scenarios must be established. The scenarios are required to identify those conditions that will be modeled and to identify the emissions, meteorology, and air chemistry inputs required by the models.

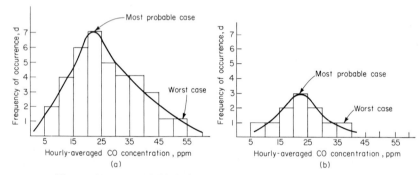

Figure 3.20   Most probable and worst case impact. (a) Winter; (b) summer.

The following two impact scenarios are typical of those employed in an air quality impact analysis:

    • *Most probable case:* That combination of emissions and atmospheric interactions that produces the most frequently encountered air quality impact.

    • *Worst case:* That combination of emissions and atmospheric interactions that produces the worst air quality impact.

The most probable case is required to evaluate the prevailing air quality impact, whereas the worst case is required to evaluate the maximum impact.

The two cases are illustrated in Fig. 3.20. The maximum hourly concentrations of carbon monoxide and the frequency of occurrence are plotted. The most probable and worst case impacts are noted on the plot.

The impact shown in Fig. 3.20 is caused by a combination of emissions and atmospheric interaction. But what causes the distribution shown? Meteorology is the culprit. From day to day, during the work week, the spatial and temporal variation in emissions remains relatively unchanged. The meteorology, however, changes. Daily variations in wind speed and direction, atmospheric stability, and mixing height are responsible for the distribution.

Despite the daily variation in meteorology, seasonal trends can be established. This is illustrated by comparing Fig. 3.20a to Fig. 3.20b. The frequency distribution shown in Fig. 3.20a is for winter. The distribution in Fig. 3.20b is for summer. Why is there a difference? The reason rests with the seasonal variation in atmospheric stability and insolation. The most unfavorable meteorological conditions for primary pollutant impact exist in early morning and late evening due to the presence of radiation inversions. Radiation inversions are most persistent in winter. These surface-based inversions last until mid-morning when the incoming solar radiation warms the ground surface and burns off the inversion (Fig. 3.12). During the winter months, high concentrations of primary pollutants (e.g., carbon monoxide) can develop. The net result is that the most probable and worst case impact scenarios must be evaluated for each season. Analysis on a monthly basis is proportionately more expensive and is normally unnecessary except for the most unusual projects (e.g., a project for which substantial changes in monthly emission rate are expected).

Methodologies for analyzing emissions, meteorology, and air chemistry, and methodologies for determining the most probable and worst case impact scenarios for both primary pollutant and secondary pollutant impact are presented in the section on Impact Scenarios later in this chapter.

Summary

Air quality models are used to describe atmospheric interactions. The models are designed to relate emissions of primary pollutants to the resulting downwind air quality. The models appropriate for air quality impact analyses vary in sophistication from direct numerical solution of the basic dispersion equation to the box model.

A word of caution is necessary at this point. Models are not absolute. The performance of a model depends upon the user, the input data, the model, and the application. The user needs to (1) be experienced (or, in the absence of experience, solicit guidance and perspective from an experienced user) and (2) recognize that the air quality models are only indicative of the cause-effect relationship between air quality and the emissions of primary pollutants—despite the apparent sophistication. The complexity of atmospheric meteorology, the mechanics of atmospheric stability, and the uncertainty of atmospheric chemistry clearly demonstrate the rationale of the preceding cautionary advice.

LEGAL ASPECTS

Till more effectual methods can take place, it would be of great service, to oblige all those Trades who make use of large Fires, to carry their Chimneys much higher into the air than they are at present; this expedient would frequently help to convey the Smoke away above the buildings, and in a great measure disperse it into distant parts without its falling on the houses below.

Workmen should be consulted, and encouraged to make experiments, whether a particular construction of the Chimneys would not assist in conveying off the Smoke, and in sending it higher into the air before it is dispersed.

A method of charring sea-coal, so as to divest it of its Smoke, and yet leave it serviceable for many purposes, should be made the object of a very strict enquiry; and Premiums should be given to those that were successful in it.[27]

The legal history of air pollution control spans centuries and reflects a long-standing sensitivity of the populace to the quality of the air. The first regulations in the United States were introduced in the late 1800s as local ordinances and addressed smoke emission from fossil fuels, primarily coal. The Chicago and Cincinnati smoke control laws adopted in 1881 are prime examples.

Many programs to control air pollution were established in the ensuing years and focused on individual sources of stationary source emissions. The focus remained as such until the pervasive and complex nature of additional air pollution problems gained recognition in the 1940s. An article in the *Los Angeles Times* on July 27, 1943, introduced a new and formidable dimension of air pollution: oxidant. The story beneath the page-one headline read: "Yesterday's annoyance was at least the fourth such 'attack' of recent date."

The "attack" was a new phenomenon in the city's booming wartime economy, something people had begun to call "the smoke nuisance." First, a butadiene plant making synthetic rubber was blamed. The plant was nearly idled. But a newspaper account reported no reduction of the nuisance: "Motorists ... declared the various tunnels approaching the downtown area yesterday were so filled with the car burning fumes (SIC) as to resemble a fog." The account variously referred to the nuisance as "smoke and fumes," "a daylight dimout," "manmade hayfever," and "the lachrymatory haze in the air." Then, in a story appearing on September 18, 1944, an old label from Pittsburgh was borrowed and adopted: "smog." The nuisance had a permanent nickname, if not a cause.

Gradually, the automobile was recognized as a potential contributor to the precursors of the oxidant. However, it was not until the early 1950s that a Stanford Research Institute study formally suggested that the automobile was a major factor.

Thus, the total air pollution picture was revealed. In addition to the persistent emission of particulate and sulfur oxide from stationary sources, oxidant air pollution was recognized and the major cause, the automobile, was identified. The latter of these has since proven elusive and intimately interwoven in the fabric of society.

In the following discussion, the legal history of air pollution is reviewed and developed from the local to the federal level. The section concludes with a detailed overview of the two federal legislative mandates—the 1970 and 1977 Clean Air Amendments—that set the framework for the modern air pollution control strategies and programs and accelerated the use and development of the air quality impact analysis.

Legal History

Air pollution control began at the local level and has since evolved into a three-tiered structure of programs at the local, state, and federal levels. The evolution of air pollution control from local, city-based programs to the broader, more comprehensive and multidimensional local, state, and federal programs was first experienced in California. The California history is a useful chronicle of the increased sensitivity of government to the regional nature of air pollution and the evolution of state and federal programs required to control the automobile and to advance control technology for major stationary sources.

Local Regulation In 1945, the city of Los Angeles established a Bureau of Smoke Control in its health department. In 1946, Los Angeles County created the Office of Air Pollution Control. However, incorporated cities within the county adopted and enforced similar control efforts that reduced the effectiveness of the county control efforts. As a result, the California legislature passed legislation in 1947 providing for the establishment of special countywide agencies to control air pollution sources. Los Angeles County was the first to take advantage of the new legislation and formed an Air Pollution Control District in 1947.

In 1949, the California legislature provided a mechanism whereby two or more contiguous county air pollution control districts could merge into unified air pollution control districts. However, few of the county air pollution control districts formed since 1949 have elected to unify.

In the early 1950s, citizens groups and legislative committees explored the type of district appropriate for the San Francisco Bay Area and concluded that both the county and the unified approaches were unsuitable. Special legislation was proposed and passed in 1955 that created the Bay Area Pollution Control District which covers all or parts of the nine Bay Area counties. Enabling legislation for the formation of similar regional districts was passed by the state legislature in 1967. As a result, a regional district can now be formed by an agreement between two or more counties whose boundaries are within one air basin. (Note that the counties have simply to be in the same air basin for a regional district and need not be contiguous as for unified air pollution control districts. Another important distinction is that regional districts, including the Bay Area APCD, have the power to tax.) Despite the enabling legislation, the Bay Area APCD remained the only regional district in California until 1977.

On February 1, 1977, the South Coast Air Quality Management District (SCAQMD) was formed by an act of the California legislature. The formation of the regional district terminated an 18-month cooperative program by four counties in the Basin (Orange, Los Angeles, Riverside, and San Bernardino) to operate a unified district. Under pressure from the California legislature to regionalize, the four counties formed the unified district in July, 1975 with the objectives to (1) discourage the legislature from requiring a regional district, and (2) retain county control over the regulatory agency. The legislature was not persuaded and formed the regional district with the stipulation that cities as well as counties be represented on the controlling board.

The emphasis in control agency structure in California has evolved from city, to county (without authority in incorporated areas), to air pollution control districts (with authority in all of the county), to unified air pollution control districts (without consideration of air basin boundaries), and finally to regional air pollution control districts (with consideration of air basin boundaries).

The evolution of control agency structure evidences the recognition of air pollution as a regional problem that requires comprehensive and coordinated control strategies developed on the air basin level. The air basin is and will continue to be the basic building block for control strategy development and implementation.

State Regulation In the early 1950s, the mechanism, pollutant precursors, and pollutant sources involved in the formation of oxidant in the Los Angeles area were identified.

Further research indicated that the problem was not unique to the Los Angeles area. The newly discovered regional nature of air pollution precipitated a variety of programs at the state level.

In 1959, the California Department of Public Health was mandated by the legislature to identify pollutants of potential impact to the public health and to set ambient air quality standards. The California Motor Vehicle Pollution Control Board was established in 1960 to establish and implement emission standards for new motor vehicles. Crankcase emission standards were implemented in 1963, and exhaust emission standards for hydrocarbons and carbon monoxide were implemented in 1966. In 1967, the California Air Resources Board (CARB) was formed with the responsibility to divide the state into air basins, compile emission inventories for the air basins, conduct air quality monitoring surveys, provide technical assistance to local and regulatory agencies, continue the review of air quality standards, and continue the motor vehicle emissions control program.

In 1968, the California Pure Air Act established increasingly stringent emission standards for 1970 through 1974 model-year automobiles for hydrocarbons and carbon monoxide, and for 1972 through 1975 vehicles for oxides of nitrogen. More stringent emission standards for 1975 and later model-year vehicles were first adopted by the CARB in 1970 and have been subsequently strengthened, extended, and delayed in keeping with technological advances and the realities of the political process.

Federal Regulation   The federal activity is summarized in Table 3.8. The first federal legislation dealing exclusively with air pollution was enacted in 1955 and provided monies for the Department of Health, Education, and Welfare (HEW) for research, data col-

TABLE 3.8 Summary of Federal Activity

1955—*Untitled (commonly referred to as the Air Pollution Control Act).* Enacted to provide the Department of Health, Education, and Welfare (HEW) with monies to support research and provide technical assistance to state agencies for abating air pollution.

1960—Congress directed Surgeon General of the Public Health Service to report on the effects of motor vehicle exhaust on human health through the pollution of air.

1962—*Untitled.* Amended Air Pollution Control Act of 1955 with a directive to study the relationship between motor vehicle exhaust and human health.

1963—*Clean Air Act of 1963.* Enacted to (1) provide grants-in-aid for research, development, and planning activities, (2) authorize Secretary of HEW to publish nonmandatory air quality criteria, and (3) authorize the Secretary of HEW to intervene directly in situations where air pollution endangers the public health and a state fails to act.

1965—*Motor Vehicle Air Pollution Control Act.* Amended the 1963 Clean Air Act to require Secretary of HEW to set emission standards for new motor vehicles (only prototype vehicles submitted by the manufacturers were subject to test).

1967—*Air Quality Act of 1967.* Amended the 1963 Clean Air Act to (1) require the Secretary of HEW to issue "criteria of air quality," (2) require the Secretary of HEW to designate broad atmospheric areas and establish more specific air quality control regions, (3) require states to set air quality standards for such regions based on the criteria set by the Secretary, and to submit plans for their implementation, and (4) provide for federal abatement action on 180 days' notice in cases of interstate air pollution.

1970—*1970 Clean Air Act Amendments.* Amended 1963 Clean Air Act to (1) establish National Ambient Air Quality Standards (NAAQS), (2) require state implementation plans to achieve and maintain the NAAQS, and (3) require 90% reductions in automotive emissions of hydrocarbons, carbon monoxide, and nitrogen oxides by certain dates.

1977—*1977 Clean Air Act Amendments.* Amended 1963 Clean Air Act to extend the deadline for achieving the NAAQS for those areas experiencing difficulty with the previous deadlines.

---

lection, and technical assistance to local and state agencies. In 1963, the Clean Air Act was passed and represented the first major federal air pollution control legislation. Through this legislation, Congress recognized the contributions of industrial and vehicular sources but retained the posture that air pollution was a state and local problem. The legislation provided for grants to state and local air pollution control agencies and established federal enforcement authority in interstate air pollution problems.

In 1965, the Clean Air Act was amended to establish federal authority to regulate emissions from new motor vehicles. These amendments also established the National Air Pollution Control Administration (NAPCA) within HEW. As a result, federal automobile emission standards for hydrocarbons and carbon monoxide for 1969 model-year automobiles were promulgated. In 1968, more stringent standards were set for the 1970 and 1971 model-year vehicles.

In 1967, the Air Quality Act was passed. This legislation identified the following tasks, responsibilities, and authority in constructing a regional approach for air quality control programs:

- Federal designation of air quality control regions.
- Federal publication of air quality criteria documents and control technique documents.
- State adoption through public hearings of air quality standards based on air quality criteria documents.
- State adoption of implementation plans for achieving and maintaining adopted air quality standards.
- Federal review of state standards and implementation plans. If accepted, the state standards and plans would become the federal standards and plans for the region. If the states took no action or if the adopted standards and plans were inconsistent with the act, the federal government would promulgate standards.
- If enforcement of adopted standards was insufficient or nonexistent, the federal government could initiate further action through the federal courts.

The general inadequacies of the 1967 Air Quality Act and the time lags associated with its implementation prompted Congress to significantly revise the Clean Air Act in 1970 through amendments. At the same time, the Environmental Protection Agency (EPA) was created by executive reorganization and was given, among other obligations, the responsibility for federal air pollution control programs. The air quality related activities and staff of the NAPCA and Public Health Service of HEW were transferred to the EPA.

The simultaneous passage of 1970 Clean Air Amendments, creation of the EPA, and delegation of responsibility to the EPA to implement the 1970 Clean Air Amendments had a profound impact on the regulatory process at the federal, state, and local level. In contrast to earlier legislative and regulatory action, these three factors combined to provide the authority and direction to create programs with potential impact on the general lifestyle of the populace.

1970 Clean Air Amendments[18]

The 1970 Clean Air Amendments contained two major titles. Title I pertained to the attainment and maintenance of the ambient air quality. Also included were sections pertaining to emissions from stationary sources. Title II pertained to the restriction of emissions from mobile sources.

A brief summary of the major sections of each title is presented here, followed by a more detailed discussion of the state implementation plans and new source performance standards.

Title I Overview   The major sections of this title and their requirements are:

- *Section 107—Air Quality Control Regions.* Required the designation of "air quality control regions." An air quality control region (AQCR) is a geographical area that shares a common resource of air. The significance of the AQCR concept is the federal position that each is a geographical region upon which air pollution control regulation may be imposed regardless of political demarcations.
- *Section 108—Air Quality Criteria.* Required that the EPA establish "air quality criteria" for individual pollutants and identify air pollution control techniques acceptable for controlling the pollutant emission. Intended for use in establishing ambient air standards and preparing implementation plans.
- *Section 109—National Primary and Secondary Ambient Air Quality Standards.* Required the establishment of "national primary and secondary ambient air quality

standards." The primary ambient air quality standards are intended to protect public health with an adequate margin of safety, while the secondary ambient air quality standards are intended to protect the public welfare.

- *Section 110—State Implementation Plan.* Required each state to develop an "implementation plan" outlining the state program for attaining the national primary ambient air quality standards. A set goal was to improve air quality throughout the United States such that by May 31, 1975, no one air quality standard would be exceeded more than one day a year. Extension of the deadline to May 31, 1977 was allowed for a given region if requested by the appropriate state and approved by the EPA.

- *Section 111—New Source Performance Standards.* Required the establishment of "new source performance standards" to limit the emission of primary pollutants from a select list of new stationary sources.

- *Section 112—Hazardous Air Pollutants.* Required the promulgation of a list of "hazardous air pollutants." Hazardous air pollutants are substances other than those which have a previously adopted national primary or secondary ambient air quality standard and have a recognized hazardous nature. Asbestos, beryllium, and mercury were initially classified hazardous air pollutants. Source emission standards and continuous monitoring were required for sources of hazardous air pollutants. In addition, permits were required from the EPA to construct, modify, and operate. The present status of hazardous air pollutant limits may be assessed by contacting the regional office of the EPA and the local air pollution control district.

Title II Overview   The major sections of this title and their requirements are:

- *Sections 202 through 210—Motor Vehicle Emission Standards.* Section 202 directed the EPA Administrator to establish "motor vehicle emission standards" for any pollutant which endangers the public health or welfare, and required a percent reduction and compliance date (90 percent reduction from 1970-model hydrocarbon and carbon monoxide levels by 1975, and a 90 percent reduction from 1971-model oxides of nitrogen levels by 1976. NOTE: These deadlines have since been extended via procedures specified in Section 202.)

Sections 203, 204, and 205 identified prohibited acts (e.g., sale of vehicle not in compliance with applicable standards, tampering with control devices, denying access to or failing to submit required reports) and detailed injunction proceedings and penalties.

Section 206 required that the EPA Administrator test new vehicles submitted to manufacturer's compliance with the applicable standards and, upon satisfactory testing results, issue a certificate of compliance for that control system/test vehicle combination when installed in corresponding production vehicles. Section 206 also detailed the administrative procedures for the certification testing.

Section 207 required that manufacturers warrant emission control systems for the useful life of the vehicle (normally taken as 50,000 miles). Further, if after the requisite surveillance testing the administrator finds that a properly-maintained control system/vehicle class combination is in violation of applicable standards, Section 207 required the manufacturer to remedy the situation at company expense.

Section 208 detailed the requirements for documenting a manufacturer's actions taken to achieve compliance with the sections governing motor vehicle emissions.

Section 209 prohibited states from adopting motor vehicle emission standards but contained a waiver for California.

Section 210 provided for federal grants to support state efforts initiated pursuant to these sections.

- *Section 211—Regulation of Fuels.* Empowered the EPA Administrator to control the manufacture and sale of fuels and fuel additives for use in motor vehicles, and detailed the conditions upon which the administrator could exercise this control (e.g., examination of available medical and scientific evidence, cost-benefit analysis). Allowed states which received a waiver for emissions standards to adopt regulations controlling fuels and fuel additives.

- *Section 212—Low-Emission Vehicles.* Outlined the procedure by which low-emissions vehicles were to be certified and required that the federal government purchase only the low-emissions vehicles if such vehicles are "certified substitutes" and procurement costs are no more than 150 percent of the costs of comparable vehicles which are not certified as low-emissions vehicles. Section 212 also provided for surveillance of "low-emissions" vehicles for compliance.

- *Sections 231 through 233—Aircraft Emission Standards.* Directed the EPA

Administrator to investigate emission from aircraft and to establish emissions standards and a compliance schedule for those pollutants which are found to endanger the public health or welfare. Required that the Secretary of Transportation be consulted when any standard is established or revised under this section in order to insure appropriate consideration of aircraft safety. Section 232 gave the Secretary of Transportation responsibility, after consulting the EPA Administrator, for promulgating regulations to insure compliance with the standards and schedules prescribed under Section 231. Section 233 prohibited states from adopting or attempting to enforce any standard different from those adopted by the EPA.

Other provisions of the 1970 Clean Air Act amendments provided for federal enforcement, penalties, inspections, and monitoring for the control of air pollutant emissions, and the federal procedures associated with achieving and maintaining the National Ambient Air Quality Standard. Further descriptions of those sections especially pertinent to air quality impact analyses follow.

State Implementation Plans   One of the more important requirements of the Clean Air Act amendments of 1970 was the requirement for State Implementation Plans (SIP) for achieving and maintaining the National Ambient Air Quality Standards. The act required each state to prepare a SIP. Several notable categories of regulations—general provisions, emergency episode plans, and nondegradation regulations—have resulted from the preparation, review, and approval process, and litigation at the federal and state level that has accompanied the evolution of the state implementation plans.

General Provisions.   The general provisions outlined the procedure required for the preparation of the SIP. First, the Air Quality Control Regions (AQCR) were established. Second, each region was classified according to the degree of air pollution that existed in the particular region. The classification criteria are given in Table 3.9. The more polluted a region, the higher the "priority." In each AQCR where the measured ambient air quality exceeded a standard, a control strategy to reduce emissions to a level which will meet the ambient standards was required. Candidate strategies suggested for consideration included the following:

▪ Emission standards to regulate the release of pollutants from stationary sources. (Table 3.10 summarizes the EPA proposal for effective air pollution control regulations.)

▪ Economic incentives/disincentives for control of pollutants.

▪ Closure of commercial or industrial facilities with excessive emissions.

▪ Inspection and maintenance of motor vehicles.

▪ Installation of motor vehicle pollution control measures (i.e., mandatory maintenance).

▪ Reduction of motor vehicle usage by commuter taxes, gasoline rationing, parking restrictions, etc.

▪ Land use controls.

▪ Prohibition of new or modified stationary sources which would prevent attaining or maintaining an ambient standard either by direct or indirect means. "Indirect" means that the emissions caused by associated activities (i.e., traffic or satellite plants) may be sufficient grounds for denial of construction, modification, or operation. To show that such impacts would not result from the proposed action, detailed data gathering and air quality modeling by the owner or operator are implied. For a new source, these data could be included in the EIS, although this provision alone does not require an EIS.

Those regions which faced difficulty in attaining and maintaining the National Ambient Air Quality Standards were classified by the EPA as an "air quality maintenance area" (AQMA). Initially, over 160 areas were classified as an AQMA for at least one pollutant. Of these, 159 were listed for particulate matter, 61 for sulfur dioxide, 24 for carbon monoxide, 49 for photochemical oxidants, and 5 for nitrogen dioxide. States were required to include an air quality maintenance plan (AQMP) in the SIP for each AQMA. The following were minimum requirements of the AQMP:

▪ Projections of emissions into the future.

▪ Estimation of air quality resulting from the future emissions.

▪ Development of a control strategy to maintain the National Ambient Air Quality Standards.

▪ Adoption of regulations to make the control strategy enforceable.

In addition, the states were required to reassess the AQMA classifications at least every five years and to classify additional regions as appropriate.

TABLE 3.9 Implementation Plans—Region Classification

| Pollutant | Priority | | | Primary ambient air quality standard | Secondary ambient air quality standard |
	I (greater than)	II (from–to)	III (less than)		
Sulfur dioxide:					
Annual arithmetic mean	100(0.04)	60–100(0.02–0.04)	60(0.02)	80(0.03)	60(0.02)
24-hour maximum	455(0.17)	260–455(0.10–0.17)	. . .	365(0.14)	. . .
3-hour maximum	. . .	1300(0.50)	1300(0.50)	. . .	1300(0.50)
Particulate matter:					
Annual geometric mean	95	60–95	60	75	60
24-hour maximum	325	150–325	150	260	150
Carbon monoxide:					
8-hour maximum	14(12)	. . .	14(12)	10(9)	10(9)
1-hour maximum	55(48)	. . .	55(48)	40(35)	40(25)
Nitrogen dioxide:					
Annual arithmetic mean	110(0.06)	. . .	110(0.06)	100(0.05)	100(0.05)
Oxidants:					
1-hour maximum	195(0.10)	. . .	195(0.10)	160(0.08)*	160(0.08)*
Hydrocarbons:					
3-hour maximum	. . .	. . .	. . .	160(0.24)	160(0.24)

Units are expressed as $\mu g/m^3$ except CO which is mg/m^3. The ppm equivalent is given in parentheses.
*Standard has since been modified (see Table 3.7) and specified in terms of ozone.
SOURCE: Reference 30, p. 15486.

Emergency Episode Plans.   As part of the general provisions for the preparation of the state implementation plans, each "Priority I region" (see Table 3.9) was required to have an "emergency episode plan." An emergency episode plan is a shutdown contingency plan to prevent ambient pollution levels from reaching harmful levels.

TABLE 3.10   Federally Recommended Air Pollution Control Regulations

Particulates:
Limit visible emissions to Ringelman No. 1 (20% opacity) except for brief periods, (e.g., soot blowing).

Initiate fugitive dust control, (e.g., wetting of earth-moving operations).

Limit incinerators to 0.2 lb of particulates per 100 lb charged.

Limit fuel-burning equipment to 0.30 lb of particulates per million Btu. (Higher Heating Value, HHV) released.

Limit processing industries by a schedule of 1 lb of particulates per pound of material processed.

Sulfur dioxide:
Use low-sulfur fuels, if available.

Limit refinery-fuel gas and flare gas to 10 grains of H_2S per 100 standard cubic feet.

Limit sulfuric acid plants to 6.5 lb of SO_2 per ton of 100% H_2SO_4 produced.

Limit sulfur-recovery plants to 0.01 lb of SO_2 per pound of sulfur recovered.

Hydrocarbons:
Equip storage tanks with floating roofs or vapor recovery systems.

Equip oil/water separators with either solid fixed or floating roofs.

Equip pumps and compressors with mechanical seals.

Incinerate ethylene plant-waste gases.

Initiate organic-solvent controls.

Carbon monoxide:
Employ CO boilers for fluid cracking and coking.

Nitrogen oxides (NO_x as NO_2):
Employ equipment designed to limit NO_x to (a) 0.2 lb per million Btu (HHV) fired for gaseous fuels, and (b) 0.3 lb per million Btu (HHV) fired for liquid fuels.

Limit nitric-acid plants to 5.5 lb of NO_x per ton of 100% HNO_3 produced.

SOURCE: Reference 30, p. 15486.

The emergency episode plans called for different actions depending upon the measured or predicted level of air pollution. These levels were "alert," "warning," and "emergency." An alert is the lowest level for an air pollution episode and has the least severe requirement. An emergency is the highest level for an air pollution episode and calls for essentially a full shutdown. The federally suggested numerical levels for these stages of alert are given in Table 3.11.

Nondegradation Regulations.   As a result of a successful court case filed by the Sierra Club against the EPA, nondegradation regulations became a necessary part of the SIP. (The nondegradation regulations were renamed Prevention of Significant Deterioration under the 1977 Clean Air Act Amendments and were revised. As discussed in the section of this chapter devoted to those 1977 Amendments, many of the nondegradation regulations were retained.) Nondegradation regulations applied only to areas which had better SO_2 and particulate air quality than the national secondary standards.

Development of the nondegradation criteria required first establishing a deterioration classification for the affected areas (not to be confused with the regional classification of Table 3.9) and the allowable degradation in ambient concentration of sulfur dioxide and particulates for each class. Three deterioration classifications were established by the EPA:

- *Class I*—Permitted only very minor increases in concentrations of SO_2 and particulate.
- *Class II*—Permitted an increase in the ambient concentrations greater than those permitted for Class I, but less than for Class III.
- *Class III*—Permitted an increase in the ambient concentrations of SO_2 and particulate up to the national *secondary* ambient air quality standards.

The EPA initially classified all areas subject to the nondegradation regulations as Class II. It was the responsibility of each state to reclassify these areas to Class I or Class III as

TABLE 3.11 Emergency Episode Plans—Episode Criteria

Pollutant	Averaging period	Primary ambient air quality standard	Alert	Warning	Episode	Never to be reached
Sulfur dioxide	24-hour	365(0.14)	800(0.3)	1600(0.6)	2100(0.8)	2620(1.0)
Particulates	24-hour	260	375	625	875	1000
SO_2 + Particulate	24-hour	...	$65 \times 10^{3*}$	$261 \times 10^{3*}$	$393 \times 10^{3*}$	$490 \times 10^{3*}$
Carbon monoxide	1-hour	40(35)	...	...	...	144(125)
	4-hour	...	...	...	...	86.3(75)
	8-hour	10(9)	17(15)	34(30)	46(40)	57.5(50)
Oxidant	1-hour	160(0.08)‡	200(0.1)	800(0.4)	1200(0.5)†	1400(0.7)
	2-hour	...	...	...	...	1200(0.6)
	4-hour	...	...	...	...	800(0.4)
Nitrogen dioxide	1-hour	...	1130(0.6)	2260(1.2)	3000(1.6)	3750(2.0)
	24-hour	...	282(0.15)	565(0.3)	750(0.4)	938(5.0)

Units are expressed in $\mu g/m^3$ except CO which is mg/m^3. The ppm equivalent is given in parentheses.
*Equal to $0.2 \times$ 24-hour SO_2 $\mu g/m^3 . \times$ 24-hour particulate $\mu g/m^3$.
†1000 $\mu g/m^3$—emergency shutdown
‡Standard has since been modified (see Table 3.7) and specified in terms of ozone.
SOURCES: References 31, 33, 34, 36, and 41.

necessary to be consistent with the intended use of that area. The degradation in ambient concentration of sulfur dioxide and particulate allowable for each class under the regulations is given in Table 3.12.

In order to construct a new source or modify an existing one in a Class I or II area, an application to the EPA was required when the proposed facility included one of the stationary sources listed below and was projected to emit more that 25 lb/h of SO_2 or particulates:

- Fossil-fueled steam electric power plants 1 million Btu or larger
- Coal cleaning plants
- Kraft pulp mill recovery furnaces
- Portland cement plants
- Primary zinc smelters
- Iron and steel mills
- Primary aluminum ore reduction plants

TABLE 3.12 Nondegradation—Allowable Increase in Ambient Concentration

Deterioration classification	Allowable increase in concentration, $\mu g/m^3$				
	Sulfur dioxide			Particulates	
	3-hr	24-hr	AAM	24-hr	AGM
Class I	25	5	2	10	5
Class II	700	100	15	30	10

SOURCES: References 37–40 and 42.

- Primary copper smelters
- Municipal incinerators 250 ton/day or larger
- Sulfuric acid plants
- Petroleum refineries
- Lime plants
- Phosphate rock processing plants
- Byproduct coke oven batteries
- Sulfur recovery plants
- Carbon black plants (furnace process)
- Fuel conversion plants
- Primary lead smelters
- Ferroalloy production facilities

New Source Review Regulations. As regards proposed new or modified sources of air pollution, the EPA and states are required under the Clean Air Act to impose restrictions which are designed to achieve the goal of attainment and maintenance of the ambient air quality standards. The traditional form of air pollution regulation is the "emission limitation" (a restraint on the amount of pollution escaping from a particular air pollution source). The act explicitly recognizes that emission limitations alone may not be sufficient in some areas to attain and maintain ambient standards. In particular, the act requires that implementation plans contain ". . . emission limitations . . . and such other measures as may be necessary to insure attainment and maintenance of such . . . (ambient) standard(s), including but not limited to, land use and transportation controls."

In formulating its basic guidelines for the content of state implementation plans in 1971, the EPA determined that ". . . other measures as may be necessary to insure . . ." maintenance of the ambient standards would be *new source review regulations.* In particular, the EPA required each state to develop a regulation which would prohibit the construction or require the modification of a new source which would interfere with the attainment and maintenance of the air quality standards.

Although most states included their own preconstruction review regulations in the original implementation plans which EPA approved in 1972, the regulations of many states were found unacceptable. EPA was therefore required to promulgate such a regulation for those states not submitting an acceptable regulation.

In general, the EPA-developed regulation requires any proposed stationary source of air pollution (except certain small sources specifically exempted) to obtain the approval of EPA before commencing construction. The source proponent has the burden of showing that the source "will not prevent or interfere with attainment or maintenance" of any National Ambient Air Quality Standard. In a nonattainment area, emission reductions from existing sources must be found to "trade-off" against the new emissions, and the "trade-off" must be greater than merely one-for-one.

Summary. The "state implementation plan" (SIP) for each state will be subject to continuous revision, much like a community general plan. The federal government is forcing compliance with stated air quality goals, while community interests act to force flexibility in the SIP process to avoid economic demise.

It is also safe to say that each SIP has a unique personality that reflects the particular climate (meteorological, political, social, and economic) of the state and regions within the state.

New Source Performance Standards   Section 111 of the Clean Air Act requires the EPA to promulgate regulations called "New Source Performance Standards" (NSPS). These standards establish the maximum emissions of primary pollutants allowed from *new* stationary sources. (A new source includes those newly constructed or a modified existing source wherein the modifications increase the emissions of pollutants or cause an additional pollutant to be emitted.) The NSPS should not be confused with the "new source review regulations." The latter address the impact of a new source on the ambient air and go beyond emission limitation. The NSPS are concerned only with emission regulation.

New source performance standards were initially adopted for the following types of sources or operations:

- Fossil fuel–fired steam generators*
- Incinerators
- Portland cement plants
- Nitric acid plants*

- Sulfuric acid plants*
- Asphalt concrete plants
- Petroleum refineries*
- Storage vessels
- Secondary lead smelters and refineries
- Brass and bronze ingot processing plants
- Iron and steel plants
- Sewage treatment plants
- Primary copper smelters*
- Primary zinc smelters*
- Primary lead smelters*
- Primary aluminum reduction plants
- Phosphate fertilizer plants
- Coal preparation plants
- Ferroalloy production facilities .

The sources identified with an asterisk (*) received special attention. First, the states were required to revise their SIP to include legally enforceable procedures which would require emission monitoring, recording, and reporting for those sources noted by asterisks. (Sources scheduled for retirement within five years of approval of SIP were exempted from the continuous monitoring requirement.) Second, the sources with asterisks were required to submit quarterly reports on any occurrence of excess emissions, the nature and cause of these excess emissions, and a listing of all repairs and maintenance performed on monitoring equipment.

The new source performance standards are summarized in Table 3.13. Note that the effective date of the summary is November 1976. The first NSPS were promulgated in 1971. A number of revisions and additions occurred in the ensuing five-year period.

The new source performance standards identify the maximum emissions allowable from these major stationary sources of air pollution. The current status of the NSPS may be assessed by contacting both the regional office of the EPA and the office of the local air pollution control district. In July of each year, all regulations that have been published in the *Federal Register* are codified for inclusion in the *Code of Federal Regulation* (CFR). As a result, the latest revision of the CFR, Title 40, Part 66, provides a convenient reference for the status of the NSPS update through July of the current year.

1977 Clean Air Act Amendments[19]

On August 7, 1977, President Carter signed into law the Clean Air Amendments of 1977. The 1977 Amendments uphold the basic goals and strategy of the 1970 Clean Air Act Amendments while benefitting from the experience gained in implementing these earlier amendments. Deadlines for attainment of air quality standards and for meeting automobile emission goals were pushed back, and a realistic approach was adopted for (1) those areas that have not yet attained ambient standards ("nonattainment areas") and (2) those areas which are cleaner than required by ambient standards ("prevention of significant deterioration areas").

Brief summaries of those sections pertinent to air quality impact analyses follow:

- *Section 306—Sewage Treatment Grants.* Provides that in certain areas, EPA may not condition sewage treatment grants on air quality grounds. However, this section contains very strict requirements for establishing the circumstances under which sewage treatment grants cannot be conditioned. By implication, it suggests that EPA may condition grants in areas which do not attain standards or have adequate plans.

- *Section 307—Economic Impact Assessment.* Requires that EPA prepare an economic assessment when setting new performance standards, promulgating regulations for prevention of significant deterioration, establishing motor vehicle emissions standards and aircraft emission standards, or certain other functions. This assessment is to contain an analysis of the costs of compliance, inflationary or recessionary effects, impacts on small businesses, consumer costs, and energy impacts of the proposed action.

- *Section 103—Air Quality Control Regions.* Requires the states to provide EPA with a list that classifies *all* areas (AQCRs or smaller) of the state with respect to: (1) attainment status for primary standards, (2) attainment status for secondary standards, and (3) regions which enjoy air quality better than that required by the standards. This designation by the state is required to determine which areas are subject to the nonattain-

TABLE 3.13 New Source Performance Standards (June 1976)

Source category	Affected facility	Pollutant	Emission level	Monitoring requirement
Subpart D*				
Steam generators (>250 million Btu/hr)	Coal-fired boilers	Particulate	0.10 lb/10^6 Btu	No requirement
		Opacity	20%	Continuous
		SO_2	1.2 lb/10^6 Btu	Continuous
Promulgated		NO_x	0.70 lb/10^6 Btu	Continuous
12/23/71 (36 FR 24876)†		(except lignite and coal refuse)		
Revised				
7/26/72 (37 FR 14877)	Oil-fired boilers	Particulate	0.10 lb/10^6 Btu	No requirement
6/14/74 (39 FR 20790)		Opacity	20%; 40% 2 min/h	Continuous
1/16/75 (40 FR 2803)		SO_2	0.80 lb/10^6 Btu	Continuous
10/6/75 (40 FR 46250)		NO_x	0.30 lb/10^6 Btu	Continuous
	Gas-fired boilers	Particulate	0.10 lb/10^6 Btu	No requirement
		Opacity	20%	No requirement
		NO_x	0.20 lb/10^6 Btu	Continuous
Subpart E				
Incinerators (>50 tons/day)	Incinerators	Particulate	0.80 g/dscf corrected to 12% CO	No requirement
Promulgated				
12/23/71 (36 FR 24876)				
Revised				
6/14/74 (39 FR 20790)				
Subpart F				
Portland-cement plants	Kiln	Particulate	0.30 lb/ton	No requirement
		Opacity	20%	No requirement
Promulgated				
12/23/71 (36 FR 24876)	Clinker cooler	Particulate	0.10 lb/ton	No requirement
		Opacity	10%	No requirement
Revised				
6/14/74 (39 FR 20790)	Fugitive emission points	Opacity	10%	No requirement
11/12/74 (39 FR 39874)				
10/6/75 (40 FR 46250)				

Subpart / Source category	Affected facility	Pollutant	Emission limit	Monitoring requirement
Subpart G Nitric acid plants *Promulgated* 12/23/71 (36 FR 24876) *Revised* 5/23/73 (38 FR 13562) 6/14/74 (39 FR 20790) 10/6/75 (40 FR 46250)	Process equipment	Opacity NO_x	10% 3.0 lb/ton	No requirement Continuous
Subpart H Sulfuric acid plants *Promulgated* 12/23/71 (36 FR 24876) *Revised* 5/23/73 (38 FR 13562) 6/14/74 (39 FR 20790) 10/6/75 (40 FR 46250)	Process equipment	SO_2 Acid mist Opacity	4.0 lb/ton 0.15 lb/ton 10%	Continuous No requirement No requirement
Subpart I Asphalt-concrete plants *Promulgated* 3/8/74 (39 FR 9308) *Revised* 10/6/75 (40 FR 46250)	Dryers; screening and weighing systems; storage, transfer, and loading systems; and dust-handling equipment	Particulate Opacity	0.04 g/dscf (90 mg/dscm) 20%	No requirement No requirement
Subpart J Petroleum refineries *Promulgated* 3/8/74 (39 FR 9308) *Revised* 10/6/75 (40 FR 46250)	Catalytic cracker Fuel-gas combination	Particulate Opacity CO SO_2	1.0 lb/1000 lb 30% (3 min. exemption) 0.05% 0.1 g H_2S/dscf (230 mg/dscm)	No requirement Continuous Continuous Continuous

TABLE 3.13 New Source Performance Standards (June 1976) (Continued)

Source category	Affected facility	Pollutant	Emission level	Monitoring requirement
Subpart K				
Storage vessels for petroleum liquids	Storage tanks >40,000-gal capacity	Hydrocarbons	For vapor pressure 78-570 mm Hg, equip with floating roof, vapor-recovery system, or equivalent; for vapor pressure >570 mm Hg, equip with vapor-recovery system or equivalent	No requirement
Promulgated 3/8/74 (39 FR 9308)				
Revised 4/17/74 (39 FR 13776) 6/14/74 (39 FR 20790)				
Subpart L				
Secondary lead smelters	Reverberatory and blast furnaces	Particulate	0.022 g/dscf (50 mg/dscm)	No requirement
		Opacity	20%	No requirement
Promulgated 3/8/74 (39 FR 9308)	Pot furnaces	Opacity	10%	No requirement
Revised 4/17/74 (39 FR 13776) 10/6/75 (40 FR 46250)				
Subpart M				
Secondary brass and bronze plants	Reverberatory furnace	Particulate	0.022 g/dscf (50 mg/dscm)	No requirement
		Opacity	20%	No requirement
Promulgated 3/8/74 (39 FR 9308)	Blast and electric furnaces	Opacity	10%	No requirement
Revised 10/6/75 (40 FR 46250)				
Subpart N				
Iron and steel plants	Basic oxygen-process furnace	Particulate	0.022 g/dscf (50 mg/dscm)	No requirement
Promulgated 3/8/74 (39 FR 9308)				

Subpart / Source	Pollutant	Standard	Monitoring requirement
Subpart O			
Sewage treatment plants			
Promulgated 3/8/74 (39 FR 9308) *Revised* 4/17/74 (39 FR 13776) 5/3/74 (39 FR 15396) 10/6/75 (40 FR 46250)			
Sludge incinerators	Particulate	1.30 lb/ton	Mass or volume of sludge
	Opacity	20%	No requirement
Subpart P			
Primary copper smelters			
Promulgated 1/15/76 (41 FR 2331)	Dryer	Particulate	0.022 g/dscf (50 mg/dscm)	No requirement
		Opacity	20%	Continuous
	Roaster, smelting furnace,[a] copper converter	SO_2	0.065%	Continuous
		Opacity	20%	No requirement
Revised 2/26/76 (41 FR 8346)	[a]Reverberatory furnaces that process high-impurity feed materials are exempt from SO_2 standard			
Subpart Q			
Primary zinc smelters			
Promulgated 1/15/76 (41 FR 2331)	Sintering machine	Particulate	0.022 g/dscf (50 mg/dscm)	No requirement
		Opacity	20%	Continuous
	Roaster	SO_2	0.065%	Continuous
		Opacity	20%	No requirement

TABLE 3.13 New Source Performance Standards (June 1976) (Continued)

Source category	Affected facility	Pollutant	Emission level	Monitoring requirement
Subpart R Primary lead smelters	Blast or reverberatory furnace, sintering machine-discharge end	Particulate	0.022 g/dscf (50 mg/dscm)	No requirement
		Opacity	20%	Continuous
Promulgated 1/15/76 (41 FR 2331)	Sintering machine, electric smelting furnace, converter	SO₂ Opacity	0.065% 20%	Continuous No requirement
Subpart S Primary aluminum-reduction plants	Potroom group	(a) Total fluorides Opacity	2.0 lb/ton 10%	No requirement No requirement
	(a) Soderberg plant			
Promulgated 1/26/76 (41 FR 3825)	(b) Prebake plant	(b) Total fluorides Opacity	1.9 lb/ton 10%	No requirement No requirement
	Anode-bake plants	Total fluorides Opacity	0.1 lb/ton 20%	No requirement No requirement
Subpart T Phosphate fertilizer	Wet-process phosphoric acid	Total fluorides	0.02 lb/ton	Total pressure drop across process scrubbing system
Promulgated 8/6/75 (40 FR 33152)				
Subpart U	Superphosphoric acid	Total fluorides	0.01 lb/ton	Total pressure drop across process scrubbing system

Subpart	Source/facility	Pollutant	Standard	Monitoring requirement
Subpart V	Diammonium phosphate	Total fluorides	0.06 lb/ton	Total pressure drop across process scrubbing system
Subpart W	Triple superphosphate	Total fluorides	0.2 lb/ton	Total pressure drop across process scrubbing system
Subpart X	Granular triple superphosphate	Total fluorides	5.0×10^{-4} lb/h/ton	Total pressure drop across process scrubbing system
Subpart Y Coal preparation plants	Thermal dryer	Particulate	0.031 g/dscf (0.070 mg/dscm)	Temperature Scrubber pressure loss
Promulgated 1/15/76 (41 FR 2232)	Pneumatic coal cleaning equipment	Opacity	20%	Water pressure No requirement
		Particulate	0.018 g/dscf (0.040 mg/dscm)	No requirement
	Processing and conveying equipment, storage systems, transfer and loading systems	Opacity	10%	No requirement
		Opacity	20%	No requirement
Subpart Z Ferroalloy production facilities	Electric submerged arc furnaces	Particulate	0.99 lb/mW·h (0.45 kg/mW·h ("high-silicon alloys") 0.51 lb/mW·h (0.23 kg/mW·h) (chrome and manganese alloys)	No requirement
Promulgated 5/4/76 (41 FR 18497)				

TABLE 3.13 New Source Performance Standards (June 1976) (Continued)

Source category	Affected facility	Pollutant	Emission level	Monitoring requirement
Revised 5/20/76 (41 FR 20659)			No visible emissions may escape furnace capture system	Flowrate monitoring in hood
			No visible emission may escape tapping system for >40% of each tapping period	Flowrate monitoring in hood
		Opacity	15%	Continuous
		CO	20% volume basis	No requirement
		Opacity	10%	No requirement
	Dust-handling equipment			
Subpart AA Iron and steel plants	Electric-arc furnaces	Particulate	0.0052 g/dscf (12 mg/dscm)	No requirement
		Opacity		
		(a) control device	3%	Continuous
		(b) shop roof	0, except 20%— charging 40%— tapping	Flowrate monitoring in capture hood Pressure monitoring in DSE system
Promulgated 9/23/75 (40 FR 43850)	Dust-handling equipment	Opacity	10%	No requirement

*Code of Federal Regulations, Title 40, Part 60.
†Federal Register (FR) Citation: Date (Volume FR Page number).
SOURCE: Reference 16, pp. 1055–1065.

ment plan requirements of Section 129 of the amendments and which areas are subject to prevention of significant deterioration requirements under Section 127.

- *Section 105—Transportation Planning and Guidelines.* Requires EPA to prepare numerous guideline documents to assist state and local governments in the preparation of nonattainment plans. EPA has to act within six months or one year (depending on the specific guidance document) to publish these guidelines.

- *Section 107—Energy or Economic Emergency Authority.* Allows (under specified circumstances in which an energy or economic emergency exists) a waiver by the President of certain portions of the SIP. For example, this section would allow for a temporary SIP suspension (no more than four months) in an area where temporary shortages of SIP-required natural gas would cause severe unemployment unless plants were allowed to burn alternative fuels.

- *Section 108—Implementation Plans.* Incorporates two concepts into the act: (1) new source review for attainment *and* maintenance, and (2) prevention of significant deterioration. It provides that after June 30, 1979, no major (greater than 100 tons per year) stationary source shall be constructed or modified in any nonattainment area if (1) it would violate standards, or (2) an adequate nonattainment plan has not been developed. Section 108 also includes (1) the requirement that the SIP contain permit procedures for new source review and for prevention of significant deterioration, (2) the requirement that the SIP show that source proponents are required to reimburse permitting agencies for the full cost of processing and enforcement of permits, and (3) the provision that EPA may delegate SIP responsibility directly to general-purpose local governments.

The inclusion in the act of an outright ban on new or modified sources after June 30, 1979 may be interpreted as a congressionally mandated sanction to prod states into developing adequate SIPs. All new permits are banned, *regardless of trade-off potential.* Also, the concept that the polluter must pay the cost of regulation has been mandated and may require a change in the revenue-raising practices of most local regulatory agencies.

- *Section 119—Consultation.* Requires involvement of local government officials in carrying out requirements of the act which are related to: (1) development of AQMPs or transportation controls, (2) preconstruction review of direct sources, and (3) any requirements relating to prevention of significant deterioration of nonattainment planning. EPA was required to promulgate within six months regulations determining how this coordination and cooperation will occur.

- *Section 122—Assurance of Plan Adequacy.* Provides that states prepare a study by August, 1978, to determine the extent to which control strategies in the SIPs are dependent upon the use of coal, petroleum products, or natural gas to comply with requirements of the act. EPA must review the state's program and modify it if it is inadequate. Section 125 includes a mechanism to provide for the use of coal in an area where it is locally or regionally available if the use of petroleum products or natural gas would cause significant local or regional economic disruption or umemployment.

- *Section 127—Prevention of Significant Deterioration (PSD).* Upholds the basic goals and strategy of the nondegradation regulations promulgated as a result of court action under the 1970 Clean Air Act Amendments, and formally establishes a congressional mandate that large areas of the country with air quality levels which exceed required standards should be protected. Requires a program for the control of emissions of particulate matter and sulfur dioxide and a three-tier-area classification scheme using the nondegradation regulations as a basis. Provides that certain areas (such as national parks and wilderness areas) receive the maximum level of protection (Class I). Most other areas are established Class II (mid level of protection). Retains the list of major stationary sources promulgated under the nondegradation regulations as requiring a preconstruction review and permit to construct should emissions be projected to exceed 100 tons per year. In addition, requires that *any* source with a potential of emissions in excess of 250 tons per year be subject to a preconstruction review. Prohibits construction if the source violates any of the provisions of the PSD requirements. Requires the states to establish a permit procedure to accomplish this review.

Allows states to redesignate areas with the provision that participation and meaningful involvement of local government be included in making any redesignations. Because the initial program contains only two pollutants, the EPA was required to develop within two years the framework for a program which would prevent significant deterioration for all criteria pollutants. State actions which accomplished the above goals could be recognized as complying with the intent of the act.

- *Section 128—Visibility Protection.* Provides for a program to protect visibility in Federal Class I prevention of significant deterioration areas (such as national parks and wilderness areas.) Requires the EPA to develop a program to ensure that new and existing sources (less than 15 years old) are controlled with respect to pollutants which impair visibility.

- *Section 129—Nonattainment Area.* Addresses the requirements for the state and local preparation of nonattainment plans. Specifies that existing EPA new source review rules remain in effect until 1979 unless very stringent requirements are met to qualify for a waiver. December 31, 1982 was set as a final date for attainment of all National Ambient Air Primary Standards, with provision for an extension to December 31, 1987 for carbon monoxide and oxidant. By January 1, 1979, a SIP revision must be submitted which demonstrates standards will be attained for nonattainment areas by the 1982 date. If implementation of all reasonable controls cannot achieve oxidant and/or carbon monoxide standards, a waiver to December 31, 1987 for achievement of these standards may be granted if: (1) all reasonable controls are being implemented, (2) a schedule for Motor Vehicle Inspection Program implementation has been established, (3) a program for rigorous alternative-site study and environmental analysis for industrial sources has been established, and (4) other measures needed to achieve standards have been identified. In the event of an extension, a second SIP revision is due on July 1, 1982.

Plans are to be prepared by state and local governments. Planning funds (75 million dollars authorized in Section 315 of the Amendments) will be made available to defray the costs incurred by local governments in developing the plan.

If an adequate plan is not developed, Section 176A provides that certain sanctions will be applied. These include the withholding of EPA grant funds, the withholding of most transportation project funds, and the denial of permits for new or modified major pollution sources. After plan adoption, Section 176B requires all federal agencies not to fund, engage in, support, license, or approve any activities not in conformance with the plan.

Section 177 relates to motor vehicle standards. It allows other states to adopt California motor vehicle standards, which are more stringent than national standards.

This section of the act provides for a nonattainment planning process. It provides congressional deadlines for standard achievement and criteria to use in judging AQMP acceptability. It also extends the planning process (until 1982 in areas where waivers for CO or oxidant are given) and provides substantial federal funding to ensure adequate resources to develop the plan. A major feature is the emphasis on the participation of local, general-purpose government in the planning process and the development of a mutually acceptable state-local role in preparing air quality plans.

Section 2: Assessment Methodology

An air quality impact analysis must include the following five elements:

1. *Existing environment*—A description of the existing environment in the area of the proposed project.

2. *Environmental impact*—A description of the future year impact on the air quality as a result of the completion and use of the proposed project.

3. *Mitigating procedures*—A description of procedures that may be implemented to reduce degradation of the air quality associated with the proposed project.

4. *Alternatives*—A description of changes in the design of the project that may be adopted to reduce the degradation of the air quality associated with the proposed project.

5. *Growth-inducing considerations*—A description of the growth-inducing potential of the proposed project and the secondary impact on air quality resulting from the induced growth.

Alternative report formats which present these elements are given in Figs. 3.21 and 3.22. The actual reporting format will depend on the preference of the author, the nature of the project, the desires of the client, and the requirements of the regulatory agencies whose approval must be obtained.

One of the most important decisions in the conduct of an air quality impact analysis is the selection of the methodologies for use in addressing each of the five elements just

1. Existing environment
   a. General project description
   b. Description of project area
      (1) Location/setting
      (2) Air pollution area/regulatory programs
   c. Meteorology
      (1) Wind rose summary
      (2) Atmospheric surface stability
      (3) Inversion height
      (4) Episode potential
   d. Existing air quality
      (1) Regional air quality
      (2) Subregional air quality
      (3) Local air quality
2. Environmental impact
   a. Detailed project description
   b. Analysis years
   c. Description of surrounding areas
   d. Traffic study
   e. Impact scenarios
      (1) Emission scenarios
      (2) Meteorological scenarios
      (3) Air chemistry scenarios
      (4) Primary and secondary impact scenarios
      (5) Visibility scenarios
   f. Air quality modeling
      (1) Existing
      (2) No-build
      (3) Construction
      (4) Build-out
      (5) Visibility
   g. Assessment
3. Mitigation procedures
4. Alternatives
5. Growth-inducing considerations

Figure 3.21 Air quality impact analysis reporting structure—example 1.

1. Existing air quality environment
   a. Overview of project area in air basin
   b. Monitored air quality and standards
   c. Topography
   d. Meteorology
   e. Climate
   f. Sources affecting project area
   g. Sources in project area
   h. Summary of transport: Upwind → downwind
   i. Air quality management agencies
2. The potential air quality environment
   a. Projected emission inventories for alternatives
   b. Predictions of regional air quality
      (1) Methodology
      (2) Results vs. standards
      (3) Model limitations/assumptions
   c. Local air quality
   d. Effects downwind
   e. Integration of project with AQMP
3. Mitigation measures
4. Unavoidable adverse impacts
5. Growth-inducing considerations

Figure 3.22 Air quality impact analysis reporting structure—example 2.

identified. A variety of methodologies are available to fulfill the requirements of each element, and the methodologies selected will depend on the source type and location, experience and engineering judgment, and the requirements of the regulatory and review process.

In addition to the use of an appropriate and adequate methodology, an air quality impact analysis must be founded upon appropriate and adequate information. However, information is not always readily available. As a result, air quality impact analyses are often not founded upon "ideal" information. More frequently, analyses are based upon "available" information. Every effort must be made to locate and utilize the best available information.

Examples of air quality impact analyses are presented later in this chapter in the section on Case Studies for the following two cases:

- A highway
- A power plant

The cases are selected to demonstrate two major source types that are commonly encountered—line (highway) and point (power plant)—and to describe a variety of methodologies available to the practitioner. These sample analyses are highly condensed, and an actual air quality impact analysis would contain considerably greater detail.

The assessment methodologies are presented here for each of the five required elements. The organizational structure of the presentation follows that of Fig. 3.21.

THE EXISTING ENVIRONMENT

General Project Description

A "general" description of the project is appropriate to place into perspective the size and purpose of the project. A "detailed" description is provided in the next section preparatory to the analysis of the air quality impact.

Description of Project Area

Location/Setting A general description of the project area is necessary to identify the major features of the site and surrounding area. The following spatial scales should be presented:

- Site plan
- Local area
- Subregional area
- Regional area

The four scales are required to adequately identify (1) existing sensitive receptors (schools, hospitals, etc.) and (2) sources surrounding the project site (so-called exogenous sources) that contribute to the existing local, subregional, and regional degradation in air quality.

Air Pollution Area/Regulatory Programs A description of the air pollution regulatory areas and regulatory agencies is necessary to identify the guidelines, criteria, and standards applicable to the proposed project. Care must be exercised to insure that accurate and current information on the air pollution regulatory area, active air pollution programs, and applicable emissions and air quality standards is utilized. The local, regional, and state regulatory agencies and the regional office of the EPA should each be consulted to establish the following:

Air Pollution Area

- *Air Basin*—the air basin within which the project is proposed
- *Air Quality Control Region (AQCR)*—the AQCR within which the project is proposed
- *Air Quality Maintenance Area (AQMA)*—the AQMA, if any, within which the project is proposed
- *Regulatory District*—the air pollution regulatory district within which the project is proposed

Air Pollution Programs

- *State Implementation Plan (SIP)*—The status of the SIP and the features of the SIP relevant to the proposed project
- *Air Quality Maintenance Plan (AQMP)*—the status of the AQMP, if any, and the consistency of the proposed project with the AQMP

Emission Standards
- *Federal Emission Standards*—the federal emission standards for stationary sources (New Source Performance Standards) and mobile sources, and applicability to the proposed project
- *State Emission Standards*—the state emission standards for mobile sources and stationary sources, if any, and applicability to the proposed project
- *Regulatory District Emission Standards*—the local emission standards for stationary sources and applicability to the proposed project

Ambient Air Quality Standards
- *National Ambient Air Quality Standards (NAAQS)*—the federal ambient air quality standards and applicability to the proposed project
- *State Ambient Air Quality Standards*—the state ambient air quality standards and applicability to the proposed project

Meteorology

A general description of the weather which includes diurnal and seasonal wind patterns (wind direction and wind speed), atmospheric stability, inversion heights, and frequency of occurrence is necessary to identify those features of the local, subregional, and regional meteorology that influence the air quality in those areas. The description is also helpful in identifying the major sources external to the project site that contribute to the local, subregional, and regional degradation in air quality. In a later section of the analysis (meteorological scenarios), the information is critically assessed to determine those conditions conducive to primary and secondary pollutant impact.

The description begins with a review of meteorological records. Wind speed and direction, atmospheric stability, inversion heights, and topographic influences each require detailed analyses as the following sections describe.

The information outlined in this section is useful in describing the general meteorological picture. Additional analyses may be required in the construction of the impact scenarios.

Wind Rose Summary   The wind rose is used to describe the local wind speeds and directions. A flowchart for constructing a wind rose summary is presented in Fig. 3.23. The chart begins with the identification of information sources. A critical point is the evaluation of whether the meteorological sources are representative of the project area. If not, field measurements may be required to obtain representative wind data.

Data Sources.   The first step in the construction of a wind rose is to collect the meteorological data available for the project area. Sources of meteorological data include:
- Local airports
- Local air pollution control regulatory agencies
- National Climatic Center (NCC), Asheville, North Carolina

The location of the meteorological observation stations should be identified on a topographic map along with the terrain features. If the decision is made that available data are representative of the local meteorology, then construction of the wind rose may proceed without a local monitoring program.

Wind Rose Construction.   An example of a radial bar-type wind rose is presented in Fig. 3.24 and an example of a tabulated wind rose summary is presented in Table 3.14. Note that the 16 compass points correspond to the mariner's compass rose with each direction equivalent to a 22½° sector of a 360° circle.

Prior to January 1964, surface wind data were reported by the U.S. Weather Bureau as one of 16 compass points. On the first day of 1964, the U.S. Weather Bureau changed the wind-direction reporting procedure from sixteen 22½° intervals to thirty-six 10° intervals. The reference (0° mark) for the 36-point wind-direction reporting procedure is the north pole. All degrees are read in a clockwise direction from north. A reading of 09 (90°) indicates a wind direction from due east. Each direction (01, 02, 03, etc.) refers to an angle (10°, 20°, 30°, etc.). The reported wind speed is in miles per hour or knots.

Problems have developed with the 36-point wind system. First, a 36-point system tends to spread tabulated frequencies and obscure directional significance. Second, a list of 36 directions is often too lengthy for convenience. Third, it is almost impossible to construct the standard radial bar-type wind rose with 36 bars. With so many bars in a wind-rose diagram, the bars crowd together at the center of the wind rose and the variation in bar length is minimized. To offset this disadvantage, Table 3.15 provides a methodology for converting 36-point readings to 16-point readings.

Table 3.14 illustrates a 16-point wind rose summary in the form of frequency table of wind direction versus wind speed groups. The highest directional frequency is from the ENE, and the highest speed frequency is 8 to 12 mi/h. The most probable wind speed and direction is the range 8 to 12 mi/h from the ENE. Average wind speeds are also shown for each direction. The average wind speed, $\overline{X}$, is found by multiplying the sum of frequencies times the class midpoint value for each class and dividing by the total number of observations:

$$\overline{X} = \frac{\Sigma f_i X_i}{\Sigma f_i} \tag{3.12}$$

where f_i = frequency of any value
      X_i = midpoint value for each speed class
The average speed should fall within the highest speed frequency range, provided sufficient data are available for a statistical analysis. In the aforementioned case, the average velocity from ENE is 7.3 mi/h, slightly lower than the highest speed frequency

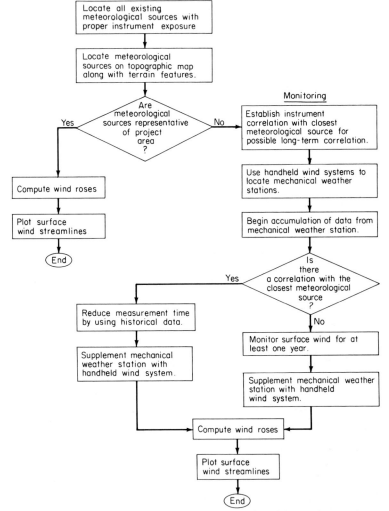

Figure 3.23   Flow chart for meteorological study. *(Adapted from Reference 4.)*

range. Computer programs for wind roses are identified and described at the conclusion of the following section under the heading "Computer Programs for Wind Rose and Stability."

The wind roses should be plotted on a topographic map of the proposed project area. This will allow the evaluation of the applicability of the available wind data for the description of surface wind streamlines within the project area.

Wind speeds and wind directions will vary during the day, and frequency distributions for various periods of the day (e.g., daytime and nighttime) are often necessary for air quality impact analyses. This additional detail, if required, is usually presented in the construction of the impact scenarios (see the section entitled Impact Scenarios later in this

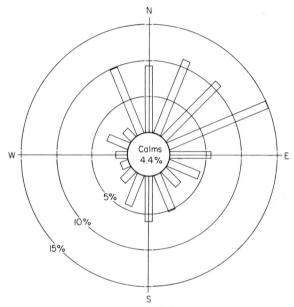

Figure 3.24   Wind rose.

chapter). In the present section, the prevailing meteorology is described in general terms, so the construction of wind rose diagrams for each season are sufficient.

Topography Review.   An evaluation of the topographic map for the area is useful in identifying terrain features that may influence local meteorology. Hills, mountains, canyons, and valleys close to the project site should be identified, and the potential should be established for the following conditions:

- "Channeling" of winds
- Promotion of "drainage" winds
- Promotion of "valley" winds

Large bodies of water must be identified, and the potential must be established for the following conditions:

- Promotion of "sea" or "lake" breezes
- Promotion of "land" breezes

Such an evaluation will (1) facilitate the interpretation of the meteorological data, (2) determine whether sufficient meteorological data are available to describe the local flow, and (3) guide the placement of mechanical weather stations that may be necessary to fill the information gaps.

Monitoring.   In some instances, the available meteorological information is insufficient to describe the surface wind streamlines within the project area, and actual field measurements are required.

Field data are collected using meteorological instrumentation referred to as "mechanical weather stations." The best location for a mechanical weather station can be estab-

TABLE 3.14 Summary of Wind Direction and Speed Data—16-Point Reporting System

Direction	Hourly observations of wind — Knots									Total	Average Speed — Knots	Average Speed — Miles per hour
	Miles per hour											
	0–3 / 0–3	4–6 / 4–7	7–10 / 8–12	11–16 / 13–18	17–21 / 19–24	22–27 / 25–31	28–33 / 32–38	34–40 / 39–46	41 and Over / 47 and Over			
N	8	13	15	18	12	3				69	10.8	12.4
NNE	1	16	28	30	7	1				83	10.2	11.7
NE	7	34	36	5						82	6.7	7.7
ENE	11	46	51	5						113	6.3	7.3
E	6	19	14	4						43	6.4	7.3
ESE	4	15	13	3						35	6.5	7.5
SE	1	13	4	2						20	6.3	7.2
SSE	2	6	20	11						39	8.3	9.6
S	3	11	21	10	1					46	8.2	9.4
SSW	3	9	9	9	4					34	9.3	10.6
SW	1	8	7							16	6.3	7.2
WSW		4	3	1						8	6.9	7.9
W	1	5	7							13	6.5	7.4
WNW	1	16	6	1						24	6.0	6.9
NW	2	3	6	1						12	7.2	8.2
NNW	1	11	29	26	6	1				74	10.6	12.2
CALM	33									33	0.0	0.0
Total	85	229	269	126	30	5				744	7.7	8.9

lished by a preliminary meteorological survey with hand-held wind systems. At least two persons with some form of communication (e.g., CB radios, walkie-talkies) are required. These hand-held wind systems should be located at various points within the project area to allow the spatial and temporal distribution of the surface winds to be evaluated. The communication system will allow each person to discuss the influences of topography on the surface winds. Measurements of wind speed and direction should be made for morning, midday, and evening conditions. This will provide data on the diurnal variation of surface winds. All of the measurements should be made in the absence of frontal activity. The preliminary survey may last from a few days to a month. Mapping the general distribution of surface winds will help to locate the mechanical weather stations in the most sensitive areas (areas where wind speeds appear to be low).

TABLE 3.15   Wind Direction and Speed—Conversion from 36-point to 16-point Reporting System

	Direction	Degree–10
1	North	35–36–01
2	North-Northeast	02–03
3	Northeast	04–05
4	East-Northeast	06–07
5	East	08–09–10
6	East-Southeast	11–12
7	Southeast	13–14
8	South-Southeast	15–16
9	South	17–18–19
10	South-Southwest	20–21
11	Southwest	22–23
12	West-Southwest	24–25
13	West	26–27–28
14	West-Northwest	29–30
15	Northwest	31–32
16	North-Northwest	33–34

SOURCE: U.S. Weather Bureau, *Tape Reference Manual.* (Reference 69).

The number of mechanical weather stations required will depend on topographic features. In some cases where level, open terrain exists, one mechanical weather station may be sufficient. In other cases where the project is located in hilly areas, two or more mechanical weather stations may be required. Once the exact location of the mechanical weather station is selected, the general guidelines for proper instrument exposure should be followed.[116]

Data should be collected for a minimum of one year to completely cover the annual meteorological cycle. The length of time required for the field measurements can sometimes be reduced to a period of a few months if a correlation between the actual field measurements and the closest meteorological monitoring station with a historical data base can be made. It should be stressed that before any correlation procedures are attempted, a calibration of the mechanical weather stations against the closest meteorological source must be made. The reasons for this calibration are as follows:[116]

- The response characteristics of instruments of various manufacture are seldom similar, and, although their directional indications are usually comparable, there may be only general agreement in wind speed.
- The exposure of wind instruments varies, with some being completely shielded from the natural air flow whereas others have excellent exposure.
- Instrument maintenance techniques can vary considerably. Some agencies service the instruments routinely, while others attend to an instrument only after it has broken down.

When performing a meteorological survey or an instrument correlation, the following realities of meteorological processes must be considered:

1. Operational practices for determining "representative" hourly wind direction and

speed range vary from an observation duration of ten seconds to an hour. At some stations, the hourly observations represent a period which includes one-half hour on each side of the stated hour, whereas others represent the preceding hour.

2. Depending on the threshold sensitivity of the instrument, some stations record the occurrence of calms (no wind flow) while other stations do not.

3. Some stations record wind direction to sixteen points of the compass, whereas others record to only eight or to as many as thirty-six points.

If the project area is located in a valley adjacent to hills, localized drainage winds may be channeled through small canyons and gullies between the hills. The surface wind streamlines may intersect the project area at different angles depending on the location of the axes of the canyons or gullies with respect to the project. Under these conditions, it is recommended that at least one mechanical weather station be located to measure the localized drainage winds for the most sensitive areas. To supplement this, measurements of wind speed and direction can be made at other locations with the hand-held wind systems. These measurements are not required each day. They can be made perhaps two or three times a week. These measurements should continue until general drainage wind streamlines can be evaluated.

Atmospheric Surface Stability   The atmospheric surface stability controls the vertical dispersion of pollutants in the vicinity of a source. This factor is especially important to the assessment of primary pollutant impact. Surface stability is governed by the atmospheric lapse rate close to the ground, which in turn is controlled in the following manner by insolation, nocturnal radiation loss, and wind speed:

- *Insolation.* During the day, the sun heats the ground by radiative heat transfer. The ground in turn heats the adjacent air by conduction. Such heating will decrease the stability close to the ground. If clouds are present, however, the radiative heat gain will be less intense and the reduction of stability less rapid.
- *Nocturnal radiation loss.* At night, the ground loses heat due to radiative heat transfer to the universe. The ground in turn cools the adjacent air by conduction. Such cooling will increase the stability close to the ground. If clouds are present, however, the radiative heat loss will be less intense and the increase in surface stability less rapid.
- *Wind speed.* The greater the wind, the more effectively the heat is transferred from the ground to the adjacent air mass. The net effect of high wind speeds is to produce a neutrally buoyant atmosphere at the surface.

The following studies have been conducted to formalize the relationships between atmospheric surface stability and those factors controlling stability insolation, nocturnal radiation, and meteorology.

- *Pasquill Stability Categories.*[108] A classification of stability in accordance with the wind speed and incoming solar radiation for day, or cloud cover for night.
- *Brookhaven Stability Categories.*[92] A classification of stability in accordance with the wind direction fluctuations.
- *TVA Stability Categories.*[15] A classification of stability in accordance with the temperature.

Of the three, the Pasquill method is the most commonly used for the determination of stability class. The selection procedure is shown in Table 3.16. Insolation is estimated by solar altitude and is modified for existing conditions of total cloud cover and ceiling height. At night, estimates of outgoing radiation are made by considering cloud cover. "Night" refers to the period from one hour before sunset to one hour after sunrise. Stability Class D should be used for all overcast conditions during day or night, regardless of wind speed.

Computer Programs for Wind Roses and Stability.   The California Department of Transportation has developed the following computer programs to analyze meteorological records for (1) the surface stability of the atmosphere and (2) wind rose construction:

- STAROS—determines the probability of occurrence of each of the six stability classes (A through F) and their associated wind roses based on meteorological records stored on microfilm by the National Climatic Center (NCC) or compiled from accessible meteorological records that exist at local airports, etc.
- WINDROS—determines the probability of occurrence of each of the six stability classes (A through F) and their associated wind roses based on meteorological records stored on magnetic tapes by the National Climatic Center (NCC).
- WIND—determines wind roses only compiled from any meteorological record without regard to stability class.

A complete user's manual for these computer programs is available.[4] The method employed to estimate the surface stability classes and associated probability of occurrence uses existing meteorological records and is based on the method of Turner[108] for estimating the Pasquill surface stability.

Inversion and Mixing Height   A description of the inversion height and mixing depth is necessary to assess the potential for oxidant formation. If oxidant modeling is to be conducted, a rather extensive quantitative description is required. For example, a description of the hour-by-hour variation in mixing depth for the entire region may be necessary.

The "inversion height" is the base of an elevated inversion as illustrated in Fig. 3.25a. The height may be determined directly from radiosonde data; the local air pollution

TABLE 3.16   Pasquill Chart for Determining Atmospheric Stability Class

| Surface wind speed (at 10 m), m/s | Day | | | Night | |
| | Incoming solar radiation | | | Thinly overcast or 4/8 low cloud | 3/8 cloud |
	Strong	Moderate	Slight		
2	A	A–B	B		
2–3	A–B	B	C	E	F
3–5	B	B–C	C	D	E
5–6	C	C–D	D	D	D
>6	C	D	D	D	D

Neutral class D should be assumed for overcast conditions during day or night.

Stability class	Class description
A	Extremely unstable
B	Unstable
C	Slightly unstable
D	Neutral
E	Slightly stable
F	Stable to extremely stable

SOURCE: Reference 108.

control agency usually compiles the data which is available for use in air quality impact analyses.

The mixing height is the height to which the pollutant mass mixes. Mixing height carries a variety of definitions. The morning mixing height has been most fully characterized by Holzworth[54] though Duckworth and Sandberg,[22] Summers,[102] and Clark[17] provide models and applications of significant value. As Holzworth states:

The morning mixing height is calculated as the height above ground [level] at which the dry adiabatic extension of the morning minimum surface temperature plus 5°C ... [intersects] ... the vertical temperature profile observed at 1200 Greenwich Median Time (GMT).* The minimum temperature is determined from the regular hourly airways reports from 0200 through 0600 Local Standard Time (LST). The 'plus 5°C' ... [allows] ... for the ... effects of the nocturnal and early morning urban heat island since ... the [weather] ... stations are located in rural or suburban surroundings.[54]

The methodology for determining the morning mixing height is based on the scenario illustrated in Fig. 3.25b. The air mass adjacent to the ground is envisaged as one large urban air parcel at the minimum surface temperature plus 5°C. The positive buoyancy of the urban air mass relative to the surroundings forces the air mass to rise as a whole. The mass rises, following the temperature history prescribed by the dry adiabat, to the height of the mixing layer. The air displaced at ground level is replenished by air from the surroundings. As the replenishing air enters the urban area, its temperature rises the arbitrarily assumed 5°C due to the urban activity (e.g., injection of hot combustion products from mobile and point sources into the air mass, heating of the air mass by the hot asphalt pavement, and rejection of waste heat into the air mass from air conditioning systems.)

*1200 Greenwich Median Time (GMT) is a standard time selected by the U.S. Weather Service for temperature soundings.

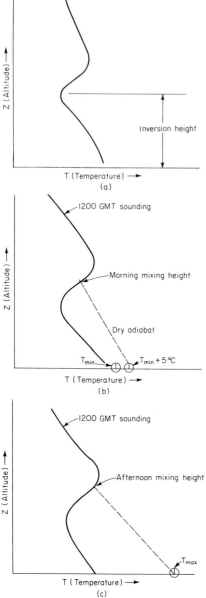

Methods available to calculate the *afternoon* mixing heights have been reviewed by Hanna,[49] who found that a procedure similar to the Holzworth approach was the most practical. Instead of using the minimum surface temperature plus 5°C, the maximum surface temperature observed from 1200 through 1600 LST is used. The methodology is illustrated in Fig. 3.25*c*.

Using these methodologies, isopleths for the seasonal morning mixing height and the afternoon mixing height for the contiguous United States have been calculated by Holzworth.[54] Examples are shown in Fig. 3.26*a* and 3.26*b*, respectively.

Existing Air Quality

An assessment of the existing air quality is required to establish the reference level to which the future year impact of the project can be compared. A full characterization of the existing air quality requires that the analysis be conducted for the regional, subregional, and local scales.

Regional Air Quality An analysis of recent ambient-air-monitoring data collected by the local air pollution regulatory agency throughout the air basin is used to describe the existing regional air quality. The records of previous years are analyzed to establish yearly and seasonal trends and to establish the spatial and temporal variation of air quality in the region. Care should be taken to assess any bias in the ambient-monitoring data due to major sources of air pollutant emissions near the air monitoring stations. On the regional scale, emphasis will usually be focused on secondary pollutants. Most local regulatory agencies conduct analyses of required air quality data as a regular practice and provide the information in a format preferred by the agency for air quality impact analyses.

A convenient general source for the analysis of meteorological data relative to secondary pollutant impact is provided by Holzworth.[54] The example presented in Fig. 3.27 illustrates isopleths that describe the propensity of meteorological conditions conducive to secondary pollutant impact. The parameter adopted by Holzworth is the "number of episode days in a five-year period." An "episode" is defined as a period lasting at least 2 days for which the mixing height is less than 1500 m, wind speeds are less than 6 m/s, and no significant precipitation has occurred.

Figure 3.25 Inversion and mixing heights. (*a*) Inversion heights; (*b*) morning mixing height; (*c*) afternoon mixing height.

Subregional Air Quality The description of subregional air quality is obtained from data collected at air monitoring stations located in the vicinity of the project area.

Explanations are appropriate for trends in air quality data that are anomalous to the region as a whole. Examples of factors that influence subregional air quality include the spatial distribution of sources and source strength, topography, and meteorology.

In the event air monitoring data are not available, or the closest air monitoring station to the project site (1) is too distant to be representative of the subregional air quality, (2) is relatively new (hence provides limited data) or (3) is not fully equipped to monitor all the pollutants of interest, then modeling, monitoring, or a combination of both may be necessary.

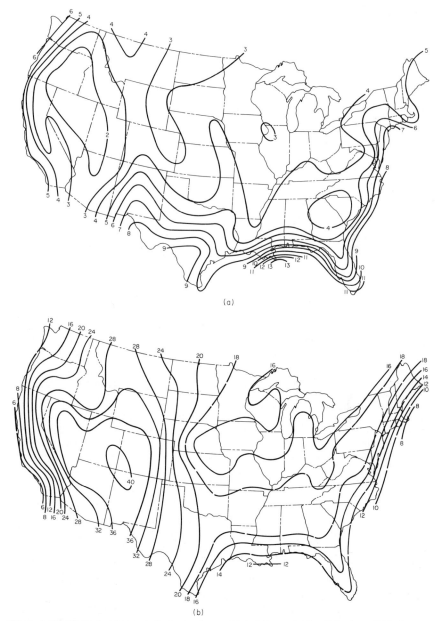

(a)

(b)

Figure 3.26   Mixing height isopleths for contiguous United States. (a) Isopleths (m × 10²) of mean summer morning mixing height; (b) isopleths (m × 10²) of mean summer afternoon mixing heights. (*From Reference 54.*)

Modeling. Two approaches are available to describe subregional air quality in the absence of sufficient data: interpolative mapping and climatogical dispersion modeling.

The first approach, Interpolative Mapping, uses measured data from individual air-monitoring stations to estimate pollutant concentration between the stations by dimensional interpolation.[6,47] A second approach to modeling subregional air quality is the Climatological Dispersion Model QC (CDMQC) developed by EPA.[110] In this approach, the seasonal or annual concentrations of conservative pollutants are calculated using

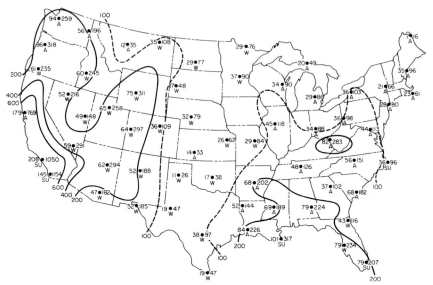

Figure 3.27 Episode day isopleths. Episode days in 5 years. Episode day: mixing heights ≤ 1500 m, wind speeds ≤ 6.0 m/s, and no significant precipitation—minimum of two consecutive days. Numerals on left and right give total number of episodes and episode days, respectively. Season with greatest number of episode days indicated as W (winter), SP (spring), SU (summer), or A (autumn). *(From Reference 54.)*

Gaussian plume modeling, average emission rates from point and area sources, and a joint frequency distribution of wind direction, wind speed, and stability for the same period. Not only is identification of the sources within the subregional area required, but the emission factors and usage rates for each must be determined as well. As a result, use of the CDM is justified for major projects only. However, the CDM is also valuable for describing the future-year impact. Once the analysis of the meteorological data is complete, only the existing year and future year spatial and temporal distributions of the pollutant emissions need be determined.

Monitoring. Should the available air quality data be insufficient to adequately describe the subregional air quality, then ambient air quality monitoring may be required. The major disadvantages of monitoring when compared to modeling are the time and expense involved. The major advantage is the collection of actual air quality data (as compared to predicted values).

Monitoring should be conducted at a height of 2 m, and at a site at which pollutant concentrations are representative of the area. Pollutant concentrations should not be subject to local influences such as nearby emission sources (e.g., major traffic arteries or point sources), high walls or buildings, trees, or highly localized meteorological conditions.

The selection and operation of air quality monitoring instruments require experience and care to insure the collection of data which are credible, representative, and technically defensible. Methods range from those that are inexpensive, cursory, and of short duration[87] to those that are expensive, comprehensive, and of long duration.[90] The local air pollution regulatory agency and regional EPA Office should be contacted to establish the need to monitor, the extent of monitoring necessary, and for recommendations of local consultants experienced in air monitoring. Morgan[67] reviews procedures applicable to ambient monitoring.

Local Air Quality The air quality at the project site may be locally degraded due to a major pollutant source near the project site. Notable examples include major roadways and large point sources such as fossil-fueled electric generating plants.

In cases where local sources impact the existing local air quality, modeling is required to estimate the degree of impact. Monitoring may be required to evaluate the accuracy of the model, but modeling is useful in (1) providing concentration isopleths, and (2) estimating the impact on air quality due to temporal variation (daily, monthly, seasonal) in the metorology and source strength. Modeling methodologies are presented in the section on Air Quality Modeling later in this chapter.

If monitoring is employed to check the accuracy of the modeling, the precautions outlined in the preceding section are applicable.

ENVIRONMENTAL IMPACT

Determination of the impact of the proposed project on air quality is the major challenge and goal of an analysis. An air quality analysis, to be meaningful, must indicate in quantitative terms whether the project in question will cause pollutant levels to exceed the ambient air quality standards established by the federal and state governments. Although emissions from the project in and by themselves may not exceed the standards, the total emissions in the project environs may produce a degradation in air quality sufficient to exceed the standards. In regions where Prevention of Significant Deterioration (PSD) regulations prevail, degradation in air quality will be limited to set increments even though the project impact itself does not violate the ambient air quality standards.

Modeling air quality impact requires the construction of emission, meteorological, air chemistry, and impact scenarios, as well as the calculation of impacts using the necessary models. In contrast to an analysis of the existing air quality, an analysis of the future-year impact of the proposed project must rely upon estimates. As a result, a thorough review and understanding of a preceding section of this chapter entitled Environmental Impacts is recommended as an appropriate background to the following sections.

For each future year analyzed, the following two alternatives must be addressed:

- *No-build*—The conditions that will occur should the project *not* be undertaken and completed
- *Build-out*—The conditions that will occur should the project be undertaken and completed

In addition to the build-out alternative, an analysis of the construction activity must be included.

The comparison of the two alternatives, no-build and build-out, is necessary to relate the estimated impact of the project to the future year air quality that would occur in the absence of the project.

Detailed Project Description

Build-out A detailed description of the project must be provided with emphasis on those features of the project that are associated with pollutant emissions. The mobile and stationary source activity associated with the project should be identified and significant variations in the diurnal, weekly, monthly, and seasonal use of the sources should be presented. If a project is phased or is part of a larger project, the impact of all phases combined or the larger project must be considered *in toto*.

Identified here are the major primary pollutants emitted by the project, the associated impacts expected and the scale on which the impacts are expected, and the applicable ambient air quality standards.

No-build A detailed description of the conditions that will exist in the absence of the project must be provided to facilitate an analysis of the no-build alternative.

Analysis Years

The estimated impact of the project will depend upon the specific years selected for the future-year impact analyses. The reason for this concerns the future-year emission factors for mobile and stationary sources. In general, the future-year emission factors decrease yearly as a result of increasingly more stringent promulgated or projected emission standards. Serving to offset the yearly reduction in future-year emission standards is the year-by-year increase in source activity. Because of the effect of the choice of analysis years upon the estimated impact, the selection of the years must be conducted with care.

The following guidelines may be used in the selection of the years for the future-year impact analyses:

• An analysis for the year of construction to account for the impact of heavy road equipment.

• An analysis for the first year of operation of the project (estimated time of completion—ETC).

• For phased projects, an analysis for each year during which a phase becomes operational.

• If deemed appropriate, an analysis for any time period for which there is (1) a substantial change in project utilization or activity in the adjacent or affected area, or (2) a substantial change in future-year emission factors associated with the dominant sources.

• Analyses to meet requirements established by the lead agency responsible for the preparation of the air quality impact analysis.

• An analysis for the existing year should existing air-monitoring data prove insufficient in describing the local air quality.

Description of Surrounding Area

A description of anticipated future development in the surrounding area is necessary to establish the existing and expected character of the project environs. Such a description should provide the information necessary to determine the existing and future-year impact of the project on the surroundings (project impact) and also provide the information necessary to determine the existing and future-year impact of the surroundings on the project area.

The existing and future-year impact of the project on the surroundings requires identifying the *sensitive receptors* likely to be impacted by the project. The existing and future year impact of the surroundings on the project area requires identifying the sources surrounding the project site (so-called exogenous sources) that impact the project site. This information is obtained using the meteorological scenarios for primary and secondary pollutants described later in this section. Any differences in the exogenous sources between the two alternatives (no-build and build-out) must be identified and described.

Traffic Study

The automobile is currently the major source of air pollutant emission in many air quality impact analyses. Documenting the use of the automobile is often elusive, yet the quantification of vehicular use is the key to an adequate and representative analysis of both the existing and future-year air quality.

For the existing and each analysis year, a traffic study is required for the no-build alternative as well as the build-out alternative. The former information is required to assess the air quality for the analysis years should the project not be implemented, whereas the latter information is required to assess the air quality for the analysis years should the project be completed.

Existing Traffic As a basis for developing detailed data on the existing vehicular emissions, a comprehensive traffic analysis must be prepared. In particular, an analysis of the existing traffic conditions in and around the site of the proposed project is an input necessary to modeling the existing local air quality.

Many local public works agencies maintain current data on vehicular usage, and many have detailed data regarding such usage in and about the site of the proposed project. Should the local public works agency not have sufficient data, a qualified traffic engineering group must be retained to provide the roadway use data. For proposed sites in the proximity of major parking and traffic use facilities, the retention of a traffic engineering group may be necessary to quantify the traffic flow (e.g., idle time, signalization).

Future-Year Traffic The traffic study required for the future-year impact analysis will be more comprehensive than that required for the existing air quality analysis, and it will rely upon estimates of projected future-year mobile source use. For each future-year analysis, the traffic study should contain a roadway analysis which should include the following information for all major roadways in the vicinity of the project and within the project itself:

- Roadway capacity
- Levels of service
- Traffic and average speed for the time period from 6:00 to 9:00 A.M., the peak 8-hour, the peak hour, and the daily average

The 6:00 to 9:00 A.M. period is used for projections of oxidant impact, the peak hour and

REAL WORLD

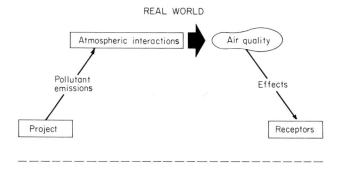

- -

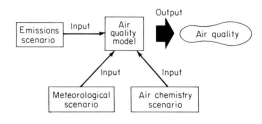

IMPACT MODELING

Figure 3.28   Relation between real world and impact modeling.

peak 8-hour are used for projections of CO impact, and the daily average is a basic reference adopted for traffic analysis.

For a project of appreciable size, it may be appropriate to develop a geocoded traffic network using the Universal Transverse Mercator (UTM) grid system (e.g., determine the x-y coordinate values for the major nodes of the traffic network using the UTM grid system). This is appropriate should geocoded emission inventories be desired for regional air quality modeling.

Impact Scenarios

The future-year air quality impact of a project is determined by modeling.* As shown in Fig. 3.28, air quality models require input information that characterizes the emissions, meteorology, and air chemistry of the project area. The purpose of this section is to identify the required information.

The approach taken is to establish impact scenarios. By definition, an "impact scenario" is that combination of emissions and atmospheric interactions (meteorology and air chemistry) which occur simultaneously to produce an air quality impact. Once constructed, the impact scenarios provide the information required to identify the appropriate input data required by the air quality models.

---

*Modeling may also be required to determine the impact of an established source on existing air quality. (See the earlier section of this chapter on Existing Air Quality.)

Impact scenarios are constructed for a minimum of two cases for each analysis:
- *Most probable impact*—That combination of emissions and atmospheric interaction that produces the most frequently encountered air quality impact
- *Worst case impact*—That combination of emissions and atmospheric interaction that produces the worst air quality impact

The most probable case provides a basis for estimating whether an air quality standard is exceeded and the level to which the standard is generally exceeded. As a result, the most probable case provides the basis for comparing the project impact to the goal of the Clean Air Act Amendments which limit the exceedance of any one air quality standard to one day a year. The worst case is useful in identifying the highest impact that will occur in a given year.

The first step toward the construction of the two impact scenarios is to separately characterize the emissions, the meteorology, and the relevant air chemistry, and to separately determine the most probable and worst case emissions and meteorology. Second, the emissions, meteorology, and air chemistry are evaluated in *combination* to establish the most probable and worst case impact scenarios. Sources at the project site and sources exogenous to the project site may be sufficiently different to warrant separate scenarios.

Pollutant	kg/h (average)
HC	—
CO	—
SO$_2$	2327
NO$_x$ (as NO$_2$)	1362
particulates	189

(a)

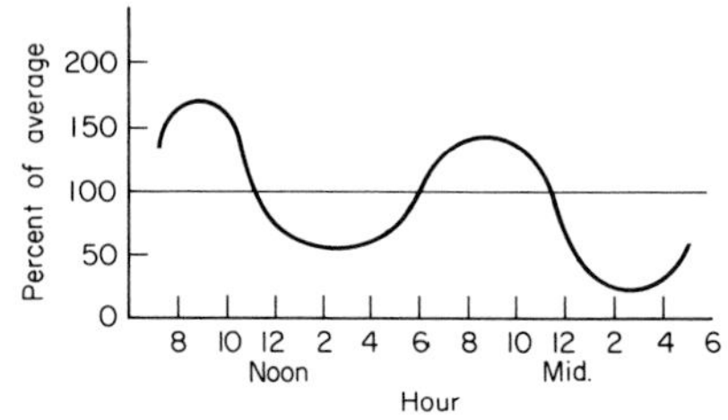

(b)

Figure 3.29 Emissions scenario—an example for a point source. (*a*) Emission inventory; (*b*) temporal variation.

Emissions Scenarios An "emissions scenario" identifies the spatial distribution of sources, the type of pollutants emitted, the emission rate, and the temporal variation of emissions.

Governing Factors. An example of an emissions scenario for a single, stationary source is presented in Fig. 3.29. The pollutants emitted, the average emission rate, and the diurnal variation of emissions are shown. From the emissions scenarios, the most probable and worst case emissions conditions can be determined. The most probable condition will generally correspond to a peak in the "emissions." The worst case condition will correspond to an unusually high emission rate generally caused by an abnormally high usage rate.

The construction of such a scenario relies upon the following four governing factors: source type, usage rate, scenario types, and spatial distribution of the sources.

SOURCE TYPE. If both stationary and mobile sources are present, emission inventories will be prepared for both. For compound sources, mobile source inventories may be further divided. For example, an airport project may have four mobile source inventories—one for aircraft operation (approach, idle and taxi, and takeoff), one for ground support vehicles, one for commercial vehicles, and one for noncommercial vehicles.

USAGE RATE. The usage rate of the sources is a major determinant of the total pollutant emission. The higher the usage rate, the greater the source impact. For some projects, the usage rate varies by season, month, week, or even day of the week. Should the variation be significant, temporal distributions will be necessary for these time periods as well.

SCENARIO TYPES. A minimum of four types of scenarios is required for an air quality impact analysis: existing, no-build, construction, and build-out.

The existing scenario provides emissions data for sources existing *at the project site* and sources *exogenous to the project site*. The construction scenario provides emissions data for sources associated with construction of the project. The no-build scenario provides emissions data assuming the project is not implemented. The build-out scenario provides emissions data for completion of the project as planned.

SPATIAL DISTRIBUTION. For some projects a fourth variable—spatial distribution of sources—may be important. In such cases, the distribution of sources is identified. The identification can vary in sophistication. For example, those sources local to the project (which may impact the ambient levels of primary pollutants in and about the project site) can be separated from those sources nonlocal to the project (which may impact the ambient levels of secondary pollutants in and about the project site). A more formal approach is to subdivide the project and surrounding areas into cells and develop an individual emission scenario for each cell. Cell dimensions of 2 km are standard, but further division may be necessary to improve resolution. To facilitate data handling, location coordinates are assigned to each cell using a standard grid system such as the Universal Transverse Mercator (UTM). The inventory is then said to be geocoded.

The formulation of geocoded emission inventories, in general, and temporally distributed geocoded emission inventories, in particular, is tedious, time consuming, and is usually justified only in select circumstances. The select circumstances may include the use of sophisticated air quality oxidant models. Such models require a geocoded emission inventory as input. The standardization of geocoding to the UTM system has the advantage of allowing the data base to be used for new oxidant models as well as current models.

The select circumstances may also include the use of the geocoded inventory itself for a qualitative assessment of oxidant impact. The geocoded inventory could be adopted as a tool for identifying the location of the high emission of hydrocarbon and oxides of nitrogen. Then, by overlaying the prevailing daytime meteorology, likely areas of impact from the resulting production of oxidant could be ascertained.

Scenario Construction. The development of an emission scenario begins with the calculation of the emission rate associated with each pollutant species. Emission rates are estimated using the following expression:

$$\overset{\text{Tabulated}}{\overbrace{\hspace{2cm}}} \quad \overset{\text{Project data}}{\overbrace{\hspace{2cm}}}$$
$$\text{Emission rate (kg/h)} = \text{emission factor} \times \text{usage rate} \tag{3.13}$$

As noted, emission factors are tabulated, and the usage rates are obtained from project data. The Environmental Protection Agency (EPA) tabulates emission factors for the commonly encountered source types.[24] The tabulation is called "AP-42." An example of the EPA tabulation of emission factors for fuel oil combustion is shown in Table 3.17. AP-42 is updated regularly as new information becomes available.

Emission factors for motor vehicles are also provided in AP-42 but in a form different from stationary sources. The many variables involved require the use of a "composite" emission factor defined by

$$e_{npstw} = \sum_{i=n-12}^{n} c_{ipn} m_{in} v_{ips} z_{ipt} r_{iptw} \tag{3.14}$$

where e_{npstw} = composite emission factor (g/mi) for calendar year n, pollutant p, average speed s, ambient temperature t, and percentage cold operation w.

c_{ipn} = the mean emission factor for the ith model year light-duty vehicles during calendar year n and for pollutant p.

m_{in} = the fraction of annual travel by the ith model year light-duty vehicles during calendar year n.

v_{ips} = the speed correction factor for the ith model year light-duty vehicles for pollutant p and average speed s.

z_{ipt} = the temperature correction factor for the ith model year light-duty vehicles for pollutant p, and ambient temperature t.

r_{iptw} = the hot/cold vehicle operation correction factor for the ith model year light-duty vehicles for pollutant p, ambient temperature t, and percentage cold operation w.

Pertinent data for each of the variables is provided in AP-42.[24] Similar expressions are also provided for light-duty trucks and heavy-duty trucks.

The AP-42 document identifies the current and most appropriate expressions to use for all sources. Care must be taken, however, to insure that the latest edition or supplement to AP-42 is used.

Care must also be exercised to insure that the AP-42 factor truly represents the emissions expected from the source in question. Frequently, the AP-42 factors underesti-

TABLE 3.17   Emission Factors—Example for Fuel Oil Combustion

	Type of boiler[a]							
			Industrial and commercial				Domestic distillate oil	
	Power plant residual oil		Residual oil		Distillate oil			
Pollutant	lb/10³ gal	kg/10³ liter	lb/10³ gal	kg/10³ liter	lb/10³ gal	kg/10³ liter	lb/10³ gal	kg/10³ liter
Particulate[b]	c	c	c	c	2	0.25	2.5	0.31
Sulfur dioxide[d]	157S	19S	157S	19S	142S	17S	142S	17S
Sulfur trioxide[d]	2S	0.25S	2S	0.25S	2S	0.25S	2S	0.25S
Carbon monoxide[e]	5	0.63	5	0.63	5	0.63	5	0.63
Hydrocarbons (total, as CH₄)[f]	1	0.12	1	0.12	1	0.12	1	0.12
Nitrogen oxides (total, as NO₂)[g]	105(50)[h,i]	12.6(6.25)[h,i]	60[j]	7.5[j]	22	2.8	18	2.3

[a]Boilers can be classified, roughly, according to their gross (higher) heat input rate, as shown below.
Power-plant (utility) boilers: $>250 \times 10^6$ Btu/h
$$(>63 \times 10^6 \text{ kg-cal/h})$$
Industrial boilers: $>15 \times 10^6$ but $<250 \times 10^6$ Btu/h
$$(>3.7 \times 10^6 \text{ but } <63 \times 10^6 \text{ kg-cal/h})$$
Commercial boilers: $>0.5 \times 10^6$, but $<15 \times 10^6$ Btu/h
$$(>0.13 \times 10^6, \text{ but } <3.7 \times 10^6 \text{ kg-cal/h})$$
Domestic (residential) boilers: $<0.5 \times 10^6$ Btu/h
$$(<0.13 \times 10^6 \text{ kg-cal/h})$$
[b]Based on References 3 through 6. Particulate is defined in this section as that material collected by EPA Method 5 (front half catch)[7].
[c]Particulate emission factors for residual oil combustion are best described, on the average, as a function of fuel oil grade and sulfur content, as shown below.
Grade 6 oil: lb/10³ gal $= 10 (S) + 3$
$$[\text{kg/10}^3 \text{ liter} = 1.25 (S) + 0.38]$$
where S is the percentage, by weight, of sulfur in the oil
Grade 5 oil: 10 lb/10³ gal (1.25 kg/10³ liter)
Grade 4 oil: 7 lb/10³ gal (0.88 kg/10³ liter)
[d]Based on References 1 through 5. S is the percentage, by weight, of sulfur in the oil.
[e]Based on References 3 through 5 and 8 through 10. Carbon monoxide emissions may increase by a factor of 10 to 100 if a unit is improperly operated or not well maintained.
[f]Based on References 1, 3 through 5, and 10. Hydrocarbon emissions are generally negligible unless unit is improperly operated or not well maintained, in which case emissions may increase by several orders of magnitude.
[g]Based on References 1 through 5 and 8 through 11.
[h]Use 50 lb/10³ gal (6.25 kg/10³ liter) for tangentially fired boilers and 105 lb/10³ gal (12.6 kg/10³ liter) for all others, at full load, and normal (>15 percent) excess air. At reduced loads, NO_x emissions are reduced by 0.5 to 1 percent, on the average, for every percentage reduction in boiler load.
[i]Several combustion modifications can be employed for NO_x reduction: (1) limited excess air firing can reduce NO_x emissions by 5 to 30 percent, (2) stage combustion can reduce NO_x emissions by 20 to 45 percent, and (3) flue gas recirculation can reduce NO_x emissions by 10 to 45 percent. Combinations of the modifications have been employed to reduce NO_x emissions by as much as 60 percent in certain boilers. See section 1.4 for a discussion of these NO_x-reducing techniques.
[j]Nitrogen oxides emissions from residual oil combustion in industrial and commercial boilers are strongly dependent on the fuel-nitrogen content and can be estimated more accurately by the following empirical relationship:
lb NO₂/10³ gal $= 22 + 400 (N)^2$
$$[\text{kg NO}_2\text{/10}^3 \text{ liters} = 2.75 + 50 (N)^2]$$
where N is the nitrogen content, by weight, in the oil.
SOURCE: Reference 24. (NOTE: References cited above are those contained in this source.)

mate or overestimate the expected emission. The reason is that the AP-42 factors are developed from source-monitoring surveys based on existing installations. As a result, the factors represent a "family of existing sources" and are often biased to the region in the county selected for the survey. To emphasize this, the following disclaimer appears in the introduction of AP-42:

The reader must be herein cautioned not to use these emission factors indiscriminately. That is, the factors generally will not permit the calculation of accurate emissions measurements from an individual installation. Only an on-site source test can provide data sufficiently accurate and precise to use in such undertakings as the design and purchase of control equipment or the initiation of a legal action. Factors are more valid when applied to a large number of processes, as, for example, when emission inventories are conducted as part of community or nationwide air pollution studies.[24]

In addition to AP-42, the other sources available for emission rate data include:

- *The references cited in each AP-42 summary.* The raw source-monitoring data used to develop the AP-42 emission factor may provide emission factors more representative of the equipment in question.
- *The state and local air pollution control agencies.* The local agency will usually have source-monitoring and operational data for the equipment class in question. In addition, local regulations may restrict emissions preferentially for the geographical area of interest and, as a result, determine explicitly the maximum emission allowed.
- *The equipment manufacturers.* Equipment manufacturers often conduct source monitoring surveys or are aware of past surveys that should be considered. The manufacturers will also be aware of any tested design changes geared to improve future-year emission performance.

The emission factor that is most representative of the proposed project should be selected only after a survey of the available information sources. Should any doubt remain, source monitoring can be conducted on equipment (if available) identical to or similar to that proposed.

Meteorological Scenarios   The success of air quality models depends to a great extent upon the meteorological parameters selected for input and the values assigned to each parameter. The number of meteorological inputs will depend upon the air quality model used. Proportional models and statistical models require no meteorological input. The Gaussian models require:

1. Local surface wind speed and direction
2. Local surface atmospheric stability

By contrast, the three-dimensional grid models used for oxidant formation require, for each hour that is to be modeled:

1. Spatially (horizontal and vertical) distributed (geocoded) wind speed and direction
2. Spatially (horizontal and vertical) distributed (geocoded) atmospheric stability
3. Spatially (horizontal) distributed mixing depth
4. Solar insolation

Values for the input parameters are assigned by undertaking a systematic analysis of the local and regional meteorological data, as well as local and regional aerodynamic and topographical influences. Included in the analyses are the construction of streamlines, assessment of aerodynamic effects of obstructions to wind flow, distortion of wind flow near natural terrain irregularities, and an assessment of those factors (e.g., sea or lake breeze, drainage winds, valley winds, global and macroscale motion, land breezes, channeling, wind shear, and weather) that influence the input parameters (wind speed and direction, stability, mixing height, and insolation).

To identify those meteorological conditions for input to the air quality models, it is necessary to establish the meteorological scenarios that are conducive to producing the most probable and worst case air quality impact—first for primary pollutants and next for secondary pollutants. As with the emissions scenarios, a minimum of four separate primary pollutant and secondary pollutant meteorological scenarios: (existing, no-build, construction, and build-out) may be required should variations in source types so warrant.

Meteorological Scenarios—Primary Pollutants.   A "meteorological scenario" for primary pollutants identifies those meteorological conditions most conducive to local air quality and the time of occurrence.

GOVERNING FACTORS.   For primary pollutants, attention is directed to the atmospheric interaction in the near vicinity of the source. The atmospheric variables that control primary pollutant dispersion near the source include wind speed and direction, mixing height, and atmospheric surface stability. "Wind speed" controls how fast a pollutant is transported horizontally away from the source, as well as the amount of diluent air mixed with the pollutant mass (the lower the wind speed, the less the dilution and the higher the impact). "Wind direction" controls the receptor area impacted relative to the source (a persistent wind direction with a minimum of variation will maximize impact, especially if directed toward a sensitive receptor). The "mixing height" controls the volume within which the pollutant is dispersed (the lower the height, the greater the impact). The "atmospheric surface stability" controls the rate at which the pollutant is transported vertically away from the source (the more stable the atmosphere, the lower the rate of vertical transport). Whether a lower rate of vertical transport increases the impact depends on the height of the source above ground level.

GROUND LEVEL SOURCE. For a ground level source, a stable atmosphere produces the maximum impact as illustrated in Fig. 3.30. As a result, the atmospheric conditions most conducive to air quality impact from a ground level source are:

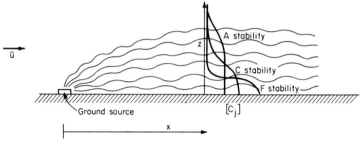

Figure 3.30 Ground source—stable atmosphere. Pollutant concentration profiles at a distance x from a ground source, calculated as a function of height above ground level and atmospheric stability.

- Wind speed—calm to low (≈ 1 m/s)
- Direction—persistent and toward populated areas and/or sensitive receptors
- Mixing height—low and persistent
- Stability—very stable (F stability)

Whether or not these conditions produce the worst case and/or most probable air quality impacts requires the construction of an impact scenario (i.e., a comparison of the emissions and meteorological scenarios).

ELEVATED SOURCE. For an elevated source, a stable atmosphere also restricts vertical transport as shown in Fig. 3.31. However, in this case the pollutant mass remains concentrated and stratified in an elevated layer, and the impact at ground level is a minimum. The maximum impact occurs when surface stability is not stable and the vertical rise of the plume is limited. The case is illustrated in Fig. 3.32 and is called a trapping plume. As a result, the atmospheric conditions most conducive to air quality impact for an elevated source are:

- Wind speed—light to strong (2 to 15 m/s)
- Direction—persistent and toward populated areas and/or sensitive receptors
- Mixing height—low, persistent, and sufficiently high to entrap elevated source emissions
- Stability—trapping conditions

Whether or not these conditions produce the worst case and/or most probable air quality impacts requires the construction of an impact scenario (i.e., a comparison of the emissions and meteorological scenarios).

SCENARIO CONSTRUCTION. For use in modeling, the meteorological scenarios must be described in more than subjective terms. The actual wind speed and direction, mixing height, and surface stability must all be identified. The task here is to construct the meteorological scenarios by analyzing actual meteorological data. For primary pollutant impact, attention is directed to meteorological data in the near vicinity of the source.

The National Climatic Center (NCC) maintains magnetic tapes containing meteorological data for many stations in the United States. For data recorded prior to 1965, hourly observations are available on magnetic tape. For data recorded after 1965, only every third hourly observation is available on magnetic tape (with the exception of most Air Force and

Navy stations for which hourly data continue to be available). If possible, tapes prior to 1965 should be used. Should a National Weather Bureau Station not be located in the near vicinity of the project, meteorological data from airports and air pollution control districts that operate meteorological stations must be used.

This analysis is generally grouped by season, though other divisions may at times be appropriate (e.g., months). For each day of each season, the hourly observation of wind speed and direction, mixing height, and surface stability may be tabulated, then analyzed to identify the most probable condition for each hour of each season. Though sufficient, this approach is tedious, and the following alternative methods may be employed to ease the burden.

COMPUTER ANALYSIS. The computer program introduced previously, WIN-DROS, uses the hourly meteorological data from magnetic tape to calculate a joint frequency distribution, ϕ (k, l, m), of a wind direction sector k, and wind-speed class l, and a stability category index m^*. The data must still be manually reviewed to identify the hour of the day associated with the various conditions. Identification of the hour is necessary to establish the *impact* scenarios since time-of-day is the one variable common to both the emissions and meteorological scenarios.

ENGINEERING JUDGMENT. Engineering judgment is often sufficient to identify the meteorological scenario most conducive to pollutant impact. Consider a ground level source for example. The

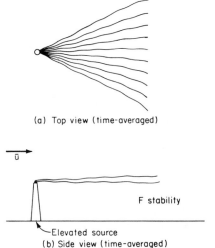

(a) Top view (time-averaged)

F stability

Elevated source
(b) Side view (time-averaged)

Figure 3.31   Elevated source—stable atmosphere.

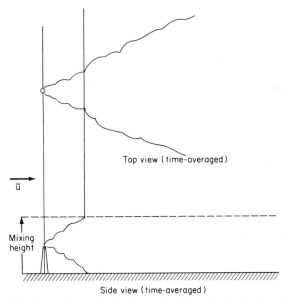

Top view (time-averaged)

Mixing height

Side view (time-averaged)

Figure 3.32   Elevated source—unstable atmosphere and limited mixing height.

---

*The computer program, STAROS, conducts the same calculations but uses as data input meteorological data compiled from accessible meteorological records of local airports, APCD's, etc.

most unfavorable meteorological conditions will usually exist in early morning and late evenings as surface based inversions. These occur generally with clear skies and light winds. The winter season inversions are usually the most persistent. These surface-based inversions generally last until midmorning when the incoming solar radiation warms the ground surface and, through convective mixing, breaks the inversions. As a result, ambient concentrations of primary pollutants emitted from ground level sources can be expected to be high in the early morning hours following long winter nights. The worst case and most probable wind speed and wind direction must be deduced from the meteorological data.

Meteorological Scenarios—Secondary Pollutants. A "meteorological scenario" for secondary pollutants identifies those meteorological conditions most conducive to subregional air quality impact and the time period of occurrence.

GOVERNING FACTORS. For secondary pollutants, attention is directed to interaction between the receptor site of interest and the sources upwind. For most sources, this will encompass an area of subregional scale. The atmospheric conditions conducive to impact include wind speed and direction, mixing height, and insolation. In addition, the length of time over which these conditions persist is important in establishing the time available for secondary pollutants to form. "Wind speed" controls how fast the secondary pollutant precursors are transported away from the source and, more importantly, the amount of diluent air mixed with the pollutant mass (the lower wind speed, the higher the impact).* "Wind direction" determines which receptor area is impacted and also the amount of fresh primary pollutants injected into the air mass while in transport. The "mixing height" controls the volume within which the secondary pollutant precursors are dispersed (the lower the height, the higher the concentration of the secondary pollutant precursors, the more rapid the rate of secondary pollutant formation, and the greater the impact). "Insolation" controls the rate of reaction when photochemical reactions (e.g., ozone formation) are involved (the more intense the insolation, the more rapid the formation of photochemically produced secondary pollutants).

As a result, the atmospheric conditions most conductive to secondary pollutant air quality impact are:

- Wind speed—low
- Direction—persistent, toward a populated area, and over a source area rich in secondary pollutant precursors
- Mixing height—low and persistent, but sufficiently high to entrap elevated source emissions
- Insolation—high and coincident with the conditions above

SCENARIO CONSTRUCTION. To describe the meteorological scenarios for secondary pollutants in more than subjective terms, meteorological data must be analyzed to establish the actual wind speeds and directions, mixing heights, and insolation associated with impact. An ideal methodology would provide a joint frequency distribution of occurrence of wind sector k, and wind speed class l, and mixing height range n, and an insolation class i—all as a function of location and time within the subregional area. (It is this spatial and temporal resolution that distinguishes secondary pollutant meteorological scenarios from primary meteorological scenarios.)

Methodology development is in its infancy. As secondary pollutant models become more sophisticated, the need increases for improved methodologies for the identification of secondary pollutant meteorological scenarios. However, the demand for improved methodologies is created only by unusually large projects such as new major airports, new major highways, new major power plants, planning for the development of large areas, or evaluating alternative transportation strategies for major urban areas. The majority of air quality impact analyses do not require sophisticated secondary pollutant analyses. Oxidant impact, for example, is produced by emissions of reactive hydrocarbons and nitrogen oxides over a large source area. One project will not significantly affect oxidant unless the emission of either reactive hydrocarbons or nitrogen oxides approaches a percent or so of the total emissions in the subregional area. As a result, a subjective discussion of the meteorology in the area of secondary impact is usually sufficient.

The state of the art notwithstanding, many agencies require an evaluation of the project's consistency with the basinwide control strategy when applicable (e.g., an

---

*The rate of secondary pollutant formation increases as the concentration of the secondary pollutant precursors increases.

approved Air Quality Maintenance Plan). In such cases, the oxidant impact of individual projects is considered using a coordinated, basin-wide modeling effort. Modeling efforts of this magnitude are beyond the scope of most individual projects and, if not coordinated, are liable to produce inconsistent results.

Air Chemistry Scenarios   An "air chemistry scenario" is a chemical description of the formation of secondary pollutants.

Governing Factors.   Chemical descriptions of the formation of secondary pollutants are based on laboratory simulations of atmospheric conditions. The controlled atmospheres are necessary to accurately maintain and monitor changes in species concentrations. The measurements obtained from these experiments are then used to construct the chemical descriptions.

Two factors govern the accuracy with which such simulations represent the actual atmospheric conditions: the understanding we have of the conditions that occur in the atmosphere, and the ability to simulate those conditions in the laboratory. The characterization of the chemical and physical properties requires expensive monitoring programs. Both aircraft monitoring and ground monitoring are required to determine the temporal and spatial variation of species concentration in the vertical direction as well as at ground level. Tracers must be released to assess the direction and rate of transport of both the secondary pollutant precursors and the secondary pollutants themselves.

The effectiveness of the simulation in the laboratory depends not only on our knowledge of the events we are attempting to simulate but also on our ability to accurately simulate the sun spectrum, to prevent the boundaries of the vessel from interacting with the reaction, and to measure the low concentrations and small changes in concentrations. Even if these challenges are resolved, there still remain features of the real world that are important to the chemical changes in the atmosphere but which are not amenable to simulation. For example, the actual atmospheric processes are complicated by several variables which include: (1) the role of trace pollutants not accounted for in the laboratory simulation, (2) the nonuniform addition of reacting pollutants into the mixture as secondary pollutants are undergoing formation, (3) the residual concentrations of secondary pollutants present in the atmosphere from the previous days, and (4) the role of meteorology in transporting the air parcel in which the secondary pollutants are being formed and in dispersing the air parcel once the secondary pollutants have formed.

Scenario Construction.   The modeling of secondary pollutants requires a model to describe the complex series of chemical reactions taking place in the atmosphere. The several reviews of atmospheric chemistry describe the complexity of the participating reactions and the questions that remain unanswered.[1,48,58,98]

Photochemical oxidant is the most common secondary pollutant encountered in modeling. A number of kinetic mechanisms applicable to modeling photochemical oxidant have been proposed.[26,45,52] Although the mechanisms proposed produce reasonable agreement with laboratory studies, the nature of the multitude of reactions occurring in the atmosphere continues to be the subject of intense inquiry. A polluted urban atmosphere typically contains upward of 100 hydrocarbon species, each of which may undergo a variety of reactions. In addition, the very low concentrations of many of these species challenge the limits of available instruments. A review of the mechanisms' application to modeling is available.[73]

For use with an air quality model, the air chemistry kinetic mechanism must be compact to avoid excessive computing times in the numerical integration of the model. This requirement necessitates the use of a lumped-parameter approach, whereby a class of compounds or reactions is assumed to be described by a single compound or reaction with an "average-rate constant" assigned. The method by which such models are constructed involves the fitting of the air chemistry model results to laboratory data.

Of the kinetic mechanisms published to date, the 15-step model of Hecht and Seinfeld[52] best combines replication of laboratory data with the goal of being relatively compact. The 15 steps are summarized in Table 3.18.

The first three steps involving nitric oxide, nitrogen dioxide, ozone, and sunlight $(h\nu)$ describe the formation and destruction of ozone in the absence of hydrocarbons. These steps are common to all of the kinetic mechanisms which have been proposed. The mechanisms diverge when it comes to describing how the presence of hydrocarbons disrupts these equilibrium conditions.

Impact Scenarios   Once the emissions, meteorological, and air chemistry scenarios are constructed, the scenarios must be evaluated in combination to establish which conditions lead to an air quality impact. For example, a meteorological condition most conducive to air quality impact may not actually correspond to such an impact if, at the same time, the emissions are low. Similarly, a peak in emissions may not result in air quality impact if the meteorological conditions are especially conducive to pollutant dispersal. Only through an evaluation of the scenarios in combination will the impact scenarios for the most probable and worst cases be identified. Fortunately for the air quality impact

TABLE 3.18   Oxidant Formation—Detailed 15-step Reaction Mechanism

1	9
$NO_2 + h\nu \rightarrow NO + O$	$HO_2 + NO \rightarrow OH + NO_2$
2	**10**
$O + O_2 + M \rightarrow O_3 + M$	$HO_2 + NO_2 \rightarrow HNO_2 + O_2$
3	**11**
$O_3 + NO \rightarrow NO_2 + O_2$	$HC + O \rightarrow \alpha RO_2$
4	**12**
$O_3 + NO_2 \rightarrow NO_3 + O_2$	$HC + OH \rightarrow \beta RO_2$
5	**13**
$NO_3 + NO_2 \xrightarrow[H_2O]{} 2\ HNO_3$	$HC + O_3 \rightarrow \gamma RO_2$
6	**14**
$NO + NO_2 \xrightarrow[H_2O]{} 2\ HNO_2$	$RO_2 + NO \rightarrow NO_2 + \epsilon OH$
7	**15**
$HNO_2 + h\nu \rightarrow OH + NO$	$RO_2 + NO_2 \rightarrow PaN$
8	
$CO + OH \xrightarrow[O_2]{} CO_2 + HO_2$	

R: Radical
$\alpha, \beta, \gamma, \epsilon$: Adjustable coefficients
PaN: Peroxyacyl nitrates
SOURCE: Reference 52.

analyst (and unfortunately for the environment), the peak in emissions often coincides with the meteorological and air chemistry conditions most conducive to air quality impact.

The four types of scenarios developed for the emissions and meteorological scenarios (existing, no-build, construction, and build-out) are combined and analyzed for temporal compatibility. For each scenario type, impact scenarios are constructed for (1) the worst case and the most probable case, and (2) for primary and secondary pollutant impact. Often the source types at the project site and those exogenous to the project site are sufficiently different to warrant separate impact scenarios as well.

If a major point source is being evaluated, an impact scenario may be required for visibility impact as well. As a result, the number of separate impact scenarios required can be substantial. However, similarities between the different types can be and are being used to reduce the number of scenarios actually constructed.

In constructing the scenarios, it is important to keep in mind the location of the existing and projected future year receptors. It can be frustrating (indeed embarrassing) to construct an impact scenario and model the impact, prepare to plot the result on a map of the project area, and find that the impact location is 1000 m offshore.

Primary Pollutant Impact.   For primary pollutant impact, an air chemistry model is not necessary, and only the emissions and meteorological scenarios must be considered.* The ease with which the primary pollutant impact scenarios are established depends upon whether or not the peak in emissions corresponds to the meteorological conditions most conducive to impact.

CASES FOR WHICH THE EMISSION PEAK AND METEOROLOGICAL CONDITIONS MOST CONDUCIVE TO IMPACT CORRESPOND. As just mentioned, for many common sources, the maximum emission coincides with the meteorological condition most conducive to impact. For example, automobile traffic peaks during the early morning hours when the radiation inversion is most intense and winds are usually calm. As a result, a typical scenario for the "most probable primary pollutant impact" from automobile traffic encompasses the following conditions:

- Time period—early morning
- Emissions—peak commuting hour
- Wind speed—calm to low (≈ 1 m/s)
- Direction—persistent and toward populated areas and/or sensitive receptors
- Mixing height—low and persistent
- Stability—very stable (F stability)
- Season—winter

Another common example is a large point source such as power plant. Power plant emissions are usually high during the morning hours. The most probable stability class depends upon the prevailing conditions. A neutral stability class is most common. As a result, a typical scenario for the "most probable primary pollutant impact" from a power plant encompasses the following conditions:

- Time period—early morning
- Emissions—peak morning loads
- Wind speed—most probable corresponding to the stability class below
- Direction—persistent and toward populated areas and/or sensitive receptors
- Stability—trapping conditions, neutral (class D)
- Mixing height—low, persistent, and sufficiently high to entrap elevated source emissions
- Season—winter

The worst case impact scenario corresponds to a condition which occurs infrequently but with predictable regularity, a condition which produces an unusually high impact due to an extreme in either emissions or meteorology. For primary pollutants, an extreme in emissions is the more common. An example for automobile traffic is a condition of obstructed flow wherein traffic is backed-up and the stop-and-go frequency is unusually high. As a result, a typical scenario for the "worst case primary pollutant impact" from automobile traffic might well correspond to the following conditions:

- Time period—early morning
- Emissions—unusually high due to an obstruction in traffic flow
- Wind speed—calm to low (≈ 1 m/s)
- Direction—persistent and toward populated areas and/or sensitive receptors
- Mixing height—low and persistent
- Stability—very stable (F stability)
- Season—winter

For the power plant, the worst case might occur on an unusually cold morning when electric loads peak higher than usual due to correspondingly heavy demand for electrical heating energy. As a result, a typical scenario for the "worst case primary pollutant impact" from a power plant might well correspond to the following conditions:

- Time period—early morning
- Emissions—unusually high due to an unusually high demand for electricity
- Wind speed—light to strong (2–15 m/s) (modeling required to determine the worst case wind speed)
- Direction—persistent and toward populated areas and/or sensitive receptors
- Mixing height—low, persistent, and sufficiently high to just entrap source emissions (i.e., critical inversion height)
- Stability—trapping conditions
- Season—winter

*An air chemistry scenario is necessary for primary pollutants that may undergo significant chemical transformation close to the source prior to receptor encounter (i.e., nonconservative primary pollutants). However, it is generally assumed that primary pollutants do not undergo sufficient change to warrant special consideration.

CASES FOR WHICH THE EMISSION PEAK AND METEOROLOGICAL CONDITIONS MOST CONDUCIVE TO IMPACT DO NOT CORRESPOND. For projects where the peak in emission does not coincide with the meteorological conditions most conducive to impact, either engineering judgment or hour-by-hour modeling are required to establish the combination of emissions and meteorology that produces the most probable and worst case scenarios. Engineering judgment is preferred in order to avoid the cost and time associated with hourly analyses. The essence of engineering judgment is to identify the time of day and the combination of emissions and meteorology that will create the most probable and worst case impacts.

Secondary Pollutant Impact. For secondary pollutant impact, the air chemistry must be considered in addition to the emissions and meteorological scenarios. The most common form of secondary pollutant impact is the impact experienced from photochemical oxidant.

Presently, the major emission sources of oxidant precursors are motor vehicles. Power plant emissions contribute as well. Since time and insolation are required for oxidant formation, attention is drawn to the emissions that occur during the morning hours. Coincidentially, this is the time period during which motor vehicle emissions and power plant emissions are high, mixing is limited, and winds are calm. As a result, a typical scenario for the "most probable secondary pollutant impact" encompasses the following conditions:

- Time period—early morning to afternoon
- Emissions—early morning to afternoon emissions from sources of reactive hydrocarbons and nitrogen oxides* (i.e., motor vehicles and powerplants)
- Wind speed—most probable during daylight hours
- Direction—toward populated areas and over a source area rich in secondary pollutant precursors
- Stability—very stable in the early morning (E, F stability)
- Mixing height—persistent and sufficiently high to entrap elevated source emissions
- Insolation—strong
- Season—summer

In the summer, the days are long, the sun is more directly overhead, and insolation is maximum. A cloudless night will create a low-lying radiation inversion into which the motor vehicle emissions will be injected. This limited mixing will maintain high concentrations and will accelerate reactions. The morning heating of the ground will break the radiation inversion and allow the power plant and motor vehicle secondary pollutant precursors to mix to the height of the mixing layer. Low winds will maintain high concentrations and accelerate oxidant formation.

The worst case impact scenario corresponds to a condition which occurs infrequently but with predictable regularity, a condition which produces an unusually high impact due to an extreme in emissions, meteorology, or a combination of both. An example of an extreme in emissions is a condition of heavy morning motor vehicle traffic that might occur after a long holiday weekend. An extreme in meteorology is a more common reason for high secondary pollutant levels. For example, consider the case of unusually low winds (bordering on stagnation) combined with a strong subsidence inversion and high temperatures that persist over a few days. This condition would likely prevent removal of oxidant and secondary pollutant precursors formed and emitted, respectively, during the previous 24-hour period and will produce worst case impacts. As a result, a typical scenario for the "worst case secondary pollutant impact" might well correspond to the following conditions:

- Time period—early morning to afternoon
- Emissions—early morning to afternoon emissions from sources of hydrocarbons and nitrogen oxides (e.g., motor vehicles and power plants)
- Wind—stagnated
- Direction—if applicable, persistent, toward populated areas, and over a source area rich in secondary pollutant precursors
- Stability—very stable (E, F stability)

---

*These are the time periods and pollutant species appropriate for use in numerical modeling. For proportional modeling, the emissions scenario is simplified to the 6:00 to 9:00 A.M. emissions of reactive hydrocarbons (see the section Air Quality Modeling—Proportional Modeling later in this chapter).

- Mixing height—unusually low and persistent
- Insolation—unusually strong and prolonged
- Season—summer

Air Quality Modeling

Air quality modeling is the heart of the air quality impact analysis. Unfortunately, it is more an art than a science. As such, air quality modeling combines intrigue with challenge and requires an effective and often imaginative use of the best engineering judgment. An EPA document designed to provide consistency in the selection and use of models provides a cogent summary:

> It would be advantageous . . . to apply a *designated* model to each proposed source . . . However, the diversity of the nation in topography and climate, and variations in source configurations and operating characteristics dictate against a "cookbook" analysis. One particular model which is capable of analyzing all conceivable situations is impractical. In addition many meteorological phenomena that are associated with threats to air quality standards are not amenable to simple mathematical treatment. *The judgment of well-trained professional analysts must always be applied.* [Emphasis added][25]

Training results from the application of models. But the application requires an understanding of the models, an understanding of the subtleties associated with the use of models, and a continuing review of the literature and government policy.

The modeling methods are introduced in order of increasing complexity. The models are listed in Table 3.19 with information on accessibility and applicability, major references, and sources for user information. At the end of the section, the subject of model validation is addressed.

Box Model The box model, introduced in an earlier section of this chapter, is the simplest model, yet it considers most of the variables essential to air quality modeling. The box model, shown in Fig. 3.33a, assumes that the pollutants are emitted into a fixed volume bounded on the top and sides. Air is transported through the volume with a face velocity of $\bar{u}$, and the pollutants are assumed to be instantaneously and uniformly mixed throughout the box. The steady state concentration is related to the emission rate Q by the following expression for the box model:

$$C_j = \frac{Q_j}{\bar{u}WD} \tag{3.15}$$

where $\bar{u}$ = wind velocity assumed constant, m/s
  W = width of box normal to the wind direction, m
  D = depth of box normal to the wind direction, m
  Q_j = emission rate of species j, g/s

The box model is most appropriate for ground sources that are uniformly distributed across a small area. Examples include roadways, parking lots, and residential tracts. The vulnerability of the box model to critical attack stems from the arbitrary choices available for the values of W and D. Engineering judgment is the best guide. For ambient concentration calculations close to the source, the mixing height D will be limited to a few hundred meters or lower for stable atmospheres. The width W will correspond to the width of the source. Far downwind of the source, D will correspond to the inversion height and W to the width of the subregional air basin.

The box model graphically demonstrates that pollutants are injected into and mixed with a limited resource of air. Although the box model is generally not acceptable for a formal air quality impact analysis, it is useful for preliminary calculations to develop qualitative estimates of source impact.

A limiting case of the box model is the case for which the wind is null and pollutant concentrations are allowed to build within the box. The dispersion of pollutants is now prevented in all directions. The box becomes a contained "closed box model," as shown in Fig. 3.33b. The corresponding equation is:

$$C_j = \frac{Q_j t}{xWD} \tag{3.16}$$

where x = length of box, m
  t = time from initiation of emission, s

The closed box model is applicable to order-of-magnitude estimates for situations where winds are null and dispersion is limited. An example is a small canyon that is axially

TABLE 3.19 Air Quality Models

Model type	Users information — Hand calculation	Users information — Computer program	Assessibility — General	Assessibility — Limited	Assessibility — Restricted	Applicability — Primary pollutant	Applicability — Secondary pollutant	Applicability — Local	Applicability — Subregional	Applicability — Regional	Representative validation studies (Reference)
Box Model:											
Open	Equation 3.15		x			x	x	x	x	x	
Closed	Equation 3.16		x			x	x	x	x	x	
Gaussian Models:											
Point sources		(Refer to Table 3.21)									
	Turner[108]		x			x		x	x		
	ASME[3]		x			x		x	x		
	TVA[65]		x			x		x	x		
		PTMAX[109]	x			x		x	x		
		PTDIS[109]	x			x		x	x		
		PTMTP[109]	x			x		x	x		
		VALLEY[12]	x			x		x	x		
		CRSTER[57]	x			x		x	x		(64)
Line Sources	Turner[108]		x			x		x	x	x	(60)
	ASME[3]		x			x		x	x	x	(60)
		HIWAY[118]	x			x		x			
		CALINE2[112]	x			x		x			
Area Sources: Urban	Turner[108]										
		APRAC1A[61]	x			x		x	x		(79)
		APRAC2[97]	x			x		x	x		
		CDMQC[110]	x			x		x	x		
Airport		AVAP[107]	x			x		x			(111)
		AQAM[84]	x			x		x			
Shopping center		ISMAP[88]	x			x		x			
Numerical Models		(Refer to Table 3.22)									
		DIFKIN[62]		x		x	x	x	x	x	(103)
		SAI80–83			x	x	x	x	x	x	
		LIRAQ-2[59]			x	x	x	x	x	x	
		REM[114]			x	x	x	x	x	x	
		IMPACT[28]			x	x	x	x	x	x	(115)

bisected by a road. In the morning, a radiation inversion may prevent vertical escape and a null wind may prevent horizontal escape.

Proportional Modeling   The proportional model is based on the principle that air quality is directly proportional to emissions, (i.e., air quality will deteriorate in proportion to an

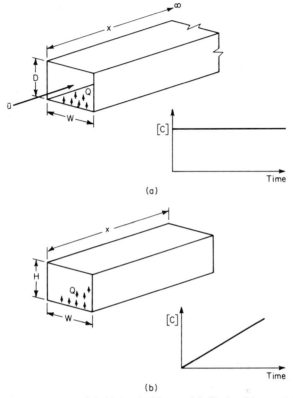

Figure 3.33   Box model. (*a*) Standard box model; (*b*) closed box model.

increase in emissions and improve in proportion to a decrease in emissions).

Several assumptions are inherent in this approach:

1. The meterology leading to the elevated pollutant concentration is similar for the base year and future year.

2. Changes in the temporal distribution of emissions are insignificant between the base year and future year.

3. Any changes in the quantity of emission occur proportionally throughout the region, (i.e., all emissions are changed by the same percentages).

4. The maximum concentration measured at the monitoring station is representative of the maximum concentration which actually occurred in the air quality control region.

Two different proportional models will be considered: (1) a model that does not consider wind speed and (2) a model that does consider wind speed.

The box model equation serves as the basis for the proportional model which does not consider wind speed. Consider, for example, the change in concentration due to an increase in emissions at a given location between some arbitrarily selected base year BY and a future year FY. For the base year:

$$C_{BY} = \frac{Q_{BY}}{\bar{u}DW} \tag{3.17a}$$

and for the future year:

$$C_{FY} = \frac{Q_{FY}}{uDW} \qquad (3.17b)$$

where C_{BY} = base-year air quality measurement, g/m^3
Q_{BY} = base-year emission rate, g/s
C_{FY} = projected future-year air quality, g/m^3
Q_{FY} = projected future-year emission rate, g/s
Dividing, we obtain an expression for the proportional model:

$$C_{FY} = C_{BY} \frac{Q_{FY}}{Q_{BY}} \qquad (3.18)$$

The proportional model is used for estimating primary and secondary pollutant concentrations or required emission reductions. In fact, the model became popular not as a predictor of future-year air quality, but as a predictor of the reduction in emissions required to meet the NAAQS. For example, if the ambient air quality standard is selected for C_{FY}, and the maximum value *measured* for the base year is selected for C_{BY}, solving for the emission ratio will identify the extent to which emissions must be "rolled back" to achieve the NAAQS.

For conserved species, a case can be argued for assuming a linear relationship between emissions and air quality. For nonconserved species, however, such a case is difficult to pursue and is not supported in fact.

In the case of the photochemical oxidants, the Environmental Protection Agency has developed the "Appendix J" rollback model.[30] The model is an attempt to account for the nonlinearities between the emissions of secondary pollutant precursors and the formation of oxidant.* The procedure involves plotting peak 1-hour oxidant measurements against the 6:00 to 9:00 A.M. average nonmethane-hydrocarbon measurement for the same day. A curve is then drawn, shown in Fig. 3.34a, such that all points are plotted below it, thus representing an upper limit to possible oxidant concentrations for a given level of emissions. This curve is then used to construct a second curve, shown in Fig. 3.34b, which relates peak oxidant measurement to percent emissions reductions required to meet the standard.

In addition to the assumptions of the proportional model already identified, the use of the proportional model for oxidant involves two further assumptions:

1. The monitoring stations selected for the analysis measure the peak oxidant produced by the 6:00 to 9:00 A.M. emission of nonmethane hydrocarbon.

2. The 6:00 to 9:00 A.M. hydrocarbon measurement is directly proportional to total regional emissions.

In short, the method ignores the space and time variations of processes which likely influence the emissions/air quality relationship.

On the local scale, the wind velocity is an important determinant in predicting the maximum concentration, and neither linear rollback nor Appendix J are appropriate. Instead, the Hanna-Gifford model may be applied for estimating concentrations of nonreactive pollutants. The Hanna-Gifford model is similar to linear rollback with the exception that the local wind speed is included as a variable as follows:

$$C_{FY} = K \frac{q_{FY}}{\bar{u}} \qquad (3.19)$$

where C_{FY} = projected future air quality, g/m^3
K = calibration constant
q_{FY} = projected future year local area emission density, $g/(s \cdot m^2)$
$\bar{u}$ = local average wind speed, m/s

and K is determined as follows:

*Although the discussion here is presented in terms of photochemical oxidant and a NAAQS of 0.08 ppm/1-hour (for which the "Appendix J" rollback model was originally developed), application of the model must be in terms of ozone and 0.12 ppm/1-hour to reflect the subsequent change of the NAAQS from photochemical oxidant to ozone and the value of the standard from 0.08 ppm/1-hour to 0.12 ppm/1-hour.[44]

$$K = \frac{C_{BY}\,\overline{u}}{q_{BY}} \qquad (3.20)$$

where C_{BY} = base year air quality measurement, g/m^3
$\quad q_{BY}$ = base-year local area emission density, $g/(s \cdot m^2)$
$\quad \overline{u}$ = local average wind speed, m/s

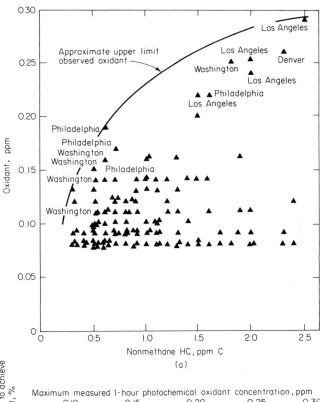

(a)

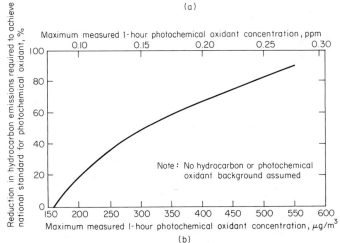

(b)

Figure 3.34 Appendix J—rollback model. (a) Maximum daily one-hour oxidant as a function of 6 A.M. to 9 A.M. nonmethane hydrocarbon concentration. (*From Reference 78*); (b) hydrocarbon control required to meet oxidant air quality standard. (*From Reference 30, p. 15502.*)

The average wind speed should be determined on a time scale consistent with the averaging time of the air quality measurement being used to define the base-year condition (e.g., 24-hour average, annual average, etc.).

Gaussian Modeling   *General Considerations.*   What the box models and proportional models feature in simplicity, they lack in reality. Most notable in this regard is the failure of the models to account for dispersion of the pollutants in the atmosphere.

The most frequently used modeling method for local air quality impact analyses is Gaussian plume modeling. Some of the many reasons for the popularity of this method are:

- Gaussian modeling accounts for turbulent dispersion and atmospheric stability.
- A computer is not necessary for calculating pollutant concentrations.
- Gaussian modeling can be understood and used by the novice and specialist alike.

As a result, Gaussian plume modeling is able to account for some of the important aspects of the dispersion processes (e.g., turbulent mixing and atmospheric stability). At the same time, it is explainable to a spectrum of those participating in the community decision-making process.

Gaussian modeling is described in major air resources textbooks such as Williamson[117] and Wark and Warner.[113] The standard workbook for Gaussian modeling is authored by Turner.[108] The American Society of Mechanical Engineers (ASME) also publishes an instructive Gaussian-modeling workbook.[3]

The Gaussian models introduced in the earlier section of this chapter for ground level and elevated point sources, and line sources are as follows:

- Gaussian dispersion model—ground level point source

$$C_j(x, y, z) = \frac{Q_j}{\pi \overline{u} \sigma_y \sigma_z} \left[\exp\left(-\frac{y^2}{2\sigma_y^2} - \frac{z^2}{2\sigma_z^2} \right) \right] \tag{3.21a}$$

- Gaussian dispersion model—elevated point source

$$C_j(x, y, z) = \frac{Q_j}{2\pi \overline{u} \sigma_y \sigma_z} \left[\exp\left(-\frac{y^2}{2\sigma_y^2} \right) \right] \left[\exp\left(-\frac{(z-H)^2}{2\sigma_z^2} \right) + \exp\left(-\frac{(z+H)^2}{2\sigma_z^2} \right) \right] \tag{3.21b}$$

- Gaussian dispersion model—ground level line source

$$C_j(x, z) = \frac{2Q_j/L}{(2\pi)^{1/2} \overline{u} \sigma_z} \left[\exp\left(-\frac{z^2}{2\sigma_z^2} \right) \right] \tag{3.21c}$$

where Q_j = emission rate of species j, g/s
σ_y = horizontal Gaussian dispersion coefficient, m
σ_z = vertical Gaussian dispersion coefficient, m
$\overline{u}$ = wind velocity in x-direction, m/s
H = height of elevated point source, m
L = length of line source, m

Examples of each source type are shown in Fig. 3.35.

Dispersion Coefficients.   The Gaussian models require information on the values of the dispersion coefficients (σ_y and σ_z) and the variation of these coefficients with atmospheric stability class and downwind distance. As a result of a variety of investigations, different sets of the σ_y and σ_z variables have been developed. The selection of the particular σ_y, σ_z set to use is important since the predicted impact will vary according to the dispersion coefficients selected. An example of this variation is shown in Fig. 3.36 for the three sets of coefficients commonly employed—Pasquill-Gifford, TVA, and ASME. Both the value of the peak concentration and the location at which the peak concentration occurs are dependent upon the coefficient selected.

The Pasquill-Gifford Gaussian coefficients are employed in the EPA workbook[108] and, as a result, are the most popular. Other sets in common use include those of the TVA[15] and ASME.[3] The selection of an alternative set requires justification that relates the conditions associated with the project to those encountered in the studies from which the dispersion coefficients are derived. It is important to note that the field data upon which the empirical coefficients are based stem from point sources only. Use of these data for line sources and area sources (as well as point sources) should be accompanied by a validation study, if possible (see the section on Model Validation later in this chapter).

PASQUILL-GIFFORD. The Pasquill-Gifford diffusion coefficients, shown in Figs. 3.37a and 3.37b, are based on a combination of empirical data for ground level point sources compiled in England by Hay and Pasquill[50,51] and a model developed by Pasquill.[63] In the model, a plume height h and an angular plume width ϕ are defined such that the concentration at the edge of the plume equals 10 percent of its value on the plume axis. Pasquill estimated values of h and ϕ as functions of meteorological conditions using meteorological principles and existing observations of dispersion. The result is the plot presented in Fig. 3.38.

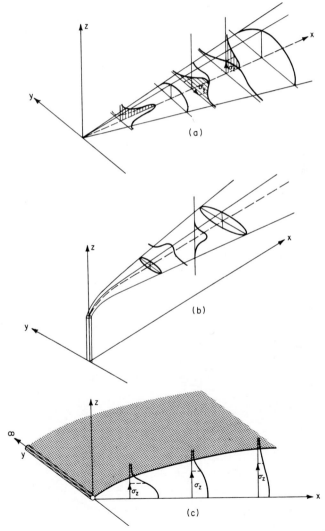

Figure 3.35 Gaussian dispersion—examples. (a) Ground level point source; (b) elevated point source; (c) line source.

Gifford later used the Pasquill results to plot the families of σ_y and σ_z curves shown in Figs. 3.37a and 3.37b. Integral to Gifford's approach was the assumption that the concentration profile follows a Gaussian distribution. This assumption provides the following relationships between Pasquill plume-geometry parameters h and ϕ and the Gaussian coefficients:

$$h = 2.15\sigma_z$$
$$\tan\frac{\phi}{2} = 2.15\,\frac{\sigma_y}{x}$$

(3.22)

Gifford believed that use of σ_y and σ_z as standard deviations rather than ϕ and h was desirable in view of their assumption of normally distributed plume concentrations and of their increasing use in summarizing atmospheric dispersion data.

The field data used to construct the Pasquill-Gifford plots are summarized by Pedri.[72] Data acquired 1 km from the source under neutral conditions are the most plentiful. In general, all data were obtained for fairly level and open country.

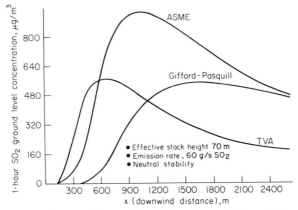

Figure 3.36 Gaussian plume modeling—effect of dispersion coefficient selection. (*Adapted from Reference 7.*)

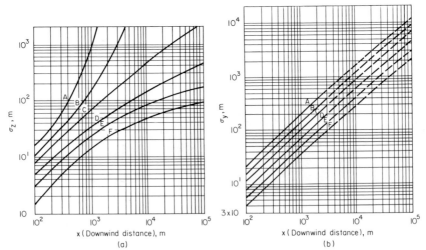

Figure 3.37 Gaussian plume modeling—Pasquill-Gifford coefficients. (*a*) σ_z, vertical dispersion coefficient; (*b*) σ_y, horizontal dispersion coefficient. (*From Reference 46.*)

The tentative statistical estimates of h and ϕ are the basis of a statement attributed to Pasquill[71] that the dispersion coefficients can give only very approximate estimates of the magnitudes of concentration. In more difficult cases of unstable and stable situations, errors in h may approach "several fold" when applied to longer distances of travel. Cases where the estimates of vertical spread may be expected to be correct within a factor of two include:

  ▪ All stabilities except at the extremes, within of a few hundred meters of the source, and in open terrain
  ▪ Neutral to moderately unstable conditions, and within a few kilometers of the source
  ▪ Unstable conditions for the first 1000 m above ground level with a marked inversion thereafter, and at distances greater than 20 km

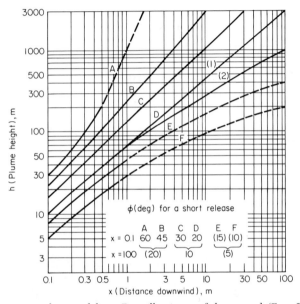

Figure 3.38   Gaussian plume modeling—Pasquill estimates of plume speed. *(From Reference 46.)*

TENNESSEE VALLEY AUTHORITY. In 1951, the TVA began operating its first major coal-fired power plant at Jacksonville. Simultaneously, a program of plume observation and surface and aerial SO_2 monitoring was initiated. Data for this and subsequent plants have since been compiled and classified according to meteorological conditions, stack height, and unit sizes.[15] From these data, the TVA developed the Gaussian coefficients presented in Fig. 3.39.

AMERICAN SOCIETY OF MECHANICAL ENGINEERS. The ASME dispersion coefficients shown in Fig. 3.40 are based on atmospheric diffusion experiments conducted over a period of 15 years at the Brookhaven National Laboratory. The Brookhaven site is situated in central Long Island, New York, a region characterized by terrain rising no more than 100 m above sea level with most relief being less than 50 m.

The stability classification shown in the figures is related to wind gustiness. The original separation of wind gustiness into categories was based on limited observations.[91] Four classes were recognized and designated A, B, C, and D in order of the decreasing amplitude of fluctuation of horizontal wind direction.

The ASME plots for σ_z do not exhibit the curvature of the TVA and Pasquill-Gifford plots. Two features which may account for the lack of curvature include: (1) in stable cases, σ_z is a very gentle function of distance, and the Brookhaven data are not detailed enough to reveal such curvature, and (2) in unstable cases, the assumption of a constant

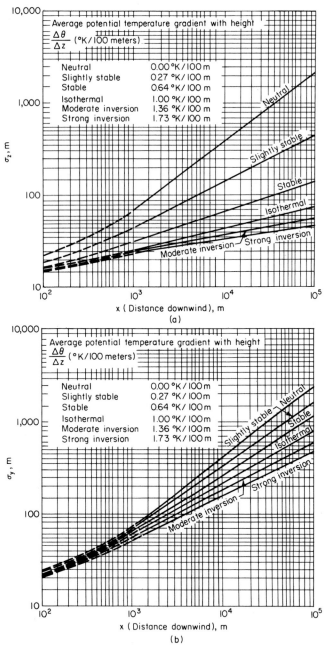

Figure 3.39 Gaussian plume modeling—TVA coefficients. (a) σ^z, vertical dispersion coefficient; (b) σ_y, horizontal dispersion coefficient. (*From Reference 15.*)

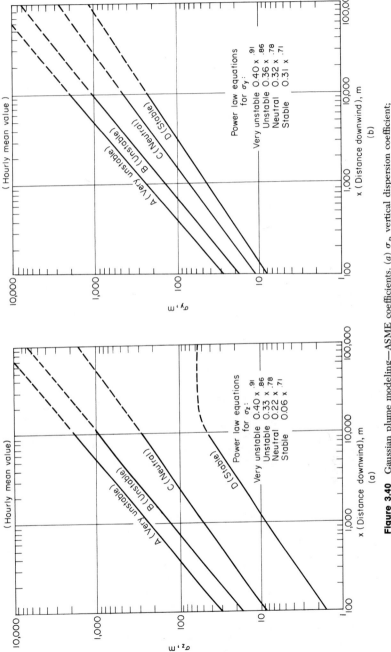

Figure 3.40 Gaussian plume modeling—ASME coefficients. (a) σ_z, vertical dispersion coefficient; (b) σ_y, horizontal dispersion coefficient. (*From Reference 3.*)

reference wind speed affects the determination of σ_z. Due to the fact that the wind speed increases with height, the σ_z plot must curve upward with increased wind speed. When reviewing data from an elevated source such as the Brookhaven data, one finds that this tendency is much less pronounced since the plume initially disperses both upward and downward. Also, the elevated reference wind speed is valid for a considerable distance.[91]

Averaging Time Selected. The averaging time selected for the analysis depends upon the applicable air quality standard to which the prediction is to be compared. Usually this is 1 hour. However, the averaging time for the prediction is determined by the averaging time base for the Gaussian coefficients (σ_y, σ_z) selected for the analysis.

The averaging times used to empirically establish the coefficients are shown in Table 3.20. Once the coefficients are selected, the remaining input data (Q_j, $\bar{u}$) must be averaged

TABLE 3.20 Averaging Time for Dispersion Coefficients

Dispersion coefficient	Averaging time (min)	References
Pasquill-Gifford	10	(108)
ASME	60	(3)
TVA	Instantaneous	(15)

over the same time period. The concentration then calculated will have a time base which corresponds to that of the coefficients. It is necessary to convert the calculated downwind concentrations into concentrations desired for comparison to the appropriate NAAQS. There is no universally accepted approach for time-averaging conversion. A major part of the problem is to account for the variation in local meteorology that can and does occur as the time base is extended. Turner[108] suggests that the following equation be used for conversion:

$$\frac{C_{j_1}}{C_{j_2}} = \left(\frac{\Delta t_2}{\Delta t_1}\right)^p \qquad 0.17 \le p \le 0.20 \qquad (3.23)$$

where C_{j_1} is the concentration of a pollutant at a specific location averaged over a time Δt_1. The desired averaging time Δt_2 will result in concentration C_{j_2}. The exponential factor p is empirical.

An exception to the use of Eq. 3.23 for conversion involves the TVA coefficients. Montgomery and Coleman[66] evaluate relationships between time-average SO_2 concentrations using data derived from monitoring networks in the Tennessee Valley. They conclude that the following equation be used to convert the calculated concentration:

$$\frac{C_{j_{1-\text{hour}}}}{C_{j_{\text{calculated}}}} = 0.5 \qquad (3.24)$$

where $C_{j_{1-\text{hour}}}$ = one-hour averaged concentration
$C_{j_{\text{calculated}}}$ = calculated concentration using TVA coefficients

Plume Rise. The vertical motion of a plume from an elevated source to the height where it becomes horizontal is known as the "plume rise." The rise is a result of momentum and thermal forces. The momentum force results from the vertical velocity of the stack gases that give an upwardly directed momentum. The thermal force results from the positive buoyancy of the effluent gases when stack gas exhaust temperatures exceed surrounding ambient temperatures.

Because of the plume rise, the physical height of the stack cannot be used for the origin of the Gaussian plume model. Instead, as illustrated in Fig. 3.41, the origin for the Gaussian model must be located at distance Δh above the stack. The term Δh is referred to as the plume rise, and the height of the origin H is called the effective stack height ($H = h + \Delta h$).

Empirically derived equations are used to predict plume rise. Dozens are available, but four have received the most use:

- Brigg's plume rise equation
- Holland's plume rise equation
- ASME plume rise equation
- TVA plume rise equation

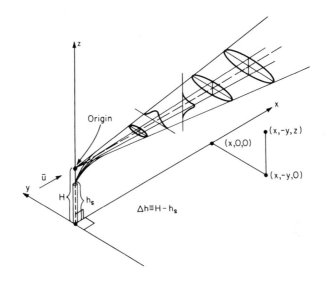

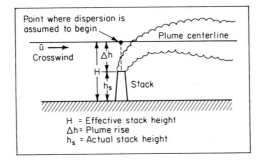

Figure 3.41 Gaussian plume modeling—plume rise.

The conditions under which the plume rise equations were developed are summarized in Pedri.[72]

BRIGGS. Briggs first used dimensional analysis to develop a plume rise model.[8] He later expanded the model to include the effect of stability[9] and used field and laboratory simulation measurements and other models to refine the models to the following set:[10]

$$x < 2\overline{u}\,S^{-1/2} \qquad\qquad x > 2\overline{u}S^{-1/2}$$

$$\Delta h = 1.6\,F^{1/3}\,\overline{u}^{-1}\,x^{2/3} \qquad \Delta h = 2.9\left(\frac{F}{\overline{u}\,S}\right)^{1/3} \qquad Stable \qquad (3.25a)$$

$$Neutral \qquad (3.25b)$$

$x/x^* \leqslant 3.5$ $\qquad$ $x/x^* \geqslant 3.5$ $\qquad$ *or*

$\Delta h = 1.6F^{1/3}\,\overline{u}^{-1}x^{2/3}$ $\qquad$ $\Delta h = 1.6F^{1/3}\overline{u}^{-1}(3.5x^*)^{2/3}$ $\qquad$ *unstable*

where $x^* = 14F^{5/8}$, m $(F < 55)$
$\qquad x^* = 34F^{2/5}$, m $(F > 55)$
$\qquad \Delta h =$ the plume rise, m
$\qquad D =$ stack diameter, m
$\qquad V_s =$ stack gas exit velocity, m/s
$\qquad \overline{u} =$ mean wind speed at height of stack, m/s
$\qquad T_s =$ stack gas exit temperature, K
$\qquad T_a =$ ambient temperature, K
$\qquad x =$ distance from stack to receptor, m
$\qquad h_s =$ height of stack, m

$\qquad F =$ buoyancy factor, m^4/s^3 $= gV_s\,D^2/4\left(\dfrac{T_s - T_a}{T_s}\right)$

$$S = \frac{9.8}{T_a}\left(\frac{\Delta T}{\Delta z} + 0.98\right)$$

HOLLAND. Albert H. Holland developed his plume rise formula primarily by observation of plumes from two sources during the period from 1948 to 1952 in the Oak Ridge Tennessee region,[53] and from observations by F. W. Thomas of the TVA at the Watts Bar Steam Plant, about 40 miles southwest of the Oak Ridge area.[105] An analysis of the data indicates that plume height decreased with increasing wind speed. This observation proved consistent with a model advanced by Rupp[85] to explain the amount of rise due to momentum:

$$\Delta h = \frac{1.5V_sD}{\overline{u}} \qquad (3.26a)$$

Holland adopted the Rupp model as the basis of his own model. His goal was to incorporate the additional effect of the thermal contribution.

Holland proceeded to plot the difference $(\overline{u}\Delta h - 1.5V_sD)$ against the emission rate Q_h for neutral conditions. The buoyancy component was found to be $4 \times 10^{-5} Q_h$. Thus, the following formula was obtained as an average plume rise during neutral conditions:

$$h = \frac{1.5V_sD + 4 \times 10^{-5}Q_h}{\overline{u}} \qquad (3.26b)$$

Further analysis of the data for nonneutral conditions indicated that the factor $\overline{u}\Delta h$ increased 10 to 20 percent in unstable atmospheres and decreased 10 to 20 percent in stable atmospheres. Holland adopted a 15 percent factor to account for the variation for the two stability classes:

$$h = \frac{1.15(1.5V_sD + 4 \times 10^{-5}\ Q_h)}{\overline{u}} \qquad Unstable \qquad (3.26c)$$

$$h = \frac{0.85(1.5V_sD + 4 \times 10^{-5}\ Q_h)}{\overline{u}} \qquad Stable \qquad (3.26d)$$

The Tennessee Valley Authority compared Holland's plume rise formula to data measured for six utility stacks owned by TVA (Paradise, Gallatin, Shawnee, Johnsonville, Colbert, and Widows Creek). It was concluded that the model exhibits fairly good agreement, with a tendency to slightly underestimate plume rise.[104]

ASME. The ASME plume rise equations are divided between the forces that dominate the rise, namely momentum and thermal.[3]

TENNESSEE VALLEY AUTHORITY. Since 1963, the Tennessee Valley Authority (TVA) has conducted comprehensive studies of plume rise at coal-fired power plants for a variety

- Momentum controlled:

$$V_s \geq 10 \text{ m/s and } T_s - T_a < 50 \text{ K}$$ (3.27a)
$$\Delta h = \frac{V_s}{\bar{u}}$$

- Thermal controlled:

$$T_s - T_a > 50 \text{ K}$$ (3.27b)

$$\Delta h = 2.9 \left(\frac{F}{\bar{u} G} \right)^{1/3} \qquad Stable$$ (3.27c)

$$\Delta h = \frac{7.4 h_s^{2/3} F^{1/3}}{\bar{u}} \qquad Neutral \text{ or unstable}$$

where $G = \frac{9.8}{T_a} \left(\frac{\Delta T}{\Delta z} + 0.98 \right)$

$\frac{\Delta T}{\Delta z}$ = the atmospheric temperature gradient, K/100 m

of meteorological and operational conditions, plant sizes, and stack heights. A substantial portion of the work has been directed towards the development of plume rise models.

The plume rise formulas developed by TVA are based on data obtained during single-stack observations at three coal-fired power plants encompassing a range of stack heights.[15] The plume rise formulas are categorized by stability class according to the following criteria:

$$\Delta h = 2.50 \ x^{0.56} \ F^{1/3} \ \bar{u}^{-1} \qquad Neutral$$ (3.28a)

$$\Delta h = 3.75 \ x^{0.49} \ F^{1/3} \ \bar{u}^{-1} \qquad Moderately \ stable$$ (3.28b)

$$\Delta h = 13.8 \ x^{0.26} \ F^{1/3} \ \bar{u}^{-1} \qquad Very \ stable$$ (3.28c)

Nonconing Plume. The Gaussian plume model for an elevated source describes a coning plume as shown in Fig. 3.42a. Coning plumes occur for environmental lapse rates that are neutral. The Gaussian plume is also applicable to the fanning plume shown in Fig. 3.42b, a limiting case of the coning plume which occurs where environmental lapse rates are stable (See also Fig. 3.31).

Field studies indicate that the coning conditions are associated with maximum surface concentrations for small units.[15] However, for larger units and taller stacks, maximum concentrations are more frequently associated with fumigation and trapping plumes. Fumigation is observed to produce maximum concentrations for stack heights less than 75 m, and trapping is responsible for maximum concentrations for stack heights in excess of 75 m.[65]

Fumigation plumes are also called inversion breakup plumes and occur when a fanning plume is initially produced in the stable air of a surface-based inversion. The surface-based inversion is normally the result of the nighttime radiational cooling of the ground. The sunrise heating of the ground initiates vertical mixing of the air, and a mixing layer develops upward from the ground surface. When the top of the surface-based mixing layer reaches the bottom of the plume embedded in the stable air layer, the plume is rapidly dispersed to the ground as shown in Fig. 3.42c. As a result, relatively high ground level plume concentrations can occur as compared to those occurring under neutral conditions.

Trapping plumes (also termed "limited mixing plumes") occur when an elevated stable layer of air limits the vertical dispersion of the plume as shown in Fig. 3.42d. This elevated stable layer is normally the result of subsidence. Trapping plumes represent a steady-state condition, whereas fumigation plumes (though otherwise similar) represent a transient condition.

The Gaussian plume model can be modified to address fumigation and trapping plumes. For example, the method-of-images approach can be employed to account for an elevated inversion.[95] An alternative approach is to use empirical data for developing modified forms of the Gaussian equations. For example, the Tennessee Valley Authority has conducted assessments of plume rise for a variety of meteorological conditions, plant sizes, stack heights, and stack configurations.[106] Based on the data collected, the following

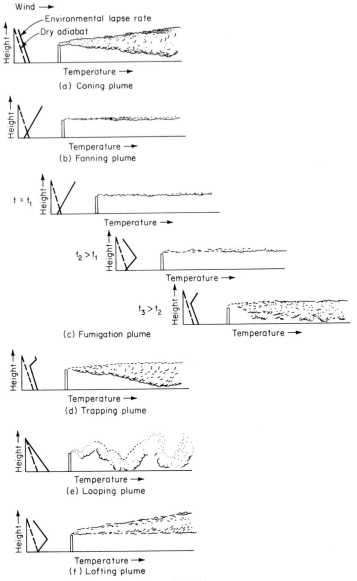

Figure 3.42 Plume types.

expressions were identified as suitable for the prediction of maximum surface SO_2 concentration when the SO_2 distribution within the plume is Gaussian:[15]

$$C_{max} = \frac{10^6 \, Q_{SO_2}}{\sqrt{2\pi}\sigma_{y_f} \rho \bar{u} \, H_f} \qquad \textit{Fumigation plume} \qquad (3.29)$$

where H_f = altitude of plume top, m = $(H + 2.14\sigma_z)(1.1)$
     H = effective source height, m

Q_{SO_2} = SO_2 emission, g/s
ρ = density of SO_2 effluent, g_{SO_2} /m^3
$\bar{u}$ = mean horizontal wind speed, m/s
σ_{y_t} = σ_y + $(0.47)H$, m
σ_y = TVA horizontal Gaussian coefficient, m

$$C_{max} = \frac{10^6 \, Q_{SO_2}}{\sqrt{2\pi} \, \sigma_{y_t} \, \bar{u} \, H_t \rho} \qquad Trapping \; plume \qquad (3.30)$$

where σ_{y_t} = σ_y + $\left[\dfrac{H_t}{1.1} - 2.15\sigma_z\right]$ (0.47), m

H_t = height from the surface to the top of the mixing layer, m
σ_z = TVA vertical Gaussian coefficient, m

The TVA Gaussian coefficients are applicable to peak concentrations, and the estimated concentrations using TVA expressions must be reduced by a factor of 0.5 to obtain a 1-hour averaged equivalent (Eq. 3.24).

Two other nonconing plume types are frequently encountered. However, neither represents an air quality impact potential. The looping plume shown in Fig. 3.42e occurs in unstable atmospheres. Although the full thrust of the plume hits the surface, the encounter is momentary, and the plume lifts only to impact another location. As a result, the duration of impact at any one location is not sufficient to produce a significant averaged concentration over 1 hour. The lofting plume shown in Fig. 3.42f occurs for plumes that are injected into elevated unstable atmospheres with a stable atmosphere below. In such cases, the surface is protected from plume encounter.

Critical Inversion Height. A trapping condition will produce impacts that increase in severity as the inversion height decreases. If the inversion is too low, the emitted flue gases may have sufficient buoyancy and momentum to rise through the elevated inversion. As shown in Fig. 3.43a, there is a critical inversion height that traps the emitted flue gas and causes the highest ground level concentrations. An inversion base below the critical value will be penetrated by the plume. An inversion base above the critical height traps the plume but provides a deeper mixing volume than the critical inversion. The effect of the limited mixing produced by an inversion at the critical height is shown in Fig. 3.43b. The critical height is taken as the height above the ground to which the plume rises in the absence of an inversion.

The critical inversion height is a function of the stack height and the plume rise. The critical inversion height can be increased and ground level concentrations can be correspondingly reduced by increasing the stack height and/or by enhancing plume rise. Since plume rise decreases with increasing wind speed, the critical inversion height also decreases with wind speed. The effect is illustrated in Fig. 3.43c.

Also shown in Fig. 3.43c is the offsetting effect of wind speed on plume rise and plume dilution. In contrast to the effect of wind speed on plume rise, the concentration downwind will decrease with increased wind speed due to the increased dilution of the source emissions. This effect, first identified in the box model (Eq. 3.15) and later in the Gaussian model (Eq. 3.21) by the appearance of the wind velocity $\bar{u}$ in the denominator, is illustrated in Figure 3.43c for the case in which critical height is *not* taken into account. As the wind speed increases, the maximum one-hour concentration increases. This indicates that the decrease of plume rise with increasing wind speed more than offsets the corresponding increase in dilution. The same does not hold for the case in which critical height *is* taken into account. The decrease of plume rise with increasing wind speed dominates to approximately 7 m/s, while the increased dilution of the plume dominates for wind speeds in excess of 7 m/s.

Terrain. Nonlevel terrain may have a pronounced effect on the predicted ground level concentrations. Figure 3.44 shows approaches for including terrain effects. While difficult to model, two approaches are available to account for this effect.

The first approach is to modify the dispersion equations. For a terrain height less than the effective stack height, the terrain height is subtracted from the effective stack height, resulting in the following equation:

$$C_j(x, y, z) = \frac{Q_j}{\pi \sigma_y \, \sigma_z \bar{u}} \left[\exp\left(\frac{-y^2}{2\sigma_y^2}\right) \right] \left[\exp\left(-\frac{(H-z)^2}{2\sigma_z^2}\right) \right] \qquad (3.31a)$$

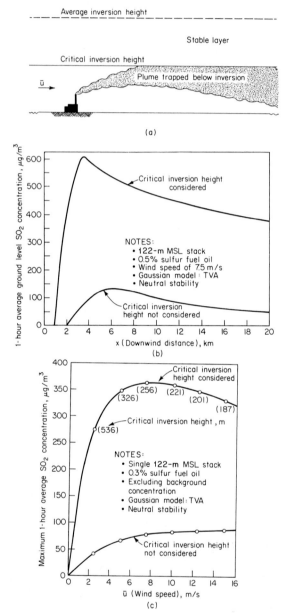

Figure 3.43 Gaussian plume modeling—effect of limited mixing. (a) Critical inversion condition; (b) effect of critical inversion condition; (c) effect of wind speed.

If the terrain height is greater than or equal to the effective stack height, then it can be assumed that the ground level concentration equals the plume centerline concentration, or:

$$C_j(x, y, z) = \frac{Q_j}{\pi \sigma_y \sigma_z \bar{u}} \exp\left(-\frac{y^2}{2\sigma_y^2}\right) \tag{3.31b}$$

This approach results in worst case or maximum possible ground level concentrations.

The second approach is to modify the plume flow for all terrain. That is, assume the plume centerline maintains its height above the ground as it would for level terrain. The basis for this is the observation that plumes will often rise as they pass over mountains, due to the "chimney effect." (The chimney effect occurs when a mountain, warmed by the sun, transfers its heat to the air passing over it. The warmed air then has a greater buoyancy than the surrounding air, and it rises.) This approach results in lower concentrations than the first. The approach may be modified by reducing the effective stack height by a fraction of the total terrain height. The appropriate fractional reduction must be determined by observing plumes in the area.

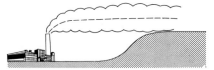

Approach 1: Modify dispersion equations, terrain height less than effective stack height.

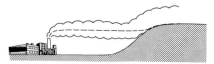

Approach 1: Modify dispersion equations, terrain height greater than effective stack height.

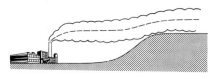

Approach 2: Modify plume flow.

Figure 3.44   Gaussian plume modeling—effect of terrain.

Downwash.   A limitation of the Gaussian model is that it cannot account for the effects of downwash caused by buildings. Moderate to high winds create pressure gradients around structures that act to draw the plume downward. When air flow encounters a sharp discontinuity in the geometry of a structure, the air flow separates from the obstruction. The separation of flow creates a low-pressure region in the wake of the stucture. The low-pressure region induces the formation of a recirculating flow behind the structure. If a stack is not sufficiently high to allow the plume to be carried above the recirculating region in the structure wake, the plume will be drawn into the turbulent region and conducted downward as shown in Fig. 3.45. This may cause the plume to be mixed to the ground near the stack, resulting in high ground level concentrations.

The probability of building downwash is reduced for stacks built at least 2.5 times the height of surrounding buildings.[10,65,94] For stack heights lower than this criterion, excessively high ground level concentrations may occur as a result of the plume being caught

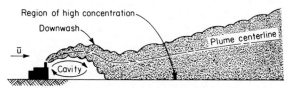

Figure 3.45   Gaussian plume modeling—effect of downwash.

in the downwash. Numerical modeling is required to describe downwash behavior in detail. However, both empirical models[11] and combination numerical/empirical models have been proposed.[56]

Plume Visibility. Recent interest in and concern for plume visibility impact is generating a need to consider visibility effects in impact analyses from major point sources. The visual impact can be categorized in terms of the distance from the source of interest. These categories are:

- Near the stack
- Intermediate zone
- Well-mixed, homogeneous atmosphere

The first category is of historical interest. The oldest emission regulations concern plume opacity near the stack. In this region, particulate matter is the major contributor to visual effects. Today, regulations limit the percent opacity (Ringelmann number) and control the visual effects for all practical purposes within this zone.

In the intermediate zone, the plume is still well-defined. Not only fly ash, but also secondary aerosols and nitrogen dioxide, can cause visual effects. Coloration of the plume is produced primarily by nitrogen dioxide. The nitrogen dioxide may be generated as a primary pollutant but, in general, is produced as a secondary pollutant by a variety of gas-phase homogeneous and photochemical reactions. As the flue gas is transported downwind, chemical reactions in the atmosphere form sufficient quantities of NO_2 from NO to cause coloration. The ratio of NO_2 to NO in the flue gas affects the amount of conversion of NO to NO_2, as well as the ambient level of ozone (O_3). A major reaction in the conversion process is the reaction between nitric oxide (NO) and ozone to form nitrogen dioxide and diatomic oxygen. The higher the background level of ozone, the faster the reaction proceeds.

Particles suspended within the plume reduce visibility, or visual range, by scattering and absorbing light. Due to the behavior of micron-sized particles in the multidirectional scattering of light into and out of the line of viewing, particles in the narrow range of 0.1 to 1.0 μm in radius have the greatest effect on visibility. Primary-pollutant particulate emissions, or secondary aerosols such as sulfates and nitrates, also influence plume visibility. The sulfate aerosol may be a significant fraction of the total particulate in a plume, especially at large distances from the source.

Plume visibility will be intensified by the increased use of fuel oil in the west and the use of higher sulfur-content fuels (oil and coal) throughout the country. Higher sulfur-content fuels result in a higher emission of sulfur oxides. Although sulfur dioxide, the dominant sulfur oxide emitted, is colorless, further oxidation in the stack or within the atmosphere can transform the species into sulfur trioxide. The sulfur trioxide, in turn, can combine with water (a major combustion product) to produce sulfate aerosols. The increased sulfate aerosol concentrations may then produce degraded visibility, depending upon conditions. In addition, high-sulfur fuel oils and coal may contain high percentages of fuel-bound nitrogen which will increase the oxides of nitrogen emission from the combustion process. The increased oxides of nitrogen may affect the coloration of the plume, again depending upon conditions.

Modeling techniques are being developed to meet the need to assess plume visibility. The initial contribution to this area is provided by Latimer and Samuelsen.[55] Models must account for light scatter from total suspended particulates, the conversion of sulfur oxides emissions to sulfates and nitrogen oxides emissions to nitrates, the gas-phase conversion of nitric oxide to nitrogen dioxide, the photochemical dissociation of nitrogen dioxide, and the oxidation of nitric oxide by ozone mixed into the plume from the ambient air. Various observer locations and elevation view angles must be considered as shown in Figs. 3.46a and 3.46b, respectively.

Computer Codes. A representative listing of computer codes is provided in Table 3.21. The increasing use of these codes is forcing improvements, and the increasing importance of code results in the decision-making process is forcing practitioners to carefully select and apply the codes.

The Environmental Protection Agency (EPA) maintains a library of codes called UNAMAP (Users Network for Applied Modeling of Air Pollution). The UNAMAP models are available on 9-track or 7-track magnetic tape from the National Technical Information Service (NTIS).* Models are added to and deleted from the UNAMAP series as the state

*NTIS Accession Number PB 229-771.

of the art dictates. The NTIS accession numbers for magnetic tapes for all the Gaussian models listed in Table 3.21 are provided in the reference list.

When selecting and applying a model, it is important that the following steps be taken to ensure (1) that the code is the most appropriate for the problem, (2) that the limitations and range of applicability of the code are established, and (3) that the latest version of the code is used:

- Review publications in the literature
- Review reports of agency-supported studies
- Contact the groups in the local, state, and federal agencies responsible for modeling for advice and guidance
- Contact the group responsible for the code development and validation

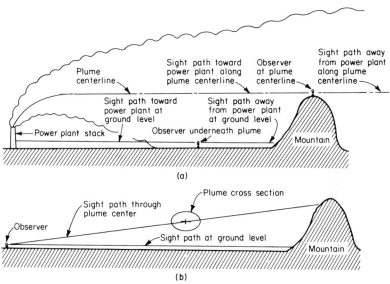

Figure 3.46   Gaussian plume modeling—plume visibility. (*a*) Observer locations; (*b*) elevation view angles.

Numerical Models   The increasing need for more precise representations of air quality (including chemical reaction) and the limitations of the Gaussian-based models have prompted development of numerical solutions of the conservation-of-mass equation on a fixed three-dimensional grid. Numerical solutions are required to account for the spatial and temporal variation of those processes which determine air quality (i.e., emissions, meteorology, and air chemistry). In addition, since only simple air chemistry processes can be modeled with the Gaussian models, numerical models are required when photochemical oxidants and other complex air chemistry processes are of interest.

Numerical models require a substantial data base. Emissions, for example, must be geocoded. Meteorological variables such as wind speed, wind direction, mixing height, and stability must also be geocoded. The atmospheric chemistry as well as the insolation must be prescribed. If successful, however, the model can be applied to a variety of adverse situations occurring in the atmosphere over urban centers.

Above-average technical expertise is required to operate numerical models, and the necessary computer time is costly. The use of the models is most appropriate for regional analyses, though application to individual line and point sources is possible for complex geometries not appropriate for the less sophisticated models. A representative listing of numerical models is provided in Table 3.22.

Model Validation   In all cases of model application, it is necessary to perform a validation or calibration analysis to verify the extent to which the model can reproduce certain conditions. In the process of performing such an analysis, certain variables within the model such as wind speed, direction, dispersion coefficients, reaction rates, or source emission rates may be adjusted to provide the "best fit" of model results to the observed

TABLE 3.21   Gaussian-based Air Quality Models

Point sources	PTMAX[109]	An interactive program that performs an analysis of the maximum short-term concentrations from a single point source as a function of stability and wind speed. Uses Briggs plume rise methods and Pasquill-Gifford dispersion methods as described in Turner.[108] An EPA UNAMAP program.
	PTDIS[109]	An interactive program that estimates short-term concentrations directly downwind of a point source at distances specified by the user. The effect of limiting vertical dispersion by a mixing height can be included, and gradual plume rise to the point of final plume rise is also considered. An option allows the calculation of isopleth half-widths for specific concentrations at each downwind distance. Uses Briggs plume rise methods and Pasquill-Gifford dispersion methods as described in Turner.[108] An EPA UNAMAP program.
	PTMTP[109]	An interactive program that estimates short-term concentrations for a number of arbitrarily located receptor points at or above ground level due to a number of point sources. Plume rise is determined for each source. Downwind and crosswind distances are determined for each source-receptor pair. Concentrations at a receptor from various sources are assumed additive. Hourly meteorological data are used; both hourly concentrations and averages over any averaging time from 1 to 24 hours can be obtained. Uses Briggs plume rise methods and Pasquill-Gifford dispersion methods as described in Turner.[108] An EPA UNAMAP program.
	VALLEY[12]	A program designed for multiple point and area source applications. It calculated pollutant concentrations for each frequency designated in an array defined by 6 stabilities, 16 wind directions, and 6 wind classes for 112 program-designated receptor sites on a radial grid of variable scale. The output concentrations are appropriate for either a 24-hour or an annual period, as designated by the user. The model contains the concentration equations, the Pasquill-Gifford dispersion coefficients, and the Pasquill stability classes, as given by Turner.[108] Plume rise is calculated according to Briggs. Plume height is adjusted according to terrain elevation for stable cases. An EPA UNAMAP program.
	CRSTER[57]	The Single Source (CRSTER) Model is a steady-state, Gaussian plume dispersion model designed for point source applications. It calculates pollutant concentrations for each hour of a year, at 180 receptor sites on a radial grid. The hourly concentrations are averaged

		to obtain concentration estimates for time increments of specifed length, such as 3-hour, 8-hour, 24-hour, and annual. The model contains the concentration equations, the Pasquill-Gifford dispersion coefficients, and the Pasquill stability classes, as given by Turner.[108] Plume rise is calculated according to Briggs.
Line sources	HIWAY[118]	An interactive program which computes the hourly concentrations of nonreactive pollutants downwind of roadways. It is applicable for at-grade highways, but can also be applied to depressed roadways. Uses Pasquill-Gifford dispersion coefficients. An EPA UNAMAP program.
	CALINE2[112]	A program which computes the hourly concentrations of conservative pollutants downwind of roadways. It is applicable for at-grade highways, but can also be applied to depressed roadways. Uses Pasquill-Gifford dispersion coefficients and differs from HIWAY primarily in the modeling of winds parallel and at small angles to the roadway.
Area sources: Urban	APRAC2[97]	An urban model developed by Stanford Research Institute that computes hourly averages of carbon monoxide for any urban location. The program requires an extensive traffic inventory for the area modeled. The computer program is a revised version of the APRAC-1A diffusion model.[61] The code uses EPA's emission calculation methodology from Supplement No. 5 to AP-42. Gridded and link-by-link emissions can be output for hydrocarbons, carbon monoxide, or oxides of nitrogen. Diffusion calculations make use of a receptor-oriented Gaussian plume model. Local winds at the receptor can be used; they are interpolated from multiple wind inputs. Mixing heights may be calculated from sounding data or directly from input. Two local source models are available, one treating pollutant behavior in a street canyon, the other treating vehicle and pollutant effects at a signalized intersection. A small program is included for decoding Federal Highway Administration data tapes. An EPA UNAMAP program.
	CDMQC[110]	The Climatological Dispersion Model QC (CDMQC) is an expanded version of the CDM program[13]; CDM determines the long-term concentrations of quasistable pollutants at any ground level receptor using average emission rates from point and area sources and a joint frequency distribution of wind

TABLE 3.21   Gaussian-based Air Quality Models (*Continued*)

		direction, wind speed, and stability for the same period. CDMQC includes three new features: (1) source contribution table, (2) internal calibration, and (3) statistical conversion of averaging times.
Airport	AVAP[107]	A program developed by the Argonne National Laboratory for the Federal Aviation Administration. Meteorological data is reduced to a series of 24-hour distributed stability wind roses on a monthly and annual basis. An emissions submodel classifies sources into category types with a time resolution up to 1 hour. Annual-averaged emissions are also calculated which are allocated in time using temporal distribution techniques. A dispersion submodel applies the Gaussian formulation to point, line, and area sources with a virtual source technique being used for sources of finite initial volume. Long-term concentrations are calculated using a predecessor of CDM.
	AQAM[84]	A program based on AVAP but expanded and specialized to Air Force operations. The improvements and refinements incorporated into AQAM, as well as a continuing series of validation studies, make the program suitable for use in general aviation projects as well.
Shopping Center	ISMAP[88]	A program developed by Stanford Research Institute specifically for complex sources. Submodels are incorporated for the simulation of traffic flow, vehicle emissions, and pollutant dispersion. Extensive traffic information is required.

TABLE 3.22   Numerical-based Air Quality Models

DIFKIN[62] (*DIFfusion/ KINetics*)	A three-dimensional trajectory (Lagrangian) model for photochemical pollutants. Uses a multiple-layered column of air transported by mesoscale windfield. A gridded emissions inventory with hourly variation is required. Horizontal dispersion is neglected. Expensive to run many trajectories to provide regional description of air quality. Meteorological input includes hourly values for regional windfield (wind speed and direction), inversion height, and stability class. Principal areas of application have been Los Angeles and San Francisco metropolitan areas.

TABLE 3.22   Numerical-based Air Quality Models   (Continued)

SAI[80-83] (Urban Airshed Model)	A three-dimensional grid (Eulerian) model for regional photochemical oxidant; also applicable to regional simulation of conservative pollutants. The primary area of application has been the Los Angeles area with some applications in the Denver, Colorado area. This model has been modified frequently, and there are several configurations of the meteorology, air chemistry, and emissions submodels. The required meteorological input includes hourly values for regional windfield (wind speed and direction), inversion height, and stability class. Hourly values for the regional emissions distribution are also required. Several options are available to treat major point and line sources on a subgrid cell basis.
	The species resolution of the hydrocarbon emissions depends on which configuration of the model is used. The model output includes estimates of hourly averages for each grid cell for carbon monoxide, oxides of nitrogen, hydrocarbons, and photochemical oxidant.
LIRAQ-2[59] (LIveRmore Air Quality model)	A grid (Eulerian) model for regional photochemical oxidant. The current computer code is specific for the San Francisco Bay Area. There is only one layer between the surface and inversion base; a mean transport wind is calculated by integrating the wind profile function. Consequently, the model is in essence a two-dimensional model. The required meteorological input includes hourly values for regional windfield (wind speed and direction), inversion height, and turbulent-eddy mixing coefficients. Hourly values for the regional emissions distribution are also required.
REM[114] (Reactive Environmental simulation Model)	A two-dimensional trajectory (Lagrangian) model for photochemical oxidant with a fairly complex air chemistry submodel. One layer between surface and inversion height. Meteorological input includes hourly values for regional windfield, inversion height, and stability class. Hourly values of regional emissions distribution are also required. Expensive to run many trajectories to provide regional description of air quality.
IMPACT[28] (Integrated Model for Plumes and Atmospherics in Complex Terrain)	A three-dimensional grid (Eulerian) model for photochemical oxidant. The air chemistry submodel can also simulate sulfate/SO_2 processes. IMPACT utilizes the DIFKIN photochemical air chemistry. Options are available for the detailed treatment of the air chemistry of major point source plumes. Meteorological input includes hourly values for regional windfield (wind speed and direction) and atmospheric stability. The inversion height is input implicitly via the stability parameter; up to 20 layers between the surface and the inversion height can be specified by the user. Hourly values for the regional emissions distribution are also required. Model output includes hourly average values for each grid cell and layer in the region.

ambient air quality. Any adjustment of these parameters must be closely scrutinized and thoroughly justified.

The danger in adjusting model parameters to fit observed data is that in any complex model there are a sufficient number of empirically derived values to manipulate the model for almost *any* desired result. The ability of a model thus modified to predict ambient air quality under a different set of meteorological or emission input conditions becomes extremely questionable if the adjustments are made either inconsistently or arbitrarily during the validation phase of model application.

Once the underlying assumptions and the calibration adjustments are verified as being reasonable, the result of the calibration analysis should be examined. A standard method for reporting calibration results involves plotting the computed pollutant concentrations against the observed concentrations as shown in Fig. 3.47. Note that the observed and

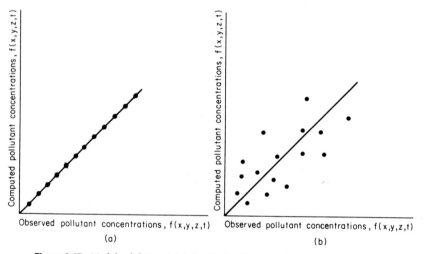

Figure 3.47   Model validation. (*a*) A "perfect" calibration; (*b*) a typical calibration.

computed concentrations are in general functions of space and time. A perfect model would produce results whereby the points fall on a straight 45°-angle line. A typical calibration analysis is presented in Fig. 3.47*b*. The calibration analysis provides the information necessary to establish confidence limits on model predictions (e.g., if the model predicts a certain concentration C_j, the uncertainty in that prediction may be estimated, $C_j \pm \Delta C_j$). It is also important to note whether any data were not used in the calibration analysis and why, since by not plotting unfavorable data a model may be made to appear more accurate than it really is. Table 3.19 lists representative validation studies conducted for the various models.

Summary   Air quality models are used to quantitatively simulate atmospheric interactions. The models are designed to relate emissions of pollutants to the resulting air quality. The models appropriate for air quality impact analyses vary in sophistication from direct numerical solution of the basic dispersion equation (including the effect of chemistry) to proportional models. The simple box model is a tool for initial, rough cut estimates of the cause-effect relationship between a source and the downwind air quality.

A word of caution is necessary here. Models are not absolute. The performance of a model depends upon the user, the model, and the application. The user needs to (1) be experienced (or, in the absence of experience, solicit guidance and perspective from an experienced user) and (2) recognize that air quality models are only indicative of the cause-effect relationship between air quality and the emissions of pollutants—despite the apparent sophistication. The accuracy of the foregoing warning is demonstrated by the complexity of atmospheric meterology, the mechanics of atmospheric stability, and the uncertainty of atmospheric chemistry.

Assessment

An assessment of the environmental impact is made by comparing the calculated pollutant concentrations to the National Ambient Air Quality Standards (NAAQS), and determining the extent to which the calculated impacts approach or exceed the NAAQS. Caution must be exercised in considering background concentrations to which the impacts must be added. Examples of assessments are provided in Section 3 for the case studies.

MITIGATION PROCEDURES

Mitigation procedures are activities that can be adopted to reduce the air quality degradation associated with the proposed project. Where mobile sources are significant contributors to the degradation, mitigation procedures will address (1) reduction in vehicle use (i.e., by introduction or improvement of alternative transit) and (2) spreading vehicular use over longer periods of time (i.e., to avoid heavy peaks during adverse meteorological conditions). Where stationary sources are significant contributors to the degradation, mitigation procedures will address (1) application of control technology to further reduce the emission of primary pollutants, (2) operation during periods of favorable meteorological conditions, and (3) reduction of emissions from other sources in the vicinity of the project (i.e., emission "trade-off" control strategy). Realistic mitigation procedures are those that are socially, politically, and economically acceptable. To meet air quality goals, conventional criteria for politically, socially, and economically acceptable procedures are being pushed to new frontiers and tested by the democratic process.

Mitigation measures which minimize significant adverse impacts could involve land use and transportation planning strategies. Examples of such mitigation measures for motor vehicle emissions are:

1. Provision of increased mass transit
2. Development of carpool systems
3. Development of bicycle lanes
4. Implementation of transit and/or carpool incentives such as favorable transit rates, preferential parking for carpools, carpool and bus lanes, employer subsidized carpools or buses
5. Implementation of penalties for use of automobile such as increased parking fees or establishment of auto free zones
6. Implementation of land use design programs that decrease the usage of and dependency on the automobile, such as (a) changes in zoning and the general plan to reduce holding capacity of the jurisdiction, (b) localized high density near transit stops, and (c) facilitating the development of work, shopping, and services nodes within easy access of homes by transit, bicycle, or walking.

For airport-related emissions, examples of mitigation measures are:

1. Eliminate the taxi mode of aircraft operation, which is a high-emission mode, by towing aircraft
2. During ground operations, have aircraft use one engine at a higher power setting rather than two engines at a lower power setting.
3. Establish satellite terminals for motor vehicle parking and provide mass transit to actual terminal site.

For stationary sources, the mitigation measures would depend on the operation and emission characteristics of the specific project. Some specific examples are:

1. For refineries, the composition of the crude oil could be adjusted to have a higher percentage of "lighter ends." This will produce more refinery gas which could then be more cleanly burned than the fuel oil.
2. For any source which burns a significant amount of fuel, switching to a less polluting fuel during adverse meteorological conditions or when fuel consumption rates increase would mitigate the air quality impact. A specific example is a power plant switching to low-sulfur fuel oil when a stagnation period is forecast.
3. For any source, the reduction of emissions from other sources in the vicinity of the proposed project would mitigate the air quality impact. This approach is referred to as an emission "trade-off" strategy and to be acceptable, will often require a reduction in emissions in excess of the increase in emissions projected for the proposed project.

ALTERNATIVES

Alternatives include options in the *design* of the project that can be adopted to reduce the degradation to the quality of air associated with the proposed project. The options range from cancellation of the project to reduced capacity. Alternatives differ from mitigating procedures in that the latter do not pertain to the design of the project per se. Alternatives are identified for each of the cases presented in the section on Case Studies later in this chapter.

GROWTH-INDUCING CONSIDERATIONS

The term "growth-inducing impact" communicates different concepts to different professional disciplines. To establish a common starting point, "growth-inducing impact" is defined, from a planner's perspective, as the degree to which a project promotes, facilitates, or provides for the increased urbanization and development of the project environs. (It is emphasized that this is not the *only* definition, and even with this definition, different perceptions will persist.)

The distinction between which projects promote, facilitate, or provide for development is also subject to discussion/debate. The following are presented only as representative of one possible interpretation.

- Non-retail employment centers are considered to promote growth
- Ancillary development such as residential, retail, and service centers facilitate growth
- Basic infrastructure such as water supply systems, wastewater treatment facilities, and major transportation facilities provide for growth

The induced-growth process can be described as follows:

Construction of a large source of employment like an industrial/office complex generates jobs which result in the nearby construction of dwelling units: these induce retail development to locate near them and generate demand for community, cultural, and religious facilities (schools, recreation areas, libraries, churches, theaters, fire and police stations, etc.). All of this requires the construction of streets and highways that then improve accessibility to the area. Better access fosters continued urban development, particularly highway-oriented commercial and office land uses. Additional sources of employment come into the area as secondary (and tertiary) industry or services located near the original major project, spurring on another round of residential development, and so forth.[5]

Positioning the specific project of concern on this cycle is one of the first steps in analyzing its growth-inducing impact and determining what measures are available for mitigating this impact. However, before proceeding, a very critical point must be made. The "growth" which is of concern in an air quality analysis is *not* growth per se but emission growth. The relationships between "growth" (be it economic, population, industrial, or whatever) and emission growth are functions of lifestyles and technology. As the effectiveness of purely technology-based air quality strategies diminishes, strategies which impact on lifestyles are becoming more important. The major example here is the increasing need to decrease the dependence on the automobile. This has significant energy resource implications as well as air quality resource implications.

Thus, returning to the air quality analysis for the specific project of concern, both the emissions/lifestyle relationships and the emissions/technology relationships associated with the project must be examined in order to identify candidate measures for mitigating the growth-inducing impact of the project. It is the *emissions* growth which is of concern.

Section 3: Case Examples

In this section, examples of air quality impact and assessment studies are presented for the two following cases:

Example	Source type
Highway	line
Power plant	point

The examples are condensed and do not contain the level of detail normally provided in an air quality analysis. In addition, both of the examples are located in the same geographical region.

The structure of the presentations is shown in Table 3.23. The highway example is presented with all sections. Redundant sections of the power plant example (e.g., meteo-

TABLE 3.23   Presentation Structure

Section	Highway	Power plant
SUMMARY	▪	▪
EXISTING ENVIRONMENT:		
General description of project	▪	▪
Description of project area	▪	▪
Meteorology	▪	←
Existing air quality		
Regional	▪	←
Subregional	▪	←
Local	▪	←
ENVIRONMENTAL IMPACT:		
Detailed project description	▪	▪
Analysis year	▪	▪
Description of surrounding area	▪	▪
Traffic study	▪	▪
Impact scenarios		
Emissions scenarios	▪	▪
Meteorology scenarios	▪	▪
Air chemistry scenarios	▪	▪
Primary impact scenarios	▪	▪
Secondary impact scenarios	▪	▪
Visibility impact scenarios	NA	▪
Modeling		
Primary pollutant impact		
Existing	▪	▪
No-build	▪	▪
Construction	▪	▪
Build-out	▪	▪
Secondary pollutant impact		
Existing	▪	▪
No-build	▪	▪
Construction	▪	▪
Build-out	▪	▪
Visibility impact	NA	▪
Assessment	▪	▪
MITIGATION	▪	▪
ALTERNATIVES	▪	▪
GROWTH INDUCING	▪	▪

▪ Covered
← Reference made to redundant section in the *Highway Case Study*
NA Non-applicable

rology, existing air quality) are not repeated but are included by reference to the highway example.

Referencing of data and information sources is essential to establishing the credibility of the report and to facilitate follow-up for those responsible for the review, use, or later expansion of the document.Where referencing is appropriate, the notation "(reference)" is used in the examples.

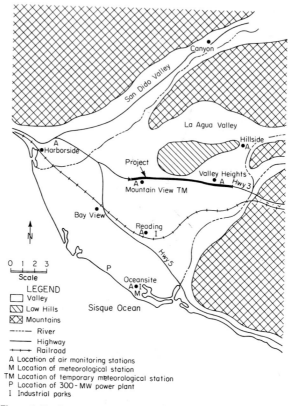

Figure 3.48   Project site—subregional perspective. (Scale in miles.)

Because the examples are condensed, the assumptions and deficiencies introduced in Section 2 which are associated with the analytical techniques are not discussed. Such a discussion is appropriate in a formal air quality analysis and must be included.

HIGHWAY CASE STUDY

SUMMARY

The proposed six-lane highway will increase local levels of carbon monoxide and subregional levels of oxidant compared to the no-build alternative. This is primarily a consequence of the growth-inducing impacts associated with the project. Although the local levels of carbon monoxide and subregional levels of oxidant at build-out (1995) will be lower than existing levels, the proposed project will prevent the attainment of National Ambient Air Quality Standards. As a result, the project is inconsistent with the State Implementation Plan.

EXISTING ENVIRONMENT

General Project Description

The project is a 9-mile highway widening of an existing two-lane roadway, shown in Fig. 3.48, to a six-lane highway. The project is required to facilitate planned development and will promote future development. Candidates for major facilities to be located in the area include a power plant.

Description of the Project Area

Location/Setting The project is located in a valley of gentle slope from ocean to low-lying hills. A map describing the setting is presented in Fig. 3.49.

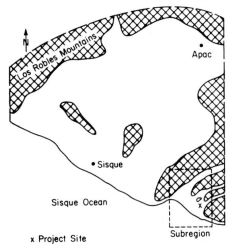

Figure 3.49 Project site—regional perspective.

Air Pollution Area/Regulatory Programs As shown in Fig. 3.49, the site is located in the Los Robles Air Basin which has been classified as the Los Robles Air Quality Control Region (AQCR) (reference). The Los Robles Regional Air Pollution Control District is the local regulatory agency responsible for air pollution control. The air basin has been classified as an Air Quality Management Area (AQMA) since ozone, sulfur dioxide, and carbon monoxide National Ambient Air Quality Standards (NAAQS) are presently exceeded and will likely be exceeded in 1985 (reference).

Air Pollution Programs. An Air Quality Maintenance Plan (AQMP) is presently being developed to establish control strategies and a policy for achieving and maintaining the NAAQS (reference). As such, the plan will deal with factors controlling all future emission sources in the AQMA, including general population growth and vehicle miles traveled. The impact of the AQMP on the proposed project is not presently known. This will depend upon the success achieved in the development of the AQMP and the timing of the project approval relative to completion of the AQMP.

When completed and approved, the AQMP will be integrated into the State Implementation Plan (SIP). The existing SIP prohibits construction or modification of facilities which will delay the attainment or prevent the maintenance of the NAAQS (reference). An assessment of the consistency of the project with the SIP is addressed in a later section, Assessment.

Emission Standards. Emission standards for the motor vehicles associated with the existing two-lane roadway and proposed project are established by the federal government. The state has not established motor vehicle emission standards separate from those of the federal government.

Ambient Air Quality Standards. The National Ambient Air Quality Standards (NAAQS) are presented in Table 3.24. The state has not promulgated separate ambient air quality standards.

Meteorology

The area has a mean annual temperature of 71.5°F (reference). The basin receives a substantial amount of rainfall, approximately 22 in/yr, and occasional snowfalls in the mountain regions (reference).

Winds The most dominant topographical features which affect the regional meteorological patterns of the basin are the Sique Ocean and the Los Robles Mountains. Winds exhibit a diurnal shift between the daytime, shore-directed sea breezes and the nighttime, lower-speed drainage winds. Within the subregional area (shown in Fig. 3.48), the sea

TABLE 3.24   National Ambient Air Quality Standards

Pollutant	Averaging time	Primary* standard	Secondary† standard	General objectives
Ozone	1 hr	240 μg/m³ (0.12 ppm)	240 μg/m³ (0.12 ppm)	To prevent eye irritation and possible impairment of lung functions in persons with chronic pulmonary disease, and to prevent damage to vegetation.
Carbon monoxide	8 hr	10 mg/m³ (9 ppm)	10 mg/m³ (9 ppm)	To prevent interference with the capacity to transport oxygen to the blood.
	1 hr	40 mg/m³ (35 ppm)	40 mg/m³ (35 ppm)	
Nitrogen dioxide	Annual average	100 μg/m³ (0.05 ppm)	100 μg/m³ (0.05 ppm)	To prevent possible risk to public health and atmospheric discoloration.
Sulfur dioxide	Annual average	80 μg/m³ (0.03 ppm)	. . .	To prevent pulmonary irritation.
	24 hr	365 μg/m³ (0.14 ppm)	. . .	
	3 hr	. . .	1300 μg/m³ (0.5 ppm)	To prevent odor.
Suspended particulate matter	Annual geometric mean	75 μg/m³	60 μg/m³	To prevent health effects attributable to long continued exposures.
	24 hr	260 μg/m³	150 μg/m³	
Hydrocarbons (corrected for methane)	3 hr	160 μg/m³ (0.24 ppm)	160 μg/m³ (0.24 ppm)	To reduce oxidant formation.

*National Primary Standards: The levels of air quality necessary, with an adequate margin of safety, to protect the public health.

†National Secondary Standards: The levels of air quality necessary to protect the public welfare from any known or anticipated adverse effects of a pollutant.

SOURCE: Reference 29, p. 8186, and Reference 44.

breeze develops an almost uniform daytime surface wind on the flat plain areas near the coast and is channeled by the mountains into the San Dido and La Agua valleys.

Meteorological data sources for the project area include a U.S. Weather Bureau station at Oceansite, a temporary meteorological monitoring station located midway between Mountain View and Valley Heights, and the ambient air quality stations operated by the Los Robles Air Pollution Control District. The temporary station was operated by the State Department of Transportation to record wind speed and direction for a twelve-month period in 1978 (reference).

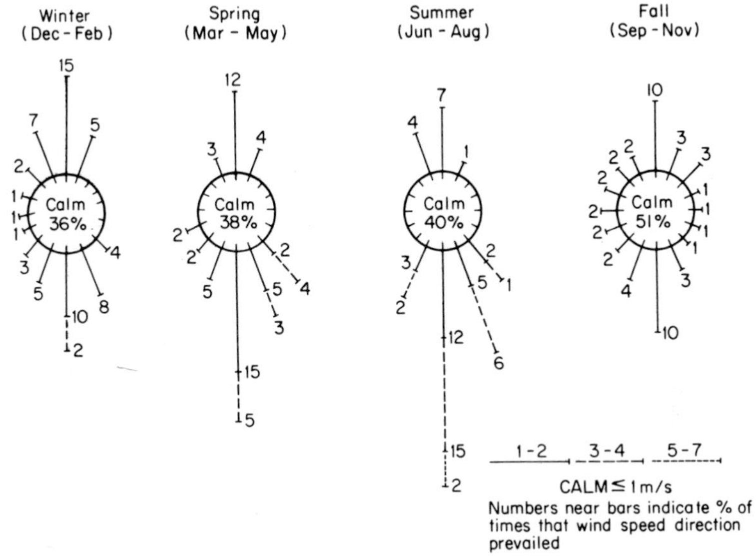

Figure 3.50   Meteorology—wind roses.

Seasonal wind roses, shown in Fig. 3.50, were developed from the data collected at the temporary meteorological station. Similarly, seasonal wind roses were developed using the National Climatic Center (NCC) program WINDROS and a five-year meteorological summary (1974–78) for the Oceansite Station. The results were basically identical to those obtained at the project site.

Streamline information was derived using meteorological data from the Oceansite station, the temporary meteorological station, and the six air monitoring stations shown in Fig. 3.48. Typical daytime streamline information is presented in Fig. 3.51.

Stability   The frequency of occurrence of stability classes and a joint relative frequency distribution of wind direction and atmospheric stability are given in Fig. 3.52 and Table 3.25, respectively. The data were obtained using the National Climatic Center (NCC) WINDROS computer program (reference) and a five-year meteorological summary (1974–78) for the Oceansite station. Periods of relatively high stability (classes E, F) are most frequent in the early morning (0700–0900 LST)* and early evening (1600–1800 LST), and are especially prominent in the fall and winter. Winds travel most often from the south (ocean breeze) or north (land breeze). The frequency of occurrence of high stability periods also coincides with these two wind directions. The seasonal variations in these stability/wind direction combinations are insignificant.

Inversion Height   Figs. 3.53a and 3.53b summarize historical inversion measurements south of the project at Oceansite (reference). The graphs show that ground-based inversions occur more often in the mornings of late fall and winter months than at any other time. By afternoon, the ground-based inversion is normally absent. Inversion bases ranging from 150 to 450 m (492 to 1476 ft) occur throughout the year but are most frequent and persistent in the summer and early fall months.

*LST = local standard time.

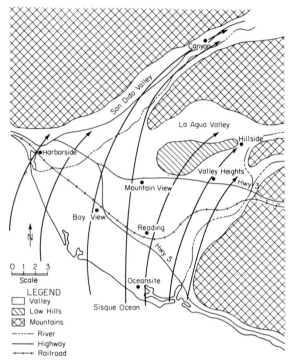

Figure 3.51 Meteorology—wind streamlines (daytime flow). (Scale in miles.)

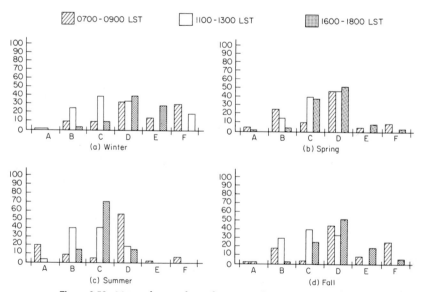

Figure 3.52 Meteorology—relative frequency of occurrence of stability class.

Episode Potential   The data from Holzworth[54] indicate that the region experiences 226 episode days in a five-year period (mixing heights < 1500 m, wind speeds < 6.0 m/s, and no significant precipitation over a minimum of two consecutive days). Thus the propensity for air pollution impact is moderate. By way of comparison, the highest number of episode

TABLE 3.25   Meteorology—Annual Frequency of Wind Direction and Stability Class and Frequency of Occurrence
(Percent)

Wind direction	Stability class						Total
	A	B	C	D	E	F	
NE	0.00	0.02	0.03	0.16	0.09	0.45	0.75
ENE	0.00	0.01	0.01	0.07	0.02	0.13	0.25
E	0.00	0.01	0.02	0.07	0.02	0.13	0.25
ESE	0.00	0.02	0.02	0.06	0.02	0.13	0.25
SE	0.03	0.28	0.15	0.67	0.53	1.84	3.50
SSE	0.09	0.64	0.69	1.91	1.66	2.51	7.50
S	0.38	2.33	1.93	6.04	3.21	3.88	17.75
SSW	0.07	0.86	1.10	1.72	0.40	0.60	4.75
SW	0.02	0.28	0.48	0.61	0.14	0.23	1.75
WSW	0.03	0.26	0.30	0.38	0.06	0.21	1.25
W	0.02	0.16	0.11	0.20	0.04	0.22	0.75
WNW	0.02	0.15	0.12	0.15	0.05	0.26	0.75
NW	0.02	0.12	0.13	0.20	0.10	0.44	1.00
NNW	0.03	0.30	0.40	0.91	0.57	1.80	4.00
N	0.04	0.55	0.77	2.56	1.56	5.52	11.00
NNE	0.01	0.10	0.14	0.95	0.53	1.53	3.25
Calm	1.12	3.58	2.60	3.58	0.00	28.88	41.25

STABILITY CLASS DESCRIPTIONS:
A–Extremely unstable   D–Neutral
B–Moderately unstable   E–Slightly stable
C–Slightly unstable   F–Moderately stable
DATA SOURCE: Oceansite, 1974–1978.
SOURCE: National Climatic Center, WINDROS Computer Program.

days in the contiguous United States over a five-year period is 1,154 experienced in southern California.

Existing Air Quality

Regional   Air quality data for a five-year period are presented in Fig. 3.54. The following three indices are used to characterize the quality of air in the region:
- Days per year for which the National Ambient Quality Standards are exceeded
- Maximum reading for year in terms consistent with the applicable air quality standard
- Seasonal average of the daily maximum readings

For each index, the value plotted in Fig. 3.54 is the maximum value which occurred anywhere in the basin. With few exceptions, the station exhibiting the most severely degraded quality of air in the basin is located in Apac (Fig. 3.48). The improvement in air quality for 1977 is attributed to unusually unstable meteorological conditions that persisted throughout the year (reference).

The regional air quality is generally degraded by particulate, ozone, and sulfur dioxide. Ozone impact, attributed to the motor vehicle activity in the basin, is increasing at a modest rate. Sulfur dioxide and particulate impact, attributed to the electrical generating fossil fuel–fired plants in the basin, is also increasing.

The air quality degradation (CO and ozone) associated with automobile-related pollutants is expected to reverse as the yearly increase in basin vehicle miles traveled (VMT) is offset by the increasingly restrictive motor vehicle emission standards. However, in the absence of a significant technological breakthrough, further reductions in emission stan-

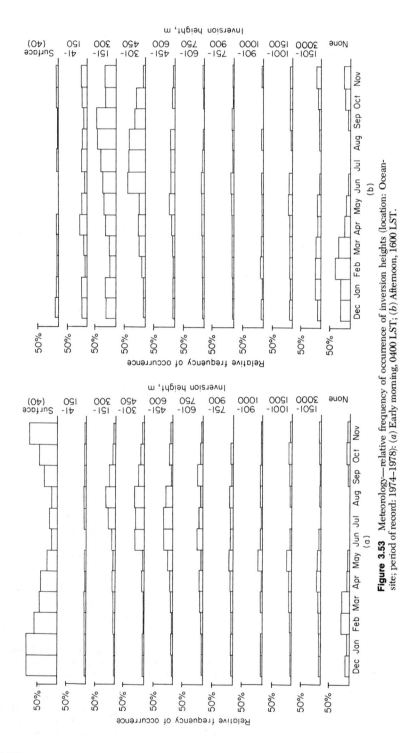

Figure 3.53 Meteorology—relative frequency of occurrence of inversion heights (location: Ocean-site; period of record: 1974–1978): (*a*) Early morning, 0400 LST; (*b*) Afternoon, 1600 LST.

dards are not expected beyond 1982 (reference). As a result, automobile use is expected to accelerate air quality degradation at the end of the century unless VMT is controlled.

The increase in air quality degradation (particulates and SO_2) due to stationary combustion and other sources is attributed to the increased generation of electrical energy in the basin and the increasing reliance on lower grade fuel oils and coal for power generation.

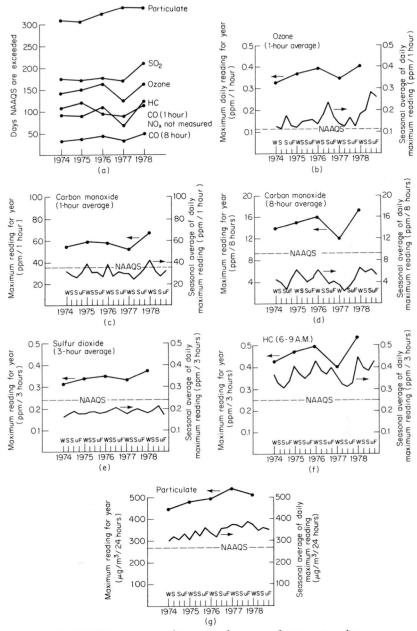

Figure 3.54 Existing air quality—regional summary of existing air quality.

Although the increasing use of energy is not expected to be abated, technological break-throughs are actively being sought to reduce the impact from the use of lower-grade fuels and coals (reference).

Subregional Four air monitoring stations are maintained by the Los Robles Air Pollution Control District in the subregional area surrounding the project site as shown in Fig. 3.48 (Reading, Mountain View, Valley Heights, and Hillside). Data from these stations for 1978 have been analyzed by season for the following three air quality indices:

- Days per year for which the National Ambient Air Quality Standards are exceeded
- Maximum readings by season in terms consistent with the applicable air quality standard
- Seasonal average of daily maximum readings

The results are summarized in Table 3.26. The maximum readings are an indication of the *worst case* conditions presently occurring in the area, and the seasonal averages are an indication of the *most probable* conditions presently occurring in the area. (The values presented are the highest that were recorded during the season and did not necessarily occur on the same day. As a result, the figures reflect a conservative estimate of the worst case and most probable conditions.)

The air quality in the subregion is degraded with respect to all criteria pollutants. Ozone and SO_2 values are especially in excess of the National Ambient Air Quality Standards. The ozone and SO_2 degradation increases with distance from the ocean and is highest at Hillside.

Local An estimate of local air quality requires modeling of the existing two-lane roadway in the rural area between the two communities, and modeling of the communities at both ends of the 9-mile project. The modeling approach and results of the modeling are presented with the construction and future-year modeling in the sections on Impact Scenarios and Air Quality Modeling later in this analysis.

ENVIRONMENTAL IMPACT

Detailed Project Description

Build-Out The project proposed is a 9-mile highway-widening project from two to six lanes. The expansion is required to support the residential and industrial development of the valley area through which the two-lane road now travels. The project is expected to commence in 1984, with construction completed in twelve months and operation initiated in 1985 (reference).

The major pollutants emitted, the impacts modeled, and the modeling methods employed to analyze the impacts are presented in Table 3.27. Gaussian modeling is used to analyze the primary pollutant impact, and proportional modeling is used to analyze the secondary pollutant impact.

No-Build In the event the project is not built, the existing two-lane highway will be saturated and growth in the area will be restricted in the absence of alternative modes of transportation.

Analysis Years

The years selected for analysis include the existing year, construction year, and two build-out years—the year of completion (1985) and ten years after the date of completion (1995). The latter is selected to account for the effect of (1) an automobile population that will be 100 percent regulated by the 1982 model year emission standards (the strictest envisioned barring a technical breakthrough), and (2) a highway usage that represents full development of the valley area. For each build-out year, an analysis is made for the no-build alternative (the case for which the project is not built).

Description of Surrounding Areas

General The project is located in a valley of gentle slope from the ocean to low-lying hills. The area is presently rural but destined for both residential and industrial development (reference). Based upon this projected development, the highway widening project has been proposed.

The area presently has 26,000 residents divided equally among two unincorporated

TABLE 3.26a Existing Air Quality—Air Quality Data (Reading)

Pollutant	NAAQS (Table 3.24) Conc.	Avg. time	Summer Days exceeded	Avg. daily max.	Seasonal max.	Fall Days exceeded	Avg. daily max.	Seasonal max.	Winter Days exceeded	Avg. daily max.	Seasonal max.	Spring Days exceeded	Avg. daily max.	Seasonal max.	Annual Max.
Ozone	0.12 ppm	1 hr	5	0.09	0.12	3	0.07	0.10	2	0.06	0.11	0	0.05	0.07	0.12
CO	9 ppm	8 hr	1	6	11	1	6	10	5	6	9	2	5	12	12
	35 ppm	1 hr	0	10	15	0	12	21	3	21	41	0	10	23	11
NO_2	0.05 ppm	Annual avg.		...			...			...			...		0.04
	0.5 ppm	3 hr	62	0.48	0.91	69	0.45	0.97	73	0.52	0.87	63	0.47	0.73	0.97
SO_2	0.14 ppm	24 hr	55	0.16	0.22	71	0.71	0.28	67	0.16	0.25	57	0.15	0.20	0.28
	0.03 ppm	Annual avg.		...			...			...			...		0.04
Total HC	0.24 ppm	3 hrs (6–9 AM)	14	0.19	0.30	21	0.18	0.29	31	0.20	0.35	13	0.19	0.32	0.32
Particulates	260 $\mu g/m^3$	24 hr	23	289	347	15	271	320	26	298	360	24	282	335	360
	75 $\mu g/m^3$	Annual geometric mean		...			...			...			...		135

SOURCE: Los Robles APCD.

TABLE 3.26b Existing Air Quality—Air Quality Data (Valley Heights)

Pollutant	NAAQS (Table 3.24) Conc.	Avg. time	Summer Days exceeded	Summer Avg. daily max.	Summer Seasonal max.	Fall Days exceeded	Fall Avg. daily max.	Fall Seasonal max.	Winter Days exceeded	Winter Avg. daily max.	Winter Seasonal max.	Spring Days exceeded	Spring Avg. daily max.	Spring Seasonal max.	Annual Max.
Ozone	0.12 ppm	1 hr	15	0.11	0.16	4	0.07	0.12	2	0.06	0.10	2	0.07	0.09	0.16
CO	9 ppm	8 hr	0	5	7	1	5	10	4	5	9	1	3	10	10
CO	35 ppm	1 hr	0	9	12	0	11	15	2	19	39	0	6	13	39
NO$_2$	0.05 ppm	Annual avg.		...			...			...			...		0.03
SO$_2$	0.5 ppm	3 hr	3	0.31	0.60	11	0.39	0.66	9	0.42	0.62	8	0.34	0.63	0.66
SO$_2$	0.14 ppm	24 hr	2	0.08	0.17	8	0.09	0.20	12	0.10	0.23	7	0.88	0.25	0.25
SO$_2$	0.03 ppm	Annual avg.		...			...			...			...		0.01
Total HC	0.24 ppm	3 hr (6–9 AM)	16	0.17	0.28	9	0.18	0.27	14	0.19	0.35	6	0.17	0.29	0.35
Particulates	260 µg/m³	24 hr	35	285	410	6	241	295	15	252	360	28	271	390	410
Particulates	75 µg/m³	Annual geometric mean		...			...			...			...		120

SOURCE: Los Robles APCD.

TABLE 3.26c Existing Air Quality—Air Quality Data (Mountain View)

Pollutant	NAAQS (Table 3.24) Conc.	Avg. time	Summer Days exceeded	Summer Avg. daily max.	Summer Seasonal max.	Fall Days exceeded	Fall Avg. daily max.	Fall Seasonal max.	Winter Days exceeded	Winter Avg. daily max.	Winter Seasonal max.	Spring Days exceeded	Spring Avg. daily max.	Spring Seasonal max.	Annual Max.
Ozone	0.12 ppm	1 hr	20	0.12	0.20	5	0.08	0.17	3	0.07	0.13	1	0.06	0.11	0.20
CO	9 ppm	8 hr	2	3	10	12	4	12	28	7	11	4	4	11	12
	35 ppm	1 hr	1	15	37	2	15	38	7	26	42	1	13	37	42
NO₂	0.05 ppm	Annual avg.		⋯			⋯			⋯			⋯		0.03
SO₂	0.5 ppm	3 hr	30	0.45	0.62	42	0.47	0.75	41	0.42	0.71	36	0.39	0.61	0.75
	0.14 ppm	24 hr	25	0.10	0.16	37	0.11	0.19	32	0.11	0.19	22	0.10	0.17	0.19
	0.03 ppm	Annual avg.		⋯			⋯			⋯			⋯		0.03
Total HC	0.24 ppm	3 hr (6–9 AM)	31	0.20	0.31	42	0.22	0.38	62	0.39	0.50	37	0.26	0.40	0.50
Particulates	260 µg/m³	24 hr	79	311	410	59	282	340	68	291	362	72	298	397	410
	75 µg/m³	Annual geometric mean		⋯			⋯			⋯			⋯		130

SOURCE: Los Robles APCD.

TABLE 3.26d Existing Air Quality—Air Quality Data (Hillside)

Pollutant	NAAQS (Table 3.24) Conc.	Avg. time	Summer Days exceeded	Summer Avg. daily max.	Summer Seasonal max.	Fall Days exceeded	Fall Avg. daily max.	Fall Seasonal max.	Winter Days exceeded	Winter Avg. daily max.	Winter Seasonal max.	Spring Days exceeded	Spring Avg. daily max.	Spring Seasonal max.	Annual Max.
Ozone	0.12 ppm	1 hr	37	0.15	0.23	12	0.11	0.17	4	0.08	0.12	7	0.08	0.8	0.23
CO	9 ppm	8 hr	0	1	1	0	1	4	0	2	4	0	1	3	4
	35 ppm	1 hr	0	6	12	0	7	15	0	7	18	0	4	13	18
NO$_2$	0.05 ppm	Annual avg.		⋯			⋯			⋯			⋯		0.04
SO$_2$	0.5 ppm	3 hr	0	0.13	0.42	0	0.17	0.48	2	0.18	0.57	0	0.15	0.38	0.57
	0.14 ppm	24 hr	0	0.03	0.08	0	0.04	0.12	1	0.04	0.15	0	0.03	0.09	0.15
	0.03 ppm	Annual avg.		⋯			⋯			⋯			⋯		0.01
Total HC	0.24 ppm	3 hr (6–9 AM)	0	0.09	0.17	0	0.07	0.13	0	0.11	0.19	0	0.10	0.18	0.19
Particulates	260 µg/m³	24 hr	12	241	321	8	235	293	13	251	309	11	248	317	321
	75 µg/m³	Annual geometric mean		⋯			⋯			⋯			⋯		110

SOURCE: Los Robles APCD.

communities, Valley Heights and Mountain View. The general plan for the area shows Mountain View growing to 62,000 and Valley Heights growing to 43,000 by 1995 (reference). Each community is being developed with integrated, low-polluting industrial complexes (reference).

TABLE 3.27   Proposed Project—Air Quality Impacts

Primary pollutants emitted	Impact modeled*		Scale	Modeling methods employed
	Primary pollutant	Secondary pollutant		
CO	CO	. . .	Local	Gaussian
HC ⎫ NO$_x$ ⎭	. . .	Ozone	Subregional, Regional	Proportional

*Local impact from HC and NO$_x$ as primary pollutants is not modeled. HC is classified as a pollutant precursor to oxidant formation. NO$_x$ is both a primary pollutant and a precursor to oxidant formation. As a primary pollutant, however, NO$_x$ is primarily NO. By the time NO$_2$ is formed, the pollutant is considered to be beyond the primary impact area.

Pollutant Sources   The pollutant sources in the surrounding area (sources exogenous to the project site) include those in the region, subregion, and local to the project area.

On the regional perspective, the pollutant sources of interest are those responsible for ozone impact. These are taken to be the sources within the region that emit the secondary pollutant precursors of reactive hydrocarbons and nitrogen oxides.

On the subregional perspective, the pollutant sources of interest are those responsible for ozone impact. These are taken to be the sources within the subregion that emit the secondary pollutant precursors of reactive hydrocarbons and nitrogen oxides.

Proportional modeling is used for the subregional and regional impact of ozone, with the 6–9 A.M. reactive hydrocarbon emissions as the proportioning parameter. For this method, identification and geocoding of the individual sources is not required.

In addition to proportional modeling, a second method is used to describe the impact on ozone at the subregional level. Subregional isopleths of the 6–9 A.M. reactive hydrocarbon emissions are presented for each analysis. For this approach, identification and geocoding of the 6–9 A.M. reactive hydrocarbon emissions within the subregion is required.

As shown in Fig. 3.48, the major sources in the subregion include the mobile source activity associated with the two communities (Mountain View and Valley Heights), Highways 3 and 5, and the communities upwind (Reading, Oceansite, and Bay View); stationary sources located in industrial complexes at Roadway and Oceansite; the Oceansite Harbor; and a 300-MW power-generating facility northwest of Oceansite. Major sources formally planned for the future are limited to a refinery and LNG terminal at Oceansite (reference).* With respect to the emission of NO$_x$, the major sources include the mobile source activity and the 300-MW power plant. With respect to the emission of reactive hydrocarbons, the major sources include the mobile source activity, the future-year LNG terminal (1985), and the future-year refinery (1985).

Areal Gaussian modeling is conducted for the local impact of CO in the present study. As a result, the emissions from the pollutant sources surrounding the project site that contribute to local impact must be individually identified and geocoded. The major sources responsible for CO impact are the mobile source activities associated with the communities of Mountain View and Valley Heights.

Receptors   The receptors that will be susceptible to primary pollutant (CO) impact of the project include the general population on both sides of the project alignment in the two cities, Mountain View and Valley Heights, and the existing and planned institutions

---

*A major source being *considered* but *not yet formally planned* is a 1500 MW power plant to be located south of the existing two-lane roadway and between the communities of Mountain View and Valley Heights. Because of its tenuous status, the contribution of this source is not considered in the present study.

(e.g., schools, hospitals) shown in Fig. 3.55 that are considered to be especially sensitive to impact.

The receptors that will be susceptible to the secondary pollutant (ozone) impact of the project include the population of the subregion in general, and the population of the downwind communities located in the La Agua and San Dido Valleys in particular.

Traffic Studies

Existing   A traffic study conducted by the State Department of Transportation (reference) addressed both the existing and build-out conditions. The area covered by the study is that area which will experience a change in traffic due to build-out of the project. Included in the study is an analysis for the existing two-lane roadway and secondary traffic in the two communities of Mountain View and Valley Heights. The traffic estimates consist of the average hourly roadway volumes, speeds, and percentage of heavy-duty

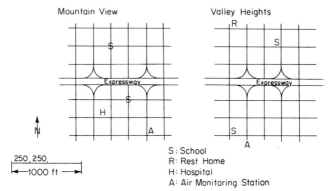

Figure 3.55   Sensitive receptors—locations.

vehicles associated with the two periods of maximum traffic—A.M. peak (0600–0900) and P.M. peak (1600–1800). The information for the two-lane roadway is presented in Fig. 3.56. The traffic data are geocoded in the report into one-kilometer calls to facilitate modeling the local air quality. It is noteworthy that the traffic on the two-lane roadway is near saturation level.

No-Build   A traffic study by the Los Robles County Road Department (reference) considers 1985 and 1995 for the case in which the project is not built and the roadway remains two lanes. The area covered by the study is that area which will experience a change in traffic should the project be built. Included in the study is an analysis for the two-lane roadway and secondary traffic in the two communities of Mountain View and Valley Heights. The traffic estimates consist of the average hourly roadway volumes, speeds, and percentages of heavy-duty vehicles associated with the two periods of maximum traffic—A.M. peak (0600–0900) and P.M. peak (1600–1800). The information for the two-lane roadway is presented in Fig. 3.56. The traffic information is geocoded to facilitate modeling of the local air quality. It is noteworthy that the effect on the two-lane roadway is to effectively reduce the average speed for the existing near-saturation conditions.

Construction   The construction-year traffic is assumed, for the purpose of analysis, to be that for the existing year.

Build-out   The State Department of Transportation study cited above includes a traffic study for the analysis years (reference). The area covered is that area which will experience a change in traffic due to completion of the project. Included in the study is an analysis for the six-lane roadway and secondary traffic in the two communities of Mountain View and Valley Heights. The traffic estimates consist of the average hourly roadway volumes, speeds, and percentage of heavy-duty vehicles associated with the two periods of maximum traffic—A.M. peak (0600–0900) and P.M. peak (1600–1800). The information for the six-lane roadway is presented in Fig. 3.56. The traffic information is geocoded into 1-km cells to facilitate modeling the local air quality.

Impact Scenarios

Emissions Scenarios Emission factors for the mobile sources were obtained from the latest supplement to AP-42 (reference). Usage data for the mobile sources were obtained from the traffic studies cited in the foregoing section.

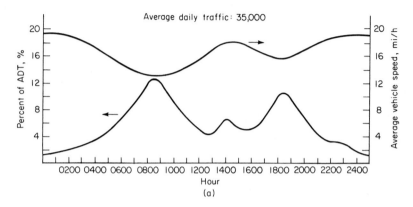

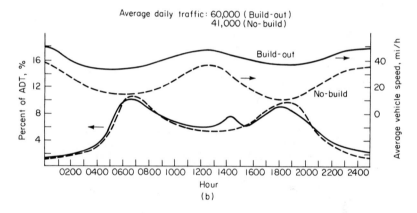

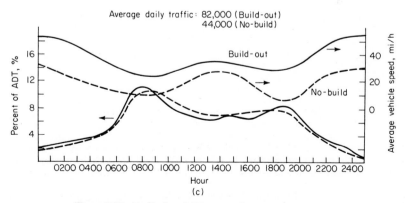

Figure 3.56 Traffic data. (*a*) Existing; (*b*) 1985; (*c*) 1995.

For the stationary sources, emission factors and usage data were obtained from the Los Robles Air Pollution Control District (reference). The data were confirmed by the state air regulatory agency (reference) and the Los Robles Regional Air Pollution Control District (reference).

Scenarios are presented first for sources at the site of the project, and second for sources exogenous to the project site.

Sources at Project Site

Existing. Emissions scenarios are presented in Table 3.28 for the existing, two-lane roadway. (No stationary sources are present at the project site.) To facilitate development of the impact scenarios for local impact modeling, temporal distributions are provided.

No-build. Emissions scenarios are presented in Table 3.28 for the two-lane roadway in the two future analysis years (1985, 1995). These future-year emissions are required to analyze the air quality impact in the event the project is not built.

Construction. Emissions scenarios are presented in Table 3.29 for the construction year.

Build-out. Emissions scenarios for the proposed six-lane roadway are presented in Table 3.29 for the two analysis years.

Sources Exogenous to Project Site

Existing. Emissions scenarios are presented for the existing pollutant sources in Table 3.28. Identification of emissions *local* to the site are necessary for modeling local CO impact, whereas the *subregional* and *regional* emissions are necessary for assessing oxidant impacts. For the purposes of modeling, the local CO emissions and the subregional 6–9 A.M. reactive hydrocarbon emissions were geocoded. The analyses are presented in the section on Air Quality Modeling.

No-build. Emissions scenarios are presented in Table 3.28 for the two future years in the event the project is not built. For the purposes of modeling, the local CO emissions and the subregional 6–9 A.M. reactive hydrocarbon emissions were geocoded. The analyses are presented in the section on Air Quality Modeling.

Construction. Emission scenarios are presented in Table 3.29 for the construction year.

Build-out. Emissions scenarios are presented in Table 3.29 for the build-out alternative. Identification of emissions local to the site are necessary for modeling local CO impact, whereas the total emissions in the subregion and region are necessary for assessing oxidant impacts. For the purposes of modeling, the local CO emissions and the subregional 6–9 A.M. reactive hydrocarbon emissions were geocoded. The analyses are presented in the section on Air Quality Modeling.

Meteorological Scenarios For primary and secondary pollutants, meteorological scenarios are developed using wind rose and stability data presented in Figs. 3.50 to 3.53, and Table 3.25. The meteorological scenarios which follow are applicable to both the sources at the project site and sources exogenous to the project site.

Primary Pollutants. The atmospheric conditions most conducive to primary pollutant impact from motor vehicle activity are:

- Wind speed—calm to low (≈ 1 m/s)
- Direction—persistent and toward populated areas and/or sensitive receptors
- Mixing heights—low and persistent
- Stability—very stable (F stability)

The winds are calm in the project area in the early morning and late evening hours during the transition between the ocean and land breezes. As shown in Fig. 3.52, periods of relatively high stability (classes E, F) are most frequent in the early morning hours (0600–0900 LST) and early evening hours (1600–1800 LST), and are especially frequent in the fall and winter. The inversion base is shown in Fig. 3.53 to be especially low, 40 m (131 ft) at the early morning hour (0400 LST), and the frequency of occurrence is maximum during the winter months. In the evening (1600 LST), the inversion base is generally higher, 150 to 450 m (494 to 1476 ft), and is likely controlled by a subsidence condition.

A review of Fig. 3.50 and Table 3.25 indicates that the winds are more frequent from the south (ocean breeze) or north (land breeze). The frequency of occurrence with periods of high stability also coincide with these two wind directions.

TABLE 3.28a Emission Scenarios—Existing Year and No-build Alternative (Average Daily Emissions)

Emissions (tons/day)

Year	Pollutant	Sources at project site Mobile†	Sources at project site Stationary	Sources at project site Total	Local Mobile†	Local Stationary	Local Total	Subregional* Mobile	Subregional* Stationary	Subregional* Total	Regional* Mobile	Regional* Stationary	Total
Existing	NMHC 1. Total	3.75	NA†	3.75	6.24	2.12	8.36	85.3	78.7	164	786	726	1512
	NMHC 2. 6–9 A.M.	0.45	NA	0.45	0.81	0.08	0.89	8.84	8.16	17	83.7	77.3	161
	NO$_x$	1.70	NA	1.70	3.21	2.42	5.63	117	37.0	154	1085	342	1427
	CO	26.5	NA	26.5	54.1	1.13	55.2	830	82	912	7681	760	8441
	SO$_x$	0.08	NA	0.08	0.21	6.2	6.41	7.06	32.1	39.2	74.6	358	433
	Particulate	0.21	NA	0.21	0.58	4.5	5.08	5.48	21.9	27.4	57.0	228	285
No-build (1985)	NMHC 1. Total	1.6	NA	1.6	4.9	3.0	6.32	54.0	50.0	104	533	493	1026
	NMHC 2. 6–9 A.M.	0.21	NA	0.21	0.63	0.13	0.76	7.44	6.86	14.3	73	68	141
	NO$_x$	1.4	NA	1.4	2.8	2.9	5.7	70.4	28.7	99.1	745	235	980
	CO	9.3	NA	9.3	23.2	1.1	24.3	310	30.7	341	3083	305	3388
	SO$_x$	0.12	NA	0.12	0.45	8.2	8.65	10.2	50.0	60.2	105	514	619
	Particulate	0.24	NA	0.24	0.54	4.9	5.44	6.2	24.8	31.0	62	249	311
No-build (1995)	NMHC 1. Total	0.81	NA	0.81	4.32	3.6	7.92	53.1	48.9	102	594	548	1142
	NMHC 2. 6–9 A.M.	0.09	NA	0.09	0.58	0.17	0.75	7.07	6.53	13.6	79	73	152
	NO$_x$	1.3	NA	1.3	2.1	3.2	5.3	65.4	26.7	92.1	752	238	990
	CO	6.5	NA	6.5	19.3	1.3	20.6	315	31.1	346	3196	316	3512
	SO$_x$	0.16	NA	0.16	0.65	10.1	10.75	12.1	59.1	71.2	160	783	943
	Particulate	0.33	NA	0.33	0.79	5.2	5.99	10.2	40.8	51.0	98	392	490

*SOURCE: Los Robles Air Pollution Control District.
†Heavy-duty vehicle mix: existing—4%, 1985—10%, 1995—14%
‡Not applicable

TABLE 3.28*b*   Temporal Distribution of Average Daily Emissions (Local Mobile Sources)

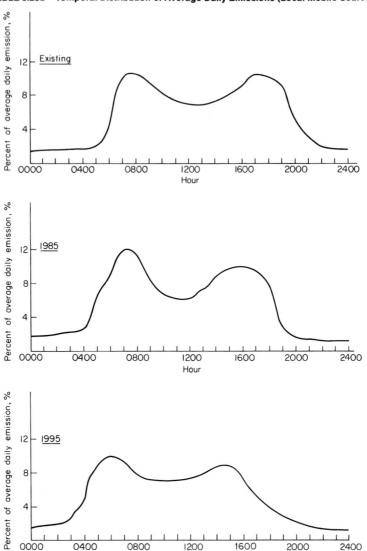

Secondary Pollutants.   A meteorological scenario is not required for secondary pollu-tants since proportional modeling is used to analyze the secondary impact. Nevertheless, a meteorological scenario is instructive in identifying (1) the conditions under which secondary pollutant impact occurs and (2) the receptor area that will experience secondary pollutant impact. For these reasons, meteorological scenarios are presented for secondary pollutants.

The atmospheric conditions conducive to the most probable secondary pollutant impact are:

    ■ Wind speed—most probable

    ■ Direction—toward a populated area, and over a source area rich in secondary pollutant precursors

TABLE 3.29a Emission Scenarios—Construction Year and Build-out Alternative (Average Daily Emissions)

Emissions (tons/day)

| Year | Pollutant | Sources at project site | | | Sources exogenous to project site | | | | | | | | | Total |
| | | Mobile† | Stationary | Total | Local | | | Subregional* | | | Regional* | | | |
					Mobile†	Stationary	Total	Mobile	Stationary	Total	Mobile	Stationary	Total	
Construction (1983)	NMHC 1. Total	0.01	NA	0.01	5.6	3.3	8.9	68	63	131	625	577	1202	1202
	2. 6–9 A.M.		NA		0.74	0.16	0.90							
	NO$_x$	0.14	NA	0.14	3.5	3.1	6.6	88	36	124	863	273	1136	1136
	CO	0.03	NA	0.03	28.2	1.6	29.8	340	34	374	3287	325	3612	3612
	SO$_x$	0.01	NA	0.01	0.48	8.1	8.58	8.4	40.8	49.2	92	451	543	543
	Particulate 1. Exhaust	0.01	NA	0.01	0.59	4.9	5.49	3.6	10.1	12.6	26	106	132	132
	2. Fugitive	0.05	NA	0.05	0.01	0.02	0.02	1.8	7.4	9.2	22	90	112	112
Build-out (1985)	NMHC 1. Total	3.1	NA	3.1	5.4	3.2	8.6	66.0	60.0	126	545	503	1048	1048
	2. 6–9 A.M.	0.11	NA	0.11	0.71	0.14	0.85	9.0	8.3	17.3	75	69.0	144	144
	NO$_x$	0.72	NA	0.72	3.1	3.2	6.3	80.9	33.1	114	706	289	995	995
	CO	4.8	NA	4.8	26.4	1.4	27.8	376	37.2	413	3149	311	3460	3460
	SO$_x$	0.06	NA	0.06	0.51	8.3	8.81	12.8	62.6	75.4	108	526	634	634
	Particulate	0.12	NA	0.12	0.62	5.1	5.72	7.0	28.2	35.2	63.0	252	315	315
Build-out (1995)	NMHC 1. Total	1.9	NA	1.9	4.68	3.8	8.48	78.5	72.5	151	619	572	1191	1191
	2. 6–9 A.M.	0.04	NA	0.04	0.61	0.18	0.79	10.2	9.4	19.6	82.2	75.8	158	158
	NO$_x$	0.55	NA	0.55	2.5	3.4	5.9	101	41.2	142	738	302	1040	1040
	CO	2.77	NA	2.77	21.3	1.3	22.6	466	46	512	3347	331	3678	3678
	SO$_x$	0.07	NA	0.07	0.72	10.5	11.2	17.7	86.3	104	166	810	976	976
	Particulate	0.14	NA	0.14	0.86	5.9	6.76	15.6	62.5	78.1	103	414	517	517

*SOURCE: Adapted from Los Robles Air Pollution Control District. Values from Table 3.28 increased by the projected increase in emissions from sources exogenous to the project site.

†Heavy Duty Vehicle Mix: Construction—90%; heavy-duty const. 1985—15%, 1995—20%

‡Vehicle mix: motor graders—3; scrapers—3; off highways trucks—3; wheeled loaders—3; miscellaneous—3
Duration of construction: 1 year

TABLE 3.29*b*   Temporal Distribution of Average Daily Emissions (Local Mobile Sources)

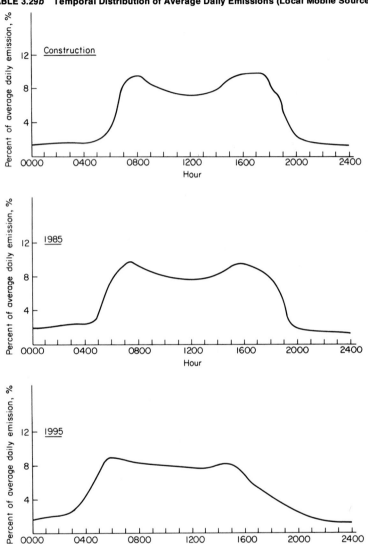

- Mixing height—low, persistent, and sufficiently high to entrap emissions from major elevated sources
- Insolation—high and coincident with conditions above

The low mixing heights are most persistent during the summer months. As demonstrated by Fig. 3.28, the 0400 LST and 1600 LST mixing heights between 150 and 600 m (492 and 1968 ft) occur frequently, with the majority less than 450 m (1476 ft). The 0400 LST readings are biased by the early morning, low-lying radiation inversion. The limited mixing height coincides with the peak insolation associated with the long daylight hours of the summer months.

The transport associated with oxidant formation begins near daybreak. The winds are initially directed toward the ocean but reverse after daybreak to an inland direction toward population centers. This is also the time of greatest ground level stability due to the radiation inversion. Hence, the meteorological scenario most conducive to secondary pollutant impact occurs during the summer, beginning in the morning when the radiation inversion and low winds allow primary pollutant concentrations to build and, along with the high levels of summer insolation, accelerate the formation of photochemical oxidant. As the radiation inversion breaks, the pollutants disperse vertically but are trapped below the inversion at 150 to 600 m, which allows the oxidant formation to continue.

The atmospheric conditions most conducive to the worst case air quality impact are:

- Wind speed—low over a prolonged period
- Direction—persistent, toward a populated area, and over a source area rich in secondary pollutant precursors
- Mixing height—low, persistent, and sufficiently high to entrap emissions from the major elevated sources
- Insolation—high and coincident with conditions above

The primary factor that distinguishes the worst case from the most probable is the persistence of the low wind speed throughout the day. Otherwise, the meteorological scenario for the worst case reads identically with that for the most probable.

Air Chemistry Scenario   An air chemistry scenario is not required since numerical modeling for secondary pollutants is not conducted in the present analysis.

Primary Pollutant Impact Scenarios   The primary pollutant impact scenarios which follow are applicable to both the sources at the project site and sources exogenous to the project site. The primary pollutant considered is carbon monoxide (Table 3.27).

Existing.   The *most probable* 1-hour primary impact occurs in the morning hours (0800–0900) when the commuter traffic on the existing two-lane roadway and within the two surrounding communities peaks. The atmospheric stability is strong (E, F stability) and the winds are low (1 m/s) and from the north. This is a case for which the peak in emissions coincides with the meteorological conditions most conducive to primary pollutant impact.

The *worst case* 1-hour primary pollutant impact occurs in the early morning hours when commuter traffic is unusually congested, the atmospheric stability is unusually strong (F stability), and the winds are low (1 m/s) and from the north. For this analysis, the net effect of this congestion on the emissions is accounted for by an arbitrarily selected 30 percent increase in traffic and a 50 percent decrease in the average speed from that experienced for the most probable condition. This is also a case where the peak in emissions coincides with the meteorological conditions most conducive to primary pollutant impact.

Because the period most conducive to 1-hour impact occurs during the time of wind reversal from a land to sea breeze, a southerly or northerly wind can occur. An examination of the hourly meteorological data, hourly emissions data, and location of most sensitive receptors indicates that a northerly wind is most likely for both the 1-hour most probable and worst case condition.

The *most probable* 8-hour primary pollutant impact occurs in the morning hours (0400–1200). During this period, the morning commuter traffic occurs, the stability is restricted and varies from F to C stability, the inversion height is restricted, and the winds change from northerly to southerly. This is a case for which the peak in emissions occurs within the time period, but engineering judgment must be exercised to identify the inclusive hours.

The *worst case* 8-hour primary pollutant impact occurs in the morning hours (0400–1200) for that condition when commuter traffic is unusually congested, the atmospheric stability is persistently strong (F stability), and the wind is persistently calm (1 m/s). The net effect on the emissions is accounted for by a 50 percent decrease in the average speed from that experienced for the most probable condition. This is also a condition where the peak in emissions occurs within the time period, but engineering judgment must be exercised to identify the inclusive hours.

No-build.   (See Existing Primary Pollutant Impact Scenario)

Construction.   The *most probable* primary pollutant impact will occur in the morning hours (0800–0900) when the construction activity is fully developed, the commuter traffic on the existing two-lane roadway peaks, the atmospheric stability is strong (E, F stability), and the winds are low (1 m/s) and from the north. This is a case in which the peak in emissions coincides with the meteorological conditions most conducive to primary impact.

The *worst case* 1-hour primary pollutant impact will occur in the early morning hours when the commuter traffic flow is unusually congested, construction activity is unusually high, the atmospheric stability is unusually strong (F stability), and the winds are low (1 m/s) and from the north. The net effect on the emissions scenario is accounted for by an arbitrarily selected 30 percent increase in traffic, 30 percent increase in construction activity, and 50 percent decrease in average vehicular speed from the most probable. This is also a case where the peak in emissions coincides with the meteorological conditions most conducive to primary pollutant impact.

The *most probable* and *worst case* 8-hour primary impact scenarios follow those presented for the existing conditions. The occurrence of the impacts will be spread across the 9-mile project length and depend upon the specific location of the construction activity.

Build-out. (See Existing Primary Pollutant Impact Scenario)

Secondary Impact Scenarios The secondary pollutant impact scenarios which follow are applicable to both the sources at the project site and the sources exogenous to the project site. Because proportional modeling is employed to analyze secondary pollutant impact, the emissions scenarios alone are sufficient. However, as discussed previously, it is instructive to include the meteorological scenarios in order to identify (1) the conditions for which secondary impact occurs, and (2) the areas susceptible to secondary pollutant impact.

Existing. The *most probable* secondary pollutant impact occurs in the late morning or early afternoon hours north of the project area as a result of primary pollutant emissions from the morning commuter traffic and stationary sources throughout the subregion.

The emissions are first transported to the south by the weakening evening northerly land breeze, then returned to the project area by the prevailing daytime southerly ocean breeze, and transported over the low-lying hills and into the greater basin. This is a case in which the peak in emissions coincides with the meteorological conditions most conducive to secondary pollutant impact.

The *worst case* secondary pollutant impact will occur during the summer months when insolation is high, traffic volume is unusually high, and low-speed winds transport the polluted air mass into the San Dido and La Agua valleys due to a prolongation in restricted vertical transport throughout the morning hours. This is a case in which the peak in emissions coincides with the meteorological conditions most conducive to secondary pollutant impact.

No-build. (See Existing Secondary Pollutant Impact Scenario)

Construction. The *most probable* secondary pollutant impact will occur in the late morning or early afternoon hours northwest of the project area as a result of primary pollutant emissions from stationary sources and traffic and construction activity during and subsequent to the commute hours.

Emissions will first be transported to the south by the weakening evening northerly land breeze, then returned to the project area by the prevailing daytime southerly ocean breeze, and transported over the low-lying hills and into the greater basin. This is a case in which the peak in emissions coincides with the meteorological conditions most conducive to secondary pollutant impact.

The *worst case* secondary pollutant impact will occur during the summer months when insolation is high, traffic volume is unusually high, and low-speed winds transport the polluted air mass into the San Dido and La Agua valleys due to a prolongation in restricted vertical transport throughout the morning hours. This is a case in which the peak in emissions coincides with the meteorological conditions most conducive to secondary pollutant impact.

Build-out. (See Existing Secondary Pollutant Impact Scenario)

Modeling

Primary Pollutant Impact The primary pollutant impact associated with sources at and exogenous to the project site and influenced by project site activities is limited to *local* air quality. As demonstrated by the modeling presented herein, the primary pollutant emissions are insufficient to produce a significant primary pollutant impact at the subregional or regional level.

Existing Year. The contribution of the existing two-lane roadway to local air quality was modeled using the most probable case and worst case impact scenarios developed and described in the previous section. The impact is described in terms of carbon monoxide (CO). Isopleths of the worst and most probable 1-hour and 8-hour CO impact are presented in Figs. 3.57 and 3.58, respectively.

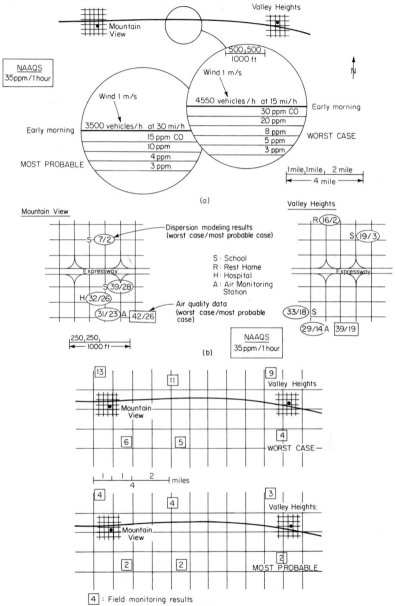

Figure 3.57 Modeling results—primary pollutant impact (existing year 1-hour CO). (*a*) Impact of existing roadway (ppm CO/1-hour), background CO concentrations not considered. (*b*) Existing impact with respect to sensitive receptors (ppm CO/1-hour), background CO concentrations not considered. (*c*) Existing background CO concentrations as determined by field monitoring programs (ppm CO/1-hour).

First, the estimated impact associated with the existing roadway between the townships is presented in each figure in the absence of background levels of CO. The values shown were estimated using the CALINE2 computer code (reference). For the 8-hour impact, calculations were made of the hourly emissions and meteorological data, and then

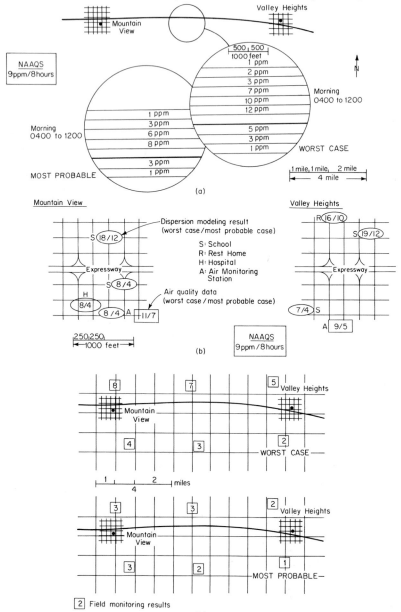

Figure 3.58 Modeling results—primary pollutant impact (existing year 8-hour CO). (*a*) Impact of existing roadway (ppm CO/8-hour), background concentrations not considered. (*b*) Existing impact with respect to sensitive receptors (ppm CO/8-hour), background CO concentration not considered. (*c*) Existing background CO concentrations as determined by field monitoring programs (ppm CO/8-hour).

averaged. Since the wind changes from a southerly to a northerly flow during this period, impacts are shown on both sides of the line source.

Second, the impact on sensitive receptors located within the townships is shown in the absence of background levels of CO. Sources considered in the local impact analysis include the two-lane roadway and sources local to the project site. Values shown were calculated using the APRAC computer code (reference). Actual air monitoring data from Table 3.26 also presented are to assess the validity of the results. The difference between the air monitoring data and calculated results is attributed to the background concentrations not considered in the calculations.

Third, background data are presented based on a field monitoring study conducted by the State Department of Transportation (reference). The total impact is the sum of the background and the calculated source impacts. The background data are of the magnitude required to explain the difference between the APRAC results and air monitoring data.

No-build.   The no-build modeling results are presented later with the build-out modeling results to facilitate comparison.

Construction Year.   The construction year primary pollutant impact is produced as a result of the combined roadway and construction activity. The description of the impact is basically identical to the existing condition. Emissions from the construction activity represent a small fraction of the emissions from the existing source* with one notable exception, particulate. The fugitive particulate (dust) emission is higher (on a per mile basis). The emission is not sufficient to warrant a dispersion analysis. However, it is sufficient to produce a dust nuisance to local residents and, as a result, warrant mitigation consideration (see the section on Mitigation later in this analysis).

Build-out/No-build (1985).   Modeling was conducted for the worst case and most probable primary pollutant impact scenarios developed and described in the section Impact Scenarios earlier in this chapter. The impact is described in terms of carbon monoxide (CO). Isopleths of the most probable and worst case 1-hour and 8-hour CO impact are presented in Figs. 3.59 and 3.60, respectively. Results are presented for both the build-out and no-build options with the format (*No-build, Build-out*).

The estimated impact associated with the proposed project between the townships is first presented in each figure in the absence of background CO levels. The values shown were estimated using the CALINE2 computer code (reference). The format follows that presented for the existing air quality modeling except the projected background concentrations were estimated for these future years using proportional modeling and subregional emissions of CO.

Build-out/No-build (1995).   Isopleths of the most probable and worst case 1-hour and 8-hour CO impacts for 1995 are presented in Figs. 3.61 and 3.62, respectively. Results are presented for both the no-build and build-out options with the format (*No-build, Build-out*). The format follows that presented for the 1985 analysis year.

Secondary Pollutant Impact   The secondary pollutant considered is ozone. The ozone impacts associated with sources at and exogenous to the project site that are influenced by project site activities occur on the subregional and regional scales. The ozone precursors associated with the project are reactive hydrocarbons and nitrogen oxides.

Existing Year.   The contribution of the existing two-lane roadway to regional air quality is represented by the emissions from the roadway relative to the region. As shown in Table 3.28, the existing 9-mile link contributes a small fraction to the total regional burden. As a result, dispersion modeling on a regional scale was considered not necessary for secondary pollutants.

The contribution of the existing two-lane roadway to the subregional air quality is represented by the emissions of the roadway relative to the subregion. As shown in Table 3.28, the existing 9-mile link contributes a small percentage (~3 percent) to the total subregional burden. As a result dispersion modeling on a subregional scale was not considered necessary for secondary pollutants.

However, contours of a major ozone precursor were generated to identify the spatial distribution of those emissions that lead to ozone formation within the subregion. This will also provide a basis for assessing future-year ozone formation potential within the subregion.

---

*Compare Tables 3.28 and 3.29 and note that the construction activity will be concentrated over only a portion (1-mile) of the 9-mile link.

Isopleths of the 6–9 A.M. emissions of reactive hydrocarbons are presented in Fig. 3.63. The plots were constructed as part of the development of the Air Quality Maintenance Plan and were provided by the Los Robles Air Pollution Control District (reference). The isopleths indicate that the ozone precursor is produced along the major road arteries, with

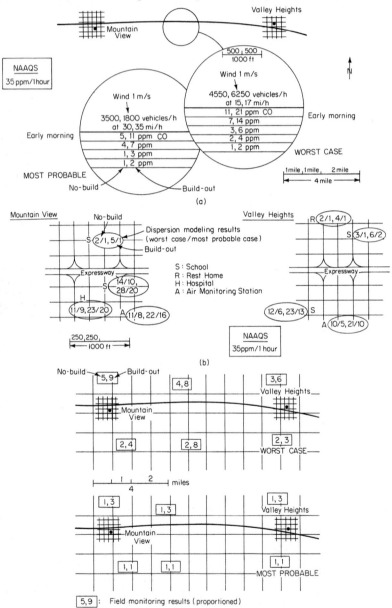

Figure 3.59   Modeling results—primary pollutant impact (1985 1-hour CO). (*a*) Impact of proposed project (ppm CO 1-hour), background CO concentrations not considered. (*b*) Impact with respect to sensitive receptors, (ppm CO/1-hour), background CO concentrations not considered. (*c*) Background CO concentrations (ppm CO/1-hour) based on proportional model applied to existing year background concentrations.

enlarged emission nodes at each community due to concentrated motor vehicle traffic and reactive-hydrocarbon-emitting stationary sources. The existing two-lane line source is shown to play a significant role in the ozone formation of the subregion, even though the contribution to the total emission of primary pollutants is low.

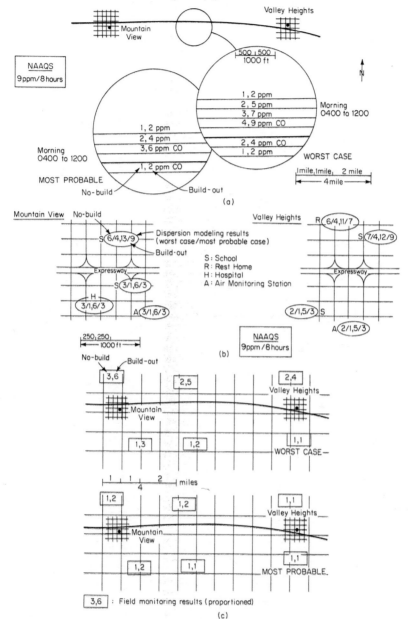

Figure 3.60 Modeling results—primary pollutant impact (1985 8-hour CO). (*a*) Impact of proposed project (ppm CO/8-hour), background CO concentrations not considered. (*b*) Impact with respect to sensitive receptors (ppm CO/8-hour), background CO concentrations not considered. (*c*) Background CO concentrations (ppm CO/8-hour) based on proportional model applied to existing year background concentrations.

No-build.   The no-build modeling results are presented with the build-out modeling results to facilitate comparison.

Construction Year.   The construction year secondary pollutant impact is produced by the combined emission of secondary pollutant precursors from the roadway and construc-

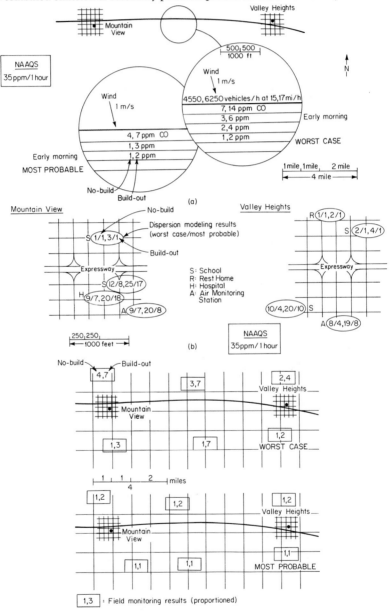

(a)

(b)

(c)

Figure 3.61   Modeling results—primary pollutant impact (1995 1-hour CO). (*a*) Impact of proposed project (ppm CO/1-hour), background concentrations not considered. (*b*) Impact with respect to sensitive receptors (ppm CO/1-hour), background CO concentrations not considered. (*c*) Background CO concentrations (ppm CO/1-hour) based on proportional model applied to existing year background concentration.

tion activity. The description of the impact is basically identical to the existing condition. Emissions from the construction activity represent a small fraction of the emissions from the existing source (see Tables 3.28 and 3.29).

Build-out/No-build (1985). The region is expected to experience modest growth by

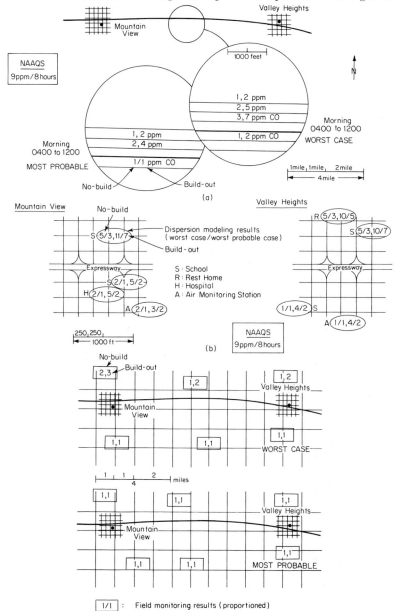

Figure 3.62 Modeling results—primary pollutant impact (1995 8-hour CO). (*a*) Impact of proposed project (ppm CO/8-hour), background CO concentrations not considered. (*b*) Impact with respect to sensitive receptors (ppm CO/8-hour), background CO concentrations not considered. (*c*) Background CO concentrations (ppm CO/8-hour) based on proportional model applied to existing year background concentrations.

1985. The impact of the basic growth will affect future-year regional levels of secondary pollutants, although projected reductions in emissions due to improved control technology will act to offset the growth.

Proportional modeling is used to estimate the impact of the basin growth on regional levels of peak ozone as shown in Table 3.30a. The results are based on data provided from the figures and tables identified. The proportional modeling projects the change in the *basin's* maximum ozone reading for the year (worst case) and the maximum seasonal

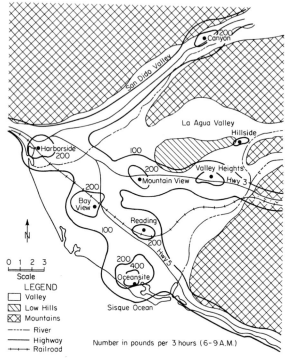

Figure 3.63   Seconday pollutant impact—isopleths of 6–9 A.M. reactive hydrocarbon emission for existing year. (Scale in miles.)

average of the daily maximum ozone reading (most probable) as a result of a change in the 6–9 A.M. reactive hydrocarbon emissions. The two indices of ozone potential are shown to decrease slightly by 1985 (indicating that the decrease in controlled HC emissions offset the increase in source activity). Note that for this regional analysis, the build-out ozone impact exceeds the no-build by 0.01 ppm for both indices.

It is noteworthy that the regional peak occurs in the inland areas of the basin (e.g., Apac) as a result of the transport of the ozone precursors. Peak ozone in and about the project area is more than 50 percent lower. This can be deduced by comparing the existing peak ozone readings in the basin (Table 3.30) to the peak ozone readings at air monitoring stations near the *project* area (Table 3.26).

By 1985, the subregion is expected to experience substantial growth that will be largely stimulated by the project. The impact of the growth will affect subregional levels of secondary pollutants, although projected reductions in emissions due to improved control technology will act to offset the growth.

Proportional modeling may be used to analyze the influence of subregional growth on subregional levels of peak ozone as shown in Table 3.30b.* The results are based on data provided from the figures and tables identified. The proportional modeling projects the

---

*It is assumed that no intrusion into the subregion of ozone or ozone precursors occurs from sources upwind. For the present project, the ocean precludes the presence of sources upwind.

TABLE 3.30a Modeling Results—Secondary Pollutant Impact (Future-Year Ozone, Regional)

Year	Data source	Hydrocarbon emissions* (tons/day)		Maximum reading for year (ppm/1-hr)		Maximum seasonal average of daily maximum readings (ppm/1-hr)	
		No-build	Build-out	No-build	Build-out	No-build	Build-out
Existing	Table 3.28	161	161	…	…	…	…
	Figure 3.54b	…	…	0.40 (Apac)	0.40 (Apac)	0.29 (Apac)	0.29 (Apac)
Projected (1985)	Table 3.28	141	…	…	…	…	…
	Table 3.29		144	…	…	…	…
	Proportional modeling result	…	…	0.35	0.36	0.25	0.26
Projected (1995)	Table 3.28	152	…	…	…	…	…
	Table 3.29	…	158	…	…	…	…
	Proportional modeling result	…	…	0.38	0.39	0.27	0.28

*Regional exogenous emissions plus project emissions, 6–9 A. M.

TABLE 3.30b Modeling Results—Secondary Pollutant Impact (Future-Year Ozone, Subregional)

Year	Data source	Hydrocarbon emissions* (tons/day)		Maximum reading for year (ppm/1-hr)		Maximum seasonal average of daily maximum readings (ppm/1-hr)	
		No-build	Build-out	No-build	Build-out	No-build	Build-out
Existing	Table 3.28	17.5	17.5	⋯	⋯	⋯	⋯
	Figure 3.54b	⋯	⋯	0.23 (Hillside)	0.23 (Hillside)	0.15 (Hillside)	0.15 (Hillside)
Projected (1985)	Table 3.28	14.5	⋯	⋯	⋯	⋯	⋯
	Table 3.29	⋯	17.4				
	Proportional Modeling Result	⋯	⋯	0.19	0.23	0.12	0.15
Projected (1995)	Table 3.28	13.7	⋯	⋯	⋯	⋯	⋯
	Table 3.29	⋯	19.7	⋯	⋯	⋯	⋯
	Proportional Modeling Result	⋯	⋯	0.18	0.26	0.12	0.17

*Subregional exogenous emissions plus project emissions, 6–9 A.M.

change in *subregional* maximum ozone readings for the year (worst case) and the maximum seasonal average of the daily maximum ozone (most probable) as a result of a change in the 6–9 A.M. reactive hydrocarbon emissions. The two ozone indices are shown to decrease for both alternatives.

Isopleths of the 6–9 A.M. reactive hydrocarbon emissions are presented in Fig. 3.64. For the no-build alternative, the emission pattern is basically similar to the existing year (Fig. 3.63) with an overall total reduction in emissions. For the build-out alternative, the total

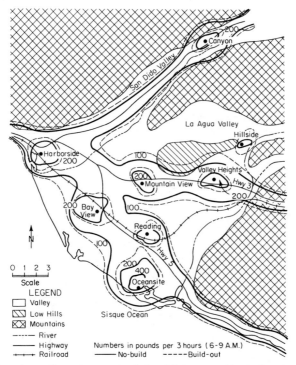

Figure 3.64   Secondary pollutant impact—isopleths of 6–9 A.M. reactive hydrocarbon emission for 1985. (Scale in miles.)

emissions pattern is also similar to the existing year (Fig. 3.63). However, there is a marked difference which reflects the development promoted by the completion of the six-lane roadway.

Build-out/No-build (1995).   The basin is expected to experience significant growth by 1995. The approach to assessing the impact of the growth on ozone follows that presented for 1985. However, the impact at the subregional level is more dramatic. As shown in Table 3.30b, the two indices are shown to decrease slightly with the no-build alternative and to increase for the build-out condition.

Isopleths of the 6–9 A.M. reactive hydrocarbon emissions are presented in Fig. 3.65. For the no-build alternative, the emission pattern is similar to the existing year (Fig. 3.63), with an overall total reduction in emissions due to implementation of mobile source emission control. For the build-out condition, the total emission is 10 percent higher than the existing year. However, there is a marked difference in the emission pattern which reflects the development promoted by the completion of the six-lane roadway.

Assessment

Primary Pollutant Impact   The existing levels of primary pollutants exceed National Ambient Air Quality Standards in the project area. For example, the 8-hour air quality standard for carbon monoxide is presently exceeded 46 days a year at Mountain View. In addition, modeling results indicate that both the most probable and worst case 1-hour and 8-hour carbon monoxide levels exceed the NAAQS adjacent to the two-lane roadway.

The modeling results for the 1985 build-out alternative indicate that both the most probable and worst case 1-hour and 8-hour carbon monoxide levels adjacent to the six-lane roadway will exceed the NAAQS. For the no-build alternative, the levels will be 50 percent less and will generally not exceed the NAAQS.

The modeling results for the 1995 build-out alternative indicate that both the most probable and worst case 1-hour and 8-hour carbon monoxide levels will generally not

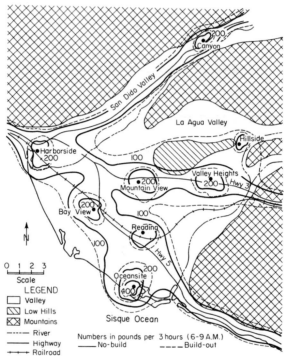

Figure 3.65   Secondary pollutant impact—isopleths of 6–9 A.M. reactive hydrocarbon emission for 1995. (Scale in miles.)

exceed the NAAQS. The improvement over the 1985 results is produced by the higher percentage of low-emitting vehicles.

It is noteworthy, as mentioned previously, that the 1995 results represent the optimum in automobile emission control technology, barring a technological breakthrough. To prevent the projected 1995 air quality from further degradation will require (1) no increase in VMT and (2) maintenance of automobile emission performance. To improve air quality over that projected for 1995 will require a reduction in VMT.

Secondary Pollutant Impact   The existing levels of secondary pollutants exceed the National Ambient Air Quality Standards in the project area. For example, the air quality standard for ozone is presently exceeded 29 days per year in Mountain View and 60 days per year in the downwind community of Hillside.

The proportional modeling results indicate that the impact of the project will be significant in the subregion. By comparison to existing air quality, the build-out alternative will result in an increase in ozone by 1995, whereas the no-build alternative will result in a net decrease in ozone.

Section 109j of the Federal Highway Act of 1972 requires that highway projects of this scope be consistent with the State Implementation Plan (SIP) (reference).

The project is inconsistent with the State Implementation Plan, which requires a zero net increase in emissions that will delay the attainment or prevent the maintenance of the ozone air quality standard (reference).

MITIGATION

Construction.   The major construction impact amenable to mitigation is dust emission. Carefully controlled and continuously implemented soil wetting procedures will reduce this impact.

Build-out.   The following mitigating procedures would serve to reduce the degradation to air quality. Evaluation of the reduction in impacts associated with each is not considered:

1. Provision for increased mass transit
2. Development of car pool systems

ALTERNATIVES

Alternative 1—No-build.   The first alternative is not to construct the project. The existing two-lane roadway would reach saturation, and the development of the area would be limited.

Alternative 2—Exclusive Busway.   One lane of the project can be devoted to an exclusive busway. The motor vehicles on the remaining four lanes would reach a saturation (there would be five rather than six lanes), whereas the total number of people moved would be retained.

Alternative 3—Ramp Metering.   Any steps taken to maintain constant flow of traffic on the six-lane highway will reduce the degradation to air quality. Ramp metering is one method that can be employed.

Of the three alternatives, only the reduction in impacts associated with the first (no-build) is considered.

GROWTH-INDUCING ASPECTS

The areas which would experience added growth pressure due to the proposed project are the communities of Valley Heights and Mountain View. The growth will be substantial, with the two communities growing from a combined present-year population of 26,000 residents to 62,000 in Mountain View and 43,000 in Valley Heights by 1995. The projected air quality impacts are primarily a result of this growth.

With respect to the urban sprawl implications of the alternatives, the following conclusions have been reached:

1. The *no-build alternative* will discourage urban sprawl.
2. The *exclusive busway* will have negligible impact on the urban sprawl in comparison to the project as proposed.
3. The *ramp metering* will have negligible impact on the urban sprawl in comparison to the project as proposed.

POWER PLANT CASE STUDY

SUMMARY

The proposed project will increase the emissions of SO_2, NO_x, and particulate in a basin presently experiencing exceedances of air quality standards for SO_2, particulate, and ozone. Although the incremental changes to the existing air quality will be modest, the State Implementation Plan (SIP) prohibits the construction of new facilities which will delay the attainment or prevent the maintenance of the National Ambient Air Quality Standards. Adoption of mitigation procedures will be necessary to meet the requirements of the SIP.

EXISTING ENVIRONMENT

General Project Description

The project is a new 1500-megawatt (MW) coal-fired power plant station. The site selected for the plant is shown in Fig. 3.66. The facility is *required* to meet the electrical energy demands of growing populations in the region.

Description of the Project Area

Location/Setting (See *Highway Case Study*)
Air Pollution Area/Regulatory Programs (See *Highway Case Study*)
Air Pollution Programs. (See *Highway Case Study*)
Emission Standards. The federal emission standards (New Source Performance Stan-

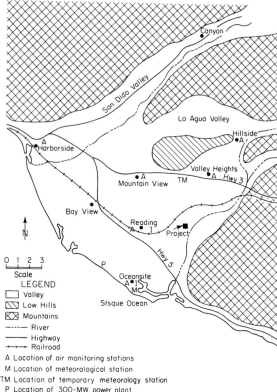

Figure 3.66 Project site—subregional perspective. (Scale in miles.)

dards—NSPS) applicable to the project are presented in Table 3.31. The state has not established stationary source standards, delegating the responsibility instead to local agencies. The Los Robles Air Pollution Control District has set more restrictive emission

TABLE 3.31   Emission Standards—Federal Emission Standards for New Coal-fuel Fired Steam Generators*

Particulate matter†	Sulfur dioxide†	Nitrogen oxides (expressed as NO$_2$)†
0.10 lb/10^6 Btu heat input (1.44 tons/hour)‡ 20 percent opacity§	1.20 lb/10^6 Btu heat input for solid fuel (17.3 tons/hour)	0.70 lb/10^6 Btu heat input (10.1 tons/hour)

†Maximum 2-hour average.
‡Numbers in parenthesis are emission standards applied to proposed project.
§Except that 40 percent opacity shall be permissible for not more than two minutes each hour.
*SOURCE: *Federal Register*, 23 December 1971, p. 24876.

standards for two pollutants. SO_2 emissions are limited to a level no greater than 1.0 lb/ million Btu heat input. Particulate emissions are limited to a level no greater than 0.05 lb/ million Btu heat input (reference).

In addition, the district is presently considering a new source review regulation. This will require that the project proponent reduce the pollutant emissions from small sources (e.g., gas stations, package boiler operators) a factor of two or more greater than the increase in emissions charged to the proposed project (reference). The proponent must arrange and finance this so-called "emissions offset" (or "emissions trade-off").

Ambient Air Quality Standards. (See *Highway Case Study*)

Meteorology

(See *Highway Case Study*)

Existing Air Quality

Regional (See *Highway Case Study*)
Subregional (See *Highway Case Study*)
Local No sources are located in the area of the proposed project. As a result, the description of subregional air quality applies to the local air quality.

ENVIRONMENTAL IMPACT

Detailed Project Description

Build-out The power plant station will have 1500 MW of generating capacity and will cover an area of about 0.7 mi². Three 500-MW units will be installed, each with a 152 m stack. The project is expected to commence in 1985, with construction completed in 36 months and operation of all three units initiated by 1988.

Coal will be the fuel with a total heat input to the plant of $28,750 \times 10^6$ Btu/h at 100 percent load. Natural gas or oil will be used only for ignition. The coal will be withdrawn from active onsite storage as required to meet the burn rate of the station. Coal dust will be controlled throughout the coal handling systems by using collection or suppression systems. Typical coal compositions are shown in Table 3.32. Two values must be

TABLE 3.32 Coal Composition

	Grade	
Coal value	Average	Worst
Heating value (Btu/lb)	11,800	11,000
Sulfur content (%)	0.57	0.84
Ash content (%)	9.7	11.6
Moisture content (%)	9.1	12.6
Nitrogen content (%)	1.8	1.8

considered—the average-grade coal burned during the lifetime operation of the station and the worst-grade coal that could be burned for short durations. The use of worst-grade coal is important for sizing air pollution abatement systems and analyzing worst case conditions, but average-grade coal must be used for analysis of the most probable conditions and long-term averaged concentration.

Coal will be transported to the site using a unit train concept consisting of sixty 100-ton cars. Diesel-electric units will be used. A major railroad trunk line exists adjacent to the project site.

Furnace bottom ash particles, which are heavier than fly ash, will fall to the water-impounded hopper located beneath the furnace. Bottom ash will be periodically removed from the hopper and pumped in the form of an ash-water slurry to dewatering bins. There, the water will be decanted and the ash off-loaded to belt conveyors for transport to the ash disposal area.

Fly ash will be removed from the flue gas by the high-efficiency particulate removal system and will be transported to storage silos, where it will be held for later disposition and transport to the ash disposal area.

After passing through the particulate removal system, the flue gas will be passed through SO_2 absorbers. The collected material will be disposed of subsequently.

The major pollutants emitted, the impacts modeled, and the modeling methods employed to analyze the impacts are presented in Table 3.33. Gaussian modeling is used

TABLE 3.33   Proposed Project—Air Quality Impacts

Primary pollutants emitted	Impacts modeled			Scale	Modeling methods employed
	Primary pollutant	Secondary pollutant	Visibility*		
SO_2	SO_2	. . .	. . .	Subregional Regional	Gaussian
	. . .	. . .	Visual range	Subregional	Hybrid Gaussian
Particulate	Particulate	. . .	. . .	Subregional	Gaussian
	. . .	. . .	Visual range	Subregional	Hybrid Gaussian
NO_x	. . .	. . .	Visual range, plume appearance	Subregional	Hybrid Gaussian
	NO_x	. . .	. . .	Subregional Regional	Gaussian
	. . .	Ozone	. . .	Subregional Regional	Numerical

*The 1977 Clean Air Amendments require that visibility be analyzed only in Federal Class II prevention of significant deterioration areas (reference). Even though the proposed project is not in a Federal Class II area, the Los Robles Air Pollution Control District has requested that visibility impacts be analyzed.

to analyze the primary pollutant impact associated with the project, numerical modeling is used to analyze the secondary pollutant impact, and a hybrid Gaussian model is used to analyze the visibility impact.

No-build   Growth in the area will not be substantially changed should the project not be built as other means will be required to meet the projected energy needs. In addition, development of the project site for residential, industrial, or other use is not envisioned in the absence of the project.

Analysis Years

The years selected for analysis include the existing year, construction years (1985–1988), and a typical year of operation. The year selected as a typical year of operation is 1995 for the following reasons:

- The power plant will be in full operation.
- The emission characteristics of the motor vehicle population will reflect the stringent emissions standards for new motor vehicles.*
- The data required to describe the mobile source emissions for 1995 are readily available from the AQMP process.

Description of Surrounding Areas

General   The project is located in a valley of gentle slope from the ocean to low-lying hills. The area is rural but is destined for both residential and industrial growth (reference). Mountain View and Valley Heights, for example, are anticipated to grow from a combined population of 26,000 today to population of 62,000 and 43,000 by 1995 (reference). Similar growth is projected for other portions of the subregion with less intense

*Mobile sources contribute to the total NO_x emission in the subregion and must be considered in the modeling of oxidant.

growth anticipated in the region as a whole (reference). Based on this projected growth, the power plant project has been proposed.

Pollutant Sources The pollutant sources in the surrounding area (sources exogenous to the project site) include those in the region, subregion, and local to the project area.

On the regional perspective, the pollutant sources of interest are those responsible for ozone impact: the sources within the region that emit the secondary pollutant precursors of reactive hydrocarbons and nitrogen oxides.

On the subregional perspective, the pollutant sources of interest are (1) those responsible for ozone impact: the sources within the subregion that emit the secondary pollutant precursors of reactive hydrocarbon and nitrogen oxides, and (2) those responsible for the primary pollutant emission of sulfur oxides, nitrogen oxides, and particulate.

As shown in Fig. 3.66, the major exogenous sources within the subregion include the mobile source activity associated with Highway 5, the two communities of Mountain View and Valley Heights, and the communities upwind (Reading, Oceansite, and Bay View); stationary sources located in industrial complexes at Reading and Oceansite; the Oceansite Harbor; and a coal-fired 300-MW power-generating facility northwest of Oceansite. Major sources formally approved for construction include a refinery and LNG terminal at Oceansite, and a six-lane highway widening project between the communities of Valley Heights and Mountain View (reference).* With respect to the emission of SO_2 and particulate, the major source impacting the project area is the existing 300-MW power facility. With respect to the emission of NO_X, the major sources include the mobile source activity and existing 300-MW power facility. With respect to the emission of reactive hydrocarbons, the major sources include the mobile source activity, the future-year LNG terminal (1990), and the future-year refinery (1992).

Proportional modeling is used to estimate the background levels of SO_2, NO_X, and particulate associated with sources exogenous to the project and upon which the power plant impact is imposed.† As a result, emissions from the sources exogenous to the project site need not be separately identified or geocoded.

For future-year ozone impacts on the subregional scale, numerical modeling is conducted. The modeling requires geocoding (e.g., identification of the temporal and spatial distributions) of the sources exogenous to the project that emit NO_X and reactive hydrocarbons.

Receptors The receptors that will be susceptible to primary pollutant (SO_2, NO_X, particulate), secondary pollutant (ozone), and visibility impact from the project include the population in the regional and subregional basins in general, and the populations in the downwind communities (Valley Heights and Hillside) and valleys (La Agua and San Dido) in particular. The existing institutions considered to be especially sensitive receptors include two schools and one rest home in Valley Heights. By the analysis year of 1995, a general hospital will be built, the schools are expected to increase to twelve, and the rest homes to four.

Traffic Study

Traffic information is required to establish the annual average NO_X impact and ozone impact resulting from the combined emissions from the power plant and mobile source activity. As part of the AQMP process, a geocoded traffic study has been developed by the state for the existing year and 1995 (reference).

Impact Scenarios

Emissions Scenarios Emission factors for the mobile sources were obtained from the latest supplement to AP-42 (reference). Usage data for the mobile sources were obtained from the traffic study cited in the immediately preceding section.

For the stationary sources exogenous to the project, emission factors and usage data were obtained from the Los Robles Air Pollution Control District (reference). The

---

*Major sources being *considered* but *not yet formally planned* include a regional shopping center west of Valley Heights. Because of its tenuous status, the contribution of this source is not considered in the present study.

†An alternative method for describing background concentrations would be to use a numerical or Gaussian dispersion method (e.g., CDMQC[110]). The identification and geocoding required of source emissions for such methods was considered not to be commensurate with the benefit returned.

emissions and usage information required for the proposed power plant are discussed hereafter.

The data used are confirmed to be appropriate by the regional office of the EPA (reference), the State Air Regulatory Agency (reference), and the Los Robles Regional Air Pollution Control District (reference). Scenarios are presented first for sources at the site of the project and second for sources exogenous to the project site.

Sources at Project Site

Existing.   There are no sources existing at the project site.

No-build.   There are no sources envisioned for the project site through 1995 should the project not be constructed.

Construction.   Emissions resulting from construction are not sufficient to warrant a separate accounting. This assessment is based on the isolation of the site from populated areas that might otherwise be impacted by dust during grading operations.

Build-out.   Emissions scenarios for the proposed project are presented in Table 3.34.

TABLE 3.34   Emissions Scenario—Build-out Alternative (Project Site Source), 1995

Pollutant	Unabated emission* (tons/hour)		Emission standards (tons/hour)		Abatement required to meet standard		Actual estimated emission (tons/hour)	
	Average coal	Worst coal	Federal†	Los Robles APCD	Average coal (%)	Worst coal (%)	Average coal	Worst coal
SO_2	13.8	22.3	17.3	17.3	0	22	1.4‡	2.2‡
NO_x	15.8	15.8	10.1	10.1	36	36	10.1	10.1
Particulate	98.5	128.6	1.44	1.44	98.5	98.9	0.49§	0.64§

*$28,750 \times 10^6$ Btu/h at 100% load.
†From Table 3.31.
‡Based on predicted removal efficiency of 90% using state-of-the-art technology.
§Based on predicted removal efficiency of 99.5% using state-of-the-art technology.

The amounts of particulate and SO_2 produced as unabated emissions are determined by mass balance based on the fuel composition and the assumption that 5 percent of the sulfur entering the boiler is retained with the particulate collected. The amount of nitrogen oxide emission is based on AP-42 emission factors for wall-firing in the absence of combustion modification to control emissions (reference).

The applicable emission standards and the abatement required to meet them are also shown in Table 3.34. In the case of NO_x, combustion modification techniques will be applied to meet the standard. In the cases of SO_2 and particulate, state-of-the-art technology will allow emissions to be reduced to levels below the applicable emission standards.

Due to the interdependency of the proposed plant and other plants in the total utility system, the hourly temporal variation in load of the proposed power plant is not firmly established, nor is it likely to be. Although the *system* may experience predictable temporal trends, any *individual plant* can vary in the contribution made to the system. The policy of the utility is to place the greatest share of the burden on the most energy-efficient units (reference). This would suggest that the three generating units associated with the proposed project, because of the technology employed to enhance efficiency, will generally follow the temporal system demand. Negating this, however, is increasing pressure on the utility company to place the greatest share of the system load on the units that produce the least amount of nitrogen oxides per kilowatt. This would place less of a burden on the new coal-fired units since 80 percent of the other system units are either gas or oil fired. As a result of the uncertainty in temporal distribution, conservative estimates are necessary to establish the load curves for use in the build-out impact scenario.

In calculating the annual average ground level concentrations for NO_x, SO_2, and particulate, average-grade coal with a sulfur content of 0.5 percent is used for the most probable case. For the worst case, worst-grade coal with a sulfur content of 0.8 percent is used. It is assumed that the load factor for both cases will be 85 percent.

For the most probable 3-hour and 24-hour SO_2 and 1-hour ozone calculations, average-grade coal with a sulfur content of 0.5 percent is used, together with an 85 percent load factor. For the worst case 3-hour and 24-hour SO_2 and 1-hour ozone calculations, worst-grade coal with a sulfur content of 0.8 percent is used with a 90 percent load factor.

Visibility. Ninety percent load operation is assumed as a conservative estimate for the conditions most conducive to visibility impact and operation on worst-grade coal. Only worst case impact is evaluated as a conservative approach. A most probable case is not appropriate because of (1) the uncertainty in the available modeling techniques and (2) the absence of an air quality standard for visibility.

Sources Exogenous to Project Site

Existing. Emissions scenarios are presented in Table 3.35 for the existing sources surrounding the project site.

No-build. Emissions scenarios are presented in Table 3.35 for the sources surrounding the project site in the event the project is not built.

Construction. As previously noted, a separate analysis is not considered necessary for the construction years.

Build-out. Emission scenarios are presented in Table 3.36 for the sources surrounding the project site in the event the project is built.

Meteorological Scenarios Meteorological scenarios for primary and secondary pollutants, and for visibility impact are developed using wind rose and stability data presented in Figs. 3.50–3.53 and Table 3.25. Scenarios are presented first for sources at the project site and second for sources exogenous to the project site.

Sources at the Project Site

The only source of major consequence at the project site is an elevated source.

Primary Pollutants. In calculating the *annual average* ground level concentrations for NO_x and particulate, the wind rose data from Figure 3.50 are used together with certain assumptions concerning stability. The assumptions are made to simplify the calculation procedures and are based on conservative estimates using the data from Table 3.25. The two stability conditions considered are (1) stable and (2) neutral with limited mixing. It is assumed that half the time the plume will be in stable air, and half of the time in neutrally stable air with an average mixing depth of 450 m (1476 ft). The data from Figure 3.53 were used to establish the average mixing depth.

For the *3-hour* and *24-hour* most probable case calculations, the atmosphere conditions most conducive to the most probable case air quality impact are:
- Wind speed—most probable (10 m/s)
- Wind direction—persistent and toward populated areas and/or sensitive receptors
- Mixing height—the height coincident with the most probable wind speed and stability which simultaneously exceeds or equals the height (critical height) sufficient to entrap the elevated source emission
- Stability—most probable (neutral, Class D)

For the *3-hour* and *24-hour* worst case calculations, the atmospheric conditions most conducive to the worst case air quality impact are:
- Wind speed—2 to 15 m/s
- Wind direction—persistent and toward populated areas and/or sensitive receptors
- Mixing height—critical height
- Stability—trapping conditions

Trapping conditions occur most frequently in the late morning or afternoon hours. This coincides with the time that winds are moderate in strength. A review of Fig. 3.50 and Table 3.25 indicates that the winds are most frequent from the south (ocean breeze) during this period. The critical mixing height must be determined by modeling and then compared to Fig. 3.53 for frequency of occurrence.

Secondary Pollutants. For secondary pollutants, the atmospheric conditions most conducive to the most probable air quality impact are:
- Wind speed—most probable (10 m/s)
- Wind direction—toward a populated area and over a source area rich in secondary pollutant precursors

TABLE 3.35 Emissions Scenarios—Existing Year and No-build Alternative (Average Daily Emissions)

Emissions (tons/day)

| | | Sources at project site | | | Local | | | Sources exogenous to project site | | | | | |
| | | | | | | | | Subregional* | | | Regional* | | |
Year	Pollutant	Mobile	Stationary	Total	Mobile	Stationary	Total	Mobile	Stationary	Total	Mobile	Stationary	Total
Existing	NMHC												
	1. Total	0.0	0.0	0.0	0.0	0.0	0.0	66.0	60.0	126	545	503	1049
	2. 6–9AM	0.0	0.0	0.0	0.0	0.0	0.0	9.0	8.3	17.3	75	69.0	144
	NOx	0.0	0.0	0.0	0.0	0.0	0.0	80.9	33.1	114	706	289	995
	CO	0.0	0.0	0.0	0.0	0.0	0.0	376	37.2	413	3149	311	3460
	SOx	0.0	0.0	0.0	0.0	0.0	0.0	12.8	62.6	75.4	108	526	634
	Particulate	0.0	0.0	0.0	0.0	0.0	0.0	7.0	28.2	35.2	63.0	252	315
No-build (1995)	NMHC												
	1. Total	0.0	0.0	0.0	0.0	0.0	0.0	78.5	75.2	151	619	572	1191
	2. 6–9AM	0.0	0.0	0.0	0.0	0.0	0.0	10.2	9.4	19.6	82.2	75.8	158
	NOx	0.0	0.0	0.0	0.0	0.0	0.0	101	41.2	142	738	302	1040
	CO	0.0	0.0	0.0	0.0	0.0	0.0	466	46	512	3347	331	3678
	SOx	0.0	0.0	0.0	0.0	0.0	0.0	17.7	86.3	104	166	810	976
	Particulate	0.0	0.0	0.0	0.0	0.0	0.0	15.6	62.5	78.1	103	414	517

*SOURCE: Los Robles Air Pollution Control District.

TABLE 3.36 Emissions Scenarios—Construction Year and Build-out Alternative (Average Daily Emissions)

Emissions (tons/day)

| Year | Pollutant | Sources at project site[*] | | | Sources exogenous to project site | | | | | | | | |
| | | | | | Local | | | Subregional[†] | | | Regional[†] | | |
		Mobile	Stationary	Total	Mobile	Stationary	Total	Mobile	Stationary	Total	Mobile	Stationary	Total
Build-out (1995)	NMHC												
	1. Total	NA[†]	Neg.	Neg.	0.0	0.0	0.0	78.5	75.2	151	619	572	1191
	2. 6–9AM	NA	Neg.	Neg.	0.0	0.0	0.0	10.2	9.4	19.6	82.2	75.8	158
	NO$_x$	NA	242/242	242/242	0.0	0.0	0.0	101	41.2	142	738	302	1040
	CO	NA	Neg.	Neg.	0.0	0.0	0.0	466	46	512	3347	331	3678
	SO$_x$	NA	33.6/52.8	33.6/52.8	0.0	0.0	0.0	17.7	86.3	104	166	810	976
	Particulate	NA	11.8/15.4	11.8/15.4	0.0	0.0	0.0	15.6	62.5	78	103	414	517

*From Table 3.34 (Most probable/worst case).
†From Table 3.35.
‡Not applicable.

- Mixing height—low, persistent, and sufficiently high to entrap the elevated source emissions
- Insolation—high and coincident with conditions above

The low mixing heights are most persistent during the summer months. As demonstrated by Fig. 3.53, 0400 LST and 1600 LST mixing heights between 150 and 600 m (492 and 1,968 ft) occur frequently, with the majority less than 450 m (1,476 ft). The 0400 LST readings are biased by the early morning, low-lying radiation inversion. The limited mixing height coincides with the peak insolation associated with the long daylight hours that accompany the summer months.

The transport of oxidant begins near daybreak. The winds are initially directed toward

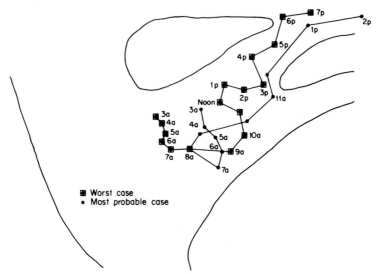

Figure 3.67   Meteorology—parcel trajectory.

the ocean, but reverse in an inland flow toward population centers after daybreak. This is also the time of high ground level stability due to the radiation inversion. Hence, the meteorological scenario most conducive to secondary pollutant impact occurs during the summer, beginning in the morning when the radiation inversion and low winds allow primary pollutant concentration to build and, along with the insolation, accelerate the formation of photochemical oxidant. As the radiation inversion breaks, the pollutants disperse vertically but are trapped below the inversion at 150 to 600 m, which allows the oxidant formation to continue.

Air parcel trajectories based upon wind data are presented in Fig. 3.67 for the most probable case. The trajectory was chosen to pass over the project area at 8:00 A.M.* The forward portion of the trajectory was extended to 3:00 P.M.

The atmospheric conditions most conducive to the worst case air quality impact are:

- Wind speed—low over a prolonged period
- Wind direction—persistent, toward a populated area, and over a source area rich in secondary pollutant precursors
- Mixing height—low, persistent, and sufficiently high to entrap the elevated source emissions
- Insolation—high and coincident with conditions above

The primary factor that distinguishes the worst case from the most probable is the persistence of the low wind speed throughout the day. Otherwise, the meteorological scenario for the worst case reads identically with that for the most probable case. A typical wind trajectory for the worst case is presented in Fig. 3.67.

*Analysis for other times of parcel passage over the site (e.g., 6:00, 7:00, and 9:00 A.M.) could be selected for analysis as well. The time of 8:00 A.M. was selected as reasonable and sufficient in view of the size of the source and uncertainty in this modeling approach.

Visibility.   Only the worst case scenario is considered for visibility impact (see Emissions Scenarios for explanation).

The atmospheric conditions most conducive to the worst case visibility impact are:[55]

- Wind speed—low (~2 m/s)
- Wind direction—toward populated areas (severity of impact depends upon location of observer relative to plume)
- Stability—neutral to visual range (Class D), stable for plume appearance (Class F)

Sources Exogenous to Project Site

The sources of major consequence surrounding the project site are motor vehicles and other ground level sources. The sole exception is the 300-MW power plant at Oceansite. Because proportional modeling is used in the present analysis to describe the background concentrations of primary pollutants, a separate and detailed meteorological scenario is required only for secondary pollutant impact.

Secondary Pollutants.   The meteorological scenarios for sources exogenous to the project site follow identically the scenarios presented for sources at the project site.

Air Chemistry Scenarios   Two scenarios are required, one for oxidant formation and one for visibility impact.

The oxidant kinetic set utilized is the set incorporated in the current version of the DIFKIN Model (reference).

The sulfate, nitrate, nitrogen oxides, and ozone chemistry utilized in the visibility model is that described by Latimer and Samuelsen.[55] Various percentages of NO to NO_2 conversion in the stack are evaluated.

Primary Pollutant Impact Scenarios   Primary pollutant impact scenarios are presented first for sources at the project site and second for sources exogenous to the project site. The primary pollutants considered include NO_X, SO_2, and particulate.

Sources at Project Site

Existing.   No sources either exist at the site or directly impact the site. The existing air quality monitoring data are sufficient in describing the local air quality at the project site.

No-build.   No sources are projected to be located at the project site in the event the proposed project is not built. Proportional modeling is used to project future-year air quality in the event the project is not built. As a result, the emissions scenario alone suffices for the primary pollutant impact scenario.

Construction.   Construction is not analyzed.

Build-out.   The absence of temporal information with respect to the emissions precludes the opportunity to temporally match the emissions and meteorological scenarios. As a result, the primary pollutants impact scenarios are simple combinations of the emissions and meteorological scenarios. For completeness, the primary pollutant scenarios for each impact considered are summarized in the following paragraphs.

For the *3-hour* and *24-hour* most probable calculations for SO_2, average-grade coal is used together with a 85 percent load factor. Neutral conditions are the most common meteorological conditions. For the worst case, trapping conditions and worst-grade coal are used together with a 90 percent load factor. The wind direction is taken along the prevailing daytime streamlines with a wind speed of 10 m/s for the most probable case and a wind speed range of 2–15 m/s for the worst case in order to identify the speed associated with the highest impact. The worst case mixing depth is determined by modeling.

For the *annual average* NO_X, SO_2, and particulate ground level calculations, average-grade coal and an 85 percent load factor is used. For the meteorology, the wind rose data of Fig. 3.50 are used together with the following assumptions regarding stability. Two stability conditions are considered: (1) stable and (2) neutral with limited mixing. It is assumed that half of the time the plume will be in stable air and half of the time in neutral stability air with an average mixing depth of 450 m (1476 ft).

Sources Exogenous to Project Site

Existing.   A primary pollutant impact scenario is not necessary since the existing ambient air quality data adequately describe the existing air quality in the absence of modeling.

No-build.   The emissions scenario suffices for the primary pollutant scenario since proportional modeling is used to describe the future-year impact in the event the project is not built.

Construction. The construction-year impacts are not analyzed.

Build-out. The emissions scenario suffices for the primary pollutant scenario since proportional modeling is used to describe the future-year impacts of sources exogenous to the project. It is noteworthy that since the project will not have a significant impact on the emissions growth of sources exogenous to the project, the future-year impacts associated with the no-build alternative will be identical to the background concentrations associated with the build-out alternative.

Secondary Pollutant Impact Scenarios   The secondary pollutant considered is ozone. The secondary impact scenarios are considered simultaneously for sources both at the project site and exogenous to the project site.

Existing. (See *Highway Case Study*)

No-build. (See *Highway Case Study*)

Construction. The construction-year impacts are not analyzed.

Build-out. The absence of temporal information with respect to the emissions associated with the project precludes the opportunity to temporally match the emissions and meteorological scenarios. As a result, the secondary pollutant impact scenarios are simple extentions of the emissions and meteorological scenarios. For completeness, the secondary impact scenarios are summarized in the following paragraphs.

The most probable secondary pollutant impact occurs in the late morning or early afternoon hours north of the project area as a result of the emission and subsequent transport of the secondary pollutant precursor NO_x during the early morning peak load condition. The emissions from the power plant are first transported to the south by the weakening evening northerly land breeze, then returned to the project area by the prevailing daytime southerly ocean breeze, and transported over the low-lying hills and into the greater basin.

The worst case secondary pollutant impact will occur during the summer months when insolation is high, power loads are unusually high, low-speed winds transport the polluted air mass into the San Dido and La Agua Valleys, and a persistent inversion restricts vertical transport throughout the morning hours.

Wind trajectories for both the most probable case and worst case were presented in Fig. 3.67.

For the most probable case, an 85 percent load is assumed with operation on average-grade coal for the sources at the project site. For the worst case, a 90 percent load is assumed with operation on worst-grade coal at the project site. For the sources exogenous to the project site over which the wind trajectory passes, proportional geocoded emissions are used. The temporal distribution selected is that documented for the existing year.

Visibility Impact Scenario   The visibility impact is calculated for the sources at the project site only. The impact of sources exogenous to the project site is taken into account by comparing the calculated impact to the existing average visual range of 55 km (34 mi) (reference).

Existing. No sources exist at the project site.

No-build. No sources are projected to be located at the project site in the event the proposed project is not built.

Build-out. The absence of temporal information with respect to the emissions precludes the opportunity to temporally compare the emissions and meteorological scenarios. As a result, the visibility impact scenarios simply combine the meteorological and emissions scenarios. For completeness, the visibility impact scenario is stated in the following paragraph.

Visibility is evaluated for the worst case condition only (see Emissions Scenarios for explanation). A 90 percent load is assumed with operation on worst-grade coal. Neutral stability (Class D) is used to evaluate visibility range, and a stable atmosphere (Class F) is used to evaluate plume appearance. The wind speed is taken to be low at 2 m/s and directed along the prevailing daytime streamlines.

Modeling

Primary Pollutant Impact   The primary pollutant emissions from the elevated point source will impact air quality at the subregional level and, because of the height of the stack and total emission rate, the project will impact air quality at the regional level as well.

Existing.   No sources exist at the site. The existing air quality monitoring data are sufficient to describe the local, subregional, and regional air quality with respect to primary pollutants.

No-build.   The no-build modeling results are presented later with the build-out modeling results to facilitate comparison.

Construction.   The construction years are not analyzed.

Build-out/No-build (1995).   The maximum concentrations of emissions at ground level were calculated for build-out using mathematical models developed by TVA from data collected at large coal-fired power plants in the southeast United States.[15] These models were used because the TVA power stations are similar in size and type to the proposed power plant and the meteorology and topography of the two regions are comparable.

The TVA models predict the incremental changes in air quality due to the project emissions. These changes must be added to the appropriate background pollutant concentrations. Because the power plant will not have a significant impact on the emissions growth of the sources exogenous to the project, the future-year air quality calculated for the no-build alternative is applicable as background concentrations for the build-out alternative. The background concentrations, presented in Table 3.37, were calculated by

TABLE 3.37   Modeling Results—Primary Pollutant Impact (1995 Background Concentrations)

Air quality	3-hr SO_2 ($\mu g/m^3$)	24-hr SO_2 ($\mu g/m^3$)	Annual SO_2 ($\mu g/m^3$)	Annual NO_x ($\mu g/m^3$)	Annual particulate ($\mu g/m^3$)
Reading	1872/3484	600/1017	120	100	300
Oceansite	insufficient data				
Mountain View	1690/2678	391/678	80	80	288
Valley Heights	1508/2366	365/913	20	80	266
Hillside	468/1482	104/391	27	80	110
NAAQS (Table 3.24)	1300	365	80	100	75

a. Values obtained by proportional modeling using air quality data from Table 3.26 and emission data from Table 3.35.

b. Format: most probable (average daily max.)/worst case (yearly max.).

proportioning the existing air quality data (Table 3.26) with the change in subregional emission between the existing year and 1995.

The TVA models determine the peak concentrations (3 to 5 minute average) under specified atmospheric conditions. Since wind speed and direction are constantly changing, the time-averaged concentrations for a given location will be less when averaged over a period longer than the 3 to 5 minute period than the peak concentration predicted by the mathematical models. To account for this variation, TVA recommends that the peak concentrations be corrected by a factor of ⅔ for the 3-hour average and ⅕ for the 24-hour average in order to be conservative.[65] To facilitate comparison to air quality standards, which refer to the worst 3-hour and 24-hour periods in a year, these conservative TVA findings were used to determine the concentrations for averaging periods of 3 and 24 hours.

To account for the fact that the emissions from the plant originate from three stacks separated from one another, a stack separation correction factor recommended by TVA was applied to the calculations.[65]

The 3-hour and 24-hour average SO_2 concentrations during neutral conditions downwind of the power plant station are shown in Figure 3.68. Calculations were made assuming the plume altitude above ground is unaffected by changing terrain elevation (i.e., the plume "follows" the terrain).

For the worst case conditions, the maximum ground level concentration under trapping conditions was calculated as a function of wind speed. Figure 3.69 shows the downwind 3-hour and 24-hour average SO_2 concentrations based on the assumption that the stable layer aloft is at the critical height (shown in the figures). Thus, the indicated concentrations represent the theoretical maximum concentrations during limited mixing conditions.

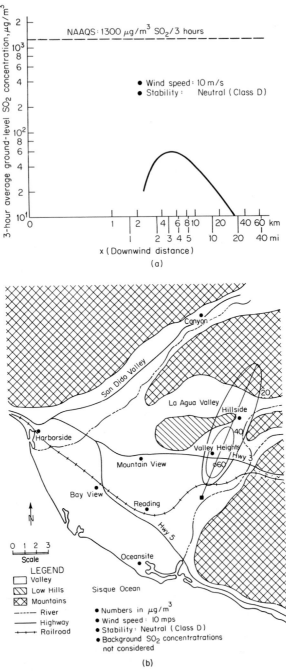

Figure 3.68 Modeling results—primary pollutant impact (1995 most probable case 3-hour and 24-hour SO₂). (*a*) 3-hour average ground level SO₂ concentrations, background SO₂ concentrations not considered. (*b*) 3-hour average ground level SO₂ concentration contours, background SO₂ concentrations not considered (scale in miles).

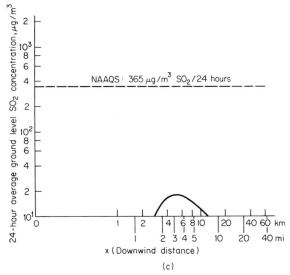

Figure 3.68 (cont.)   (*c*) 24-hour average SO_2 ground level concentrations, background SO_2 concentrations not considered.

A computerized mathematical model based on Turner[108] that calculates annual average concentrations as a function of direction and distance from a point source was used to calculate annual average concentrations of particulate, NO_X, and SO_2.

The increments in the annual average background concentrations of particulate matter, sulfur dioxide, and nitrogen oxides are shown in Figs. 3.70, 3.71, and 3.72. The maximum increases in annual average air pollutant concentrations resulting from station operation are less than 0.3 $\mu g/m^3$ for particulate matter, 0.6 $\mu g/m^3$ for sulfur dioxide, and 4 $\mu g/m^3$ for nitrogen oxides.

Secondary Pollutant Impact   The secondary pollutant impact produced by the elevated point source will occur on the subregional and regional scale due to the emission of NO_X. Power plants emit relatively little reactive hydrocarbon.

Existing.   There are no existing sources at the site. The existing air quality monitoring data are sufficient to describe the local, subregional, and regional air quality with respect to secondary pollutants.

No-build.   (See *Highway Case Study,* Build-out Condition.)

Construction.   Construction is not evisioned to produce a secondary pollutant air quality impact.

Build-out.   To predict the impact of the project on secondary pollutants, it was necessary to utilize the DIFKIN trajectory air quality model (reference). Computations were conducted for the wind trajectories presented in Figure 3.67. Results are presented in Table 3.38 for O_3 at hourly intervals for both the no-build and build-out alternatives.

Visibility Impact   Although no standards for visibility have been promulgated by the federal government or the state for urban areas, the project participants are very much aware of the importance of visibility. The visibility impact was evaluated using the method of Latimer and Samuelsen.[55] Both the visual range impact and plume appearance impact were considered.

Figure 3.73 shows visual range for 2 m/s as a function of the viewer's location downwind from the plant for three cases: (1) looking across (perpendicular to) the plume, (2) along the plume in the direction of the plant, and (3) along the plume in the direction away from the plant. All cases are at ground level with neutral atmospheric stability conditions.

The existing average visual range in the area is about 55 km (34 mi) (reference). The results show that visibility could be reduced to 6.2 km (10 mi) at certain locations.

The brown atmospheric haze characteristic of NO_2 will be visible whenever the apparent intensity of light with wave length of 4000 Å (angstroms) has been reduced by 70 percent or more.[100] Brown discoloration would be visible over that distance in which the attenuation of light intensity (I/I_o) is at or below 0.3.

The model was used to calculate this intensity ratio (I/I_o) for the following points of views: (1) along the plume axis at ground level looking both toward and away from the stack, and (2) across the plume axis at the height of the plume centerline.

During strong inversion conditions when the plume is a relatively narrow "ribbon," brown discoloration will be observed for NO to NO_2 conversion ratios greater than 15–20 percent (see Fig. 3.74). If the viewer is situated beneath the plume centerline (either toward or away from the plant), the atmosphere will appear brownish at certain downwind distances for a realistic NO to NO_2 conversion ratio of 15 percent (see Fig. 3.75).

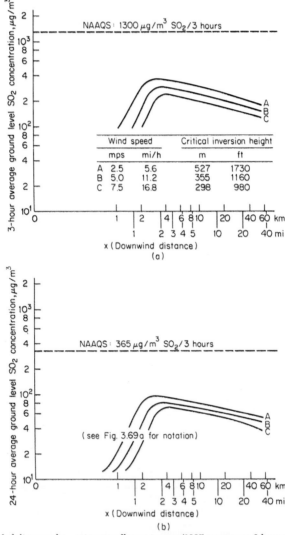

Figure 3.69 Modeling results—primary pollutant impact (1995 worst case 3-hour and 24-hour SO_2). (a) 3-hour average ground level SO_2 concentrations, background SO_2 concentrations not considered. (b) 24-hour average ground level SO_2 concentrations, background SO_2 concentrations not considered.

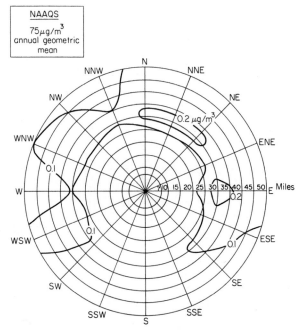

Figure 3.70 Modeling results—primary pollutant impact (1995 maximum increase in annual average particulate).

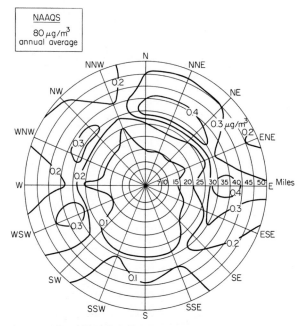

Figure 3.71 Modeling results—primary pollutant impact (1995 maximum increase in annual average SO₂).

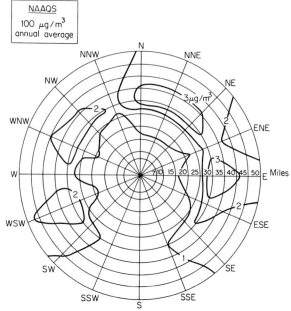

Figure 3.72   Modeling results—primary pollutant impact (1995 maximum increase in annual average NO_x).

TABLE 3.38   Modeling Results—Secondary Pollutant Impact (1995 Ozone Production)

	1-hr average ozone, ppm*	
Time	No-build	Build-out
6 AM	0.11/0.18	0.11/0.18
7	0.11/0.18	0.11/0.18
8	0.11/0.20	0.11/0.18
9	0.11/0.22	0.11/0.19
10	0.13/0.22	0.12/0.18
11	0.14/0.21	0.13/0.19
Noon	0.14/0.22	0.14/0.20
1 PM	0.15/0.22	0.16/0.23
2	0.16/0.24	0.18/0.25
3	0.18/0.26	0.20/0.28
4	0.18/0.28	0.20/0.29
5	0.18/0.26	0.20/0.28

*Most probable/worst case.

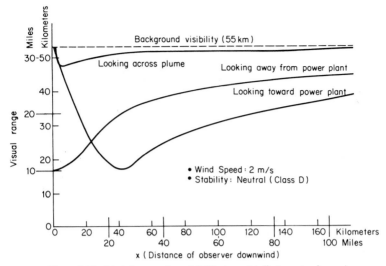

Figure 3.73 Modeling results—visibility impact (effect on visual range).

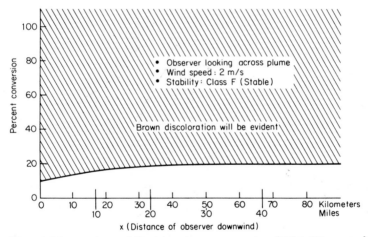

Figure 3.74 Modeling results—visibility impact (percent conversion of NO to NO_2 required to produce brown atmospheric discoloration when viewed *across* plume).

Assessment

The project will produce an incremental degradation in air quality with respect to SO_2, NO_x, particulate, and ozone. Since the facility is designed to meet the restrictive New Source Performance Emission Standards, the incremental degradation will be modest and, in and by itself, will not cause an exceedance of air quality standards. However, when superimposed upon the projected background levels of SO_2, NO_x, particulate, and ozone, the project is shown to delay the attainment of and possibly prevent the maintenance of the National Ambient Air Quality Standards. The State Implementation Plan (SIP) specifically prohibits the construction of such a facility without the adoption of mitigation procedures to meet the requirements of the SIP.

MITIGATION

Reduction in Energy Demand

A public information program could be implemented to encourage energy conservation. Fee schedules could be restructured as well to encourage energy conservation.

System Biasing

Except for high-load periods, the utility could increase the burden to the power stations (1) removed from population centers and/or (2) operating on natural gas or low-sulfur oil.

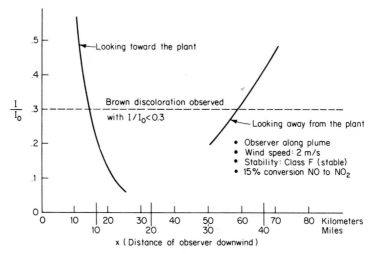

Figure 3.75 Modeling results—visibility impact (locations at which brown atmospheric column is produced for observer *along* plume).

Low-sulfur Coal

Importation of low-sulfur and low-nitrogen coal for use under meteorological conditions conducive to worst case air quality impacts could be encouraged.

Emissions Trade-Off

Implementation of the new source regulations proposed by the Los Robles Air Pollution Control District will mitigate the impacts projected. Under the plan, emissions from existing sources in the proximity of the project area must be reduced in an amount up to double the emissions estimated for the proposed project.

ALTERNATIVES

Project Not Constructed

The project participants have appraised the future power demands and the projected power production for their service areas and determined that without the project the forecasted peak demands in 1990 and thereafter cannot be consistently met. Any delays in the project will shorten the period before service interruptions are experienced in the service area.

Alternate Power Sources and Generating Methods

Any alternate power source must be available by 1990 and of sufficient capacity to meet the projected increases in demand. Alternate sources of electrical power are shown in Table 3.39. This table evaluates all possible sources of electrical power against selection criteria.

An evaluation of power alternatives for an electric utility must be performed in the context of requirements for the total generating system of the utility. These evaluations must recognize the general objective of maintaining a reasonable balance among various primary resource forms used for generating electricity. The precise resource mix necessary to achieve such a balance is subject to a number of variables, with resource availabil-

TABLE 3.39   Alternatives—Alternate Electrical Power Sources

Power sources	Sufficient lead time (available by 1990)	Sufficient generating capacity	Sufficient technology	Sufficient available resources (fuel)	Comments
			Selection criteria		
Purchased power	No	No	Yes	No	. . .
Nuclear power	No	Yes	Yes	Yes	. . .
Gas (steam generator)	Yes	Yes	Yes	No	. . .
Oil (steam generator)	Yes	Yes	Yes	No	. . .
Gas turbine	Yes	Yes	Yes	No	. . .
Hydro	No	No	Yes	No	. . .
Solar	No	Yes	No	Yes	Technical feasibility undetermined
Geothermal	No	No	No	No	Needs research
Fuel Cells	No	No	No	No	Development of 25 MW size underway
Tidal	No	No	No	No	No suitable West Coast sites
Magnetohydrodynamics	No	No	No	Yes	Needs research

ity being one of the most important. During the operation time frame of the project, the availability of massive supplies of coal as a primary energy source for producing electricity cannot be ignored as a major factor in evaluating alternatives.

Purchase Power   One of the potential alternatives to the building of new generation facilities is to purchase the necessary power and energy from a neighboring utility that has an excess. The advantages of such a procedure are (1) the energy is available to a region without the concomitant environmental side effects and (2) the adjacent utility is able to operate at a higher capacity factor, resulting in more efficient system operation. The disadvantages of purchase power include the need for longer transmission lines, ordinarily resulting in line losses equal to 5 to 10 percent of the total energy generated. Also, contracts for such power are, of necessity, written so that the energy and/or the capacity are only available on an interruptible basis, with only excess energy or capacity available to the purchaser. Since adjacent utilities customarily have largely coincident peak loadings, the energy is often unavailable when it is most needed. No reliable source of purchase power is available to the utilities participating in this project.

Nuclear   Utilization of nuclear energy for electrical generation is an existing, highly developed technology. Its applicability in a given situation is determined by its availability and the economics relative to other feasible methods of generation. The participating utilities have recognized the benefits of nuclear power and are involved in either existing or planned nuclear projects. However, to achieve and maintain a balanced energy resource mix, and in consideration of a longer lead time, nuclear projects can only be regarded as complementary to power projects using coal resources, rather than as an alternative.

Utilities experience indicates that about ten years are required from initiation of a new nuclear project to start-up. This period includes more than six years for actual construction. Therefore, it is not presently possible to meet the capacity requirement of 1990 with a nuclear power plant.

Steam Generation with Alternate Fossil Fuels

Natural Gas. The supply of natural gas is low and is decreasing rapidly. Curtailment of deliveries for electrical generation is growing so that no existing unit can be expected to receive an uninterrupted supply of gas. Therefore, it is not considered feasible to depend upon natural gas for a new plant.

Oil. The supply of oil in the world is more limited than that of coal. Furthermore, there is much more competition for that supply for uses other than electrical generation, such as domestic space heating and transportation. Low-sulfur oil, which is being required increasingly for environmental reasons, is only a fraction of the total. Most low-sulfur oil is currently imported from abroad. The cost and availability of this fuel is subject to many uncertainties beyond the control of the utility. In general, even before the competition for this small supply of preferred oils began, the cost of oil was much higher than that of coal for a given energy output.

Gas Turbine   Gas turbine technology is highly developed. These units are characterized by low capital cost but relatively high operating cost, resulting in a high energy cost per kilowatthour. They require large quantities of fuels that are in short supply—natural gas or high-grade diesel oil. Consequently, gas turbines are typically used as peaking units and operate only a few hours per week during periods of maximum demand.

An additional disadvantage of gas turbines is that during hot weather, their power output can be reduced as much as 20 percent. Extra units would be needed to offset hot-weather capacity reduction.

Combined Cycle   A combined cycle plant uses a gas turbine to produce energy from the flue gas, which in turn passes into a steam generating boiler to drive a steam turbine. Such plants are being built in the medium-size range, but thus far are only developed for natural gas or distillate fuels. Experimental work is under way to develop gas turbines to burn residual fuel, but the technique has not yet been perfected. In addition, indications are that if residual fuel is burned, the efficiency of the combined cycle unit may be reduced well below that attainable with a conventional steam unit due to the need for reduced turbine inlet temperature.

Hydroelectric   Further development of hydroelectric power is limited by the small number of remaining feasible sites. In addition, there is insufficient lead time remaining to develop a hydroelectric site. The lead time requirements have been lengthened considerably due to required environmental studies.

Solar   Two types of solar-energy conversion are currently the goal of some research and development work. The first would have large "farms" of solar collectors that convert energy into a heated fluid from which mechanical energy would be extracted in a conventional manner.

The second is a more exotic technique in which geostatic satellites would collect the sun's energy, converting it to electricity by direct solid-state conversion. The electrical energy would then be beamed to centralized collectors on earth by microwave antennae.

At the present time, these concepts are merely in the proposal stage and do not provide a viable alternative to proven power sources.

Geothermal   Thermal potential energy exists in the earth in the form of steam, hot brine, and hot rock. In some areas this supply of energy is located within several thousand feet of the surface, and it is feasible to drill well holes at these points. Successful utilization of this energy for electrical generation has been made with several such steam heat sources. However, hot steam turns out to be a relatively rare and localized resource. Efforts are currently being made by the participants to develop technology for use of hot brine resources, which are more widespread than steam, but as yet this technology has not beem demonstrated. Consequently, geothermal power will not be able to provide the energy needed by 1990 to offset that which would otherwise be supplied by the proposed project.

Fuel Cells   These electrochemical devices convert the chemical energy of conventional fuel oxidation directly into low-voltage, direct-current electricity. The oxidation of hydrogen in such cells, for example, could have an electrical conversion efficiency of 60 to 80 percent, higher than the Carnot efficiency limitation of conventional steam plants. At present, the costs and lack of experience in the use of fuel cells limit their applicability.

Tidal   The technological feasibility of harnessing the tidal variations in a natural coastal basin has been proven, but it is economically viable in only a very few locations.

Such a facility, in addition to requiring the impoundment of large areas of coastline, would necessitate long transmission lines with consequent impact upon land utilization.

Magnetohydrodynamic (MHD)   Magnetohydrodynamic generation of power relies on the same principle as conventional generation. However, instead of a solid conductor, such as a turbine rotor moving across a magnetic field, a jet of ionized fluid flows past it. By placing electrodes in this fluid stream, one can obtain direct current electricity at a relatively high potential (2000 V or more).

The higher flame temperature associated with the MHD process means emission of more nitrogen oxides, but the higher energy conversion efficiency of a combined steam-MHD plant would mean less waste heat transmitted to the environment. Before MHD can be utilized for central power stations, there are many significant technical problems that must be solved.

Site

Alternative sites require the availability of cooling water and transportation systems to insure the availability of fuel. Outside urban areas, the siting of power plants is restricted by regulations for the prevention of significant deterioration. These factors entered into the decision to locate the power plant at the proposed site.

GROWTH-INDUCING IMPACTS

No significant growth-inducing impacts are associated with the proposed project. This is based on the assumption that, should the project not be built, an alternative site will be necessary to supply the energy demand. Alternative sites have the disadvantage of higher transmission costs, higher fuel transportation costs, and low availability of cooling water.

References

1. Altshuller, A. P., and J. J. Bufalini: "Photochemical Aspects of Air Pollution: A Review," *Environmental Science and Technology,* vol. 5, no. 1, January 1971, p. 39.
2. American Chemical Society: *Cleaning Our Environment: The Chemical Basis for Action,* Washington, D.C., September 1969.
3. American Society of Mechanical Engineers: *Recommended Guide for the Prediction of the Dispersion of Airborne Effluents,* May 1968.
4. Beaton, J. L., A. J. Ranzieri, and J. B. Skog: *Air Quality Manual, Vol. 1: Meteorology and Its Influence on the Disperion of Pollutants from Line Sources,* Federal Highway Administration Report No. FHWA-RD-72-33, April 1972.
5. Benesh, F., P. Guldberg, and R. D'Agostino: *Growth Effects of Major Land Use Projects, Volume II: Summary,* Environmental Protection Agency Report No. EPA-450/3-76-012-C, September 1976.
6. Boone, D., and G. S. Samuelsen: "Computer Mapping of Air Quality," *Journal of the Environmental Engineering Division,* ASCE, vol. 103, no. EE6, December 1977, p. 969.
7. Brewer, D. H., S. L. Ahmed, and S. J. Johnson: *Air Dispersion Modeling for Economic Stack Design,* presented at the 70th Annual Meeting of the Air Pollution Control Association, APCA Paper No. 77-43.4, June 1977.
8. Briggs, G. A.: "A Plume Rise Model Compared with Observations," *Journal of the Air Pollution Control Association,* vol. 15, no. 9, September 1965.
9. Briggs, G. A.: *Penetration of Inversions by Plumes,* paper presented at the 48th Annual Meeting of The American Meteorological Society, San Francisco, 1968.
10. Briggs, G. A.: *Some Recent Analyses of Plume Rise Observations,* Proceedings of the Second International Clean Air Congress, Academic Press, New York, 1971.
11. Briggs, G. A.: *Diffusion Estimation for Small Emissions,* ATDL Contribution File No. (Draft) 79, Turbulence and Diffusion Laboratory, NOAA, Oak Ridge, Tenn., May 1973.
12. Burt, E., and J. Mersch: *Valley Model Computer Program,* Environmental Protection Agency Source Magnetic Tape EPA/DF-78/002, September (NTIS PB 275-700), 1977.
13. Busse, A. D., and J. R. Zimmerman: *User's Guide for the Climatological Dispersion Model,* Environmental Protection Agency Report No. EPA-R4-73-024. (NTIS PB 227-346), 1973.

14. California Air Resources Board: *ARB Bulletin*, August 1976, p. 3.
15. Carpenter, S. B., T. L. Montgomery, J. M. Leavitt, W. C. Colbaugh, and F. W. Thomas: "Principal Plume Dispersion Models: TVA Power Plants," *Journal of the Air Pollution Control Association*, vol. 21, no. 8, August 1971, p. 491.
16. Chaput, L. S.: "Federal Standards of Performance for New Stationary Sources of Air Pollution, A Summary of Regulations," *Journal of the Air Pollution Control Association*, vol. 26, no. 11, November 1976, p. 1055.
17. Clark, F. J.: "Nocturnal Urban Boundary Layer Over Cincinnati," Ohio, *Monthly Weather Review*, vol. 97, p. 582, 1969.
18. Clean Air Amendments of 1970, 42 USC 1857 et. seq. as amended by the Clean Air Amendments of 1970, PL 91-604.
19. Clean Air Amendments of 1977, 42 USC 1857 et. seq. as amended by the Clean Air Amendments of 1977, PL 95-95, as amended PL 95-190.
20. Committee on Public Works: *Proceedings of the Conference on Health Effects of Air Pollutants*, Serial No. 93-15, November 1973.
21. Dobson, G. M. B.: *Exploring the Atmosphere*, Oxford Press, England, 1968.
22. Duckworth, F. S., and J. S. Sandberg: "The Effect of Cities Upon Horizontal and Vertical Temperature Gradients," *Bulletin of the American Meteorological Society*, vol. 35, p. 198, 1954.
23. Environmental Protection Agency: *Air Quality Criteria for Nitrogen Oxides*, AP-84, January 1971.
24. Environmental Protection Agency: *Compilation of Air Pollutant Emission Factors*, 2d ed., AP-42, February 1976.
25. Environmental Protection Agency: *Guidelines on Air Quality Models*, Environmental Protection Agency Report No. EPA 450-2-78-027, April 1978.
26. Eschenroeder, A. Q.: *Validation of Simplified Kinetics for Photochemical Smog Modeling*, General Research Corporation, September 1966.
27. Evelyn, J.: *Fumifugium: or The Inconvenience of the Aer and Smoake of London Dissipated. Together with Some Remedies Humbly Proposed*, 1661 (second printing, 1772).
28. Fabrick, A., R. Sklrew, and J. Wilson: *Point Source Model Evaluation and Development Study*, Report to the California Air Resources Board and the California Energy Resources Conservation and Development Commission, Contract No. A5-058-87, March 1977.
29. Federal Register, vol. 36, no. 84, pp. 8186–8201, April 30, 1971.
30. Federal Register, vol. 36, no. 158, pp. 15486–15506, August 14, 1971.
31. Federal Register, vol. 36, no. 206, p. 20513, October 23, 1971.
32. Federal Register, vol. 36, no. 228, pp. 22421–22448, November 25, 1971.
33. Federal Register, vol. 36, no. 243, pp. 24002–24003, December 17, 1971.
34. Federal Register, vol. 37, no. 238, pp. 26310–26312, December 9, 1972.
35. Federal Register, vol. 38, no. 116, pp. 15834–15837, June 18, 1973.
36. Federal Register, vol. 39, no. 50, pp. 9672–9675, March 13, 1974.
37. Federal Register, vol. 39, no. 235, pp. 42510–42517, December 5, 1974.
38. Federal Register, vol. 40, no. 11, p. 2802, January 16, 1975.
39. Federal Register, vol. 40, no. 111, pp. 24534–24535, June 9, 1975.
40. Federal Register, vol. 40, no. 114, pp. 25004–25011, June 12, 1975.
41. Federal Register, vol. 40, no. 162, pp. 36330–36335, August 20, 1975.
42. Federal Register, vol. 40, no. 176, pp. 42011–42012, September 10, 1975.
43. Federal Register, vol. 43, no. 21, pp. 26962–26986, June 22, 1975.
44. Federal Register, vol. 44, no. 28, pp. 8202–8237, February 8, 1979.
45. Friedlander, S. K., and J. H. Seinfeld: "A Dynamic Model of Photochemical Smog," *Environmental Science and Technology*, vol. 3, no. 11, November 1969, p. 1175.
46. Gifford, F. A.: "Use of Routine Meteorological Observations for Estimating Atmospheric Dispersion," *Nuclear Safety*, vol. 2, no. 4, 1961, p. 47.
47. Gustafson, S-Å, K. O. Kortanek, and J. R. Sweigart: *Numerical Optimization Techniques in Air Quality Modeling, Objective Interpolation Formulae for the Spatial Distribution of Pollutant Concentration*, Environmental Protection Agency Report No. EPA 600/4-76-058, December 1976.
48. Hagen-Smith, A. J., and L. G. Wayne: In Stern, A. C.(ed.) *Air Pollution, Volume 1*, Academic Press, New York, 1967.
49. Hanna, S. R.: "The Thickness of the Planetary Boundary Layer," *Atmospheric Environment*, vol. 3, p. 519, 1969.
50. Hay, J. S., and F. Pasquill: "Diffusion from a Fixed Source at a Height of a Few Hundred Feet in the Atmosphere," *Journal of Fluid Mechanics*, vol. 2, p. 299, 1957.
51. Hay, J. S., and F. Pasquill: "Dispersion from a Continuous Source in Relation to the Spectrum and Scale of Turbulence," in F. N. Frenkiel and P. A. Sheppard (eds.), *Advances in Geophysics*, vol. 6, Academic Press, New York, 1957, p. 345.
52. Hecht, T. A., and J. H. Seinfeld: "Development and Validation of a Generalized Mechanism for Photochemical Smog," *Environmental Science and Technology*, vol. 6, no. 1, January 1972, p. 47.

53. Holland, J. Z.: *A Meteorological Survey of the Oak Ridge Area: Final Report Covering the Period 1948–1952*, Atomic Energy Commission Report Number ORO-99, p. 554, 1953.
54. Holzworth, G. C.: *Mixing Heights, Wind Speeds, and Potential for Urban Air Pollution Throughout the Contiguous United States*, Environmental Protection Agency, AP-101, January 1972.
55. Latimer, D. A., and G. S. Samuelsen: "Modeling Plume Visibility From Major Power Plants," *Atmospheric Environment*, vol. 21, p. 1445, 1978.
56. Lavery, T. F., B. A. Egan, and R. M. Iwanchuk: *The Numerical Simulation of the Advection and Diffusion of a Plume Under Aerodynamic Downwash Conditions*, presented at the 67th Annual Meeting of the Air Pollution Control Association, APCA Paper #74-215, June 1974.
57. Lee, R., and J. Mersch: *Single Source (CRSTER) Model Computer Programs*, Environmental Protection Agency Source Magnetic Tape EPA/DF-78/004 1977. (NTIS PB 275-701).
58. Leighton, P. A.: *Photochemistry of Air Pollution*, Academic Press, New York, 1961.
59. MacCracken, M. C., and G. D. Sauter (eds.): *Development of an Air Pollution Model for the San Francisco Bay Area*, Final Report to the National Science Foundation, Lawrence Livermore Laboratory Reports UCRL-51920, vols. 1 and 2, 1975.
60. Maldonado, C., and J. A. Bullin: "Modeling Carbon Monoxide Dispersion from Roadways," *Environmental Science and Technology*, vol. 11, no. 12, November 1977, p. 1071.
61. Mancuso, R. L., and F. L. Ludwig: *User's Manual for the APRAC-1A Urban Diffusion Model*, Coordinating Research Council Report Number CRC-CAPA-3-4, New York (NTIS PB 213-091), 1972.
62. Martinez, J. R., R. A. Nordsick, and M. A. Hirshberg: *Users Guide to Diffusion/Kinetics (DIFKIN) Code*, Final Report EPA Contract No. 68-02-0336, General Research Corporation, Santa Barbara, Calif., 1973.
63. Meade, P. J.: "*Rassegna internazionale elettronica, nucleare e teleradio. Cinematogra Sezione nucleare, 6th*," Attidel congreso scientifico, II. Roma, comitato nazionale recherche nucleari, p. 107, 1959.
64. Mills, M. T., and Stern: *Improvements to Single-Source Model, Volume 2: Testing and Evaluation of Model Improvements*, Environmental Protection Agency Report No. EPA-450/3-77-003G, January 1977.
65. Montgomery, T. L., W. B. Norris, F. W. Thomas, and S. B. Carpenter: "A Simplified Technique Used to Evaluate Atmospheric Dispersion of Emissions from Large Power Plants," *Journal of the Air Pollution Control Association*, vol. 23, no. 5, May 1973, p. 388.
66. Montgomery, T. L., and J. H. Coleman: "Empirical Relationships Between Time-Averaged SO_2 Concentrations," *Environmental Science and Technology*, vol. 9, no. 10, October 1975, p. 953.
67. Morgan, G. B.: "Monitoring the Quality of Ambient Air," *Environmental Science and Technology*, vol. 11, no. 4, April 1977, p. 352.
68. National Academy of Sciences: *Ozone and Other Photochemical Oxidants*, Washington, D.C., 1977.
69. National Climatic Center: *Surface Observations, Magnetic Tape Reference Manual*, National Climatic Center Report No. TDF-14, October 1975.
70. National Environmental Pollution Act: 42 USC 4321 et. seq., PL 91-190, 1969.
71. Pasquill, F.: "The Estimation of the Description of Windborne Material," *The Meteorological Magazine*, vol. 90, no. 1063, February 1961, p. 35.
72. Pedri, C.: *Comparative Tests of Plume Rise Models and Dispersion Coefficients*, UCI Combustion Laboratory, ARTR-79-3, University of California, Mechanical Engineering, Irvine, 1979.
73. Pitts, J. N., Jr., W. P. Carter, K. R. Darnell, G. J. Doyle, W. Kuby, A. C. Lloyd, J. M. McAfee, C. Pate, J. P. Smith, J. L. Sprung, and A. M. Winer: "Development of Experimentally Validated Models of Photochemical Air Pollution," in *Proceedings of the Second Annual NSF-RANN Trace Contaminants Conference*, August 1975.
74. Public Health Service, U.S. Department of Health, Education, and Welfare: *Air Quality Criteria for Particulate Matter*, AP-49, January 1969.
75. Public Health Service, U.S. Department of Health, Education, and Welfare: *Air Quality Criteria for Sulfur Oxides*, AP-50, January 1969.
76. Public Health Service, U.S. Department of Health, Education, and Welfare: *Air Quality Criteria for Carbon Monoxide*, AP-62, March 1970.
77. Public Health Service, U.S. Department of Health, Education, and Welfare: *Air Quality Criteria for Photochemical Oxidants*, AP-63, March 1970.
78. Public Health Service, U.S. Department of Health, Education, and Welfare: *Air Quality Criteria for Hydrocarbons*, AP-64, March 1970.
79. Racin, J. A., P. D. Allen, and A. J. Ranzieri: *Transportation Systems and Regional Air Quality— Evaluation of a Modified APRAC-1A Carbon Monoxide Diffusion Model for the Sacramento Region*, Federal Highway Administration Report FHWA/CA-76-30, California State Department of Transportation, Sacramento, Calif., April 1976 (NTIS PB 272-477).
80. Reynolds, S. D., P. M. Roth, and J. H. Seinfeld: "Mathematical Modeling of Photochemical Air Pollution: I. Formulation of the Model," *Atmospheric Environment*, vol. 7, p. 1033, 1973.
81. Reynolds, S. D., M. K. Lin, T. A. Hecht, P. M. Roth, and J. H. Seinfeld: *Urban Airshed*

Photochemical Simulation Study: Vol. I—Development and Evaluation, Environmental Protection Agency Reports EPA-R4-73-020a,b,c,d,e, Systems Applications, Inc., San Rafael, Calif., 1973.

82. Reynolds, S. D., M. K. Liu, T. A. Hecht, P. M. Roth, and J. H. Seinfeld: "Mathematical Modeling of Photochemical Air Pollution: III. Evaluation of the Model," *Atmospheric Environment,* vol. 8, p. 563, 1974.

83. Reynolds, S. D., J. Ames, T. A. Hecht, J. P. Meyer, D. C. Whitney, and M. A. Yocke: *Continued Research in Mesoscale Air Pollution Simulation Modeling: Volume II, Model Development and Refinement,* Environmental Protection Agency Report No. EPA 600/4-76-016b, Systems Applications, Inc., San Rafael, Calif., 1976.

84. Rote, D. M., and L. E. Wangen: *A Generalized Air Quality Assessment Model for Air Force Operations,* Air Force Weapons Laboratory Report AFWL-TR-74-304, February 1975 (NTIS AD/A-0006-807).

85. Rupp, A. F., S. E. Beall, L. P. Borwasser, and D. F. Johnson: *Dilution of Stack Gases in Crosswinds,* Atomic Energy Commission Report No. AECD-1811 (CE-1620), 1948.

86. Samuelsen, G. S.: "Air Quality: Who is Responsible," *California Air Environment,* October/December 1971, p. 1.

87. Samuelsen, G. S., G. L. Guymon, F. Greve, and N. Frager: *Carbon Monoxide Transport from Freeways: Evaluation of Major Models,* UCI Combustion Laboratory Report UCI-ARTR-78-6, Mechanical Engineering, University of California, Irvine, 1978.

88. Sandys, R. C., P. A. Buder, and W. F. Dabberdt: *ISMAP—A Traffic/Emissions/Dispersion Model for Mobile Pollution Sources, User's Manual,* SRI Report No. 3628, Stanford Research Institute, Menlo Park, Calif., January 1975.

89. Section 109(j) Title 13, 23 USC 109(j), Added by Section 136(B) of Federal Aid Highway Act, PL 91-605, 1970.

90. Shy, C. M., and J. F. Finklea: "Air Pollution Affects Community Health," *Environmental Science and Technology,* vol. 7, no. 3, March 1973, p. 204.

91. Singer, I. A., and M. E. Smith: "Relation of Gustiness to Other Meteorological Parameters," *Journal of Meteorology,* vol. 10, p. 121, 1953.

92. Singer, I. A., and G. S. Raynor: *Analysis of Meteorological Tower Data, April 1950 to March 1952,* Brookhaven National Laboratory, Report Number AFCRC TR-57-220, June 1957.

93. Slade, D. H. (ed.): *Meteorology and Atomic Energy,* U.S. Atomic Energy Commission, 1968.

94. Snyder, W. H., and R. E. Lawson, Jr.: *Determination of Height for Stacks near Buildings—Wind Tunnel Study,* Environmental Protection Agency Report No. EPA/600/4-76/001, February 1976.

95. Somers, E. V.: "Dispersion of Pollutants Emitted into the Atmosphere," in W. Strauss (ed.), *Air Pollution Control,* Part One, Wiley-Interscience, New York, 1971.

96. South Coast Air Quality Management District: *Source Receptor Areas,* El Monte, Calif., 1977.

97. Stanford Research Institute: *Users' Manual for the APRAC-2 Emissions and Diffusion Model,* Menlo Park, Calif., June 1977 (NTIS PB 275-459).

98. Stephens, E. R.: "Reactions of Oxygen Atoms and Ozone in Air Pollution," *International Journal of Air and Water Pollution,* vol. 10, no. 10, October 1966, p. 649.

99. Stephens, E. R.: In Atkisson, A. and R. S. Gaines, (eds.). *Development of Air Quality Standards,* Merrill, Columbus, Ohio, 1970, p. 204.

100. Stern, C. A.: *Air Pollution, Volume 1,* Academic Press, New York, 1968.

101. Stern, A. C., H. C. Wohlers, R. W. Boubel, and W. P. Lowry: *Fundamentals of Air Pollution,* Academic Press, New York, 1973.

102. Summers, P. W.: "An Urban Heat Island Model; Its Role in Air Pollution Problems with Application to Montreal," *Proceedings of the First Canadian Conference on Micrometeorology,* Toronto, Ontario, Department of Transportation, Canada, April 1967.

103. Taylor, G. H., and A. Q. Eschenroeder: *Tests of the DIFKIN Photochemical/Diffusion Model Using LARPP Data,* Paper 17-2, International Conference on Photochemical Oxidant Pollution and Its Control, Environmental Protection Agency, Raleigh, N.C., 1976.

104. Tennessee Valley Authority: *Full Scale Study of Plume Rise at Large Electric Generating Stations,* Muscle Shoals, Ala., 1968.

105. Thomas, F. W.: "TVA Air Pollution Studies Program," *Air Repair,* vol. 4, 1954, p. 59.

106. Thomas, F. W., Carpenter, S. B., and Colbaugh, W. C.: "Plume Rise Estimates for Electric Generating Stations," *Journal of the Air Pollution Control Association,* vol. 20, no. 3, March 1970, p. 120.

107. Tigue, J.: *Airport Vicinity Air Pollution Model Computer Source Code,* Federal Aviation Administration Source Magnetic Tape FAA/DF-76/001, December 1975 (NTIS AD-A031 027).

108. Turner, D. B.: *Workbook of Atmospheric Dispersion Estimates,* Public Health Service, U.S. Department of Health, Education, and Welfare, AP-26, 1970.

109. Turner, D. B., and A. D. Busse: *Users' Guide to the Interactive Versions of Three Point Source Dispersion Programs: PTMAX, PTDIS, and PTMTP,* Environmental Protection Agency, June 1973 (NTIS PB 229-771).

110. Turner, B., and J. Mersch: *Climatological Dispersion Model QC (CDMQC) Computer Program,* Environmental Protection Agency Source Magnetic Tape EPA/DF-78/003, May 1977 (NTIS PB 276-516).

111. Wang, I. T., D. M. Rote, and L. A. Conley: *Airport Vicinity Air Pollution Study, Model Application and Validation and Air Quality Impact Analysis at Washington National Airport,* Argonne National Laboratory, Illinois, July 1974 (NTIS AD/A-001-564).

112. Ward, C. E., Jr., A. J. Ranzieri, and E. C. Shirley: *CALINE 2—An Improved Microscale Model for the Dispersion of Air Pollutants from a Line Source,* Federal Highway Administration Report FHWA/RD-77-74, California State Department of Transportation, Report CA-DOT-TL-7218-1-76-23, Sacramento, Calif., 1977 (NTIS PB 275-683).

113. Wark, K., and C. F. Warner: *Air Pollution, Its Origin and Control,* IEP, New York, 1976.

114. Wayne, L. G., A. Kokin, and M. I. Weirburd: *Controlled Evaluation of the Reactive Environmental Simulation Model (REM), Vol. III,* Final Report, Environmental Protection Agency Contract 68-02-0345, Pacific Environmental Services, Inc., Santa Monica, Calif., 1973.

115. Wayne, L. G., and P. J. Drivas: *Validation of an Improved Photochemical Air Quality Simulation Model,* Pacific Environmental Services Report PES TP-014, Santa Monica, Calif., March 1977.

116. Weather Bureau, U.S. Department of Commerce: *Meteorological Summaries Pertinent to Atmospheric Transport and Dispersion over Southern California,* Technical Bulletin No. 54, 1965.

117. Williamson, S. J.: *Fundamentals of Air Pollution,* Addison-Wesley, Philippines, 1973.

118. Zimmerman, J. R., and R. S. Thompson: *Users' Guide for HIWAY, A Highway Air Pollution Model,* Environmental Protection Agency Report No. EPA-650/4-74-008, June 1975 (NTIS PB 229-771).

Chapter 4

Noise Impact Analysis

VINCENT E. MESTRE AND
DAVID C. WOOTEN

The noise levels to which the urban population is exposed have been increasing at a substantial rate over the past several decades. This has led to increasing numbers of complaints received by public agencies and to the recent flood of noise-related litigation now taking place in the courts. Also, federal, state and local government agencies have enacted numerous laws, ordinances, and regulations to control noise. The question of community noise control has, therefore, become an important consideration in urban planning, in construction practices, and in public administration.

One means for incorporating noise control considerations into land use planning is the Environmental Impact Statement (EIS). The following sections of this chapter on the environmental impact of noise address the relevant topics related to writing the noise portions of an EIS. These include various environmental assessment topics such as description of existing noise environment, physical description of noise, assessing noise impacts for various noise-sensitive land uses, designing mitigation measures to alleviate potential community noise problems, and the analysis of various project alternatives which may avoid noise conflicts. Detailed assessment methodologies for the more technical reader include the fundamentals and equations necessary to project noise levels for a variety of sources under a variety of conditions. Sources include aircraft, helicopters, automobiles, trucks, construction equipment, and stationary sources. Finally, illustrative applications are shown for a variety of projects including a residential project, a shopping center, and an airport.

INTRODUCTION

Typical Considerations

Typical considerations in environmental noise assessment can be divided into two separate categories: those relating to the noise source and those relating to potential receivers. For example, in describing the noise impact characteristics of a source (whether it be aircraft, roadway traffic, or stationary source), it is necessary to have a physical description of the sound itself, a description of how the loudness varies with time, when the noise occurs, and the location of the noise source. The noise environment the receiver will hear must then be determined and an assessment must be made of the effects of the noise as it relates to sleep disturbance, interference with speech communication, or interference with other human activities. A judgment must then be made as to whether such noise levels are acceptable. All environmental noise problems include these basic elements.

The physical description of sound concerns its loudness as a function of frequency. Noise, in general, is sound which is composed of many frequency components of various loudness distributed over the audible frequency range. Various noise scales have been introduced to describe, in a single number, the response of an average human to a complex sound made up of various frequencies at different loudness levels. The most common and heavily favored of these scales is the A-weighted decibel (dBA). This scale has been designed to weight the various components of noise according to the response of the human ear; that is, the ear does not perceive low frequency or high frequency sound as well as the middle frequencies. Therefore, in the dBA scale noise with predominant middle frequencies is given a much higher loudness value than noises which are predominantly low or high frequency in nature. This assignment of a noise level to complex sound in order to relate the sound's "loudness" to actual human response is the first step in determining the "impact" of a noise. A more detailed and technical treatment of sound levels and various scales is given in a later section of this chapter.

Another characteristic of the noise source which affects community response is how the noise level varies with time. This includes such parameters as whether the noise is impulsive in nature (a gunshot) or continuous in nature (traffic noise near a freeway). The time-varying nature of noise is an important factor in answering two questions: "How long is the noise at a given loudness?" and "Is the noise steady (constant loudness such as a steady hum) or does it vary radically with time (intermittent traffic noise)?" This is important because, first, it is well known that a steady noise is not as annoying as a noise that is continually varying in loudness, and second, the less amount of time a noise source generates loud noise levels the less potential it has for adversely impacting a community. These qualities of a noise are usually described by reporting the nature of the noise loudness level versus time. Methods of describing the time-varying nature of noise are described in subsequent parts of this chapter.

The time of day that a noise occurs is also an important factor in determining the acceptance of a noise source. For example, loud noise levels at night in residential areas are not acceptable because of sleep disturbance, while nighttime noise near professional buildings is not as important as noise during the working portion of the day. Various schemes have been developed for weighting noise levels for time-of-day of occurrence. In the Day-Night Noise Level scale (L_{dn}), one event at night is equivalent to ten identical noise events during the day. The Community Noise Equivalent Level (CNEL) scale further weights evenings such that one noise event occurring between 7:00 and 10:00 P.M. is equivalent to three identical noise events during the day. The section of this chapter on Sound Propagation presents the technical methodologies for weighting noise for time of occurrence.

The location of a noise source as it relates to the location of noise-sensitive land use is an important factor in determining the impact of the noise; that is, how loud and how long the noise exposure at a receiver will be. In planning as well as project design, determining the location of noise-sensitive land use in relation to noise sources is one of the most effective ways of controlling noise in our communities. A later section of this chapter presents the theories of the propagation of sound. For more technical and complete treatment of the propagation of sound, acoustics, and architectural acoustics, see one of the more traditional books on noise control listed in the References.

Once the noise source is fully characterized and the noise at a receiver's location is described, the impact must be assessed in terms of noise impact on human activities. One of the most common and effective ways of doing this is to classify land uses affected by noise sources according to their sensitivity to noise. Table 4.1 presents a list of land uses according to their sensitivity to noise. Note that this table is general and the sensitivity is somewhat arbitrary and must be determined by considering the specific characteristics of land use. In the section of this chapter on Effects of Noise on People these sensitivities are correlated with actual noise levels using various noise scales.

Environmental Impacts and Effects

The environmental impact of noise can have several effects varying from hearing loss to annoyance. A noise problem is said to exist when noise interferes with human activities.

TABLE 4.1   Noise Sensitivity of Various Land Uses

Land use	Sensitivity
Educational facilities Hospitals Convalescent homes Theaters Wildlife sanctuaries Churches Mobile-home parks	Very sensitive
Single-family (detached) dwellings Single-family (attached) dwellings Multifamily (low-rise) dwellings Multifamily (mid-rise) dwellings Multifamily (high-rise) dwellings Dormitories Resort hotels Outpatient clinics Preschools	Sensitive
Cemeteries Country clubs Scientific testing Professional research Government services Restaurants and bars Motor inns General merchandising Professional offices Recreational vehicle parks	Moderately sensitive
Agriculture Mining and extraction Water areas Natural open space Undeveloped land Motor vehicle transport Auto parking Raceways and drag strips	Insensitive

Specifically, sufficiently loud noise may:
1. Damage hearing or health
2. Interfere with work tasks
3. Interfere with speech communication
4. Affect interroom privacy
5. Interfere with sleep
6. Cause annoyance

Physical measurements of noise can produce data that may be used to determine ratings specific to each of the six effects listed above. However, the choice of criteria for assessing noise impacts is difficult because each human activity is influenced differently by noise.

The damage risk criteria for hearing as enforced by the Occupational Safety and Health Administration (OSHA) are listed in Table 4.2. It must be emphasized that these levels were established to reduce hearing loss. Community noise acceptability criteria are generally in terms of much lower sound levels. Any community noise exposure comparable to those shown in Table 4.2 is clearly unacceptable and, in fact, constitutes a community health hazard. Table 4.2 indicates that noise levels up to 90 dBA are acceptable for eight hours per day. In community noise control, annoyance and physiological damage can occur at much lower noise levels. For example, sleep interference begins to occur at around 45 dBA. Table 4.3 lists the trends in public reaction to peak noise in residential areas. Adverse community response comes at much lower levels than does the risk of hearing loss. Land use acceptability criteria for noise is based on preventing both hearing loss and adverse physiological responses, as well as preventing interference with human activities. These land use criteria are discussed fully in a later section.

Legal Aspects

The legal aspects of community noise control range from the general laws against public nuisances to detailed laws and regulations which limit the noise that a given source may make to land use restriction designed to protect noise-sensitive land uses.

In general, there have been laws for a number of years which indirectly imposed noise control in our communities. Laws against public nuisances and damages due to negligence as well as inverse condemnation procedures are classic examples of measures which have been taken to control noise. Public nuisance laws require that local authorities, rather arbitrarily, use their own discretion in determining what constitutes a public

TABLE 4.2   Damage Risk Criteria for Hearing Loss

Maximum allowable duration per day, h	Noise level, dBA, slow response
8	90
6	92
4	95
3	97
2	100
1½	102
1	105
½	110
¼ or less	115 (maximum)

nuisance. Laws against damages due to negligence require that people exposed to excessive noise show damages in order to obtain relief. Damages from noise have always been difficult to demonstrate and, furthermore, such proof requires that the damage has been done rather than prevented.

Inverse condemnation occurs when an action interferes with the intended use of the

TABLE 4.3   Trend of Public Reaction to Peak Noise Near Residences

dBA	
90	Local committee activity with influential or legal action
	Petition of protest
80	
	Letters of protest Complaints likely
70	Complaints possible
	Complaints rare
60	
	Acceptance
50	

property of another. With regard to noise, inverse condemnation occurs when the noise from a given source prevents the normal use of another's property. Lawsuits around airports generally take the form of inverse condemnation suits.

As a result of growing public concern over noise control, numerous laws have been written to deal specifically with noise control. These laws have been enacted on the federal, state, and local levels.

On the federal level, several agencies of the federal government are charged with enforcing noise-related legislation. The Federal Aviation Administration (FAA) establishes noise standards for aircraft, implements aircraft noise abatement procedures, and sets measures for noise standards around airports. The Federal Highway Administration (FHWA) sets noise standards for location of new highways. The Department of Housing and Urban Development (HUD) establishes noise standards for residential construction. The Department of Labor (DOL) sets and enforces occupational noise limits through the Occupational Safety and Health Administration (OSHA). The Environmental Protection Agency (EPA), with responsibility for federal noise control on an overall basis, sets vehicle noise limits, requires noise labels on appliances, and reviews all federal noise activities for adequacy and compliance.

On the state level, it is beyond the scope of this discussion to summarize the noise activity in all states; however, it is meaningful to examine the noise laws and regulations in the state of California because California has been a leader in establishing environmental legislation including noise control. The California Vehicle Code establishes maximum permissible noise limits for new and operating vehicles. The Noise Regulations for California Airports establish criteria for compatibility of various land uses in the vicinity of airports, require noise monitoring in the vicinity of airports, and require airports to reduce the noise impact area if land use conflicts exist. The Division of Industrial Safety regulates occupational noise exposure. The state of California also requires that all cities and counties develop a noise element in their general plan which will introduce noise considerations in land use planning. The California Noise Control Act also establishes an Office of Noise Control in the State Department of Health. Further, the Noise Insulation Standards establish requirements for sound and impact insulation in multifamily dwellings which share common walls and floor/ceiling combinations and, in addition, establishes indoor noise limits for outdoor noise sources such as highways and airports.

On the local level in California, noise control has been introduced in many ways. Land use control has been achieved by general plan land use controls through the Environmental Impact Report (EIR) approval requirements. Also, many cities and counties have adopted noise ordinances, which outlaw noise nuisances and intrusions. Varying levels of detail can exist in a noise ordinance, including very detailed descriptions of allowable noise levels as a function of duration of the noise as well as the establishment of both daytime and nighttime allowable noise levels.

In carrying out a noise assessment, therefore, one must consider laws and regulations from all levels of government that may be applicable.

Noise assessment in an environmental impact statement can take many forms, including consideration of the potential damage to hearing, potential physiological responses, annoyance, and general community responses. Typical noise assessments in the EIS will take the form of comparisons of existing and anticipated noise levels with the criteria established by various regulatory agencies of the federal, state, and local governments. As a conclusion to this introductory section and a prelude to the remainder of this chapter on environmental noise assessment, the following typical outline of the approach to be followed is provided:

1. Existing noise
   a. Describe existing noise levels
      (1) Measurement results
      (2) Modeling results
   b. Determine current noise regulations
   c. Assess the existing noise environment
2. Noise impact
   a. Determine the potential noise sources associated with the project
      (1) Motor vehicle noise
      (2) Airport and aircraft noise
      (3) Train noise

      (4) Stationary noise sources
    b. Describe the noise characteristics of the project
      (1) Through measurement of an existing similar project
      (2) Through the acoustical modeling of the noise sources associated with the project
    c. Determine the noise sensitivity of the surrounding land uses as well as the noise sensitivity of the project
    d. Assess the impact of the project on surrounding land uses
    e. Assess the impact of existing or future noise levels on the project itself if the project is noise-sensitive
  3. Develop mitigation measures for potential impacts
  4. Discuss the noise considerations of alternatives to the project
The following sections are provided to assist in completing these analyses.

NOISE AND SOUND

Noise is most often and most simply defined as unwanted sound. In considering environmental noise, the analyst studies how noise affects the health or interferes with the activities of people. We are principally concerned with determining quantitatively the amount and type of noise to which people will be (or already are) exposed in a given situation. Based upon this, we assess the possible effects on health and human activities such as sleep, conversation, and work. We are also concerned with the potential effectiveness of alternative mitigation measures in reducing noise and its effects upon people.

    Almost all noise problems involve three basic elements: the noise source, the transmission path, and the receiver. To analyze a noise problem, to develop the appropriate environmental assessment, and to evaluate possible mitigation measures the dominant noise sources must be identified and characterized as to their loudness, pitch or frequency, duration, and other characteristics. Further, the transmission paths from the sources to receivers must be established, the amount and characteristics of the noise reaching people (receivers) must be determined, and criteria for the amount of noise that is considered acceptable must be used to assess the noise impact. The basic physical characteristics of sound which are needed to start developing the analysis of environmental noise are discussed in this section. A later section deals with the detailed analysis of sound; that is, methods used to quantify the basic properties of sound and noise.

The Nature of Sound

The technical definition of sound is that it is a disturbance that propagates through an elastic medium at a speed that is determined by the properties of that medium. In environmental noise, we are usually concerned with sound propagation through air but in some cases are interested in sound propagation through solids, such as the walls of buildings. In dealing with the practical aspects of environmental noise, it is not necessary to study the detailed physics of sound—acoustics. Instead, we will draw upon the science of acoustics as needed. A review of several basic properties of sound is needed to develop an understanding of the methods of dealing with environmental noise.

    Sound as a disturbance propagating through a medium (air) is a wave propagation phenomenon. Similar propagation of waves can be readily observed in other situations. For example, a rock dropped into a pond causes ripples of waves to propagate radially outward on the pond's surface. Or if a suspended rope is snapped at one end, we observe a wave traveling down the rope to the opposite end. In the first example, the propagating disturbance is a ripple on the water's surface and, in the second, it is a displacement of the rope about its original or equilibrium position. Sound is a similar type of wavelike disturbance in air which we can hear but cannot see. The disturbance in air, which we call sound, can be sensed by measurement of some physical quantity in the air which is disturbed from its equilibrium value. The physical quantity of interest here is sound pressure which is the pressure disturbance or variation in pressure above and below ambient atmospheric pressure.

    Sound waves are generated by a vibrating body or air turbulence. The sound generating mechanism, called the sound source, acts like an oscillating force pushing on the air which causes alternating regions of compression and rarefaction to travel from the source outward through the air. This can be illustrated by considering a piston moving in a tube as shown in Fig. 4.1. The piston is attached by a crank to a flywheel so that it moves back and forth in the long tube. The piston displacement plotted as a function of time is called sinusoidal, or harmonic, motion. The action of the piston causes the air adjacent to it to undergo periodic compression and rarefaction. Due to the elastic nature of the air, the

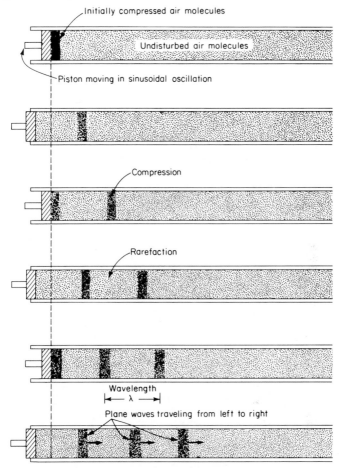

Figure 4.1   Sound waves in a tube generated by a piston moving in sinusoidal motion.

alternating regions of compression and rarefaction propagate down the tube to the right as shown. A plot of the pressure disturbances in the tube at a given instant of time is shown as the solid line in Fig. 4.2. At a time Δt later, the disturbance (or wave), indicated by the dotted line, has moved down the tube a distance Δx. The speed of propagation of the wave c is given by:

$$c = \frac{\Delta x}{\Delta t}$$

If an observer located at some position down the tube, say at X_o, were to measure the passing pressure disturbance, he could plot the pressure versus time at his location as shown in Fig. 4.3. The time between successive peaks is the period τ of the harmonic wave. The period τ represents the time it takes for the wave to travel a distance of one wavelength λ. That is,

$$\lambda = c\tau \tag{4.1}$$

Since the period τ is the reciprocal of the frequency f in cycles per second, or hertz, Eq. 4.1 may be rewritten as

$$c = \lambda f \qquad (4.2)$$

In the units used in this chapter,

$$c = \text{speed of propagation, m/s}$$
$$\lambda = \text{wavelength, m}$$
$$f = \text{frequency, Hz}$$

From the theory of acoustics, the speed of sound in air is a function only of the absolute temperature. The equations for the speed of sound in both metric and English units are:

$$c = 20.05\sqrt{T} \qquad \text{(m/s)}$$
$$c = 49.03\sqrt{R} \qquad \text{(ft/s)}$$

where T is the absolute temperature in kelvin (that is, 273.2° plus the temperature in degrees centigrade) and R is the absolute temperature in degrees Rankine (or 459.7° plus the temperature in degrees Fahrenheit).

From the preceding, we can calculate the speed of sound in air and use it to relate the frequency and wavelength of sound in air.

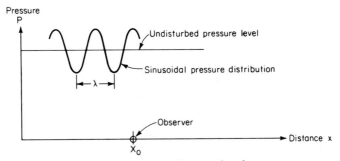

Figure 4.2   Pressure disturbance in the tube.

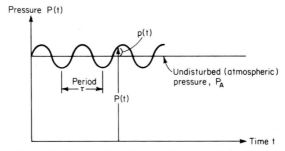

Figure 4.3   Pressure disturbance at observer plotted as function of time.

EXAMPLE 4.1   Calculate the speed of sound in air at 70°F (21.1°C) in both metric and English units.

$$T = 273.2 + 21.1 = 294.3 \text{ K}$$

and

$$R = 459.7 + 70 = 529.7°R$$

Then

$$c = 20.05\sqrt{T} = 20.05\sqrt{294.3} = 344 \text{ m/s}$$

and

$$c = 49.03\sqrt{R} = 49.03\sqrt{529.7} = 1128 \text{ ft/s}$$

EXAMPLE 4.2 Middle C on the musical scale corresponds to a frequency of 254 Hz. Calculate the wavelength in meters and feet assuming a temperature of 70°F.

$$\lambda = c/f = 344/254 = 1.35 \text{ m}$$

or
$$\lambda = 1128/254 = 4.4 \text{ ft}$$

In discussing the piston-tube experiment shown in Fig. 4.1, we must assume that the tube is either infinitely long or has a nonreflecting (anechoic) end so that no sound is reflected from the end opposite the piston. The sound waves generated by the piston then propagate down the tube, from left to right, without interference from the side walls or waves reflected from the end. These waves are called one-dimensional, plane, free-progressive waves— one-dimensional since they can be fully described in terms of distance down the tube and time, plane because the wavefronts are planar, and free-progressive because of the absence of interferences with solid objects, reflected waves, or changes in the medium.

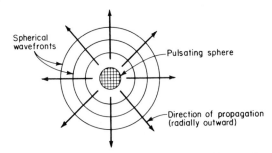

Figure 4.4 Spherical sound waves generated by pulsating sphere.

A sound source similar to a vibrating piston which we may consider is a pulsating sphere. As the surface of a sphere oscillates, the air adjacent to it is alternately compressed and rarefacted so that spherical waves are radiated outward from the sphere's surface as shown in Fig. 4.4. The wavefronts are no longer planar. Spherical waves are assumed to be generated by sound sources other than spheres, provided the dimensions of the source are small compared to the wavelength of the sound generated. Such sources are called point sources. As we shall see later, the study of such idealized sources provides a useful means of analyzing more complex noise sources.

Sound Pressure As noted in Fig. 4.3, at any point along the tube an observer can measure the pressure disturbance which is the variation of pressure relative to ambient or atmospheric pressure P_A. The amount by which the pressure varies about atmospheric pressure is called the sound pressure. We will denote this by $p(t)$ indicating that the pressure disturbance is a function of time. The time-varying pressure $P(t)$ at the observer's location is therefore given by

$$P(t) = P_A + p(t)$$

where $P(t)$ is defined as the sound pressure. Sound pressure is measured in newtons per meter squared (N/m²), called pascals (Pa) or, in English units, pounds per square inch (lb/in²). Atmospheric pressure at standard conditions is about 1.01×10^5 Pa or 14.7 lb/in².

Sound pressures are extremely small; that is, sound represents a very small disturbance relative to atmospheric pressure. For normal speech, an observer located a distance of one meter from the talker would measure a sound pressure varying approximately 0.1 Pa above and below atmospheric pressure. The sound pressure in this case would be about one-millionth of atmospheric pressure.

Thus, we see that sound can be represented by a pressure wave propagating away from a sound source with speed c. For a sinusoidal or harmonic sound wave, frequency and wavelength are related to the sound speed. The sound pressure, the time-varying magnitude of the pressure disturbance, is very small relative to atmospheric pressure.

The Environmental Noise Problem

Environmental noise can be considered most simply in terms of its basic elements: the sound source, the sound propagation path, and the receiver. The source and the propagation path can be analyzed in terms of basic physics and engineering principles, but analysis of the receiver deals with the physiology of human hearing and how noise affects health as well as the psychological mechanisms by which noise otherwise interferes with human activities. In this chapter, we assume that the receivers of interest are people. On occasion, however, noise problems involve animals as receivers or delicate instrumentation which sound pressure disturbances may upset. After discussing how sound is analyzed or quantified in the next section, we will deal with sound propagation in the following section, the effects of noise on people in the section after that, and methods used to measure environmental noise and assessment of several types of environmental noise in the subsequent sections.

ANALYSIS OF SOUND

Sound or noise may be thought of as energy that is generated by a sound source (or sources) and which propagates away from the source through a medium (air) until it reaches a receiver. The noise source may be characterized by the amount of sound energy it generates as a function of frequency (that is the amount of sound energy generated in various frequency intervals) and the times of day over which this sound is generated. The transmission path over which the sound travels from the source to receiver may be characterized in terms of the properties of the medium, usually air, and the presence of any objects in the transmission path which may interfere with the free propagation of sound such as trees, vegetation, walls, berms, or other barriers. The characteristics of the noise generated by the source and its propagation over the transmission path to the receiver will determine the noise exposure of the receiver. Methods used to describe the sound generated by noise sources are discussed in this section.

Sound Power and Sound Intensity

Sound waves propagating away from a moving body or sound source transmit energy through the elastic medium. That is, energy from the moving body exerting pressure back and forth on the medium is propagated as sound energy through the medium. The energy transmitted to the medium per unit time by the sound source is defined as the sound power emitted by the source. In the piston-tube model shown in Fig. 4.1, for example, the moving piston (source) emits sound energy which propagates from left to right down the tube at a speed c, the speed of sound in the air in the tube. The amount of energy generated and propagated per unit time is called the sound power of the piston source. The units of sound power are watts (one watt is equal to one newton-meter per second, and 746 watts is equal to one horsepower).

The amount of sound energy per unit time passing through a unit area is defined as the sound intensity. Now consider two piston-tube models like that in the previous example, but each of different diameters. Suppose that both pistons are driven so that they put out the same amount of sound power. Then obviously the sound power per unit area, or sound intensity, in the smaller tube will be greater than that in the larger tube because the same amount of energy is distributed over a smaller area. That is, the sound intensity in the smaller tube is greater. This implies that the sound pressure disturbances in the smaller tube are larger because the sound intensity is higher. This is in fact the case, that sound intensity and sound pressure are related.

If the sound source generates sound power W which is propagated through an area A, then the sound intensity I is given by

$$I = \frac{W}{A} \qquad \text{W/m}^2 \qquad (4.3)$$

At a distance r from a point source, the surface area A of the spherical wave front is $4\pi r^2$, so the sound intensity at this distance is

$$I = \frac{W}{4\pi r^2} \qquad (4.4)$$

Note that the intensity decreases inversely as the square of the distance from a point source generating a fixed sound power. This relationship is called the "inverse square law" for sound intensity.

As implied by the previous discussion, sound pressure and sound intensity are related by the equation

$$I = \frac{p^2}{\rho c} \quad \text{W/m}^2 \tag{4.5}$$

where p = root-mean-square[1] sound pressure, Pa
ρ = air density, kg/m^3
c = sound propagation speed, m/s

The quantity ρc is called the "specific acoustic impedance" of the medium in units of kg/(m$^2 \cdot$s) or Mks rayls. At standard conditions, $T = 22°C$ and $P_A = 1.013 \times 10^5$ N/m^2, ρc = 412 Mks rayls.

In the remainder of this chapter, any reference to sound pressure will mean the root-mean-square (rms) sound pressure. That is, when we say that the sound pressure is "so many pascals," we will be referring to the rms value which can be directly substituted into Eq. 4.5 to calculate sound intensity. The following examples illustrate calculations relating sound power, pressure, and intensity.

EXAMPLE 4.3 The sound power from a point source is 100 W. What is the sound intensity measured by observers at distances of (*a*) 10 m and (*b*) 100 m from the source?

Assuming spherical waves and using Eq. 4.4,

$$(a)\ I = W/4\pi r^2 = 100/4\pi(10)^2 = 7.9 \times 10^2\ \text{W/m}^2$$
$$(b)\ I = 100/4\pi(100)^2 = 7.9 \times 10^{-4}\ \text{W/m}^2$$

EXAMPLE 4.4 Calculate the rms sound pressure that would be measured by the observer for (*a*) in Example 4.3. Assume standard atmospheric conditions (that is, $\rho c = 412$ Mks rayls). From Eq. 4.5,

$$p^2 = \rho c I^2$$

or

$$p^2 = 412 \times 7.9 \times 10^2 = 32.5\ [\text{Pa}]^2$$

Then

$$p = 5.7\ \text{Pa}$$

NOTE: In this equation for p^2, the units are

$$[p^2] = \frac{\text{kg}}{\text{m}^2 \cdot \text{s}} \times \frac{\text{W}}{\text{m}^2} = \frac{\text{W} \cdot \text{kg}}{\text{m}^4 \cdot \text{s}}$$

Since

$$1\ \text{W} = 1\ \frac{\text{N} \cdot \text{m}}{\text{s}}\ \text{and}\ 1\ \text{kg} = 1\ \frac{\text{N} \cdot \text{s}^2}{\text{m}}$$

then

$$[p^2] = \frac{\dfrac{\text{N} \cdot \text{m}}{\text{s}} \times \dfrac{\text{N} \cdot \text{s}^2}{\text{m}}}{\text{m}^4 \cdot \text{s}} = \frac{\text{N}^2}{\text{m}^4} = [\text{Pa}]^2$$

where the brackets [·] stand for "units of."

Decibels and Levels

The sound pressure disturbances that we are capable of hearing are very small relative to atmospheric pressure, since normal speech communications are on the order of one-millionth of atmospheric pressure. The range of pressures from the faintest sounds we can hear to the loudest is, however, very large. For example, based upon hearing tests, the lowest audible pressure for a 1000 Hz pure tone is 2×10^{-5} Pa while the sound pressure which will cause pain is approximately 200 Pa. This represents a range of pressure by a factor of 10^7 (ten million) or a corresponding range of sound intensity by a factor of 10^{14} (one hundred trillion). The decibel scale used to describe sound power, intensity, and pressure compresses this rather large range of numbers into a more meaningful scale. We will define the decibel scale and then review the methods of analysis using decibels.

[1]The root-mean-square (rms) of a time-varying quantity is a type of average.

In terms of decibels, a quantity Q is related to a reference quantity Q_o by the equation

$$\text{decibel } (Q) = 10 \log\left(\frac{Q}{Q_o}\right) \text{ dB re } Q_o \qquad (4.6)$$

The notation "dB re Q_o" means that Q is expressed in terms of decibels relative (re) to the reference quantity Q_o. In environmental noise sound power, intensity and pressure expressed in terms of decibels are called sound power level, sound intensity level, and sound pressure level. These are defined in the following paragraphs.

Sound Power Level   Sound power level is defined as

$$L_w = 10 \log \frac{W}{W_o} \text{ dB re } W_o \qquad (4.7)$$

where W is the sound power emitted by the source and where the reference sound power is

$$W_o = 10^{-12} \qquad (4.8)$$

Sound Intensity Level   Sound intensity level is defined as

$$L_I = 10 \log \left(\frac{I}{I_o}\right) \text{ dB re } I_o \qquad (4.9)$$

where I is the sound intensity and the reference intensity is

$$I_o = 10^{-12} \text{ W/m}^2 \qquad (4.10)$$

Sound Pressure Level   Most sound measuring instruments measure the sound pressure, and the word "decibel" usually refers to sound pressure level. Decibel scales are usually associated with ratios of power-like quantities. So the sound pressure level is expressed in terms of the sound pressure squared since, from Eq. 4.5, sound pressure squared is proportional to sound intensity (power). Sound pressure level is defined as

$$L_p = 10 \log \left(\frac{p^2}{p_o^2}\right) \text{ dB re } p_o \qquad (4.11a)$$

$$= 20 \log \left(\frac{p}{p_o}\right) \text{ dB re } p_o \qquad (4.11b)$$

where the reference pressure is

$$p_o = 2 \times 10^{-5} \text{ N/m}^2 \qquad (4.12)$$

Both p and p_o are expressed as rms values. Table 4.4 lists sound pressures, the corresponding sound pressure levels, and typical sounds associated with each. Note that here and in all subsequent discussions the term "level" used with a quantity means that the quantity is expressed in terms of decibels.

Since sound meters measure sound pressure (level) to determine sound power or intensity, we must establish the relationships between these and sound pressure.

TABLE 4.4   Sound Pressure Levels and Corresponding Sound Pressures

Sound pressure, Pa	Sound pressure level, dB	Typical source
20.0	120	Jet aircraft takeoff at 100 ft
6.32	110	Same aircraft at 400 ft
0.632	90	Motorcycle at 25 ft
0.200	80	Garbage disposal
0.0632	70	City street corner
0.0200	60	Conversational speech
0.00632	50	Typical office
0.00200	40	Living room (without TV)
0.000632	30	Quiet bedroom at night

Relationships Between Sound Power Level, Sound Intensity Level, and Sound Pressure Level

Substitution of Eq. 4.5 into Eq. 4.9 gives the following for a free progressive wave

$$L_I = 10 \log \left(\frac{I}{I_o}\right) = 10 \log \left(\frac{p^2}{\rho c I_o}\right)$$

$$= 10 \log \left(\frac{p^2}{p_o^2}\right) + 10 \log \left(\frac{p_o^2}{\rho c I_o}\right)$$

For standard atmospheric conditions in air, the quantity $p_o^2/\rho c I_o$ is almost equal to unity and, hence, its logarithm is approximately zero. For most practical conditions, therefore, we may assume that

$$L_I = L_p \tag{4.13}$$

EXAMPLE 4.5 Calculate $10 \log (p_o^2/\rho c I_o)$ for standard atmospheric conditions.
From Example 4.4 we know that $\rho c = 412$ Mks rayls at standard conditions. Hence

$$10 \log \left(\frac{p_o^2}{\rho c I_o}\right) = \frac{(2 \times 10^{-5})^2 \text{ N/m}^2}{(412)(10^{-12})}$$

$$= 10 \log \left(\frac{4}{4.12}\right) = -0.13 \text{ dB}$$

Thus, Eq. 4.13 is accurate to within 0.13 dB for standard atmospheric conditions.

If the intensity is uniform over an area, then Eq. 4.3 relates sound power and sound intensity. Hence the sound power level can be related to the sound intensity level as follows

$$L_I = 10 \log \left(\frac{I}{I_o}\right) = 10 \log \left(\frac{W/A}{W_o/A_o}\right)$$

or

$$L_w = L_I - 10 \log A \tag{4.14}$$

where $A_o = 1$ m². Clearly, L_I will equal L_w only when A is equal to 1 m². If the source can be represented as a point source, the surface area A is equal to $4\pi r^2$. L_w and L_I are related only by the distance from the source, which is, of course, implied directly from Eq. 4.4.

Multiple Sound Sources

In many environmental noise control problems, we are interested in the total noise at an observer's location due to more than one source. For example, suppose that one source produces a sound pressure level L_p of 70 dB and we ask what L_p two such sources will produce, or N sources? The answer for two sources is *not* 140 dB but 73 dB because it is not correct to add decibels directly. Rather, we must add the sound intensities (or sound pressures squared) first and *then* convert to decibels. For a single sound source

$$L_{p_1} = 10 \log \left(\frac{p_1}{p_0}\right)^2$$

Taking the antilog of both sides after dividing by 10 we have

$$\frac{p_1^2}{p_o^2} = \text{antilog} \left(\frac{L_{p_1}}{10}\right) = 10^{L_{p_1}/10}$$

Similarly, for the second sound source

$$\frac{p_2^2}{p_o^2} = \text{antilog} \left(\frac{L_{p_2}}{10}\right) = 10^{L_{p_2}/10}$$

The combined sound pressure squared of these two waves is obtained by adding as follows:

$$\frac{p_{\text{total}}^2}{p_o^2} = \frac{p_1^2}{p_o^2} + \frac{p_2^2}{p_o^2} \tag{4.15}$$

This holds true as long as the two waves are *not* pure tones of the same frequency. If they are, then interference between the two waves introduces an additional term that must be included in Eq. 4.15. The sound pressure level due to the two sources combined is

$$L_{p_{\text{total}}} = 10 \log \left(\frac{p_{\text{total}}^2}{p_0^2} \right) = 10 \log \left(\frac{p_1^2}{p_0^2} + \frac{p_2^2}{p_0^2} \right) \tag{4.16}$$

If the sound pressures from the two sources are equal, then the total sound pressure level is

$$L_{p_{\text{total}}} = 10 \log \left(\frac{2p_1^2}{p_0^2} \right) = L_{p_1} + 10 \log 2$$
$$= L_{p_1} + 3 \text{ dB}$$

For N sources, the same procedure gives

$$\frac{p_{\text{total}}^2}{p_0^2} = \sum_{i=1}^{N} \left(\frac{p_i^2}{p_a^2} \right) = \sum_{i=1}^{N} (10^{L_{p_i}/10})$$

Substituted into Eq. 4.16 this gives the logarithmic sum (log. sum)

$$L_{p_{\text{total}}} = 10 \log \left(\sum_{i=1}^{N} 10^{L_{p_i}/10} \right) \tag{4.17}$$

which is the total sound pressure level produced by N separate sources. If the L_p for all the sources are equal, say to L_1, then this becomes

$$L_{p_{\text{total}}} = 10 \log (N \, 10^{L_1/10}) = L_1 + 10 \log N \tag{4.18}$$

In adding sound pressure levels two at a time, it is convenient to use the curve shown in Fig. 4.5. To add two sound pressure levels L_1 and L_2, for example, locate the difference in the sound pressure levels $(L_2 - L_1)$ on the abscissa and read the corresponding ordinate value of ΔL from the curve. The value ΔL (dB) is then added to the larger of L_1 and L_2 to obtain the combined sound pressure level. Note that if the difference between the two sound pressure levels being added is greater than 10 dB, the effect of the lower sound source is less than 0.5 dB. It can generally be neglected so that the combined L_p approximately equals that of the larger of the two sources.

Sound Spectra

In general, environmental noise is not a pure tone but a combination of a wide band of frequencies. We are often interested in knowing how sound energy is distributed as a function of frequency. There are several reasons for this. The dominant noise generation mechanisms in a sound source may often be identified by the sound frequencies they produce. The effects of air absorption and attenuation by barriers upon sound transmission paths are strong functions of the sound frequency. In addition, human hearing and overall response to noise are also strongly dependent upon the sound frequency. Thus, the sound (or noise) spectrum is often of great interest in dealing with environmental noise problems.

A sound spectrum plot is a graphical representation of the frequency characteristic of the sound. Specifically, the sound spectrum is a graph of the sound energy or intensity (usually represented as sound pressure squared) in various frequency bands versus frequency. For example, a pure tone of a single frequency would have the sound spectrum shown in Fig. 4.6a. Similarly, the spectrum for a sound consisting of several pure tones is shown in Fig. 4.6b. These are called "line spectra."

Most sounds in environmental noise contain a large number of frequencies and,

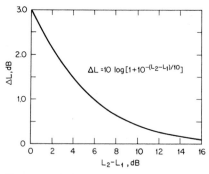

$$\Delta L = 10 \log[1 + 10^{-(L_2 - L_1)/10}]$$

Figure 4.5 Curve used to add sound pressure levels two at a time.

in fact, a continuum of frequencies. That is, these sounds contain energy at essentially all frequencies within a certain range. Examples of such sounds include the noise from an automobile, the hiss of air escaping from a valve, the splash of water from a faucet, and, in fact, human speech. An example of the spectrum for such sounds is shown in Fig. 4.6c.

In order to specify the spectrum of such a sound, we measure the acoustic energy, or the equivalent mean-squared pressure, in contiguous frequency bands and plot the results. One way to describe a continuous-spectrum noise, for example, would be to measure the sound pressure in increments of frequency 1 Hz in width along the frequency scale. The

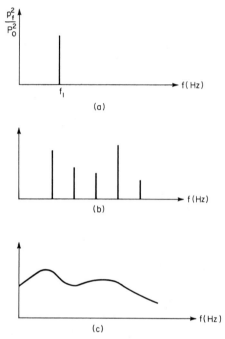

Figure 4.6   Sound spectra for (a) a single pure tone, (b) a sound consisting of several pure tones, and (c) a sound with a continuous spectrum.

spectrum so plotted would be the sound pressure (squared) per unit frequency band. Generally, however, instruments are not available which measure sound pressure in 1-Hz bands across a wide range of frequencies. Instead, measurements are made with a wider bandwidth and converted to the 1-Hz bandwidth method of presenting the data. Such spectral plots, sound pressure squared versus frequency for 1-Hz bandwidths, are called "power spectral density plots." Even if the instrumentation were available, such measurements would be very cumbersome. For example, to measure in 1-Hz bands, the sound pressure across the range of human hearing (20 to 20,000 Hz) would require almost 20,000 measurements. For this reason, and because it is usually not necessary to know the spectrum in such detail, measurements using wideband filter sets called octave filters are most often used in acoustics and in environmental noise assessment.

Octave Band Analysis

The octave bands are a contiguous set of frequency bands (or intervals) that span the audible range. Sound measuring instruments electronically filter the measured noise into these frequency bands and measure the overall sound pressure level of the filtered sound within the band. For an octave band, the upper frequency limit f_u is twice the lower frequency limit f_l of the band, that is,

$$f_u = 2f_l \tag{4.19}$$

Each octave band is referred to in terms of its center frequency f_c, which is defined as follows:

$$f_c = \sqrt{2}f_l = f_u/\sqrt{2} \qquad (4.20)$$

The set of octave bands currently used is shown in Table 4.5. Note that the center frequency for each higher octave band increases by a factor of two.

TABLE 4.5   Center and Approximate Cutoff Frequencies for Standard Set of Contiguous-octave and One-third-octave Bands Covering the Audio Frequency Range

	Frequency, Hz					
	Octave			One-third octave		
Band	Lower band limit	Center	Upper band limit	Lower band limit	Center	Upper band limit
12	11	16	22	14.1	16	17.8
13				17.8	20	22.4
14				22.4	25	28.2
15	22	31.5	44	28.2	31.5	35.5
16				35.5	40	44.7
17				44.7	50	56.2
18	44	63	88	56.2	63	70.8
19				70.8	80	89.1
20				89.1	100	112
21	88	125	177	112	125	141
22				141	160	178
23				178	200	224
24	177	250	355	224	250	282
25				282	315	355
26				355	400	447
27	355	500	710	447	500	562
28				562	630	708
29				708	800	891
30	710	1,000	1,420	891	1,000	1,122
31				1,122	1,250	1,413
32				1,413	1,600	1,778
33	1,420	2,000	2,840	1,778	2,000	2,239
34				2,239	2,500	2,818
35				2,818	3,150	3,548
36	2,840	4,000	5,680	3,548	4,000	4,467
37				4,467	5,000	5,623
38				5,623	6,300	7,079
39	5,680	8,000	11,360	7,079	8,000	8,913
40				8,913	10,000	11,220
41				11,220	12,500	14,130
42	11,360	16,000	22,720	14,130	16,000	17,780
43				17,780	20,000	22,390

The frequency is plotted logarithmically on the frequency axis for octave spectrum plots. The sound pressure measured in each band is usually plotted in terms of the octave-band sound pressure level (in dB re 2×10^{-5} Pa) as a point for each center frequency. The plotted points are usually connected by straight lines as shown in the octave band spectrum shown in Fig. 4.7.

Octave filter data can be combined to obtain the overall sound pressure level by the methods discussed earlier in the section Multiple Sound Sources. Most sound measuring instruments, however, have the means of doing this electronically.

EXAMPLE 4.6 Calculate the overall sound pressure level for the octave band data given in the following table:

Center frequency f_c, Hz	63	125	250	500	1000	2000	4000	8000
Octave band L_p, dB	79	86	89	88	89.5	78	68	59

This corresponds approximately to the spectrum for the automobile shown in Fig. 4.7.
To solve this example problem, construct the following table and combine the decibel levels by twos using Fig. 4.6:

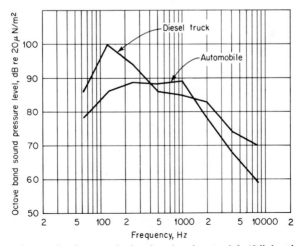

f_c, Hz	Octave band L_p, dB
63	79
125	86
250	89
500	88
1000	89.5
2000	78
4000	68
8000	59

The answer is 94.6 dB. That is, 79 dB and 80 dB added logarithmically are equal to 86.8 dB. Then, 86.8 dB added logarithmically to 89 dB gives 91.0 dB, and so on. Adding 78 dB to 94.5 dB illustrates the fact that when two levels are more than 10 dB apart, their sum is approximately 0.1 dB larger than the higher value.

One-third-Octave Spectra

In some cases, the spectrum of a noise of interest may contain details that are not brought out by the relatively wideband octave spectrum. For a more detailed analysis of sounds, finer sets are available such as $\frac{1}{3}$- and $\frac{1}{10}$-octave bands. The center frequencies of the $\frac{1}{3}$-octave bands along with their respective upper and lower cutoff frequencies are also shown in Table 4.5. The $\frac{1}{3}$-octave filter provides three times as many data points as an octave filter set over a given frequency range; hence, a finer filtering analysis of the sound results.

SOUND PROPAGATION

Sound propagation from a source to a receiver depends upon the properties of the atmosphere and the presence of any objects or barriers in the transmission path. For

Figure 4.7 Typical octave band spectra for diesel truck and automobile (full throttle acceleration at 35 mi/h at 50 ft).

example, sound is absorbed by the atmosphere and is diffracted and scattered by atmospheric inhomogeneities. Objects in or very near the transmission path reflect and diffract sound. Hence, we must consider the characteristics of the sound transmission path in order to be able to analyze the noise environment at a receiver's location.

Outdoor Sound Propagation

Sound propagation outdoors, that is outside of any building or structure, is usually a part of any environmental noise problem. Even though a receiver may be located indoors, sound transmission from an outdoor source first involves propagation through the air and then through the building structure or its openings to the receiver. Since the indoor noise for a given outdoor level can be characterized for typical structures of various types, noise standards are often specified in terms of the allowable outdoor noise levels. Then, the exterior-to-interior attenuation provided by the structure is depended upon to reduce the

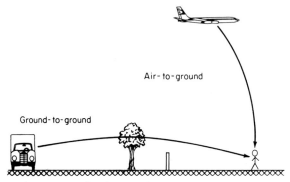

Figure 4.8   Two types of outdoor sound propagation from a source to a receiver.

outdoor level to an acceptable indoor level. We will first review outdoor sound propagation and then consider the effects of structural attenuation upon the indoor sound level.

Outdoor sound propagation is influenced by a number of factors, including the weather. Local micrometeorological conditions, for example, strongly affect sound propagation outdoors. The interactions between these factors can be quite complex, and research is currently underway on sound reduction by barriers, ground absorption, and atmospheric conditions to provide a better understanding of outdoor sound propagation. The following information is presented to enable one to estimate the sound reduction that takes place along the transmission path from an outdoor source to a receiver.

The generalized outdoor noise propagation problem may be thought of in terms of the two cases shown in Fig. 4.8, ground-to-ground propagation and air-to-ground propagation. In the first case, propagation of sound from the source to the receiver may be influenced by objects in the transmission path as well as by atmospheric effects. In the second case, sound transmission is usually only affected by atmospheric factors. This may not be the case, however, if the observer is on the back side of a hill or behind a building or other barrier relative to low-flying aircraft.

Wave Divergence

The sound pressure level generated by a noise source decreases with increasing distance from the source due to wave divergence. By considering the sound energy radiated from a symmetric spherical or point source such as that shown in Fig. 4.4, for example, we can see that the sound energy radiated by the source is distributed over larger and larger spherical surfaces as it radiates away from the source. Thus, the sound intensity (power per unit area) decreases with distance from the source. This is expressed quantitatively in Eq. 4.4. Substitution of Eq. 4.13 into Eq. 4.14 gives a relation between sound power level (which is assumed constant for a given steady source) and sound pressure level (which varies with distance from the source) given by

$$L_w = L_p + 10 \log A \qquad \text{(dB)} \qquad (4.21)$$

where for a symmetric or nondirectional source

$$A = 4\pi r^2 \tag{4.22}$$

and where r is the distance of the source from the receiver. If we substitute Eq. 4.22 into Eq. 4.21, we obtain an equation for L_p in terms of L_w and r as follows:

$$L_p = L_w - 10 \log 4\pi r^2 \tag{4.23a}$$

or

$$L_p = L_w - 20 \log r - 11 \quad \text{(dB)} \tag{4.23b}$$

where $20 \log r$ represents the decrease in sound pressure level with distance due to wave divergence and $10 \log 4$ is approximately 11 dB.

In an actual noise propagation problem such as shown in Fig. 4.8, however, there is usually an additional decrease in sound pressure level with distance from the source due to atmospheric effects or interaction with objects in the transmission path. This additional decrease in L_p, called excess attenuation, may be represented by an additional term A_e in Eq. 4.23b as follows:

$$L_p = L_w - 20 \log r - A_e - 11 \quad \text{(dB)} \tag{4.24}$$

where L_p is the sound pressure level at an observer located a distance r from a source with a sound power level L_w emitting a spherically symmetric radiation pattern.

For a sound source that can be approximated as a point source (a source that is small compared to the wavelength of the radiated sound) located above a flat rigid surface, the radiation pattern is approximately hemispherical and the area A is therefore approximately equal to $2\pi r^2$. The equation for sound pressure level corresponding to Eq. 4.24 is

$$L_p = L_w - 20 \log r - A_e - 8 \quad \text{(dB)} \tag{4.25}$$

where $8 \cong 10 \log 2\pi$.

Often, the sound power of a source is not known, but the sound pressure level L_{p_1} at a distance r_1 from the source is known. The sound pressure level L_{p_2} at a distance r_2 from the source can then be calculated from the equation

$$L_{p_2} = L_{p_1} - 20 \log \frac{r_2}{r_1} - A_{e_{1,2}} \quad \text{(dB)} \tag{4.26}$$

where $A_{e_{1,2}}$ is the excess attenuation along the path $r_2 - r_1$ between observers 1 and 2. In environmental noise assessment, Eq. 4.26 is of more general use than Eqs. 4.24 or 4.25 since the sound power of a source is seldom known. Rather, the sound pressure level of a source is usually given at a reference distance and Eq. 4.26 is then used to calculate sound pressure levels at other distances from the source.

All three of these equations indicate that for a spherical or hemispherical sound wave propagating through a homogeneous loss-free atmosphere $(A_e = 0)$, the sound pressure level decreases by 6 dB for each doubling of the distance from the source. The following examples illustrate the use of the above equations when excess attenuation is zero, that is, where the decrease in sound pressure level with distance from the source is due to wave divergence only.

EXAMPLE 4.7 A small airplane radiates noise with a sound power of 100 W. Assuming spherical radiation and no excess attenuation, what will be the peak sound pressure level when the airplane flies directly over an observer at an altitude of 300 m?

From Eqs. 4.24 and 4.25,

$$\begin{aligned}
L_p &= L_w - 20 \log r - 11 \\
&= 10 \log W/W_o - 20 \log r - 11 \text{ dB} \\
&= 10 \log 100/10^{-12} - 20 \log 300 - 11 \\
&= 140 - 49.5 - 11 \\
&= 79.5 \text{ dB}
\end{aligned}$$

Or using the procedure illustrated in Example 4.3:

$$I = W/4\pi r^2 = 100/4\pi(300)^2 = 8.84 \times 10^{-5} \text{ W/m}^2$$

and
$$L_p = L_I = 10 \log \frac{8.84 \times 10^{-5}}{10^{-12}}$$
$$= 10 \log (8.84 \times 10^7)$$
$$= 79.5 \text{ dB}$$

EXAMPLE 4.8 A truck produces a sound pressure level L_{p_1} of 85 dB at a distance of $r_1 = 15$ m. Assuming no excess attenuation, calculate the sound pressure level at a distance $r_2 = 100$ m from the truck.

Use Eq. 4.26 to calculate the sound pressure level L_{p_2} at the distance r_2.

$$L_{p_2} = L_{p_1} - 20 \log r_2/r_1$$
$$= 85 - 20 \log (100/15)$$
$$= 68.5 \text{ dB}$$

Excess Attenuation

Excess attenuation is the additional reduction in sound pressure level with increasing distance from a source beyond that caused by wave divergence. Excess attenuation may include one or more of the following:

A_{e_1} = attenuation by air absorption, dB

A_{e_2} = attenuation by rain, snow, sleet, or fog, dB

A_{e_3} = attenuation by barriers, dB

A_{e_4} = attenuation by vegetation, dB

A_{e_5} = attenuation due to atmosphere inhomogeneities and atmospheric turbulence

We will discuss each of these factors in enough detail so that its importance in an environmental noise assessment can be determined. If a more detailed analysis is required, additional information can be found in the references.

Attenuation by Air Absorption As sound propagates through air, a small part of the sound energy is extracted by the air and converted into heat. The amount of sound energy

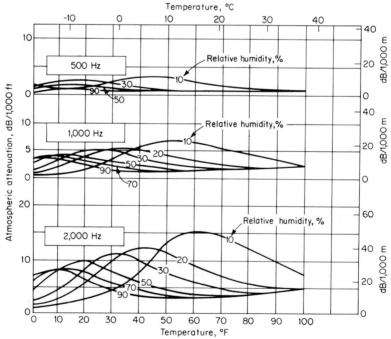

Figure 4.9 Atmospheric excess attenuation for aircraft-to-ground sound propagation for octave bands with center frequencies of 500, 1,000, 2,000, 4,000, and 8,000 Hz.

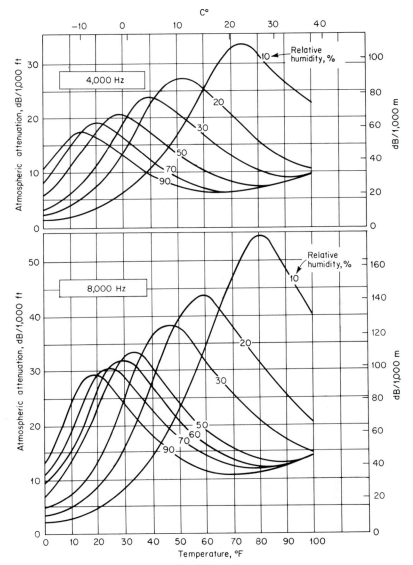

Fig. 4.9 (cont.)

absorbed depends upon frequency, temperature, and humidity. A number of tables and curves of atmospheric sound attenuation are published in the references listed at the end of this chapter. See, for example, Beranek (1971) and Crocker and Price (1975). Samples of these curves for aircraft-to-ground propagation are shown in Fig. 4.9.

Atmospheric attenuation at a temperature of 20°C (68°F) may also be calculated from the equation

$$A_{e_1} = 7.4 \frac{f^2 r}{\phi} 10^{-8} \quad \text{(dB)} \tag{4.27}$$

where f = frequency, Hz, or octave or ⅓-octave-band center frequency
  r = distance between source and receiver, m
  ϕ = relative humidity, %

This equation and the curves shown in Fig. 4.9 can be used for estimates of atmospheric absorption. More accurate data can be obtained from the bibliographic literature, such as Harris (1966). In using Eq. 4.27, note that ϕ is relative humidity in percent and not the fractional humidity.

The excess attenuation due to atmospheric absorption previously discussed is for still, homogeneous air. Atmospheric turbulence and wind, temperature, and humidity gradients can cause considerable deviation from these values of absorption. The mechanisms for these latter effects, due to scattering and diffraction of sound waves, are different from the conversion of sound to heat in the absorption mechanisms already mentioned. Attenuation due to atmospheric scattering and diffraction may be quite significant, especially for propagation distances greater than several thousand feet.

While atmospheric absorption is certainly significant for propagation, in distances greater than several hundred feet, especially for the higher frequencies, absorption is rarely counted upon to reduce noise levels in environmental noise problems. This is because of the high sensitivity of absorption to temperature and humidity which, of course, cannot be predicted except on an average basis. For example, even in areas where low humidity and high temperatures often combine to foster greater absorption, especially of higher frequencies, these conditions usually last only during certain parts of the day and during limited seasons. Thus, atmospheric attenuation due to absorption of sound cannot generally be included as an effective noise mitigation factor.

EXAMPLE 4.9 A sound of frequency 1,000 Hz propagates over a distance of 1,000 ft (305 m) when the temperature is 68°F (20°C) and relative humidity is 10 percent. Calculate the excess attenuation due to atmospheric absorption.
From Eq. 4.27,

$$A_e = \frac{7.4f^2r}{\phi} \times 10^{-8} \text{ dB}$$
$$= \frac{7.4(10^3)^2\, 305}{10} \times 10^{-8}$$
$$= 2.2 \text{ dB}$$

NOTE: This is not the same value of attenuation shown in Fig. 4.9 which is for octave bands versus single frequencies.

Attenuation by Rain, Fog, or Snow Atmospheric precipitation such as rain, snow, fog, or suspended dust particles causes only slight attenuation of sound compared to atmospheric absorption. In environmental noise assessment, we neglect excess attenuation due to this effect because it is so small relative to other effects.

Attenuation by Barriers Walls or barriers located in the transmission path between a noise source and receiver can provide significant noise reduction. Such barriers are often used to protect noise sources.

Typical geometry for a noise barrier is shown in Fig. 4.10. The barrier could consist of a solid wall, an earth berm, or other solid, nonporous object interrupting the direct path or line of sight between the source and receiver. For purposes of analysis, the barrier is considered to be infinitely long and perpendicular to the line of sight. Two types of idealized sources are considered: a point source and an infinitely long incoherent line source located parallel to the barrier. An example of the first type would be a single vehicle on a roadway, and an example of the second type would be dense traffic on a roadway.

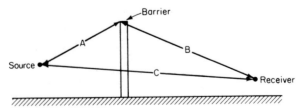

Figure 4.10 Noise Barrier Geometry. The path C is called the line of sight between the source and receiver.

The noise reduction for these idealized cases can be determined by first calculating the path length difference δ defined by the equation

$$\delta = A + B - C \quad \text{(m)} \qquad (4.28)$$

and then determining the dimensionless Fresnel number N given by

$$N = \frac{\delta}{\lambda} = \frac{A + B - C}{\lambda} \qquad (4.29)$$

where λ = the wavelength of the sound. Note that we assume that the barrier interrupts, or protrudes above, the line of sight. Having N, the sound attenuation A_b caused by the barrier can be obtained from the curves shown in Fig. 4.11 for a point source or a line source. From experiment, an attenuation of 24 dB is found to be a practical upper limit for barrier attenuation, and for barriers barely intersecting the line of sight, attenuation of 5 to 7 dB is obtained.

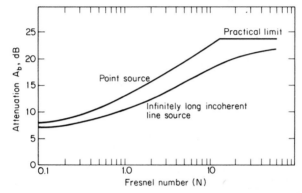

Figure 4.11   Sound attenuation of an infinite barrier for a point source and line source as a function of Fresnel number.

This procedure for calculating barrier attenuation can be used to approximate the excess attenuation for walls, berms, and the like. A more detailed discussion of barrier attenuation for traffic noise is given in a later section on Transportation Noise. Deviation from the predicted attenuation values just outlined can be caused by atmospheric effects, ground absorption, and differences between the actual situation and the idealized geometry.

Buildings located between a noise source and a receiver can also provide a shielding or barrier effect. Measurements of the propagation of traffic noise into residential areas indicate that if the highway is visually blocked by buildings, excess attenuation of 15 to 20 dB can be obtained.

EXAMPLE 4.10   For the geometry which follows, calculate the barrier attenuation for the automobile using Eqs. 4.27 and 4.28 and Fig. 4.11. Assume that the auto has a noise spectrum shown in Fig. 4.7 and the speed of sound c is 340 m/s.

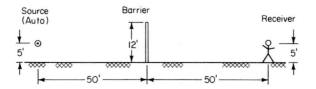

To solve this problem, construct the following table:

A	B	C	D	E	F	G
$(f_c)_{\Delta B}$ (Hz)	$(L_p)_{\Delta B}$ at 50' (dB)	$(L_p)_{\Delta B}$ at 100' (dB)	λ (m)	N	A_{e_b} (dB)	L_p (dB)
63	78	72	5.44	0.11	8	64
125	86	80	2.72	0.22	9	71
250	88	82	1.36	0.44	10	72
500	88	82	0.68	0.87	13	69
1000	89	83	0.34	1.75	15	68
2000	78	72	0.17	3.5	18	54
4000	68	62	0.085	7.0	21	41
8000	58	52	0.042	14.0	24	28
Log. sum	94.1	88.1			27.1	76.6

Column B shows the octave-band sound pressure levels at a 50-ft distance taken from Fig. 4.7 for the automobile. Column C shows the octave-band sound pressure levels that would exist at 100 ft from the automobile with no barrier. Column D shows the wavelength of sound at the octave-band center frequencies shown in column A, calculated from the equation

$$\lambda = c/f = 340/f \quad \text{(m)}$$

Column E is the Fresnel number, and column F is the excess barrier attenuation taken from Fig. 4.11. Column G, equal to column C minus F, is then the octave band sound-pressure level produced at the receiver by the automobile with the barrier present. The logarithmic sum of column C is the overall sound pressure level at the receiver without the barrier, and the logarithmic sum of column G is that with the barrier, the difference being the overall barrier attenuation of 88.1 − 76.6 = 11.5 dB. Note that this is *not* the same as the logarithmic sum of the barrier attenuations for each octave band (column F). Care must be exercised in calculations of this type. To avoid mistakes, it is best to calculate the sound pressure levels at the observer both with and without excess attenuation in each octave (or ⅓ octave) band and then combine the band levels as previously illustrated. The overall attenuation by the barrier or other excess attenuation factor is then just the difference between these two.

Attenuation by Grass, Shrubs, or Trees   Attenuation by vegetation is highly variable and depends upon the nature of the ground surface, the type and structure of vegetation, and the heights of the source and receiver above ground. Excess attenuation due to shrubbery or grass can be roughly estimated from the empirical formula

$$A_{e_s} = (0.18 \log f - 0.31)r \quad \text{(dB)} \tag{4.30}$$

where f = frequency of sound, Hz
r = path length through shrubbery or over grass, m
Attenuation by forests can be approximated by use of the equation

$$A_{e_f} = 0.01 f^{1/3} r \quad \text{(dB)} \tag{4.31}$$

This is for an average U.S. forest. Depending upon the type of trees and, for deciduous forests, whether the trees are bare, values of excess attenuation significantly differ from that given by Eq. 4.31 may be obtained. Even though excess attenuation by vegetation is highly variable, the above equations can still be used to obtain approximate expected values of attenuation to determine whether such air effect is worthy of more detailed analysis.

Atmospheric Inhomogeneities   Sound waves are refracted by wind and temperature gradients in the atmosphere and are scattered by turbulence. The effects due to wind and temperature gradients can be quite large, 10 to 30 dB or more. However, since constant atmospheric conditions cannot be counted upon to provide protection of a noise-sensitive area from a noise source, the excess attenuation due to these is usually assumed to be zero in environmental noise assessments.

Outdoor-to-Indoor Sound Transmission

The preceding discussion is directed toward determination of the outdoor noise levels for a given noise source–transmission path configuration. In some instances, we may be interested in the indoor noise level that will occur for a given outdoor noise level. Measurements indicate that for a typical southern California wood-frame house the outdoor-to-indoor attenuation is approximately that given in Table 4.6. The values of

TABLE 4.6   Approximate Outdoor-to-Indoor Attenuation (Pre-energy insulation standards) for Typical Southern California Wood-Frame Houses

	Windows closed	Windows partially open
Highway noise	20 dBA	15 dBA
Aircraft noise	28 dBA	20 dBA

attenuation shown are actually based upon A-weighted decibels (dBA) which will be discussed in a later section. The actual outdoor-to-indoor attenuation of a given building is a strong function of the sound frequency and details of construction. If it is necessary to determine accurately the outdoor-to-indoor attenuation, it usually can only be done by direct measurement.

THE EFFECTS OF NOISE ON PEOPLE

If sufficiently loud, noise may adversely affect people in a number of ways. For example, noise may interfere with human activities such as sleep, speech communication, and tasks requiring concentration or coordination. It may also cause annoyance, hearing damage, and other physiological problems. While it is possible to study these effects upon people on an average or statistical basis, it must be remembered that all the stated effects of noise upon people vary greatly with the individual. The effects discussed in this section, therefore, apply to people on a statistical basis only, and the effects upon a given individual may vary significantly in some cases from those indicated.

The effects of noise upon people depend upon the physical characteristics of the sound being classified as noise. Generally, the effects of noise upon people depend, of course, upon the loudness of the sound which is related to sound pressure level and frequency content of the sound. Other factors, such as duration of the noise, time of occurrence during the day, the presence of pure tones, and changes of sound pressure level with time, influence the degree to which noise affects people. These effects will be discussed in this section, and several noise scales and rating methods used to quantify the effects of noise upon people will be discussed in detail in a later section of this chapter devoted to that subject.

Human Hearing

Hearing is certainly one of the more vital human senses. Most human communication is by speech and hearing. Try, for example, to watch television without the sound; in most cases a much better understanding of the program can be gained by listening to the sound without the picture. Not only is hearing very important for human development, but the human ear is a remarkably sensitive and versatile instrument. For example, the dynamic range of the ear (the range between the weakest sound that can be heard and the loudest sound before experiencing pain) is about 140 dB; most good sound measuring instruments have a dynamic range of no more than about 50 dB. Also, the audible frequencies for a very healthy human ear range from approximately 20 to 20,000 Hz.

A cross section of the human ear is shown in Fig. 4.12, and Fig. 4.13 is a schematic diagram of the ear in which the cochlea, a spiral-shaped organ, is shown straightened out for convenience. The functions of the parts of the ear can be described in terms of the three main components: (1) the outer ear, consisting of the pinna, auditory canal, and eardrum or tympanic membrane; (2) the middle ear, which is an air-filled cavity containing three small bones called the ossicles that transmit vibrations from the eardrum to the oval window; and (3) the inner ear, consisting of the semicircular canals, which provide our sense of balance, and the cochlea which contains the organ of Corti, the sensory organ for hearing.

The outer ear serves to gather sound and transmit it to the eardrum where the sound pressure disturbances in the air are converted to vibrations of the eardrum. The tissue and bony structure surrounding the outer ear also serve to protect the sensitive parts of the middle and inner ears from mechanical damage. Vibrations of the eardrum are transmitted to the inner ear by the ossicles, which are the three small bones in the middle ear called the malleus (hammer), incus (anvil), and stapes (stirrup). These small bones provide a large mechanical advantage between the pressure vibrations in the air and the vibrations transmitted to the liquid-filled cochlea and organ of Corti in the inner ear. Also, small muscles connected to the ossicles help protect the inner ear from intense sounds.

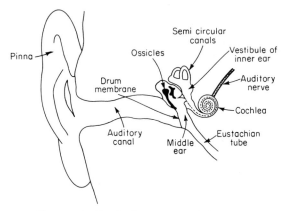

Figure 4.12   Sketch of cross section of a human ear.

Vibrations in the organ of Corti are converted into nerve impulses and are transmitted to the brain along the auditory nerve.

The functions of the outer, middle, and inner ears are strongly frequency dependent. For example, the auditory canal provides a filtering action that depends upon the sound frequency, and the inertial properties of the ossicles are frequency dependent. For these reasons, hearing and the resulting perception of loudness of sounds are frequency dependent.

Loudness of Sounds

Researchers have established the relative loudness of sounds as a function of frequency by having test subjects alternately listen to a pure reference tone at 1,000 Hz and a second tone at some other frequency. The person being tested is asked to adjust the second tone

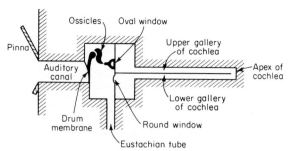

Figure 4.13   Schematic diagram of human ear with cochlea straightened.

until the two seem equally loud. Curves of equal loudness are drawn for sound pressure levels in 10-dB increments at a reference frequency (1,000 Hz) as shown in Fig. 4.14. The minimum audible field (MAF) is just the minimum audible sound pressure level plotted as a function of frequency. The loudness levels associated with these curves are called "phons." By examination of these curves, we see that the ear is most sensitive to sounds with frequencies of around 4,000 Hz, and sensitivity diminishes at lower and higher frequencies. Note, for example, that a 40-dB tone at 1,000 Hz appears equally as loud as a 51-dB tone at 100 Hz and a 32-dB tone at 4,000 Hz.

Other experiments have been performed to determine the increase in sound pressure level necessary for a pure tone to double the apparent loudness. The result is that for pure tones at 1,000 Hz an increase in sound pressure level of about 10 dB represents an

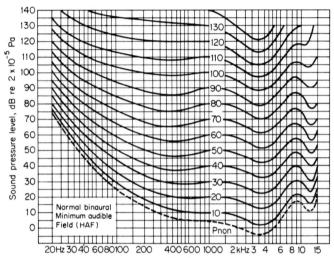

Figure 4.14 Equal loudness contours.

apparent doubling of loudness. At other frequencies, an increase of about 10 dB also approximately corresponds to a doubling of loudness of sounds.

A-weighted Sound Pressure Level

Since loudness of sounds is important to the effects of noise on people, the dependence of loudness upon frequency must be taken into account in environmental noise assessments. Several methods have been developed by researchers using the frequency spectrum of the sound and the curves shown in Fig. 4.14 to arrive at a loudness index for the given sound. These methods are more complicated and time-consuming than required for most situations. Therefore, simplified techniques have been developed to account for the dependence of perceived loudness upon frequency. This is done by the use of weighting filters in noise measuring instruments which give a direct reading of approximate loudness.

The most common weighting filters are called the A, B, and C frequency weightings shown in Fig. 4.15. These curves are smoothed approximations of the inverse of the 40, 70, and 100 phon equal-loudness curves, respectively. The A-weighting is most commonly used for environmental noise, and measurements of sound pressure level made using the A-weighting filter are reported in units of dBA (A-weighted decibels). Sometimes these units are reported as dB(A) or dB-A, but the dBA notation is preferred.

It should be noted that although the dBA scale is not as accurate as the more complex loudness-rating procedures, it is adequate for most noise assessments. Spectral data are still needed in many noise studies, however, for determining excess attenuation, transmission properties through structures, and source radiation mechanisms.

The A-weighting factors for octave band sound-pressure levels corresponding to the curve in Fig. 4.14 are given in Table 4.7.

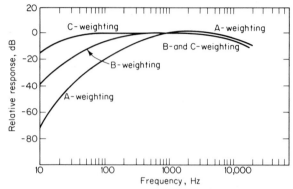

Figure 4.15   Relative responses of the A-, B-, and C-weighting networks.

TABLE 4.7   A-weighting Correction Factors for Octave Band Sound Pressure Levels

Octave band center frequency, Hz	31.5	63	125	250	500	1000	2000	4000	8000
A-weighting correction, dB	−42	−28	−18	−9	−3	0	+1.5	+0.5	+2.0

EXAMPLE 4.11   Calculate the A-weighted sound pressure level at 50 ft from the automobile whose spectrum is shown in Fig. 4.7.
Construct the following table:

Octave band center frequency, Hz	Octave band levels at 50 ft, dB	A-weighting correction, dB	A-weighted octave band levels,* dB
63	78	−28	50
125	86	−18	68
250	88	− 9	79
500	88	− 3	85
1000	89	0	89
2000	78	+ 1.5	79.5
4000	68	+ 0.5	68.5
8000	59	+ 2.0	61

*Add the right-hand column logarithmically to obtain 91.1 dBA.

Hearing Loss

It is clearly established that continued exposure to loud noise causes hearing loss which can lead to hearing handicap and deafness. In most cases, hearing loss is caused by a gradual shifting of the threshold of hearing due to repeated exposure to noise. For instance, a factory worker who is exposed to high noise levels over a period of years might experience some hearing loss. It has been demonstrated that the length of exposure is as important as the magnitude of the noise level in producing hearing loss.

Hearing loss due to noise exposure is caused by disruption and damage to the nerve cells in the organ of Corti. Damage to these cells caused by continued noise exposure is permanent, and regeneration or surgical repair is not presently possible. Therefore, noise-induced hearing loss is a serious handicap and is one of the more significant effects of noise upon people.

Various criteria have been established to prevent hearing loss. The Occupational Safety and Health Administration (OSHA) of the Department of Labor has established noise standards to protect industrial workers from hearing loss. Table 4.8 shows the maximum permissible noise exposure times for industrial workers. The Environmental Protection Agency has recommended that 85 dBA be the level not to be exceeded when an individual is exposed to noise for an 8-hour work day in the working environment.

The EPA has also recommended an equal energy criterion which would lower the exposure time by a factor of two for each increase of 3 dBA in noise level.

The noise exposures (level and duration) indicated in Table 4.8 are rarely encountered in nonoccupational noise situations. However, when noise levels in the environment exceed about 85 dBA, the potential for hearing damage exists and a careful analysis is required.

The EPA has adopted hearing-conservation criteria for noise exposure of the general public as opposed to the occupational exposure criteria published by OSHA. The data on hearing loss of industrial workers was adjusted to account for the intermittent nature of environmental noise: a yearly dose of 365 days versus 250 working days, and continuous versus 8 hours per day exposure. The result was a recommended maximum of 70 dBA average equivalent sound level to protect the public from hearing damage by environmental noise pollution. The yearly average equivalent sound level of 70 dBA may

TABLE 4.8   Permissible Noise Exposure Times for Occupational Noise Levels

Maximum duration per day, h	Sound level, dBA
8	90
6	92
4	95
2	100
1	105
½	110
¼ (or less)	115

therefore be used as one of the criteria for environmental noise exposure. The technical definition of and methods of calculating equivalent sound levels will be discussed in the section on Noise Scales and Rating Schemes.

Speech Interference

As indicated earlier, speech is one of the most important and essential forms of human communication. Background noise in the environment may interrupt speech communication whether it be person-to-person conversation, classroom communication, or listening to television or radio. The interference of unwanted sounds, or noise, with verbal communications is called masking. The degree of masking of speech by noise depends upon the characteristics of the speaker, the complexity level or familiarity of the material being spoken, the loudness of the speech or vocal effort, the spatial locations of the noise source, speaker, and listener, and the acuity of the listener's hearing. The overall degree of masking is usually measured in terms of the percentage of spoken messages or phrases understood by the listener.

Based upon laboratory measurements, several methods of rating speech interference by background noise have been developed. These include the articulation index (AI), speech interference level (SIL), and preferred speech interference level (PSIL). These methods all involve analysis of the background spectrum to rate the level of speech interference. For example, the simpler PSIL method uses the average of octave band sound pressure levels with center frequencies of 500, 1,000, and 2,000 Hz to enter a table of distance between talker and listener versus talker's voice effort. This determines the steady noise level at which reliable speech communication is barely possible. A simpler method, based upon conversion of the laboratory speech interference data to dBA, has been presented in the graph of Fig. 4.16 by the EPA which can be used to gauge speech interference. It should be noted that the dBA and other indices are not accurate measures of speech interference by noise containing intense low-frequency components. It has been shown that intense low-frequency noise can mask speech completely. For example, a sound pressure level of 115 dB at 50 Hz will provide a 10- to 30-dB masking effect through 3,000 Hz.

Figure 4.16 should only be used for the outdoor environment. For example, with a steady outdoor noise level of 70 dBA, relaxed conversation is virtually impossible and, even with a raised voice, satisfactory conversation is not possible for communicating distances greater than about 1.5 m.

Indoor speech interference criteria have also been developed from octave-band noise data. These can also be converted to criteria in terms of dBA and are presented in Table 4.9. The dBA levels shown are those at which speech interference begins to occur thereby impairing use of the room for the stated purpose.

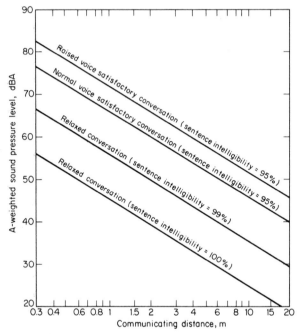

Figure 4.16 Maximum distances outdoors at which conversation is considered to be satisfactorily intelligible over steady noise.

Speech interference is often used as one of the criteria for the acceptability of environmental noise. However, since noise in many communities is fluctuating, the problem of establishing acceptability criteria in terms of the degree of speech interference is quite complex and imprecise. The noise rating schemes discussed in the next section provide methods of averaging fluctuating noise exposure in the environment. The criteria presented for levels of acceptability are related to speech interference for general outdoor and indoor environments. For example, the EPA has concluded that to avoid outdoor activity (speech) interference in areas such as schoolyards and playgrounds, the daily equivalent noise level should not exceed 55 dBA. This is the level that would be obtained if the total daily average sound intensity is expressed in terms of sound intensity level (which is approximately equal to sound pressure level).

Interference with Sleep

Everyone is familiar with the fact that noise can cause awakening from sleep or keep one from falling asleep. This is clearly one area in which noise affects people. However, in addition to this, noise can also affect the quality of sleep. This is because there are several stages of sleep ranging from fully awake to deep sleep, and noise can cause a shifting from one stage to another. There is considerable research effort devoted to developing an understanding of the nature of sleep, its beneficial effects, and the relationships between the various stages of sleep. From this research, it has been established that noise can adversely affect sleep patterns without fully awakening the subject.

Sleep interference is highly individual and depends upon a number of factors including the motivation to wake, loudness and duration of the noise, fluctuation of the noise, sleep deprivation, differences between men and women, differences between age groups, meaning and familiarity of the noise, and a whole host of other factors. While data on sleep disturbance are dependent upon the aforementioned factors, several general conclusions

TABLE 4.9   Speech Interference Criteria for Indoor Rooms for Various Uses

Type or use of space	Approximate A-weighted sound level, dBA
Concert halls, opera houses, recital halls	21 to 30
Large auditoriums, large drama theaters, churches (for excellent listening conditions)	Not above 30
Broadcast, television, and recording studios	Not above 34
Small auditoriums, small theaters, small churches, music rehearsal rooms, large meeting and conference rooms (for good listening)	Not above 42
Bedrooms, sleeping quarters, hospitals, residences, apartments, hotels, motels (for sleeping, resting, relaxing)	34 to 47
Private or semiprivate offices, small conference rooms, classrooms, libraries, etc. (for good listening conditions)	38 to 47
Living rooms and similar spaces in dwellings (for conversing or listening to radio and television)	38 to 47
Large offices, reception areas, retail shops and stores, cafeterias, restaurants, etc. (moderately good listening)	42 to 52
Lobbies, laboratory work spaces, drafting and engineering rooms, general secretarial areas (for fair listening conditions)	47 to 56
Light maintenance shops, office and computer equipment rooms, kitchens, laundries (moderately fair listening conditions)	52 to 61
Shops, garages, power-plant control rooms, etc. (for just-acceptable speech and telephone communication)	56 to 66

SOURCE: L. L. Beranek, W. E. Blazier, and J. J. Figwer, "Preferred Noise Criteria (PNC) Curves and their Application to Rooms," *J. Acoust. Soc. Am.*, **50**, 1223–1228 (1971).

can be drawn. Awakening from light stages of sleep can be caused by noise levels as low as 30 to 40 dBA. Overall awakening by a single noise occurs about 10 percent of the time for a 40 dBA noise increasing to about 90 percent for an 80 dBA noise. The choice of a noise criterion for avoiding sleep interference is at best a rough approximation involving still unresolved questions relating to long-term health effects of sleep loss. The EPA has tentatively concluded, however, that a daily average (equivalent) level of 45 dBA should be sufficient to protect the majority of the population from the adverse health effects of sleep loss due to noise.

Other Effects

People's response to noise, that is, the degree of disapproval of sound, depends upon several physical factors. The overall degree of disapproval or annoyance caused by a sound is termed the "noisiness." The judgment of noisiness of a sound can be predicted in terms of physical measurements of the following factors:

- Intensity and frequency content—noisiness increases with sound intensity as does loudness, and sounds with frequencies in the 2 to 8 kHz range are judged noisier than sounds with equal sound pressure level outside this interval

- Spectrum complexity—complex spectra containing pure tones are generally judged noisier than similar noncomplex sounds
- Duration—noisiness increases approximately with the logarithm of the duration of a sound
- Duration of period of rising sound level—sounds that are increasing in level are judged noisier than sounds of decreasing level with time
- Impulsive noise—sudden noises are judged to be very noisy

These factors are combined in a scale called perceived noisiness levels (PNL) expressed in units of PNdB. While the PNL scale may more accurately correlate with people's judgments of annoyance produced by sounds than, say, the dBA scale, the use of the PNL scale is complex and too unwieldy for most environmental noise assessments. However, in some situations where rapidly changing noise levels or sounds with large

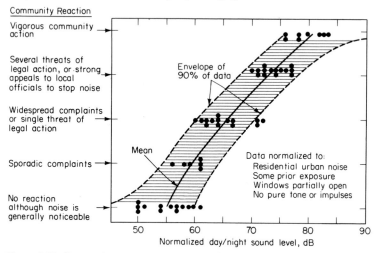

Figure 4.17 Community reaction to environmental noise function of normalized L_{dn}.

pure tone energy content are involved, the PNL scale may be a useful noise assessment tool.

Noise may also interfere with the performance of tasks by people. It is difficult to assess quantitatively the effects of noise upon task performance. Complex tasks and those requiring a high degree of concentration are most adversely affected by noise, and irregular bursts of noise seem to be more disruptive than steady noises. Also, high-frequency noise also appears to cause more disruption than low-frequency noise (below 2 kHz).

There is some evidence that noise causes other adverse psychological and physiological effects upon people. For example, there is speculation that noise causes circulatory problems, heart disease, and mental illness. There is no conclusive scientific evidence, however, that noise alone causes these effects. Rather, noise is one of many causes of stress in people's daily lives, and while the elimination of noise may reduce the incidence of such problems in the population as a whole, the efforts and costs required for such noise reduction must be weighed against the control of other known causes of these diseases.

Community Reaction

We have discussed how noise affects people in terms of the reaction of individuals to specific effects such as interference with speech, sleep, and other activities. It is also possible to characterize the effects of noise on people by studying the aggregate response of people in communities to various noise environments. This subject has received considerable study in the past, and the results show a significant correlation with the interference with human activities by noise levels.

One such result is illustrated in Fig. 4.17. The normalized day-night noise level, L_{dn} is just the day-night noise level corrected for seasonal differences, outdoor residual noise level, and previous community exposure and attitudes. The L_{dn} rating is discussed in detail in a later section of this chapter. With these corrections, the (normalized) L_{dn} correlates closely with overall community response or reaction. A normalized outdoor L_{dn} of about 50 to 60 dB with a mean of 55 dB would result in no community reaction. The data shown in Fig. 4.17 suggest that widespread complaints may be expected when the normalized L_{dn} reaches about 65 dB. (For a detailed description of the normalization procedure, see "Community Noise," published by the EPA in December of 1971.) These results are useful in establishing criteria for noise in residential areas.

NOISE SCALES AND RATING METHODS

As we have previously discussed, the reaction of people to noise depends upon a large number of factors including loudness, duration, and frequency content of the sound. Also, a noise problem, said to exist when noise interferes with human activities, certainly depends upon the activity being interfered with as well as the physical properties of the intruding noise. These together with the fact that most community noise environments are highly variable and are caused by a large number of different types of sources indicate the complexity of carrying out a comprehensive community noise assessment. Because of this complexity, noise scales and rating methods have been developed which account for the more significant of these factors. Also, criteria of acceptability in terms of these scales and ratings have been adopted based upon the degree of annoyance or interference with human activities. These scales or rating methods greatly simplify community noise assessment, provide for general classification of the noise environment, serve as a means of evaluating noise control measures, and are a basis for laws and regulations for noise control.

We will distinguish between noise scales and rating methods or procedures. A "noise scale" describes only the characteristics of the noise itself such as the simple A-weighted sound pressure level which can be read directly at any instant from a sound level meter. More complex noise scales may take into account frequency or duration of the sound as well as factors that relate the scale to some subjective response such as loudness, annoyance, speech interference, or other perceived effect. In other words, a noise scale is constructed from the fundamental physical properties of the sound and, in itself, is independent of the characteristics of the observer or the effects the sound may have upon the person hearing the sound.

On the other hand, a noise rating method or procedure (which we will call a "noise rating") attempts to account for the context in which the noise is heard and usually includes corrections for not only the physical properties of the sound but also the situation into which the noise intrudes. That is, the rating procedure may include corrections which consider whether the noise occurs in the daytime or night, the overall noise exposure, and the type of activities (land use) that occur in the area exposed to the noise. As opposed to a noise scale, therefore, which is derived solely from the physical properties of the sound itself, a noise rating includes factors that relate to the context or environment in which the sound is heard. For example, in environmental noise studies we are most often concerned with noise exposure in residential and recreational areas and less often in commercial, industrial, or other land use areas. This is, of course, because residential areas are more noise sensitive than most other areas.

Community Noise

Most environmental noise assessments involve community noise usually in an urban or suburban setting. Figure 4.18 shows an example of community noise measured in a suburban neighborhood near a street. The ambient sound pressure level is characterized by significant variations above a base or residual noise level caused by auto traffic, an aircraft overflight, and sources that are not as loud as these such as doors slamming, people talking, and others common to residential areas. The definitions of several terms used to describe such a noise record are as follows. The ambient sound pressure level is defined as the total noise level at a given receiver location due to all sources. (The ambient level is the noise plotted in Fig. 4.18.) The residual noise level is that level below which the ambient noise does not seem to drop during a given time interval and is due to more distant, generally unidentified sources. The residual noise level in Fig. 4.18, for example, is about 44 dBA. Residual noise may be due to traffic on more distant streets, wind blowing through trees, or other sources which, when combined, provide a relatively steady residual or floor noise level. The term "background noise" is sometimes confused with the residual noise level; background noise is the noise level without a particular noise source in operation and is not the same as the residual noise level.

If noise were recorded at night at the same location in a suburban area, the residual noise level would usually be 10 to 20 dB lower due to less traffic on nearby streets and less general activity. The peaks due to automobiles, trucks, or aircraft flyovers, however, generally have the same maximum noise levels but are fewer in number. Usually, night-time noise is characterized by a lower residual noise level and fewer noise peaks or noise events. Since the annoyance that people experience depends upon the number of noise events that occur during a time interval, the peak levels reached during the events, and the degree to which these peak levels exceed the residual noise level, the nighttime noise environment may be more or less significant relative to overall annoyance than the day-time noise. Any noise rating, to be useful, must account for these factors.

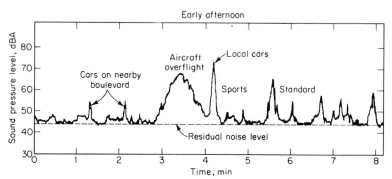

Figure 4.18 Outdoor community noise level measured in a suburban neighborhood in the early afternoon.

A recording of the ambient noise such as that shown in Fig. 4.18 can be made over an entire day, week, or year and can, of course, be used to describe the noise in a community. However, it is clear that to use such a record to describe noise over a period of time of more than a few minutes is impractical. Because such a record contains so much detailed information, some way is needed to describe the noise environment in a simpler, more concise way. Several of the more important and widely used noise scales and ratings that

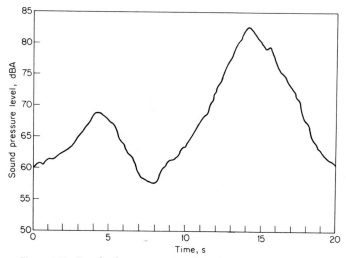

Figure 4.19 Sample of community noise level during 20-second interval.

are used to describe community noise in an average or overall sense are discussed in the following sections.

Statistical Noise Levels

If we examine the recorded noise level such as that shown in Fig. 4.18 over a given time interval, it is possible to describe several important features of the noise using statistical quantities. These, in a way, are a type of averaging of the detailed noise environment. Consider a recording such as that shown in Fig. 4.19 over a 20-second time period. The noise level at each second can be determined and classified into intervals of, for example, 5-dBA increments. That is, the noise can be treated as twenty distinct measurements and classified into the number of measurements or number of times that the noise level falls between 50 and 55 dBA, 56 and 60 dBA, 61 and 65 dBA, and so on. These can be plotted on a histogram that shows the percent of the time during the interval that the noise level falls within these 5-dBA intervals. To illustrate this, consider the simple example which follows.

EXAMPLE 4.12   Derive a histogram for the noise level recording shown in Fig. 4.19.

The noise levels taken from the figure at times 1 through 20 seconds are 61, 62, 65, 69, 68, 64, 59, 58, 61, 64, 67, 72, 77, 82, 80, 76, 72, 67, 62, and 61 dBA, respectively. The number of times the noise level falls within the 5-dBA intervals from 55 to 85 dBA are shown in Table 4.10. The percentage of the time is just the number of times the noise level falls within a given interval. From this table, the histogram shown in Fig. 4.20 can be drawn.

TABLE 4.10   Number of Times and Percentages of the Time the Noise Level Shown in Fig. 4.19 Falls Within the 5-dBA Intervals Shown

Noise interval, dBA	Number of occurrences	Percentage of time
56–60	2	10%
61–65	8	40%
66–70	4	20%
71–75	2	10%
76–80	3	15%
81–85	1	5%

The "histogram" is simply a means of statistically summarizing the noise levels that occur during a given time period. However, a more meaningful and more widely used measure, obtained from the histogram, is to calculate the percent of the time certain noise

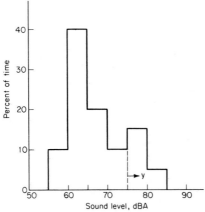

Figure 4.20   Histogram showing the percent of the time the sound pressure level shown in Fig. 4.19 is within 5 dBA intervals.

levels are exceeded during the time interval. The notation for these statistical quantities is:

L_{90} = the noise level exceeded 90 percent of the time
L_{50} = the noise level exceeded 50 percent of the time
L_{10} = the noise level exceeded 10 percent of the time

L_{90} is a measure of the residual noise level, L_{50} is the median noise, and L_{10} is a measure of the peak noise levels observed during a given time period. The calculation of these is illustrated in the following example.

EXAMPLE 4.13   Calculate the L_{10}, L_{50}, and L_{90} levels for the noise record shown in Fig. 4.19.

First, calculate the area under the histogram for this noise record, as shown in Fig. 4.20.

$$\text{Area} = 5\,(10 + 40 + 20 + 10 + 15 + 5) = 500 \text{ units}$$

where "units" refers to percent times dBA.

Calculate L_{90}—Ten percent of the area under the histogram would correspond to 50 units. Measuring from 55 dBA to the right, an area of 50 units is reached at 60 dBA. Therefore, the noise level exceeded 90 percent of the time (and, correspondingly, 90 percent of the area of the histogram is greater than) 60 dBA.

Calculate L_{50}—Calculate the area, starting from the left of the diagram (at 55 dBA) corresponding to one-half of the total area under the histogram or 250 units.

$$\text{Area (from left)} = 5 \times 10 + 5 \times 40 = 250 \text{ units}$$

That is, half the area is below 65 dBA and half is above. Therefore, 65 dBA is the value of L_{50}.

Calculate L_{10}—The area under the histogram corresponding to 90 percent of the area is 450 units. Starting with 55 dBA, the area under the curve is

$$(5 \times 10) + (5 \times 40) + (5 \times 20) + (5 \times 10) + (y \times 15) = 450$$

where y is measured from 75 dBA. This may be written as

$$y \times 15 = 450 - 400 = 50 \text{ units}$$
or
$$y \cong 3.3 \text{ dBA}$$

The value of L_{10} is therefore $75 + 3.3 \cong 78.3$ dBA. That is, the noise level exceeds 78.3 dBA 10 percent of the time during the period measured, or 10 percent of the area under the histogram is greater than (to the right of) 78.3 dBA.

Usually, the statistical noise levels are described in terms of a longer time interval than 20 seconds, and more than 20 data points are used. Also, one can use a sound level meter to observe the noise level at, perhaps, 10-second intervals over a period of time, say from one-quarter hour up to perhaps one hour. Using this technique, data can be recorded as shown in Fig. 4.21. For example, each x in Fig. 4.21 corresponds to a noise measurement

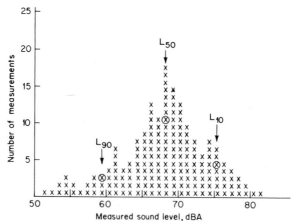

Figure 4.21   Record of sound measurements taken at equal intervals over a period of time.

taken from an observation by a sound level meter on a record such as that shown in Fig. 4.18. This is a type of histogram and represents 180 measurements. Thus, L_{90}, L_{50}, and L_{10} would correspond to the 18th, 90th, and 162nd x, respectively, counted from the left-hand end of the graph. These are shown circled in the figure and correspond to 59, 68, and 75 dBA, respectively. Sound recording instruments are also available which give the statistical noise levels measured over a period of time.

Figure 4.22 shows examples of the statistical noise levels for a number of different locations. Note that near airports, the peak noise levels indicated by L_{10} and L_1 (the noise level exceeded only one percent of the time) are much higher than the median noise

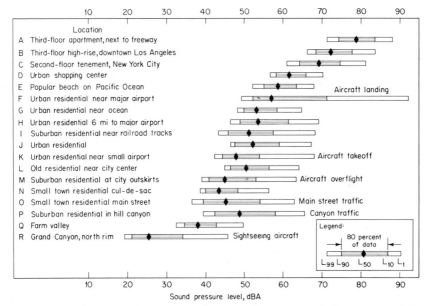

Figure 4.22 Outdoor statistical noise measured in various locations over a 24-hour time period. Data show levels exceeded 99, 50, 10, and 1 percent of the time over the 24-hour time period.

level L_{50}. That is, the distribution is skewed by the very loud noise events associated with aircraft overflights. These wide distributions generally lead to more annoyance partly because they represent high peak noise levels above a relatively low residual level.

Noise criteria for various land use categories have been published by the Federal Highway Administration (Policy and Procedure Memorandum 90-2, April 1972) in terms of L_{10}. The exterior design (maximum) noise level for areas where serenity and quiet are needed such as amphitheaters or certain areas of parks is $L_{10} = 60$ dBA. For residential areas, schools, and the like, the criterion exterior level is $L_{10} = 70$ dBA and the interior level $L_{10} = 55$ dBA. For the design of roadways requiring federal approval, these criteria must be met.

Equivalent Sound Pressure Level

A record of sound level such as that shown in Fig. 4.18 is equivalent to a plot of sound-intensity level, since $L_I \cong L_p$, versus time. That is, it is a measure of the sound energy reaching an observer per unit time per unit area. This intensity is given by

$$I(t) = I_o 10^{L_I(t)/10} \qquad \text{W/m}^2 \qquad (4.32)$$

where I_o is the reference sound intensity and $L_I(t)$ is the sound intensity level as a function of time. In terms of the sound pressure level, this is

$$I(t) = I_o 10^{L_p(t)/10} \qquad \text{W/m}^2 \qquad (4.33)$$

Over a period of time T, we may average the sound energy as follows

$$\bar{I} = \frac{1}{T} \int_0^T I(t) \, dt = \frac{I_0}{T} \int_0^T 10^{L_p(t)/10} \, dt$$

or

$$\frac{\bar{I}}{I_0} = \frac{1}{T} \int_0^T 10^{L_p(t)/10} \, dt \tag{4.34}$$

We may express this in terms of decibels as follows

$$L_{eq} = 10 \log \frac{\bar{I}}{I_0} = 10 \log \left[\frac{1}{T} \int_0^T 10^{L_p(t)/10} \, dt \right] \text{dBA} \tag{4.35}$$

This energy-average sound level is defined as the "equivalent sound level" scale, and if $L_p(t)$ is expressed in dBA then the units of L_{eq} are also expressed as dBA. This scale, the average sound energy at an observer expressed in decibels, is a useful measure of noise exposure and forms the basis of several of the noise ratings in current use. The equivalent sound level can be calculated from a noise record such as that shown in Fig. 4.18 and there are also noise measurement systems which give a direct reading of L_{eq}.

Although in some cases it may be possible to evaluate the integral in Eq. 4.35 directly, more often it is necessary to numerically evaluate the integral. If we divide the noise level scale into n 1-dB intervals with a value at the middle of each interval of L_i, an approximate expression for the equivalent noise level is

$$L_{eq} = 10 \log \left(\sum_{i=1}^n f_i 10^{L_i/10} \right) \text{dBA} \tag{4.36}$$

where f_i is the fraction of time that the sound pressure level is in the ith interval. For example, assume that the sound pressure level is 60 dBA for 10 minutes and 70 dBA for 10 minutes. Over the total 20-minute time period, the equivalent sound pressure level is

$$L_{eq} = 10 \log (0.5 \times 10^{60/10} + 0.5 \times 10^{70/10})$$
$$= 10 \log (0.5 \times 1.1 \times 10^7)$$
$$= 67.4 \text{ dBA}$$

The L_{eq} over a priod of time may easily be calculated from a histogram as illustrated by the following example.

EXAMPLE 4.14 Calculate the equivalent sound pressure level for the measurements shown in Fig. 4.21.

In this case, for simplicity, use 5-dBA intervals (51–55, 56–60, 61–65 dBA, etc.), with L_1 = 53 dBA, L_2 = 58 dBA, . . . , and construct the following table.

L_i, dBA	Number of measurements in interval	f_i, %
53	9	5
58	13	7.2
63	33	18.3
68	70	38.9
73	39	21.7
78	15	8.3
83	1	0.6

Carrying out the logarithmic addition in Eq. 4.36, we obtain

$$L_{eq} = 73.9 \text{ dBA}$$

Note that this is not the same as L_{50} which was 68 dBA.

Single Event Noise Level

Sometimes it is desirable to describe the noise due to a single event such as an aircraft flyover or a noisy vehicle driving by an observer. Such an event may have a noise level versus time plot such as that shown in Fig. 4.23. The annoyance caused by such an event depends upon both the peak noise level and duration. A single event noise equivalent level (SENEL) is used in the California Airport Noise Regulations, for example, to quantify a single aircraft flyover noise. The SENEL is defined as

$$\text{SENEL} = 10 \log \int_{T_1}^{T_2} 10^{L_p(t)/10} \, dt \tag{4.37}$$

This quantity is similar to the equivalent noise level, Eq. 4.35, except that the integral is not divided by the time interval. T_1 and T_2 are the times before and after the peak at

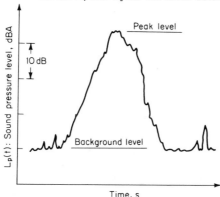

Figure 4.23   Noise recording of a single event such as an aircraft flyover.

which the noise level is Δ dBA below the peak level. Δ is usually taken as 10, 20, or 30 dBA; actually, the contribution to the SENEL value for those portions of the curve less than 10 dBA below the peak is generally very small.

Day-Night Sound Level

A noise rating developed by the EPA for specification of community noise from all sources is the day-night sound level, L_{dn}. It is similar to a 24-hour equivalent sound level except that during the nighttime period, which extends from 10:00 P.M. to 7:00 A.M., a 10 dBA weighting penalty is added to the instantaneous sound level before computing the 24-hour average. This nighttime penalty is added to account for the fact that noise at night when people are trying to sleep is judged more annoying than the same noise during the daytime.

The EPA has adopted L_{dn} as the rating method used to describe community noise exposure. To arrive at a recommended outdoor L_{dn} criterion, the EPA recommends that an L_{eq} of 45 dBA should not be exceeded indoors in residential buildings; this allows freedom from speech interference. If the outdoor-to-indoor attenuation is taken, on the average, as 15 dBA with the windows partially open, this implies an outdoor criterion of L_{eq} = 60 dBA. The EPA then applies corrections to this to account for other annoyance factors, including the 10 dBA nighttime penalty and a 5 dBA margin of safety, to arrive at a recommended criterion of L_{dn} = 55 dBA as a desirable maximum outdoor noise level for residential areas. It should be noted, however, that the majority of the urban and a large fraction of the suburban areas in the United States are currently subjected to noise levels greater than this value.

A very similar rating used in the California Airport Noise Regulations is the community noise equivalent level (CNEL). This rating uses the same weighting at night as the L_{dn} rating but also adds a penalty of 5 dBA to the noise measured during the evening time period, 7:00 P.M. to 10:00 P.M. Numerically, these two ratings, L_{dn} and CNEL, are about equal for most community noise environments. The California law uses a CNEL of 65 dBA as the criterion for maximum outdoor noise levels in residential areas. This corresponds to an L_{dn} of also about 65 dBA. A different noise control law in California, however, relating to noise-insulation standards for residential structures, specifies that when the outdoor noise level exceeds CNEL ($\cong L_{dn}$) = 60 dB for residential areas in the vicinity of transportation noise sources, a noise study shall be carried out to assure that the indoor level does not exceed a CNEL of 45 dB.

Although based upon a daily noise exposure, the CNEL used in the California airport noise law is based upon an annual (energy) average. This is intended to account for changes of noise with day of the week and seasonal variations that may exist in some areas.

The L_{dn} for a given location in a community may be calculated from the hourly L_{eq}'s. That is, the L_{eq} for each hour of the day, with a 10 dBA correction added to the nighttime values, may be added on a logarithmically averaged basis to obtain L_{dn}. An equation expressing this method is as follows:

$$L_{dn} = 10 \log\left\{\frac{1}{24}\left[\sum_{i=1}^{15} 10^{(L_{eq})_i/10} + \sum_{j=1}^{9} 10^{((L_{eq})_j+10)/10}\right]\right\} \qquad (4.38)$$

where i denotes the sum over the 15 hours during the daytime and j denotes the sum over the 9 hours during the nighttime. $(L_{eq})_i$ denotes the equivalent noise level for the ith hour during the day and similarly $(L_{eq})_j$ for the jth hour during the night. A simpler expression may be written as

$$L_{dn} = 10 \log\left\{\frac{1}{24}\left[15(10^{L_d/10}) + 9\,(10^{(L_n+10)/10})\right]\right\} \qquad (4.39)$$

where L_d is the equivalent noise level during the daytime (7:00 A.M. to 10:00 P.M.) and L_n is the equivalent noise level during the nighttime (10:00 P.M. to 7:00 A.M.).

EXAMPLE 4.15   The hourly equivalent noise levels L_{eq} measured in a community over a 24-hour time period are given in the table below. Calculate the L_{dn} corresponding to this noise environment.

Time	L_{eq}, dBA	Adjusted L_{eq} (L_i), dBA	$10^{L_i/10}$ ($\times 10^4$)
10 P.M.	50	60	100
11	50	60	100
12 midnight	50	60	100
1 A.M.	40	50	10
2	30	40	1
3	30	40	1
4	30	40	1
5	40	50	10
6	50	60	100
7	60	60	100
8	70	70	1000
9	60	60	100
10	60	60	100
11	60	60	100
12 noon	70	70	1000
1 P.M.	60	60	100
2	60	60	100
3	60	60	100
4	60	60	100
5	70	70	1000
6	70	70	1000
7	60	60	100
8	60	60	100
9	50	50	10

First correct the nighttime levels by adding the 10 dBA adjustment to obtain L_i. Then add these logarithmically as indicated in the right-hand column to obtain the sum S given by

$$S = \sum_{i=1}^{24} 10^{L_i/10} = 5433 \times 10^4$$

Then, from Eq. 4.37,

$$L_{dn} = 10 \log (5433 \times 10^4) = 77.3 \text{ dBA} \tag{4.40}$$

Other Noise Scales and Ratings

There are a large number of noise scales and rating methods that have been developed over the years for various purposes. Since these are usually some sort of average or weighted-average quantity derived from the detailed noise characteristics, it is not usually possible to directly relate one to the other. However, for specific types of noise environments approximate comparisons among some of these scales or ratings do exist. For example, we have already pointed out that CNEL $\cong L_{dn}$ for most situations; one would expect this approximation to be in error only when the noise environment is dominated by noise sources during the evening time period (7:00 P.M. to 10:00 P.M.). For example, suppose that a given residential area is relatively quiet except when several trains pass through during the evening hours. In such an instance, CNEL would be expected to be about 5 dB greater than L_{dn}. However, this type of situation is not often encountered in noise assessment.

The equation

$$L_{eq} \cong L_{50} + \frac{(L_{10} - L_{90})^2}{60} \tag{4.41}$$

is valid when the noise level histogram is approximately Gaussian—that is, a relatively smooth bell-shaped curve. This may also be written as

$$L_{eq} \cong L_{50} + \frac{(L_{10} - L_{50})^2}{15} \tag{4.42}$$

subject to the same condition.

One of the problems in carrying out a comprehensive noise assessment is that different noise criteria using different noise scales or ratings must often be examined. If a complete description of the noise environment is available for a representative set of conditions, then each applicable scale or rating may be calculated from the same data base. Often, however, data are available only in terms of one scale or rating method. In these cases, relationships such as those given in Eqs. 4.40 through 4.42 are useful.

The use of these noise scales and ratings and the noise criteria discussed earlier are illustrated in the next section on transportation noise and in the examples in the following section.

ESTIMATING TRANSPORTATION NOISE IMPACTS

Highway Noise

Estimating highway noise impact can be an involved and complex process which may require a large computer analysis to model in detail. However, the methodology presented here is a general highway noise model that predicts the equivalent noise level L_{eq} and is adequate for most noise assessment requirements. The advantage of a model which predicts L_{eq} is that L_{eq} is the "energy average" noise level and as such is not dependent on the statistics of the traffic flow; that is, it does not matter whether the traffic is flowing freely with even spacing or is subject to queuing such that the traffic passes in clusters. In contrast, the noise descriptors L_{10}, L_{50}, and L_{90} can be very sensitive to the flow characteristics of the traffic and can, therefore, be difficult to model. Furthermore, the model presented here can be applied equally well to high traffic volume and low traffic volume roadways.

For our purposes, we will only consider the case of an infinitely long roadway. The model does have the capability of handling finite road segments, but that is beyond the scope of this handbook. (For finite roadways, see Ref. 14.)

The Highway Noise Model   The highway model presented here is for calculating the one hour L_{eq}. Predicting the day-night noise level L_{dn} will be discussed later. The model was recently developed by the Federal Highway Administration (FHWA), and all units are metric. The model is based upon calculating the hourly L_{eq} for automobiles, medium trucks, and heavy trucks separately and then adding these logarithmically to obtain the overall hourly L_{eq} as follows:[2]

$$L_{eq}(h)_i = \overline{L_{0_E}} + 10 \log\left(\frac{N_i}{S_i T}\right) + 10 \log\left(\frac{15}{d}\right)^{1+\alpha} + \Delta_s - 13 \qquad (4.43)$$

where

- $L_{eq}(h)_i$ is the L_{eq} at hour h for the ith vehicle type, i.e., autos, medium trucks, or heavy trucks,[3] h is the hour at which the noise is measured, i.e., $L_{eq}(0700)_{\text{auto}}$ refers to the automobile L_{eq} at 7:00 A.M.
- $\overline{L_{0_{E,i}}}$ is the reference mean energy level for the ith vehicle type. This is the noise emission level for a given vehicle type and is found by measurement or published data (see Fig. 4.24).
- N_i is the number of class i vehicles passing during the time T. For now, this would be the one-hour traffic flow but, as shall be shown later, can be for a different period.
- S_i is the average speed for the ith vehicle class in km/h.
- T is the duration for which the L_{eq} is desired and must correspond to N_i, the count of vehicles during the time T. Normally, T would be one hour but can be longer or shorter.
- D is the perpendicular distance, in meters, from the centerline of the traffic lane to the location of the observer, i.e., the location where the noise level is desired.
- α is a factor which relates to the absorption characteristics of the ground cover between the roadway and the observer.
- Δ_s is the shielding factor such as provided by a noise barrier. Each of these factors is discussed in detail later in this section.

The preceding equation is used three times; once for automobiles, once for medium trucks, and once for heavy trucks to obtain three values, L_{eq_A} for autos, $L_{eq_{MT}}$ for medium trucks, and $L_{eq_{HT}}$ for heavy trucks. The final total L_{eq} can be calculated by logarithmically adding the three L_{eq} values as follows:

$$L_{eq(h)_{\text{total}}} = 10 \log\left[10^{L_{eq_A}/10} + 10^{L_{eq_{MT}}/10} + 10^{L_{eq_{HT}}/10}\right] \qquad (4.44)$$

EXAMPLE 4.16   Consider a highway for which the automobile L_{eq} equals 55 dBA, the medium truck L_{eq} equals 57 dBA, and the heavy truck L_{eq} equals 60 dBA. Find the total L_{eq}.

$$L_{eq} = 10 \log\left[10^{55/10} + 10^{57/10} + 10^{60/10}\right]$$
$$= 62.6 \text{ dBA}$$

Considering Eq. 4.43, it is important to discuss some of the terms in more detail. For example, the term $\overline{L_{0_{E,i}}}$ is the reference energy mean emission level for the population of vehicles being evaluated. The level is the energy average[4] maximum noise level for a

---

[2]The Highway Noise Model developed by the FHWA contains a term for finite roadway adjustments that is not included here. The finite roadway adjustment is for cases where the roadway does not meet the requirements for infinite roadways; i.e., the observer does not have an unobstructed view of the road in both directions. A peculiarity of the FHWA model is that the finite roadway adjustment factor is not zero for infinite roadways but, in fact, results in a 1 to 1.5 dB noise reduction for infinite roadways. This factor was not included here and, therefore, the results given here can be considered conservative (high) by 1.5 dB, or the reader can adjust the values given here by 1.5 dB to match FHWA computer models.

[3]A heavy truck is one that has three or more axles and weighs more than 26,000 lb. A medium truck can have two or three axles and must weigh between 10,000 and 26,000 lb.

[4]Energy average as opposed to arithmetic average is defined as follows:

$$\text{Arithmetic average} = \frac{1}{n}\sum_{i=1}^{n} L_i$$

$$\text{Energy average} = 10 \log \frac{1}{n}\sum_{i=1}^{n} 10^{L_i/10}$$

single vehicle passby at constant speed at a distance of 15 m. This value can be measured and used in the model or estimated from data taken from a national sampling of vehicles. Figure 4.24 shows the reference energy mean emission levels as a function of vehicle speed for the national average population of automobiles, medium trucks, and heavy trucks.

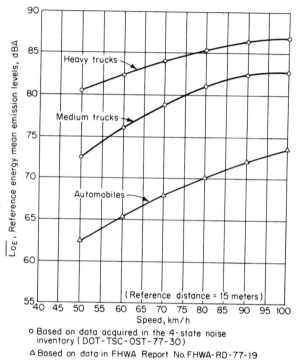

o Based on data acquired in the 4-state noise inventory (DOT-TSC-OST-77-30)

△ Based on data in FHWA Report No. FHWA-RD-77-19

Figure 4.24 Reference energy mean emission levels as a function of speed.

EXAMPLE 4.17 Determine the reference energy mean emission levels for automobiles, medium trucks, and heavy trucks for an average vehicle speed of 75 km/h at a distance of 15 m. Answer from Fig. 4.24:

$$\overline{L_{O_{E,A}}} = 69 \text{ dBA}$$

$$\overline{L_{O_{E,MT}}} = 80 \text{ dBA}$$

$$\overline{L_{O_{E,HT}}} = 84.6 \text{ dBA}$$

The distance D in meters is the perpendicular distance between the observer and the centerline of the traffic lane. (See Fig. 4.25.) Note that this assumes all traffic in one lane. This assumption is valid for multiple lane roadways if an "effective center lane" is used and the traffic is relatively evenly distributed among all the traffic lanes. The effective center lane should be located in the center of the road. If the traffic flow is heavy in one direction and light in the other direction, then the roadway can be modeled as two separate roadways. The total L_{eq} can then be obtained by logarithmically adding the noise from these "two" roadways together. In reality, satisfactory results can usually be obtained using the centerline of the roadway for estimating D.

The absorption factor α is an empirical number which varies with ground surface. If the ground cover is vegetated or has a soft texture, sound will decrease at the rate of 4.5 dB every time the distance between the roadway and the observer is doubled. This results in an α of 0.5. If the ground between the roadway and observer is paved or hard, then the dropoff rate is only approximately 3 dB every time the distance is doubled. This results in an α of 0.

Table 4.11 lists the rules for when to use each α. Note that combinations of soft and hard patches of ground between the roadway and the observer can be handled by calculating the noise level in a series of steps corresponding to each patch. In most cases, a doubling rate 4.5 dB ($\alpha = 0.5$) should be used.

The shielding factor Δ_s is another empirical number which takes into account the environment surrounding the roadway and the observer. Figure 4.26 shows the various physical characteristics that should be considered. Figure 4.26a and Fig. 4.26b illustrate the cases just mentioned. Figure 4.26c shows that for every 30 meters of very dense landscape vegetation (at least 15 ft high), 5 dB of additional attenuation can be obtained up to a maximum of 10 dB. A word of caution is warranted here. The landscape must be very dense with absolutely no line-of-sight visibility between the source and the observer. This can be likened to a dense tropical forest. Note that if the landscape attenuation is used, then the dropoff rate used for every doubling of distance should be 3 dB ($\alpha = 0$).

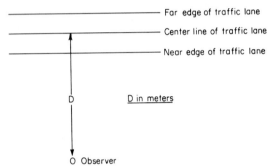

Figure 4.25 Schematic sketch of observer distance D for a one-lane roadway.

Additional attenuation caused by intervening rows of houses is depicted in Fig. 4.26d. If the first row of houses covers at least 40 to 60 percent of the area, then approximately 3 dBA additional attenuation will result. If the area of coverage is 70 to 90 percent, then the attenuation will be about 5 dBA. An additional 1.5 dBA for each additional row of houses may be taken, but the total attenuation by intervening rows of houses should not exceed 10 dBA. Note that these are estimates but are not important in many cases since the first row of houses is usually subjected to the design criteria for noise from roadways.

TABLE 4.11   Dropoff Rate per Doubling of Distance and Corresponding α

Situation	Dropoff rate, dB	α
1. All situations when the sound or the receiver is located 3 m above the ground or whenever line-of-sight* averages more than 3 m above the ground.	3	0
2. All situations involving propagation over the top of a barrier 3 m or more in height.	3	0
3. Where the height of the line-of-sight is less than 3 m and		
(a) There is a clear, unobstructed view of the highway, the ground is hard, and there are no intervening structures.	3	0
(b) The view of the roadway is interrupted by isolated buildings, clumps of bushes, scattered trees, or the intervening ground is soft or covered with vegetation.	4.5	0.5

*The line-of-sight (L/S) is a direct line between the noise source and the observer.

Also, as shown in Fig. 4.26e, walls and berms can be very effective in attenuating highway noise. These are discussed for highways in the next section. It is curious to note, however, that for a solid wall barrier and earthen berm of the same height, the berm has been observed to provide about 3 dB more attenuation than the solid wall. This is

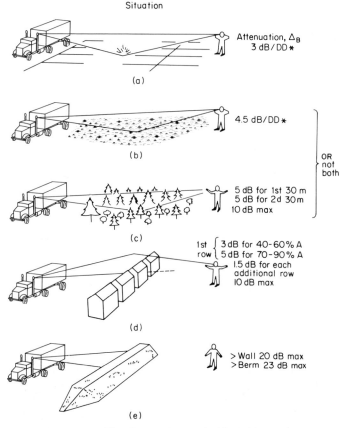

Situation

(a)   Attenuation, Δ_B
          3 dB/DD *

(b)   4.5 dB/DD *

(c)   5 dB for 1st 30 m
          5 dB for 2d 30 m
          10 dB max

OR not both

(d)   1st { 3 dB for 40-60% A
          row { 5 dB for 70-90% A
          1.5 dB for each
          additional row
          10 dB max

(e)   > Wall 20 dB max
          > Berm 23 dB max

Figure 4.26 The shielding factor Δ_s.(*DD = doubling of distance)

probably due to absorption of sound by the earth and a shape factor. In any case, the maximum attenuation of a solid wall is taken as 20 dBA and for an earthen berm, 23 dBA.

EXAMPLE 4.18 Consider a house located near a major roadway and calculate L_{eq} in the yard next to the road. The distance from the house to the edge of the roadway is 50 m. The roadway has four traffic lanes and is 20 m wide. The average daily traffic (ADT) is 40,000 vehicles of which 3 percent are medium trucks and 1 percent are heavy trucks. The average vehicle speed is 75 km/h. The ground is grassy between the roadway and observer, and there are no intervening structures. Find the L_{eq} for the peak hour when the hourly traffic flow volume is 10 percent of the ADT.

1. Find the $\overline{L_{0_{E,i}}}$ for each vehicle type from Fig. 4.25:

$$\overline{L_{0_{E,A}}} = 69 \text{ dBA}$$

$$\overline{L_{0_{E,MT}}} = 80 \text{ dBA}$$

$$\overline{L_{0_{E,HT}}} = 84.6 \text{ dBA}$$

2. Find the peak hour traffic flow N_i for each vehicle type:

$$N_{HT} = (1\%)(10\%)(40{,}000) = (0.01)(0.10)(40{,}000)$$
$$= 40$$
$$N_{MT} = (3\%)(10\%)(40{,}000) = (0.03)(0.10)(40{,}000)$$
$$= 120$$
$$N_A = (96\%)(10\%)(40{,}000) = (0.96)(0.10)(40{,}000)$$
$$= 3840$$

3. $S_{AUTO} = S_{MT} = S_{HT} = 75$ km/h
4. $T = 1$ h
5. Find the distance D between the observer and the centerline of the roadway:

$$D = 50 + (1/2)(20) = 60 \text{ m}$$

6. Use $\alpha = 0.5$ for grassy ground cover (Fig. 4.26)
7. Find the automobile L_{eq}, L_{eq_A}, using Eq. 4.43:

$$L_{eq_A} = 69 + 10 \log \left[\frac{3840}{75(1)} \right] + 10 \log \left(\frac{15}{60} \right)^{1+(0.5)} - 13$$
$$= 64.1 \text{ dBA}$$

8. Find the medium truck $L_{eq}, L_{eq_{MT}}$:

$$L_{eq_{MT}} = 80 + 10 \log \left[\frac{120}{75(1)} \right] + 10 \log \left(\frac{15}{60} \right)^{1+(0.5)} - 13$$
$$= 60.0 \text{ dBA}$$

9. Find the heavy truck $L_{eq}, L_{eq_{HT}}$:

$$L_{eq_{HT}} = 84.6 + 10 \log \left[\frac{40}{75(1)} \right] + 10 \log \left(\frac{15}{60} \right)^{1+(0.5)} - 13$$
$$= 59.8 \text{ dBA}$$

10. Find the total L_{eq} using equation (2):

$$L_{eq_{total}} = 10 \log [10^{64.1/10} + 10^{60/10} + 10^{59.8/10}]$$
$$= 66.6 \text{ dBA}$$

Day-Night Noise Level, L_{dn}   The L_{eq} model presented can easily be used to calculate the day-night noise level, L_{dn}. Recall that L_{dn} is a 24-hour time-weighted annual average noise level based on 1-hour L_{eq}'s. The L_{dn} noise descriptor divides every 24-hour period into a daytime period (0700–2200 hours) and a nighttime period (2200–0700 hours). Noise levels during the nighttime period are penalized 10 dB for occurrences during this noise-sensitive period. Using this penalty correction, then, the L_{dn} is the 24-hour energy average of the twenty-four 1-hour L_{eq}'s that make up a day. That is,

$$L_{dn} = 10 \log \frac{1}{24} \left[\sum_{\text{daytime}} 10^{Leq_i/10} + \sum_{\text{nighttime}} 10^{(Leq_i + 10)/10} \right] \qquad (4.45)$$

where the summation $\sum_{\text{daytime}}$ refers to the sum of all the daytime L_{eq} (0700 to 2200 hours) and the summation $\sum_{\text{nighttime}}$ refers to the sum of all the nighttime L_{eq} (2200 to 0700 hours).

Examination of Eq. 4.45 shows that L_{dn} can be calculated by logarithmically adding the daytime and corrected nighttime L_{eq} values where each of the 24 L_{eq} values is determined from Eqs. 4.43 and 4.44. Since Equation 4.43 must be computed three times for every hour, the number of equations that must be calculated to compute one L_{dn} value is almost 100, a very tedious and time consuming process. A look at Eq. 4.43 suggests that the process can be greatly simplified by not calculating each 24-hour L_{eq} but instead by directly calculating only two L_{eq} values—a daytime "L_{eq} sum" and a nighttime "L_{eq} sum." To do this, simply change the N values in Eq. 4.43 to correspond to the daytime and nighttime traffic volumes. The time factor T would remain 1 hour just as used in the hourly L_{eq}. The resulting "L_{eq} sum" from this substitution is the energy sum of the daytime L_{eq}'s and nighttime L_{eq}'s respectively without having to compute all twenty-four 1-hour L_{eq}'s—this results in a tremendous saving of time. The traffic volumes should then consist of the 15-hour daytime volume and the 9-hour nighttime volume instead of the hourly volume used in the hourly L_{eq}.

In general, 90 percent of the average daily traffic (ADT) flows during the daytime period, and 10 percent flows at night. Using these factors, the L_{dn} can easily be computed.

EXAMPLE 4.19   Using all the data in the previous example, compute the L_{dn} value assuming 90 percent of the ADT flows during the daytime period.

Solution: Proceed just as in the earlier example, except this time keep track of two L_{eq} values, the daytime L_{eq} sum, $L_{eq_{day}}$, and the nighttime L_{eq} sum, $L_{eq_{night}}$.

1. The reference energy mean emission levels are the same as the previous example:

$$\overline{L_{O_{E,A}}} = 69 \text{ dBA}$$

$$\overline{L_{O_{E,MT}}} = 80 \text{ dBA}$$

$$\overline{L_{O_{E,HT}}} = 84.6 \text{ dBA}$$

2. Find the daytime and nighttime traffic flow N_i for each vehicle class:

Daytime:

$$N_{HT_{day}} = (1\%)(90\%)(40{,}000) = (0.01)(0.9)(40{,}000)$$
$$= 360$$
$$N_{MT_{day}} = (3\%)(90\%)(40{,}000) = (0.03)(0.9)(40{,}000)$$
$$= 1{,}080$$
$$N_{A_{day}} = (96\%)(90\%)(40{,}000) = (0.96)(0.9)(40{,}000)$$
$$= 34{,}560$$

Nighttime:

$$N_{HT_{night}} = (1\%)(10\%)(40{,}000) = (0.01)(0.1)(40{,}000)$$
$$= 40$$
$$N_{MT_{night}} = (3\%)(10\%)(40{,}000) = (0.03)(0.1)(40{,}000)$$
$$= 120$$
$$N_{A_{night}} = (96\%)(10\%)(40{,}000) = (0.96)(0.1)(40{,}000)$$
$$= 3{,}840$$

3. $S_A = S_{MT} = S_{HT} = 75$ km/h
4. $D = 60$ m
5. $\alpha = 0.5$
6. Find the daytime L_{eq} sum, $L_{eq_{day}}$, from the energy sum of the daytime L_{eq} for autos, medium trucks, and heavy trucks using Equation 4.42:

$$L_{eqA_{day}} = 69 + 10 \log \left[\frac{54560}{75}\right]$$
$$+ 10 \log \left(\frac{15}{60}\right)^{1.5} - 13 = 73.6 \text{ dB}$$

$$L_{eqMT_{day}} = 80 + 10 \log \left[\frac{1080}{75}\right] + 10 \log \left(\frac{15}{60}\right)^{1.5}$$
$$- 13 = 69.6 \text{ dB}$$

$$L_{eqHT_{day}} = 84.6 + 10 \log \left[\frac{360}{75}\right] + 10 \log \left(\frac{15}{60}\right)^{1.5}$$
$$- 13 = 69.4 \text{ dB}$$

therefore

$$L_{eq_{day}} = 10 \log \left[10^{73.6/10} + 10^{69.6/10} + 10^{69.4/10}\right] = 76.1 \text{ dB}$$

7. Find the nighttime L_{eq} sum, $L_{eq_{night}}$, from the energy sum of the nighttime L_{eq} for autos, medium trucks, and heavy trucks just as above:

$$L_{eqA_{night}} = 69 + 10 \log \left[\frac{3840}{75}\right] + 10 \log \left(\frac{15}{60}\right)^{1.5} - 13$$
$$= 64.1 \text{ dB}$$

$$L_{eqMT_{night}} = 80 + 10 \log \left[\frac{120}{75}\right] + 10 \log \left(\frac{15}{60}\right)^{1.5} - 13$$
$$= 60.0 \text{ dB}$$

$$L_{eqHT_{night}} = 84.6 + 10 \log \left[\frac{40}{75} \right] + 10 \log \left(\frac{15}{60} \right)^{1.5} - 13$$
$$= 59.8 \text{ dB}$$

therefore

$$L_{eq_{night}} = 10 \log \left[10^{64.1/10} + 10^{60.0/10} + 10^{59.8/10} \right] = 66.6 \text{ dB}$$

8. Calculate the total L_{dn} by taking the energy sum of the daytime L_{eq} sum and the penalized nighttime L_{eq} sum and averaging over 24 hours:

$$L_{dn} = 10 \log 1/24 \left[10^{76.1/10} + 10^{(66.6+10)/10} \right] = 65.6 \text{ dB}$$

Highway Noise Barriers Barriers adjacent to roadways can be very effective noise attenuators if they are designed properly. Barriers can include the following:

- Walls
- Berms
- Buildings
- Terrain (such as hills)
- Combination of the above (such as earth berm and wall)

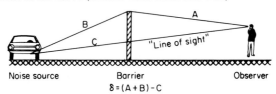

Figure 4.27 Sketch of barrier geometry and path-length difference, δ.

Some of the more important design criteria of barriers include the surface density which should be at least 4 lb/ft², that the barrier have no holes or cracks, that the barrier be long enough such that leaks do not occur around the edges, and that the barrier be high enough to obtain the necessary amount of noise attenuation. The following section describes the method of determining the necessary barrier or wall height for an infinitely long barrier. The special case of a finite-length barrier can be found in several references.

As discussed earlier in the section on Sound Propagation, the attenuation by a given barrier design can be estimated from the path-length difference, δ (meters), due to the barrier. The path-length difference is the additional distance that sound must travel due to the barrier as opposed to the "line-of-sight" distance. The key design factor is that the barrier must break the line of sight between the observer and noise source. This is shown in Fig. 4.27. As the path-length difference increases, the barrier attenuation increases. For traffic noise, the amount of attenuation achieved can be estimated from the Fresnel number N_o where

$$N_o = 3.21 \, \delta$$

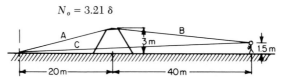

Figure 4.28 shows the relationship between the Fresnel number N_O and barrier attenuation. The curve and Eq. 4.46 are based on the assumption that the "effective frequency" of traffic noise is 550 Hz. This curve (Fig. 4.28) applies to solid walls and solid walls on top of earthen berms. For earthen berms, 3 dB additional attenuation is obtained.

Data required for a barrier analysis are the source height for motor vehicles and the observer height. The following source heights should be used for motor vehicles:

- Automobiles: 0 m
- Medium Trucks: 0.7 m
- Heavy Trucks: 2.44 m

A typical observer height is 1.5 m.

For example, calculate the barrier attenuation for automobile noise of a 3 m high earthen berm located 20 m from the centerline of the roadway for an observer 40 m from the berm. In this case

$$A = \sqrt{20^2 + 3^2} = 20.224 \text{ m}$$

$$B = \sqrt{(3-1.5)^2 + 40^2} = 40.028 \text{ m}$$

$$C = \sqrt{(1.5)^2 + 60^2} = 60.019 \text{ m}$$

$$\delta = (20.224 + 40.028) - 60.019$$

$$= 0.233 \text{ m}$$

$$N_o = 3.21 \, (0.233)$$

$$= 0.75$$

(4.46)

Using Fig. 4.28, the barrier attenuation for a solid wall is 9 dB. Therefore, the barrier attenuation for an earthen berm is

$$\Delta_B = 9 + 3 = 12 \text{ dB}$$

Noise Reduction for Future Motor Vehicle Populations An important consideration for projects which involve long-range projections and planning for future traffic volumes is the reduction in vehicle noise mandated by federal law. Regulations limiting motor vehicle noise at the time of manufacture will result in a continuing decrease in motor vehicle noise levels as new vehicles enter the population and replace the older, noisier vehicles.

Estimates of the noise reduction that will be realized are very tentative; however, automobile and motorcycle noise could be reduced by up to 10 to 20 dB. Truck noise reduction is more uncertain because of technical problems in reducing tire noise.

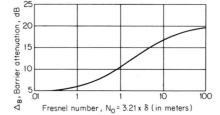

Figure 4.28 Traffic noise barrier attenuation vs. Fresnel number N_o for infinitely long barriers.

A 5 to 10 dB overall noise reduction could be realized from the truck population. Recognizing the fact that motor vehicle noise is being reduced and at the same time taking care not to overestimate the reduction, a very conservative assumption can be made that future motor vehicle populations will be 3 dB quieter than today's motor vehicles.

Aircraft Noise

Aircraft noise is one of the most significant noise problems today, and one of the most controversial. Assessing the noise impact of aircraft noise is the most complex of the noise problems that will be discussed in this handbook. In general, noise predictions for airports involve the use of complex computer programs that have been specifically written to model noise around airports. The intent here is to present the basic model methodology and consider some simple cases to demonstrate the technique.

Jet aircraft are certainly the most significant cause of the aircraft noise problem. Therefore, a brief discussion of jet noise is in order. The main sources of noise from jet aircraft can be categorized as fan, combustion, and jet noise. These are shown schematically in Fig. 4.29. Fan noise is associated with the aerodynamic and machinery noise in

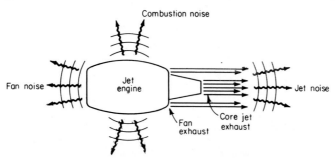

Figure 4.29 Main sources of jet aircraft noise.

the intake portion of the jet engine; it is one of the main sources of jet engine noise and can be recognized by its high-pitched whine. Combustion noise is associated with combustion processes in the engine and is caused by the high-velocity flow of high-temperature gases in the engine. Jet noise is caused by the high-velocity exhaust gases exiting the engine into the atmosphere and creating high levels of turbulence and noise. Jet noise is characterized by the lower frequency rumble or roar observed especially during takeoff.

It should be recognized that in the time since commercial jet aircraft were introduced in the 1960s, a great deal of technological improvements have occurred in jet-engine design. Figure 4.30 compares the relative magnitude of jet-engine noise sources for older engines that have been retrofitted with quiet engine technology and those that have not been treated. New technology engines are significantly quieter than the older technology engines.

New technology engines used on the newer wide-body aircraft (Boeing 747, DC-10, and Lockheed L-1011) are high bypass ratio engines that are much quieter than the older engines. "Bypass ratio" is the ratio of the mass flow through the fan discharge to the mass flow through the core discharge. The older jet aircraft use a much lower bypass ratio (the oldest turbojet engines have zero bypass) and include such aircraft as Boeing 737, 727, DC-9, and DC-8. Also, most business-jet engines are low-bypass ratio engines. Simplistically, high-bypass ratio engines are quieter primarily because exhaust gases are released at lower velocities than the low-bypass ratio engines and noise is critically proportional to velocity (noise varies as the eighth power of velocity). Data presented later in this section allow comparison of the noise levels of various aircraft and engine technologies.

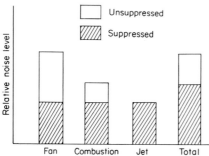

Figure 4.30 Comparison of jet engine noise sources for suppressed and unsuppressed engines.

Characterizing Aircraft Noise Aircraft noise is unique compared to most noise sources because that is, it is not a continuous constant noise. Consider Fig. 4.31 which depicts the time history of an aircraft flyover. The important parameters for describing aircraft noise are the maximum noise level, the number of noise events (flyovers), the time of occurrence of the noise event, and the duration of the noise event.

The noise scale day-night noise level L_{dn} is one of the most common methods of combining these aircraft flyover characteristics into a single number that can be used for assessing the impact of the noise. The L_{dn} used for aircraft noise is the same as was described earlier in the section on Noise Propagation and is also used for highway noise assessment. The calculation of the L_{dn} around an airport is much more complex than for highways, as is shown later. Recalling that L_{dn} is a "time-weighted annual energy average

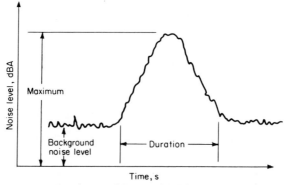

Figure 4.31 Time history of the noise level during an aircraft flyover.

noise level," it can be seen that a means for quantifying the sound energy of a flyover is needed. The sound pressure level in dBA as shown in Fig. 4.31 is only an instantaneous description of the pressure characteristics of the noise. The Sound Exposure Level (SEL) is the scale used to describe the energy content of flyover noise. Recall that for highway noise, the equivalent sound level L_{eq} accomplished this for the more continuous type of noise from highways. The energy content of the flyover shown in Fig. 4.31 is the logarithmic integral of the curve of sound pressure level versus time. The SEL is the energy of the flyover normalized to one second. For practical purposes, only the energy contained within 10 dB of the peak is used to compute SEL. For smooth curves that look similar to triangular pulses, the following approximate equation can be used to compute SEL for a flyover:

$$SEL \cong L_{max} + 10 \log t_{1/2} \tag{4.47}$$

where L_{max} is the maximum A-weighted noise level, and $t_{1/2}$ is one-half the 10 dB downtime, t_{10} (that is, one-half of the total time that the noise is within 10 dB of the peak.)

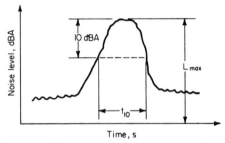

Figure 4.32   Quantities used to compute the approximate SEL for an aircraft flyover.

Figure 4.32 depicts the important quantities for computing the SEL of a flyover. Referring to the figure, $t_{1/2} = \frac{1}{2}t_{10}$.

Computing Noise Contours for Airports   The following algorithms and examples for computing noise levels around airports are presented to demonstrate the technique of generating noise contours. Because of the complexity of the problem and the need to perform many computations for even a small airport, the digital computer is ideally suited for computing noise contours. Several computer programs exist for this, including programs written by the Federal Aviation Administration, Environmental Protection Agency, and the United States Air Force.

Currently, only the state of California has laws requiring the establishment of noise contours around airports (California Administrative Code, Title 4, Subchapter 6). In California, the Community Noise Equivalent Level (CNEL) scale is used and is very closely related to L_{dn}. The only difference is that CNEL has a weighting for the evening time period (7:00 P.M. to 10:00 P.M.) while L_{dn} does not. In practice, the difference is only slight (generally less than 1 dB). California law specifies the following techniques to establish noise contours around an airport:

1. Estimate the location of the contour analytically.
2. Measure the noise for a short period.
3. Adjust the contours based on the short measurements and initiate long-term monitoring.

Simply stated, the L_{dn} at a particular location is computed by summing on an energy basis the L_{dn} due to each individual aircraft type and operation. The following equation expresses this mathematically:

$$L_{dn} = 10 \log \left[\sum_{i=1}^{a} \sum_{j=1}^{b} 10^{L_{dn_{ij}}/10} \right] \tag{4.48}$$

where a is the number of aircraft types, b is the number of operation types, and $L_{dn_{ij}}$ is the L_{dn} for the ith aircraft performing jth type of operation. $L_{dn_{ij}}$ is computed as follows:

$$L_{dn_{ij}} = \overline{SEL}_{ij} + 10 \log N_{t_{ij}} - 49.4 \tag{4.49}$$

where $\overline{SEL_{ij}}$ is the energy average SEL for the ith aircraft type performing the jth type of operation, and $N_{t_{ij}}$ is the *equivalent* number, N_t, of aircraft type i performing the jth type of operation.

N_t is computed as follows:

$$N_t = N_{\text{day}} + 10\,(N_{\text{night}})$$

where N_{day} is the number of daytime operations (7:00 A.M. to 10:00 P.M.) and N_{night} is the number of nighttime operations (10:00 P.M. to 7:00 A.M.).

From the equation for N_t, the nighttime penalty is included; that is, one flight at night is equivalent to ten flights during the day.

The following information is required to generate noise contours for an airport:

1. Number of operations by runway according to aircraft type, time of day, flight track, and profile.

2. SEL versus Slant Range Distance (SRD) by aircraft type.

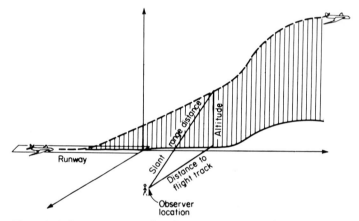

Figure 4.33 Important terms in describing airport operations for noise purposes.

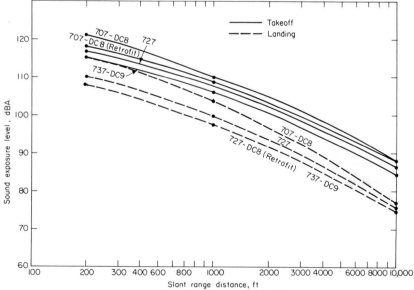

Figure 4.34 Sound exposure level vs. slant range distance for 707, DC8, 727, 737, and DC9 aircraft.

Some important terms need to be defined at this point. Referring to Fig. 4.33, "flight path" is the actual path of the aircraft in three-dimensional space. The "flight track" is the projection of the flight path on the ground. The "profile" is the altitude of the aircraft as a function of distance from the runway. Every unique combination of aircraft type, flight track, and flight profile needs to be treated separately in computing the noise level at specified observer locations. Noise contours are constructed by calculating the L_{dn} for a grid of locations, and the contour is established by curve-fitting techniques. Numerical methods for solving the preceding equations for a specified airport's operations using the digital computer have been developed but are beyond the scope of this analysis.

The "slant range distance" (SRD) is the line-of-sight distance between the observer and the closest point on flight path to the observer as shown in Fig. 4.33.

Given SRD, the SEL for a given operation can be obtained from SEL versus SRD curves for various aircraft shown in Figs. 4.34 through 4.37.

Sample Airport Noise Problem Consider an observer located near an airport which has only one runway and where all flights are straight in and straight out. The observer location is depicted in Figure 4.38, and the aircraft takeoff and landing profiles are shown in Fig. 4.39. Aircraft take off at a 10° angle and land at a 3° angle (glide slope).

As an example, calculate the L_{dn} at the observer location for Boeing 737 aircraft which make the following (annual average) number of takeoffs and landings per day.

| | Number of daily operations (B737) | |
Time period	Takeoff	Landing
Daytime (7 A.M. to 10 P.M.)	50	50
Nighttime (10 P.M. to 7 A.M.)	4	4

Solution:
  1. Find N_t:

$$N_{t_{\text{takeoff}}} = N_{t_{\text{landing}}} = N_{\text{day}} + 10(N_{\text{night}})$$
$$= 50 + 10(4)$$
$$= 90 \text{ equivalent operations per day}$$

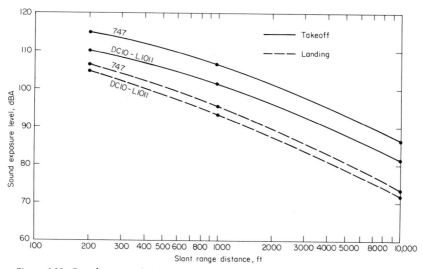

Figure 4.35 Sound exposure level vs. slant range distance for 747, DC-10, and L-1011 aircraft.

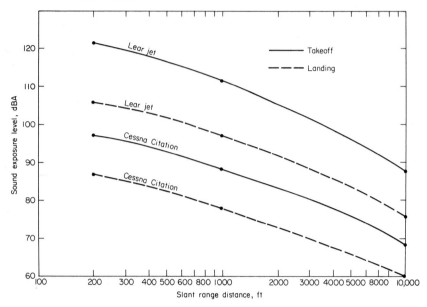

Figure 4.36   Sound exposure level vs. slant range distance for two business jets.

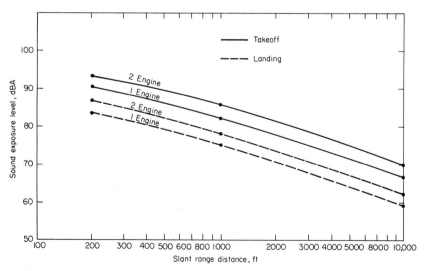

Figure 4.37   Sound exposure level vs. slant range distance for general aviation propeller aircraft.

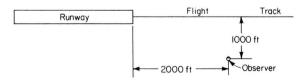

Figure 4.38   Sample airport configuration.

2. Find $L_{dn_{\text{Landing}}}$:

  a. Find the SEL for landing B737:

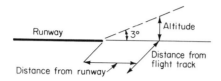

$$\text{SRD} = \frac{\text{Find the slant range distance (SRD):}}{\sqrt{(\text{altitude})^2 + (\text{distance to flight track})^2}}$$
$$\text{Altitude} = 2,000 \tan 3°$$
$$= 104 \text{ ft}$$

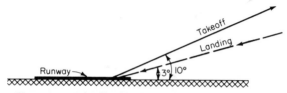

Figure 4.39   Takeoff and landing profiles for sample aircraft.

$$\text{SRD} = \sqrt{(104)^2 + (1000)^2}$$
$$= 1,006 \text{ ft}$$

Use Figure 4.34 to find the SEL for a landing B737 at a slant-range distance of 1006 ft.

$$\text{SEL}_{\text{landing}} = 97.5 \text{ dB}$$

  b.
$$L_{dn_{\text{landing}}} = \text{SEL} + 10 \log N_t - 49.4$$
$$= 97.5 + 10 \log (90) - 49.4$$
$$= 67.6 \text{ dB}$$

3. Find $L_{dn_{\text{takeoff}}}$

  a. Find the SRD for B737 takeoff:

$$\text{SRD} = \sqrt{(\text{altitude})^2 + (\text{distance to flight track})^2}$$
$$\text{altitude} = 200 \tan 10°$$
$$= 352 \text{ ft}$$
$$\text{SRD} = \sqrt{(352)^2 + (1,000)^2}$$
$$= 1,060$$

From Figure 4.34:

$$\text{SEL}_{\text{takeoff(B 737)}} \text{ at } 1,060 \text{ ft} = 106 \text{ dB}$$

  b.
$$L_{dn_{\text{takeoff}}} = \text{SEL} + 10 \log N_t - 49.4$$
$$= 106 + 10 \log (90) - 49.4$$
$$= 76.1 \text{ dB}$$

4. Find total L_{dn}:

$$L_{dn} = 10 \log \left(10^{L_{dn_{\text{landing}}}/10} + 10^{L_{dn_{\text{takeoff}}}/10}\right)$$
$$= 10 \log \left(10^{67.6/10} + 10^{76.1/10}\right)$$
$$= 76.7 \text{ dB}$$

Validating Noise Contours Around Airports   The best way to validate calculated airport noise contours (that is, to make sure they are correct) is to continuously monitor the noise at several locations around the airport 24 hours a day, 365 days per year. This, however, is very expensive and time consuming. If statistical sampling techniques are used, short-term monitoring can produce meaningful comparisons of analytical noise predictions and the actual noise environment.

Consider the equation used to analytically predict L_{dn}:

$$L_{dn} = \overline{SEL} + 10 \log N_t - 49.4 \tag{4.50}$$

In the analytical approach, the SEL value for a particular aircraft type was taken from published data. This number could very well have been measured in the field. There is a significant uncertainty if one measures too small a sample and thereby inaccurately estimates the true energy average SEL for a given aircraft type. Typically, at least 20 samples of a given aircraft type should be averaged at a given location.[5] However, cases around atypical airports (military, for example) where flight operations are highly variable have shown that as many as 1,000 flyover measurements may be necessary to get a reasonable estimate of the true average. Statistical techniques may be used to test the adequacy of the sample size.

SEL measurements are most easily made using an integrating sound level meter. Generally, integration is done digitally, and there are several low-cost meters on the market which will directly measure SEL.

Accuracy of Noise Contours Around Airports   An important note to make in this section is about the relative accuracy of noise contours around airports. In California, state law requires monitoring to be within an accuracy of plus or minus 1.5 dB. This may appear to be quite accurate, but when this range is translated to error bands of contours projected on the ground, the inaccuracy of noise contours becomes apparent. For example, for an airport with an average of 30 jet operations per day, the 1.5 dB accuracy translates to a distance accuracy of plus or minus 360 ft on the sideline of the 65 L_{dn} contour and plus or minus 3,900 ft on the closure point of the contour. The closure point is the location on the contour usually directly under the flight path. For 300 jet operations per day, these bands become plus or minus 1,300 ft on the sideline and plus or minus 14,200 ft on the closure point—almost 3 mi. Clearly, noise contours should not be considered as accurate lines on the ground across which the noise environment may go from acceptable to unacceptable. They should be used with other factors related to land use planning to provide valuable input to making land use decisions relative to the acceptability of the noise environment around airports for various land uses and human activities.

Railroad Noise

Railroad noise prediction can be relatively simple or highly complex depending on the type of operations. The case of single track with normal train operations will be addressed briefly here to give the reader an insight into the nature of railroad noise. However, more complex operations such as switching yards, railroad crossings, and booster engines will not be addressed here. Reference 20 contains a complete methodology for railroad noise prediction.

Railroad Noise Prediction Methodology   The basic equation for predicting railroad day-night noise levels (L_{dn}) is the one that was used for aircraft noise and is as follows:

$$L_{dn} = \overline{SEL} + 10 \log N_t - 49.4 \tag{4.51}$$

where $\overline{SEL}$ is the energy average sound exposure level for a train passby and N_t is the equivalent number of operations per day.

The equivalent number of operations per day is the time-weighted average number of operations per day and is calculated as follows:

$$N_t = N_{day} + 10 \, (N_{night}) \tag{4.52}$$

where N_{day} is the number of daytime operations (7:00 A.M. to 10:00 P.M.) and N_{night} is the number of nighttime operations (10:00 P.M. to 7:00 A.M.).

[5]In noise calculation, "average" is not taken literally to mean arithmetic average but is the average on an energy basis. On an energy basis, the SEL average is calculated from individual SEL measurements as follows:

$$\overline{SEL} = 10 \log \frac{1}{n} \left(\sum_{i=1}^{n} 10^{SEL_i/10} \right)$$

The key to predicting railroad noise is determining the $\overline{\text{SEL}}$ for the operations on the track. A simplified approximate methodology is presented here. (For a more precise methodology, consult Ref. 28.) Even in this simple methodology, locomotive noise must be considered separately from car noise. The following methodology should be used to determine the $\overline{\text{SEL}}$ for trains.

1. Calculate the effective time duration (half the 10 dB downtime) for the passby using the following equation:

$$t_{1/2} = 0.68 \times L/V \qquad \text{(seconds)} \tag{4.53}$$

where L is the train length in ft and V is the train speed in mi/h.

2. Calculate the $\overline{\text{SEL}}$ for the cars at 100 ft from the track using the following equation:

$$\overline{\text{SEL}}_{\text{cars}} = L_{\text{max}} + 10 \log t_{1/2} \tag{4.54}$$

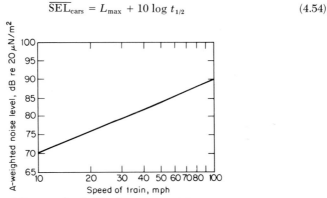

Figure 4.40   Noise level of freight cars at 100 ft from the track.

where L_{max} is the maximum noise level during the passby at 100 ft and can be found in Fig. 4.40 as a function of train speed.

This $\overline{\text{SEL}}$ value for train cars must now be adjusted for track characteristics. Table 4.12 presents the correction factor to be added to the car $\overline{\text{SEL}}$.

The $\overline{\text{SEL}}$ for locomotives at 100 ft can be found directly in Figure 4.41 as a function of speed and track grade.

The total $\overline{\text{SEL}}$ for locomotives and cars at 100 ft can be found by logarithmically summing the individual $\overline{\text{SEL}}$'s as follows:

TABLE 4.12   Corrections to SEL for Track Characteristics

Track characteristic	Correction, dB
1. Mainline welded or jointed track	0
2. Low-speed classified jointed track	8
3. Presence of switching frogs or grade crossing	8
4. Tight radius curve	
a. Radius less than 600 ft	8*
b. Radius 600 to 900 ft	2*
c. Radius greater than 900 ft	0
5. Presence of bridgework	
a. Light-steel trestle	20
b. Heavy-steel trestle	10
c. Concrete structure	0

*Interpolate between values for additional refinement.
NOTE: In case of simultaneous occurrence for these factors, the single largest correction is to be applied.
SOURCE: "Assessment of Noise Environments Around Railroad Operations," Wyle Labs, July 1973.

$$\overline{SEL}_{total} = 10 \log \left[10^{\overline{SEL}_{locomotive}/10} + 10^{\overline{SEL}_{cars}/10} \right] \tag{4.55}$$

The L_{dn} at 100 ft can now be calculated as follows:

$$L_{dn} = \overline{SEL}_{total} + 10 \log N_t - 49.4 \tag{4.56}$$

To find the L_{dn} at some distance other than 100 ft, use the following correction factor:

$$\Delta L_{dn} = 15 \log \frac{100}{D} \tag{4.57}$$

where ΔL_{dn} is the change in L_{dn} as a result of the observer being located some distance D in feet from the tracks instead of 100 ft.

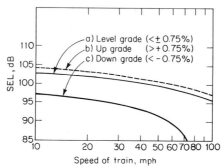

Figure 4.41   Engine SEL contribution at 100 ft from the track for a train traveling at the grade condition indicated.

Sample Railroad Noise Problem   Calculate the L_{dn} for an observer located 400 ft from a single track (welded, no grade, no crossing) which handles the following operations.[6]

		Number of operations	
Speed, mi/h	Length, ft	Day	Night
35	3,000	10	2

Solution:
1. Find the equivalent number of operations:

$$N_t = 10 + 10(2) = 30$$

2. Calculate the car $\overline{SEL}$:
   a. Calculate the effective time duration:

$$t_{1/2} = 0.68 \times \frac{3000}{35} = 58 \text{ seconds}$$

   b. Find L_{max} from Fig. 4.40 at 100 ft:

$$L_{max} = 81 \text{ dBA}$$

   c. Calculate the $\overline{SEL}_{car}$ at 100 ft

$$\overline{SEL}_{car} = 81 + 10 \log 58 = 98.6 \text{ dB}$$

   d. Track correction = 0 dB (Table 4.12)
3. Find the locomotive $\overline{SEL}$ from Fig. 4.41 at 100 ft:

$$\overline{SEL}_{locomotive} = 102 \text{ dB}$$

[6]If there had been more than one unique train speed and length, each unique case would be treated individually and the component L_{dn}'s would be summed logarithmically.

4. Calculate the total $\overline{\text{SEL}}$ at 100 ft:

$$\overline{\text{SEL}}_{\text{total}} = 10 \log (10^{98.6/10} + 10^{102/10}) = 103.6 \text{ dB}$$

5. Calculate the L_{dn} at 100 ft:

$$L_{dn} = 103.6 + 10 \log 30 - 49.4 = 69.0 \text{ dB}$$

6. Correct the L_{dn} at 100 to 400 ft by finding the correction factor:

$$\Delta L_{dn} = 15 \log \frac{100}{400} = -9 \text{ dB}$$

Therefore the total L_{dn} at 400 ft is

$$L_{dn} = 69.0 - 9$$
$$= 60.0 \text{ dB}$$

EXAMPLES OF EIS NOISE SECTIONS

Sample noise sections from Environmental Impact Statements are presented in this section. These examples have been extracted from real EIS's but have been modified considerably for inclusion here. Basically, the examples have been edited to remove background material, definitions, and discussions that have been presented earlier and are not necessary to repeat in the examples. The examples have been modified to show the essential parts of a noise analysis of an EIS. Supporting calculations for numbers presented are not shown but can be derived using the techniques presented in the preceding sections of this chapter.

The three examples are for the following types of projects:
1. Residential development
2. Shopping center development
3. Airport modification

The names of projects and cities have been changed to prevent the edited EIS sections presented here from being confused with the actual EIS prepared for the projects.

Example 1: Residential Development

Introduction The purpose of this analysis is to determine the noise impact of roadways in and near the project in the city. Roadways of concern include Canyon Road and Rock Drive. Of the two roads, only Rock Drive is currently in existence. The noise impact analysis was carried out using the noise criteria specified in the General Plan of the city. These criteria are presented and described in the next section. Existing noise levels are also briefly described in this report, followed by a detailed discussion of the noise emanating from and impacting on proposed residences in the project adjacent to the critical roads.

Noise Criteria and Noise Projections The noise criterion used in this analysis, from the General Plan for the city, is a maximum of 65 L_{dn} exterior noise level, where L_{dn} stands for day-night noise level. L_{dn} is a time-weighted annual average noise level based on the A-weighted decibel (a measure of loudness that corresponds closely to the human perception of loudness). "Time-weighted" refers to the fact that nighttime noise levels are penalized 10 dB if the noise occurs between the hours of 10:00 P.M. and 7:00 A.M. The most common criteria for determining the acceptability of noise in residential communities is an outdoor level of 65 L_{dn} and an indoor level of 45 L_{dn}.

Noise projections made in this report are based upon anticipated ultimate traffic volumes for the two roads. Projections were made using the basic noise model described in the section on Highway Noise at the beginning of this chapter. (No calculations are presented here but would usually be included as an appendix to the EIS. For the purposes of this handbook, the reader is referred to the aforementioned section for details of calculations.)

Existing Noise Levels The existing noise levels in the vicinity of the project are now quite low since Rock Drive is the only existing roadway adjacent to the project and current traffic volumes are low. No recent traffic counts are currently available for this section of

Rock Drive. A traffic study, now one year old, identified a volume of 300 Average Daily Trips (ADT) existing at that time. In view of the limited number of new residences occupied in nearby developments since that time, an existing ADT of 500 has been conservatively assumed adjacent to the project. Using the estimated ADT on Rock Drive of 500 vehicles (no trucks and an average vehicle speed of 45 mi/h), the 65 L_{dn} contour lies within the right of way of the road. Because of the low existing noise levels in the community, noise measurements are not warranted.

Noise Impact

Exterior Noise   The traffic data used to project roadway noise levels due to ultimate traffic volumes are shown in Table 4.13. Ultimate volumes for Rock Drive were obtained

TABLE 4.13   Roadway Traffic Data

Roadway	ADT Existing	ADT Ultimate	Auto speed, mi/h	Truck speed, mi/h	Percent heavy trucks	Percent grade
Canyon Road	$\cdots$	16,000	55	55	3.5	1
Rock Drive	500	5,600	40	$\cdots$	Neg.	2

from the city and are based upon a recent traffic study. Ultimate volumes for Canyon Road were based on the city's Traffic Analysis Program (TAP). Typical sections were taken for Canyon Road and Rock Drive and were used for the noise analysis.

Each road is discussed separately as follows. The projected ultimate traffic for Canyon Road is 16,000 ADT. Based on this volume, the approximate distance from the centerline of the road to the 65 L_{dn} noise contour is 182 ft. An examination of the proposed project shows that the buildings will be located 108 ft from the centerline of Canyon Road. The L_{dn} at this location is 68.4 dB. Mitigation measures are required and are discussed later in this report.

The projected ultimate traffic volume for Rock Drive is 5,600 ADT. For the ultimate traffic volume, the 65L_{dn} contour lies 31 ft from the approximate centerline of Rock Drive. The proposed setback along Rock Drive is 100 ft from the road centerline. Clearly, no residence is impacted by the 65 L_{dn} contour. In fact, for observers located at the nearest wall of the residence to the roadway, the projected exterior noise is 57.4 L_{dn}. No mitigation is required for noise purposes along Rock Drive.

A summary of the exterior noise impact analysis is presented in Table 4.14.

TABLE 4.14   Summary of Exterior Noise Impact Analysis

Roadway	Effective distance from C_L to observer,* ft	Effective distance to 65 CNEL from C_L, ft	CNEL at observer,† dB	Noise reduction needed, dB	Required noise barrier height, ft
Canyon Road	108	182	68.4	3.4	6
Rock Drive	100	31	57.4	0	None

*Effective distance from the centerline (C_L) of the roadway to the observer is equal to the square root of the distance to the centerline of the nearest lane times the distance to the centerline of the far lane and is usually not significantly different from the distance to the actual centerline.
†In all cases, the observer was located at the nearest wall of the residence to the roadway.

Interior Noise   Interior noise levels have not been specifically addressed in this analysis. However, in light of anticipated exterior noise levels and assuming standard building construction, there should be no difficulty complying with the interior noise standard of 45 L_{dn}. For construction of the type to be used for the proposed residences, indoor noise levels are generally attenuated about 20 dB for highway noise when all windows and doors are closed.

Mitigation Measures   Homes located along Canyon Road will be exposed to L_{dn} levels of approximately 68.4 dB unless some mitigation is designed into the project. Practical alternatives include placing the home outside the 65 L_{dn} contour by increasing the setback of homes an additional 74 ft or constructing a noise barrier between the road and the homes. The homes are at grade with the roadway, and a barrier constructed at the property line would be located 70 ft from the centerline of Canyon Road. Using the barrier-analysis technique described in an earlier section of this chapter, the required barrier height for this project is approximately 6 ft.[7]

Noise Reduction Due to Quieter Vehicles   Both the federal government and the state have enacted legislation establishing noise emission standards for automobiles and trucks. As this legislation takes effect (dependent on the turnover from old to newer vehicles), vehicular noise levels are expected to decrease. Estimates of the noise reduction achieved by this legislation have been from 5 to 15 dB for some vehicles. But, because of uncertainties in the rate of change of the vehicle population (for trucks, in particular) and enforcement of the noise emission standards, a more conservative estimate of the noise reduction is common. It is quite common to assume a 3 dB noise reduction for future-year vehicle populations, i.e., the 1985 to 1990 timeframe when ultimate traffic volumes are reached. In the case of this study, *no* such noise reduction was accounted for in the noise projections, and thus the levels presented here represent a conservative analysis. The effect of a 3 dB noise reduction due to technological improvements in vehicular design is that future volumes at twice the traffic level of presently assumed ultimate traffic volumes would produce the same noise level.[8] Consequently, it can be concluded that ultimate traffic volumes could reach twice the volumes projected here and still meet the noise criterion specified. For example, the residences along Canyon Road would not be adversely impacted by ultimate traffic volumes of up to 32,000 ADT.

Example 2: Shopping Center Noise Analysis

Existing Noise   The existing noise environment around the proposed shopping center is dominated by motor vehicle noise, and the freeway is the major noise source in the vicinity of the project. The existing noise levels in the community are established in terms of the day-night noise level (L_{dn}) by modeling the roadways for the current traffic and speed characteristics. A description of L_{dn} is provided in the Noise Impact section of this report. The traffic characteristics used in the model are given in Table 4.15.

TABLE 4.15   Traffic Data Used to Generate Existing Noise Levels

Roadway	ADT	Average speed, mi/h*
Freeway	139,000	50
Gabriel Blvd.	17,600	35

*24-hour average speed; 3 percent heavy trucks.

The distances to the L_{dn} contours for the roadways affected by the proposed shopping center are given in Table 4.16. These distances represent the distance from the centerline of the road to the contour value shown.

TABLE 4.16 L_{dn} Contours for Current Conditions

Roadway	Distance to CNEL contour from centerline of road, ft	
	60 L_{dn}	65 L_{dn}
Freeway	1,359	631
Gabriel Blvd.	205	95

Note that the values given in Table 4.16 do not take into account the effect of any noise barriers (such as the wall on the north side of the freeway) or topography that may affect ambient noise levels. Noise level criteria are given in the Noise Impact section of this report, but it should be pointed out that $65L_{dn}$ is the usual criterion for determining the land uses.

[7]Assumptions include a source height of 8 feet (heavy trucks) and an observer height of 5 feet for a single story structure.
[8]A 3 dB noise reduction is the same as dividing the traffic volume by one-half.

Noise Impact The impact of the proposed shopping center will be analyzed and quantified by comparing community noise with and without the project for the year 1982. The comparison will be based on the noise scale L_{dn}. The same scale was used in the Existing Noise section of this report.

The Noise Model The noise model used to project the noise levels for 1982 (as well as the noise levels shown under existing conditions) was based on the "Federal Highway Administration Highway Traffic Noise Prediction Model" developed by the FHWA in 1977. The programming and all calculations for this model were done by computer. The model uses the projected roadway average daily traffic (ADT), vehicle speed, and traffic distribution to project noise levels. Noise levels for the year 1982 with and without the project are given in the next section.

The traffic data used to project 1982 noise levels are shown in Table 4.17:

TABLE 4.17   Traffic Data Used to Project 1982 Noise Levels

| | 1982 Traffic conditions | | | |
| | Without project | | With project | |
Roadway	ADT	Speed*	ADT	Speed*
Freeway	174,000	50	178,000	50
Gabriel Blvd.	22,000	30	27,200	30

*24-hour average speed in miles per hour; 3% heavy trucks.

Noise Impact of the Project The distances to the 60 and 65 L_{dn} contours from the centerline of the roadway for each roadway, with and without the project, are shown in Table 4.18 for the year 1982. Note that Table 4.18 does not take into account the effect of any walls or barriers that may affect ambient noise levels.

TABLE 4.18   Distances to L_{dn} Contours for 1982 With and Without Project

| | Distances to contour from centerline (ft) | | | |
| | Without project | | With project | |
Roadway	60 L_{dn}	65 L_{dn}	60 L_{dn}	65 L_{dn}
Freeway	1,561	724	1,609	747
Gabriel Blvd.	203	94	236	110

An interesting observation should be noted about Table 4.18. If Table 4.18 is compared with Table 4.16 (Existing Conditions), it can be seen that the contours for Gabriel Boulevard are smaller for 1982 than those for 1978 even though the traffic volume increased. This is caused by the assumption that from 1978 to 1982 the average speed decreased from 35 to 30 mi/h. Noise levels increase with increasing traffic volume and decrease with decreasing traffic speed. In the case of Gabriel Boulevard, the decrease in the traffic speed had a more dominant effect than the increase in traffic volume, and, therefore, the noise is reduced. In reality, it is difficult to say if this projected reduction will actually occur. In any case, the difference between existing and future noise levels will be quite small.

Table 4.18 can be used to directly compare the project with the no-project alternative in terms of area of impact. Sometimes it is valuable to compare the alternatives in terms of a change in noise level in decibels. Table 4.19 presents the increase in noise levels due to the project for each roadway.

TABLE 4.19   Increase in Noise Levels Due to the Project

Roadway	Increase in noise level due to the project, dBA
Freeway	0.1
Gabriel Blvd.	1.0

The projected increase in noise level shown in Table 4.19 is insignificant when one considers that the human ear is just barely able to discern a noise increase of 3 dBA. The conclusion, therefore, is that the project will not adversely impact community noise levels in a significant way.

Mitigation Measures   Mitigation measures to reduce the impact of the proposed shopping center on the noise environment include reducing both automobile traffic and automobile noise. Current legislation in both the state and on the federal level requires new cars to meet even stricter noise emission standards. Conservatively, the effect of this legislation has been estimated to account for 3 to 5 dB reductions in motor vehicle noise levels. This reduction was not taken into account in the noise level projections made in this report and may further reduce the impact of the shopping center.

Example 3: Airport Noise Analysis

Environmental Setting   The sample airport currently consists of a single runway 8,600 ft long and 150 ft wide and is oriented in a north-south direction. The airport serves five major airlines and several third-level carriers with limited operations, and it also serves as a major pilot training facility for airlines in addition to their normal operations.

Based upon traffic, the sample airport is officially classified as a small hub airport by the Civil Aeronautics Board. The number of airline passengers serviced by the airport has grown substantially in the past few years. From 1968 to 1975 the number of passengers has doubled from 554,140 to an estimated 1,100,000. However, while the scheduled air carrier operations have fluctuated from year to year, they have not increased from 1968 to 1975 and have remained at approximately 40,500 annually.

Environmental Impact Analysis

Environmental Impact of the Proposed Action   Airport noise has been an environmental problem since the introduction of commercial jet aircraft in 1959. This problem has increased in magnitude since that time due to increased aircraft operations, related increases in surface traffic, and ineffective land use planning in the surrounding environs. Only in recent years have those decision makers associated with both rapid urbanization and increased mobility considered the environmental aspects of their actions.

This section presents the effects of noise and the noise impact analysis. The noise impacts were evaluated for three study years of 1975, 1985, and 1995. The two future years were selected to compare the noise generated by the existing runway configuration and airport operation with noise impacts from the project and alternative runway configurations.

In summary:

- Noise may be a potential health hazard.
- Federal laws and regulations and local government ordinances are designed to minimize potential noise-induced handicaps.
- Federal and local government regulations also provide some protection to citizens from annoying or disturbing noise.
- Additional regulations are being considered to provide more desirable living conditions.

Using the appropriate input data, the L_{dn} computer program is able to calculate the noise generated by aircraft operations. The graphical depiction of the mathematical results of an L_{dn} analysis is a set of noise contours. The program includes standard performance data for all aircraft types expected to operate at sample airports. Standard performance data include flight profiles (aircraft altitude during takeoff as a function of distance from the start of takeoff roll) and sound profiles as a function of the distance from the aircraft.

To generate specific noise contours for the sample airport, the following additional information was used in the computer analysis:

1. A landing glide slope of 3 degrees
2. The existing and future runway lengths at the sample airport
3. For each separate flight path segment, the volume of operations during a typical 24-hour period, segregated by day and night and by trip length

The L_{dn} methodology can be combined with projections for airport operations to provide anticipated noise contours for future operations. In general, variations in aircraft flight paths tend to increase as distance from the airport increases. Thus, the accuracy of

TABLE 4.20 Estimated Volume of Operations at Sample Airport in 1975

Trip length, mi	Average departures per day by aircraft type							
	4-eng. HBRF	4-eng. turbofan	4-eng. turbojet	3- and 2- engine HBRF*	3-eng. stretched turbofan	3-eng. turbofan	2-eng. turbofan	Total
Training	5.5	11.6	0.7	3.5	2.5	1.1	3.5	28.4
0 to 500		3.1	3.0		19.0		16.5	41.6
500 to 1,000								
1,000 to 1,500					2.0			2.0
1,500 to 2,500		1.0						1.0
2,500 to 3,500								
3,500 to 4,500								
4,500 +								
Total	5.5	15.7	3.7	3.5	23.5	1.1	20.0	73.0

Average departure per day for general aviation and other operations

Aircraft type	Air taxi	Business jet	Military jets	Nonjet general aviation		Total
				1-eng	2-eng	
Average daily departures	9.3	4.9	3.6	30.2	12.1	60.1

TOTAL 133.1

*HBRF—High-Bypass-Ratio Turbofan

TABLE 4.21 Estimated Volume of Operations at Sample Airport in 1995

Trip length (miles)	4-eng. HBRF*	4-eng. turbofan	4-eng. turbojet	3 and 2-engine HBRF*	3-eng. stretched turbofan	3-eng. turbofan	2-eng. turbofan	Total
Training	12.0			19.0	6.0		2.0	39.0
0 to 500				5.2	36.9		32.9	75.0
500 to 1,000				5.0	8.5			13.5
1,000 to 1,500				10.0				10.0
1,500 to 2,500				8.0				8.0
2,500 to 3,500	5.2			.9				6.1
3,500 to 4,500	4.0							4.0
4,500 +	2.0							2.0
Total	23.2			48.1	51.4		34.9	157.6

Average departures per day by aircraft type

Average departure per day for general aviation and other operations

Aircraft type	Air taxi	Business jet	Military jets	Nonjet general aviation		Total
				1-eng.	2-eng.	
Average daily departures	32.9	26.4	4.1	52.7	38.1	154.2

TOTAL 311.8

*HBRF—High-Bypass-Ratio Turbofan

the L_{dn} contour tends to decrease as distances from the airport increase. The contour presentation is useful in displaying the general impact of aircraft noise in the neighboring communities. The contours were based on the best available projections of airport activity. Tables 4.20 and 4.21 present operations data for the sample airport.

In order to quantify and assess the net change in noise exposure, L_{dn} contours were developed for study cases. These cases included the existing runway configuration and the project. The two cases are shown in Table 4.22.

TABLE 4.22 L_{dn} Noise Study Cases

Case	Description
1	Existing runway in 1975
2	The project in 1995

The basic case parameters of runway length, allocation of operations to runway(s), daily volumes of operations, and day, evening, and night splits for each case are shown in Table 4.23.

TABLE 4.23   Noise Study Case Parameters

Case	Runway length	Alloc.	Vol. of ops. (daily)	Time splits
Case 1: 1975 existing runway	8,600 ft	100%	133.1	Training: 100% day Air carrier: 71.9% day 20.8% eve. 7.3% night
Case 2: 1995 The project (existing runway and second runway*)	12,000 ft	50%	311.8	Same as above

*The separation between runways for the project is 6,700 ft.

The time allocations for day, evening, and night were derived from FAA data for a sample day in 1975 (May 2). The sample day showed a time split for air carriers of 71.9 percent day (7:00 A.M. to 7:00 P.M.), 20.8 percent evening (7:00 P.M. to 10:00 P.M.), 7.3 percent night (10:00 P.M. to 7:00 A.M.). The time split used for air taxi was 88 percent day (7:00 A.M. to 7:00 P.M.) and 12 percent evening (7:00 P.M. to 10:00 P.M.). The same day, evening, and nighttime splits used for air carriers were used for general aviation operations. Training operations were allocated 100 percent to the day period (7:00 A.M. to 7:00 P.M.) based on tower conversations. The same time allocation percentages were assumed for all study years.

Areas currently exposed to aircraft noise can be defined and the exposure can be established by the direct measurement of sound pressure levels. If this approach were followed, however, it would not only be time-consuming and expensive, but it would provide data only on current operations and not about anticipated future levels of noise. More importantly, the noise observations would have to be translated into some expression of degree of interference or annoyance in order to be of use in the community planning and noise control processes.

The L_{dn} noise exposure for the two study cases and the 60, 65, 70, and 75 L_{dn} values are shown in Figs. 4.42 and 4.43. These contours were compared to determine whether the noise impact as a result of airport operations in the project runway configuration would increase or decrease.

The noise contours generally show an increase in total area impact in future years. While the number of aircraft departures increases from 133.1 in 1975 to 311.8 in 1995, the noise contours do not increase by a similar proportion. This is a result of the increased use of FAR Part 36 certified aircraft (aircraft with noise generation restrictions) from 7 percent of the fleet mix in 1975 to almost 100 percent of the fleet mix in 1985 and 1995. Therefore, by 1985 the fleet mix is stabilized to the extent that further reductions in the L_{dn} contours will not occur without additional noise abatement measures. Some of the benefits, however, will not be fully attained in 1985 since a greater number of future aircraft will be traveling greater trip lengths, which will result in larger gross weights and increased noise.

Land use incompatible with the proposed airport project is defined by the Noise Standards as single-family dwelling units, multiple-family dwelling units, schools, hospitals, and trailer parks. One objective of this standard is to create an urban development pattern in which all the land included within the criterion L_{dn} contour is devoted to either airport or nonresidential purposes. If the land uses surrounding the airport do not comply with the standards, the airport operator is required to obtain a variance to continue airport operations. To obtain this variance, an airport operator must file a plan showing how full compliance will be attained. Further, an airport operator cannot cause an increase in the size of the impact area. The incompatible acres surrounding the sample airport that will result from the project are shown in Table 4.24. For comparative purposes, the acreage

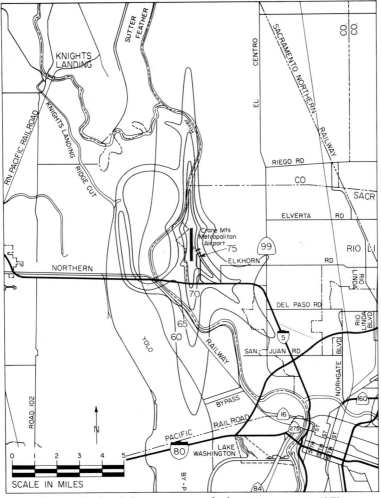

Figure 4.42   Case 1: L_{dn} noise exposure for the existing runway in 1975.

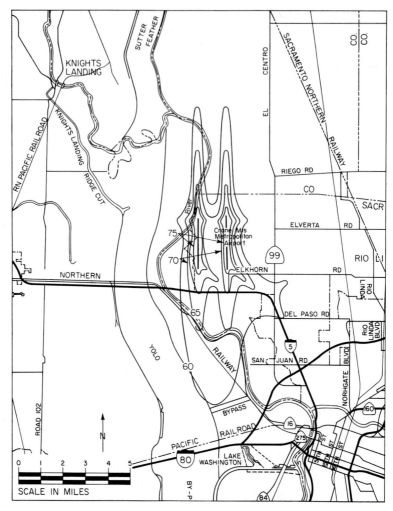

Figure 4.43   Case 2: L_{dn} noise exposure for the project in 1995.

TABLE 4.24   Incompatible Acres Associated with the Project

		Incompatible acres	
Year	L_{dn} standard	Project	Do nothing
1995	65	34	32

TABLE 4.25   Population and Area Impacted for Comparison Runway Configuration*

		60 L_{dn}		65 L_{dn}		70 L_{dn}		75 L_{dn}	
Case	Description	Pop.	Acres	Pop.	Acres	Pop.	Acres	Pop.	Acres
1	1975 Existing runway	235	15,850	100	4,750	60	1,050	30	210
2	1995 Project	250	11,350	120	1,800	60	350	25	40

*Population projections are based on estimates from the county. Acres impacted consist of off-airport area and projected airport acquisition of property by 1985 and by 1995.

that would exist under continuation of the single runway or "do nothing" case are also shown.

The population and area impacted for the two study cases and the 60, 65, 70, and 75 L_{dn} noise contours are shown in Table 4.25. The population projections for future-case years are based upon estimates from the county. The number of acres impacted consists of off-airport area and projected airport acquisition of property either through air easements or on a purchase and leaseback basis. By 1985, additional property is projected to be acquired to the north and east, and by 1995, further acquisition is projected to take place to the south and west. The total area projected to be acquired by 1995 is estimated to be 3,869 acres.

Mitigation It is recommended that the Department of Airports, in cooperation with local and regional governmental agencies, undertake one or more of the following mitigation measures to minimize airport noise impact to incompatible land uses:

- Acquire incompatible land uses within the 1995 estimated 65 L_{dn} and convert to compatible uses
- Purchase and rent-back noise impacted homes within the 1995 estimated 65 to 75 L_{dn} contours
- Guarantee future resale price to the landowner
- Purchase aviation easements in noise-sensitive areas
- Assure that future incompatible land uses are not constructed within the 1995 estimated 65 L_{dn} contour, and that all new development (within the Airport 60 L_{dn} contour) follow the noise insulation standards
- Soundproof residential units within the 1995 estimated 65 L_{dn} contour, to achieve an indoor sound level of 45 L_{dn} in exchange for aviation easements.

REFERENCES

1. Beranek, L. L., *Noise and Vibration Control*, rev. ed., McGraw-Hill, New York, 1971.
2. Bragdon, C. R., *Noise Pollution: The Unquiet Crisis*, University of Pennsylvania, 1970.
3. Broch, J. T., *Acoustic Noise Measurement*, Bruel & Kjaer, 1971.
4. Crocker, M. J. and A. J. Price, *Noise and Noise Control*, vol. I, CRC Press, Cleveland, 1975.
5. Cunniff, P. F., *Environmental Noise Pollution*, Wiley, New York, 1977.
6. Doelle, L. L., *Environmental Acoustics*, McGraw-Hill, New York, 1972.
7. Environmental Protection Agency, "Public Health and Welfare Criteria for Noise," July 1973.
8. Environmental Protection Agency, "Information on Levels of Environmental Noise Requisite to Protect Public Health and Welfare with an Adequate Margin of Safety," March, 1974.
9. Environmental Protection Agency, "Sound Exposure Level versus Distance Curves for Civil Aircraft," October 1974.
10. Federal Highway Administration, "Highway Noise Barrier: Selection, Design, and Construction Experiences," Implementation Package 76-8, Washington, D.C., 20590, 1976.
11. Federal Highway Administration, "Noise Barrier Design Handbook," Research Report FHWA-RD-76-58, Washington, D.C., 20590, February 1976.
12. Federal Highway Administration, "Noise Barrier Attenuation: Field Experience," Research Report FHWA-RD-76-54, Washington, D.C., 20590, February 1976.
13. Federal Highway Administration, "Manual for Highway Noise Prediction," Report No. OST-TSC-FHWA-72-1, Washington, D.C., 20590.
14. Federal Highway Administration, "FHWA Highway Noise Model," (Draft Report), December 1977.
15. Highway Research Board, "Highway Noise—Measurement, Simulation and Mixed Reactions," National Cooperative Highway Research Program Report 78, National Academy of Sciences, National Academy of Engineering, 1969.
16. Highway Research Board, "Highway Noise—A Design Guide for Engineers," National Cooperative Highway Research Program Report 117, National Academy of Sciences, National Academy of Engineering, 1971.
17. Highway Research Board, "Highway Capacity Manual," 1965.
18. U.S. Department of Housing and Urban Development, "Policy Circular No. 1390.2," August 1971.
19. Shultz, T. J., U.S. Department of Housing and Urban Development, Shultz, T. J., and McMahon, N. M., "Noise Assessment Guidelines," August 1971.
20. Kinsler, L. E., and A. R. Frey, *Fundamentals of Acoustics*, 2d ed., Wiley, New York, 1962.
21. Kryter, K., *Noise and Vibration Control*, McGraw-Hill, New York, 1971.
22. Miller, J. D., "Effects of Noise on People," *Journal of the Acoustical Society of America*, **56**:729, 1974.
23. Peterson, A. P. G., and E. E. Gross, *Handbook of Noise Measurement*, 7th ed., General Radio, 1974.
24. Shultz, T. J., *Community Noise Ratings*, Applied Science, London, 1972.

25. Swing, J. W., "Simplified Procedure for Developing Railroad Noise Exposure Contours," *Sound and Vibration*, February 1975.
26. Transportation Research Board, "Highway Noise—Generation and Control," National Cooperative Highway Research Program Report 173, Washington, D.C., 1976.
27. Transportation Research Board, "Highway Noise—A Design Guide for Prediction and Control," National Cooperative Highway Research Program Report 174, Washington, D.C., 1976.
28. Wyle Laboratories, "Assessment of Noise Environments Around Railroad Operations," July 1973.

Chapter 5

Energy Impact Analysis

BLAIR FOLSOM

This chapter addresses the analysis and preparation of an energy element for an environmental impact statement (EIS). Since the legal requirements are in a state of flux, the discussion focuses on the underlying principles of energy impact assessment and the organization of comprehensive energy elements for all project types. The Introduction outlines existing requirements on the national and state levels, defines energy impacts, and describes typical energy impact considerations. The second section is an overview of energy-related issues important to energy impact analysis and discusses the interrelationship between energy and environmental problems. The third section describes the organization of an energy element focusing on the methodology of energy impact assessment, supply/demand analysis, and energy conservation measures. The final section lists energy-related data useful in the preparation of energy elements and gives data sources for additional information.

INTRODUCTION

Legal Requirements

At the time that most environmental legislation was promulgated, energy and its interrelationship with the environment were not considered critically important issues. Consequently, energy impact assessments and energy elements are not currently required in the majority of environmental impact statements.

This scenario is rapidly changing. The current interest in energy problems has brought with it a barrage of new legal requirements demanding consideration of energy impact and dictating energy-efficient design and conservation standards. Energy impact analysis is now required in certain federal EIS's, and some states (most notably California) require an energy impact assessment as part of every environmental impact report (EIR) produced under their jurisdiction.

Chapter 1 lists the recent federal legislation affecting the preparation of EIS's. The Department of Energy (DOE) is the federal agency with authority in this field.[102] This responsibility includes:

- Performing environmental review from an energy perspective of Environmental Protection Agency (EPA) programs and projects
- Providing guidance for preparing EIS's
- Coordinating comments on EIS's prepared by others

The EIS's reviewed by DOE deal either directly or indirectly with energy resource development, production, distribution, and end use. The review process is performed to assure that adequate consideration is given to energy-efficient alternatives, that the proposed action is in keeping with national energy policy, and that adequate energy supplies are maintained on a national level.

Legislation requiring energy impact assessment has also been promulgated at the state level. In particular, the state of California has established very specific requirements for energy elements in the preparation of EIR's,[27] such as:

(a) Describe avoidable adverse impacts, including *inefficient and unnecessary consumption of energy,* and the measures proposed to minimize these impacts. This discussion shall include an identification of the acceptable levels to which such impacts will be reduced, and the basis upon which such levels were identified. Where alternative measures are available to mitigate an impact, each should be discussed and the basis for selecting one alternative should be identified.

(b) *Energy conservation measures,* including both the available alternatives and those incorporated into the design and operation of the proposed project, shall be discussed as mitigation measures. There are many ways in which a project may be designed or operated to cause less energy to be consumed both directly and indirectly. Examples include but are not limited to:

    (i) Insulation and other protection from heat loss or heat gain to conserve fuel used to heat or cool buildings and mobile homes.

    (ii) Use of resource conserving forms of energy such as solar energy for water and space heating, wind for operating pumps, and falling water for generating electricity.

    (iii) Energy efficient building design including such features as orientation of structures to summer and winter sunlight to absorb winter solar heat and reflect or avoid summer solar heat.

    (iv) Measures to reduce energy consumption in transportation such as access to alternate means of transportation, use of small cars, alternate means of shipping, energy conserving lighting and construction practices, use of alternate power sources, waste heat recovery and recycling and use of recycled materials.[27]

It is expected that most states will specifically require consideration of energy impact issues in environmental impact statements (or related documents) in the near future.

The task of performing an energy analysis for an EIS has been made significantly easier by the passage of legislation dictating energy design and conservation standards. Prime examples of such standards are the regulations of the California Energy Resource Conservation and Development Commission. This recently established (1975) state agency has supported energy research and development and promulgated many very specific regulations and standards including insulation requirements and minimum design requirements for heated and/or cooled buildings to be constructed within the state.[20] The value of energy design and conservation standards, from the viewpoint of the preparation of an energy element for an EIS, lies not only in the standards themselves but also in studies demonstrating effectiveness of the standards and in information provided by cognizant government authorities to aid in compliance.

The standards themselves provide a simple and meaningful mechanism for comparing a specific project with others within the jurisdiction. For example, the California standards for energy insulation specify maximum overall thermal conductance for walls, ceilings, and floors between heated and unheated areas. Without these standards a technical analysis must be conducted to determine the thermal conductance required to minimize energy impact. If a project meets the standards, it has by governmental definition acceptably minimized the energy impact.

Supportive reports and other data generated by the cognizant governmental jurisdiction to demonstrate the validity and effectiveness of energy design and conservation standards can also aid in the preparation of the energy element for an EIS. For example, to determine the effectiveness of proposed nonresidential energy conservation standards, the California Energy Resources Conservation and Development Commission contracted for an economic and energy effectiveness study.[68] This study compared the cost and energy consumption of six building types as designed under present practice and with modifications conforming to the proposed standards. The results and conclusions from this study can be used to assess the energy impact of similar designs.

Since energy design and conservation standards are complex technical requirements, government agencies usually provide supportive design information to aid in compliance. Again using California as an example, an energy design manual for residential buildings has been prepared in connection with energy insulation standards.[22] This manual explains the regulations and describes methods of applying them to specific structures. A public domain computer program for calculating building energy consumption and a manual for preparing energy elements for EIR's within California are also being prepared.

Typical Energy Impact Considerations

The energy impact of a proposed project deals with the amounts and types of energy consumed, produced, or conserved by the project in the context of the supply and demand for those types of energy. Since essentially all projects consume, produce, or conserve measurable amounts of energy in one form or another, an energy impact assessment is appropriate for all projects.

An energy impact assessment includes the determination of:

- The amount of energy consumed, produced, or conserved by the project
- The sources of energy supply available to the project
- Other demands for these energy supplies
- The net effect on supply/demand
- Effects on resource bases of nonrenewable fuels
- Alternatives and their energy impacts
- Conservation measures

An energy impact assessment is not the environmental impact of an energy-intensive project. For example, if a proposed industrial project included the combustion of coal in a boiler, the energy impact assessment would determine the amount of coal burned, the supply of coal to the plant, other potential uses for coal in the surrounding area, anticipated changes in coal supply/demand and associated changes in coal price, alternative fuels such as fuel oil, and energy conservation measures, planned or possible, to reduce coal consumption. The energy impact assessment would not include an assessment of the air pollution from coal combustion, noise produced by the boiler, or other environmental effects related to the project not involving energy.

For the purposes of energy impact analysis, projects are grouped into three categories depending upon the energy flow associated with the project's major function, namely:

1. Energy-consuming projects
2. Energy-producing projects
3. Energy-conserving projects

Most projects are the energy-consuming type. Examples include residential, commercial, and most industrial developments and transportation systems. The energy impact analysis for these projects focuses on determining the types and amounts of energy consumed by the project, translating this energy consumption into increased demand, and comparing the increased demand with the available supply. The analysis should answer this question: "Why does it matter if this project consumes energy?" Possible answers include:

- Reduction of nonrenewable resources
- Elimination of energy available for other potential purposes
- Limited supply problem
- Indirect detrimental environmental effects

In contrast to energy-consuming projects, the major function of energy-producing or energy-conserving projects is increasing the supply of usable energy or reducing the amount of energy normally consumed in the performance of specific tasks. Examples of energy-producing projects include refineries, coal mines, oil fields, nuclear fuel processing plants, thermal power plants, and hydroelectric installations (dams). Examples of energy-conserving projects include thermally insulating walls and ceilings, reducing the speed limits on highways, and tuning boilers for higher efficiency.

Energy-producing and energy-conserving projects also differ from energy-consuming projects in that their purpose and major function is energy-related and the amount of energy produced or conserved must, by definition, be greater than zero, even when all indirect energy flows are added together. If this were not true, the project would be pointless from an energy point of view.

The key issues in energy impact analysis for producing/conserving projects are the amount of energy produced or conserved by the project, the translation of this energy into increased supply or reduced demand, and the ramifications associated with the changes in energy supply/demand. The analysis should answer this question: "Why does it matter if this project produces or conserves energy?" Possible answers include:

- Reduction of use rate of nonrenewable resources
- Availability of energy resources for other uses
- Increased supply of scarce fuels
- Reduction in indirect detrimental environmental effects

ENERGY OVERVIEW

Importance of Energy Impact Analysis

There are three primary reasons for including energy impact assessment in an environmental analysis:

- Adequate energy supply is important to practically every project
- Energy resources are limited
- Energy use causes environmental problems

Energy is a necessary ingredient for a high-technology society. It is used in the production of raw materials, in agriculture, in manufacturing, in hospitals, for transportation, lighting, cooking, heating, and so on. Figure 5.1 shows the relationships between gross national product and total energy consumption for several countries on a per capita basis. While there are some exceptions, high gross national product generally corresponds to high energy consumption.

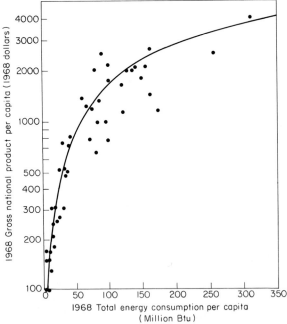

Figure 5.1   Gross national product as a function of energy consumption.

The resource bases for most energy sources are limited. In the past, this has not been a problem because the resource bases were sufficient for many years of consumption. However, the greatly increased energy consumption in recent years is beginning to cause energy supply problems for fuels such as natural gas and petroleum. With limited supplies, energy must be allocated to energy consumers on a priority basis. Any new project consuming, producing, or conserving energy will influence the supply/demand balance. An energy impact analysis should address these effects.

Energy and environmental problems are closely related, since it is nearly impossible to produce, transport, or consume energy without significant environmental impact.[13] Pressures for low environmental impact affect energy cost, supply, and supply/demand balance. The environmental problems directly related to energy production and consumption include air pollution, water pollution, thermal pollution, and solid waste disposal.

The emission of air pollutants from fossil fuel combustion is the major cause of urban air pollution. Although recent research and development in this area has led to methods for significantly reduced emissions, the large quantities of fuel consumed make even trace pollutant concentrations in exhaust gases a significant problem. Chapter 3 discusses this problem in more detail.

Diverse water pollution problems are associated with energy usage. One major problem is oil spills. In all petroleum-handling operations, there is a finite probability of spilling oil either on the earth or in a body of water. The average spill rate is about one barrel for each million barrels handled. At current production rates, about 500,000 barrels are spilled annually in approximately 7,500 incidents worldwide. A large fraction of this oil is never recovered: some evaporates, some is absorbed by soil particles, some remains in fresh or sea waters, and some is ingested by living organisms.

Coal mining can also pollute water. Changes in groundwater flow produced by mining operations often bring otherwise unpolluted waters into contact with certain mineral materials which are leached from the soil and produce an acid mine drainage. These and other waste pollution problems are discussed in Chapter 6.

Thermal pollution is another important energy-related environmental problem. The end result of essentially all energy use is waste heat: energy used to heat buildings eventually leaks away to the environment; energy used to power airplanes, trucks, trains, automobiles, etc., produces hot exhaust gases, warm radiator water, and rapidly mixed air, all forms of waste heat; even electrical power generated at central power plants is eventually entirely converted into heat. Although the atmosphere is a large heat sink, it is not limitless. As the amount of heat released by combustion of fuels approaches the same order of magnitude as incident solar radiation, significant changes in weather may occur. Such weather changes may not necessarily be undesirable, but no satisfactory method of predicting local and global weather changes induced by thermal pollution has been developed.

Thermal pollution is a growing problem. In 1970, the heat release in the northeastern portion of the United States was about 1 percent of the solar radiation in that same area; in the Los Angeles Basin it was 5 percent.[86] This heat release is approaching the point where measurable climatic changes are expected to occur. Although heat release is not yet critical in most areas, each new energy use adds to the problem. For example, a single 1,000-MW fossil-fueled power plant produces about as much heat as the solar radiation on 1,100 acres.

Solid waste is also a by-product of some forms of energy usage. Coal mining requires the removal of large quantities of earth as well as coal. In the past this solid waste was left to weather unattended in unsightly spoil banks. Although land reclamation for the removal of spoil banks has recently become an important element of mining technology, the generation and disposal of this solid waste is still a major problem. Ash remaining after combustion of fossil fuels also presents a solid waste problem. Ash is primarily the impurities in coal or other fossil fuels which are not consumed in the combustion process. A portion of the ash falls to the bottom of the furnace where it accumulates and must be discarded. Nuclear fuels are another important source of solid waste. Unlike the other forms of solid waste mentioned previously, nuclear wastes pose serious long-range health and safety problems and require isolation from the environment for hundreds of years.

These are only a few of the adverse environmental effects associated with energy usage. In general, environmental problems increase with energy use and this combined with the limited energy resource base is the crux of the energy crisis. An energy impact assessment should compare these costs with the benefits to be derived from energy use.

Types of Energy and Energy Utilization

The similarities and differences between heat and work are important in energy impact assessment but are often misunderstood. Heat and work are both forms of energy and are measured with the same units. Specifically, heat is the form of energy which raises the temperature of materials and is proportional to the random molecular motion in the material. Work is the form of energy resulting in orderly motion such as a rotating shaft or a moving vehicle. It is important to differentiate between heat and work because while work can be entirely converted into heat, heat can *never* be entirely converted into work. This is a statement of the second law of thermodynamics and has important consequences in energy impact analysis.

Most energy sources are harnessed as heat. For example, fuel is burned to produce heat, and geothermal energy is available as steam. If work is desired instead of heat, a heat engine must be used to convert a portion of the heat into work. The portion of input heat not converted into work is called "waste heat" and is usually discarded. The ratio of work produced to heat supplied is the efficiency and is always less than 100 percent.

It can be shown that the maximum theoretical efficiency of any heat engine is limited by the maximum temperature at which heat is supplied and the minimum temperature at which waste heat is rejected. Maximum efficiency is produced when the temperature difference is as large as possible. The maximum temperature is usually limited by materials; at high temperature, materials lose their strength, corrosion becomes a problem, and lubricants break down. The minimum temperature is set by the environment since the waste heat rejection temperature must be greater than the environment's temperature. State-of-the-art heat engines tuned for optimum thermodynamic performance usually achieve efficiencies ranging from 20 to 40 percent in practical service.

If a portion of the waste heat also serves some useful purpose, it is helpful to define another type of efficiency to express the ratio of the work plus useful heat produced to the total heat energy supplied. This "energy utilization efficiency" is always greater than or equal to the heat engine efficiency and can approach 100 percent where all waste heat is put to use. Examples of waste heat utilization include low-temperature steam from power plants used for district heating and total energy systems which produce electrical power and utilize waste heat for space and water heating and as the input to special absorption cooling units for summer space cooling.[46]

This concept of efficiency in the conversion of heat into work demonstrates that all forms of heat are not equivalent. In general, heat energy available at a temperature much higher (or lower) than ambient conditions is termed "high-grade heat" and is much more valuable than "low-grade heat" at temperatures close to ambient because the high-grade heat can be used to produce more work.

Examples of high-grade heat include the heat of combustion of fossil fuels (3000 to 5000°F), the high temperature steam produced in a power plant (1000°F), and some forms of geothermal heat (up to 7000°F). Examples of low-grade heat include hot water for domestic purposes (130 to 150°F), heat for space heating (50 to 80°F), and heat from flat-plate solar collectors (100 to 250°F). Since energy is a scarce resource, it is desirable to use equipment with a high-energy utilization efficiency and match energy sources to the most compatible end uses.

There are usually several alternative methods available to produce the same useful energy output. For example, Figure 5.2 shows four ways to supply 100 energy units* of low-grade heat for a space-heating application. Alternative A is the direct combustion of a fossil fuel such as natural gas or fuel oil in a small furnace and the direct use of the combustion heat. The less than perfect efficiency (80 percent) is due to heat losses in the flue gases and limitations of small-scale furnaces. Alternative B is electrical resistance heating. Here, electrical energy is converted to low-grade heat by passing it through a high-resistance wire. When comparing these alternatives, it is important to remember that a unit of electricity is work, and work cannot be produced from heat without some losses. Although the electrical-resistance heating process can be made to approach 100 percent efficiency with only negligible losses, the conversion of heat into work (electricity) at the power plant introduces substantial inefficiencies. The low-grade space heating is really

---

*Specific energy units and conversion factors are listed in a later section of this chapter, Measures of Energy.

produced by combustion of fossil fuel. Considering the entire process, the energy utilization efficiency is only 33 percent, less than half of Alternative A.

Alternative C uses electricity from a central power plant to power a heat pump. A heat pump is the thermodynamic equivalent of a heat engine operating in reverse "pumping" heat from a low-temperature source to a high-temperature sink. Here the low-temperature source is the outside ambient air temperature and the high-temperature sink* is the warmed inside air. Using a typical design value (twice as much heat produced as electrical power supplied) and again including the inefficiencies of the central power plant, the energy utilization efficiency is 67 percent, which is double the efficiency of the electrical-resistance heater but is not quite as high as direct combustion.

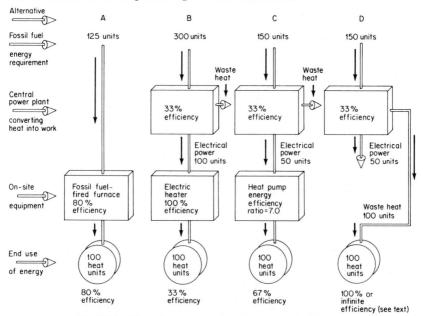

Figure 5.2 Alternative methods of producing space heating.

Alternative D is the use of existing waste heat for this low-temperature space heating application. For example, waste heat in the form of low-temperature steam or hot water from a power plant cooling system could be passed through a heat exchanger to warm air to room temperature. Under ideal conditions, no additional fuel would be consumed. From the viewpoint of the end user who desires low-temperature space heat, there is no additional fuel combustion and as a result the energy utilization efficiency is infinite. From the viewpoint of the utility, the amount of waste heat is reduced and the energy utilization efficiency is increased. The amount of the increase depends upon the amount of waste heat used, and the overall energy utilization efficiency may approach 100 percent. Regardless of viewpoint, alternative D is the most attractive from an energy utilization standpoint.

This example demonstrates that the important element in energy impact analysis is not the amount of useful energy consumed, produced, or conserved but is rather the net energy flow required to produce the desired results. The four alternatives examined here each involve the same amount of useful energy but differ substantially in their energy requirements. It is recommended that for energy impact analyses associated with an environmental impact study, all conversion processes be taken into account and that all energy usage be converted to common heat units. In most projects, the energy requirement consists of two distinct parts: heat supplied by some fossil fuel, and electricity. To

---

*High temperature here means higher than ambient, and it may be as low as 60 to 80°F, depending upon specific design.

calculate the total energy consumption, the electrical energy supplied should be divided by the electrical power plant efficiency and the result should be added to the heat supplied divided by the efficiency of producing heat from the fossil fuel. This can be expressed in equation form as follows:

$$\text{Total energy consumption} = \frac{\text{electrical energy}}{\text{power plant efficiency}} + \frac{\text{heat energy}}{\text{heat production efficiency}}$$

For example, consider an industrial process which requires these energy flows:
- 3,127 units of electrical energy produced from an oil-fired power plant with 31 percent efficiency
- 7,492 units of heat produced by burning oil with 79 percent efficiency

The total energy consumption is:

$$\text{Total energy consumption} = \frac{3127}{0.31} + \frac{7492}{0.79}$$
$$= 19{,}571 \text{ units}$$

In this example, electrical power and heat were assumed to be expressed in similar units. Conversion factors for several energy units are listed in a subsequent section of this chapter.

Energy Sources and Resources

A wide range of fuels and power-generating technologies are currently available, and recently increased research and development efforts should significantly expand the alternatives in the next few years. An energy impact assessment as part of an environmental impact study must examine those alternatives suitable for the project and determine the optimum choice. This section briefly reviews the characteristics of existing and projected alternatives. More detailed information on specific fuels for power generating technologies can be found in References 46, 58, 62, 86–88, and 101. Sources of supply and demand data are listed in a later section of this chapter.

In comparing alternative fuels and power sources, availability and magnitude of the resource base are important considerations. Many fuels such as coal, petroleum, and natural gas have limited resource bases which are depleted by removing and consuming the fuels. These fuels are termed "nonrenewable resources" since they were produced by long-term geological processes and cannot be replaced in the near future.

In contrast to nonrenewable resources, renewable resources are not depleted by use or are available in such abundance that even intensive utilization will not measurably deplete the supply. Examples include hydroelectric power, solar energy, and geothermal energy.

The uses of nonrenewable and renewable resources must be viewed differently. Since the supply of nonrenewable resources is limited, there is competition between various uses. Using energy for one application reduces the availability for other applications. As the supply dwindles, the resource becomes more valuable and the price rises, discouraging use or encouraging fuel switching. This price rise is inevitable for nonrenewable fuels and is exemplified by the doubling of petroleum prices in 1973 and again in 1979. Conservation is one way to forestall price increases and is clearly desirable for nonrenewable resources.

The price rise associated with a limited resource base does not apply to renewable resources. Conservation is only appropriate if hardware or equipment operational costs can be reduced, or if environmental, socioeconomic, or other problems can be reduced. In a mixed system with both renewable and nonrenewable resources,* conservation through reduced demand can be applied to the nonrenewable portion.

Alternative fuels and power sources are of three general types, namely:
1. Fuel and heat sources
2. Work sources
3. Power generating technologies

Fuels, heat sources, and work sources are the raw energy supplies. Power generating technologies allow conversion of heat into work.

Table 5.1 lists the characteristics of the alternative fuel and heat sources now available

---

*For example, an electrical utility with hydroelectric and fossil-fueled power plants feeding the grid.

TABLE 5.1 Alternative Fuels and Heat Sources

Description	Development status	Renewable?	Physical state	Mining method	Processing required	Refinery losses	Sulfur content	Comments
Petroleum	Commercial	No	Liquid	Drilling	Refinery	Small	Varies	Raw material is crude oil.
Natural gas	Commercial	No	Gas	Drilling	None	None	Low	Clean, easy to handle.
Coal	Commercial	No	Solid	Strip or underground	None	None	High	An inherently dirty, hard to handle fuel.
Coal products	Pilot plant	No	Liquid/gas	Strip or underground	Several energy-intensive processes are available	Large	Low	Final products are synthetic crude oil and synthetic natural gas.
Oil shale	Pilot plant	No	Liquid	Strip, underground, or in situ	Heating process	Large	Low	Final product is synthetic crude oil.
Tar sands	Small commercial	No	Liquid	Strip	Heating process	Large	Very low	Final product is synthetic crude oil.
Solar heat	Experimental and small commercial	Yes				None	None	Useful for small-scale water and space heating.
Fuel from organic materials or solid wastes	Experimental	Yes	Solid, liquid, or gas	Mechanical harvesting	Processing required to produce liquid or gaseous fuel	Large	Low	Most efficient use is direct combustion. Partial pyrolysis and other processes can produce synthetic oils and gas.
Hydrogen	Experimental	Yes	Gas		Produced by disassociating water or by processing other fuels	Large	None	Can be used with some equipment modification in place of natural gas, but production losses are very high.
Uranium	Commercial	No	Solid	Strip or underground	Refining, concentration	Small	None	Small volumes give massive amounts of energy.
Liquified natural gas	Commercial	No	Liquid		Cooling and compression	Large	Low	Same as natural gas but is cooled and compressed for storage and shipment.
Geothermal heat	Commercial	Yes		Drilling	None	None	High	Low-temperature heat source can be used directly or with a power plant to generate electricity.

and currently under development. Most of the fuels under development including coal products, oil shale, tar sands, and fuels from organic materials or solid waste will be processed into final fuels much like natural gas and petroleum products. These "synthetic fuels" will be direct replacements for fuels in short supply. Liquefied natural gas is natural gas which has been cooled and compressed to facilitate its transportation, usually by ship.

Hydrogen is the only fuel listed that is not available naturally in an energetic form. It is produced by disassociation of water or some other energy-intensive process. Since hydrogen is available in essentially limitless quantities in sea water, it is termed renewable even though its production requires energy from another (possibly nonrenewable) source. Hydrogen could be produced from solar, nuclear, or geothermal energy and could be burned as a replacement for hydrocarbon fuels in short supply.

Solar and geothermal energy can be used in a variety of heating applications or for producing electrical power. They are both renewable and are the subject of intensive development efforts.

Several energy sources directly producing work are listed in Table 5.2. Since no heat

TABLE 5.2   Work Sources

Description	Development status	Ultimate energy source	Ultimate resource base	Air pollution	Suitable for: Large scale	Small scale
Hydroelectric	Commercial	Solar	Renewable	No	Yes	Yes
Tidal	Experimental	Lunar	Renewable	No	Yes	Yes
Waves	Experimental	Solar	Renewable	No	*	Yes
Wind	Experimental	Solar	Renewable	No	*	Yes
Muscle	Commercial	Solar	Renewable	No	No	Yes

*Several small-scale units can be used to generate large-scale power.

engine is required, the percentage of the energy resource which can be converted to usable work (usually electricity) is much higher than for fuel and heat sources. All are renewable since they draw their energy from the combination of solar heat and the movement of the earth and moon. Even muscle power (primarily from humans, horses, and elephants) is renewable since the fuel to run these "engines" is organic material derived from solar energy.

Unfortunately, these clean renewable work sources do not have the capacity and technological development status to provide a large portion of the work energy required by society. Therefore, heat engines which convert heat into work must provide the base load.

Table 5.3 lists the salient characteristics of alternative power generating technologies (heat engines) now available and currently under development. Energy storage units are listed along with the other heat engines because although they do not convert heat into work, they increase the capacity of commercial power plants to handle peak loads. Pumped storage systems are hydroelectric installations where excess energy generated in off-peak periods is used to pump water from a lower reservoir to a higher reservoir. The energy "stored" can be used in peak demand periods to run turbines and to generate electrical power. Although pumped storage can be considered a power generating technology, it actually consumes energy. The losses associated with pumping water to a higher reservoir and later producing power with a turbine are considerable. If power from a fossil-fueled power plant is used directly, the overall efficiency of power generation is about 40 percent. However, if the energy is stored in a pumped storage facility and used later, the overall efficiency may drop to less than 27 percent. Furthermore, pumped storage reduces the capital cost per unit of power produced by power plants by allowing them to operate at a higher load factor; at the same time, it increases peak capacity at low additional capital cost.

Measures of Energy

The majority of the confusion over energy measurement and units stems from a misunderstanding of the difference between energy and power. For example, contrary to common

TABLE 5.3 Alternative Power Generating Technologies (Heat Engines)

Technology	Development status	Date commercial production	Fuel	Fuel relative abundance	Power plant scale	Approximate efficiency, %*	Nuclear† safety ranking (1 = Best)	Air pollution	Suitability for large-scale electrical power generation	Small-scale mechanical power generation
Nuclear:										
Conventional light-water reactor (LWR)	Commercial	Now	Uranium ore	Limited	Large	30–35	4	No	Yes	No§
High-temperature gas reactor (HTGR)	Demonstration	1980–1982	Uranium/Thorium	Slightly larger than (LWR)	Large	39–43	2	No	Yes	No§
Gas turbine—high-temperature gas reactor (GT-HTGR)	Experimental	1986	Uranium/Thorium	Slightly larger than (LWR)	Large	36–45	2	No	Yes	No§
Liquid metal fast breeder reactor (LMFBR)	Demonstration	1990	Plutonium/Uranium	10 times LWR	Large	40	3	No	Yes	No§
Fusion reactor	Early experimental	2000	Lithium/Deuterium	Almost infinite	Large	38	1	No	Yes	No§
Fossil fueled:										
Conventional steam turbine	Commercial	Now	Any fossil fuel	Limited	Large	40		Yes	Yes	Yes
Combined Cycle Gas Turbine-Steam Turbine	Commercial	Now	Liquid or gaseous	Limited	Large	40–50		Yes	Yes	No§
Combined cycle other topping and bottoming cycles	Experimental	Demonstration 1981	Varies	Limited	Large	Up to 60		Yes	Yes	No§
Gasoline engines	Commercial	Now	Gasoline	Limited	Small	20–25		Yes	No	Yes
Diesel engines	Commercial	Now	Diesel oil	Limited	Small	25–29		Yes	No‡	Yes

TABLE 5.3 Alternative Power Generating Technologies (Heat Engines) (Continued)

Technology	Development status	Date commercial production	Fuel	Fuel relative abundance	Power plant scale	Approximate efficiency, %*	Nuclear† safety ranking (1 = Best)	Air pollution	Suitability for large-scale electrical power generation	Small-scale mechanical power generation
Gas turbine engines	Commercial	Now	Nat. gas or oil	Limited	Small	20–27		Yes	Not‡	Yes
Total energy systems	Commercial	Now	Nat. gas or oil	Limited	Small	Overall energy usage up to 70		Yes	Not‡	Yes
Fuel cell	Experimental	1990	Hydrogen	Produced indirectly	Small	60–80		No	Not‡	
Energy storage:										
Pumped	Commercial	Now	Electricity		Large	67		No	Yes	No
Other (electrical energy storage)	Large-scale experimental / Small-scale commercial	1985+ / Now	Electricity		Variable	50–70		No	Yes	No

*Efficiency: Power plant: $\dfrac{\text{electrical energy output}}{\text{heat energy input}}$

Storage: $\dfrac{\text{electrical energy output}}{\text{electrical energy input}}$

†Relative based on assessment of current data.

‡Several small-scale units can be used to generate larger quantities of electrical power.

§These technologies may be used for generating mechanical power for ships.

belief, the terms horsepower and kilowatt are not energy units but power units. Power is the rate of energy utilization.

Table 5.4 lists thirteen common energy units and factors for converting any one to any of the others.

One of the most basic and widely used energy units is the British thermal unit (Btu). It is defined as the amount of energy required to heat 1 pound of water 1°F under standard conditions. The metric equivalent of the Btu is the calorie and is defined as the amount of energy required to heat 1 gram of water 1°C under standard conditions.

While useful, these basic heat units are quite small. For example, the energy required to make a single cup of coffee is 68 Btu or 17,080 calories.*

Larger energy units are clearly desirable to avoid large numbers when referring to sizable energy quantities. The barrel of oil is an energy unit equivalent to energy released in the combustion of 42 gal of crude oil. While all crude oils are not the same, an average value for the purposes of defining units is 5.8×10^6 Btu/barrel of oil. The barrel of oil is a convenient size for expressing energy consumption of many projects. For example, a typical residence consumes heating oil at a rate of about 2 barrels of oil per month.

For much larger projects, an even larger energy unit is required. In 1970 the total U.S. energy consumption was approximately 10^{10} barrels per year. The "quad" is a large energy unit suitable for measuring such large amounts of energy and is equivalent to 10^{15} Btu. The 1970 U.S. energy consumption thus represents 63 quads.

Power, the rate of energy utilization, has other units. Table 5.5 lists ten common power units and conversion factors. The "horsepower," a common power unit, is the amount of strenuous work that a horse can perform per unit time. It is equivalent to 0.7073 Btu/s. A horsepower-hour is not a power unit but is the amount of energy produced by 1 horsepower operating for 1 hour and is equivalent to 2,547 Btu.

A "kilowatt" is another convenient power unit and is commonly used for measuring electrical power. A kilowatt is equal to 1.341 hp. This unit can be used, like the horsepower unit, to derive an energy unit, where 1 kilowatthour is equal to 3,415 Btu.

It is recommended that energy usage be presented in a unit system that produces numbers ranging from 0.1 to 10,000 for easy reading. All similar energy quantities listed in an environmental impact assessment should be listed with the same units so the reader may compare figures without difficulty. A conversion table listing all energy and power units used should also be provided.

Energy impact analysis often requires several quantities of energy expressed in different power and energy units to be added. In calculating the sum, the following rules must be observed:

- Quantities of power and energy cannot be added directly
- All energy quantities must be converted to common units before addition

Power units can be converted to energy units by multiplying by the time over which the power acts. This can be expressed as follows:

$$\text{Energy} = \text{Power} \times \text{Time}$$

For example, consider a suburban shopping center which uses electricity for lighting and air conditioning. The lighting load is 300 kW during the 8 hours per day, 6 days per week that the shops are open. The remainder of the time, the load is reduced to 75 kW. The air conditioning load is 700 kW for 10 hours a day during the four summer months. The total electrical energy can be calculated for any time period; however, a yearly basis is convenient here because of the seasonal air conditioning load. The yearly electrical energy required is:

Lighting energy while shops are open $= 300 \text{ kW} \times 8 \dfrac{\text{hours}}{\text{day}} \times 6 \dfrac{\text{days}}{\text{week}} \times 52 \dfrac{\text{weeks}}{\text{year}}$ $= 748{,}800 \text{ kWh}$

Lighting energy while shops are closed $= 75 \text{ kW} \left(16 \dfrac{\text{hours}}{\text{day}} \times 6 \dfrac{\text{days}}{\text{week}} + 24 \dfrac{\text{hours}}{\text{day}} \times 1 \dfrac{\text{day}}{\text{week}} \right) \times 52 \dfrac{\text{weeks}}{\text{year}} = 468{,}000 \text{ kWh}$

Total Lighting Energy $= 1{,}216{,}800 \text{ kWh}$

*Heating 1 cup (0.52 pounds) of water from 60 to 190°F.

TABLE 5.4 Energy Units Conversion Table

a \ b	quad	Btu	kWh	mWh	Bbl oil	hp-hr	ft-lb	joules	ergs	cal	kcal	liter-atm	therm
quad	1.0	1.0×10^{15}	2.928×10^{11}	2.928×10^{8}	1.724×10^{8}	3.926×10^{11}	7.775×10^{17}	1.054×10^{18}	1.054×10^{25}	2.520×10^{17}	2.520×10^{14}	1.040×10^{16}	1.0×10^{10}
Btu	1.0×10^{-15}	1.0	2.928×10^{-4}	2.928×10^{-7}	1.724×10^{-7}	3.926×10^{-4}	777.5	1054.	1.054×10^{10}	252.0	0.2520	10.40	1.0×10^{-5}
kWh	3.415×10^{-12}	3415	1.0	1.0×10^{-3}	5.889×10^{-4}	1.341	2.655×10^{6}	3.600×10^{6}	3.600×10^{13}	8.605×10^{5}	860.5	3.553×10^{4}	3.415×10^{-2}
mWh	3.415×10^{-9}	3.415×10^{6}	1000	1.0	0.5889	1341	2.655×10^{9}	3.600×10^{9}	3.600×10^{16}	8.605×10^{8}	8.605×10^{5}	3.553×10^{7}	34.15
Bbl oil	5.800×10^{-9}	5.800×10^{6}	1698	1.698	1.0	2277	4.510×10^{9}	6.113×10^{9}	6.113×10^{16}	1.462×10^{9}	1.462×10^{6}	6.032×10^{7}	58.00
hp-hr	2.547×10^{-12}	2547	0.7457	7.457×10^{-4}	4.392×10^{-4}	1.0	1.980×10^{6}	2.684×10^{6}	2.684×10^{13}	6.417×10^{5}	641.7	2.649×10^{4}	2.547×10^{-2}
ft-lb	1.286×10^{-18}	1.286×10^{-3}	3.766×10^{-7}	3.766×10^{-10}	2.217×10^{-10}	5.051×10^{-7}	1.0	1.356	1.356×10^{7}	0.3240	3.240×10^{-4}	1.338×10^{-2}	1.286×10^{-8}
joules	9.488×10^{-19}	9.488×10^{-4}	2.778×10^{-7}	2.778×10^{-10}	1.636×10^{-10}	3.725×10^{-7}	0.7376	1.0	1.0×10^{7}	0.2391	2.391×10^{-4}	9.869×10^{-3}	9.488×10^{-7}
ergs	9.488×10^{-26}	9.488×10^{-11}	2.778×10^{-14}	2.778×10^{-17}	1.636×10^{-17}	3.725×10^{-14}	7.376×10^{-8}	1.0×10^{-7}	1.0	2.391×10^{-8}	2.391×10^{-11}	9.869×10^{-10}	9.488×10^{-16}
cal	3.968×10^{-18}	3.968×10^{-3}	1.162×10^{-6}	1.162×10^{-9}	6.842×10^{-10}	1.558×10^{-6}	3.086	4.183	4.183×10^{7}	1.0	1.0×10^{-3}	4.129×10^{-2}	3.968×10^{-8}
kcal	3.968×10^{-15}	3.968	1.162×10^{-3}	1.162×10^{-6}	6.842×10^{-7}	1.558×10^{-3}	3086	4183	4.183×10^{10}	1000	1.0	41.29	3.968×10^{-5}
liter-atm	9.612×10^{-17}	9.612×10^{-2}	2.815×10^{-5}	2.815×10^{-8}	1.658×10^{-8}	3.774×10^{-5}	74.73	101.3	1.013×10^{9}	24.22	2.422×10^{-2}	1.0	9.615×10^{-7}
therm	1.0×10^{-10}	1.0×10^{5}	29.28	2.928×10^{-2}	1.724×10^{-2}	39.26	7.775×10^{7}	1.054×10^{8}	1.054×10^{15}	2.520×10^{7}	2.520×10^{4}	1.040×10^{6}	1.0

To convert a quantity with units of a to units of b, multiply by the number shown.
EXAMPLE: 1 quad = 10^{15} Btu

TABLE 5.5 Power Units Conversion Table

a \ b	Horsepower	Watt	Kilowatt	Megawatt	Bbl/day	Bbl/yr	Quad/yr	Btu/s	Btu/h	ft·lb/s
horsepower	1.0	745.7	0.7457	7.457×10^{-4}	1.054×10^{-2}	3.850	2.233×10^{-10}	0.7073	2.547	550.0
watt	1.341×10^{-3}	1.0	1.0×10^{-3}	1.0×10^{-6}	1.413×10^{-5}	5.161×10^{-3}	2.994×10^{-11}	9.486×10^{-4}	3.415	0.7376
kilowatt	1.341	1000	1.0	1.0×10^{-3}	1.413×10^{-2}	5.161	2.994×10^{-8}	0.9486	3415	737.6
megawatt	1341	1.0×10^{6}	1000	1.0	14.13	5161	2.994×10^{-5}	948.6	3.415×10^{6}	7.376×10^{5}
bbl/day	94.88	7.077×10^{4}	70.77	7.077×10^{-2}	1.0	365.3	2.119×10^{-6}	67.13	2.417×10^{5}	5.220×10^{4}
bbl/yr	0.2598	193.7	0.1937	1.937×10^{-4}	2.738×10^{-3}	1.0	5.800×10^{-9}	0.1838	661.6	142.9
quad/yr	4.479×10^{7}	3.340×10^{10}	3.340×10^{7}	3.340×10^{4}	4.720×10^{5}	1.724×10^{8}	1.0	3.169×10^{7}	1.141×10^{11}	2.464×10^{10}
Btu/s	1.414	1054	1.054	1.054×10^{-3}	1.490×10^{-2}	5.441	3.156×10^{-8}	1.0	3600	777.5
Btu/h	3.927×10^{-4}	0.2928	2.928×10^{-4}	2.928×10^{-7}	4.138×10^{-6}	1.511×10^{-3}	8.766×10^{-12}	2.778×10^{-4}	1.0	0.2160
ft·lb/s	1.818×10^{-3}	1.356	1.356×10^{-3}	1.356×10^{-6}	1.916×10^{-5}	6.997×10^{-3}	4.059×10^{-11}	1.286×10^{-3}	4.630	1.0

To convert a quantity with units of a to units of b, multiply by the number shown.
EXAMPLE: 1.0 kilowatt = 5.161 Bbl/yr

$$\frac{\text{Air conditioning}}{\text{energy}} = 700 \text{ kW} \times 10 \frac{\text{hours}}{\text{day}} \times 7 \frac{\text{days}}{\text{week}} \times 4 \frac{\text{weeks}}{\text{month}} \times 4 \frac{\text{months}}{\text{year}} \qquad = 784{,}000 \text{ kWh}$$

Total Electrical Energy $\qquad\qquad = 2{,}000{,}800 \text{ kWh}$

The electrical energy required is only part of the total energy consumed at the shopping center. Additional energy is required in the form of natural gas for space heating (130,000 therms) and fuel oil for water heating (650 million Btu). To calculate the total energy required, these quantities must first be converted to similar units. The barrel of oil is a convenient unit to use because of the size of this project. The total yearly energy consumption is:

$$\text{Electrical energy} = 2{,}000{,}800 \text{ kW} \frac{1^*}{0.31} \times 5.889 \times 10^{-4} \frac{\text{Bbl oil}}{\text{kWh}} = 3{,}800.9 \text{ Bbl oil}$$

$$\text{Natural gas} \quad = 130{,}000 \text{ therms} \times 1.724 \times 10^{-2} \frac{\text{Bbl oil}}{\text{therm}} \quad = 2{,}241.2 \text{ Bbl oil}$$

$$\text{Fuel oil} \quad = 650 \times 10^6 \text{ Btu} \times 1.724 \times 10^{-7} \frac{\text{Bbl oil}}{\text{Btu}} \quad = \underline{ 112.1 \text{ Bbl oil}}$$

Total Energy Consumption $\qquad\qquad = 6{,}154.2 \text{ Bbl oil}$

ENERGY IMPACT ASSESSMENT—ORGANIZATION AND METHODOLOGY

Although the legal requirements for evaluation of energy in an environmental impact statement are considerably less stringent than for evaluation of other impacts, it is expected that future legislation will place at least as much emphasis on energy as on air quality (for example) and will require a very detailed energy impact assessment. It is recommended that the energy element of an EIS be structured to contain the same information required by the National Environmental Policy Act (NEPA) for environmental areas. Specifically:

(i)   The environmental impact of the proposed action
(ii) Any adverse environmental effects which cannot be avoided should the proposal be implemented
(iii) Alternatives to the proposed action
(iv) The relationship between local short-term uses of man's environment and the maintenance and enhancement of long-term productivity
(v)   Any irreversible and irretrievable commitments of resources which will be involved in the proposed action should it be implemented.[82]

Section 102 of the Act directs the preparation of environmental impact statements and describes their content in general terms. Each item listed can be applied to energy impact assessment. For example, the environmental impact is analogous to the energy impact defined here as the amounts and types of energy consumed, produced, or conserved by the project in the context of the supply and demand for those types of energy. Unavoidable adverse environmental effects are analogous to the consumption of nonrenewable energy resources; this is intimately related to short-term versus long-term productivity, and irreversible and irretrievable commitments of resources. Alternatives includes both alternatives in the normal sense and any energy conservation measures applicable to the project.

The project type (consuming, producing, or conserving) determines the depth of energy impact assessment required. The recommended organization of energy impact elements for these project types is shown in Table 5.6.

For a consuming project, energy is not directly related to the project's major function and the energy impact assessment should include the same level of detail as other environmental issues such as air quality. The tone of the energy impact discussion should reflect the relative importance of the project's energy flows in relation to the local and

---

*This factor is the estimated electrical power plant efficiency and converts the electrical energy to equivalent heat input to the power plant. See the earlier section, Types of Energy and Energy Utilization.

national energy supply and demand scenarios. Excessive energy consumption should be treated in the same manner as excessive air pollutant emissions, noise, or solid waste generation. The energy element should begin with an energy inventory—the amounts and types of energy flows created by the project. Next, conservation measures employed in the project to minimize energy consumption and their effects should be described. Within this framework of anticipated energy flows, the supply and demand scenarios impacted by the project can be discussed.

The next section should discuss the energy impacts of alternatives to the proposed action as well as energy alternatives and energy conservation measures. The discussion should highlight the energy-related aspects and justify why they are not recommended, particularly if they conserve energy. The background information section should be writ-

TABLE 5.6   Energy Impact Organization

Consuming Project
   A. Summary
   B. Energy inventory—consumption
   C. Conservation measures employed
   D. Supply/demand scenarios
   E. Alternatives
   F. Background information
Producing Project
   A. Summary
   B. Energy inventory—production, consumption
   C. Supply/demand scenarios
   D. Alternatives
   E. Background information
Conserving Project
   A. Summary
   B. Energy inventory—conservation, consumption
   C. Supply/demand scenarios
   D. Alternatives
   E. Background information

ten as an appendix amplifying the discussion, listing energy units, and generally supporting the text.

Unlike an energy-consuming project, the major function of an energy-producing project is directly energy related. The energy element should demonstrate that the project's major purpose is the production of usable energy, that it will be produced as efficiently as possible, and that, as a result of the project, energy supplies will be increased to meet anticipated demands. Since energy is the key element in the project, the energy impact assessment should form a major portion of the EIS.

Although the outline suggested in Table 5.6 for a producing project is similar to that for a consuming project, there are significant differences. The energy inventory section for a producing project should receive major emphasis and include an inventory of (1) energy production through the project's major function, (2) the consumption of energy due to indirect affects, and (3) the net energy balance. Conservation measures employed to reduce indirect energy consumption should also be discussed.

The supply/demand section should be greatly amplified over the equivalent section for a consuming project since the changes in supply/demand scenarios constitute the ultimate purpose and justification for a producing project. The alternatives section should follow the same lines as for a consuming project.

As with an energy-producing project, the energy element for an energy-conserving project should receive major emphasis. It should demonstrate that the project's major purpose is to conserve energy, reduce demand, and produce a more favorable supply/demand balance. The energy inventory section should tabulate the reduction in demand produced directly by conservation and the indirect consumption required to operate the project.

The following sections discuss the energy assessment topics in Table 5.6 in more detail.

Energy Inventory

An energy inventory is a detailed tabulation of the amounts and types of energy consumed, produced, and conserved by the project. This inventory of energy flows combined with supply/demand data gives a complete description of the project's energy impact.

Data gathering is the most laborious task in the calculation of an energy inventory. All energy flows produced by the project must be determined by calculation or by direct measurement. Data and data sources for this purpose are listed in a later section of this chapter.

The level of detail appropriate for an energy inventory is not easily determined. It is virtually impossible to accurately total all energy flows related to a specific project. Most energy flows can only be estimated to within a few percent, and a complete energy inventory would require assessing energy flows back to the point at which raw materials are mined and processed. For example, consider the energy flows associated with construction workers commuting to a specific construction site. A partial list of the energy flows includes:

- Gasoline to power automobiles
- Oil burned by automobiles (including regular oil changes)
- Energy to operate automotive maintenance shops
- Energy to manufacture expendable replacement parts (tires, filters, shock absorbers, etc.)
- Energy to manufacture and sell automobiles
- Energy to manufacture automotive components (belts, engines, etc.)
- Energy to maintain streets, operate signals, etc.
- Energy to process raw materials for all previously listed items into gasoline, oil, iron, plastic, glass, paper, asphalt, concrete, etc.

There are large uncertainties associated with many of these items. Considering these uncertainties, energy flows which add up to only 1 to 5 percent of the total project's energy consumption can probably be neglected without significantly altering the energy inventory accuracy. Based on this criterion, the energy associated with construction workers commuting to a construction site can be approximated by the gasoline directly consumed in the commuting process.

Data gathering for an energy inventory can be facilitated by the use of a standard form such as Table 5.7. As specific energy flows are identified and calculated, they should be given arbitrary reference numbers. This provides a convenient way to keep track of all data. Table 5.7 is a working form used to summarize the results of the specific energy flow calculations. It includes columns for all items which need to be presented in an energy inventory. Energy flows can be entered in any convenient order because this form is only to be used as an initial tabulation rather than for presentation. The organization of inventory data for presentation in an energy element is discussed in the following section.

Organization of Energy Inventory Data   It is not sufficient merely to tabulate energy flows in an energy inventory. The data must be organized to demonstrate the importance of specific energy flows in relation to the project as a whole and the appropriate supply/demand scenarios. Energy inventory should be organized along the following lines:

- Fuel type
- Consumption, production, or conservation
- Time
- Construction and operational phases

Table 5.8 is a standard form for organizing and presenting energy inventory data along these lines. Form tables should be prepared as follows: three to itemize consumption, production, and conservation separately, and one to summarize all energy flows. Inventory data tabulated in standard forms such as Table 5.7 can be transferred directly onto these forms.

Each fuel or power source should be tabulated separately. The types of fuel listed should correspond to those tabulated in the energy supply and demand scenarios. For example, if local supply and demand data for gasoline, distillate heating oil, and residual oil are available, the energy flows associated with these fuels should be listed separately. However, if available supply/demand data only lists processed petroleum products, the energy flows for the three specific fuels should be summed and the totals calculated in forms such as Table 5.8.

TABLE 5.7 Energy Inventory Initial Tabulation Form

Number	Year	Item	Energy source or fuel	Consumption	Production	Conservation	Units

TABLE 5.8 Form for Presenting Energy Inventory Data

Fuel or source	Major use	Year						Total
		1	2	3	4	5	6	
Nonrenewable sources:								
Natural gas								
Electricity								
Gasoline								
Heating oil								
Other								
Total								
Renewable sources:								
Solar energy								
Hydroelectric power								
Wind energy								
Waste wood								
Total								
Total (all sources)								

(Construction ← → Operation)

The main reasons for concern over energy impact are the adverse environmental effects (often termed "environmental residuals") produced by some forms of energy usage and the limited availability of energy sources. The difference between renewable and nonrenewable resources is a key issue. Renewable energy sources are those with such large resource bases that significant use by human beings, even for centuries, will not measurably reduce their supply. It follows that for such energy sources (including solar energy, for example), the supply/demand problem reduces the supply of equipment for converting the energy into useful forms rather than the supply of the energy source itself. The use of renewable resources instead of nonrenewable resources has several advantages (see the section on Energy Sources and Resources) and should be encouraged. Consequently, if renewable resources will be utilized in the project, the associated energy flows should be tabulated separately as in Table 5.8.

Production, consumption, and conservation energy flows should be itemized separately. All consumption and production energy flows created or changed by the project can be directly summed. However, not all energy conservation flows may be included. For the purposes of energy inventory, energy conservation must be viewed carefully. The purpose of the energy inventory is to itemize the energy flows actually caused by the project as it is proposed. Conservation in an energy inventory refers to a reduction in energy consumption *external* to the project, not *internal*. For example, if a certain residential construction project requires reducing the speed limit on an existing highway, it will produce external energy conservation since the energy consumed by vehicles which normally travel that road will be reduced. The key element is that energy consumption which would exist independent of the project will be reduced. Additional examples of external energy conservation include:

- Governmental regulations which mandate reduced energy consumption such as:
  55 mi/h speed limit
  Energy insulation standards
  Appliance efficiency standards
  Thermostat setback programs
- Fuel switching to renewable resources such as:
  Solar energy
  Hydroelectric power
  Wastewood chips
- Utilization of waste heat to replace fuel combustion
- Facility relocations which reduce commuting distances
- Improvement of efficiencies of boilers or other equipment

Internal energy conservation refers to reduction in project energy consumption due to certain design or operational features. These conservation measures should not be tabulated in an energy inventory. The reduction of energy consumption due to heavily insulating the roof of a new building is an example of internal energy conservation. Although the thermal insulation will conserve energy, it merely reduces the increased energy consumption required to heat the new building. For the purposes of energy inventory, the energy required to heat the building, as designed with insulation, should be listed as energy consumption. The key element in this example is that the energy conservation did not reduce any energy flows which would have existed without the project. Additional examples of internal energy conservation include:

- Location of a new facility to minimize commuting energy
- Building design to reduce heating and/or cooling loads
- Optimization of power plant design for high overall efficiency

Since energy impact is largely a supply/demand problem, the timing of energy flows is very important. It is usually convenient to tabulate energy flows on a yearly basis as in Table 5.8. Construction and operational phases can also be separated as shown. The construction phase generally involves much less energy than the operational phase; however, it may be concentrated in a relatively short time period. These timing effects can be presented on a graph expressing the total anticipated energy flows per year and integrated over the project's life. Similar graphs itemizing energy flows by fuel or project function could also be prepared.

The "energy balance" is the net energy consumed, produced, or conserved over the project life and should be calculated as the final energy inventory output. The project type does not necessarily determine the overall energy balance. It is possible, but certainly not desirable, for a project designed to conserve energy to actually consume energy. For example, using daylight savings time on a year-round basis has been argued as both a conservation measure and as an energy-consuming measure termed "conserving." The evaluation of energy impact should discuss the overall energy balance including all measurable effects, and compare direct and indirect energy flows.

Energy Inventory Examples This section discusses three examples of energy inventory organization:

- Single-family detached housing development (consuming project)
- New oil-fired power plant (producing project)
- Minimum insulation standard for water heaters (conserving project)

Housing Development. A residential development consisting of 200 single-family detached dwelling units has been proposed for a 50-acre site on the outskirts of a small town in the southwest. The site is currently used for grazing cattle and is adjacent to a highway with a 55 mi/h speed limit. As part of the proposed project, the speed limit on 3/4 mi of the highway will be reduced to 40 mi/h both as a safety measure and to reduce highway-generated noise within the development to acceptable levels. Grading and construction will require two years, but the first phase of 25 houses will be completed and occupied during the first year. Full occupancy will be achieved by the fourth year.

Only two basic floor plans will be constructed. However, each floor plan will be available with either the regular floor plan or as the mirror image of the regular plan for a total of four possibilities. In addition, four different exterior finishes or facades will be used on each floor plan for a total of 16 different-looking models.

Space heating, water heating, and cooling energy will be supplied by natural gas, while electricity will provide energy for appliances, lights, and air conditioning. Air conditioning will be an extra-cost option, and it is estimated that 42 percent of the units will be air conditioned initially. However, since the climate is quite warm, it is anticipated that additional air conditioning units will be installed beginning in the fifth year so that 80 percent of the units will be air conditioned by the eighth year.

The cost of natural gas in the area is rising, and solar heating is estimated to become cost-competitive with natural gas heating by the fifth year. Installation of solar heating units is estimated to total 30 percent of the units by the eighth year.

Table 5.9 is a "raw" energy inventory for the project. The information in the previous paragraphs has been used to calculate the specific energy flows. Data sources and procedures for calculating these energy flows are discussed in a later section on Energy Inventory Data. Table 5.9 is an initial compilation of energy flows which will be retabulated later. Consequently, energy quantities are entered in convenient units which will subsequently be converted to a common basis.

The first four items in Table 5.9 are related to energy consumed during construction. Gasoline used by construction workers commuting to the site and electricity used by power tools, etc. are the largest items.

Items 4 to 8 include the natural gas required by the occupants during the first 5 years. Beginning in the sixth year, part of the heating load supplied by natural gas will be replaced by solar energy (items 22 to 27). The space heating energy requirement in this project should be calculated using one of the computerized methods discussed in a later section of this chapter on Building Heating and Cooling Data. Since only one climate and two basic models with well-specified designs are involved, the calculations will be straightforward.

Items 9 to 12 are the electrical power required during the first ten years. Electrical power is required for lighting, appliances, and air conditioning. Items 9 to 12 were based on the initial 42 percent air conditioner installation estimate. The additional energy required by the retrofit air conditioners is listed in items 18 to 21.

Items 13 to 16 are the gasoline used by the occupants for transportation. The energy was estimated from the projected number of people, number of vehicles per capita, average annual mileage per vehicle, and average fuel economy.

Item 17 is the only energy conservation energy flow denoted. Reducing the speed limit on the highway will reduce the energy consumed by vehicles traveling the highway by an estimated 20 percent. Since the vehicles would travel the highway independent of the project's existence, this item is external energy conservation and is included in the energy inventory.

The conversion of some of the heating systems to solar energy is an internal energy

TABLE 5.9 Raw Energy Inventory for Residential Project

Number	Year	Item	Energy source	Consumption	Production	Conservation	Units
1	1-2	Construction workers commuting	Gasoline	60,000			Gallons
2	1	Earth-moving equipment	Diesel fuel	6,000			Gallons
3	1-2	Tools, lights, etc. for construction	Electricity	45,000			kWh
4	1-2	Space heating during construction	Kerosene	9.0			$\times 10^6$ Btu
5	1	Occupancy—space heating, etc.—25 units	Natural gas	4,430			$\times 10^6$ Btu
6	2	Occupancy—space heating, etc.—100 units	Natural gas	17,720			$\times 10^6$ Btu
7	3	Occupancy—space heating, etc.—175 units	Natural gas	31,010			$\times 10^6$ Btu
8	4-5	Occupancy—space heating, full occupancy—200 units	Natural gas	35,440			$\times 10^6$ Btu
9	1	Occupancy—appliances, air conditioning, etc.—25 units	Electricity	328			$\times 10^3$ kWh
10	2	Occupancy—applicances air conditioning, etc.—100 units	Electricity	1,314			$\times 10^3$ kWh
11	3	Occupancy—appliances air conditioning, etc.—175 units	Electricity	2,299			$\times 10^3$ kWh
12	4-10	Occupancy—applicances, air conditioning, etc.—200 units	Electricity	2,627			$\times 10^3$ kWh
13	1	Occupancy—automobiles	Gasoline	50			$\times 10^3$ gal
14	2	Occupancy—automobiles	Gasoline	200			$\times 10^3$ gal
15	3	Occupancy—automobiles	Gasoline	350			$\times 10^3$ gal
16	4-10	Occupancy—automobiles	Gasoline	400			$\times 10^3$ gal
17	2-10	Highway speed limit reduction	Gasoline			73	$\times 10^3$ gal
18	5	Additional air conditioning	Electricity	187			$\times 10^3$ kWh
19	6	Additional air conditioning	Electricity	455			$\times 10^3$ kWh
20	7	Additional air conditioning	Electricity	635			$\times 10^3$ kWh
21	8-10	Additional air conditioning	Electricity	858			$\times 10^3$ kWh
22	6	Space & water heating, etc.	Solar	1,709			$\times 10^6$ Btu
23	6	Space & water heating, etc.	Natural gas	33,731			$\times 10^6$ Btu
24	7	Space & water heating, etc.	Solar	3,418			$\times 10^6$ Btu
25	7	Space & water heating, etc.	Natural gas	32,022			$\times 10^6$ Btu
26	8-10	Space & water heating, etc.	Solar	5,127			$\times 10^6$ Btu
27	8-10	Space & water heating, etc.	Natural gas	30,313			$\times 10^6$ Btu

TABLE 5.10 Energy Inventory for Housing Project
$(\times 10^9$ Btu/yr)

Fuel or source	Major use	Construction										Total
		Year										
		1	2	3	4	5	6	7	8	9	10	
		Annual Consumption										
Nonrenewable sources:												
Natural gas	Space heat	4.4	17.7	31.0	35.4	35.4	33.7	32.0	30.3	30.3	30.3	280.5
Electricity*	Air conditioning and appliances	4.1	15.0	25.3	28.9	31.0	33.9	38.4	38.4	38.4	38.4	291.8
Gasoline	Automobiles	14.0	33.2	44.7	51.1	51.1	51.1	51.1	51.1	51.1	51.1	449.6
Diesel fuel	Earth-moving equipment	0.8										0.8
Kerosene	Space heat	0.9	0.9									1.8
Total nonrenewable		24.2	66.8	101.0	115.4	117.5	118.7	121.5	119.8	119.8	119.8	1,024.5
Renewable sources:												
Solar energy	Space heat						1.7	3.4	5.1	5.1	5.1	20.4
Total all sources		24.2	66.8	101.0	115.4	117.5	120.4	124.9	124.9	124.9	124.9	1,044.9
		Annual Conservation										
Gasoline	Lower speed limit	9.3	9.3	9.3	9.3	9.3	9.3	9.3	9.3	9.3	9.3	83.7
		Annual Energy Balance										
Gasoline		14.0	23.9	35.4	41.8	41.8	41.8	41.8	41.8	41.8	41.8	365.9
Nonrenewable sources		24.2	57.5	91.7	106.1	108.2	109.4	112.2	110.5	110.5	110.5	940.8
All sources		24.2	57.5	91.7	106.1	108.2	111.1	115.6	115.6	115.6	115.6	961.1

*Heat input to power plant assuming 31% efficiency.

conservation measure since it reduces the energy consumption of the project but does not affect external energy flows. Consequently, items 22 to 27 list the impact of conversion of solar energy as an interchange between consumption of two fuels. However, if solar energy conversion was being considered for an entirely separate project, it would be an external energy conservation measure.

After the "raw" energy inventory has been completed, the information must be organized into the format of Table 5.8. Since all energy flows in this example are "consumption" except for the energy conservation associated with the reduced speed limit, only one table is required. However, more complex projects will require separate tables for consumption, production, conservation, and the net energy balance. Table 5.10 shows the processed energy inventory data for the project. All energy flows from Table 5.9 have been converted to common units and are itemized by year and fuel source. Electrical energy was converted to heat energy units assuming a power plant efficiency of 31 percent and using the method discussed earlier in the section Types of Energy and Energy Utilization. All energy amounts have been rounded to the nearest 0.1×10^9 Btu/yr. This represents about 0.1 percent of each year's total energy flow and is probably much less than the uncertainty in the estimates.

The addition or comparison of energy quantities from different fuels or power sources is an important part of every energy inventory. In this example, the energy inventory data will be compared with local supply and demand data for specific fuels, and it is therefore appropriate to express the data in terms of the energy released by burning these fuels. However, the summation of these energy quantities to calculate an energy balance must be viewed with caution. The energy balance so calculated is an *approximate* indication of the amount of *heat energy* that would be available for other uses without the project. The heat balance does not represent the total amount of energy consumed by the project. Calculation of this quantity would be exceedingly complex and would involve tracing each energy flow to the point of extraction. It would also require the inclusion of several ancillary data such as the energy associated with: resource exploration and development, employees working to process the energy, equivalent energy content of materials, and equipment used to produce energy, etc. Table 5.30, presented later, lists factors for comparing energy quantities at various points in the fuel cycle.

Power Plant. A 225-MW capacity oil-fired electrical power generating plant has been proposed for a site 100 mi from a major northeastern metropolitan area. The power plant will serve two important needs: (1) it will provide additional electrical power to meet the growing demand in the metropolitan area, and (2) it will allow an all natural gas–fired power plant within the metropolitan area to be derated to conserve natural gas and reduce air pollutant emissions in the congested area.

The electrical power produced at the power plant will be transported to the metropolitan area over an existing 345-kV power line now operating at less than full capacity. The power plant boilers will be fired with Number 6 residual oil from a refinery 75 mi from the power plant site. Oil will be shipped by rail on unit trains.

Power plant construction will require 5 years and a total of 2,600 worker-years of labor. The work force will peak at 800 persons during the third and fourth years. Construction workers will commute primarily from the area within 40 mi of the plant. The average commute is estimated to be 55 mi/day round trip.

Power plant operation will begin in the sixth year and will initially average 25 percent capacity rising to 50 percent in the seventh year and 75 percent in subsequent years. The power plant will have a lifetime of 35 years (40 years after start of construction) but will be derated to 25 percent capacity during the last year. The power plant will operate with a heat rate of 9,000 Btu/kWh or 37.92 percent.

When the new power plant is fully operational in the eighth year, the natural gas power plant will be derated by 60 MW. This plant has a heat rate of 11,000 Btu/kWh or 31.03 percent and will be operated until the twentieth year, when it is dismantled.

This project involves consumption, production, and conservation energy flows. Energy consumption is required for construction, plant operation, and, most important, for boiler fuel. Useful electrical power is produced and natural gas is conserved by derating the natural gas–fired plant.

Table 5.11 is a raw energy inventory for the project. All energy flows were calculated from the information listed above using the data in the section Data and Data Sources

TABLE 5.11 Raw Energy Inventory for Power Plant

Number	Year	Item	Energy source	Consumption	Production	Conservation	Units
1	1	Heavy equipment operation	Diesel fuel	1,000			$\times 10^3$ gal
2	2	Heavy equipment operation	Diesel fuel	500			$\times 10^3$ gal
3	3	Heavy equipment operation	Diesel fuel	300			$\times 10^3$ gal
4	4	Heavy equipment operation	Diesel fuel	200			$\times 10^3$ gal
5	5	Heavy equipment operation	Diesel fuel	100			$\times 10^3$ gal
6	1	Tools, lights, etc. for construction	Electricity	200			$\times 10^3$ kWh
7	2	Tools, lights, etc. for construction	Electricity	600			$\times 10^3$ kWh
8	3–4	Tools, lights, etc. for construction	Electricity	800			$\times 10^3$ kWh
9	5	Tools, lights, etc. for construction	Electricity	200			$\times 10^3$ kWh
10	1	Portable space heaters during construction	Propane	300			$\times 10^6$ Btu
11	2	Portable space heaters during construction	Propane	1,800			$\times 10^6$ Btu
12	3	Portable space heaters during construction	Propane	2,400			$\times 10^6$ Btu
13	4	Portable space heaters during construction	Propane	600			$\times 10^6$ Btu
14	5	Portable space heaters during construction	Propane	300			$\times 10^6$ Btu
15	1	Temporary construction offices—space heating	Propane	363			$\times 10^6$ Btu
16	1	Temporary construction offices—cooling	Electricity				$\times 10^3$ kWh
17	1	Temporary construction offices—lights, etc.	Electricity				$\times 10^3$ kWh
18	2	Temporary construction offices—space heating	Propane	1,000			$\times 10^6$ Btu
19	2	Temporary construction offices—cooling	Electricity	11.95			$\times 10^3$ kWh
20	2	Temporary construction offices—lights, etc.	Electricity	2.40			$\times 10^3$ kWh
21	3–4	Temporary construction offices—space heating	Propane	1,451			$\times 10^6$ Btu
22	3–4	Temporary construction offices—cooling	Electricity	15.94			$\times 10^3$ kWh
23	3–4	Temporary construction offices—lights, etc.	Electricity	32			$\times 10^3$ kWh

TABLE 5.11 Raw Energy Inventory for Power Plant (Continued)

Number	Year	Item	Energy source	Consumption	Production	Conservation	Units
24	5	Temporary construction offices—space heating	Propane	363			$\times 10^6$ Btu
25	5	Temporary construction offices—cooling	Electricity	4			$\times 10^3$ kWh
26	5	Temporary construction offices—lights, etc.	Electricity	8			$\times 10^3$ kWh
27	1	Construction workers commuting	Gasoline	183.3			$\times 10^3$ gal
28	2	Construction workers commuting	Gasoline	550			$\times 10^3$ gal
29	3–4	Construction workers commuting	Gasoline	733.3			$\times 10^3$ gal
30	5	Construction workers commuting	Gasoline	183.3			$\times 10^3$ gal
31	1	Equipment deliveries	Diesel fuel	93.75			$\times 10^3$ gal
32	2–4	Equipment deliveries	Diesel fuel	187.5			$\times 10^3$ gal
33	5	Equipment deliveries	Diesel fuel	93.75			$\times 10^3$ gal
34	6–40	Equipment deliveries	Diesel fuel	18.75			$\times 10^3$ gal
35	6	Operating crew commuting	Gasoline	83.4			$\times 10^3$ gal
36	7	Operating crew commuting	Gasoline	41.7			$\times 10^3$ gal
37	8–40	Operating crew commuting	Gasoline	20.85			$\times 10^3$ gal
38	3–40	Shop & control room space heating	Natural gas	520			$\times 10^6$ Btu
39	3–40	Shop & control room cooling	Electricity	35.7			$\times 10^3$ kWh
40	1	Domestic water heating	Propane	182.5			$\times 10^6$ Btu
41	2	Domestic water heating	Propane	547.5			$\times 10^6$ Btu
42	3–4	Domestic water heating	Natural gas	730			$\times 10^6$ Btu
43	5–6	Domestic water heating	Natural gas	182.5			$\times 10^6$ Btu

44	7	Domestic water heating	Natural gas	91.25			$\times 10^6$ Btu
45	8–40	Domestic water heating	Natural gas	45.63			$\times 10^6$ Btu
46	6	Power generation	Electricity		0.493		$\times 10^9$ kWh
47	7	Power generation	Electricity		0.986		$\times 10^9$ kWh
48	8–39	Power generation	Electricity		1.478		$\times 10^9$ kWh
49	40	Power generation	Electricity		0.493		$\times 10^9$ kWh
50	6	Boiler fuel	No. 6 oil	4.437			$\times 10^{12}$ Btu
51	7	Boiler fuel	No. 6 oil	8.874			$\times 10^{12}$ Btu
52	8–39	Boiler fuel	No. 6 oil	13.311			$\times 10^{12}$ Btu
53	40	Boiler fuel	No. 6 oil	4.437			$\times 10^{12}$ Btu
54	6	Railroad transportation refinery—power plant	Diesel fuel	9.72			$\times 10^9$ Btu
55	7	Railroad transportation refinery—power plant	Diesel fuel	19.43			$\times 10^9$ Btu
56	8–39	Railroad transportation refinery—power plant	Diesel fuel	29.15			$\times 10^9$ Btu
57	40	Railroad transportation refinery—power plant	Diesel fuel	9.72			$\times 10^9$ Btu
58	6	Transmission line	Electricity		−11.83		$\times 10^6$ kWh
59	7	Transmission line	Electricity		−23.66		$\times 10^6$ kWh
60	8–39	Transmission line	Electricity		−35.50		$\times 10^6$ kWh
61	40	Transmission line	Electricity		−11.83		$\times 10^6$ kWh
62	3–40	Plant lighting	Electricity	87.6			$\times 10^3$ kWh
63	8–20	Natural gas power plant derating	Natural gas	−525.6			$\times 10^{12}$ Btu
64	8–20	Natural gas power plant derating	Electricity			5.78	$\times 10^6$ kWh

later in this chapter. Items 1 to 33, 41, and 42 are related entirely to construction and include energy flows similar to those in the previous example, such as:
- Heavy equipment operation and equipment deliveries—diesel fuel
- Portable space heaters and space and water heating for temporary construction offices—propane
- Tool operation, lighting, and cooling for temporary construction offices—electricity
- Construction workers commuting—gasoline

The remaining items are related to power plant operation. Items 34 to 45 and 62 are indirect energy flows, i.e., those required for plant operation but not directly related to the production of electrical power from oil. These include:
- Equipment deliveries (maintenance)—diesel fuel
- Operating crew commuting—gasoline
- Space and water heating—natural gas
- Space cooling and lighting—electricity

The generation of electrical power is represented by items 46 to 49. This production energy flow may be calculated from the plant capacity, the capacity factor, and the number of hours per year:

$$\begin{matrix} \text{kWh electricity} \\ \text{produced per} \\ \text{year} \end{matrix} = \left(\begin{matrix} \text{plant capacity,} \\ \text{MW} \times 10^3 \end{matrix} \right) \left(\begin{matrix} \text{capacity factor,} \\ \div 100 \end{matrix} \right) \left(\begin{matrix} 8{,}760 \\ \text{h/yr} \end{matrix} \right)$$

For example, the electrical power produced in years 8–39 may be calculated as follows:

$$\begin{matrix} \text{kWh electricity} \\ \text{produced per} \\ \text{year} \end{matrix} = (225 \times 10^3) \left(\frac{75}{100} \right) (8{,}760)$$

$$= 1.478 \times 10^9 \text{ kWh/yr}$$

Items 50 to 53 are the oil required at the power plant to produce electrical power. The heating value of the oil is the product of the plant heat rate in Btu/kWh and the number of kilowatthours of electrical power generated, i.e.,

$$\text{Oil heating value, Btu/yr} = \left(\begin{matrix} \text{heat rate,} \\ \text{Btu/kWh} \end{matrix} \right) \left(\begin{matrix} \text{power generated,} \\ \text{kWh/yr} \end{matrix} \right)$$

For years 8–39, this may be calculated as follows:

$$\text{Oil heating value, Btu/yr} = (9{,}000) (1.478 \times 10^9)$$

$$= 13.3 \times 10^{12} \text{ Btu/h}$$

Items 54 to 57 are the diesel fuel consumed in transporting the oil from refinery to power plant. The method for calculating this energy flow is discussed later in an example in the Energy Inventory Data section.

Items 58 to 61 are the energy losses in transmitting the electrical power to the metropolitan area. While this could be viewed as energy consumption, it is convenient to represent it as negative production since it directly impacts the amount of electrical power available at the urban center. Note, however, that the other items of electrical power consumption are treated as consumption.

Items 63 and 64 relate to the derating of the natural gas power plant. The derating will reduce the electrical power available from the project and is thus treated as negative production similar to transmission line losses. The derating will also conserve natural gas as shown.

Table 5.12 shows the energy consumption data converted to common units and organized by fuel type and year. The energy consumption in years 8–39 is constant and is thus represented by a simple column. Since this project uses no renewable resources, it is not necessary to differentiate between renewable and nonrenewable fuels. However, the amount of Number 6 oil used as boiler fuel is much greater than the consumption of other fuels, and it is appropriate to provide a subtotal as shown.

Consumption of electrical power must be viewed cautiously. In some power plants the power required for plant operation (lights, space cooling, instrument operation, etc.) is *not* supplied from the main generators but from a separate line attached to the local grid. This allows the entire power plant to be controlled and operated independent of the electrical

TABLE 5.12 Power Plant Energy Consumption
($\times 10^9$ Btu/yr)

Fuel or source	Major use	Construction years						Operation years			Total
		1	2	3	4	5	6	7	8–39	40	
Diesel fuel	Fuel shipping on railroad	151.2	95.0	67.4	53.6	16.8	12.3	22.0	31.7	12.3	1,455.0
Electricity	Plant lighting	2.3	7.0	19.4	19.4	12.4	10.0	10.0	10.0	10.0	410.5
Gasoline	Commuting	23.4	70.2	93.6	93.6	23.4	10.7	5.3	2.7	2.7	409.3
Natural gas	Space heating			1.3	1.3	0.7	0.7	0.6	0.6	0.6	24.4
Propane	Space heating	0.5	3.4	3.9	2.1	0.7					10.6
Subtotal		177.4	175.6	185.6	170.0	64.0	33.7	37.9	45.0	25.6	2,309.8
No. 6 oil	Boiler fuel						4,437	8,874	13,311	4,437	443,700
Total		177.4	175.4	185.6	170.0	64.0	4,470.7	8,911.9	13,356.0	4,462.8	446,009.8

power generators, transformers, etc. While it could be argued that the electrical power consumed at the plant decreases the net electrical power available from the plant, the power consumed is supplied from a separate power plant and may not impact the net power available to the metropolitan area.

Another way of considering the electrical power consumption is to assume the power is generated by a small power plant totally dedicated to the major power plant. In this case the energy consumption would be the heating value of the fuel supplied to the small power plant. Thus, in Table 5.12 the electrical power consumed was converted to heat energy by dividing by the assumed efficiency of the local power plant (31.0%).*

In large power generating projects such as power plants, the energy consumed for construction and operation of the plant is usually much smaller than the main energy flow utilized in energy production. The following energy flow totals from Table 5.12 illustrate this point:

Energy consumption	× 10⁹ Btu	% of total
Plant construction	773	0.17
Plant operation	1,537	0.34
Boiler fuel	443,700	99.49
Total	446,010	100.00

In this example the entire energy consumption associated with plant construction and operation is only 0.51 percent of the total. Thus, uncertainties in these energy flows are not important in relation to the plant's overall energy impact. However, these energy flows may be important when considered in terms of the local supply and demand for specific fuels.

Table 5.13 summarizes all energy flows including production, conservation, and consumption. Transmission line losses and the derating of the natural gas power plant are treated as negative production so that the net electricity produced represents the additional electrical power available in the metropolitan area as a result of this project.

Natural gas conservation through derating of the natural gas power plant is constant in years 8–20 and is shown as a single item. The total energy consumption for each year from Table 5.12 is also shown. The net energy consumed is calculated as follows:

$$\begin{pmatrix} \text{Net energy} \\ \text{consumed} \end{pmatrix} = \begin{pmatrix} \text{total energy} \\ \text{consumed} \end{pmatrix} - \begin{pmatrix} \text{net} \\ \text{electricity} \\ \text{produced} \end{pmatrix} - \begin{pmatrix} \text{natural gas} \\ \text{conserved} \end{pmatrix}$$

where all amounts are in 10^{12} Btu energy units. Even though this project is termed "energy-producing," its net effect is energy consumption because more energy is consumed than produced and conserved.

Data from Table 5.13 may be used to calculate the overall heat rate and efficiency of the project. The overall heat rate is the amount of heat energy required per unit of electrical power produced, given as follows:

$$\begin{aligned} \begin{pmatrix} \text{Overall heat} \\ \text{rate, Btu/kWh} \end{pmatrix} &= \frac{\begin{pmatrix} \text{total energy} \\ \text{consumed, Btu} \end{pmatrix} - \begin{pmatrix} \text{natural gas} \\ \text{conservation, Btu} \end{pmatrix}}{\text{net electricity produced, kWh}} \\[2mm] &= \frac{(446.010 \times 10^{12}\ \text{Btu}) - (75.14 \times 10^{12}\ \text{Btu})}{41,230 \times 10^6\ \text{kWh}} \\[2mm] &= 8{,}995\ \text{Btu/kWh} \end{aligned}$$

Thus, the overall efficiency is:

$$\begin{aligned} \text{Overall efficiency, \%} &= \frac{3{,}413\ \text{Btu/kWh}}{\text{overall heat rate, Btu/kWh}} \times 100 \\[2mm] &= \frac{3{,}413\ \text{Btu/kWh}}{8{,}995\ \text{Btu/kWh}} \times 100 \\[2mm] &= 37.94\% \end{aligned}$$

*See the example in the earlier section, Types of Energy and Energy Utilization.

TABLE 5.13 Power Plant Energy Flows

Energy flow	Construction years					Production years					Total
	1	2	3	4	5	6	7	8–20	21–39	40	
Electricity production ($\times 10^6$ kWh)											
Electricity produced at power plant						493	986	1478	1478	493	49,268
Transmission line losses						−12	−24	−36	−36	−12	−1,200
Net electricity available						481	962	1442	1442	481	48,068
City power plant derating								−526			−6,838
Net electricity produced						481	962	916	1442	481	41,230
Net electricity produced ($\times 10^{12}$ Btu)						1.64	3.28	3.13	4.92	1.64	140.72
Natural gas conservation ($\times 10^{12}$ Btu)								5.78			75.14
Total energy consumption, all fuels ($\times 10^{12}$ Btu)	0.177	0.175	0.186	0.170	0.064	4.471	8.912	13.356	13.356	4.463	446.01
Net energy consumed ($\times 10^{12}$ Btu)	0.177	0.175	0.186	0.170	0.064	4.471	8.912	7.576	13.356	4.463	370.87

The following table compares heat rates and efficiencies for the oil-fired power plant, the natural gas–fired power plant, and the entire project.

Plant type	Heat rate, Btu/kWh	Efficiency, %
Oil-fired	9,000	37.92
Natural gas–fired	11,000	31.03
Project	8,995	37.94

The oil-fired plant has higher efficiency than the natural gas–fired plant, and the derating conserves energy thus raising the net efficiency. However, this efficiency improvement is balanced by the energy loss in the transmission line and the energy required for plant construction and operation resulting in a net heat rate and overall efficiency essentially the same as for the oil-fired plant.

Insulation Standard.   An insulation standard has been proposed for residential water heaters. A minimum of 3 inches of thermal insulation would be required on all exposed surfaces of water heaters installed after 1980. The standard would be applied throughout the United States, including Alaska and Hawaii.

The previous two examples discussed projects with a relatively small number of energy-related components. This example, however, affects millions of water heaters. The approach of the previous examples involving itemizing each individual component is clearly inappropriate. Instead, the effect of the insulation standard on four representative water heater designs will be estimated, and the results will then be extrapolated to the entire United States. Reference 78 discusses residential water heating and fuel conservation achievable with varying amounts of insulation. The data in this example were obtained from this source.

Residential water heaters are manufactured in a variety of sizes and are heated primarily by four fuels: fuel oil, natural gas, liquid petroleum gas (LPG), and electricity. The fuel-fired units average 40 gal capacity and 1.0-in insulation thickness. Electrically heated units average 50 gal capacity with 2.0 in insulation.

Table 5.14 shows the energy consumption of these average units as currently designed

TABLE 5.14   Energy Consumption of Typical Residential Water Heaters

Fuel and capacity	Current design annual energy consumption		Project design annual energy consumption		Percent reduction
	Units	×10⁶ Btu	Units	×10⁶ Btu	
Fuel oil, 40 gal	271.9 therms	27.19	213.1 therms	21.13	21.6
Natural gas, 40 gal	260.3 therms	26.03	204.0 therms	20.40	21.6
LPG, 40 gal	260.3 therms	26.03	204.0 therms	20.40	21.6
Electricity	4400 kWh	48.44*	4158 kWh	45.78*	5.5

*Heat required to produce electricity assuming 31.0% power plant efficiency.

and with 3.0 in of insulation. Energy consumption varies with usage, and the amounts listed are based on average values. The insulation standard would reduce energy consumption of fuel-fired water heaters by 21.6 percent and electric water heaters by 5.5 percent.

The energy consumption for domestic water heating without the insulation standard is shown in Table 5.15 for 1970 and 1990. The number of units listed refers to units of average design shown in Table 5.14. The 1970 data is historical, and the 1990 data is based on projections assuming electricity doubling in price relative to fuel costs.

Figure 5.3 shows the number of water heaters of each type by year assuming linear growth between 1970 and 1990. The effect of the insulation standard is also shown. Since residential water heaters have average lifetimes of 10 years, it was assumed that the number of water heaters of old design without the extra insulation would decrease at 10 percent per year beginning in 1980. All new water heaters would meet the insulation standard, and the number in service would rise from zero in 1980 to 100 percent in 1990.

Based on these assumptions, the projected energy consumption with and without the insulation standard were calculated. The energy consumption in each year is as follows:

$$\begin{matrix} \text{Energy consumption,} \\ \text{all units} \end{matrix} = \begin{pmatrix} \text{number of} \\ \text{old units} \end{pmatrix} \times \begin{pmatrix} \text{energy consumption} \\ \text{per old unit} \end{pmatrix} + \\ \begin{pmatrix} \text{number of} \\ \text{new units} \end{pmatrix} \times \begin{pmatrix} \text{energy consumption} \\ \text{per new unit} \end{pmatrix}$$

For example, the energy consumption of natural gas–fired water heaters in 1985 with and without the insulation standard is:

$$\begin{matrix} \text{Energy consumption} \\ \text{without standard} \end{matrix} = \begin{pmatrix} 59.43 \times 10^6 \\ \text{old units} \end{pmatrix} \begin{pmatrix} 26.03 \times 10^6 \\ \text{Btu/unit} \end{pmatrix}$$

$$= 1547 \times 10^{12} \text{ Btu}$$

TABLE 5.15   United States Residential Water Heating Energy Consumption Estimate

	1970			1990		
	Number of	Total annual energy consumption		Number of	Total annual energy consumption	
Fuel	units, $\times 10^6$	Units	$\times 10^{12}$ Btu	units, $\times 10^6$	Units	$\times 10^{12}$ Btu
Fuel oil	6.179	29.090×10^6 bbl	168	2.427	11.430×10^6 bbl	66
Natural gas	34.883	908×10^9 ft³	908	67.653	1.761×10^9 ft³	1,761
LPG	3.150	20.440×10^9 bbl	82	2.151	14.000×10^9 bbl	56
Electricity	16.114	70.9×10^9 kWh	781*	13.636	60.0×10^9 kWh	668*
Total	60.326		1,939	85.867		2,551

*Heat required to produce electricity assuming 31% power plant efficiency.

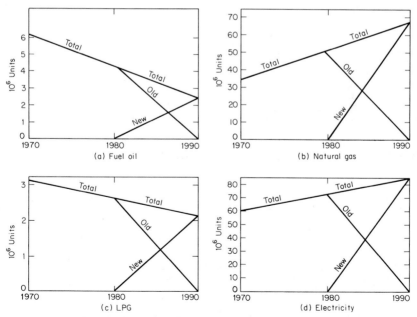

Figure 5.3 Number of water heaters by year and fuel used.

and

$$\text{Energy consumption} \atop \text{with standard} = \left({25.62 \times 10^6 \atop \text{old units}}\right)\left({26.03 \times 10^6 \atop \text{Btu/unit}}\right) + \left({33.82 \times 10^6 \atop \text{new units}}\right)\left({20.40 \times 10^6 \atop \text{Btu/unit}}\right)$$

$$= \left({667 \times 10^{12}\text{ Btu} \atop \text{old units}}\right) + \left({690 \times 10^{12}\text{ Btu} \atop \text{new units}}\right)$$

$$= 1357 \times 10^{12}\text{ Btu}$$

Tables 5.16 and 5.17 show the results for each year from 1981 to 1990.

The energy conserved as a result of the insulation is the difference between consumption with and without the standard. The energy consumed for each fuel and year is thus calculated as follows:

$$\text{Energy conserved} = \left({\text{energy consumed} \atop \text{without standard}}\right) - \left({\text{energy consumed} \atop \text{with standard}}\right)$$

TABLE 5.16   Energy Consumption Without Project
($\times 10^{12}$ Btu/yr)

						Year					
Fuel	1981	1982	1983	1984	1985	1986	1987	1988	1989	1990	Total
Fuel oil	112	107	102	97	92	86	81	76	71	66	890
Natural gas	1,377	1,420	1,462	1,505	1,547	1,590	1,633	1,676	1,718	1,761	15,689
LPG	68	66	65	64	63	61	60	59	57	56	619
Electricity	719	713	708	702	696	691	685	679	674	668	6,935
Total, all fuels	2,276	2,306	2,337	2,368	2,398	2,428	2,459	2,490	2,520	2,551	24,133

For natural gas in 1985, this becomes:

$$\text{Natural gas conserved in 1985} = 1547 \times 10^{12}\text{ Btu} - 1357 \times 10^{12}\text{ Btu}$$

$$= 190 \times 10^{12}\text{ Btu}$$

Table 5.18 shows the results for all fuels and years considered. Totals and percentages have also been calculated.

Additional factors which could be included in a more complete analysis are:
- Alternative fuel cost scenarios
- Other water heater designs
- Variable insulation composition
- Different assumed attrition rate of old units
- Ancillary energy required to manufacture insulation and/or administer the program.

Supply/Demand Scenarios

The changes in energy supply/demand scenarios produced by a project are the key reasons for assessing energy impact. This section discusses the organization of supply/demand data to illustrate energy impact. The following subsection discusses the relationship between the components of supply and demand, and an accounting method for presenting the data to interface with energy inventory data. Potential adverse impacts are also addressed. The subsection entitled Supply/Demand Examples considers three examples of the organization of supply/demand data to interface with the example energy inventories previously discussed. Sources for supply/demand data are listed in a later section, Supply and Demand Data Sources.

Supply/Demand Accounting and Organization   There are several ways to organize supply and demand data, each with a different accounting method. If the data are to be used for

TABLE 5.17 Energy Consumption With Project
($\times 10^{12}$ Btu/yr)

Fuel	Type	1981	1982	1983	1984	1985	1986	1987	1988	1989	1990	Total
Fuel oil	Old	105	94	82	70	58	47	35	23	12	0	526
	New	5	10	15	21	26	31	36	41	46	51	282
	Total	110	104	97	91	84	78	71	64	58	51	808
Natural gas	Old	1,201	1,068	934	801	667	534	400	267	133	0	6,005
	New	138	276	414	552	690	828	966	1,104	1,242	1,380	7,590
	Total	1,339	1,344	1,348	1,353	1,357	1,362	1,366	1,371	1,375	1,380	13,595
LPG	Old	62	55	48	41	34	28	21	14	7	0	310
	New	4	9	13	18	22	26	31	35	39	44	241
	Total	66	64	61	59	56	54	52	49	46	44	551
Electricity	Old	648	576	504	432	360	288	216	144	72	0	3,240
	New	62	125	187	250	312	375	437	499	562	624	3,433
	Total	710	701	691	682	672	663	653	643	634	624	6,673
Total, all fuels		2,225	2,213	2,197	2,185	2,169	2,157	2,142	2,127	2,113	2,099	21,627

energy impact assessment, selecting a particular method is not as important as maintaining consistency and carefully defining terms such as production, consumption, supply, and demand. The accounting method should be chosen on the basis of convenience in organizing the specific data available. This section discusses one systematic accounting method for organizing supply and demand data for an energy impact assessment.

Strictly speaking, supply and demand are economic terms quantifying the relationships between the amounts of energy suppliers will sell and consumers will buy as a function of price. However, there are two reasons for using more generalized definitions for energy accounting. First, energy is intrinsically related to most activities of society and is often viewed as a necessity. This is particularly true over short time periods where consumers are committed to using nearly fixed amounts of energy independent of price. For example,

TABLE 5.18 Net Energy Conservation
[× 10^{12} Btu/yr (% of Consumption Without Project)]

Fuel	1981	1982	1983	1984	1985	1986	1987	1988	1989	1990	Total
Fuel oil	2	3	5	6	8	8	10	12	13	15	82
	(2)	(3)	(5)	(6)	(9)	(12)	(12)	(16)	(18)	(23)	(9)
Natural gas	38	76	114	152	190	228	267	305	343	381	2,094
	(3)	(5)	(8)	(10)	(12)	(14)	(16)	(18)	(20)	(22)	(13)
LPG	2	2	4	5	7	7	8	10	11	12	68
	(3)	(3)	(6)	(8)	(11)	(12)	(13)	(17)	(19)	(21)	(11)
Electricity	9	12	17	20	24	28	32	36	40	44	262
	(1)	(2)	(2)	(3)	(3)	(4)	(5)	(5)	(6)	(7)	(4)
Total	51	93	140	183	229	271	317	363	407	452	2506
	(2)	(4)	(6)	(8)	(10)	(11)	(13)	(15)	(16)	(18)	(10)

a recent study of the demand for gasoline as a function of price found that a large percentage change in price produced a negligible change in the amount of gasoline consumed. This behavior is termed "inelastic demand." When the demand is inelastic, the price of energy and the amount consumed are effectively uncoupled.

The second reason for using more generalized definitions is governmental control. Most energy-producing industries are under varying degrees of governmental regulation and/or control. In some areas, the price, supply, and allocation of fuels are determined by the government; in others, the production and distribution of energy are handled by regulated utilities. Thus, the balance between supply and demand is not determined solely by price.

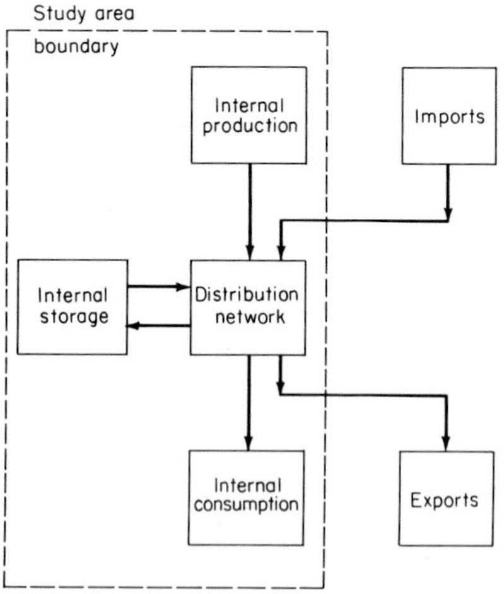

For the purposes of energy impact assessment, "energy supply" will be defined as the total energy flowing into an area including internal energy production, energy supply from storage, and energy imported into the area from external sources. "Energy demand" will be defined as the total energy required by the area and includes energy consumed and stored in the area and energy exported to other areas.

These energy flows and the recommended energy accounting method can best be described in terms of an example study area. This study area could be the world, a nation, a city, a county or any other clearly described entity such as an electric utility's distribution area. The importance of choosing the area properly will be discussed later.

Figure 5.4 Energy flows in a study area.

Figure 5.4 describes the energy flows in the area schematically. The study area boundary separates energy flows within the study area from those outside. Within the study area, the heart of the energy system is the distribution network. This complex combination of businesses and governmental agencies includes all mechanisms for distributing energy from producers to consumers. Energy flows into the distribution network from three sources, namely:

1. Internal production
2. Internal storage
3. Imports

These three sources together represent the energy supply to the study area. Internal production is that portion of the energy supply generated within the study area. For natural gas, it is the gas drawn from wells within the study area, and for electrical power, it is the electricity generated at power plants within the study area.

Unlike internal production, internal storage is an amount of energy rather than a rate of energy flow. Internal storage represents energy stored within the study area, such as natural gas in utility tanks and gasoline in automobile gasoline tanks and industrial facilities. When these energy stores are reduced, energy is available for consumption or export.

The third source of energy supply is energy imported from outside the study area. For natural gas, it is gas flowing into the study area through pipelines, and for electricity, it is electricity generated at power plants outside of the study area supplying power over the powerline grid.

Energy flows out of the distribution system in three ways:

1. Internal consumption
2. Internal storage
3. Exports

These three energy flows constitute the energy demand. Internal consumption is that portion of the energy demand consumed within the study area and is equated with the entire demand in some accounting methods. Examples of internal consumption include natural gas burned to heat homes and the electricity to operate lights, motors, etc., within the study area. Internal storage contributes to demand if the amount stored increases. Changes in internal storage can represent a large portion of the total demand. For example, the energy stored in the form of gasoline in an average full-size car can fluctuate from 0 to 20 gal, depending on the size of the tank, and variations in this amount can be equivalent to a month's (or more) consumption. Exports are that portion of the energy demand transported outside of the study area through powerlines, pipelines, etc.

The energy flows associated with the study area at any point in time represent the energy supply and demand scenario for that time frame. The purpose of the supply/demand section of an energy impact assessment is to evaluate these scenarios for present and future time frames.

The supply/demand scenarios should be organized to present the following key points:

- Supply/demand scenarios significantly impacted by the project
- Comparison of project energy flows with supply/demand scenarios
- Evaluation of results

Supply/demand scenarios can be formulated on many levels ranging from a single dwelling to the world. However, the scenarios presented in an energy impact assessment for a specific project should be limited to those significantly impacted by the project. A "significant impact" is defined here as a projected change in any of the energy flows described above of more than 1 percent of the existing or projected amount without the project. The number of scenarios to be presented depends upon the project's size. For example, a small project may affect a city's energy flow by less than 1 percent, and in this case, one scenario—the city—may be sufficient. However, a large project which changes the nation's energy flow by more than 1 percent would require several levels—for example, the United States and specific states, counties, and cities. The scenarios presented should effectively cover the time period during which the project affects energy flows, and the fuel and energy sources actually employed in the project. The data presented in these scenarios should be factual for present and historical scenarios and should be the best estimates for future scenarios.

Table 5.19 shows one suitable way to present these supply/demand data. The effects of the project on the supply/demand scenario can be illustrated by preparing similar tables with and without the project energy flows.

This type of presentation compares energy flow rates and is appropriate for most projects. However, for some projects the total amount of energy conserved, produced, or consumed may represent a significant portion of the resource base for certain fuels. In these cases, the totals over the project life should be compared to existing and projected resource bases for those fuels as shown in Table 5.20. Note that energy types such as electricity, gasoline, heating oil, etc., are not listed because they have no readily definable resource base. Only basic fuels and energy sources can be accurately compared.

The presentation of this supply/demand data should both utilize the same system of energy units as the energy impact section and discuss all data sources and assumptions in enough detail to enable the reader to verify the results. It is particularly important to explain the energy accounting method and the reasons it was chosen over other methods.

Analyzing the supply/demand data in this fashion demonstrates that regardless of project size or type, there is a study area level with energy flows significantly impacted by

TABLE 5.19   Illustrative Table: Supply/Demand Scenario

Energy flow	Year					
	1	2	3	4	5	6
Demand: Internal consumption Exports Storage increase						
Total demand						
Supply: Internal production Imports Storage decrease						
Total supply						

the project. The evaluation of the supply/demand data which follows should recognize this and discuss impacts on the appropriate levels. For a consuming project, it should be shown that the additional energy required by the project will be available within the time constraints imposed by the project. It should be further demonstrated that any anticipated energy shortages are either acceptable or can be mitigated to the satisfaction of other energy consumers and suppliers within the study area. For a producing or conserving project, it must be shown that the additional energy made available by the project will be utilized effectively and that any energy surpluses are either acceptable or can be adjusted to minimize oscillations in the energy supply cycle. The determination of these shortages or surpluses is straightforward based on the use of the tables described previously. However, the potential problems and their solutions require careful analysis.

TABLE 5.20   Illustrative Table: Total Resource Base and Project Energy Consumption, Production, or Conservation

Basic fuel type	Existing resource base	Projected resource base	Project lifetime consumption, production, or conservation
Natural gas Petroleum Coal Geothermal heat Uranium Other			
Total			

There are three types of potential adverse impacts and problems which must be addressed, namely:

1. Power demand problems—short term
2. Energy demand problems—long term
3. Economic problems

Power demand problems stem from restrictions or bottlenecks in the energy supply chain between the raw fuel and the final processed energy supply. These bottlenecks are not shortages of energy per se, but result from the inability to process energy fast enough to meet demand. Most energy supply/demand problems are strictly demand problems. For example, the petroleum shortage in 1973 was not due to the limited supply of petroleum. Even conservative estimates show that the United States has enough domestic oil to meet demand for at least several decades. The shortage was a result of the inability of domestic suppliers to pump, process, and deliver petroleum products at a sufficient rate to satisfy the total United States demand without foreign imports.

Power demand problems are usually facility limited. If facilities have already been constructed and are operating at less than peak capacity, the power produced (rate of energy production or processing) can be increased rapidly. However, if new facilities must be constructed, increased power production may require several years. Consequently, if the energy supply/demand analysis shows that a consuming project will require rapid construction of new energy processing facilities, there is a potential power demand problem.

The electrical power generation industry is particularly plagued with power demand problems. Since electricity travels almost instantaneously and most electrical power networks have negligible electrical energy storage, an increase in demand such as the starting of a motor or the turning on of a light bulb must be met with a nearly instantaneous increase in the amount of power generated. Fortunately, since a large number of users are connected to the same power grid, the small variations due to thousands of lights, motors, air conditioners, heaters, etc., cycling on and off tend to cancel each other; however, there is a significant daily variation in average power. This diurnal variation peaks on summer days in midafternoon when air conditioners contribute heavily and drops (to roughly half the peaks) in early morning hours when most people sleep. Since the electric utility must meet peak demand, the installed generating capacity must be much higher than the average load. Thus, in evaluating supply/demand scenarios, time of day could be an important parameter.

Power demand problems begin when the peak power demand approaches the peak generating capacity. When the power demand momentarily exceeds the peak generating capacity, there are several possibilities. The system may overload, causing one or more generating units to cut out. Since demand is not lessened, the system becomes more overloaded and additional failures occur progressively until the entire system fails and no power is produced. This is called a "blackout." Another possibility is that the electric utility sets up a priority list for customers termed interruptible. In times when demand exceeds supply, the interruptible customers with low priority are cut off to maintain full service to the remainder. This process is termed "service interruption." The final possibility is to program the generators to reduce the voltage supply to all customers equally. Since the power supplied is roughly proportional to the square of the voltage, a 20 percent overload can be handled with only a 10 percent voltage reduction. This action is termed a "brownout" and causes problems with certain electrical equipment designed to operate only on well-controlled voltage.

Similar problems occur in other energy supply systems. Some notable examples include gasoline rationing, fuel oil rationing, and interruption of natural gas supply to certain industrial customers and power plants during peak demand periods. The key to evaluating and mitigating potential power demand problems is accurate demand forecasting and long-range planning.

Energy demand problems stem from the limited resource bases of certain nonrenewable energy sources and are long-range problems. While power demand problems are associated with the rate of energy supply and demand, energy demand problems relate to the total projected energy consumption for *all* times in the future compared with known and projected resource bases. Since all nonrenewable resources have finite resource bases, eventually the total consumption for each nonrenewable fuel will approach the total resource base. The resulting fuel shortages are energy demand problems and are inevitable for all nonrenewable energy sources.

For specific fuels, the magnitude of the problem can be expressed in terms of the projected date when the total consumption and resource base will be equal using conservative demand assumptions. If this date is in the distant future (500 years, for example), energy demand is not a current problem. However, if the date is within 25 years, it is a significant problem and changeover to other fuels should be initiated as soon as possible. If the date is within 5 years, it is a critical problem and emergency steps must be taken to curtail the use of the "endangered" fuel through conservation or switching to other fuels. These dates are approximate, but are intended to give a relative indication of potential energy demand problems.

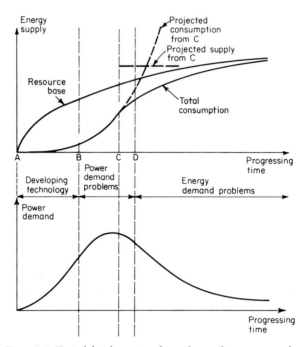

Figure 5.5   Typical development and curtailment of an energy supply.

A typical fuel cycle from development to curtailment is shown in Figure 5.5. The upper graph shows total consumption and known resource base as functions of time, and the lower graph shows the power demand over the same time interval. Point A marks the "practical" discovery of the new fuel when its potential for large-scale energy generation is first realized. Since the fuel was not previously in demand, little exploratory development had been done and the known resource base was essentially zero. However, the realization that large-scale use is feasible encourages resource exploration, and the known resource base grows rapidly. After a short period, all of the obvious resources have been discovered and resource discoveries taper off to moderate growth. At the same time, technology for using the new resource is developed and large-scale processing facilities are constructed. During this period of developing technology, the business community views the exploitation of this energy resource as a "golden opportunity" and invests heavily in capital equipment. As a result, demand follows supply and expands as more uses for the energy are developed.

At point B the demand outstrips the supply, and power demand problems arise as the energy industry is unable to supply enough energy to meet mushrooming demand. During this entire process, projections of total supply and demand are continuously being made. When projections, such as the one shown at point C, indicate that total consumption will equal the resource base at some time in the near future, industry starts to get "cold feet." Why should industry continue to invest in capital equipment with a lifetime of say 50 years if the resource base is projected to be only 25 years? As a result, the construction of new processing facilities is stopped, causing further power demand problems.

Fortunately, due to secondary and tertiary recovery methods, the resource base continues to expand slowly, and at point D the supply and demand become influenced by the total resource base. This is the beginning of energy demand problems. As the available supply dwindles, the fuel becomes more precious and its price rises, reducing demand. This interaction of supply and demand continues to reduce the resource base ad infinitum.

These considerations apply directly to energy impact assessment. The supply/demand section should describe the resource base considerations important to the project and point out any energy demand problems. If energy demand problems are inevitable, the relative benefits of the short-term energy demand or supply associated with the project should be compared with the long-term effects on the resource base. For a consuming project, this analysis must show that the project does not contribute materially to any shortage which may develop. However, for producing or consuming projects, these energy demand problems are the key reasons for the project's desirability and should be emphasized. The potential solutions to energy demand problems offered by the project should be described in detail.

Economics are intrinsically related to the supply and demand for energy. The following four types of potential economic problems should be addressed:

1. Price changes associated with power demand
2. Price changes associated with energy demand
3. Energy industry solvency
4. Governmental intervention regulations

Price changes associated with power demand are those resulting from the rate of energy supply and reflect the traditional balance between supply and demand. Price changes associated with energy demand are due to the increased cost of secondary, tertiary, and the other recovery methods which increase the resource base. These price changes are intrinsically related to the financial solvency of businesses in the energy industry. For example, low prices tend to reduce profits which discourage further investment in new equipment to expand supply. Each of these factors is also affected by governmental intervention and regulation, and this can perpetuate further economic problems.

Supply/Demand Examples This section discusses three examples of energy supply/demand data organization, each using energy inventory data extracted from corresponding examples stated previously. These are:

1. Single-family detached housing development—county gasoline scenario
2. New oil-fired power plant—electric utility distribution area scenario
3. Water heater insulation standard—natural gas resource base: United States and world.

Housing Project. The supply/demand data for a consuming project should be organized to show that the additional energy consumption will not adversely impact any supply/demand scenarios. If adverse impacts cannot be avoided, mitigation measures for reducing the impact should be discussed. The single-family detached housing development discussed in the earlier section is a small project and, as a result, only local scenarios are expected to be impacted. The local supply and demand for each fuel consumed by the project should be examined. Supply and demand for gasoline are discussed in the following paragraphs.

Most states tax gasoline purchases at the point of final sale to the consumer. Thus, tax records provide an excellent source of gasoline consumption data.[23,26] At the time that the energy impact for this project was assessed (three years prior to the planned start of construction date), the gasoline consumption in the county was 7.159×10^6 gal/yr. The county planning department estimates that the county's population will grow at a rate of 7.0 to 9.5 percent for the next 15 years. Assuming that gasoline consumption per capita will be constant, the future county gasoline consumption can be calculated by multiplying the current rate by the growth rate as follows:

$$\text{Projected gasoline consumption} = \begin{pmatrix} \text{current} \\ \text{gasoline} \\ \text{consumption} \end{pmatrix} \left(1 + \frac{\text{growth rate, \%}}{100}\right)^N$$

The exponent N is the number of years in the future. For example, the county gasoline consumption for the fifth year after the start of construction (seven years in the future), assuming the 9.5 percent growth rate, is:

$$\text{Projected gasoline consumption, year 5} = (7.159 \times 10^6 \text{ gal}) \left(1 + \frac{9.5}{100}\right)^7$$

$$= 13.51 \times 10^6 \text{ gal}$$

Using this method, gasoline consumption has been projected for the first 10 years of the housing development based on the high and low growth rates. The results are shown in Table 5.21. The net consumption of gasoline as a result of the project is also shown. This

TABLE 5.21   Projected Gasoline Demand in County
($\times 10^6$ gal/yr)

Energy flow	Year										Total
	1	2	3	4	5	6	7	8	9	10	
Projected demand:											
High growth rate—9.5%	9.40	10.29	11.27	12.34	13.51	14.80	16.20	17.74	19.43	21.27	146.25
Low growth rate—7.0%	8.77	9.39	10.04	10.75	11.50	12.30	13.16	14.08	15.07	16.13	121.19
Project consumption:											
Net consumption	0.110	0.187	0.277	0.327	0.327	0.327	0.327	0.327	0.327	0.327	2.863
Percent of demand:											
High growth rate	1.17	1.81	2.46	2.65	2.42	2.21	2.01	1.84	1.68	1.54	1.96
Low growth rate	1.25	1.99	2.76	3.04	2.84	2.66	2.48	2.32	2.17	2.03	2.36

was obtained from the data in Table 5.10 by converting the net gasoline energy consumption in Btu's to gallons. The net consumption of gasoline is also listed in Table 5.21 as a percentage of the high and low projected gasoline demand. Figure 5.6 plots these percentages as functions of time.

These results demonstrate that the housing project will have a negligible impact on gasoline supply and demand in the county. The project's gasoline consumption will not require more than 3 percent of the projected county demand, and the uncertainty in projected demand is several times greater then the entire project's gasoline consumption.

Since the project has a negligible impact at the county level, larger scenarios need not be examined.

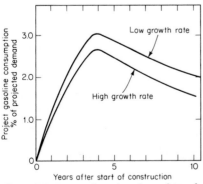

Figure 5.6 Projected gasoline demand in relation to county demand.

Power Plant Project. Supply and demand data for a producing project must be organized to demonstrate that the energy made available by the project is necessary and will be utilized effectively, and that any energy surplus will be acceptable. It is of utmost importance to demonstrate these points since they are the basic reasons for the project's existence.

An electrical power generation and distribution system must be capable of generating enough power to meet the maximum demands of its customers at any point in time. It must also be capable of meeting the average annual power demands without unduly taxing any other interconnected power generating systems.

To meet peak demand, the installed generating capacity must equal the peak anticipated demand plus a factor of safety to account for potential failures. Peak demand usually occurs in hot humid weather when air conditioners contribute heavily to power demand. Since such weather conditions often cover wide areas, adjacent interconnected power generating systems cannot rely upon mutual support in times of peak demand. Each must have sufficient capacity to meet its own peak power demand.

The average power demand is usually substantially less than peak demand. Common practice is to design a power generating system with large high-efficiency power plants to meet the base load requirements and smaller low-cost peaking units to meet peak demand. This allows the base load plants to operate efficiently at design point for a majority of the time.

In reference to the oil-fired power plant project discussed in an earlier section, the supply and demand data must be organized to show that:

- Without the project, the electrical power available will be insufficient to meet the projected demand.
- With the project, capacity will be increased enough to meet projected demand.
- The additional capacity made available by the project will not be excessive.

Although this project concerns the construction of a single oil-fired power plant, the operation of all power plants serving the same utility distribution area must be considered. Table 5.22 lists the power plants relevant to this project. The first plant is the natural

TABLE 5.22   Power Plants Serving Utility Distribution Area

Number	Name	Capacity, MW	Capacity factor, %	Service life*	Effects of project
1	Natural gas	300	75	−10 to +20	Derate 60 MW in year 9
2	Hydroelectric/peaking	100	26	−10 to +40	None
3	Nuclear plant	500	65	+20 to +50	None
4	Oil fired	225	75	+6 to +40	Construction

*Years after start of construction of oil-fired plant.

gas–fired base load power plant serving the area. The project will result in altering this plant to meet local air quality regulations and conserve natural gas. The second plant is a hydroelectric peaking unit. It has a capacity of 100 MW, but based on average rainfall data, there will be only enough water to operate at an annual average of 26 percent of full load. This plant will not be affected by the project. The third plant is a nuclear unit designed to replace the natural gas plant and increase capacity to meet previous demand projections. However because of greater than anticipated urban growth, the oil-fired power plant (Number 4) will be required as the following discussion demonstrates.

Table 5.23 shows the annual electrical power supply and demand projections for the utility service area without the project. Internal consumption is the projected power

TABLE 5.23   Supply and Demand for Electricity without Power Plant Project
($\times 10^9$ kWh/yr)

| Energy flow | Years after start of construction | | | | | | | | | | | | | | | | |
	−10	−5	0	5	6	7	8	10	15	20	21	22	25	30	35	39	40
Demand:																	
Internal consumption	1.60	1.78	1.95	2.25	2.30	2.36	2.40	2.54	2.90	3.30	3.38	3.44	3.75	4.20	4.70	5.00	5.10
Exports	0.60	0.42	0.25	0	0	0	0	0	0	0	0	0	0	0	0	0	0
Total demand	2.20	2.20	2.20	2.25	2.30	2.36	2.40	2.54	2.90	3.30	3.38	3.44	3.75	4.20	4.70	5.00	5.10
Supply:																	
Internal production	2.20	2.20	2.20	2.20	2.20	2.20	2.20	2.20	2.20	2.20	2.50	3.05	3.05	3.05	3.05	3.05	3.05
Imports	0	0	0	0.05	0.10	0.16	0.20	0.34	0.70	1.10	0.88	0.39	0.70	1.15	1.65	1.95	2.05
Total supply	2.20	2.20	2.20	2.25	2.30	2.36	2.40	2.54	2.90	3.30	3.38	3.44	3.75	4.20	4.70	5.00	5.10

demand within the service area based on the most recent projections. Internal production is the annual power which can be generated by all power plants within the utility service area operating at their design points. These values were calculated from the data in Table 5.21, summing the contribution from each plant as follows:

$$\begin{matrix}\text{Internal}\\\text{production,}\\\times 10^9 \text{ kWh/yr}\end{matrix} = \begin{pmatrix}\text{Plant}\\\text{capacity,}\\\text{MW}\end{pmatrix}\left(\frac{1{,}000 \text{ kW}}{\text{MW}}\right)\begin{pmatrix}\text{capacity}\\\text{factor, }\%\\\overline{100}\end{pmatrix}\left(\frac{8{,}760 \text{ h}}{\text{yr}}\right) \times (10^{-9})$$

For example, in year 10:

$$\begin{matrix}\text{Natural gas}\\\text{power plant}\\\text{production}\end{matrix} = 300 \times 1{,}000 \times \frac{75}{100} \times 8{,}760 \times 10^{-9}$$
$$= 1.97 \times 10^9 \text{ kWh/yr}$$

$$\begin{matrix}\text{Peaking}\\\text{power plant}\\\text{production}\end{matrix} = 100 \times 1{,}000 \times \frac{26}{100} \times 8{,}760 \times 10^{-9}$$
$$= 0.23 \times 10^9 \text{ kWh/yr}$$

$$\begin{matrix}\text{Total}\\\text{internal}\\\text{production}\end{matrix} = 1.97 \times 10^9 + 0.23 \times 10^9$$
$$= 2.20 \times 10^9 \text{ kWh/yr}$$

In years when internal consumption is less than internal supply, the difference must be made up by imports from other power plants if they have reserve capacity. Figure 5.7 shows these data as functions of time. The utility service area changes from an exporter of energy to an importer at year 5, and the situation worsens with time. Although the nuclear plant reduces the projected energy imports, it alone is insufficient to meet projected demand.

Table 5.24 is similar to Table 5.23 but includes the changes in supply resulting from this project. These data are plotted in Figure 5.8. The increased generating capacity eliminates the energy imports required up to year 33 except for small imports in years 5 and 20.

This discussion has addressed only the average annual electrical power supply and demand. However, it is also important that the system be capable of meeting peak demand. Based upon this utility's historical electrical demand data, the peak demand has been 40 percent greater than the annual average demand. Table 5.25 lists the projected

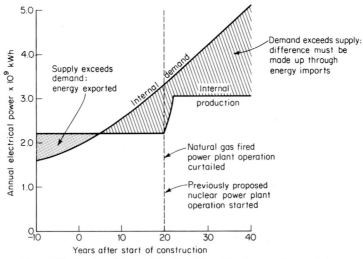

Figure 5.7 Supply and demand for electricity without power plant project.

TABLE 5.24 Supply and Demand for Electricity with Power Plant

Energy flow	Year																
	−10	−5	0	5	6	7	8	10	15	20	21	22	25	30	35	39	40
Demand:																	
Internal consumption	1.60	1.78	1.95	2.25	2.30	2.36	2.40	2.54	2.90	3.30	3.38	3.44	3.75	4.20	4.70	5.00	5.10
Exports	0.60	0.42	0.25	0	0.38	0.80	0.72	0.58	0.22	0	0.56	1.05	0.74	0.29	0	0	0
Total demand	2.20	2.20	2.20	2.25	2.68	3.16	3.12	3.12	3.12	3.30	3.94	4.49	4.49	4.49	4.70	5.00	5.10
Supply:																	
Internal production	2.20	2.20	2.20	2.20	2.68	3.16	3.12	3.12	3.12	3.12	3.94	4.49	4.49	4.49	4.49	4.49	4.49
Imports	0	0	0	0.05	0.0	0	0	0	0	0.18	0	0	0	0	0.21	0.51	0.61
Total supply	2.20	2.20	2.20	2.25	2.68	3.16	3.12	3.12	3.12	3.30	3.94	4.49	4.49	4.49	4.70	5.00	5.10

TABLE 5.25 Peak Capacity and Demand

Energy flow	Years after start of construction																
	−10	−5	0	5	6	7	8	10	15	20	21	22	25	30	35	39	40
Without oil-fired plant:																	
Peak generating capacity, MW	400	400	400	400	400	400	400	400	400	400	600	600	600	600	600	600	600
Peak power demand, MW	256	284	312	360	368	377	384	406	463	528	540	550	599	671	752	799	815
Reserve, %	+56.3	+40.8	+28.2	+11.1	+8.7	+6.1	+4.2	−1.5	−13.6	−24.2	+11.1	+9.1	+0.2	−10.6	−20.2	−24.9	−26.4
With oil-fired plant:																	
Peak generating capacity, MW	400	400	400	400	625	625	625	625	625	625	825	825	825	825	825	825	825
Peak power demand, MW	256	284	312	360	368	377	384	406	463	528	540	550	599	671	752	799	815
Reserve, %	+56.3	+40.8	+28.2	+11.1	+69.8	+65.8	+62.8	+53.9	+35.0	+18.4	+52.8	+50.0	+37.7	+23.0	+9.7	+3.3	+1.2

peak supply and demand with and without the project. The peak demand was calculated from the annual power demand as follows:

$$\text{Peak demand, MW} = \frac{\left(\begin{array}{l}\text{average annual}\\ \text{power demand,}\\ \text{kWh/yr}\end{array}\right) \times 1.4}{8{,}760 \text{ h/yr} \times 1{,}000 \text{ kW/MW}}$$

For example, in year 10:

$$\text{Peak demand} = \frac{2.54 \times 10^9 \times 1.4}{8{,}760 \times 1{,}000}$$

$$= 406 \text{ MW}$$

Peak power generating capacity is the total installed power generating capacity and represents the maximum power which can be generated with all plants operating at full

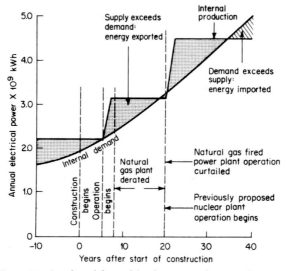

Figure 5.8 Supply and demand for electricity with power plant project.

capacity. For example, in year 10 with the project, all plants except the nuclear plant can be operated and the peak generating capacity is from Table 5.22 as follows:

$$\text{Peak generating capacity} = 300 + 100 + 225$$
$$= 625 \text{ MW}$$

Peak generating capacity should be greater than peak demand by a small margin to account for unanticipated occurrences such as equipment failures, scheduled maintenance coinciding with peak demand, and record hot weather (causing correspondingly high power demands). The percentage reserves in Table 5.25 were calculated as follows:

$$\text{Reserve} = \frac{(\text{peak capacity, MW}) - (\text{peak demand, MW})}{(\text{peak demand, MW})} \times 100$$

For example, in year 10 with the project:

$$\text{Reserve} = \frac{625 - 406}{406} \times 100$$
$$= +53.9\%$$

Figures 5.9 and 5.10 show these peak supply and demand data as functions of time. Without the project, the reserve drops to zero at year 10, and power demand problems would be inevitable. However, with the project there is always a positive reserve.

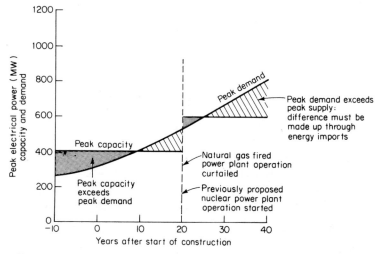

Figure 5.9 Peak capacity and demand for electricity without power plant project.

Insulation Standard. Supply and demand data for a conserving project should be organized to show (1) the need for conserving the energy sources impacted by the project and (2) the significance of the project's energy conservation for each energy source in relation to the applicable supply/demand scenarios. If no need for energy conservation can be demonstrated or the project's energy conservation is insignificant, the project is pointless and should not be pursued.

The water heater insulation standard being considered here will conserve several fuels, and the supply/demand scenarios for each should be examined. However, this section will address natural gas only.

The primary reason for conserving natural gas is the limited resource base—an energy demand problem. In relation to Figure 5.5, the typical development and curtailment of an energy supply, natural gas is approximately at point C. In recent years, the discovery of new gas fields has lagged behind a rapid increase in natural gas demand. As a result, projections based upon known resources and projected consumption rates show that all

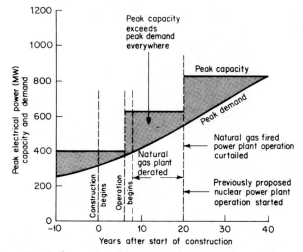

Figure 5.10 Peak capacity and demand for electricity with power plant project.

reserves will soon be exhausted. Table 5.26 shows some projections based upon 1970 data.[78] Assuming that consumption would remain at the 1970 level and that no additional resources would be discovered, the number of years until exhaustion can readily be calculated:

$$\text{Time to resource exhaustion, yr} = \frac{\text{known reserves, Btu}}{\text{consumption rate, Btu/yr}}$$

For the United States, the years to exhaustion can be calculated as follows:

$$\text{Time to exhaustion} = \frac{279 \times 10^{15}}{24{,}104 \times 10^{12}}$$
$$= 11.57 \text{ years}$$

These data clearly demonstrate the limited natural gas resource base and the need to conserve. However, they should not be interpreted as an indication that we will run out of

TABLE 5.26   Project Natural Gas Conservation in Relation to Resource Base

Natural gas	United States	World
Resource base:		
1970 known reserves ($\times 10^{15}$ Btu)	279	1,487
1970 consumption ($\times 10^{12}$ Btu)	24,104	49,014
Years remaining at 1970 consumption rate	11.57	30.34
Project conservation:		
Conserved in 1990 ($\times 10^{12}$ Btu)	381	381
Percent 1970 known reserves	0.14	0.03
Percent 1970 consumption	1.58	0.78
Conserved 1981–1990 ($\times 10^{12}$ Btu)	2,094	2,094
Percent 1970 known reserves	0.75	0.14
Percent 1970 consumption	8.69	4.27

natural gas in a few years. Additional resources will undoubtedly be discovered and consumption will decrease through fuel switching and conservation. Thus, the resource base will never be totally exhausted.

Now that the need for natural gas conservation has been demonstrated, the impact of the project's energy conservation relative to these scenarios must be addressed. Table 5.26 compares the natural gas consumed by the project in 1990 and 1981–1990 with the known resources and 1970 consumption rates. When all "old" water heaters have been replaced with new ones meeting this standard (1990), the energy conserved each year will be equivalent to 1.58 percent of the 1970 United States consumption rate. The total energy consumption over the first 10 years of the project (1981–1990) will be 8.69 percent of the 1970 consumption rate. These are significant results, particularly for such a simple energy conservation measure.

The project energy conservation is also significant in relation to the total known resource base. The first 10 years of the project will conserve 0.75 percent of the known U.S. resources.

Table 5.26 also compares the project energy conservation with world supply and demand data. The results show correspondingly smaller but still significant impact as expected.

Energy Conservation

The purposes of the energy conservation section are to describe the project's energy-conserving features, to explain why they are to be implemented, to tabulate the energy savings due to these features, and to compare the project to applicable energy standards and regulations. It should be written to reflect favorably on the project describing how and why certain aspects of the project will save energy and why this is important. This is in contrast to the alternatives section which discusses possible energy-conserving features and other alternatives *not* employed in the project.

The level of detail and the format to use in writing this section are primarily dependent upon the relative importance of energy conservation in the overall energy balance. For energy-conserving projects whose sole justification for existence is energy conservation, this section should be written as the main frame of the energy element and with considerable detail. For other projects where energy conservation is less important, detail should be suitably reduced. However, all impact reports should contain an energy conservation section regardless of project size or whether the project management has made any special attempts to conserve energy. It is hard to imagine a project which has no energy-conserving features or, stated differently, which could not decrease its energy consumption by some means. Some energy-conserving features can always be found.

Throughout this discussion the term "energy conservation feature" will be used as opposed to the more commonly used (and misused) term "energy conservation measure." An energy conservation measure is a special project whose sole justification is energy conservation. However, the energy conservation section should discuss all aspects of the project which tend to reduce energy consumption even though they may be employed for substantially different reasons. Thus, energy conservation features include energy conservation measures and all other project aspects which reduce energy consumption or increase production or conservation. For example, if a new industrial plant employing 5,000 workers were to be located near a residential area because of groundwater availability and sewage processing facilities in the immediate area, the site selection could hardly be called an energy conservation measure. However, the amount of energy indirectly expended due to the 5,000 workers commuting only short distances is probably much smaller than for other sites, and hence, the site selection conserves energy and is an energy conserving feature.

The energy conservation section should contain a discussion of the following points:

- Description of energy-conserving features
- Separation of internal and external effects
- Inventory of energy conservation
- Comparison of project with energy regulations and standards

The description of energy-conserving features should provide the following information:

- Functional description
- Why it will conserve energy
- Why it is employed

The discussion of these points is straightforward, except for why it is employed. This point must be addressed because it helps provide justification for not adopting those energy conserving features described in the alternatives section. Some suitable reasons for employing specific energy-conserving features include:

- Cost-effectiveness
- Substantial energy savings
- Mandatory nature due to regulations
- Indirect benefits

Each energy-conserving feature has either internal, external, or combined effects. Internal energy conservation features are those which reduce the energy consumption (or increase production) of the project itself independent of the project's environment. Specific examples of internal conservation measures include thermal insulation, solar energy systems, fluorescent light fixtures, low motor vehicle speed limits within the project boundaries, and energy-efficient processes within an industrial plant. In contrast, external energy conservation features are those which reduce the energy consumption (or increase production) of the project's environment rather than the project itself. All projects defined as "energy conserving" fit this classification since their purpose is to reduce energy flows in certain segments of society. Specific examples include: (1) project relocations which reduce existing transportation energy requirements and (2) reduced speed limits on existing highways required by the project.

The energy conservation produced as a result of internal conservation measures should be inventoried in the energy conservation section. Since internal conservation features do not affect the overall energy balance, they are not inventoried in the supply/demand section and the only numerical accounting of their effects will appear here. The most accurate way to inventory internal energy conservation is to first inventory the energy consumption, production, and conservation of a fictitious alternative project that could be constructed without internal energy-conserving features and then compare that inventory with the real inventory as determined in the energy impact section. Although this provides a complete assessment of internal energy conservation effects, it is usually too complex and laborious to justify here.

A simplified approach is to determine the conservation from each of the internal conservation features on an individual basis and then add the results. This sum should express the total amount of energy conservation due to implementing all conservation measures as compared to implementing none. Complications arise when two or more conservation features affect the same energy flow. Table 5.27 shows the heating energy

TABLE 5.27   Energy Conservation Summation Problems

Heating system design	Heating energy per year	
	Consumption	Conservation
Base case:	1000	0
Electric resistance heat		
No insulation		
Conservation feature 1:	500	500
Heat pump heating		
No insulation		
Conservation feature 2:	500	500
Electric heating		
Thermal insulation		
Combined conservation 1 & 2:	250	750
Heat pump		
Thermal insulation		

conservation associated with two such energy-conserving features: (1) heat pump heating as opposed to electric resistance heating, and (2) thermal insulation. In this particular project, each feature conserves 50 percent of the energy consumption without the feature. Consequently, if the total heating energy required for the base case is 1,000 units per year, each energy-conserving feature examined independently is responsible for saving 500 units. If these energy conservation numbers are summed, the result is an erroneous total conservation of 1,000 units per year where the actual total should be 750 units per year.

The correct procedure for reporting this type of energy-conserving feature is to establish an order of adding features to the base case. For example, if the first feature is the replacement of electrical resistance heat with a heat pump, the energy conservation is 500 units per year. When the thermal insulation is then added, the incremental conservation is 250 units per year for a total conservation of 750 units.

As an example of an energy conservation inventory, consider the housing development discussed in the section on Energy Inventory Examples. Two energy conservation features have already been discussed: the highway speed limit reduction and the installation of solar heating equipment. The highway speed limit reduction was an external conservation feature and was directly itemized in the energy inventory. The installation of solar heating equipment was considered an internal energy conservation feature, and although its effects on the proposed energy flows were evaluated, they did not appear directly in the inventory. This energy-conserving feature and all others which help to minimize the energy consumption of the project should be listed in the energy conservation inventory. Table 5.28 is an energy conservation inventory for the project and includes the estimated amount of energy conserved by each feature. Most of the features listed obviously conserve energy. However, some conservation features are not so obvious. Fluorescent lights (Number 2) provide almost four times as much light per unit power input as incandescent lights. Extra wide awnings and deciduous trees (Numbers 7 and 8) conserve heating and cooling energy. In the summer the wide eaves and foliage shade the walls from direct solar radiation, reducing air conditioning cooling loads. In winter, when the sun is lower on the horizon, the eaves and bare trees do not interfere with solar heat gain, thus reducing heating loads.

The location of a bus route adjacent to the site (Number 14) conserves gasoline by encouraging more efficient bus transportation. Of course, this energy conservation is partially offset by increased direct fuel consumption of buses. The energy conservation listed is net and is compensated for this effect.

High-pressure sodium street lights (Number 15) are about twice as efficient as the mercury lights more commonly installed. Note that the energy inventory in the earlier

TABLE 5.28   Energy Conservation Inventory for Housing Project

Number	Energy-conserving feature	Type of energy conserved	10-year conservation, $\times 10^9$ Btu
1	Conversion to solar space and water heating—30% of units	Natural gas	20
2	Fluorescent kitchen and bathroom lights	Electricity	10
3	Heavy insulation in walls and ceilings	Natural gas Electricity	20 10
4	Double-glazed windows	Natural gas Electricity	10 5
5	Weatherstripping around all doors and windows	Natural gas Electricity	10 5
6	Attic ventilating fans—standard	Electricity	5
7	Extra wide eaves on south and west	Natural gas Electricity	6 3
8	Landscaped with deciduous trees	Natural gas Electricity	6 3
9	High-efficiency water heaters (thick insulation)	Natural gas	5
10	High-efficiency furnaces	Natural gas	10
11	High-efficiency air conditioners (optional)	Electricity	10
12	Heat recovery fireplaces	Natural gas	4
13	Night setback of thermostats	Natural gas	20
14	Bus route adjacent to site	Net	15
15	High-pressure sodium street lights	Electricity	5
16	Natural gas space and water heating and cooking appliances	Net	50
	Total internal energy conservation		232

section on Energy Inventory Examples neglected street lighting energy because it was insignificant in comparison with total energy consumption.

Natural gas space and water heating and cooking appliances (Number 16) is listed as a conservation feature because gas-fired units of this type usually consume less energy than comparable electric units when the efficiency of power generation is included. However, natural gas is a scarce fuel, and the use of electrical appliances could be considered a natural gas conservation feature.

Table 5.28 also lists the total internal energy conservation. This is the additional amount of energy which would be consumed if none of the conservation features were implemented. The total internal conservation is nearly three times the external energy conservation (highway speed limit reduction) and is 25 percent of the project's net consumption of energy from nonrenewable sources (See Table 5.10). Additional suggestions for energy conservation are listed in the subsequent section Energy Conservation Data.

The preceding example considered an energy-consuming project. The energy conservation inventory for a producing project should be organized similarly. In the case of an energy-conserving project, it is recommended that the energy conservation associated with the project's main function be deleted from the table. The purpose of this energy conservation inventory is not to present the main energy aspects of the project but to evaluate the impact of incidental energy conservation features on the project energy balance.

The final part of the energy conservation section compares the project with all applicable energy regulations and standards. Compliance with these standards is an indication that the project's energy flows are acceptable from a governmental or regulatory point of view. The comparison should be presented point by point showing how each and every regulation or standard is met and how this maximizes energy conservation.

Energy Alternatives

The purpose of the energy alternatives section is to discuss the alternatives *not* to be employed in the project as designed and to justify the decision not to employ them from an energy point of view. If the alternatives are not to be employed, by definition, the project management has determined that the project as designed is preferable. Two types of alternatives must be discussed, namely:

1. General alternatives
2. Energy alternatives

General alternatives are possible project modifications which primarily affect energy flows. If the project is energy intensive or is either a producing or a consuming project, these modifications should have been discussed in the main alternatives section in the environmental impact statement. In this case, they should be treated similar to the general alternatives which follow. Except for these, energy alternatives will not have been discussed elsewhere in the EIS. Consequently, the discussion in the energy alternatives section should logically present all pertinent points just as if an EIS were being written for the energy alternative rather than for the project itself. The detail required in discussing energy alternatives depends upon how reasonable the alternative appears and its general desirability. The discussion should convince the reader that the project as designed is preferable either from an energy point of view or because of other overriding factors.

Two energy alternatives deserve special attention here, namely:

1. Alternative fuels
2. Energy conservation features

The shortage of certain fossil fuels has intensified interest in fuel switching. The energy alternatives section should contain a discussion justifying the project's fuel choices. The following fuels or energy supply systems should be considered:

- Petroleum and natural gas
- Coal
- Solar energy
- Electrical power—central power plant
- Total energy system

Not all of these are appropriate for all projects, and some projects may require considerations of some of the more exotic fuels and power-generating technologies discussed in the earlier sections of this chapter.

The justification for choosing among these alternatives should be based on analyses of the following factors:

- Fuel availability
- Nonrenewable resources
- Technology and equipment availability
- Certainty of supply
- Cost
- Other environmental factors (air pollution, etc.)

The features of the fuel supply systems chosen for the project should be compared to those not chosen so that reasons for the choices are obvious.

The energy conservation features employed in the project as designed are discussed in the energy conservation section. Those not employed should be discussed here. Since it would be impossible to address all conservation features not included, the discussion should center on the obvious conservation features, particularly those examined by the project management and rejected. By definition, energy conservation features tend to reduce energy consumption (or increase production or conservation) and produce a favorable energy balance. Since the measures are not to be employed, the discussion must show how these energy advantages are overridden by other factors. Suitable reasons for not employing energy conservation features include:

- High cost
- High cost/benefit ratio
- Technological uncertainties
- Insignificant energy savings
- Inconvenience
- Aesthetics
- Other environmental problems

Inconvenience and aesthetics are weak arguments for not employing energy-conserving features, but sometimes they are the overriding reasons and should be presented as such. For example, one of the primary limitations of public bus systems is their inconvenience relative to private automobiles. Also, it is well known that heating and air conditioning energy requirements increase with glass area, thus favoring small or zero window areas in all heated and cooled structures; however, aesthetics generally overrides this consideration.

The energy aspects of general alternatives to the project must also be addressed. The logical location for this discussion is the energy alternatives section. The change in energy flows associated with each alternative should be tabulated and compared with those in the project as proposed.

DATA AND DATA SOURCES

Energy Inventory Data

This section addresses the numerical aspects of calculating energy flows, given specific project functions. A complete comprehensive listing of energy usage factors is beyond the scope of this handbook. Since varying amounts of energy are used in essentially every type of building, industrial process, transportation system, and most appliances (and the amount and type used in each application varies), only average or typical values will be given here. These values may be used to estimate energy impacts unless refined and more specific data are available.

The following subsection discusses the energy content of various fuels and the incidental losses in production, refining, transportation, and conversion. Subsequent subsections address typical energy use factors for a variety of applications, organized as follows:

- Building heating and cooling
- Lighting, appliances, and industrial processes
- Transportation

Fuel Energy Data Table 5.29 lists the average energy content of common fuels and energy sources and may be used to convert fuel requirements (such as gallons of gasoline)

TABLE 5.29 Energy Equivalents of Various Fuels

Fuel type	Fuel quantity	Energy Btu	Common units	Reference
Petroleum fuels:				
Crude oil	1 barrel	5.8×10^6	138,100 Btu/gal	Definition
Diesel fuel	1 barrel	5.8×10^6	138,200 Btu/gal	Average value
Jet fuel (JP 4)	1 lb	18,400	119,200 Btu/gal	15
Kerosene	1 lb	19,810	135,100 Btu/gal	15
Gasoline	1 lb	20,750	127,650 Btu/gal	15
Gaseous fuels:				
Natural gas	1 ft³	1,000		Average value
Butane gas	1 gal	102,000		Average value
Propane gas	1 gal	91,500		Average value

TABLE 5.29   Energy Equivalents of Various Fuels   (Continued)

Fuel type	Fuel quantity	Energy Btu	Common units	Reference
Coal fuels:				
Sublignite	1 lb	<6,300		7
Lignite	1 lb	6,300–8,300		7
Subbituminous	1 lb	8,300–11,500		7
Anthracite	1 lb	14,000		101
Bituminous	1 lb	10,500–14,000		7
Power gas (low Btu)	1 ft^3	100–200		101
Intermediate Btu	1 ft^3	300–650		101
Synthetic natural				
gas (high Btu)	1 ft^3	900–1,050		101
Liquified coal	1 lb	17,000		101
Oil shale:				
Raw oil shale	1 lb	2,780		15
Shale oil—synthetic				
crude	1 lb	18,400		15
Tar sand:				
Raw tar sand	1 lb	2,100		101
Synthetic crude oil	1 barrel	6.3 × 10^6		101
Hydrodynamic				
energy:				
Water falling 1.0 ft	1 lb	1.286 × 10^{-3}		Calculated
Tidal power 1.0 ft				
water level change	1 ton	2.572		Calculated
Solar energy:				
Incident solar				
energy (per hour)	1 ft^2	442		101
Average useful solar				
energy (per day)	1 ft^2	1,450		101
Wind energy (100-				
ft-diameter rotor in				
30 mi/h wind per				
hour)		3.679 × 10^6		Calculated
Ocean thermal				
gradients from 41°				
to 77°F	1 lb H$_2$O	36		Calculated
Other organic				
materials:				
Wood (typical,				
nonresinous, and				
seasoned)	1 lb	6,300		15
Wastewood	1 lb	4,600		15
Charcoal	1 lb	13,530		15
Methyl alcohol	1 lb	9,600	63,825 Btu/gal	15
Ethyl alcohol	1 lb	12,810	84,850 Btu/gal	15
Garbage				
Solid waste refuse	1 ton	10.5 × 10^6		101
Geothermal energy:				
At "geyser #5"				
steam at 355°F,				
100 psi cooled to				
70°F	1 lb	1,151		62

into energy requirements. The energy content for fuels is the total heat of combustion released under typical combustion conditions; however, this is not necessarily the useful heat produced. In all applications, some heat is lost in stack gases and other areas so that the amount of useful heat is never greater than about 90 percent of the total, and usually is substantially less.

The energy equivalents for solar and geothermal energy are also heat values and represent the total available rather than the useful energy produced. The heat produced by these energy sources is generally at significantly lower temperatures than that produced by fuel combustion.* Consequently, if solar and geothermal energy are used for producing work, the conversion efficiency will be substantially less.

Unlike the heat sources, hydrodynamic and wind energy are work sources. The energy equivalents listed are the total work energy available, and although they have the same units as heat, a unit of work is more valuable than a unit of heat because substantial losses are incurred in converting heat into work, while work can be converted into heat with 100 percent efficiency.

The energy equivalents listed in Table 5.29 are average values rather than specific. For example, specifying coal as "bituminous" only determines its energy equivalent within ±16 percent. While this is accurate enough to determine order-of-magnitude energy impact, precise inventory requires more specific data. The best source for this data is the fuel distributor.

As a demonstration of the use of Table 5.29 to convert fuel requirements into energy requirements and vice versa, consider the following example. A large oil-fired steam power plant has been given notice that its oil supply will be curtailed. The plant requires 500 gal of oil per min when operating under full load. A nearby coal mine has the capacity to supply 160 tons of high-grade bituminous coal per hour. Will this be sufficient to supply the power plant? The answer is determined as follows:

$$\text{Energy supplied as oil (diesel fuel)} = 500\,\frac{\text{gal}}{\text{min}} \times \frac{60\,\text{min}}{\text{h}} \times 138{,}200\,\frac{\text{Btu}}{\text{gal}} = 4.146 \times 10^9\,\frac{\text{Btu}}{\text{h}}$$

$$\text{Energy available as coal} = 160\,\frac{\text{ton}}{\text{h}} \times \frac{2{,}000\,\text{lb}}{\text{ton}} \times \frac{14{,}600\,\text{Btu}}{\text{lb}} = 4.672 \times 10^9\,\frac{\text{Btu}}{\text{h}}$$

Thus, the energy content of the coal available is greater than the energy requirement. Of course, the decision to switch fuels will be based on additional factors such as environmental and economic considerations.

Table 5.30 lists the fuel cycle efficiencies for common fuels and power generating technologies, and may be used to convert the energy consumed and produced in any phase of the fuel cycle to the equivalent amount in another phase. "Extraction" refers to mining or drilling to remove the raw fuel. The efficiency is less than 100 percent because not all of the resource is recovered and some energy is consumed in the process. Very low extraction efficiencies, such as for oil, sometimes justify secondary or tertiary recovery at a later time since a substantial portion of the resource remains after initial development. "Processing" refers to preparing the fuel for its ultimate use (usually combustion). For coal, it includes in and near mine transportation, crushing, and conveying; for oil, it includes refining; and for coal gasification, it includes the gasification process. Transport efficiency includes the losses in transporting the processed fuel to the market (assumed to be a power plant) with pipelines, barges, etc. Conversion to electricity is assumed to occur in a conventional steam power plant, and transmission is by high-voltage power lines. If the ultimate fuel use is not generation of electricity, the final two columns should not be utilized. The efficiencies listed are average values based on current practice for existing fuels (such as oil and coal) and on estimates for future fuels (such as gasified coal).

The following example demonstrates the use of Table 5.30 in environmental impact analysis. A proposed 750-MW power plant will have an efficiency of 38 percent and will be fired with either a liquified coal product or with refined petroleum. The estimated energy required to operate the fuel processing plants for each alternative is:

---

*Fuel combustion can produce temperatures in excess of 3000°F while, except for concentrating solar collectors and geothermal energy from very hot materials such as hot lava, the temperatures of most solar and geothermal sources are in the range of 150 to 500°F.

$$\text{Energy required in processed fuel at power plant site} = \frac{750 \text{ MW}}{0.38} = 1973.7 \text{ MW}$$

$$\text{Energy required in processed oil at processing plant} = \frac{1973.7 \text{ MW}}{0.98} = 2014.0 \text{ MW}$$

$$\text{Energy required in processed liquified coal at processing plant} = \frac{1973.7 \text{ MW}}{0.98} = 2014.0 \text{ MW}$$

$$\text{Energy required in raw oil at processing plant} = \frac{2014.0 \text{ MW}}{0.88} = 2288.6 \text{ MW}$$

$$\text{Energy required in raw coal at processing plant} = \frac{2014.0 \text{ MW}}{0.65} = 3098.5 \text{ MW}$$

$$\text{Energy required for processing oil} = 2288.6 \text{ MW} - 2014.0 \text{ MW} = 274.6 \text{ MW}$$

$$\text{Energy required for processing coal} = 3098.5 \text{ MW} - 2014.0 \text{ MW} = 1084.5 \text{ MW}$$

Thus the coal liquefaction plant requires almost four times as much energy to process its fuel.

TABLE 5.30   Fuel Cycle Efficiencies in Percent

Fuel	Extraction	Processing	Transport (processed fuel to power plant)	Conversion to electricity	Transmission	Reference
Deep-mined coal	56 (56)	92 (52)	98 (50)	38 (19)	91 (17)	30
Surface-mined coal	79 (79)	92 (73)	98 (71)	38 (27)	91 (25)	30
Onshore oil	30 (30)	88 (26)	98 (26)	38 (10)	91 (09)	30
Offshore oil	40 (40)	88 (35)	98 (34)	38 (13)	91 (12)	30
Natural gas	73 (73)	97 (71)	95 (67)	38 (26)	91 (23)	30
Nuclear (LWR)	95 (95)	57 (54)	100 (54)	31 (17)	91 (15)	30
Coal gasification (high Btu)	79* (79)	60 (47)	95* (45)	38* (17)	91* (16)	101
Coal gasification (low Btu)	79* (79)	73 (58)	95* (55)	38* (21)	91* (19)	101
Coal liquefaction	79* (79)	65 (51)	98* (50)	38* (19)	91* (17)	101
Tar sands to synthetic crude	80 (80)	70 (56)	98* (55)	38* (21)	91* (19)	101
Oil shale	63 (63)	59 (37)	98* (36)	38* (14)	91* (13)	101

*Inferred from Reference 30 data.
NOTES: 1. Values in parentheses are net efficiency including the process; other values are process efficiency.
2. "Processing" includes all refining operations and conversion to other fuel forms.
3. "Conversion" is conversion to electricity in a conventional steam power plant.
4. Efficiencies are average values.

Table 5.31 presents the energy consumed in transporting fuels based on average haul or trip length for a variety of transportation modes. The primary efficiency expresses the percentage of fuel lost during transport. Losses occur for a variety of reasons. Coal is lost in surface transportation due to spillage and wind-blown coal dust. When it is transported as a coal and water slurry, although no fuel material is lost, the energy released in combustion

TABLE 5.31 Illustrative Losses In Transportation Of Energy*

Fuel	Transportation mode	Average haul, mi	Primary efficiency, percent	Ancillary energy required, 10^8 Btu per 10^{12} Btu transported	Ancillary energy required per mile, 10^8 Btu per 10^{12} Btu transported
				Losses per haul	
Coal	Underground conveyer	5.2		2.44	0.47
	Surface truck near mine	4.1		4.43	1.08
	Unit train	251	99	13.2	0.053
	Mixed or conventional train	259	98	10.9	0.042
	River barge	467	98	12.6	0.027
	Surface truck	10	99	1.06	0.106
	Surface conveyer	5	99	0.39	0.078
	Slurry pipeline	273	98	7.09	0.026
Oil shale	Underground conveyer	1	100	3.0	3.0
	Surface truck near mine	1	100	26.0	26.0
	Synthetic crude oil pipeline	300	100	34.0	0.113
Crude oil and petroleum products	Pipeline	300	100	36.9	0.123
	Tankers and supertankers	10,000	99.9	407	0.041
	Barge	1,500	100	257	0.171
	Tank truck	500		141	0.282
	Tank car (railroad)	500		146	0.292
Natural gas	Pipeline (Trans-Canadian) Capacity: 11.8×10^9 ft³/day	1,007	100	92.0	0.091
	LPG truck		100	51.1	
Geothermal steam	Pipeline	0.25	99		
Electricity	345 kV @ 500 MW	100	100	240	2.4
	500 kV @ 1,500 MW	100	100	147	1.47
	765 kV @ 2,500 MW	100	100	104	1.04

*All data from References 58 and 101.

is reduced because of incomplete drying after dewatering the slurry. Geothermal energy is lost due to heat transfer when hot steam is conveyed through pipes.

In addition to these direct fuel (or energy) losses, some energy is consumed in moving the fuel. Sometimes the energy required is withdrawn from the fuel being transported. For example, the pumping stations on long-distance natural gas pipelines are often driven by natural gas–powered turbines coupled to pumps or compressors. In transporting electricity through transmission lines, energy is consumed directly in the transmission process and appears as reduced current and voltage at the terminal end. In other cases, including most forms of surface transportation (conveyors, trucks, trains, etc.), the energy required comes from a different fuel because of its convenience.

Table 5.31 also presents the ancillary energy requirements in terms of 10^8 Btu per 10^{12} Btu of fuel transported. This is equivalent to 0.01 percent. The energy requirements are presented both in terms of energy per trip or haul based on average haul distances and in terms of energy consumed per mile. Comparison of these figures reveals several things. In all cases, the energy consumed is a small fraction of the total energy transported. The largest amount of energy consumed per trip is for a tanker and amounts to 4.07 percent of the total energy. The energy consumed per mile is a function of the size of the unit, the transportation mode, and the haul distance. Large units such as tankers, barges, pipelines, and trains are the most energy efficient. The highest per mile energy requirement is for oil shale moved short distances by truck. The energy consumed by this transportation mode is about 100 times the amount per mile consumed by a river barge transporting coal, the best alternative. Energy consumed by the oil shale trucks is high because of the short haul distance (1 mi), the low energy content of the fuel (2,780 Btu/lb), and the steep roads commonly traveled near oil shale mines.

All transportation modes listed, except high-voltage transmission lines, move heat energy or fuel to be combusted to produce heat energy. Transmission lines, however, transport electricity which is work rather than heat. Consequently, the energy consumed in transport must be viewed differently. Assuming a central power plant efficiency of 38 percent, the amount of fuel transported to the power plant (expressed in energy units) is $100/38 = 2.63$ times as great as the amount of electricity transmitted from the power plant (expressed in heat energy units). Consequently, in comparing the transportation energy consumption for fuel and electricity, as in power plant siting analyses, the energy consumed by transmission lines should be divided by a factor of 2.63 and then compared with the fuel transportation. On this basis, the energy consumed per mile by high-voltage transmission lines is in the same range as railroad tank cars and tank trucks moving petroleum products.

As an example of the use of Table 5.31 in an energy impact analysis, consider three possible sites for a coal-fired power plant producing 1000 MW for use in a distant city.

Power plant site	Coal transport	Electricity transport
A	200 mi by barge	75 mi using 345-kV lines
B	100 mi by unit train	150 mi using 500-kV lines
C	10 mi by truck	200 mi using 765-kV lines

The energy losses incurred in transporting the coal and electricity and the amount of coal required at the mine will now be calculated based on a 38 percent overall efficiency power plant.

$$\text{Electrical energy} = 1000 \text{ MW} \times 3.415 \times 10^6 \frac{\text{Btu/h}}{\text{MW}} = 3.415 \times 10^9 \frac{\text{Btu}}{\text{h}}$$

$$\text{Coal energy} = 1000 \text{ MW} \left(\frac{1}{0.38}\right) \times 3.415 \times 10^6 \frac{\text{Btu/h}}{\text{MW}} = 8.987 \times 10^9 \frac{\text{Btu}}{\text{h}}$$

For Site A we have:

$$\begin{array}{l}\text{Ancilliary} \\ \text{energy in} \\ \text{electricity}\end{array} = 3.415 \times 10^9 \text{ Btu/h} \times 2.4 \times 10^{-4} \frac{1.0}{\text{mi}} \times 75 \text{ mi} = 6.147 \times 10^7 \frac{\text{Btu}}{\text{h}}$$

$$\text{Ancilliary energy in coal} = 8.987 \times 10^9 \text{ Btu/h} \times 0.027 \times 10^{-4} \frac{1.0}{\text{mi}} \times 200 \text{ mi} = \underline{0.485 \times 10^7 \text{ Btu/h}}$$

$$\text{Total ancilliary energy} = 6.632 \times 10^7 \text{ Btu/h}$$

$$\text{Coal required at mine} = \frac{8.987 \times 10^9}{0.98} \text{ Btu/h} = 9.170 \times 10^9 \text{ Btu/h}$$

Similar calculations can be conducted for the other sites. The following table summarizes the results:

Power plant site	Total ancilliary energy required, Btu/h	+	Coal required at mine, Btu/h	=	Total energy required, Btu/h
A	6.632×10^7		9.170×10^9		9.236×10^9
B	8.006×10^7		9.078×10^9		9.158×10^9
C	7.198×10^7		9.078×10^9		9.150×10^9

Thus, alternative C requires the minimum total energy.

Transportation of energy materials is discussed in depth in the Reference 37. This two volume study addresses the costs and availability of alternative energy transportation modes for present and future scenarios.

The data presented here are necessarily general. A complete compilation of specific energy use factors is beyond the scope of this handbook. More detailed information can be found in the references or by contacting the distributors and manufacturers of specific fuels.

Building Heating and Cooling Data   Energy usage for building heating, ventilating, and air conditioning (HVAC) varies with building and HVAC system design, building operation, and weather conditions; however, there are three methods available to estimate building energy usage:

1. Computer programs
2. Manual methods
3. Gross energy use factors

The accurate estimation of energy usage in a building involves much more computation than the calculation required to size the HVAC system. In the latter case, the HVAC engineer examines the building construction and calculates the heat flow to and from the building for the worst case conditions (hottest and coldest days, peak occupancy, etc.). Then, an HVAC system of sufficient capacity is chosen to match these conditions. The calculation of building energy usage requires that actual occupancy and weather conditions be forecast and that heat flows be calculated at many times over a yearly cycle; then, the energy used in each increment is totaled. The accuracy of the calculation is improved by detailing the building construction and shortening the time increments. Unfortunately, this also increases the number of repetitive calculations required.

Computers are adept at repetitive calculations, and in recent years several computer programs have been written to simplify building energy usage calculations. Table 5.32 lists several available programs, their owners/distributors, and references for more specific information. Most programs are proprietary and may be used on a fee basis only. The owners/distributors supply instruction manuals and the necessary support to input data and interpret the results.

The American Society of Heating, Refrigerating, and Air Conditioning Engineers (ASHRAE) has been quite active in this field and is responsible for many of the calculating procedures and data used in current programs. A Task Group on Energy Requirements (TGER) was founded in 1967 to develop the necessary data and methodologies for computer techniques and has published several reports documenting their work.[73,94] ASHRAE and TGER have sponsored several symposiums as well and have conducted field validations of computer-calculated energy consumption.[9,95] ASHRAE also publishes the periodicals *ASHRAE Journal* and *ASHRAE Transactions*, which include reports of current work in this field.

Table 5.33 lists the typical input data required for these computer programs. In general,

TABLE 5.32 Computer Programs to Calculate Building Energy Usage

Program name				
Abbreviation	Full name	Owner/distributor	Computer system	Reference number
AXCESS	Alternative choice comparison for energy system selection	Electric Energy Association	National CSS Inc.	14, 41
CREAP	Correlated residential energy analysis program	Hittman Associates		67
ECUBE	Energy conservation using better engineering	Control Data Corp. and American Gas Association	Control Data Corp. Cybernet Network	5, 6, 14
MACE	McDonnell annual consumption of energy	McDonnell Douglas Automation Co.	McDonnell Douglas Automation-McAuto Systems	14
MEDSI	Mechanical Engineering Data Services, Inc.	J.R. McClure and Associates	United Computing Systems	14
REAP	Residential energy analysis program	U.S. Postal Service	Listing given in Reference	67
RMA	Energy system analysis series	Ross F. Meriwether and Associates	University Computing Corp.	14, 84
TRACE	Trane air conditioning economics	McDonnell Douglas Automation and Trane Co.		14, 96, 97, 98

the user must supply enough data to completely specify the heat flows to and from the building. This includes building structural details, building operation, HVAC system design, and the estimated weather environment for a year. The programs differ in the details of supplying and entering these data. For example, some programs provide built-in historical weather data for selected cities, while for others the user must supply his own

---

TABLE 5.33   Typical Data Inputs Required for Computerized Building Energy Programs

- Building design data:
  - Building size, shape, number of floors, etc.
  - Construction detail—type of walls, roof, etc.
  - Orientation
  - Glazing area, type, and shading
  - Insulation type and thickness
  - Doorway sizes and locations
  - Outside radiative heat transfer properties
  - Conditioned/unconditioned areas
- Building operation data:
  - Desired internal temperature and humidity
  - Internal heat generation:

People	Machinery
Lights	Hot water
Computers	Cooking

  - Open windows: number, location, and time
  - Doorway openings: number, location, and time
  - Elevator requirements
- HVAC system design:
  - Type of system
  - Full-load performance characteristics
  - Part-load performance characteristics
  - Number and location of zones
  - System operating schedule
  - System economics:
    - Capital cost and amortization
    - Interest rate
    - Operating cost
- Weather Data:
  - (For specified time period—usually annual)
  - Temperature
  - Dew point or humidity
  - Cloud cover
  - Wind speed
  - Wind direction

---

weather data. Some programs require the design and performance of one or more alternative HVAC systems as inputs to the program. Others select suitable HVAC system components based upon building requirements and provide an optimum HVAC system design as a final output. Since these computer programs require specific input data regarding building design and operation, they are suitable for estimating the building energy usage for well described buildings. They are not suitable for estimating the building energy usage for a city or other project involving many types of poorly defined structures.

The details of the computer calculations also differ. However, the main process consists of selecting time periods and calculating the heat flows to and from the building based on solar intensity, cloud cover, inside and outside temperatures, and internal heat generation. These flows are summed to give the net heat flow. The energy required by one or more alternative HVAC systems is then calculated, and the procedure is repeated for all time periods as occupancy and weather vary. The energy requirements for each increment are then summed and analyzed to give a range of outputs.

Table 5.34 lists typical outputs from computerized building energy programs. The data outputs vary according to program details and the user requirements. Heat balance data give the net heat flows to and from the building itemized in several different ways such as annual net, maximum, minimum, etc. HVAC system operation data give the operation and energy requirements of the alternative HVAC systems chosen by the user or designed by the program. Economic data compare the capital, operating, and life cycle costs for the HVAC systems.

TABLE 5.34 Typical Outputs from Computerized Building Energy Programs

- Heat balance data:
    Heat gain and heat loss itemized as follows:
    By zone
    Hourly
    Daily
    Monthly
    Annually
    Maximum
    Minimum
    Net
- HVAC system operation:
    Alternative system comparisons
    Hours of operation as functions of load
    Peak cooling power demand
    Peak heating power demand
    Total annual electrical power consumed
    Total annual fuel consumed
    Optimum energy-efficient system design
- Economic data:
    Alternative system comparisons
    Fuel cost
    Electrical power cost
    Capital equipment and amortization
    Total energy cost—amortized
    Optimum system design:
      Minimum capital cost
      Minimum fuel cost
      Minimum electrical power cost
      Minimum total annual cost

Manual methods for calculating building energy consumption use the same computational methods as the computer programs but are simplified to allow the calculations to be performed by hand. For example, one manual method[75] uses the BIN method where the total hourly energy requirements are calculated for each 5° outdoor temperature increment and are then multiplied by the frequency of occurrence. Another method[76] separates the year into a heating and a cooling season and calculates heating and cooling loads based on degree-hours.*

There are several manual methods in common use, and as might be expected, ASHRAE has made a sizable contribution here as well. The principles of energy usage calculation and several practical methods are discussed in the *ASHRAE Handbooks.*[10] This series of four volumes—*Fundamental, Equipment, Systems,* and *Applications*—is updated each 4 years and provides data on all aspects of HVAC system design and building energy usage. ASHRAE Standard 90P[8] also discusses manual methods and includes recommended procedures and forms for simplifying the required tabulations.

Some manual methods have been specifically developed for single-family residences. *The Household Energy Game*[89] and *Making the Most of Your Energy Dollars in Home Heating and Cooling*[71] are notable examples where the methods have been simplified

---

*A degree-hour is the product of the number of hours and the difference between the outdoor temperature and a reference temperature summed for a typical weather year.

into almost nontechnical form to enable homeowners to evaluate their home energy consumption without involved calculations.

As with computer programs, manual methods require very specific data inputs and a considerable amount of labor to determine the energy usage for buildings. Thus, they are not suitable for projects involving many buildings.

Energy usage factors are average values for energy consumption based on typical building design and construction. Since they are nonspecific, they may be applied to cities and other large projects where the detailed information required for computer and manual methods is either unknown or too complex to assimilate. However, their use provides the least accurate method for estimating building energy usage.

Tables 5.35 through 5.44 list average energy usage factors for various types of residential and nonresidential structures at four typical U.S. locations. Table 5.35 decribes these locations and the assumed weather and building operating conditions. To best describe the weather conditions within each region, weather conditions compiled by the U.S. Weather Bureau were weighted by population centers to determine the number of heating degree days for each state. These factors were further weighted according to the total number of housing units in each state to determine the regional values listed in Table 5.35.

The energy usage factors listed in Tables 5.36 through 5.44 were calculated using the Degree Day Manual Method as described in the *ASHRAE Systems Handbook 1973.*[10]

TABLE 5.35   Building Energy Usage Factors—Engineering Design Parameters

		Residential assumptions			
		Heating		Cooling Equivalent full-load	
Region	City	Degree-days	Design ΔT*	Operating hours	Design ΔT*
Northeast	Norwalk, Conn.	5,400	70	300	15
North central	Detroit, Mich.	6,200	75	500	15
South	Pine Bluff, Ark.	2,800	50	1,600	25
West	Roswell, N.M.	3,800	55	1,600	25
U.S. average		4,540			

	Nonresidential assumptions			
	Northeast (Norwalk, Conn.)	North central (Detroit, Mich.)	South (Pine Bluff, Ark.)	West (Roswell, N.M.)
Square feet/ton:				
Office building	250	250	200	300
Retail establishment	225	225	175	250
School	300	300	250	400
Hospital	200	200	150	225
Equivalent full-load operating hours	800	800	1,500	1,400

	Fuel utilization assumptions, %	
Fuel	Existing construction (<2 yr old)	New construction
Gas	60	70
Oil	50	60
Electricity—electric furnace	100	100

*Design ΔT is the temperature difference (°F) between indoor and outdoor conditions used to calculate heating and cooling loads.
SOURCE: From Reference 3.

TABLE 5.36   Mobile Home: Unit Demand, 1970
(10^6 Btu/unit/year)

	Fuel			
	Electricity	Gas	Oil	Kerosene
Space heating:				
Northeast	45.7	90.7	105.8	87.8
North central	52.0	104.7	122.0	101.3
South	24.7	50.0	58.3	48.4
West	29.8	61.0	71.1	59.0
Cooling:				
Northeast	3.4			
North central	5.9			
South	19.2			
West	18.7			

Energy usage factors based on:
- 720 ft² unit; 12 ft by 60 ft; 7 ft ceiling heights. Walls: Precoated aluminum exterior siding, 2½ in fiberglass insulation with ⅛-in plywood. Ceiling: 30 gauge galvanized steel roof, 3½ in fiberglass insulation, ¼-in softboard ceiling. Floor: ⅛-in vinyl tile, ⅝-in particleboard underlayment, 3½-in insulation. Windows: Aluminum, single glazing 12% of floor area.

SOURCE: From Reference 3.

TABLE 5.37   Single-family, Detached: Unit Demand, 1970   (Continued)
(10^6 Btu/unit/year)

		Fuel		
	Stories	Electricity	Gas	Oil
Space heating:				
Northeast	1	66.2	184.5	215.2
	2	66.3	174.5	203.5
North central	1	75.4	205.5	239.8
	2	74.9	196.3	229.2
South	1	37.2	84.7	
	2	35.2	81.2	
West	1	48.5	113.1	
	2	48.4	109.4	

		Fuel—electricity	
	Stories	Electric-designed home	Gas- or oil-designed home
Cooling:			
Northeast	1	3.5	4.0
	2	3.9	4.6
North central	1	6.0	7.0
	2	6.5	7.6
South	1	28.1	32.0
	2	28.0	30.1
West	1	29.0	31.4
	2	28.5	32.0

TABLE 5.37   Single-family, Detached: Unit Demand, 1970   (Continued)
(10^6 Btu/unit/year)

Energy usage factors based on:

Region	Size, ft²	Foundation	Predominant exterior walls	Average height, stories
Northeast	1,560	Basement	Wood and aluminum	1.5
North central	1,560	Basement	Wood and brick	1.3
South	1,560	Crawl space	Brick and wood	1.0
West	1,560	Slab	Stucco and wood	1.2

Construction details:
   Frame construction. Ceiling and walls insulation for houses using fossil fuels assumed to be R-11. For electric fuel, ceiling insulation R-19, walls R-11. Basements assumed to be unfinished; floors over basements and crawl spaces assumed insulated with R-7 when using electric heating, uninsulated when using fossil fuels. Glass assumed to be a single pane when fossil fuels are used, double pane when electricity used. In addition, certain design loads were increased 5 to 15% to account for the heat losses (or gains) within the distribution system, particularly with one-story, slab-on-grade residences.

SOURCE: From Reference 3.

Other details of this analysis are described in Reference 3. In using these tables, it must be remembered that the factors listed are average values for the particular region and that energy usage for specific structures may vary considerably. More accurate estimates of building energy usage for particular locations may be obtained from local utilities. Data

TABLE 5.38   Single-family, Attached: Unit Demand, 1970
(10^6 Btu/unit/year)

	Stories	Fuel		
		Electricity	Gas	Oil
Space heating:				
Northeast	1	45.7	125.5	146.3
	2	37.2	106.7	124.1
North central	1	52.0	156.2	182.2
	2	51.7	133.5	155.8
South	1	23.4	58.4	
	2	24.3	52.8	
West	1	30.6	70.1	
	2	34.8	71.1	

	Stories	Fuel—electricity	
		Electric-designed home	Gas- or oil-designed home
Cooling:			
Northeast	1	2.4	2.7
	2	2.2	3.2
North central	1	4.1	5.3
	2	4.5	5.2
South	1	18.3	22.1
	2	19.7	20.0
West	1	18.3	19.5
	2	20.5	20.8

Energy usage factors based on:
   Duplex consisting of two 1,100 ft² units; sharing one common wall. Insulation and materials assumed to be similar to detached single-family units.

SOURCE: From Reference 3.

TABLE 5.39   Multifamily, Low-rise: Unit Demand, 1970
(10^6 Btu/unit/year)

	Fuel		
	Electricity	Gas	Oil
Base load (hot water heating)			24.8
Space heating:			
Northeast	23.8	73.8	86.1
North central	27.1	86.3	100.7
South	12.3	30.5	
West	16.5	38.4	

	Fuel—electricity	
	Electric-designed home	Gas- or oil-designed home
Cooling:		
Northeast	1.3	1.6
North central	2.2	2.9
South	9.3	11.5
West	9.9	10.7

Energy usage factors based on:
Three-story, 24-unit garden apartment complex, 900 ft² per unit. Insulation and materials assumed to be similar to detached single-family unit.

SOURCE: From Reference 3.

TABLE 5.40   Multifamily, High-rise: Unit Demand, 1970
(10^6 Btu/unit/year)

	Fuel		
	Electricity	Gas	Oil
Base load (hot water heating)			24.8
Space heating:			
Northeast	21.2	68.3	79.6
North central	23.3	78.1	91.1
South	10.4	27.1	
West	14.6	32.8	

	Fuel—electricity	
	Electric-designed home	Gas- or oil-designed home
Cooling:		
Northeast	1.1	1.5
North central	2.7	1.9
South	7.9	10.2
West	8.7	9.1

Energy usage factors based on:
Ten-story building, six units per floor, 900 ft² per unit. Walls: 4-in common brick, 4-in concrete block, reinforced concrete framing, ½-in gypsum wallboard.

SOURCE: From Reference 3.

such as average electrical power consumption, gas consumption, and oil consumption according to building types are commonly prepared by utilities to aid in energy forecasting and planning. Various state and federal agencies also accumulate similar data.[24]

To illustrate the application of energy usage factors to large nonspecific developments, the natural gas consumption for space heating in a proposed residential development will now be calculated. This development will be located on a 640-acre site in the northeast adjacent to a metropolitan area. The development will be a residential suburb providing housing, schools, churches, and professional offices for a community of 20,000 people.

Table 5.45 lists the proposed types and numbers of buildings and the energy use factors from Tables 5.36 to 5.44. The natural gas consumed is the product of the number of units and the energy use factor given by

$$\begin{matrix} \text{Natural gas} \\ \text{consumed} \\ \text{per year} \end{matrix} = \begin{pmatrix} \text{Number of units} \\ \text{or square feet} \end{pmatrix} \begin{pmatrix} \text{unit demand factor} \\ \text{from tables} \end{pmatrix}$$

TABLE 5.41   Retail Establishments: Unit Demand,* 1970
[10^3 Btu/(ft^2)(yr)]

	Fuel		
	Electricity	Gas	Oil
Space heating:			
Northeast	22	52	63
North central	26	62	73
South	14	25	30
West	16	31	38
Space cooling:			
Northeast	12.2		
North central	12.2		
South	29.2		
West	19.0		

Factors may be applied to:
Both urban and suburban commercial stores and other merchandising activities.
Energy usage factors based on:
67,000 ft^2, single-story, suburban mall-type shopping center, 260 ft × 260 ft. Walls: 12-in concrete block, painted both sides. Roof: 4-ply built-up roofing. 2-in rigid insulation, steel decking, ½-in softboard. Windows: ¼-inch plate, 69% one wall, 0% other walls.

*Also used for "other" category.
SOURCE: From Reference 3.

TABLE 5.42   Office Buildings: Unit Demand, 1970
[10^3 Btu/(ft^2)(yr)]

	Fuel		
	Electricity	Gas	Oil
Space heating:			
Northeast	44	96	113
North central	51	113	113
South	24	59	71
West	25	61	72
Space cooling:			
Northeast	10.9		
North central	10.9		
South	25.5		
West	16.0		

Factors may be applied to:
General office space, public and administration buildings, etc.
Energy usage factors based on:
40,000 ft^2, three stories; 90 ft × 150 ft. Walls: 4-in common brick, 8-in concrete block structural steel framing, 1-in rigid insulation, ½-in gypsum wallboard. Roofing: 2-in rigid insulation steel decking, ½-in softboard. Windows: $\frac{2}{16}$-in sheet, 30% of external wall areas.

SOURCE: From Reference 3.

For example, the natural gas consumed by the mobile homes may be calculated as follows:

$$\begin{matrix}\text{Natural gas}\\\text{for mobile}\\\text{homes}\end{matrix} = \left(960\ \text{units}\right)\left(90.7 \times 10^6\ \frac{\text{Btu}}{\text{unit-year}}\right)$$

$$= 87.1 \times 10^9\ \text{Btu/yr}$$

The tables include energy usage factors for all structure types in this development except churches. Churches are generally constructed similar to schools (auditorium and classrooms), and so the school energy usage factor was assumed appropriate.

TABLE 5.43   Educational, Schools: Unit Demand, 1970
[10^3 Btu/(ft^2)(yr)]

	Fuel		
	Electricity	Gas	Oil
Space heating:			
Northeast	40	85	100
North central	46	99	117
South	18	44	52
West	23	54	64
Space cooling:			
Northeast	9.2		
North central	9.2		
South	20.4		
West	11.9		

Factors may be applied to:
Educational classrooms, laboratories, libraries, and related institutions on all levels.
Energy usage factors based on:
40,000 ft^2, single-story, 100 ft × 400 ft. Wall: 4-in common brick, 8-in concrete block, 1-in polyurethane insulation, ¾-in lath and plaster. Roof: 4-ply built-up roofing, 2-in rigid insulation, steel decking, ½-in softboard. Windows: single-strength sheet, 20% all walls.

SOURCE: From Reference 3.

TABLE 5.44   Hospitals: Unit Demand, 1970
[10^3 Btu/(ft^2)(yr)]

	Fuel		
	Electricity	Gas	Oil
Space heating:			
Northeast	46	103	122
North central	54	121	143
South	19	51	61
West	25	63	76
Space cooling:			
Northeast	13.6		
North central	13.6		
South	34.0		
West	21.1		

Factors may be applied to:
Hospitals, clinics, and other intensive medical care facilities.
Energy usage factors based on:
60,000 ft^2, four-story, 60 ft × 250 ft. Walls: 4-in common brick, 4-in concrete block, 2-in rigid insulation, ½-in gypsum drywall. Roof: 4-ply builtup roofing, 2-in rigid insulation, steel decking, ½-in softboard. Windows: double-pane insulated, 15% all walls.

SOURCE: From Reference 3.

The total natural gas consumed is the sum of the consumption of each building type, 612.5×10^9 Btu/yr. This can be converted to cubic feet of natural gas by dividing by 1,000 Btu/ft^3 (from Table 5.29), that is,

$$\text{Natural gas, ft}^3/\text{yr} = \frac{612.5 \times 10^9 \text{ Btu/yr}}{1,000 \text{ Btu/ft}^3} = 612.5 \times 10^6$$

TABLE 5.45   Natural Gas Energy Usage for Proposed Development

Building type	Number of units	Square feet	Unit demand factor	Energy usage, $\times 10^9$ Btu/yr
Residential:				
Mobile homes	960		90.7×10^6 Btu/yr	87.1
Low-rise apartments	1728		73.8×10^6 Btu/yr	127.5
Townhouses				
(1 story)	1280		125.5×10^6 Btu/yr	160.6
Single-family				
detached (1 story)	640		184.5×10^6 Btu/yr	118.1
Single-family				
detached (2 story)	560		174.5×10^6 Btu/yr	97.7
Total	5168			591.0
Nonresidential:				
Shopping center		150,000	52×10^3 Btu/(ft^2)(yr)	7.8
Professional offices		10,000	96×10^3 Btu/(ft^2)(yr)	1.0
Schools		100,000	85×10^3 Btu/(ft^2)(yr)	8.5
Churches		50,000	85×10^3 Btu/(ft^2)(yr)	4.2
Total		310,000		21.5
Grand Total				612.5

Lighting, Appliances, and Industrial Process Data   The optimum method for determining energy consumption for lighting, appliances, and industrial processes depends upon the project scale. For small projects energy usage factors are appropriate. The total energy consumption is calculated by multiplying the power required for each unit by its hours of operation per year and summing all units. For larger projects where thousands of units are involved, supply/demand data can be manipulated to give average energy utilization factors. The following discussion addresses the first method and lists energy consumption factors for several applications.

Energy consumption factors for electric lights are easily determined since lights are usually rated according to their power requirements. A 100-W lamp requires 100 W of power, and if operated for 10 hours, the energy consumed is $100 \times 10 = 1000$ Wh, or 1.0 kWh.

The light output of an electric lamp is measured in lumens where one lumen is the amount of light required to illuminate one square foot of area at a level of one foot candle. The number of foot candles required in various applications depends upon several factors. Typical values range from 3.0 for an intimate restaurant dining area to 2,500 for a hospital operating table. The *IES Lighting Handbook*[69] lists values for several other applications.

Electric lamps are not very efficient; only a small fraction of the electrical power is converted to light with the remainder going to heat. Table 5.46 lists the efficiency of several commonly available lamps in units of lumens per watt.

The power required for a specific lighting application can be calculated by applying the following formula:

$$\text{Power, W} = \frac{\left(\begin{array}{c}\text{lighting level,}\\ \text{footcandles}\end{array}\right)\left(\begin{array}{c}\text{area,}\\ \text{ft}^2\end{array}\right)}{(\text{lamp efficiency, lumens/watt})}$$

For example, a 250-ft^2 office is to be lighted with 48-in-long 40-W cool white fluorescent lamps to a level of 100 footcandles. The power required is thus:

$$Power = \frac{(250 \text{ footcandles})(100 \text{ ft}^2)}{(67 \text{ lumens/watt})}$$

$$= 373 \text{ W}$$

Since each lamp requires 40 W, the number of lamps is:

$$Number = \frac{\text{total power, W}}{\text{lamp power, W}}$$

$$= \frac{373}{40}$$

$$= 9.32$$

Thus, use 10 lamps.

TABLE 5.46   Efficiency of Alternative Lighting Sources

Lamp description	Efficiency, lumens/watt*
Incandescent:	
40-W general service	11
60-W general service	14.3
100-W extended service	14.8
100-W general service	17.4
1000-W general service	22
Fluorescent:	
24 in cool white	50
48 in cool white—40 W	67
96 in cool white	73
High intensity discharge:	
400-W mercury	46
400-W metal halide	74
400-W high-pressure sodium	100
1000-W metal halide	85

*Typical values, specific units may vary.
SOURCE: From Reference 80.

Typical energy usage factors for unitary heating, ventilating, and air conditioning equipment are listed in Table 5.47. Since a complete listing of energy usage factors for every type, size, and make of HVAC equipment is beyond the scope of this book, data on selected units are presented instead. The power requirements for these units were taken from either name plate information or catalog data. The annual hours of operations were estimated for "normal" service. For example, a room air conditioner was assumed to be operated an average of 4 hours per day during a 3-month summer cooling season. The annual energy consumption tabulated is the product of the power requirement and the annual hours.

EER is an index of air conditioner performance defined as the ratio of cooling capacity in Btu/h to power in watts. High EER corresponds to low power requirement for a unit amount of cooling. Note that the furnace fan power requirement is normally not included in central air conditioner power. Including furnace fan power lowers the EER by approximately one point.

The data in Table 5.47 can be used to translate building energy requirements as determined by one of the methods discussed in the section on Building, Heating, and Cooling Data into electrical power requirements. For example, the output of a computerized calculation of cooling requirements for a 2,000 ft² single-family detached structure specifies the following cooling energy requirements:

Maximum capacity   = 42,000 Btu/h
Total cooling energy per season = 10.1 × 10⁶ Btu

The electrical power requirements for three alternative cooling systems will now be calculated:

Alternative	Cooling system
A	42,000 Btu/h central system
B	42,000 Btu/h high-efficiency central system
C	High-efficiency room air conditioners:
	One @ 21,300 Btu/h
	One @ 12,000 Btu/h
	Two @ 5,000 Btu/h

TABLE 5.47   Unitary HVAC Equipment Energy Consumption Factors

Item and capacity		Power,* W	EER†	Annual hours‡	Annual kWh‡
Central air conditioning:					
Standard:	23,000 Btu/h	3,300	7.0	1,000	3,300
	31,000 Btu/h	4,300	7.2	1,000	4,300
	42,000 Btu/h	5,900	7.1	1,000	5,900
High efficiency:	24,000 Btu/h	2,750	8.9	1,000	2,750
	36,000 Btu/h	4,050	9.0	1,000	4,050
	42,000 Btu/h	5,000	8.4	1,000	5,000
Electrostatic filter		11		2,000	22
Evaporative cooler:	2,000 cfm	280		1,000	280
	3,100 cfm	522		1,000	522
	4,900 cfm	845		1,000	845
Attic fan, 42 in	12,000 cfm	650		500	325
Roof ventilator:	500 cfm	165		500	82
	2,000 cfm	450		500	225
Bathroom ventilator:	90 cfm	220		365	80
Same with heat:	90 cfm	1,020		365	372
Room fan	12 in	47		200	9.4
	20 in	180		200	36
Dehumidifier:	14 pint/day	240		500	120
	35 pint/day	560		500	280
Humidifier:	128 pint/day	60		500	30
Room air conditioners					
High efficiency:	4,000 Btu/h	745	5.4	500	370
	5,000 Btu/h	625	8.0	500	310
	6,000 Btu/h	860	7.0	500	430
	8,000 Btu/h	855	9.4	500	430
	10,000 Btu/h	1,375	7.3	500	690
	12,000 Btu/h	1,380	8.7	500	690
	14,000 Btu/h	1,380	10.1	500	690
	18,000 Btu/h	2,745	6.6	500	1,370
	21,300 Btu/h	2,855	7.5	500	1,420
	25,000 Btu/h	3,140	8.0	500	1,570
	29,000 Btu/h	4,170	7.0	500	2,080

*Power data from specific manufacturer's data for commonly produced units. Other models may vary.
†EER is the energy efficiency ratio for an air conditioning system and is defined as the ratio of Btu/h cooling capacity to power requirement in watts. It is an index of coefficient of performance.
‡Annual hours and annual energy vary with usage—typical values are shown.

The first two alternatives require a furnace fan (750 W) to run at the same time as the air conditioning unit. The furnace fan operation can be determined from the total cooling energy as follows:

$$\text{Operating hours} = \frac{\text{total cooling energy, Btu}}{\text{system capacity, Btu/h}}$$
$$= \frac{10.1 \times 10^6}{42,000}$$
$$= 240.5 \text{ hr}$$

The electrical power requirement for alternative A is:

$$\text{Air conditioner power, Wh} = \frac{\text{total cooling energy, Btu}}{(\text{EER, Btu/h} \cdot \text{W})}$$
$$= \frac{10.1 \times 10^6}{7.1}$$
$$= 1.42 \times 10^6$$

$$\begin{aligned}
\text{Furnace fan power, Wh} &= (\text{power, W})(\text{hours}) \\
&= 750 \times 240.5 \\
&= 0.18 \times 10^6
\end{aligned}$$

$$\begin{aligned}
\text{Total power, Wh, for A} &= 1.42 \times 10^6 + 0.18 \times 10^6 \\
&= 1.60 \times 10^6
\end{aligned}$$

The power requirement for alternative B is calculated similarly:

$$\text{Total power, Wh, for B} = 1.38 \times 10^6$$

The electrical power requirement for alternative C can be calculated assuming the same duty cycle for each of the four units:

$$\begin{aligned}
\text{Total capacity, Btu/h} &= 21,300 + 12,000 + 2 \times 5000 \\
&= 43,300
\end{aligned}$$

$$\text{Hours of operation, h} = \frac{10.1 \times 10^6}{43,300}$$
$$= 233.3$$

$$\begin{aligned}
\text{Electrical power, Wh} &= (\text{power, W})(\text{hours}) \\
&= 2855 \times 233.3 + 1380 \times 233.3 + 2 \times 625 \times 233.3 \\
&= 1.28 \times 10^6
\end{aligned}$$

Table 5.48 lists the energy requirements for typical residential water heaters itemized according to the fuel type. The energy serves two purposes: (1) heating water from the supply temperature to the delivery temperature, and (2) overcoming heat losses from the tank. Heat losses can be minimized through effective insulation, and the energy required to warm the water depends upon consumption. Based on the amount of hot water supplied as listed in Table 5.48, the minimum energy requirement with no heat loss is 121.8 therms/yr. Thus, the fired water heaters are about 45 percent efficient, and the electric

TABLE 5.48   Water Heating Energy

Fuel type	Capacity, gal	Insulation thickness, in	Annual energy consumption	
			Amount	Units
Electric	50	2	4,400	kWh
LPG	40	1	260.3	therms
Natural gas	40	1	271.9	therms

NOTES:
1. All data from Reference 78.
2. Insulation thickness is typical thickness for specific type of heater.
3. Energy consumption based on:
   - 140°F tank temperature
   - 50 gal/day hot water supplied
   - 70°F ambient temperature
   - 60°F cold water feed temperature

water heater is about 81 percent efficient. However, the electric water heater requires fuel to be burned at the power plant equivalent to about three times the electrical power. Thus, the overall electric water heater efficiency is about 27 percent.

Energy usage factors for household refrigerators and freezers are listed in Table 5.49. These values are the average energy consumption for units in typical domestic household service based on a survey of 1975 model-year refrigerators and freezers conducted by the Association of Home Applicance Manufacturers.[11] The range of values listed reflects the differences in design of similar size units and antisweat heaters. Heat leakage near the door seals of refrigerators and freezers make these areas the coldest and promotes water

TABLE 5.49   Energy Consumption of Household Refrigerators, Freezers, and Combination Units*

Ranges of total refrigerated volume, ft³				Annual electricity usage, kWh			
Rated		Actual		Partial automatic defrost		Fully automatic defrost	
Min	Max	Min	Max	Min†	Max†	Min†	Max†
				Refrigerators‡			
	2.5		3.5	330	600		
2.5	4.5	1.5	5.5	330	810		
4.5	6.5	3.5	7.5	420	810		
6.5	8.5	5.5	9.5	420	660		
8.5	10.5	7.5	11.5	540	720		
10.5	12.5	9.5	13.5	540	780		
12.5	15.5	11.5		540	870		
				Refrigerator-freezer combinations			
	10.5		11.5	900	1,320		
10.5	12.5	9.5	13.5	900	1,320	1,590	2,040
12.5	14.5	11.5	15.5	900	1,440	1,380	2,040
14.5	16.5	13.5	17.5	660	960	1,050	2,490
16.5	18.5	15.5	19.5	660	960	1,050	2,910
18.5	20.5	17.5	21.5			1,200	2,910
20.5	22.5	19.5	23.5			1,440	2,880
22.5	24.5	21.5	25.5			1,440	3,210
24.5	26.5	23.5	27.5			1,650	3,210
26.5	28.5	25.5	29.5			2,400	2,400
28.5	29.5	27.5	30.5			2,400	2,400
				Freezers			
	6.5		7.5	690	1,020		
6.5	8.5	5.5	9.5	720	960		
8.5	10.5	7.5	11.5	840	1,290		
10.5	12.5	9.5	13.5	900	1,530	1,890	2,070
12.5	14.5	11.5	15.5	1,020	2,040	1,890	2,370
14.5	16.5	13.5	17.5	1,020	2,040	1,620	2,370
16.5	18.5	15.5	19.5	1,050	1,680	1,560	2,250
18.5	20.5	17.5	21.5	1,050	1,830	1,800	2,250
20.5	22.5	19.5	23.5	1,080	1,980	1,980	2,340
22.5	24.5	21.5	25.5	1,470	2,070	1,980	1,980
24.5	26.5	23.5	27.5	1,590	2,520		
26.5	28.5	25.5	29.5	1,620	2,520		
28.5		27.5		1,920	2,040	2,970	2,970

*All data from Reference 11.
†When antisweat switches are provided, the maximum value corresponds to heaters fully energized and minimum when switched to lowest position.
‡Refrigerators have no defrost provision.

condensation under humid conditions. Antisweat heaters are small electric resistance heaters which warm these areas, preventing condensation. Some manufacturers provide a switch to deactivate these heaters in dry weather and thus reduce energy consumption.

Tables 5.50, 5.51, and 5.52 list energy consumption factors for kitchen appliances, other small appliances, and office equipment, respectively. As with previous tables, the energy consumption numbers are based on specific manufacturer's data, and annual hours are estimated. Energy usage factors for other equipment not listed here can be easily obtained by contacting specific manufacturers or examining nameplate data.

TABLE 5.50   Energy Consumption of Kitchen Appliances

	Power,* W	Annual hours†	Annual energy,† kWh
Electric range	12,000	180	2,160
Trash compactor	460	90	42
Microwave oven	1,300	180	234
Dishwasher	1,200	360	340
Garbage disposal	390	50	20
Toaster (two slice)	800	30	24
Blender	500	3	1.5
Frying pan (electric)	1,225	180	220
Toaster oven	1,200	180	216
Electric grill	1,000	52	52
Can opener/knife sharpener	100	180	18
Hand mixer	125	90	11
Stationary mixer	150	90	13
Hot plate (two burner)	1,600	180	288
Coffee pot (electric)	1,000	60	60
Slow cooker	160	520	83
Deep fryer	1,150	52	60
Waffle iron	1,000	26	26
Fruit juicer	60	60	3.6
Hot dog cooker	1,200	6	7.3

*Power data from specific manufacturer's data for commonly produced applicances. Other models may vary.
†Annual hours and annual energy vary with usage—typical values are shown.

These energy usage factors express direct energy consumption which is only one component of the total. Other components include the energy required for manufacturing, maintenance, sales, and ultimate disposal. These components are difficult to quantify due to the complex nature of the various processes. Table 5.53 lists rough estimates of the ratio of energy consumed in manufacturing to energy directly consumed by selected appliances in 1 year. Since these appliances are not large energy consumers, many years of operation are required to equal the energy used in manufacturing.

Unlike these appliances, automobiles are energy intensive and consume large amounts of energy each year. A recent study of the motor vehicle industry determined that 7,000 kWh of energy was required to manufacture the average 1974 automobile.[12] Assuming an average fuel consumption of 15 mi/gal and a usage rate of 15,000 mi/yr, the energy used in manufacturing corresponds to 2.2 months of use.

Since there are many types of industrial processes and equipment, it is not possible to provide a complete list of their energy consumption in this chapter; however, the energy consumption of many processes can be estimated from the horsepower of their electric motors and the steam, compressed air, and other energy requirements. Table 5.54 lists the energy requirements of several sizes of electric motors and materials commonly used in industrial processes.

As an example of the application of Table 5.54, consider the following industrial process. A plastic molding machine has two 30-horsepower electric motors and utilizes the following materials:

Material	Flow rate
200 psig steam	175 lb/h
100 psig compressed air	110 ft³/min
35°F chilled water	10 gal/min

The energy requirement is the product of each material's flow rate and production energy factor (from Table 5.54). Assuming a motor power factor of 0.8, the energy requirements are as follows:

$$\text{Electric motors} = 2 \times (31{,}073 \text{ Volt-Amps}) \times 0.8 \, \frac{\text{Watts}}{\text{Volt-Amps}}$$
$$= 49{,}717 \text{ W}$$

TABLE 5.51   Energy Consumption of Small Appliances

Item	Power,* W	Annual hours†	Annual energy,† kWh
Color TV, 25-in solid state	190	1,825	347
Black and white TV, 19-in solid state	68	1,825	124
Stereo receiver	50	1,460	73
Solid state radio	5	1,460	7.3
Casette tape recorder	5	180	0.9
Clock	5	8,760	44
Electric blanket (twin size)	135	730	99
Slide projector, 35 mm	400	52	21
Movie projector, 8 mm	230	52	12
Movie projector, 16 mm	600	52	31
Sewing machine	120	52	6.2
Hand iron	1,100	156	172
Curling iron	40	90	3.6
Hand held hair dryer	1,000	30	30
Stationary hair dryer	725	100	72
Electric toothbrush	2	18	0.04
Floor polisher	400	52	21
Upright vacuum cleaner	700	365	256
Canister vacuum cleaner	700	365	256
Shop vacuum cleaner	900	26	23
Washing machine (clothes)	840	365	307
Clothes dryer, electric	5,600	365	2,044
Swimming pool pump, ¾ hp	840	8,760	7,358
Garage door opener	625	6	3.8
Electric drill, hand held	360	13	4.7
Circular saw, hand held	1,320	13	17
Radial arm saw	1,200	13	16
Belt sander	900	13	12
Router	720	13	9.4
Arc welder, 300 amp	11,700	26	304
Lawnmower, electric	1,150	26	30
Edger, electric	500	26	13
Electric shaver	10	61	0.6
Auto battery charger, 10 amp	180	300	52
Drill press, ½ hp	650	13	8.5
Small electric organ	20	400	8
Large electric guitar amplifier	235	400	94

*Power data from specific manufacturer's data for commonly produced appliances. Other models may vary.
†Annual hours and annual energy vary with usage—typical values are shown.

$$\text{Steam} = \left(175\,\frac{\text{lb}}{\text{hr}}\right)\left(0.425\,\frac{\text{kWh}}{\text{lb}}\right)$$
$$= 74.375\ \text{kW}$$

$$\text{Air} = \left(\frac{110}{100}\times100\ \text{ft}^3/\text{min}\right)\left(\frac{60\ \text{min}}{\text{h}}\right)\left(\frac{0.383\ \text{kW}}{100\ \text{ft}^3/\text{min}}\right)$$
$$= 25.278\ \text{kW}$$

$$\text{Water} = \left(\frac{10\ \text{gal}}{\text{min}}\right)\left(\frac{60\ \text{min}}{\text{h}}\right)\left(0.034\,\frac{\text{kWh}}{\text{gal}}\right)$$
$$= 20.400\ \text{kW}$$

$$\text{Total energy (except steam)} = 49.717 + 25.278 + 20.400$$
$$= 95.395\ \text{kW}$$

The steam energy is not included in this sum since it is not produced directly from electricity. To convert all energy components to common heat units, the kW units should

TABLE 5.52   Energy Consumption of Office Equipment

Item	Power, W*	Annual hours†	Annual energy, kWh†	Reference
Computers:‡				
Mini	8,000	2,500	20,000	3
Small	18,000	2,000	36,000	3
Medium	35,000	3,800	133,000	3
Large	70,000	5,200	364,000	3
Giant	105,000	6,500	683,000	3
Remote terminal	1,000	1,000	1,000	3
Dry copier	700	1,000	700	
Small water cooler	250	2,920	730	
Large water cooler	550	2,920	1,600	
Coffee vending machine	20	8,760	175	
Candy vending machine	45	8,760	394	
Microfilm reader	100	1,000	100	
Adding machine	30	2,000	60	
Hand held calculator	5	2,000	10	
Electric elevator, 600 floors/day	12,000		2,600	
Photocopier§			9,000	

*Power data from specific manufacturer's data for commonly produced appliances. Other models may vary.
†Annual hours and annual energy vary with usage; typical values are shown.
‡Includes peripherals, air conditioning, etc.
§Estimate of effective energy usage.

TABLE 5.53   Manufacturing Energy for Selected Appliances

	Ratio of energy in manufacturing to energy in use per year
Carving knife, electric	20
Can opener	17
Garage door opener	50
Garbage disposal	30
Power mower	3
Roto-Tiller	25
Shredder	20
Snow blower	25
Trash compactor	20

SOURCE: From Reference 89.

TABLE 5.54   Energy Requirements of Electric Motors and Various Industrial Materials

CASE: Electric Motors

Horsepower	Volt-amperes	
1/500	23	
1/125	50	
1/10	391	
1/4	644	NOTE: "Volt-amperes" is the product of the
1/3	690	current and voltage required at full
1/2	897	load. The actual power is the product
3/4	1,344	of the volt-amperes and the power
1	1,600	factor which is nominally 0.8, but
2	2,530	varies from 0.7 to 0.95 according to
3	4,300	motor design.
5	5,750	
7.5	8,280	Data from manufacturers catalog
10	10,120	information.
20	20,795	
30	31,073	
40	40,634	
50	49,398	

CASE: Industrial Materials

Material	Amount	Energy content, kWh*	Production energy, kWh†
Ice at 32°F	1 lb	0.053	0.021
Steam at 100 psig (saturated)	1 lb	0.337	0.421
Steam at 200 psig (saturated)	1 lb	0.340	0.425
Hot water at 140°F	1 gal	0.170	0.214
Chilled water at 35°F	1 gal	0.085	0.034
Compressed air at 100 psig	100 ft³	0.164	0.383

*Energy content is energy released to produce material at atmospheric pressure and 70°F.
†Production energy is energy required to produce material with commercially available equipment assuming:
Ice and chilled water—refrigeration COP—2.5
Steam and hot water—boiler efficiency—80%
Compressed air—2-stage compressor (production)
    isothermal expansion (energy content)

be converted to Btu/hr (see Table 5.5) and the electrical power should be adjusted for power plant efficiency, assumed to be 38 percent, as follows:

$$\text{Total heat energy} = (74.375 \text{ kW})\left(\frac{3415 \text{ Btu/h}}{\text{kW}}\right) + \frac{(95.395 \text{ kW})\left(\frac{3415 \text{ Btu/h}}{\text{kW}}\right)}{0.38}$$
$$= 1.111 \times 10^6 \text{ Btu/h}$$

Transportation Energy Data   The amount of energy consumed in transportation depends upon the material transported and the transit mode. Energy usage in transporting fuels was summarized in Table 5.31. Tables 5.55 and 5.56 list energy usage factors for people and freight.

The most meaningful way to express energy use factors for transporting people is to calculate the net energy consumed (Btu) per passenger mile. The total energy consumption is then the product of this factor, the number of passengers, and the distance traveled.

One complication is that energy usage factors depend heavily upon the number of passengers actually carried in addition to the maximum capacity. For example, the energy consumed by automobiles is almost independent of the number of passengers. If a full-size automobile with only a driver uses 7 gal on a 100-mi trip, increasing the number of people carried to six would only increase the fuel consumption to approximately 7.7 gal, or 10 percent.[74] The energy usage factors for transit modes with highly variable passenger loads listed in Table 5.55 were calculated assuming full passenger capacity and list the fuel consumption and passenger capacity as well. For those transit modes which usually operate with relatively fixed passenger loads, only the energy usage factors are listed.

The following example illustrates the application of Table 5.55. A manufacturing plant has proposed the construction of a private runway to reduce the time required for company personnel to commute to an airport 75 mi away (straight line). Because of geographical considerations, the most direct highway distance to the airport is 100 mi. Seventy-five

TABLE 5.55   Energy Usage Factors for Transporting People*

Transit mode	Miles per gallon	Passengers (capacity)	Btu per passenger mile
Ground transportation:			
Ultra-compact auto	35	4	910
Small auto	22	5	1,160
Family auto	13	6	1,960
Motorcycle	75	2	850
School bus	7	30	610
Intercity bus	6.1	50	420
Train			2,000
Personal rapid transit			3,720
Air transportation:			
Small single-engine (Cessna 150)	14.3	2	4,460
Twin (Beech Baron)	8.6	6	2,470
Business jet (Airesearch 731)	122	10	10,460
Small helicopter (Rotorway 133)	10	2	6,380
Airliners:			
DC 3			2,630
DC 6			3,130
DC 7			3,030
DC 8			4,000
Electra			3,330
B-747			2,700
SST			6,250

*Data from specific manufacturers and References 50 and 99.

TABLE 5.56   Energy Usage Factors for Transporting Freight

Transit mode	Btu/ton · mi
Aircraft	42,000
Trucks	2,800
Waterway	680
Rail	670
Pipeline	450

SOURCE: Reference 46.

people make this trip weekly: one-third take a bus, and the other two-thirds drive their personal cars at an average of 1.5 people per car. The runway construction would allow the company plane, a light twin-engine, to carry all personnel to the airport at an average of 4 people per trip.

The energy consumption is calculated as follows:

$$\text{Energy} = \sum_{\text{all transportation modes}} [(\text{Number of people})(\text{number of miles})(\text{energy, Btu/passenger mile})]$$

Without the runway,

$$\text{Energy, bus} = 25 \times 200 \times 420$$
$$= 2.1 \times 10^6 \text{ Btu/week}$$

$$\text{Energy, autos} = 50 \times 200 \times \frac{6 \times 1960}{1.5}$$
$$= 78.4 \times 10^6 \text{ Btu/week}$$

$$\text{Total energy} = 2.1 \times 10^6 + 78.4 \times 10^6$$
$$= 80.5 \times 10^6 \text{ Btu/week}$$

The bus energy consumption was calculated assuming that the buses operate at full capacity. Since the automobiles operate at less than full capacity, the Btu per passenger mile factor was adjusted by multiplying by the capacity (6) and dividing by the average load (1.5).

With the runway,

$$\text{Energy} = 75 \times 150 \times \frac{6 \times 2470}{4}$$
$$= 41.7 \times 10^6 \text{ Btu/week}$$

As with the automobiles, the aircraft Btu per passenger mile factor was adjusted for load. Thus the construction of the runway conserves transportation energy.

Energy usage factors for freight transportation are listed in Table 5.56 in Btu/ton·mile. As with other energy usage factors, these numbers are approximate. The total energy consumed is the product of the factor, the number of tons moved, and the distance in miles, i.e.,

$$\text{Energy} = (\text{factor, Btu/ton} \cdot \text{mi}) (\text{weight, tons}) (\text{distance, mi})$$

For example, if 750 tons are to be shipped 3500 mi, the energy required to ship by truck and by rail are as follows:

$$\text{Truck energy} = 2800 \times 750 \times 3500$$
$$= 7.35 \times 10^9 \text{ Btu}$$

$$\text{Railroad energy} = 670 \times 750 \times 3500$$
$$= 1.76 \times 10^9 \text{ Btu}$$

Supply and Demand Data Sources

Supply and demand data are the historical, present, and projected magnitudes of specific energy flows. While historical data are relatively easy to obtain, the primary challenge for the author of an energy impact section is to obtain accurate, up-to-date information on current energy flows and meaningful projections. The time lag between occurrence and reporting of specific energy flows in the literature usually ranges from four months to a year. The time lag from initial reporting of raw data to the publication of comprehensive reports and projections is usually another one to two years.

In times of stable energy supply and demand, this time lag is not important. However, the 1970s have been very turbulent with some fuels in short supply and others doubling in price almost overnight. At the same time, there has been increased awareness of energy's critical importance. In times like these, it is important to base an energy impact assessment on the most recent and comprehensive data available. Any current energy supply and demand data presented in this section will be only as useful as historical data at the time this handbook is published, and historical data is easily obtainable elsewhere. Consequently, this section focuses on data sources rather than the data itself.

Energy supply and demand data are usually organized along geographical lines as follows:

- International (by country, groups of countries, or continents)
- U.S. National (by state or region)
- State (by county or city)
- Local (by local area or utility service area)

On national and international levels, the best data sources are recent reports from federal agencies active in the energy field. Unfortunately, federal reorganization is continuously creating or eliminating agencies, and almost every agency has produced at least one report discussing some aspect of energy supply and demand. In the past, three federal organizations have been the leading publishers of energy supply and demand data:

1. Federal Energy Administration (FEA)*
2. Bureau of Mines (BOM)
3. Federal Power Commission (FPC)

Prior to being reorganized as the Department of Energy, the FEA published several reports dealing with national and international supply and demand issues. The most comprehensive report is the *Project Independence Blueprint Report* published in 1974.[50] It was a response to the Arab oil embargo and investigated the possibility of modifying U.S. energy supply and demand to achieve self-sufficiency by 1985. The report summarized comprehensive studies by several agencies and was published as a summary report, a main body report, and several Task Force reports. The Task Force reports are excellent sources for historical supply and demand data and projections for specific fuels.

The Department of Energy (DOE) publishes the *National Energy Outlook*[48] annually during the first quarter of the year. This comprehensive document presents recent energy trends for major fuel groups (including oil, coal, natural gas, and nuclear fuel) and electrical power on national and world bases. It also presents the near and long-term forecasts and identifies critical supply and demand problems.

The Bureau of Mines is part of the Department of the Interior and is an excellent source of current national energy use data. Bureau of Mines publishes news releases containing energy use data and fuel supply data that may be as recent as three months old.[36] However, the information is essentially unprocessed and does not contain detailed correlations or projections. Bureau of Mines also publishes more comprehensive reports focusing on specific fuel types: *Minerals Yearbook*[35] and *Mineral Industry Surveys: Petroleum Statement Annual*,[34] *Natural Gas Annual*,[33] *Fuel Oil Shipments*,[31] and *LPG Shipments*.[32] In addition, Bureau of Mines publishes reports covering several fuel types such as *Fuels and Energy Data*[28] which lists reserves, production, and consumption of fuels and energy by state.

The Federal Power Commission (superseded by the Federal Energy Regulatory Commission) is primarily concerned with electrical power production. Data concerning the number and type of U.S. power plants, electrical power production itemized by utility, state, etc., and world electrical power production can be found in current FPC publications. Relevant examples include *Statistics of Publicly-Owned Utilities in the United States*,[54] *Statistics of Privately-Owned Utilities in the United States*,[53] *Hydroelectric Power Resources of the United States, Developed and Undeveloped*,[52] and *World Power Data*.[55]

Lists of current publications from these and other agencies can be obtained directly. The National Technical Information Service (NTIS), a branch of the Department of Commerce, maintains a file of all publications produced by the federal government. NTIS publishes a weekly listing of current publications categorized by technical area, and recent publications can be ordered in either printed copy or microfiche for a nominal charge. Subscribing to this weekly listing is a convenient way to keep abreast of current publications. Documents addressing energy supply and demand data are listed in the "Energy" category under the subtitle "Energy Use, Supply and Demand." Computer-aided searches of the NTIS and other data banks are available from several sources.

Energy supply and demand data on the state level are not as readily available as national data. Every state has one or more agencies with responsibilities in this area. Some publish data regularly, others publish rarely, if at all. In California, for example, there are several active energy-related agencies, namely:

1. Energy Resources Conservation and Development Commission (Energy Commission)
2. Department of Transportation
3. Division of Mines and Geology
4. Public Utilities Commission
5. Resources Agency
6. Division of Oil and Gas

The California Energy Commission is a new organization established in 1974 to provide comprehensive leadership in energy-related areas. It publishes a *Quarterly Fuel and Energy Summary*[25] listing current supply and demand in detail and comparisons with

---

*Now combined with the Energy Research and Development Administration (ERDA) as part of the Department of Energy.

previous data for at least two years. The Energy Commission also publishes forecasts of energy supply and demand such as *Electricity Forecasting and Planning Report.*[24]

The cognizant energy agencies in specific states can be determined by contacting the governor's office or by the use of a special directory. *The Energy Directory*[43] is published annually and is a comprehensive guide to energy organizations on national, regional, and state levels. It also lists professional, trade, and industry organizations active in the energy field.

In many states, specific fuels such as gasoline are taxable, and the tax records are available to the public. Knowledge of the taxing rate and a review of the tax records can give surprisingly detailed fuel consumption data. For example, in California the State Board of Equalization collects sales tax on gasoline, diesel fuel, and LPG used for highway transportation. Taxes are also collected on those fuels if used for other purposes, but the amount taxed is returned at the end of the fiscal year much as extra income tax is refunded on an income tax return. Consequently, the State Board of Equalization publishes records of both taxable[21] and nontaxable[23,26] fuel sales.

Local supply and demand data are, in general, the most difficult to find. Local governments such as cities and counties usually lack the funding and the interest to compile and publish energy data. However, there are a few exceptions. For example, San Bernardino County (California's largest county) has recently compiled energy supply/demand data for the county and Southern California as part of its Joint Utilities Management Program.[100] Another example is the Southern California Association of Governments (SCAG), a regional organization of local governments which publishes local statistics, some of which are energy-related.[91]

Local energy distributors, such as electrical power companies, gas companies, and petroleum distributors, are other sources of local supply and demand data. They compile summaries of their operations and often calculate average fuel and electrical power demand itemized by the characteristics of the end user. Much of this data is published in reports to the state regulatory agencies.[92,93] Furthermore, public relations personnel in these companies are usually very helpful and will often secure unpublished data if requested.

This discussion has focused on energy supply and demand data organized along geographical lines. Another commonly used method of reporting these data is to group energy use into sectors. The four most commonly used sectors include:

- Residential
- Commercial
- Industrial
- Transportation

References 3, 19, 29, 42, and 66 are examples of compilations of supply and demand data on a national level for specific sectors. Similar studies for portions of sectors have also been published. For example, FEA recently published a series of reports on energy consumption in the motor vehicles industry[12] and several other energy-intensive industries.

Energy Conservation Data

This section discusses the identification and evaluation of energy conservation features as applied to energy impact analysis. The following subsection addresses the potential for energy conservation in several areas and the various forms of energy conservation. Subsequent sections discuss the evaluation of conservation features and list some energy conservation measures applicable to a variety of projects.

Potential For Energy Conservation   There is no question that there is a great potential for energy conservation. In the past, with ample supplies of fossil fuels available and with expectations of continued supply, energy consumption by processes and equipment was a secondary concern and conservation was often sacrificed in favor of lower capital cost or increased comfort and convenience. As a result, we are now burdened with a large inventory of equipment which consumes much more energy than necessary and social customs requiring high energy consumption.

Examples of such equipment include:

- Automobiles with fuel economy of less than 10 mi/gal
- Electric heating systems for buildings
- Reheat air conditioning systems

- Frost-free refrigerators
- Buildings with no thermal insulation
- Buildings with large glass areas

Examples of energy-wasting customs include:

- Living in the suburbs and driving many miles to work
- Illuminating the outside of buildings for decorative effects
- Illuminating the inside of buildings 24 hours a day, regardless of use
- Warming up automobiles on cold mornings by prolonged idling
- Using open freezers in supermarkets
- Using gas lights and fireplaces for decorative purposes

The potential for energy conservation in the United States has been estimated as follows:

. . . if full application of the economically justifiable technical improvements presently available were made to equipment and practices in buildings and industry, as much as 25 percent of the total primary fuel consumption in the USA could be conserved.[16]

There are several areas in which careful attention to energy conservation can yield considerable results in the form of energy savings or substitute energy production. The following numbers are all approximate. For comparison, the total energy consumption in the U.S. during 1970 was about 70×10^{15} Btu. . . . The efficiency of electrical power generation averages about 33 percent. Thus, the "waste heat" from electrical generation amounts to 10×10^{15} Btu. Much of this wasted energy could be salvaged by extensive use of total energy systems. Topping cycles could recover about a third of this energy. Space and water heating consume 17×10^{15} Btu. More than half of this heating could ultimately be supplied by solar energy, thus conserving oil and gas.

More efficient transportation could save about 5×10^{15} Btu. The energy content of the agricultural urban and sewage wastes are about 7×10^{15} Btu. Currently, little serious effort is being made to recover this energy.

Better insulation and weather stripping could save about 3×10^{15} Btu in the area of space conditioning.

Better housekeeping could conserve 10 percent of the energy used in industry. This would save about 2.5×10^{15} Btu.[63]

The various forms of energy conservation can be grouped as follows:

- Advanced technology
- Application of existing technology
- Combined uses
- Reduced energy demand

Advanced technology is new technology which provides the same output at lower costs, or reduced energy consumption for the same end use at the same cost. Advanced technology of this type always lowers total cost, dollar cost, and energy consumed. It should be noted, however, that advanced technology which lowers the cost of energy may also result in increased consumption.

Energy may also be conserved by applying existing technology to new areas. For example, building energy consumption for heating and air conditioning could be reduced by insulating walls and ceilings, installing attic fans, etc. Such measures are already being applied where they are clearly cost-effective or where governmental standards require them, but in many cases economics and convenience do not favor energy conservation.

Another area of energy conservation deals with combined uses of energy. (For example, the combustion of fuel to provide steam to produce electricity and heat.) This conservation measure lowers energy consumption by matching the wasted energy of one consumer with the energy needs of another. Such combinations are usually difficult to identify because minimizing the cost for one customer usually requires degradation of the "quality" of the wasted energy below the required standards of others. As an example, consider the cooling of central power stations. Heat is rejected from central power stations at as low a temperature as is economically feasible—usually about 100°F—in order to maximize the thermal efficiency of the plant. Unfortunately, this temperature is too low for the majority of industrial applications. However, if the central power station could be matched to a customer who desired waste heat, at say 200°F, the central power station might consider raising its heat-rejection temperature, reducing its efficiency, but reducing its total cost by selling the waste energy to the industrial customer. This conservation measure has the potential for reducing both costs and total energy consumed. However, it requires careful energetic and economic analyses as well as a certain degree of diplomacy for success.

The preceding conservation measures involve providing the same energy services to customers but with reduced total energy consumption. Energy conservation can also be achieved by reducing the energy service. For example, the energy consumed to heat buildings could be reduced by setting thermostats to lower temperatures in the winter.

Evaluation of Energy Conservation Measures   Energy conservation measures usually involve more than just conserving energy. It is important to consider all aspects of a proposed energy conservation project and to weigh both the potential for conservation and other effects such as indirect costs, public opinion, and environmental effects. The following is a list of several items to be considered:

1. Conservation potential
   - The significance of the energy conservation relative to the total project
   - The significance of the energy conservation relative to local supply and demand for the fuels involved
   - The significance of the energy conservation relative to the relevant fuel resource bases
   - A comparison between the proposed project and other similar projects
   - A comparison of existing technology with advanced technology
   - A comparison of existing energy efficiency with theoretical maxima
   - The identification of conservation areas in which immediate substantial savings can be obtained with little deleterious effects
   - The identification of conservation areas in which existing technology can be readily applied to achieve substantial energy savings
   - The identification of technology areas in which research to expand the state of the art could produce some substantial energy savings.
2. Cost or benefits of implementation
   - Cost of administrating conservation measures
   - Cost of enforcing conservation measures
   - Cost of monitoring performance of conservation measures
   - Cost of new energy-saving equipment
   - A comparison between cost per unit energy consumed for this conservation measure and other alternatives
3. Indirect effects
   - Air pollution
   - Water pollution
   - Solid waste
   - Noise
   - Public opinion
   - Business expansion/employment
   - Convenience

Sample Energy Conservation Measures   There have been thousands of suggestions for energy conservation documented in many publications. In reviewing conservation measures discussed in the literature, it should be noted that many have never been put to practice because of cost or other constraints, and others concern only certain applications.

The interrelationships between the various energy uses in a large building are an excellent example. One almost universally recognized conservation measure is the lowering of thermostat settings in winter. The reasoning is that the difference between indoor and outdoor temperature determines heating load, and that minimizing this temperature difference also minimizes energy consumption. However, in some buildings the internal heat generation due to appliances, lighting, industrial processes, and human heat release is so great that cooling is required even when outdoor temperatures are lower than room temperature. Lowering the thermostat in this case will increase cooling load.

The following subsections list several sample conservation measures that are applicable to a variety of project types. They are arranged as follows:
   - Buildings
   - Transportation
   - Industrial processes

Building Conservation Measures.   Energy used in buildings generally means energy used for space heating and cooling, domestic water heating, lighting, and appliances. The three primary energy sources for buildings are heating oil, natural gas, and electricity. Natural gas is used for many heating applications, for air conditioning with the vapor-absorption cycle, and for decorative lighting. The short supply of natural gas, coupled with its high use in buildings, make energy conservation in buildings particularly important. The two primary means of conserving energy in buildings are (1) better design of new buildings and (2) improved operation and maintenance of existing structures. Retro-fitting older buildings to reduce energy consumption is usually a more costly alternative.

The possibilities of conserving energy through building design are numerous. In the past, minimum consumption of energy was secondary to aesthetic considerations. Now, however, building architects are under increasing pressure to design buildings for reduced energy consumption. Several recent reports have been written suggesting specific energy conservation measures which might be employed in design of new buildings and in the operation of existing buildings. Several states have also promulgated insulation standards and other regulations incorporating energy conservation into building design.

The following list contains a few suggestions for energy conservation through the design of new buildings. These suggestions are taken from Reference 40 and represent only about 20 percent of the total suggestions available there.

Site and building:

1. Cover exterior walls and/or roof with earth and planting to reduce heat transmission and solar gain.

2. Reduce paved areas and use grass or other vegetation to reduce outdoor temperature buildup.

3. Locate buildings on site to induce airflow effects for natural ventilation and cooling.

4. Locate buildings to minimize wind effects on exterior surfaces.

5. Construct building with minimum exposed surface area to minimize heat transmission for a given enclosed volume.

6. Utilize building configuration and wall arrangement (horizontal and vertical sloping walls) to provide self-shading and wind breaks.

7. Locate insulation for walls, roofs, and floors over garages at the exterior surface.

8. Construct exterior walls, roof, and floors with higher thermal mass with a goal of 100 lb/ft³.

9. To minimize heat gain in summer due to solar radiation, finish walls and roofs with a light-colored surface having a high emissivity.

10. To increase heat gain due to solar radiation on walls and roofs, use a dark-colored finish having a high absorptivity.

11. Reduce infiltration quantities by one or more of the following measures:
    - Reduce building height
    - Use impermeable exterior surface materials
    - Reduce crackage area around doors, windows, etc. to a minimum
    - Provide all external doors with weather stripping

12. Do not heat parking garages.

13. Consider the use of the insulation type which can be most efficiently applied to optimize the thermal resistance of the wall or roof; for example, some types of insulation are difficult to install without voids or shrinkage.

14. To reduce heat transfer due to windows, consider one or more of the following:
    - Use minimum ratio of window area to wall area.
    - Use double glazing.
    - Avoid window frames that form a thermal bridge.
    - Use operable thermal shutters which decrease the composite "U" value to 0.1.

15. To take advantage of natural daylight within the building and reduce electrical energy consumption, consider the following:
    - Increase window size but do not exceed the point where yearly energy consumption, due to heat gains and losses, exceeds the saving made by using natural light.
    - Locate windows high in wall to increase reflection from ceiling but reduce glare effect on occupants.
    - Control glare with translucent drapes operated by photo cells.

- Provide exterior shades that eliminate direct sunlight but reflect light into occupied spaces.

Ventilation and infiltration:

1. Where outdoor conditions are close to but less than indoor conditions for major periods of the year, and the air is clean and free from offensive odors, consider the use of natural ventilation when yearly trade-offs with other systems are favorable.

2. Provide controls to shut down all air systems at night and weekends except when used for economizer-cycle cooling.

Heating, ventilation, and air conditioning (HVAC):

1. Use outdoor air for sensible cooling whenever conditions permit and when recaptured heat cannot be stored.

2. In the summer when the outdoor air temperature at night is lower than indoor temperature, use full outdoor air ventilation to remove excess heat and pre-cool structure.

3. Design HVAC systems so that they do not heat and cool air simultaneously.

4. To reduce fan horsepower, consider the following:
   - Design duct systems for low pressure loss.
   - Use high-efficiency fans.
   - Use low-pressure loss filters concomitant with removable contaminant.
   - Use one common air coil for both heating and cooling.

5. Design piping systems for low pressure loss and select routes and locate equipment to give shortest pipe runs.

6. Consider chilled water storage systems to allow chillers to operate at night when condensing temperatures are lowest.

7. Extract waste heat from boiler flue gas by extending surface coils or heat pipes.

8. Consider the use of heat pumps both water/air and air/air if a continuing source of low-grade heat exists near the building, such as a lake, river, etc.

9. Consider the direct use of solar energy via a system of collectors for heating in winter and absorption cooling in summer.

10. If electric heating is contemplated, consider the use of heat pumps in place of direct resistance heating since by comparison they consume one-third of the energy per unit output.

11. Consider the use of spot heating and/or cooling in spaces having large volume and low occupancy.

12. Consider the use of a total energy system if the life cycle costs are favorable.

Lighting and power:

1. Use natural illumination in areas where effective when a net energy conservation gain is possible vis-á-vis heating and cooling loads.
   - Provide exterior reflectors at windows for more effective internal illumination.

2. Consider a selective lighting system in regard to the following:
   - Reduce the wattage required for each specific task by review of user needs and method of providing illumination.
   - Consider only the amount of illumination required for the specific task accounting for the duration and character and user performance required as per design criteria.
   - Group similar tasks together for optimum conservation of energy per floor.
   - Design switch circuits to permit turning off unused and unnecessary light

3. Consider the use of light colors for walls, floors, and ceilings to increase reflectance but avoid specular reflections.

4. Lower the ceilings or mounting height of luminaries to increase level of illumination with less wattage.

5. Use lamps with higher lumens per watt input such as:
   - One 8-ft fluorescent lamp versus two 4-ft lamps
   - One 4-ft fluorescent lamp versus two 2-ft lamps
   - U-tube lamps versus two individual lamps
   - Fluorescent lamps in place of all incandescent lamps except for very close task lighting, such as at a typewriter paper holder

6. Match motor sizes to equipment shaft power requirements and select to operate at the most efficient point.

7. Minimize power losses in distribution system by:

- Reducing length of cable runs
- Increasing conductor size within limits indicated by life cycle costing
- Use high-voltage distribution within the building

8. Match characteristics of electric motors to the characteristics of the driven machine.

9. In canteen kitchens, use gas for cooking rather than electricity.

10. Use conventional ovens rather than the self-cleaning type.

Transportation in major buildings:

1. Utilize a sloping site to accommodate entrances on multilevels to reduce elevator mileage.

2. Use elevators rather than escalators for vertical travel.

3. For high traffic densities through one or two floors, use staircases or ramps rather than escalators.

4. Reduce the number of elevators installed by scheduling their use for essential purposes only (for example, stops at every other floor would eliminate single-floor trips and enforce use of staircases).

Domestic hot and cold water:

1. To reduce the quantity of hot and cold water used, consider the following:
- Select kitchen equipment such as dishwashers that have minimum water requirements.
- Use a single system to meet handwashing needs in rest rooms.
- Use foot-operated, self-closing valves to control faucets.
- Select a water treatment system for cooling towers that allow high cycles of concentration (suggest target greater than 10:7) and reduces blowdown quantity.
- Schedule boiler blowdown on an "as needed" basis rather than a fixed timetable.
- Recycle wastewater for toilet flushing.

2. To reduce the yearly energy used to generate domestic hot water, consider the following:
- Reduce the generating and storage temperature to the minimum required for hand washing (suggested goal, 105°F).
- Avoid the use of straight electric heating for hot water; consider instead using a heat pump.
- Boost hot water temperature locally for kitchens, etc., rather than provide higher temperatures for the entire building.
- If boilers are used as primary heat source for domestic hot water, install a boiler to match the load rather than use an oversized heating boiler all summer. (Careful selection of modular heating boiler sizes could achieve the same end).
- Use gravity circulation for domestic hot water rather than pumps.

3. Domestic hot water generation at low temperatures is a fertile field for the use of waste energy. Consider meeting the hot water heating needs from the following sources:
- Waste heat from incinerators
- Rejected heat of compression from refrigeration units (both air conditioning and kitchen freezers and cold rooms).

4. Consider the use of solar water heaters using flat-plate collectors with heat pump boosters for winter.

Much can be done to conserve energy through improved operation and maintenance of buildings also. There are jobs for the professional, such as balancing an air conditioning system for a large building, and jobs for the layman, such as turning off lights when they are not required, lowering household thermostat settings in winter and raising them in summer, and keeping windows and doors to the outside closed while air conditioning and heating systems are operating.

The following list from Reference 8 outlines several energy conservation measures of these types:

1. Heat building to no more than 68°F when occupied.

2. Heat building to no more than 60°F when unoccupied.

3. Do not cool building when it is unoccupied.

4. Schedule morning startup in winter so that the building is at 63°F when occupants arrive and warms up to 68°F over the first hour.

TABLE 5.57 Actions to Reduce Energy Consumption and Their Institutional and Legal Considerations

Action group	Action	Regional energy reduction, %	Months to implement	Implementation cost	Implementing agency	Institutional and legal change required	Possibly new legislation	Initial public reaction	Enforcement
1. Measures to improve flow of high occupancy vehicles	Bus-actuated signals	0–0.5	6–12	L	L, S	None	No	+/–	No
	Bus-only lanes on city streets	0–2.0	2–6	L	L, S	None	No	+/–	Maybe
	Reserved freeway bus or bus/carpool lanes and ramps	1.0–3.0	2–24	L–H	L, S	None	No	+/–	Yes
	Bus priority regulations at intersections	0–0.5	3–9	L	L, S	None	Yes	+/–	Yes
2. Measures to improve total vehicular traffic flow	Improved signal systems	1.0–4.0	6–18	M	L, S	None	No	+	No
	One-way streets, reversible lanes, no on-street parking	1.0–4.0	6–12	M	L, S	None	No	+/–	Yes
	Eliminate unnecessary traffic control devices	0–2.0	3–6	L	L, S	None	No	+	No
	Widening intersection	0–1.0	6–12	M	L, S	None	No	+	No
	Driver advisory system	0–0.5	6–12	L–H	L, S	None–adapt	No	+	No
	Ramp metering, freeway surveillance, driver advisory display	0–1.0	6–18	M–H	L, S	None	No	+/–	Yes
	Staggered work hours	0	4–12	L	P, L, S	None–new	No	+/–	No
3. Measures to increase car and van occupancy	Carpool matching programs	3.0–6.0	2–6	L	P, L, S	Adapt	No	+/–	No
	Carpool public information	2.0–4.0	2–6	L	P, L, S	Adapt	No	+	No
	Carpool incentives	4.0–6.0	2–6	L–M	P, L, S	Adapt	No	+/–	Maybe
	Neighborhood ride sharing	0–1.0	3–24	L	P, L	None–new	No	+	No

TABLE 5.57 Actions to Reduce Energy Consumption and Their Institutional and Legal Considerations (Continued)

Action group	Action	Regional energy reduction, %	Months to implement	Implementation cost	Implementing agency	Institutional and legal change required	Possibly new legislation	Initial public reaction	Enforcement
4. Measures to increase transit patronage	Service improvements	1.0–3.0	3–18	M	P, L, S	None	No	+	No
	Fare reductions	4.0–6.0	2–12	M–H	L, S	None	Yes	+	No
	Traffic-related incentives	1.0–5.0	2–24	L–M	L, S	None	No	+/–	Maybe
	Park/ride with express bus service	0.5–2.5	18–24	M–H	L, S	Adapt	No	+	No
	Demand-responsive service	0–1.0	6–12	H	L, S	Adapt-new	Yes	+	No
5. Measures to encourage walk and bicycle modes	Pedestrian malls	0.5–2.5	6–12	M–H	L	Adapt	Yes	+	Maybe
	Second-level sidewalks	0–0.5	6–12	M	L	Adapt	No	+/–	No
	Bikeway system	0.5–2.0	6–12	L–M	L, S	Adapt	Yes	+	Maybe
	Bicycle storage facilities	0–1.0	2–4	L	L, S	Adapt	No	+	No
	Pedestrian-actuated signals	0–0.5	6–12	L	L, S	None	No	+/–	No
	Bicycle priority regulations at intersections	0–0.5	3–9	L	L, S	None	Yes	+/–	Yes
6. Measures to improve the efficiency of taxi service and goods movement	Improve efficiency of taxi service	0–2.0	3–18	M	P, L	None-adapt	Yes	+	Yes
	Improve efficiency of urban goods movement	0–1.5	6–18	H	P, L, S	Adapt-new	Yes	+	Yes
7. Measures to restrict traffic	Auto-free or traffic-limited zones	0.5–2.5	12–18	M–H	L	Adapt	Yes	+/–	Yes
	Limiting hours or location of travel	0–3.0	4–12	M–H	L, S	Adapt-new	Yes	–	Yes
	Limiting freeway usage	0–1.0	3–6	L–M	L, S	None-adapt	Yes	–	Yes

Category	Measure								
8. Transportation pricing measures	Bridges and highway tolls	1.0–5.0	12–24	L–M	L, S	None-new	Yes	−	No
	Congestion tolls and road cordon tolls	1.0–5.0	18–24	M–H	L, S	Adapt-new	Yes	−	Maybe
	Increased parking costs	0.5–3.0	3–12	M	L	Adapt-new	Yes	−	Maybe
	Fuel tax	2.0–6.0	2–6	L	L, S	Adapt	Yes	−	No
	Mileage tax	2.0–6.0	6–12	M	L, S	Adapt	Yes	−	Maybe
	Vehicle-related fees	2.0–10.0	6–12	M	S	Adapt	Yes	−	No
9. Measures to reduce the need to travel	Four-day work week	1.0–6.0	4–12	L	P, L, S	None-new	No	+/−	No
	Zoning	1.0–10.0	6–12	L	L, S	None-new	Yes	+/−	Maybe
	Home goods delivery	0–1.0	12–24	L	P, L	New	No	+/−	No
	Communications substitutes	0–1.0	18–24	L–H	P, L, S	None-new	No	+/−	No
10. Energy restriction measures	Gas rationing without transferable coupons	10.0–25.0	2–6	L–H	S, F	New	Yes	−	Yes
	Gas rationing with transferable coupons	10.0–25.0	2–6	L–H	S, F	New	Yes	−	Yes
	Restriction of quantity of sales on a geographic basis	5.0–20.0	0–6	L–M	P, L, S	New	Yes	−	Maybe
	Ban on Sunday and/or Saturday gas sales	2.0–10.0	1–6	L	P, L, S	New	Yes	−	Yes
	Reduced speed limits	0–2.0	1–6	L	L, S	Adapt	Yes	−	Yes

SYMBOLS:
Implementation cost: L = low, M = medium, H = high, within the low-cost constraint on type of actions considered
Implementing agency: P = private, L = local, S = state, F = federal
Initial public reaction: + = positive, − = negative, +/− = positive or negative, depending on group affected

TABLE 5.58 Actions to Reduce Energy Consumption and Their Indirect Socioeconomic Effects

Action group	Action	Regional energy reduction %	Socioeconomic				
			Travel time	Safety	Lifestyle change	Economic dislocation	Development opportunities
1. Measures to improve flow of high-occupancy vehicles	Bus-actuated signals	0–0.5	Decrease	Improve	NE	NE	NE
	Bus-only lanes on city streets	0–2.0	Decrease	Improve	Minor	NE–Minor	NE
	Reserved freeway bus or bus/carpool lanes and ramps	1.0–3.0	Decrease	Improve	Minor	NE	NE
	Bus priority regulations at intersections	0–0.5	Decrease	Improve	Minor	NE	NE
2. Measures to improve total vehicular traffic flow	Improved signal systems	1.0–4.0	Decrease	Improve	NE	NE	NE
	One-way streets, reversible lanes, no on-street parking	1.0–4.0	Decrease	Improve	NE–Minor	NE–Minor	NE
	Eliminate unnecessary traffic control devices	0–2.0	Decrease	Improve	NE	NE	NE
	Widening intersection	0–1.0	Decrease	Improve	NE	NE	NE
	Driver advisory system	0–0.5	Decrease	Improve	NE	NE	NE
	Ramp metering, freeway surveillance, driver advisory display	0–1.0	Decrease	Improve	NE	NE	NE
	Staggered work hours	0	Decrease	NE	Minor/major	Minor	Minor/major
3. Measures to increase car and van occupancy	Carpool matching programs	3.0–6.0	NE	NE	NE	NE	Major
	Carpool public information	2.0–4.0	NE	NE	NE	NE	Major
	Carpool incentives	4.0–6.0	NE	NE	NE	NE	Minor
	Neighborhood ride sharing	0–1.0	NE	NE	Minor	NE	NE
4. Measures to increase transit patronage	Service improvements	1.0–3.0	Decrease	Improve	NE	NE	Major
	Fare reductions	4.0–6.0	NE	NE	NE	NE	NE
	Traffic-related incentives	1.0–5.0	NE	NE	NE	NE	NE–Minor
	Park/ride with express bus service	0.5–2.5	Decrease	Improve	NE	NE	Major
	Demand-responsiveness service	0–1.0	Decrease	Improve	NE	NE	Major
5. Measures to encourage walk and bicycle modes	Pedestrian malls	0.5–2.5	Decrease	Improve	Minor	NE–Minor	Major
	Second-level sidewalks	0–0.5	Decrease	Improve	NE	NE	Major
	Bikeway system	0.5–2.0	Decrease	Improve	Minor	NE	Major

Category	Measure		Effect				
	Bicycle storage facilities	0–1.0	NE	Improve	NE	NE	Minor
	Pedestrian-actuated signals	0–0.5	Decrease	Improve	NE	NE	NE
	Bicycle priority regulations at intersections	0–0.5	Decrease	Improve	NE	NE	NE
6. Measures to improve the efficiency of taxi service and goods movement	Improve efficiency of taxi service	0–2.0	Decrease	NE	NE	NE	Minor
	Improve efficiency of urban goods movement	0–1.5	Decrease	NE	Minor	NE	Minor/major
7. Measures to restrict traffic	Auto-free or traffic-limited zones	0.5–2.5	Increase	Improve	Minor	NE-Minor	Major
	Limiting hours or location of travel	0–3.0	Increase	Improve	Minor/major	Minor/major	NE-Major
	Limiting freeway usage	0–1.0	Increase	Improve	Minor	NE	NE
8. Transportation pricing measures	Bridges and highway tolls	1.0–5.0	NE	NE	NE-Minor	NE-Minor	NE
	Congestion tolls and road cordon tolls	1.0–5.0	NE	NE	NE-Minor	NE-Minor	NE
	Increased parking costs	0.5–3.0	NE	NE	NE-Minor	Minor	NE
	Fuel tax	2.0–6.0	NE	NE	NE	NE-Minor	NE
	Mileage tax	2.0–6.0	NE	NE	NE	NE-Minor	NE
	Vehicle-related fees	2.0–10.0	NE	NE	NE	NE-Minor	NE
9. Measures to reduce the need to travel	Four-day work week	1.0–6.0	NE	NE	Major	Minor	Major
	Zoning	1.0–10.0	NE	NE	Major	Major	Major
	Home goods delivery	0–1.0	NE	NE	Minor	NE	Minor
	Communications substitutes	0–1.0	NE	NE	Minor	Minor	Minor/major
10. Energy restriction measures	Gas rationing without transferable coupons	10.0–25.0	NE	NE	Major	Minor/major	NE
	Gas rationing with transferable coupons	10.0–25.0	NE	NE	Major	Minor/major	NE
	Restriction of quantity of sales on a geographic basis	5.0–20.0	NE	NE	Major	Major	NE
	Ban on Sunday and/or Saturday gas sales	2.0–10.0	NE	NE	Major	Minor/major	NE
	Reduced speed limits	0–2.0	Increase	Improve	Minor	NE	NE

SYMBOL: NE = no effect

TABLE 5.59 Actions to Reduce Energy Consumption and Their Indirect Environmental Effects

Action group	Action	Regional energy reduction, %	Environmental effects			
			Air pollution	Noise	Congestion	Land use patterns
1. Measures to improve flow of high-occupancy vehicles	Bus-actuated signals	0–0.5	Decrease	Decrease	Decrease	NE
	Bus-only lanes on city streets	0–2.0	Decrease	Decrease	Decrease	NE–Minor
	Reserved freeway bus or bus/carpool lanes and ramps	1.0–3.0	Decrease	Decrease	Decrease	NE–Minor
	Bus priority regulations at intersections	0–0.5	Decrease	Decrease	Decrease	NE
2. Measures to improve total vehicular traffic flow	Improved signal systems	1.0–4.0	Decrease	Decrease	Decrease	NE
	One-way streets, reversible lanes, no on-street parking	1.0–4.0	Decrease	Decrease	Decrease	NE–Minor
	Eliminate unnecessary traffic control devices	0–2.0	Decrease	Decrease	Decrease	NE
	Widening intersection	0–1.0	Decrease	Decrease	Decrease	NE–Minor
	Driver advisory system	0–0.5	Decrease	Decrease	Decrease	NE
	Ramp metering, freeway surveillance, driver advisory display	0–1.0	Decrease	Decrease	Decrease	NE
	Staggered work hours	0	Decrease	Decrease	Decrease	NE
3. Measures to increase car and van occupancy	Carpool matching programs	3.0–6.0	Decrease	Decrease	Decrease	NE
	Carpool public information	2.0–4.0	Decrease	Decrease	Decrease	NE
	Carpool incentives	4.0–6.0	Decrease	Decrease	Decrease	NE
	Neighborhood ride sharing	0–1.0	Decrease	Decrease	Decrease	NE
4. Measures to increase transit patronage	Service improvements	1.0–3.0	Decrease	Decrease	Decrease	NE
	Fare reductions	4.0–6.0	Decrease	Decrease	NE	NE
	Traffic-related incentives	1.0–5.0	Decrease	Decrease	Decrease	NE
	Park/ride with express bus service	0.5–2.5	Decrease	Decrease	Decrease	Minor
	Demand-responsiveness service	0–1.0	Decrease	Decrease	Decrease	NE
5. Measures to encourage walk and bicycle modes	Pedestrian malls	0.5–2.5	Decrease	Decrease	Decrease	Minor/major
	Second-level sidewalks	0–0.5	Decrease	Decrease	Decrease	Minor
	Bikeway system	0.5–2.0	Decrease	Decrease	Decrease	Minor

Measure	Range				
Bicycle storage facilities	0–1.0	Decrease	NE	NE	NE
Pedestrian-actuated signals	0–0.5	NE	NE	Decrease	NE
Bicycle priority regulations at intersections	0–0.5	NE	NE	Decrease	NE
6. Measures to improve the efficiency of taxi service and goods movement					
Improve efficiency of taxi service	0–2.0	Decrease	Decrease	Decrease	NE
Improve efficiency of urban goods movement	0–1.5	Decrease	Decrease	Decrease	Minor
7. Measures to restrict traffic					
Auto-free or traffic limited zones	0.5–2.5	Decrease	Decrease	Decrease	Minor/major
Limiting hours or location of travel	0–3.0	Decrease	Decrease	Decrease	Minor/major
Limiting freeway usage	0–1.0	Decrease	Decrease	Decrease	Minor
8. Transportation pricing measures					
Bridges and highway tolls	1.0–5.0	Decrease	Decrease	Decrease	NE
Congestion tolls and road cordon tolls	1.0–5.0	Decrease	Decrease	Decrease	NE
Increased parking costs	0.5–3.0	Decrease	Decrease	Decrease	NE
Fuel tax	2.0–6.0	Decrease	Decrease	Decrease	NE
Mileage tax	2.0–6.0	Decrease	Decrease	Decrease	NE
Vehicle-related fees	2.0–10.0	Decrease	Decrease	Decrease	NE
9. Measures to reduce the need to travel					
Four-day work week	1.0–6.0	Decrease	Increase/Decrease	Increase/Decrease	NE–Minor
Zoning	1.0–10.0	Decrease	Decrease	Decrease	Major
Home goods delivery	0–1.0	Decrease	Decrease	Increase/Decrease	NE
Communications substitutes	0–1.0	Decrease	Decrease	Decrease	Major
10. Energy restriction measures					
Gas rationing without transferable coupons	10.0–25.0	Decrease	Decrease	Decrease	Minor/major
Gas rationing with transferable coupons	10.0–25.0	Decrease	Decrease	Decrease	Minor/major
Restriction of quantity of sales on a geographic basis	5.0–20.0	Decrease	Decrease	Decrease	Major
Ban on Sunday and/or Saturday gas sales	2.0–10.0	Decrease	Decrease	Decrease	Minor
Reduced speed limits	0–2.0	Decrease	Decrease	NE	NE

SYMBOL: NE = no effect

5. Limit pre-cooling startup in morning to give building a temperature of 5°F less than outdoor temperature or 80°F, whichever is highest.

6. Turn off heating or cooling 30 minutes before the end of the occupied period.

7. Allow humidity to vary naturally in the building between 20 percent RH and 65 percent RH. Only add or remove moisture when building conditions exceed those limits.

8. Use cool night air to flush building and remove heat from structure, provided energy to run fans is less than that required to run chillers.

9. Turn off lights that are not needed.

10. Draw drapes over windows or close thermal shutters when daylight is not available and when building is unoccupied.

11. Maintain equipment to retain "as new" efficiency.

12. Clean air filters on a regular maintenance schedule.

13. Clean lighting fixtures and change lamps on a regular maintenance schedule to maintain desired lighting levels.

Transportation Conservation Measures.   High reliance on the automobile for transportation, coupled with the relatively low efficiency of the automobile as an energy conversion device, implies a great untapped potential for energy conservation. References 38, 39, 44, and 45 contain excellent descriptions of energy conservation measures dealing with transportation. The reader is referred particularly to Reference 45, a report which contains a thorough treatment of several energy conservation measures dealing with automobiles. Tables 5.57 to 5.59, reproduced from this report, summarize 46 specific conservation measures in 10 categories. Each table covers the same conservation measures but examines a different set of considerations. Each measure is described in more detail in Reference 45.

Industry.   "Industry" here refers to industrial uses of energy other than for transportation and buildings as described above. Typical industrial energy uses include electrical power to run machine tools, computers, etc., and various fuels for heating appliances (in steel making, for example).

The diversity of energy uses in industry makes it difficult to cite specific conservation measures applicable to more than a limited number of industries. Some more general energy conservation measures applicable to almost all industries include:

▪ Recover flue gas heat to heat water for use in industrial processes or domestic water heating.

▪ Locate energy-intensive processes (furnaces, etc.) outside to reduce air conditioning loads.

▪ Tune combustion units for better fuel economy.

▪ Plan factories to reduce energy required for material handling.

▪ Insulate hot-process piping and equipment.

▪ Use automatic controls of process heating and scheduling.

▪ Install heat recuperators to save process heat normally wasted to heat raw materials.

REFERENCES

1. Acton, J. P., M. N. Gralbard, and D. J. Weimschectt: "Electricity Conservation Measures in the Commercial Sector: The Los Angeles Experience," Rand Corp., Santa Monica, Calif., Report No. R-1592-FEA, September 1974.

2. A. D. Little, Inc. (Cambridge, Mass.): "Energy Policy Issues for the United States During the Seventies," July 1971. Presented at the National Energy Forum, Washington, D.C., September 1971.

3. A. D. Little, Inc. (Cambridge, Mass.) "Residential and Commercial Energy Consumption," Draft Final Report, Council on Environmental Quality, August 7, 1974.

4. Altman, Manfred (University of Pennsylvania): "Conservation and Better Utilization of Electric Power By Means of Thermal Energy Storage and Solar Heating," N.S.F. Contract No. NSF GI 27976, NTIS PB-210359, October 1971.

5. American Gas Association, "Ecube Input Instruction Manual," New York, N.Y., December 1973.

6. American Gas Association (Arlington, Va.): "Ecube Application Manual," 1973.

7. American Society for Testing Materials. Standard D-388.

8. ASHRAE: "Energy Conservation in New Building Design," ASHRAE Standard 90P, 1974.

9. ASHRAE: "A Field Study of Load Profiles and Energy Requirements for Heating and Cooling Systems Selected for Field Validation Test," ASHRAE Symposium Bulletin No. 72-2, 1972.

10. ASHRAE: "Handbook and Product Directory: Handbook of Fundamentals (1972)"; 1972 Equip-

ment Volume; 1973 Systems Volume; 1974 Applications Volume, Published by ASHRAE, New York, NY.

11. Association of Home Appliance Manufacturers (AHAM), Chicago, Ill.: "1975 Directory of Certified Refrigerators and Freezers," September 1975.

12. A. T. Kearney, Inc.: "Industrial Energy Study of the Motor Vehicles Industry," Federal Energy Administration Report No. FEA-EI-1671, NTIS No. PB 236 694, July 1974.

13. Austin, A. L., B. Rubin, G. C. Werth (Lawrence Livermore Laboratory): "Energy: Uses, Sources, Issues." Prepared for U.S. Atomic Energy Commission under Contract No. W-7405-ENG-48, C.C.C. Report No. UCRL-51221.

14. Ayres and Hayakawa Energy Management: "State Energy Conservation Study Computer Program Volume I." Prepared for Calif. Dept. of Housing and Community Development, Division of Codes and Standards, April 1975.

15. Baumeister, T.: "Marks' Standard Handbook for Mechanical Engineers," 7th ed., McGraw-Hill, New York, N.Y., 1969.

16. Berg, Charles A.: "Conservation Via Effective Use of Energy at the Point of Consumption," National Bureau of Standards, NTIS No. COM-79-10479, April 1973.

17. Berg, Charles A. (National Science Foundation): "Energy Conservation Through Effective Utilization," NSF Report No. MBSIR 73-102, NTIS No. COM-73-10856, February 1973.

18. Bernstein, H. M., and P. M. McCarthy: "Analysis of Correct Uses, Incentives, Constraints, and Institutional Factors Related to Retrofit for Energy Conservation in the Commercial Sector," Federal Energy Administration Interim Report No. HIT-595, December 1974.

19. Bullard, C. W., and R. A. Herendeen (University of Illinois at Urbana-Champaign): "Energy Use in the Commercial and Industrial Sectors of the U.S. Economy–1963," National Science Foundation Report No. NSF-RANN-'74-057, NTIS Report No. PB 235 487, Nov. 21, 1973.

20. California Administrative Code: "Regulations Establishing Energy Conservation Standards for Residential Buildings," Title 20, Section 1401, Title 24, Sections T20-1401 - T20-1406, February 1977.

21. California Board of Equalization: "Taxable Sales of Gasoline, Diesel and LPG," 1969–1974.

22. California Dept. of Housing and Commercial Development, Division of Codes and Standards: "Energy Design Manual for Residential Buildings," April 1976.

23. California Dept. of Transportation: "Gasoline Sales by Counties," 1974.

24. California Energy Resources Conservation and Development Commission: "Electricity Forecasting and Planning Report, Vol. II," October 1976, Report EAD-5B:01.

25. California Energy Resources, Conservation and Development Commission: "Quarterly Fuel and Energy Summary," vol. I, no. 3, third quarter 1975.

26. California Office of the State Comptroller: "Non-Taxable Gasoline Sales 1969–1974."

27. California Resources Agency: "Guidelines for EIR's, Section IX: Contents of an EIR."

28. Crump, Lulie H.: "Fuels and Energy Data: United States by States and Census Divisions, 1973." Bureau of Mines, Div. of Interfuels Studies, Document No. BUMINES-IC-8722, NTIS No. PB 262 363/7WE, December 1976.

29. Curran, H. and R. Anderson (Hittman Associates, Inc.): "Residential Energy Consumption Multi-Family Housing Data Acquisition," Department of Housing and Urban Development Report No. HUD-HAI-3, NTIS No. PB-216 440, October 1972.

30. Department of Interior, "United States Energy—A Summary Review," 1972.

31. Department of Interior, Bureau of Mines. "Mineral Industry Surveys, Fuel Oil Shipments 1964–1973."

32. Department of Interior, Bureau of Mines: "Mineral Industry Surveys, L.P.G. Shipments, 1964–1973."

33. Department of Interior, Bureau of Mines: "Mineral Industry Surveys, Natural Gas Annual, 1964–1973."

34. Department of Interior, Bureau of Mines: "Mineral Industry Surveys, Petroleum Statement Annual, 1964–1973."

35. Department of Interior, Bureau of Mines: "Minerals Yearbook, 1960–1972."

36. Department of Interior, Bureau of Mines: "Annual U.S. Energy Use Drops Again," news release, Apr. 5, 1976.

37. Department of Transportation: "Analysis of Requirements and Constraints on the Transport of Energy Materials, Vol. I and II," FEA Project Independence Blueprint, Final Task Force Report, U.S. Govt. Printing Office Stock No. 4118-00025, November 1974.

38. Department of Transportation, Transportation Systems Center: "Research and Development Opportunities for Improved Transportation Usage," NTIS No. PB-220-612, September 1972.

39. Department of Transportation, Urban Mass Transit Administration: "Guidelines to Reduce Energy Consumption Through Transportation Actions," May 1974.

40. Dubin-Mindell-Bloome Associates: "Energy Conservation Design Guidelines for Office Buildings," General Services Administration Report, January 1974.

41. Electric Energy Association: "The Engineering Cost Section of Access: Alternate Choice Comparison for Energy System Selection," New York, N.Y.

42. Energy and Environmental Analysis, Inc.: "Energy Management in Manufacturing 1967–1990, Vol. I, Summary Report," Draft Final Report, Contract No. EQ4AC024, Council on Environmental Quality, Aug. 7, 1974.
43. Environment Information Center, Inc.: *The Energy Directory*, Library of Congress No. 74-79869, Sept. 1974.
44. Executive Office of the President, Office of Emergency Preparedness: "The Potential for Energy Conservation: A Staff Study," U.S. Government Printing Office, Stock No. 4102-00009, October 1972.
45. Exxon Research and Engineering Company (prepared for EPA): *Feasibility Study of Alternative Fuels for Automotive Transportation*, June 1974, NTIS No. PR-235-581, EPA Report #EPA-460/3-74-009a.
46. Federal Council of Science and Technology: "Total Energy Systems, Urban Energy Systems, Residential Energy Consumption," NTIS PB-221 374, October 1972.
47. Federal Energy Administration: "Final Assessment of the Environmental Impacts of the State Energy Conservation Program," FEA Report No. FEA/D-76/363, NTIS Report No. PB 256044, April 1976.
48. Federal Energy Administration; "National Energy Outlook," Report No. FEA-N-75/713, February 1976.
49. Federal Energy Administration: "A Pilot Project in Homeowner Energy Conservation," FEA Report, Oct. 31, 1974.
50. Federal Energy Administration: "Project Independence Blueprint Summary and Final Report," November 1974.
51. Federal Energy Office, Office of Energy Conservation: "Federal Energy Reduction Program, Fiscal Year 1974," Second Quarterly Report (October–December 1973) March 1974.
52. Federal Power Commission: "Hydroelectric Power Resources of the United States, Developed and Undeveloped," 1972.
53. Federal Power Commission: "Statistics of Privately Owned Electric Utilities in the United States: 1960–1972."
54. Federal Power Commission: "Statistics of Publicly Owned Electric Utilities in the United States: 1960–1972."
55. Federal Power Commission: "World Power Data," 1969.
56. Federal Power Commission, Office of the Chief Engineer: "Staff Report on Guidelines for Energy Conservation for Immediate Implementation: Small Business and Light Industries," NTIS No. PB 231840, Feb. 1974.
57. Felton, L. A. and L. R. Glocksman (Massachusetts Institute of Technology—Energy Laboratory): "Energy Conservation: A Case Study for a Large Manufacturing Plant," Report No. MIT-EL 74-010, NTIS No. PB 239302, May 1974.
58. Folsom, Blair A.: "Alternate Sources of Fuel and Electrical Power Generating Technologies," Ultrasystems, Inc., Irvine, Calif., 1976.
59. Folsom, B. A., C. W. Courtney, and M. P. Heap: "The Effects of LPG Composition and Combustor Characteristics on Fuel NO_x Formation," ASME Paper No. 79-GT-185, Presented at the ASME Gas Turbine Conference, March 12–15, 1979, San Diego, Calif.
60. Gatts, R. R., Massey, R. G., Robertson, J. C. (National Bureau of Standards): "Energy Conservation Program Guide for Industry and Commerce (EPIC)," NBS Handbook 115, Library of Congress No. 74-600153, U.S. Govt. Printing Office SD Catalog No. C13.11.115, Sept. 1974.
61. Germain, L. S.: "Energy Conservation," Lawrence Livermore Laboratory Report No. UCRL-51488, Nov. 20, 1973.
62. Goldsmith, M.: "Geothermal Resources in Calif., Potentials and Problems," Caltech Environmental Quality Lab., Report No. 5, December 1971.
63. Goldstein, D. B. and A. M. Rosenfeld: "Conservation and Peak Power—Cost and Demand," Lawrence Berkeley Laboratory, Berkeley, Calif., Report LBL-4438, 1925.
64. Harris, Jeffrey P. (California Energy Resources Conservation and Development Commission): "Interim Conservation Plan Report: Opportunities for Energy Conservation in California," Report No. CON-10A-01, Staff Draft, 1976.
65. Hirst, E. and J. C. Moyers: "Efficiency of Energy Use in the United States," *Science,* **179,** Mar. 30, 1973, pp. 1299–1304.
66. Hittman Associates, Inc.: "Residential Energy Consumption Single Family Housing, Final Report," Dept. of Housing and Urban Development Report No. HUD-HA 102, Publication No. HUD-PDR-29-2, U.S. Government Printing Office Stock No. 2300-00258, March 1973.
67. Hoffman, John D. (National Bureau of Standards): "Energy Conservation at the NBS Laboratories," NBS Report No. NBS IR 74-539, NTIS No. COM-74-11574, July 1974.
68. Hugh Carter Engineering Corp.: "Nonresidential Energy Conservation Standards Title 24 Economic and Energy Effectiveness Study." Prepared for California Energy Resources Conservation and Development Commission, RFP No. 75-CON-1, Nov. 5, 1975.
69. Illuminating Engineering Society (New York, N.Y.): *IES Lighting Handbook,* 5th ed., 1972.
70. Irving, J. H. and H. Bernstein: "The Impact of Area-Wide Personal Rapid Transit on Energy Consumption," in Caltech Series on Energy Consumption in Private Transportation, Department of Transportation, NTIS No. PB-235 348, June 30, 1974.

71. Jacobs, M., et al. (National Bureau of Standards): "Making the Most of Your Energy Dollars in Home Heating and Cooling," National Bureau of Standards, NTIS, No. COM-75-11029, August 1975.

72. Limaye, Sharke, et al., Mathematica, Inc., and Peat Marwick Mitchell and Co.: "Comprehensive Evaluation of Energy Conservation Measures, Final Report," Environmental Protection Agency Report No. EPA-230/1-75-003, NTIS PB 250 824, March 1975.

73. Lokmanhekim, M. (American Society of Heating, Refrigerating and Air Conditioning Engineers): "Procedures for Determining Heating and Cooling Loads for Computerized Energy Calculations: Algorithms for Building Heat Transfer Subroutines," ASHRAE, 1971.

74. Marks, Craig: "Which Way to Achieve Better Fuel Economy," in Caltech Seminar Series on Energy Consumption in Private Transportation, Department of Transportation, NTIS No. PB-235 348, June 30, 1974.

75. Meckler Associates (Long Beach, Calif.): "Comparison of Manual Energy Calculation Methods for Nonresidential Buildings," prepared for California Advisory Committee on Energy Conservation in New Nonresidential Structures, Feb. 17, 1973.

76. Meckler Associates (Los Angeles, Calif.): "A Manual Computation Method for Estimating Energy Requirements of Proposed Nonresidential Buildings." Prepared for California Advisory Committee on Energy Conservation in New Nonresidential Buildings, Jan. 10, 1975.

77. Miller, Harold G., et al (Department of Transportation and Environmental Protection Agency): "Study of Potential for Motor Vehicle Fuel Economy Improvement: Technology Panel Report," Department of Transportation Report No. DOT-TSC-OST-75-13, NTIS No. PB 241774, January 1975.

78. Mutch, J. J. (Rand Corp.): "Residential Water Heating: Fuel Conservation, Economics and Public Policy," Rand Corp. Report No. R-1498-NSF, May 1974.

79. National Bureau of Standards: "Eleven Ways to Reduce Energy Consumption and Increase Comfort in Household Cooling," U.S. Government Printing Office Stock No. 0303-0876, 1976.

80. National Bureau of Standards, "Technical Options for Energy Conservation in Buildings." Prepared for Joint Emergency Workshop on Energy Conservation in Buildings, Sponsored by National Conference of States on Building Codes and Standards, June 19, 1973.

81. National Petroleum Council, "Potential for Energy Conservation in the United States: 1974–1978, Industrial," Library of Congress No. 74-79107, Sept. 10, 1974.

82. Public Law 91-190: "National Environmental Policy Act (NEPA) of 1960," Section 102(2)(C), Jan. 1, 1970.

83. Resource Planning Associates, Inc.: "Energy Management Case Histories," Federal Energy Administration Report No. FEA/D-75/335, NTIS Report No. PB 246763, October 1975.

84. Ross F. Meriwether and Associates, Inc., San Antonio, Tex.: "Input Instruction Manual for the Energy System Analysis Series," 1974.

85. Salter, R. G., R. L. Petruschell, and K. A. Wolf (Rand Corp.): "Energy Conservation in Nonresidential Buildings," Rand Corp., Santa Monica, Calif., Report No. R-1623-NSF, October 1976.

86. Scego, G. C. (Inter Technology, Inc.): "The U.S. Energy Problem, Vol. I: Summary," National Science Foundation Report No. NSF-RANN 71-1-1, NTIS No. PB 207 517, November 1971.

87. Scego, G. C. (Inter Technology, Inc.): "The U.S. Energy Problem, Vol. II: Part A," National Science Foundation Report No. NSF-RANN 71-1-2, NTIS No. PB 207 518, November 1971.

88. Scego, G. C. (Inter Technology, Inc.): "The U.S. Energy Problem, Vol. II: Part B," National Science Foundation Report No. NSF-RANN 71-1-3, NTIS No. PB-207 519, November 1971.

89. Smith, J. W. and J. Jenkins (Wisconsin University): "The Household Energy Game," National Oceanic and Atmospheric Administration Report No. NOAA-75020306, NTIS No. COM-75-10304, Dec. 1974.

90. Snell, Jack E., Paul R. Achenbach, and Stephen R. Peterson: "Energy Conservation in New Housing Design," *Science*, **192**, p. 1305, June 25, 1976.

91. Southern California Association of Governments (SCAG): "Growth Forecast Selection," 1974.

92. Southern California Edison Co. (SCE): "Financial and Statistical Report," 1973.

93. Southern California Edison Co. (SCE): "Report to the Calif. Public Utilities Commission on Loads and Resources," 1975.

94. Stoeker, M. F., "Proposed Procedures for Simulating the Performance of Components and Systems for Energy Calculations," ASHRAE Bulletin, 1974.

95. Task Group on Energy Requirements, ASHRAE, "Progress Report II," ASHRAE Symposium, June 1969.

96. The Trane Company (LaCrosse, Wis.): "Output Interpretation Guide," 1973.

97. The Trane Company (LaCrosse, Wis.): "Trace/McAuto User Manual," 1975.

98. The Trane Company (LaCrosse, Wis.): "Trace Program Input," 1974.

99. Transportation System Center, "Research and Development Opportunities for Improved Transportation Energy Usage," NTIS No. PB-220 612, September 1972.

100. Ultrasystems, Inc., Irvine, Calif.: "Energy Supply and Demand in San Bernardino County," Final Report to San Bernardino County, Calif., Joint Utilities Management Plan, 1974.

101. University of Oklahoma, Science and Public Policy Program: "Energy Alternatives: A Comparative Analysis." Prepared for CEO, ERDA, EPA, FEA, FPC, Dept. of Int. and NSF. Available

from U.S. Government Printing Office, Stock No. 041-011-00025-4, Catalog No. PREX 14.2:EN2, May 1975.

102. Westinghouse Environmental Systems Department (Pittsburgh, Penn.): "Environmental Guidance Manual," Federal Energy Administration, October 1975.

103. Wheeler, J., M. Graubarp, and J. P. Acton: "How Business in Los Angeles Cut Energy Use by 20 Percent," U.S. Government Printing Office Stock No. 041-018-00042, January 1975.

Chapter 6

Water Quality Impact Analysis

DAVID YORK AND JAMES SPEAKMAN

While transportation of large quantities of water dates back to as early as 97 A.D., concern with water quality is a relatively new concept only extending over the past 150 years. The earliest documented problems of water quality degradation resulted from the use of storm drains for conveying human excrement to nearby waterways to relieve the intolerable conditions created by the privies and cesspolls which were saturating urban slums (Refs. 1,2).

The earliest noticeable effects on water quality were aesthetic; streams receiving sewage began to emit noxious odors (Ref. 1). Eventually, the role of water in transmitting disease was discovered. In London in 1854, it was proven that victims of cholera all drank from the same water source, the "Broad Street pump." This observation represents the earliest documented evidence of adverse water supply (Ref. 3). In 1892, an epidemic of cholera caused 8600 deaths in Hamburg, Germany. This death rate was nearly six times greater than that in Altona, Germany, which used the same water supply source, the Elbe River, but treated its water supply by slow sand filtration (Ref. 4).

During the nineteenth century, increased crowding and disease in United States cities resulted in the development of sanitary sewerage systems (Ref. 2). In addition, advanced technology produced industrial processes which generated a variety of effluents, most of which were discharged without treatment into nearby watercourses.

As awareness of the degradation of the nation's waterways has grown, the state-of-the-art of water pollution control has advanced. Scientific and technological advancements have led to the discovery of additional sources of water quality degradation, including mining, agriculture, forestry, construction, residential development, and stream channel modifications. Water quality impact assessments must be performed, and new methods of water pollution control must be devised to deal with the numerous sources of water quality degradation.

WATER QUALITY CRITERIA AND STANDARDS

Before conducting a water quality impact assessment, all applicable water quality criteria and standards must be known. "Water quality criteria," as distinguished from standards, are defined as the levels of specific concentrations of constituents which are expected, if not exceeded, to assure the suitability of water for specific uses (Ref. 5).

Since the early 1900s, substantial research has been devoted to the establishment of water quality criteria. In 1952, the state of California (Ref. 6) published the first summarization of criteria for eight major uses of water. In 1963, the 1952 Water Quality Criteria was expanded and published by the Resources Agency of California, State Water Quality Board (Ref. 7). In that landmark document, which contained 3827 references, criteria were identified for a multitude of constituents as they related to domestic water supply, industrial water supply, agricultural water supply (irrigation), stock and wildlife watering, fish and other aquatic and marine life, shellfish culture, swimming and bathing, boating, and even power and navigation.

In 1968, the National Technical Advisory Committee (NTAC) report to the Secretary of the Interior was published (Ref. 8). This report recommended concentrations of constituents which would insure the preservation of water for five specified uses: domestic water supply, recreation, fish and wildlife, agricultural uses, and industrial uses.

The latest report on water quality criteria (Ref. 5) departs from the identification of criteria for specific uses and arranges the criteria alphabetically. The criteria recommended in this report are designed to achieve the goals of Public Law (PL) 92-500; that is, to provide for the protection and propagation of fish and other aquatic life and for recreation in (and on) the water.

Both McKee and Wolf (Ref. 7) and the NTAC report (Ref. 8) are excellent guides for predicting the effects of a proposed action on specific downstream water uses. The *Quality Criteria for Water* was tailored to protect the water for use by aquatic life and for recreation and domestic supply, since those uses represent the highest beneficial water uses achievable (Ref. 5). That report reflects the most recent research investigations and is continuously revised as new data become available.

"Water quality standards" are legal regulations established by the states, limiting the concentrations of various constituents in water. Stream quality standards apply to ambient waterways, and effluent standards apply to discharges of liquid effluents into those waterways. Since water quality standards are continually revised, the most recent standards for waterways under consideration should be obtained from the state(s) having jurisdiction over the initiation of water quality assessment. The states which have U.S. Environmental Protection Agency (USEPA) approved water quality standards, and the documents containing those standards, are reported in the *Federal Register* (Ref. 9). Each annual reference issue of the *Water and Sewage Works Journal* (Ref. 10), publishes a directory of state and territorial water pollution control agencies, which provides the names and addresses of agencies from which water quality standards may be obtained.

For projects involving effluent discharges, state effluent standards and the effluent guidelines developed by the USEPA for the National Pollutant Discharge Elimination System (NPDES) apply. The state effluent standards are usually applicable to any discharge, while the EPA effluent guidelines are specific to an industry or type of public treatment facility. Industrial effluent guidelines have been developed for specific Standard Industrial Classification (SIC) codes and may be obtained from the Effluent Guidelines Division of the USEPA.

The Safe Drinking Water Act (PL 93-523) substantially broadened the spectrum of constituents to be considered in water quality assessment, especially in streams currently serving as sources of public water supply. Subsequent to that act, the USEPA published Interim Primary Drinking Water Regulations (Ref. 11), which specified maximum allowable concentrations of constituents, including many organic chemicals. Since those regulations are in a state of evolution, the most current regulations should be obtained from the *Federal Register*.

OVERVIEW OF WATER QUALITY IMPACT ANALYSIS

The activities that must be performed by the water quality assessment team are basically those that must be followed in any impact assessment process, as shown schematically in Figure 6.1. The general procedure includes the following steps:

1. Perform a preliminary review of the existing environment and proposed project
2. Select environmental indicators to be used for describing the environment and gauging the effects of the project

3. Describe the existing environment by providing quantitative descriptions of each indicator, using existing data sources

4. Conduct field sampling programs to complete the description of the environmental setting

5. Make predictions of the effects of the proposed project on the environment (impact assessment)

6. Propose modifications which could minimize adverse impacts resulting from the project

7. Prepare the appropriate sections dealing with water quality for the environmental impact statement or report.

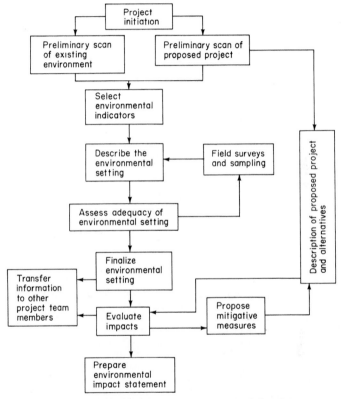

Figure 6.1 Environmental impact assessment—work flow diagram.

The activities just described and those depicted in Figure 6.1 are interrelated. If the environmental setting is not judged to be sufficiently complete, other sources of information should be investigated, or field measurements should be made to complete the setting. If the impact analysis reveals that additional environmental indicators are required to fully describe the character and magnitude of the impacts, these indicators should be developed and described. The evaluation of the impacts may reveal a means of minimizing adverse impacts. As a result, modification of the proposed project may be warranted, which would necessitate re-evaluation of impacts. This highlights the fact that the impact assessment should be an integral part of the planning process and not simply an evaluation step added onto the end of the project to satisfy the requirements of the NEPA. Incorporation of the impact assessment into the planning process can minimize adverse impacts as well as public objections, while increasing the compatibility of a project with the environment.

The water quality assessment group is a member of a multidisciplinary impact assessment team. For this reason, close coordination must be maintained between the other project groups, particularly the biological and socioeconomic groups. Water quality and water quantity have a significant influence on other areas of the environment. Much of the information generated by the water quality group will be used in the assessment of biological and socioeconomic indicators. Similarly, input from other disciplines is needed for the completion of the water quality assessment.

ENVIRONMENTAL SETTING

The basic approach to conducting an assessment of water quality impacts is shown schematically in Figure 6.1. The earliest major phase of the process is the description of the environmental setting: an inventory of existing water quality, hydrologic resources, and conditions influencing water quality and water resources. This phase of the project involves the selection of appropriate environmental indicators and descriptions of the environment through literature reviews and field investigations. This phase also includes a description of aspects of the proposed project and alternative courses of action which could influence water quality or quantity.

This section focuses on the development of the environmental setting. The selection of environmental indicators is discussed, followed by a discussion of information sources for the description of the setting. The final topic addressed concerns water quality field surveys. Field surveys may play a significant role in the development of the environmental setting, particularly for a major project.

Environmental Indicators

At the onset of the assessment, the water quality team must scan the environment in the area of the proposed project and review the characteristics of the project. The object of this preliminary review is to determine possible water quality impacts which might result from the project and to identify sensitive or critical environmental areas. Surface and groundwater resources should be identified and their general water quality described (such as severely polluted, occasional violations of stream standards, clean stream, etc.). Water requirements and wastewater discharges, along with construction requirements for the proposed project, should be defined. Possible changes in water quality and quantity resulting from the project are identified and described in qualitative terms.

The next task in detailing the environmental setting is the selection of environmental indicators. The environmental indicators selected should be described as quantitatively as possible during preparation of the environmental setting. During the impact evaluation phase, the proposed project is superimposed on the existing environmental setting, and predictions of the new values for each environmental indicator are made.

The selection of environmental indicators is extremely important since this selection significantly defines the level of detail to be developed during the environmental impact assessment. Selection of a multitude of extremely detailed indicators may serve to cloud the effort and bury the focus of the impact evaluation under a blanket of extraneous detail. Selection of only a few general indicators may minimize the potential for quantification and may be inadequate to fully describe the setting and anticipated project impacts.

Table 6.1 contains a list of possible environmental indicators in the following general categories: geophysics, hydrology, water quality, water systems, and wastewater systems. This is not a definitive list of indicators to be included in all impact assessments. Rather, the list contains indicators which should be considered for inclusion in any impact assessment. The project team must take the responsibility for selecting appropriate indicators for each individual project and geographical location. Emphasis should be placed on this reference to "project team," because it is unlikely that any one individual will have the breadth of knowledge needed to fully address all five general categories of indicators listed in Table 6.1. Assistance from a geologist/soils scientist, hydrologist or water resources engineer, and an environmental engineer may be required.

The selection of environmental indicators should be based on the scope and nature of the proposed project, as well as on the nature of the local or regional environment. A preliminary scan of the environment and review of the proposed project will be valuable to this selection. For example, a large project which might result in major impacts, such as the location of a major industrial facility, requires consideration of a large number of relatively detailed indicators. As the scope of the project decreases and as the potential for

TABLE 6.1   Environmental Indicators

Category	Sub-category	Indicators
Geophysical	Geology	Bedrock type Bedrock characteristics Depth to bedrock
	Soils	Soil type Soil characteristics Depth to water table
	Topography	Watershed description Watershed map Drainage areas Slope Relief
	Erosion/sedimentation	Locate erosion problems Erodability of soils Locate sedimentation problems Stream bed loads
Hydrology	Surface water	Inventory water sources Inventory water withdrawals Water budget Lake water surface elevations Surface area Lake stratification Depth of flow Flow velocity Discharge (average, low, peak, seasonal variation) Flood and drought records (include flood frequency analysis) Describe flood control facilities Stream order Reservoirs (purposes, operating schedule)
	Groundwater	Salt water intrusion Permeability of aquifers Porosity of aquifers Depth to groundwater Yields Seasonal variations Long-term trends Recharge areas Recharge rates Inventory withdrawals Inventory deep-well discharges
	Meteorology	Temperature (daily and seasonal variation, high, low, mean) Wind (speed, direction, wind rose) Precipitation (seasonal variations, extremes, storm frequency analysis) Snow (monthly distribution, extremes) Frost (earliest, latest) Humidity (daily and seasonal variations) Dew point (daily and seasonal variations) Solar radiation (daily and seasonal variations)
Water quality	Surface water	Classification of stream Stream standards Temperature pH

TABLE 6.1   Environmental Indicators   *(Continued)*

Category	Sub-category	Indicators
	Surface water (cont.)	Conductivity Turbidity Total dissolved solids Total suspended solids Color BOD (5-day, 20°C)* BOD—ultimate* COD† TOC‡ Dissolved oxygen Hardness Alkalinity Acidity Nitrate Nitrite Ammonia Total Kjeldahl nitrogen Organic nitrogen Phosphate Ortho-phosphate Organic phosphorus Sulfates Chloride Fluoride Iron Manganese Magnesium Potassium Sodium Calcium Silica Mercury Phenol Total coliforms Fecal coliforms Sodium adsorption ratio (SAR) Pesticides Radioactivity Surfactants Heavy metals Trace organics Carcinogens
	Groundwater	(Same indicators as for Surface Water)
Water systems	Water use	Flow (daily and seasonal variation) Residential water use Industrial water use Agricultural water use Commercial water use Municipal water use Metering systems Water importation Water diversion
	Water treatment facilities	Intake water quality (See water quality indicators above)

Category	Sub-category	Indicators
		Describe intake Describe plant Design capacity Current demand (time variation) Chemical additions Energy requirements Sludge type and quantity Sludge disposition Product water quality (See water quality indicators above) Operational difficulties
	Distribution system	Size of lines Age and conditon of lines Capacity of lines Current flows (daily and seasonal variations) Pressure Storage requirements and capacity
Wastewater systems	Collection system	Sewer sizes Sewer age and condition Capacity Current flows (daily and seasonal variations) Problems (odor, sludge, etc.) Infiltration/inflow analysis Stormwater collection (separate and combined sewers)
	Treatment system	Describe systems Locate facilities Age and condition of plants Design capacity NPDES effluent limitations§ Raw waste characteristics (See water quality indicators above) Effluent characteristics (See water quality indicators above) Flows and loads (average and time variation) Describe sludge handling systems Sludge (type, quantity, moisture content, disposition) Outfalls Operational difficulties (odor, insects, poor effluent, etc.)

*Biochemical oxygen demand
†Chemical oxygen demand
‡Total organic carbon
§NPDES–National Pollutant Discharge Elimination System, administered by the U.S. Environmental Protection Agency

relatively detailed indicators. As the scope of the project decreases and as the potential for significant adverse impacts declines, the number and detail of indicators usually can be reduced.

Another factor influencing indicator selection is the environment itself. Indicators should be chosen which will enable the assessment team to adequately describe the environment. Sensitive or unique environments must be described in great detail, particularly if the proposed action could have an impact upon such environments. If a proposed project is to have only surface wastewater discharges and no interaction with the groundwater system, and groundwater is relatively unimportant in the study area, the groundwater indicators would not be selected for detailed description. However, if an industrial facility having major wastewater discharge was proposed for location on a natural, pristine estuary, surface water quality indicators would receive detailed attention.

Sources of Information

Most of the description of the environmental setting will be derived through a search of existing sources of information. In some cases, these existing sources may be completely adequate for relatively small projects with little potential for major impacts. Major projects may require extensive field surveys to adequately define the setting. This particular section addresses existing sources of information, and a subsequent section discusses field surveys.

First, it is imperative that key members of the assessment team visit the study area. Enough emphasis cannot be placed on this point. It is virtually impossible to rationally describe the environment of an area without having seen it. The site visit should be carefully documented with photographs. A record should be kept during the field visit of each photograph taken, noting location and description of subject, time of day, date, and direction faced. This information should be transferred to the back of each print, since unlabeled prints are often worthless. Photographs are relatively inexpensive and can serve to familiarize members of the assessment team with the nature of the study area.

The next step begins with an investigation of general sources of information. Local, county, or regional planning agencies may have a wealth of information, particularly if a comprehensive plan has been prepared for the study area. Planning agencies may have information in various elements of a comprehensive plan or inventory regarding meteorology, hydrology, water quality, and water and wastewater systems. While this information may not be of sufficient detail for the environmental impact assessment, it represents an excellent starting point. These reports will often contain lists of references leading to more detailed information.

Previous environmental impact statements for projects in the general study area can provide significant information regarding water resources and water quality as well as much of the information needed for the subject environmental setting. For example, impact statements for major power plants, which may be several volumes in length, are important sources of detailed information.

An excellent source of information is a nearby college or university. Individual departments should be contacted directly for information concerning soils, geology, hydrology, and water quality. If the university is also the site of the state's water resources institute, hydrologic and water quality information may be readily available.

With regard to geophysical information, topographic maps obtainable from the U.S. Geological Survey (USGS) can serve as base maps, showing both natural and constructed features. Watersheds and drainage areas can be directly delineated on these maps. The USGS also serves as a data source for geologic information. The U.S. Soil Conservation Service (SCS) and county agents are valuable sources of data concerning soils. Erosion and sedimentation data may be available from the USGS and SCS. University programs in geology, soil science, earth sciences, water resources, or civil engineering may also furnish data pertaining to the study area.

The USGS usually can provide information regarding surface water and groundwater hydrology. University departments of water resources engineering or civil engineering may also have information regarding the local area. State water resources agencies or boards may maintain hydrologic records. Previous water quality or water resources studies in the area should be reviewed. Studies completed under requirements of Public Law 92-500, including 201 facilities plans, 208 areawide plans, and 303 basin plans, often contain valuable water quality data. The U.S. Army Corps of Engineers and local or regional flood control districts are sources of information regarding flood frequencies and flood control structures. The National Weather Service is the prime source of meteorologic/climatic information.

The state environmental control agency can provide printouts of historical water quality data contained in the EPA's STORET data base.[1] The state agency may also maintain additional historical water quality data of its own. University departments of civil or

[1]A national computerized water quality base administered by the U.S. Environmental Protection Agency. Contact the state environmental control agency or the U.S. Environmental Protection Agency regional office for detailed information.

environmental engineering may have conducted water quality surveys in the study area. In addition, USGS collects data on a large number of parameters at its water quality stations, including metals and pesticides.

The local or regional water department can provide information regarding existing water use, water treatment plants, the distribution system, and quality of raw and treated water. The city engineer's office may also furnish this information. Previous water resources studies, particularly comprehensive water supply studies, should be reviewed. Again, local university departments in civil or environmental engineering should be consulted.

Information regarding wastewater flows and loads, collection systems, treatment facilities, and wastewater characteristics (raw and treated) can usually be obtained from the local or regional wastewater control department. Effluent data filed to support an NPDES permit will be available through the state environmental control agency or the EPA regional office. Previous water quality studies, wastewater management studies, 201 facilities plans, and 208 areawide plans should be consulted. Environmental engineering departments of local universities may also furnish detailed information. In addition, the U.S. Army Corps of Engineers has developed wastewater management plans for many metropolitan areas as part of their urban studies program.

Assessment of the Adequacy of the Setting

The site visit, literature review, and interviews with agency representatives are used to complete the description of the environmental setting by describing each selected environmental indicator as quantitatively as possible. The setting can be illustrated through a series of tables or matrices presenting the values or descriptions assigned to each indicator. When the description of the existing environmental setting is completed, the water resources/water quality assessment team must assess its overall quality. The following types of questions should be addressed:

1. Is the physical condition of the environment accurately described?
2. Are sensitive and unique environmental features described in sufficient detail?
3. Is each indicator described in sufficient detail?

In addition, the assessment team must look ahead to the evaluation of impacts phase of the environmental assessment. This requires that the following questions be addressed:

1. Is the setting described in sufficient detail to allow for the prediction of impacts of the proposed project on the environment?
2. Are the proposed project and alternatives described in sufficient detail to permit prediction of impacts?

These questions are particularly critical if a modeling study is to be performed to predict water quality impacts. Sufficient data must be available to calibrate and verify the models.

If the answer to any of these and other related questions is "no," the environmental setting and/or description of the project will need to be expanded and refined. This may require a review of additional existing data sources. If additional sources are not available, field surveys must be performed. The following section addresses water quality field surveys, an important feature of many impact assessments.

Field Surveys

Water quality surveys can provide significant input for the environment setting. This is particularly true when there is not enough historical data to fully describe existing water quality. In addition, water quality surveys can serve as valuable baseline data which can be compared to conditions after the proposed project is operational. The success or failure of water quality surveys is, in large measure, dependent upon the amount of planning performed prior to taking the initial samples. A well-conceived, carefully planned sampling program has a high probability of success. Another important consideration is the quality and conscientiousness of individuals performing both the field work and the analytical work, as well as the degree of training they have received.

The overall plan for the water quality survey must contain the following elements:

1. Detailed plan for sample collection
2. Provision for laboratory analysis
3. Description of the methods to be used for data reduction and manipulation, including statistical analysis.

This plan must also address: (1) location of sampling stations; (2) parameters to be analyzed; (3) time schedules, including time of day, time of year, and frequency; (4) method of data collection; and (5) sample handling prior to analysis. These key topics are addressed in the following sections. For a more detailed discussion of water quality surveys, the reader is referred to Kittrell (Ref. 12).

Location of Sampling Points   Sampling points should be located to provide an accurate description of existing water quality. In addition, the sampling points should be selected to maximize the ease of sampling. This is facilitated by locating sampling points on good base maps, such as county highway maps or USGS topographic maps. For large navigable streams, lakes, and estuaries, navigation charts prepared by the U.S. Coast Guard or U.S. Army Corps of Engineers are useful.

The actual location of sampling points is primarily dependent on the physical situation. For example, it can be assumed that a rapidly moving, narrow, shallow stream will be completely mixed, both laterally and vertically. As a result, only one sampling point need be employed at each location along the stream. For wide rivers, canals, lakes, or estuaries, it may be necessary to collect multiple samples at each cross section along the stream. If the water body is stratified, at least two samples should be taken, one at the midpoint of the epilimnion (above the thermocline or chemocline) and one at the midpoint of the hypolimnion (below the thermocline or chemocline). For streams less than 1000 ft wide but more than 100 ft wide, three equally spaced sampling points are preferred across the cross section. For wide rivers or lakes (width greater than 1000 ft), at least five equally spaced sampling points are suggested.

In determining the location of sampling points, it is important to recognize that two general types of water quality parameters are normally considered: conservative and nonconservative materials. Concentrations of conservative materials, such as chlorides and total dissolved solids, do not change as the water moves downstream, whereas concentrations of nonconservative parameters, such as biochemical oxygen demand (BOD) and temperature, change with time in response to biological, chemical, or physical processes.

At this point, a basic decision must be made as to which water quality parameters to investigate. This decision depends primarily on the type of project proposed and the anticipated impacts of the project. In general, one should select parameters on which the proposed project may have an impact and which are expected to be of the most concern in the study area streams. If the project involves thermal discharges, the temperature must be measured along with parameters which are somewhat temperature dependent, such as dissolved oxygen and BOD. Projects involving the discharge of organic wastes require an investigation of the oxygen resources of the stream, including BOD, dissolved oxygen, and temperature. If sanitary sewage is to be discharged, total coliforms or fecal coliforms should be assessed. If there will be discharges of dissolved solids, measurements should be taken of total dissolved solids, conductivity, alkalinity, acidity, pH, and specific elements or compounds. If suspended solids are discharged or if increased erosion is anticipated, suspended solids should be measured and sediment sampling should be considered. If nutrients are discharged, the analysis must consider nitrogen and phosphorus forms as well as the BOD-dissolved oxygen system.

The nonconservative BOD-dissolved oxygen system is frequently of concern in impact assessment because organic wastes exert an oxygen demand which can deplete the dissolved oxygen concentration in the stream. This depletion and the subsequent re-establishment of dissolved oxygen are shown in Figure 6.2. Multiple cross sections along the stream should be selected for sampling. At least one point along the stream should be selected which will not be influenced by the proposed project (point A), and four downstream points should also be selected. These points could be:

    1. In the zone of degradation, immediately below proposed project or outfall (point B)

    2. At the sag point within the zone of active decomposition (point C)

    3. Within the zone of recovery (point D)

    4. Below the influence of the project (point E).

Additional points may be added for more detail if financial and/or time constraints permit. In general, the locations of the points indicated on Figure 6.2 are not known for a specific project area. However, a review of existing water quality data and some preliminary modeling, using techniques presented later in this chapter, can be employed to formulate rough predictions of the curve shown in Figure 6.2 and to locate sampling stations.

If the proposed project only involves the discharge of conservative materials, a minimum of two sampling cross sections are required—one above the discharge and one below the discharge. The sample below the discharge should be located below the mixing zone to insure mixing of waste across the cross section. Additional samples will provide further detail. In wide rivers, lakes, or estuaries, cross sections should be located so as to define the three-dimensional extent of the plume. This is particularly necessary in studies regarding new thermal discharges. In general, it is wise to collect water quality samples at the same points biological samples and sediment samples are collected.

The second factor to be considered in the location of sampling stations is the ease of obtaining samples. An "ideal" sampling point accessible only by a 10-mile hike through dense underbrush has little utility. The sampling crew must have ready access to all stations. For water bodies having minimum depths of 3 ft, access to all stations can be gained by use of a small boat. Bridges are some of the most popular locations for sampling, offering easy access and a means of readily taking samples across the cross section. However, bridges are not always located exactly where one prefers to sample.

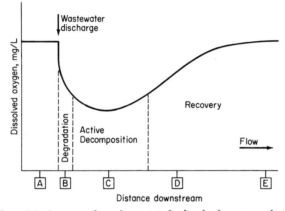

Figure 6.2 Location of sampling points for dissolved oxygen analysis.

Sample Frequency Three basic time element decisions must be made:
1. What time of day to sample
2. What season of the year to sample
3. How often to sample.

Normally, only one sample is taken at each sampling point on any one day. This is valid, however, only as long as large diurnal variations do not occur. In streams with significant algal activity, diurnal variations in dissolved oxygen of several milligrams per liter may be observed. If this is the situation, it is suggested that at least one day of round-the-clock sampling be conducted. At a minimum, two sets of samples should be obtained: one in the afternoon and one shortly before dawn.

The diurnal pattern becomes increasingly obvious as more samples are taken during the day at a single point. Four samples, at six-hour intervals during the day, will normally provide sufficient diurnal data. Other alternatives for determining diurnal patterns are using automatic samplers, which take a series of discrete samples during the day, or using continuous recording instruments (particularly for temperature, dissolved oxygen, and pH).

Sampling over a one-year period is desirable to provide good baseline data. This is normally done for large projects, such as proposed power plants. Sampling is conducted on a routine basis, such as once a week, if a one-year sampling program is conducted. An alternative is to conduct four intensive sampling programs, one during each season, five to ten days in length. For smaller projects, budgetary constraints may rule out year-round sampling. In this case, a five- to ten-day sampling program could be implemented. If the BOD-dissolved oxygen system is of primary concern, the sampling should occur during the warm weather and low-flow conditions of late summer or early autumn.

Rainfall must also be considered in establishing the sampling program. Routine baseline monitoring should be suspended during periods of rainfall. If only a short-term sampling program is conducted, then it is imperative that the program not be biased by a storm. If it is decided that the effects of rainfall will be measured, then a special field sampling program should be designed to determine the patterns of discharge and pollutant concentration resulting from the storm event.

Method of Data Collection   Two types or samples are commonly employed. They are "grab samples," which represent conditions existing at the particular location and instant the sample was obtained, and "composite samples," which are a series of samples over a period of time. Composites have the advantage of being more reflective of average conditions, while grab samples require less time to obtain. In a water body in which flows and concentrations do not change rapidly, grab samples are preferred. In a stream that is subject to rapid variation, such as those which are used by industrial discharges using batch operations, composites can be used to better define average conditions. Multiple grab samples may also be used to give a better indication of temporal variations.

Another method of sampling is to use continuous monitoring equipment with recorders and automatic composite samplers. Automatic composite samplers, which take one large sample over a period of hours or a series of discrete smaller samples, are available.

Several methods of measuring velocity and discharge in flowing streams are available.

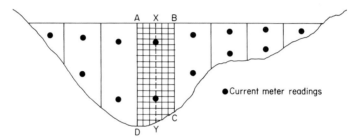

Figure 6.3   Discharge measurements with a current meter.

A simplistic approach is the use of surface floats. In this method, the time t required for a float to travel a known distance d is observed, and the average velocity $\bar{V}$ is obtained by:

$$\bar{V} = \frac{d}{1.2t}$$

The factor 1.2 accounts for the fact that surface velocities are normally about 1.2 times the mean velocity. If the cross-sectional area A is measured, the discharge Q is given by:

$$Q = \bar{V}A$$

This method is useful in small, shallow streams.

The most popular method is to use current meters. Either a propeller-type or rotating-cup (Price) meter can be used. This method requires the following steps:

1. Partition the stream into a series of vertical sections (see Figure 6.3).
2. Measure the depth at the midpoint of each section.
3. Obtain velocity measurements at 0.2 and 0.8 of the depth on the centerline of each section. In shallow sections, measure velocity at 0.6 of the depth.

Average velocity $\bar{V}_i$ in the ith section is estimated by:

$$\bar{V}_i = \frac{V_{0.2} + V_{0.8}}{2} \text{ for deep sections}$$
$$\bar{V}_i = V_{0.6} \text{ for shallow sections}$$

The total stream discharge Q is then estimated by:

$$Q = \sum_{i=1}^{n} \bar{V}_i A_i$$

where n = number of sections and A_i = cross-sectional area of ith section.

For a further discussion of this method, the reader is referred to Linsley, Kohler, and Paulhus (Ref. 13).

In lakes, estuaries, and in some stratified river situations, sensitive directional-indicating current meters must be used. This is particularly true if the flow in the bottom layer is in a different direction from the flow in the surface layer.

Tracers such as dyes or radionuclides can also be used. Two basic tracer methods are available. In the first method, a slug of dye is spiked into a stream and the concentration of the dye is monitored at a known distance d downstream. The time of travel t for the peak concentration to pass the downstream point is observed, and the average velocity $\overline{V}$ is calculated as:

$$\overline{V} = \frac{d}{t}$$

The second method involves introducing a tracer into a stream at a known rate (concentration and flow). Concentrations of the tracer upstream from the input spot and downstream below the mixing zone are monitored. When the system comes to steady state, the discharge is given by:

$$Q_R = \frac{Q_I(C_I - C_D)}{C_D - C_U}$$

where Q_R = stream discharge
       Q_I = flow rate of feeder stream
       C_I = salt concentration in feeder stream
       C_D = salt concentration at the downstream station
       C_U = salt concentration at the upstream station

Brine solutions, such as lithium chloride, are commonly used as tracers in this type of study. Kulin and Compton (Ref. 14) discuss methods for measuring velocity and discharge in streams and sewers.

Instrumentation and Equipment   A wide range of field sampling instrumentation is available. For example, reliable field instruments are available for the measurement of dissolved oxygen, temperature, conductivity, and pH. Instruments are also available with probes which can be lowered to a particular depth and the measurement made *in situ*.

If discrete samples which must be obtained from a certain depth are required, either a Van Dorn or Kemmerer sampler can be used. The Van Dorn offers the advantage of lower impedence of water into the sampler, but is difficult to use in swift streams due to its large size. Polyethylene or polyvinyl chloride samplers are generally preferred. Samples can also be pumped from preselected depths using a peristaltic pump or a vacuum system.

Several types of equipment are available for obtaining samples of bottom sediments. A wide variety of core samplers are on the market, ranging from simple hand-operated units for shallow water sampling to sophisticated devices for deep water and marine applications. Many of these core samplers can also be used to obtain soil samples.

Three types of dredges are used to take bottom samples. The Ekman dredge can be used in relatively shallow water and for soft sediments. The Ponar grab dredge can be used in deeper water, including marine applications, and in hard sediments such as clay and gravel. The Petersen grab dredge is commonly used in deep lake and marine environs and for sampling very hard sediments. The Ponar offers the widest range of application.

Sample Handling and Laboratory Analyses   The specific goal in sampling is to collect samples that are fully descriptive of the overall water quality. The field crew would generally collect as small a sample as possible and handle the sample in such a manner that the characteristics of the sample do not change prior to analysis. Both *Standard Methods for the Examination of Water and Wastewater* (Ref. 15) and *Methods for Chemical Analysis of Water and Wastes* (Ref. 16) present detailed discussions of sample-handling requirements. Some general rules for sample handling are as follows:

1. Rinse the sample bottle two or three times before filling.
2. Avoid collecting surface debris or bottom sediments in the sample.
3. Clearly label each sample bottle with waterproof ink, and record the details of each sample in a field notebook.
4. Measure temperature and pH in the field.
5. Be sure the sampler and sample bottles are clean.
6. Take a sufficient volume of sample; normally 2 liters is sufficient (See Table 6.2).
7. Do not use the same sample for chemical, bacteriological, and microscopic analysis.

8. Minimize the time delay from sampling to analysis.

9. Preserve samples as needed prior to analysis (See Table 6.2).

Immediate analysis is optimal. Refrigeration at 4°C is normally sufficient to maintain samples for less than 24 hours. Chemical preservatives should be used when required and when they will not interfere with the analysis. Table 6.2 presents recommended preservation techniques, while *Standard Methods* (Ref. 15) presents a more detailed discussion of sample preservation.

It is imperative that care be taken to insure accurate laboratory analysis of samples. This means that quality control measures must be implemented to insure valid results. The reader is referred to both *Standard Methods* (Ref. 15) and the USEPA's manual for laboratory analysis (Ref. 16) for detailed descriptions of analytical procedures for individual determinations. Three basic means of quality control are:

1. Obtain duplicate samples in the field for analysis.

2. Split the field samples into two or three smaller samples for analysis.

3. Use artificially spiked samples.

Duplicate samples are fed to the laboratory without the knowledge of lab technicians. The supervisor of the laboratory takes responsibility for comparing results of the duplicated analyses and may require re-analysis if wide discrepancies exist. "Spiked samples" must be used in which a known amount of the material to be analyzed is introduced into an actual sample. The object of spiking samples is to determine the amount of the material being recovered by the analytical procedures employed. If recovery is low, investigations for faulty techniques or interfering substances must be implemented. If severe interferences exist, the analytical procedure may have to be modified or a new procedure substituted.

Remember that careful planning and organization are the primary determinants of the success of the sampling program. Carefully train all field and laboratory personnel. The study manager must be personally involved in the field investigations. Quality control on all laboratory analyses is essential.

Each sampling crew chief should maintain a detailed field notebook. If temperature studies are conducted, weather data must be obtained since meteorological conditions are prime determinants of water temperatures. At each sampling station, estimates of wind direction and speed, cloud cover, and measurements of air temperature should be made. In addition, complete weather information should be obtained from nearby weather stations during the sampling program. If water temperature is of concern and solar-radiation data is not available from local weather stations, solar radiation measurements should be made using a pyrheliometer.

Final Comments About the Environmental Setting

The environmental setting is generally presented as a series of tables showing the numerical value of each environmental indicator with supporting descriptive text. The units used for each indicator should be clearly stated. When showing concentrations of the various forms of nitrogen or phosphorus, one should indicate whether the result is expressed in terms of the element (i.e., as N) or as the entire group (i.e., as NO_3). Be sure that diurnal and seasonal variations are shown. Water quality data can be summarized by season or by month. Means and extremes (highs and lows) should be shown. The standard deviation can also be used to present data variability. For parameters such as temperature, which vary with the season, "average" is meaningless, and, therefore, seasonal data is required.

Furthermore, because water quality standards and criteria are the guideposts for judging relative water quality, comparisons of existing water quality with standards and criteria should be presented. Violations of standards should be highlighted wherever they occur.

Finally, this section has dealt with the development of the environmental setting—a description of the environment as it exists before the proposed project is implemented. The selection of environmental indicators was discussed, and possible sources of information were outlined. A significant portion of this section consisted of a discussion of field surveys, reflecting the importance of this subject. In many instances, the results of field surveys may constitute the bulk of the description of the setting. In addition, field surveys may represent valuable baseline data that can serve as a basis for evaluation of postconstruction conditions.

TABLE 6.2   Preservation Techniques

Parameter	Container*	Sample volume (mL)	Preservative	Allowable holding time
Acidity	P,G	100	Refrigerate, 4°C	24 hr
Alkalinity	P,G	100	Refrigerate, 4°C	24 hr
BOD	P,G	1,000	Refrigerate, 4°C	6 hr
COD	P,G	50	H_2SO_4 to pH <2	7 days
Chloride	P,G	50	None	7 days
Color	P,G	50	Refrigerate, 4°C	24 hr
Cyanides	P,G	500	Refrigerate, 4°C; NaOH to pH 12	24 hr
Dissolved oxygen:				
Probe	G	300	Determine on site	No holding
Winkler	G	300	Fix on site	4–8 hr
Hardness	P,G	100	Refrigerate, 4°C	7 days
Metals	P,G	200	HNO_3, to pH <2	6 mo
Nitrogen:				
Ammonia	P,G	500	Refrigerate, 4°C; H_2SO_4 to pH <2	24 hr
Kjeldahl	P,G	500	Refrigerate, 4°C; H_2SO_4 to pH <2	7 days
Nitrate	P,G	100	Refrigerate, 4°C; H_2SO_4 to pH <2	24 hr
Nitrite	P,G	100	Refrigerate, 4°C	24 hr
Organic carbon	P,G	50	Refrigerate, 4°C; H_2SO_4 to pH <2	24 hr
pH	P,G	100	Determine on site	No holding
Phenol	G	500	Refrigerate, 4°C; H_2PO_4 to pH <4; 1.0 g. $CaSO_4$/liter	24 hr
Phosphate	P,G	50	Refrigerate, 4°C	24 hr
Residue	P,G	100	Refrigerate, 4°C	7 days
Specific conductance	P,G	100	Refrigerate, 4°C	24 hr
Sulfate	P,G	50	Refrigerate, 4°C	7 days
Temperature	P,G	1,000	Determine on site	No holding
Turbidity	P,G	100	Refrigerate, 4°C	7 days
Coliform bacteria	G	500	Refrigerate, 4°C	36 hr
Pesticides	G	1,000	Refrigerate, 4°C	24 hr
Radioactivity	P,G	1,000	None required	24 hr

*P indicates plastic; G indicates glass.

Adapted from: *Standard Methods for the Examination of Water and Wastewater*, 14th Edition, 1975 (Ref. 15). Used with permission of the American Health Association.

Methods for Chemical Analysis of Water and Wastes, EPA 625-/6-74-003a, 1976 (Ref. 16).

"Gaging and Sampling Industrial Wastewaters," by Rabosky and Koraido in *Chemical Engineering*, *80* (1): 111–120, Jan. 1973 (Ref. 17). Used with permission of McGraw-Hill Publishing Company.

IMPACT ANALYSIS

Water quality impacts are changes in the values of water quality indicators resulting from a proposed action or project. If a proposed project is expected to reduce average summer dissolved oxygen concentrations in a stream from 6.2 mg/L to about 4.5 mg/L, then this represents an adverse water quality impact. The magnitude of the impact is also of importance. For example, a change in dissolved oxygen from 6.2 to 6.0 mg/L represents a relatively minor impact. However, if dissolved oxygen were to drop from 6.0 to 4.5 mg/L, the impact would be considered very significant. Impacts should also be evaluated on the basis of water quality criteria, effluent standards, and stream standards. A decrease in dissolved oxygen of 2.0 mg/L, from 6.2 to 4.2 mg/L, is more significant if the applicable stream standard is 5.0 mg/L rather than 3.0 mg/L.

Impacts can be classified as primary or secondary. Primary impacts result directly from the proposed action. The location of a major industry has a primary impact on water quality due to the discharge of industrial wastewater. If the proposed industry is expected to promote additional urban and residential development, secondary impacts will result from increased domestic wastewater flows from the additional residential areas.

The objective of water quality impact assessment is the evaluation of the nature and magnitude of changes in water quality indicators as a result of a proposed project. This evaluation should be as quantitative as possible.

It must be remembered that the objective of water quality impact assessment is not only to define the changes in water quality and quantity, but also to provide input to other areas of the comprehensive environmental impact assessment. Changes in water quality may result in significant effects on aquatic biota, on the costs of water treatment, and on other important factors.

There are three basic methods of evaluating impacts on water quality, namely: (1) professional judgment based on relevant experience, (2) extrapolation from similar proj- ects that have been implemented, and (3) modeling. All three methods are valuable and are used to some extent in impact analysis.

While there is no substitute for sound professional judgment tempered with relevant experience, judgment can, at best, result in qualitative assessment of project impacts. Extrapolation from similar projects located in similar environments can be valuable since the action will tend to result in similar impacts. However, results of extrapolation are qualitative or, at best, semi-quantitative. It is extremely doubtful that the identical project has already been implemented in an identical environment, thus precluding direct quantitative extrapolation. The use of models, on the other hand, offers the advantage of a greater degree of quantification, as indicated in the following discussion.

Modeling

Modeling is the process by which major ecosystem interactions are identified and described. The model is a representation of the environment that can be used to simulate environmental conditions. Once a model has been constructed, the effects of the proposed project and alternatives can be predicted.

Three main types of modeling systems are normally used. The first is the development of a mathematical model using a series of equations to represent interactions in the environment. Simple mathematical models can be solved by using hand calculations, while more detailed models require the use of computers. The second type is a physical model where an actual scale model of the study area is constructed. The U.S. Army Corps of Engineers' Waterways Experiment Station at Vicksburg, Mississippi and the Tennes- see Valley Authority make extensive use of physical models. These are commonly used when the physical situation is too complex to be analyzed using mathematical models. Analog models using mechanical or electrical principles can also be used. The most common type of model is the mathematical model.

Models are classified by the number of dimensions modeled and by the type of model employed. These can be described as follows:

1. *One-dimensional Models*—Flowing streams are frequently modeled as being one-dimensional. The only direction modeled is the horizontal, along the direction of flow. Concentrations of materials in the stream are considered uniform laterally, across the width of the stream, and vertically, throughout the entire depth.

2. *Two-dimensional Models*—Wide rivers, such as the Mississippi River, may not be

uniform in concentration across the entire width at any one point. For this reason, a two-dimensional model may be needed to adequately describe water quality. In this case, both the horizontal and lateral dimensions would be described, with vertical uniformity assumed. For deep, narrow rivers, lakes, or estuaries, the horizontal and vertical dimensions may be modeled, while uniformity in the lateral is assumed. This type of model enables description of thermally stratified conditions.

3. *Three-dimensional Models*—An actual stream exhibits variation in all three dimensions. Some models describing variations in the horizontal, lateral, and vertical directions have been developed. Three-dimensional models require large amounts of computer time, computer storage, and specification of many input parameters, many of which may be difficult to measure with few published values available. As the number of dimensions modeled increases, the potential accuracy increases, but the cost of developing and running the model also increases.

A distinction is made between dynamic and steady-state models. "Dynamic models" provide information on variations in water quality as a function of both distance and time. Dynamic models could be used to model variations in dissolved oxygen during the day resulting from changes in solar radiation, temperature, and algal activity. As the time increment used in the model is reduced, the degree of accuracy increases, but the cost of running the model also increases.

"Steady-state models" assume that no variations exist over time. Therefore, water quality at a particular point at 2:00 P.M. will be identical to water quality at the same point at 2:00 A.M. In streams exhibiting significant algal activity or wide variation in stream flow, the assumption of steady state may not be applicable. Steady-state models are normally less expensive to run and easier to construct than dynamic models.

The parameters which are to be modeled must be determined at the beginning of a modeling study along with the type(s) of models to be used. The parameters considered should reflect the list of environmental indicators selected earlier in the study. The type of model selected will reflect the degree of precision required and the resources (computer, manpower, and dollars) available. Remember, as the number of dimensions and parameters increases and the time increment used decreases, the potential accuracy increases. However, the data requirements, computation requirements, and costs also increase dramatically. The project team assessing water quality impacts must make trade-offs between detail and precision versus data and resource availability.

Modeling Procedures   Five basic phases are contained within a modeling study. They are: model development, calibration, verification, model application, and sensitivity analysis.

Model Development.   Numerous "canned" programs are available to the modeller. A decision must be made as to whether these existing programs can be used or whether a new model should be developed to better fit the situation. Before this decision can be made, a thorough review of historical data should be conducted. This type of information should have been compiled and reviewed during the description of the environmental setting. Available data may enable the modeller to evaluate the utility of one-dimensional models versus two- or three-dimensional models for description of the study area. Based on this review, an assessment of available information and resources (time and money) and an assessment of the required level of detail, the type of model, and parameters to be evaluated can be selected. "Canned" programs may minimize the cost of program development. On the other hand, development of a model to fit the particular situation may result in higher initial costs but may yield better accuracy. In any event, the model must not be forced to fit the situation. If a deep, slow-moving stream is thermally stratified, it is unwise to attempt to force a one-dimensional stream model to fit the situation.

Calibration.   The model is applied to the best set of available data. The values predicted by the model are plotted, along with the values actually observed in the stream. Plotting is usually done as a function of distance for steady-state models. If a dynamic model is used, values for key points in the stream are plotted as a function of time. Differences between observed values and values predicted by the model should be reconciled as much as possible. This reconciliation process, referred to as calibration, is accomplished by adjusting model parameters including reaction rate constants, temperature-dependence factors, and other coefficients; re-running the model; and comparing results to actual values.

Various curve-fitting and optimization methods can be used, with caution, to aid in the calibration procedure. In cases where major differences remain, the model structure may have to be re-evaluated. Calibration continues until model predictions agree satisfactorily with observed values. In general, the model should predict water quality to within ±10 percent, although ±20 percent may be acceptable. The model output should be plotted and compared to the observed data as shown in Figures 6.4 and 6.5. Slopes of the model predictions and the location of maximum and minimum points should be compared to observed data. Figure 6.4 presents the results of a calibration run for a steady-state dissolved oxygen model. This model cannot be considered acceptable since dissolved oxygen predictions are too high and the sag point does not correspond to the observed data. Figure 6.5 presents an acceptable calibration run for dissolved oxygen over a 24-hour period at a single point in a stream. The dissolved oxygen levels predicted by the model, the general slope of the curve, and the timing of peaks and valleys for the model in Figure 6.5 show good agreement with observed data.

Verification.   Once calibration has been completed, the model should be applied to a second set of observed data. If the values predicted by the model agree satisfactorily with

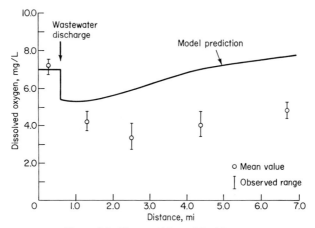

Figure 6.4   Unacceptable model calibration run.

the second set of observed data, the model is considered verified. The same general criteria used for calibration apply. If significant discrepancies exist, the general model formulation should be reviewed and the calibration/verification cycle repeated.

Model Application.   The model that has been calibrated and verified is now employed to evaluate the water quality impacts resulting from the proposed project. First, the model is applied to existing conditions without the project (the environmental setting). Next, the proposed project's wastewater streams and water intakes are added and the model is run again to predict the water quality after the project is operational. The differences in water quality between the two runs comprise the environmental impacts of the proposed project. This same procedure is used to evaluate each potential alternative course of action.

Application of the model can have significant input into the formulation of mitigative measures. For example, if dissolved oxygen currently is 8.0 mg/L, the stream standard is 5.0 mg/L, and the proposed project's wastewater discharge is predicted to lower dissolved oxygen to 3.5 mg/L, it is apparent that a higher degree of wastewater treatment is needed.

Sensitivity Analysis.   Sensitivity analysis enables the modeller to evaluate the relative importance of the basic parameters and input data in the model study. One model run is selected for use in the sensitivity analysis, generally the run depicting existing conditions. Another candidate is the run depicting the results of the proposed activity. Input parameters (such as reaction rate constants) are perturbed one at a time by an incremental amount, perhaps ±20 percent. The resulting percent changes in predicted water quality values (BOD, dissolved oxygen, temperature, etc.) are noted. If the change in water quality is small compared to the percent change in a parameter, then the model is relatively insensitive to that parameter. However, if the change in predicted water quality is great, the model is sensitive to that parameter. It is imperative that the best available values be obtained for sensitive parameters.

Sensitivity analysis should be performed throughout the model study—not as the final step in the modeling procedure. This analysis is an integral and important component of the entire model study because critical parameters are identified during model formulation and calibration. Results of the sensitivity analysis should be displayed in the final report. This enables the reviewer to evaluate the relative accuracy of the impact predictions and to determine the level of confidence merited by the results.

Individuals conducting model studies have an axiom: "garbage in, garbage out." This implies that the results of the model study are only as good as the input data. Care must be taken in the description of the environmental setting and related field sampling and analysis to provide viable input data.

If models are not used, the impact assessment team avoids the primary means for quantification of impacts. In the absence of models, professional judgment becomes the

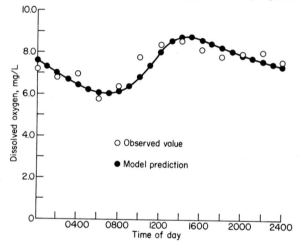

Figure 6.5   Acceptable model calibration run.

primary means of assessing impacts. Individuals using models are cautioned not to regard the results of a model study as the ultimate answer. Their results must be tempered with sound professional judgment based on an understanding of the physical, biological, and chemical processes involved in water quality interactions.

The following sections describe some basic models and model components that can be used in impact assessment. This is not intended to be a comprehensive review of models and modeling, but should provide an introduction to the subject. Emphasis is placed on relatively simple models that can be used in many, if not most, impact assessments. While the five phases of a model study may not be needed for very simple models, the basic concepts presented in the previous discussion of modeling should be kept in mind. Several existing model packages are presented briefly.

Mixing Zone Model   A very simple mass balance model can be developed for the mixing zone of a stream, as shown in Figure 6.6. A waste stream is discharged into a stream. At some point downstream, the waste discharged will be uniformly mixed across the entire cross-sectional area of the stream. The concentration of a substance at the downstream end of this reach is given by:

$$C_r = \frac{C_w Q_w + C_u Q_u}{Q_r}$$

where C_r = concentration in the river downstream, mg/L
C_w = concentration in waste discharge, mg/L
C_u = upstream river concentration, mg/L

Q_u = upstream river flow, ft³/s
Q_w = waste discharge rate, ft³/s
$Q_r = Q_u + Q_w$ = river flow at downstream station, ft³/s

Using this method, water quality impacts can quickly be estimated. The method is particularly useful for evaluating conservative substances, that is, materials that do not degrade chemically or biologically. This model is particularly applicable to relatively shallow, narrow streams.

Example 1: Mixing Zone Calculations.   An industrial plant is proposed for location on a stream. The industry would discharge wastewater containing 1300 mg/L of total dissolved solids (TDS) at a rate of 100 ft³/s. The receiving stream has an average velocity of 1.5 ft/s, an average width of 45 ft, and an average depth of 2 ft with a TDS concentration of 310 mg/L. Will this proposed plant discharge result in violation of the state stream standard of 500 mg/L?

$$C_w = 1300 \text{ mg/L} \qquad Q_w = 100 \text{ ft}^3/\text{s} \tag{1}$$
$$C_u = 310 \text{ mg/L} \tag{2}$$
$$Q_u = (1.5 \text{ ft/s})(45 \text{ ft})(2 \text{ ft}) = 135 \text{ ft}^3/\text{s} \tag{3}$$
$$Q_r = 100 \text{ ft}^3/\text{s} + 135 \text{ ft}^3/\text{s} = 235 \text{ ft}^3/\text{s} \tag{4}$$
$$C_r = \frac{(1{,}300 \text{ mg/L})(100 \text{ ft}^3/\text{s}) + (310 \text{ mg/L})(135 \text{ ft}^3/\text{s})}{(235 \text{ ft}^3/\text{s}} \tag{5}$$
$$= 731 \text{ mg/L}$$

Conclusion: The standard will be exceeded.

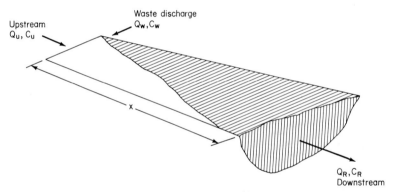

Figure 6.6 Simple mixing zone relationship.

Dissolved Oxygen Models Numerous models ranging from simple one-dimensional models to complex two-dimensional ecologic models have been developed to simulate dissolved oxygen. Before presenting several simple models, the basics of the interactions between biochemical oxygen demand (BOD), dissolved oxygen, and temperature will be discussed.

The ultimate carbonaceous BOD is the total amount of dissolved oxygen that would eventually be consumed by microorganisms in an aerobic environment while oxidizing organic material present in the water. The kinetics of the disappearance of ultimate BOD is commonly described as a first-order reaction given by:

$$\frac{dL}{dt} = -K_1 L$$

where L = ultimate BOD
    t = time
    K_1 = deoxygenation rate constant (base e)
This can be integrated to yield:

$$L_t = Le^{-K_1 t}$$

where L_t = amount of ultimate BOD remaining at time t

e = base of natural logarithms (2.718)

The amount of BOD exerted at time t is given by:

$$BOD = L - L_t = L\,(1 - e^{-K_1 t})$$

BOD is normally measured after an incubation period of five days at 20°C. In this case, the five-day, 20°C BOD is given by:

$$BOD_{5-20} = L(1 - e^{-5K_1})$$

The rate of this reaction is temperature dependent. This temperature dependence can be described using the following relationship:

$$K_T = K_{20}\,\theta^{T-20}$$

where K_T = reaction rate constant at temperature T

T = temperature, °C

θ = temperature coefficient

The value of the rate constant at 20°C is normally in the range of 0.1 to 0.8 day^{-1}, while θ is on the order of 1.02 to 1.06 (Refs. 18,19). A value of θ commonly used is 1.047 and the rate of deoxygenation (K_1) at 20°C is frequently taken as 0.23 day^{-1}.

Water that has a dissolved oxygen concentration of less than saturation will receive additional oxygen through surface reaeration. The rate of reaeration can be expressed as:

$$\frac{dDO}{dt} = K_2(DO_s - DO)$$

where DO = dissolved oxygen concentration at time t, mg/L

DO_s = dissolved oxygen concentration if the water were saturated, mg/L

K_2 = rate of reaeration (day^{-1} for base e)

The reaeration coefficient K_2 varies with temperature and with physical characteristics of the stream. This temperature dependence can be expressed as:

$$K_T = K_{20}\,\theta^{T-20}$$

where the temperature coefficient θ is normally in the range 1.015 to 1.047 (Ref. 19).

The saturation value (Refs. 18,19) can be estimated for unpolluted streams using the relationship:

$$DO_s = 14.652 - 0.41022\,T + 0.007991\,T^2 - 0.000077774\,T^3,$$

where T is the stream temperature (°C). Tables relating saturation concentration to temperature are available in many references and textbooks.

The estimation of the reaeration coefficient K_2 is an extremely important aspect of modeling dissolved oxygen. Numerous models have been developed for estimating this coefficient. Several popular equations for estimating K_2 at 20°C include the following:

1. O'Conner and Dobbins (Ref. 20) for low stream velocities and isotropic flow conditions (Chezy's coefficient greater than 17) use;

$$K_2 = \frac{(D_m \overline{U})^{0.5}}{D^{1.5}}$$

where D_m = molecular diffusion coefficient (ft^2/day)

$\overline{U}$ = mean stream velocity (ft/day)

D = mean stream depth (ft)

$$D_m = (1.91 \times 10^{-3})\,(1.037)^{T-20}$$

2. O'Conner and Dobbins (Ref. 20) for high velocities and nonisotropic conditions use:

$$K_2 = \frac{480 D_m^{0.5} S_o^{0.25}}{D^{1.25}} \times 2.31$$

where S_o = slope of the stream bed.

3. Churchill, Elmore, and Buckingham (Ref. 21) use:

$$K_2 = 5.026 \overline{U}^{0.969} D^{-1.673} \times 2.31$$

4. Baca, *et al.* (Refs. 18,19) use:

$$K_2 = \frac{r_1 \overline{U}^{r_2}}{D^{r_3}}$$

where r_1 = constant (2.0–12.0)
     r_2 = constant (0.6–1.0)
     r_3 = constant (0.85–1.85)
5. Langbien and Durum (Ref. 22) use:

$$K_2 = \frac{3.3 \overline{U}}{D^{1.33}}$$

6. Thackston and Krenkel (Ref. 23) use:

$$K_2 = [10.8(1 + F^{0.5})U^*/D] \times 2.31$$

where F = Froude number = $U^*/\sqrt{gD}$
     U^* = shear velocity (ft/s) = $\sqrt{gDS_e}$
     g = gravitation constant = 32.2 ft/s^2
     S_e = slope of energy grade line

7. Owens, Edwards, and Gibbs (Ref. 24) use (for stream velocities of 0.1–5.0 ft/s and depths of 0.4–11.0 ft):

$$K_2 = \frac{9.4 \overline{U}^{0.67}}{D^{1.85}} \times 2.31$$

8. Owens, Edwards, and Gibbs (Ref. 24) use (for stream velocities of 0.1–1.8 ft/s and depths of 0.4–2.4 ft):

$$K_2 = \frac{10.1 \overline{U}^{0.73}}{D^{1.75}} \times 2.31$$

Typical values of K_2 range from 0.05 to 16.0 day^{-1} (Refs. 18,19). Fair (Ref. 25) defines a purification constant f as:

$$f = \frac{K_2}{K_1}$$

Typical values for f are shown in Table 6.3.
    The preceding models of carbonaceous BOD and atmospheric reaeration can be com-

TABLE 6.3   Fair's Purification Factor f

Water source	Values of f* at 20°C
Small ponds and back waters	0.5–1.0
Sluggish streams, large lakes, or impounding reservoirs	1.0–1.5
Large streams of low velocity	1.5–2.0
Large streams of normal velocity	2.0–3.0
Swift streams	3.0–5.0
Rapids and waterfalls	Above 5.0

$*f = \dfrac{K_2}{K_1}$

SOURCE: "The Dissolved Oxygen Sag–An Analysis" by G. M. Fair, *Sewage Works Journal*, May 1939 (Ref. 25). Used with permission of Water Pollution Control Federation.

bined to form a simple dissolved oxygen model (referred to as the "Streeter-Phelps Model") of a stream:

$$\frac{dD}{dt} = K_1 L - K_2 D$$

where D = dissolved oxygen deficit $(DO_s - DO)$
$\qquad L$ = ultimate carbonaceous BOD
This equation can be integrated to yield the classic Streeter-Phelps model (Ref. 26):

$$D_t = \left(\frac{K_1 L_a}{K_2 - K_1}\right)[e^{-K_1 t} - e^{-K_2 t}] + D_a e^{-K_2 t}$$

where D_t = dissolved oxygen deficit t days downstream, mg/L
$\qquad L_a$ = initial ultimate BOD concentration, mg/L
$\qquad D_a$ = initial dissolved oxygen deficit, mg/L
The time t_c, in days, at which the dissolved oxygen reaches its minimum value downstream is given by:

$$t_c = \frac{1}{K_2 - K_1} \ln\left\{\frac{K_2}{K_1}\left[1 - \frac{D_a(K_2 - K_1)}{K_1 L_a}\right]\right\}$$

and the minimum dissolved oxygen is calculated by:

$$DO_{min} = DO_s - \frac{K_1 L_a}{K_2}(e^{-K_1 t_c})$$

This one-dimensional, steady-state model can be useful in impact assessment. However, numerous assumptions are included in the use of the model, including:

1. Constant K_1, K_2, and T along the stream reach
2. Channel cross section remains constant
3. Carbonaceous BOD is the only dissolved oxygen sink
4. Atmospheric reaeration is the only oxygen source

Example 2: Streeter-Phelps Model. It is proposed that an industrial facility be located on a relatively clean stream. The stream has a BOD$_{5-20}$ of 2.0 mg/L, a dissolved oxygen content of 8.0 mg/L, temperature of 22°C, and a flow of 250 ft^3/s. The industrial wastewater has BOD$_{5-20}$ of 800 mg/L, temperature of 31°C, flow of 125 ft^3/s, and will be aerated to a dissolved oxygen concentration of 6.0 mg/L. After mixing, it is anticipated that the average depth of flow will be 3 ft, and the stream width is 50 ft. The stream standard for dissolved oxygen is 5.0 mg/L. The following values have been determined for the rate constants and temperature coefficients:

$$K_1 @20°C = 0.23 \text{ day}^{-1}$$
$$K_2 @20°C = 3.0 \text{ day}^{-1}$$
$$\theta_1 = 1.05$$
$$\theta_2 = 1.02$$

Using this information, compute the following:

1. Find the maximum BOD$_{5-20}$ that can be discharged from the industrial plant
2. Plot the dissolved oxygen profile downstream
3. If $K_2@20°C$ is 2.5 day^{-1}, plot the dissolved oxygen profile

The solution is as follows for the mixed conditions indicated:

$$Q = 250 \text{ ft}^3/\text{s} + 125 \text{ ft}^3/\text{s} = 375 \text{ ft}^3/\text{s}$$
$$\text{Velocity} = \frac{375 \text{ ft}^3/\text{s}}{(3 \text{ ft})(50 \text{ ft})} = 2.5 \text{ ft/s}$$
$$T = \frac{(22°C)(250 \text{ ft}^3/\text{s}) + (31°C)(125 \text{ ft}^3/\text{s})}{375 \text{ ft}^3/\text{s}} = 25.0°C$$
$$DO = \frac{(8.0 \text{ mg/L})(250 \text{ ft}^3/\text{s}) + (6.0 \text{ mg/L})(125 \text{ ft}^3/\text{s})}{375 \text{ ft}^3/\text{s}}$$
$$= 7.33 \text{ mg/L}$$

From table, $DO_s = 8.38$ mg/L. Therefore,

$$D_a = 8.38 \text{ mg/L} - 7.33 \text{ mg/L} = 1.05 \text{ mg/L}$$
$$K_1 = 0.23(1.05)^{25-20} = 0.29 \text{ day}^{-1}$$
$$K_2 = 3.0(1.02)^{25-20} = 3.3 \text{ day}^{-1}$$

Maximum allowable deficit is given by:

$$D_c = 8.38 \text{ mg/L} - 5.00 \text{ mg/L} = 3.38 \text{ mg/L}$$

Use trial and error to find L_a (mg/L) yielding a critical deficit corresponding to dissolved oxygen of 5.0 mg/L.

$$t_c = \frac{1}{K_2 - K_1}\left(\ln\left\{\frac{K_2}{K_1}\left[1 - \frac{D_a(K_2 - K_1)}{K_1 L_a}\right]\right\}\right)$$

$$t_c = \frac{1}{3.3 \text{ day}^{-1} - 0.29 \text{ day}^{-1}}\left(\ln\left\{\frac{3.3 \text{ day}^{-1}}{0.29 \text{ day}^{-1}}\left[1 - \frac{(1.05 \text{ mg/L})(3.3 \text{ day}^{-1} - 0.29 \text{ day}^{-1})}{(0.29 \text{ day}^{-1})L_a}\right]\right\}\right)$$

$$D_c = \frac{K_1 L_a}{K_2}(e^{-K_1 t_c})$$

$$= \frac{(0.29 \text{ day}^{-1})L_a}{3.3 \text{ day}^{-1}}(e^{-0.29 t_c})$$

$$= (0.0879)L_a(e^{-0.29 t_c})$$

TABLE 6.4 Trial-and-Error Solution for Allowable Ultimate BOD Loading

Trial	Ultimate BOD loading L_a (mg/L)	Critical flow time t_c (days)	Dissolved oxygen deficit D_c (mg/L)
1	100	0.770	7.03
2	50	0.727	3.56
3	40	0.703	2.87
4	45	0.716	3.12
5	48	0.723	3.42
6	47	0.721	3.35
7	47.4	0.721	3.38 (acceptable)

Table 6.4 presents the results of this iterative procedure. In this case, the ultimate BOD concentration is:

$$L_a = 47.4 \text{ mg/L}$$

Now, find total BOD_{5-20}.

$$BOD_{5-20} = L(1 - e^{-5K_1})$$
$$= (47.4 \text{ mg/L})[1 - e^{-(0.23 \text{ day}^{-1})(5 \text{ day})}]$$
$$= 32.4 \text{ mg/L (after mixing)}$$

For waste discharge

$$BOD_{5-20} = \frac{(32.4 \text{ mg/L})(375 \text{ ft}^3/\text{s}) - (2.0 \text{ mg/L})(250 \text{ ft}^3/\text{s})}{125 \text{ ft}^3/\text{s}}$$
$$= 93.2 \text{ mg/L}$$

Maximum allowable discharge = 93.2 mg/L. Therefore, 88 percent of the influent BOD_{5-20} of 800 mg/L must be removed.

Now plot dissolved oxygen profile:

$$D_t = \frac{K_1 L_a}{K_2 - K_1}[e^{-K_1 t} - e^{-K_2 t}] + D_a e^{-K_2 t}$$
$$= \frac{(0.29 \text{ day}^{-1})(47.4 \text{ mg/L})}{3.3 \text{ day}^{-1} - 0.29 \text{ day}^{-1}}[e^{-0.29 t} - e^{-3.3 t}] + 1.05 \text{ mg/L}(e^{-3.3 t})$$
$$= 4.57[e^{-0.29 t} - e^{-3.3 t}] + 1.05 e^{-3.3 t}$$

Table 6.5 summarizes these calculations and Figure 6.7 presents the dissolved oxygen profile. Now develop a profile of K_2 @ 20°C = 2.5 day^{-1}.

K_2 @25°C $= (2.5)(1.02)^{25-20} = 2.8$ day^{-1}

$$t_c = \frac{1}{(2.8\ \text{day}^{-1} - 0.29\ \text{day}^{-1})} \left(\ln \left\{ \frac{2.8\ \text{day}^{-1}}{0.29\ \text{day}^{-1}} \left[1 - \frac{(1.05\ \text{mg/L})(2.8\ \text{day}^{-1} - 0.29\ \text{day}^{-1})}{(0.29\ \text{day}^{-1})(47.4\ \text{mg/L})} \right] \right\} \right)$$

$= 0.819$ days

$$D_c = \frac{(0.29\ \text{day}^{-1})(47.4\ \text{mg/L})}{2.8\ \text{day}^{-1}} [e^{-(0.29\ \text{day}^{-1})(0.819\ \text{days})}]$$

$= 3.87$ mg/L

$$D_t = \frac{(0.29\ \text{day}^{-1})(47.4\ \text{mg/L})}{2.8\ \text{day}^{-1} - 0.29\ \text{day}^{-1}} [e^{-0.29t} - e^{-2.8t}] + 1.05\ \text{mg/L}(e^{-2.8t})$$

$= 5.48\ [e^{-0.29t} - e^{-2.8t}] + 1.05\ e^{-2.8t}$

TABLE 6.5   Calculations for Plot of Dissolved Oxygen Profile

Time t (days)	D (mg/L) [K_2 @20 = 3.0 day^{-1}]	D (mg/L) [K_2 @20 = 2.5 day^{-1}]
0.0	1.05	1.05
0.1	1.91	1.98
0.2	2.49	2.64
0.3	2.88	3.11
0.4	3.13	3.43
0.5	3.28	3.65
0.6	3.35	3.78
0.7	3.38	3.85
0.8	3.37	3.87
0.9	3.34	3.86
1.0	3.29	3.83
1.5	2.93	3.48
2.0	2.55	3.05
2.5	2.21	2.65
3.0	1.91	2.29
4.0	1.43	1.72
5.0	1.07	1.29
6.0	0.80	0.96
7.0	0.60	0.72
8.0	0.45	0.54
10.0	0.25	0.30
12.0	0.14	0.17
15.0	0.06	0.07
20.0	0.01	0.02

Table 6.5 presents the calculations of dissolved oxygen deficit as a function of flow time. Figure 6.7 presents the dissolved oxygen profile.

The Streeter-Phelps model can be expanded to include other sources and sinks of oxygen by an equation of the form:

$$\frac{dDO}{dt} = -K_{1c}L_c - K_{1n}L_n - B - R + P + K_2(DO_s - DO)$$

where L_c = ultimate carbonaceous BOD
$\quad L_n$ = ultimate nitrogenous BOD
$\quad K_{1c}$ = rate constant for utilization of carbonaceous BOD
$\quad K_{1n}$ = rate constant for utilization of nitrogenous BOD
$\quad B$ = sink resulting from aerobic decomposition of benthic materials

R = respiration oxygen demand exerted by algae and other aquatic plants during hours of darkness

P = oxygen contributed by photosynthetic activity of algae and other aquatic plants during daylight hours

Baca (Refs. 18,19) suggests that the net photosynthesis/respiration effects of attached algae and benthic plants may be ±0.01–2.0 mg/L of dissolved oxygen over a daily cycle, and that suspended algae may contribute ±0.1–2.0 mg/L.

Sedimentation and/or adsorption can be a significant sink of BOD. The sink can be modeled as:

$$\frac{dL}{dt} = -K_3L$$

where K_3 equals the rate constant for sedimentation/adsorption (days^{-1}).

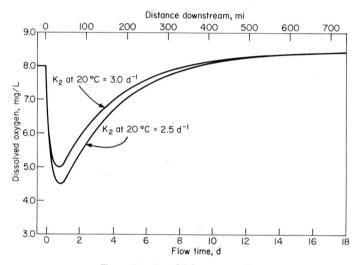

Figure 6.7 Dissolved oxygen profile.

Temperature Models Temperature is an extremely important parameter to be evaluated because temperatures affect rates of chemical and biological reactions, solubility of gases in water (particularly oxygen), and may induce stratified conditions.

One method of modeling water temperatures is by carefully adding all heat inputs into the water volume and subtracting all heat losses. This type of approach is described by Edinger, et al. (Ref. 27) and Velz (Ref. 28) and can be expressed as:

$$\Delta H = H_s - H_{sr} + H_a - H_{ar} - H_b - H_c - H_e + H_p$$

where ΔH = net change in heat content of the water
H_s = heat input from short-wave solar radiation
H_{sr} = reflected solar radiation
H_a = long-wave atmospheric radiation
H_{ar} = reflected atmospheric radiation
H_b = back radiation
H_c = heat loss or gain by conduction
H_e = loss due to evaporation
H_p = heat released by the project to the water

Edinger, et al. (Ref. 27) present methods for evaluating each of the above terms. The rate of change in water temperature is given by:

$$\frac{dT_w}{dt} = \frac{\Delta H}{\rho CD}$$

where T_w = water temperature, °C
    t = time, sec
    ΔH = change in heat content, W/m^2
    ρ = density of water, ≈ 1000 kg/m³
    C = heat capacity of water, ≈ 4186 J/kg·°C
    D = depth of water, m

Edinger, et al. (Ref. 27) present a simplified model using the equilibrium temperature concept. The equilibrium temperature is the water temperature at which the net rate of surface heat exchange would be zero. The equilibrium temperature can be approximated as:

$$T_e = T_d + \frac{H_s}{K}$$

where T_e = equilibrium temperature, °C
    T_d = dew-point temperature, °C
    H_s = gross rate of short-wave radiation, W/m^2
    K = surface heat-exchange coefficient, $W/m^2(°C)$

This expression for the equilibrium temperature demonstrates that during the night-time hours, the equilibrium temperature equals the dew-point temperature. The dew point is commonly measured at most weather stations. Since the dew point is relatively constant during the day, the major source of diurnal variations in equilibrium temperature results from changes in solar radiation.

If solar radiation observations are available from a weather station nearby, this data can be utilized. Unfortunately, relatively few weather stations record solar radiation data. Lacking solar radiation information, the impact assessment team must resort to implementation of a pyrheliometer study or the use of empirical equations to estimate radiation.

Numerous models for estimating short-wave solar radiation are available. For example, Mosby (Ref. 29) developed the following equation:

$$H_s = K(1.0 - 0.071C)\overline{h}$$

where H_s = solar radiation cal/cm² · min
    C = average cloud cover, 1/10s of sky covered
    K = a constant which is a function of latitude
    $\overline{h}$ = average altitude of the sun, degrees

Anderson (Ref. 30) evaluated the performance of the Mosby equation and found that the equation produced results that were about 15 percent less than observed values. Baca *et al.* (Ref. 31) present a detailed equation for predicting solar radiation.

Kennedy (Ref. 32) presents an approach for extrapolating solar radiation data from pyrheliometer stations having the same altitude and latitude as the study area. His method involves the development of the constants in the expression:

$$H_s = l_o a^m$$

where H_s = solar radiation
    l_o = solar radiation received at the exterior of the earth's atmosphere on a horizontal surface
    a = atmospheric transmission coefficient
    m = the solar air mass, or the ratio of the actual path length of the solar beam to the path length through the zenith

In applying Kennedy's method, the a factor is first calculated on a daily basis for the pyrheliometer station (Note: H_s, l_o, and m are assumed to be known at this station). A relationship is then developed for the a factor as a function of cloud cover. This information is then applied to the study area. Anderson (Ref. 30) found that Kennedy's method gave excellent results for clear days at Lake Hefner, but that errors of up to 50 percent were observed on cloudy days. Duffie and Beckman (Ref. 33) present a detailed discussion of various methods for measuring and empirically calculating solar radiation.

Figure 6.8 presents Edinger's (Ref. 27) graphical means for estimating the value of the surface heat exchange coefficient. This figure represents K as a function of wind speed and average temperature T_m given by:

$$T_m = \frac{T_d + T_w}{2}$$

Temperature can then be modeled as:

$$\frac{dT_w}{dt} = \frac{K(T_e - T_w)}{\rho C D}$$

For a small, well-mixed body of water receiving a constant heat input of H_p (in W) the water temperature will be given by:

$$T_s = T_n + \frac{H_p}{AK}$$

where T_s = water temperature with heated water discharge, °C
T_n = natural water temperature before heated discharge, °C
H_p = plant heat rejection rate, W
A = surface area of the water body, m²

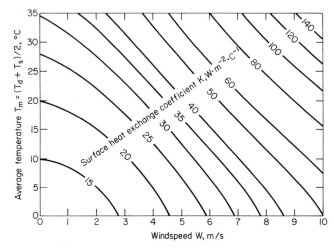

Figure 6.8 Evaluation of surface heat exchange coefficient. (SOURCE: Heat Exchange and Transport in the Environment *by Edinger, Brady, and Geyer (Ref. 27). Used with permission of the senior author.*)

For a more detailed discussion of temperature modeling and related topics, the reader is referred to Edinger, et al. (Ref. 27) and Mathur (Ref. 34).

Small Lake Model Small, shallow lakes that receive significant wind-induced mixing can be modeled as completely mixed reactors (CSTR) using a mass balance approach described by Metcalf and Eddy (Ref. 35)[1]. Figure 6.9 depicts a small lake receiving inflow from two sources. The mass balance is given by:

$$V dC = Q_1 C_1 dt + Q_2 C_2 dt - QC dt + O - VCK dt$$

[change in storage] = [inflow] − [outflow] + [sources] − [sinks]

where Q_1, Q_2 = flow rates of streams No. 1 and 2 into the lake
C_1, C_2 = concentration in streams No. 1 and 2
$Q = Q_1 + Q_2$ = flow rate out of lake
C = concentration in outflow
V = volume of lake
K = first-order reaction rate
(NOTE: No other sources present.)

This mass balance equation can be integrated to obtain:

$$C = \frac{W}{BV}(1 - e^{-\beta t}) + C_o e^{-\beta t}$$

[1]Adapted from *Wastewater Engineering* by Metcalf and Eddy, Inc. (Ref. 35), McGraw-Hill Book Company, New York, 1972. Used with permission of McGraw-Hill Book Company.

where W = mass flow rate into the lake = $Q_1C_1 + Q_2C_2$

$$\beta = \frac{1}{t_o} + K$$

C_o = concentration in the lake at time $t = 0$

t = time

$$t_o = \text{detention time in the lake} = \frac{V}{Q}$$

At steady state, the expression can be reduced to:

$$C = \frac{W}{\beta V}$$

A completely mixed model will not adequately describe large deep lakes. However, if a deep lake stratifies, this type of a model could be used to represent the epilimnion (upper layer) of the lake. Lakes having large surface areas and isolated bays could be modeled using several interconnected CSTR models.

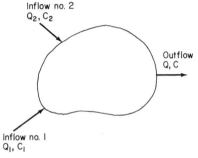

Figure 6.9 A small lake.

Additional source and sink terms could be added to the mass balance relationship to enable dissolved oxygen to be modeled. CSTR models can also be linked sequentially to form a model of a flowing stream. To represent a stream, the mass of materials being considered would be routed from CSTR to CSTR in the downstream direction.

Runoff Models Proposed projects and developments can exert a dramatic influence on the amount and temporal distribution of runoff from an area. Three basic approaches are available for modeling runoff, namely: (1) empirical relationships, (2) relating runoff to rainfall by a proportionality coefficient (Rational Method), and (3) detailed accounting of the rate of rainfall.

While empirical models were once very popular, their use has decreased since these models are only applicable to specific areas and cannot be generalized. The second and third approaches are commonly used today.

The so-called "Rational Method" is the most popular method in the United States for estimating runoff (Ref. 36). The model is given by:

$$Q = CIA$$

where Q = peak runoff rate, ft³/s

    C = runoff coefficient

    I = average rainfall intensity, in/hr

    A = drainage basin area, acres

The following assumptions are involved in the use of this model:

1. Peak runoff rate is a function of average rainfall intensity during the time of concentration.

2. Frequency of peak discharge is the same as the frequency of average rainfall intensity.

3. Time of concentration is the time required for runoff from the most remote portion of the drainage area to contribute to the point of concern.

This method is most applicable to small, urban drainage areas. However, the model does not provide definition of the hydrograph.

The runoff coefficient is selected from a table similar to Table 6.6. Normally this coefficient is assumed to be constant over the entire duration of a storm. Indeed, this coefficient varies as a function of antecedent rainfall and time of prior wetting during a storm. This is one source of error inherent in use of the rational method. Schaake, et al. (Ref. 37) compared results using the rational method to actual runoff frequency curves and found significant differences.

Hershfield (Ref. 38) has prepared a series of isopluvial maps of the United States showing rainfall intensities for durations of 30 min to 24 hr for return periods of 1 to 100 yr. These generalized isopluvial maps can be used to estimate design storms if local data is not readily available.

For a given watershed and rainfall intensity, the impact of development in a watershed area can be estimated as:

$$Q_1 = \frac{(Q_0)(C_1)}{C_0}$$

where Q_0 = runoff before development, ft³/s
Q_1 = runoff after development, ft³/s
C_0 = runoff coefficient before development
C_1 = runoff coefficient after development

TABLE 6.6   Runoff Coefficient

Land uses	Runoff coefficient (C)
Business:	
Downtown	0.70 to 0.95
Neighborhood	0.50 to 0.70
Residential:	
Single-family	0.30 to 0.50
Multi-units, detached	0.40 to 0.60
Multi-units, attached	0.60 to 0.75
Residential (suburban)	0.25 to 0.40
Apartment	0.50 to 0.70
Industrial:	
Light	0.50 to 0.80
Heavy	0.60 to 0.90
Parks, cemeteries	0.10 to 0.25
Playgrounds	0.20 to 0.35
Railroad yard	0.20 to 0.35
Unimproved	0.10 to 0.30
Surface types:	
Pavement:	
Asphalt and concrete	0.70 to 0.95
Brick	0.70 to 0.85
Roofs	0.75 to 0.95
Lawns, sandy soil:	
Flat, 2 percent	0.05 to 0.10
Average, 2 to 7 percent	0.10 to 0.15
Steep, 7 percent	0.15 to 0.20
Lawns, heavy soil:	
Flat, 2 percent	0.13 to 0.17
Average, 2 to 7 percent	0.18 to 0.22
Steep, 7 percent	0.25 to 0.35

SOURCE: *Design and Construction of Sanitary and Storm Sewers*, MOP No. 37 by a Joint Committee of ASCE and WPCF (Ref. 36). Copyright 1969 by ASCE. Used with permission of ASCE.

Example 3: Rational Method. It is proposed that a 100-acre underdeveloped watershed be developed into a light industrial park. For a design rainfall intensity of 2.0 in/hr, what impact will the proposed development have on runoff?

Given information is that $A = 100$ acres and $I = 2.0$ in/hr. Consulting Table 6.6 we find that $C_0 = 0.20$ and $C_1 = 0.65$. Therefore, the present condition can be shown by:

$$Q_o = C_o IA = (0.20)(2.0 \text{ in/hr})(100 \text{ acres}) = 40 \text{ ft}^3/\text{s}$$

Future conditions are:

$$Q = \frac{Q_o C}{C_o} = \frac{(40 \text{ ft}^3/\text{s})(0.65)}{(0.20)} = 130 \text{ ft}^3/\text{s}$$

$$\text{Increase} = \frac{(130 \text{ ft}^3/\text{s} - 40 \text{ ft}^3/\text{s})}{40 \text{ ft}^3/\text{s}} \times 100\% = 225\%$$

The U.S. Soil Conservation Service (SCS) (Ref. 39) presents an empirical method for developing synthetic hydrographs. This method involves the use of standardized curves and could be used for assessment of impacts. The appropriate curve number before development would be used to evaluate runoff. A new curve number reflecting post-project conditions would then be used to evaluate post-project runoff.

Several models have been developed for the formulation of runoff hydrographs. These models estimate the amount of rainfall going to interception, infiltration, surface detention, overland flow, and flow in channels. This type of approach has been incorporated into the Storm Water Management Model (Ref. 40), the Stanford Watershed Model (Ref. 41), and the Chicago Stormwater Model (Ref. 42). The reader is referred to these references for detailed descriptions of these model formulations. Keys (Ref. 43), Heeps, and Mein (Ref. 44) and Marsalek, et al. (Ref. 45) compare models used to estimate runoff.

With regard to flooding, there are two steps involved in assessing impacts. First, the change in the magnitude and streamflow must be established. The preceding sections have dealt with this topic for small watersheds (Rational Method). For large watersheds, the impacts on flood magnitudes normally are reduced since increased flows and longer distances tend to dampen out fluctuations. Most localized projects will not exert significant influence on flows in a large watershed. However, local projects may have a significant impact on flows in small local drainage areas. Major basin-wide developments, major channel improvement projects, and flood control reservoirs may significantly alter the flood hydrographs for large watersheds. Texts and handbooks on hydrology should be consulted for methods of predicting major changes in flood hydrographs.

The second step is the estimation of the areal extent of flooding and flood depths. This normally involves backwater computations using the principles of gradually varied open-channel flow (Refs. 46, 47, 48). These methods take into account channel and floodplain geometry as well as physical characteristics. The computations can be done by hand, but are cumbersome. Computer programs can greatly simplify the task. One of the most popular programs for computing backwater profiles is HEC-2 written by the Hydraulic Engineering Center of the U.S. Army Corps of Engineers (Ref. 49). The HEC-2 program can be run for a given flood discharge for both preproject and postproject conditions to estimate the impact of a project that would change the geometry or physical characteristics of the stream channel or floodplain. Keyes (Ref. 43) and Eichert (Ref. 50) compare models that can be used to compute backwater curves.

Erosion and Sedimentation Models   Consideration of possible changes in erosion rates and changes in sedimentation patterns are often important in assessing impacts. This may be particularly true during the construction stage when large areas of land may be cleared and left exposed. A valuable tool which can be used to estimate erosion rates is given by the universal soil loss equation which is expressed as (Refs. 51–54):

$$E = R \cdot K \cdot L \cdot S \cdot C \cdot P$$

where E = average soil loss, ton/acre $\cdot$ yr
   R = factor expressing the erosion potential of average rainfall in the area
   K = soil erodibility factor, ton/acre $\cdot$ yr
   L = length of slope factor
   S = slope factor
   C = cropping management factor
   P = erosion control practice factor

The following discussion focuses on each term in this model. Wischmeier and Smith (Ref. 53) present an excellent discussion of the equation and its use.

Rainfall Factor (R).   The rainfall factor, discussed in detail by Wischmeier and Smith (Ref. 55), is the product of the total kinetic energy of a storm and the storm's maximum 30-minute intensity. The value of R can be calculated using precipitation records from a continuous recording gauge and the energy tables presented by Wischmeier and Smith (Ref. 55). The U.S. Army Corps of Engineers (Ref. 56) suggests reducing R for snowfall events to one-third of the value corresponding to equivalent rainfall. Figure 6.10 presents an iso-erodent map that can be used to estimate the value of R for areas east of the Rocky Mountains.

Soil Erodibility Factor (K).   This factor is defined as the erosion rate per unit of rainfall factor R for a plot 72.6 ft long on a slope of 9.0 percent for a specific soil in continuous, cultivated fallow. Olson and Wischmeier (Ref. 54) reported the results of studies to evaluate K factors for about 20 soil types. Wischmeier, et al. (Ref. 57) presented a nomograph for the estimation of K, and Wischmeier and Mannering (Ref. 58) developed a regression equation relating K to 24 soil properties. Table 6.7 presents average values of K for generalized soil types.

Length of Slope Factor (L).   The length of slope factor is given by (Ref. 53):

$$L = \left(\frac{\lambda}{72.6}\right)^m$$

where λ = length of slope, ft
   m = constant

Figure 6.10 Average annual values of the rainfall factor R. [SOURCE: Predicting Rainfall—Erosion Losses from Cropland East of the Rocky Mountains, *Agricultural Handbook No. 282, by Wischmeier and Smith (Ref. 53). Published by the Agricultural Research Service, U.S. Department of Agriculture.*]

The exponent m is normally equal to 0.5. However, for slopes greater than 10 percent, it is suggested that a value of 0.6 be used for m (Ref. 53). The value of m may be reduced to 0.3 for long slopes having gradients of less than 0.5 percent.

Slope Factor (S).   The slope factor is calculated by (Refs. 53, 60):

$$S = \frac{0.43 + 0.30s + 0.043s^2}{6.613}$$

where s = slope, %

The slope and length factors are usually evaluated together. The equations for L and S can then be combined for $m = 0.5$ to yield:

$$L \cdot S = (0.00761 + 0.005325s + 0.000761s^2)\sqrt{\lambda}$$

Figure 6.11 presents a graph of this expression that can be used to estimate the $L \cdot S$ factor.

TABLE 6.7   Soil Erodibility Factor, K

Soil type	Organic matter content		
	<0.5%	2%	4%
Sand	0.05	0.03	0.02
Fine sand	0.16	0.14	0.10
Very fine sand	0.42	0.36	0.28
Loamy sand	0.12	0.10	0.08
Loamy fine sand	0.24	0.20	0.16
Loamy very fine sand	0.44	0.38	0.30
Sandy loam	0.27	0.24	0.19
Fine sandy loam	0.35	0.30	0.24
Very fine sandy loam	0.47	0.41	0.33
Loam	0.38	0.34	0.29
Silt loam	0.48	0.42	0.33
Silt	0.60	0.52	0.42
Sandy clay loam	0.27	0.25	0.21
Clay loam	0.28	0.25	0.21
Silty clay loam	0.37	0.32	0.26
Sandy clay	0.14	0.13	0.12
Silty clay	0.25	0.23	0.19
Clay		0.13–0.29	

SOURCE: *Control of Water Pollution from Cropland* (Ref. 59), published by the Agricultural Research Service, U.S. Department of Agriculture.

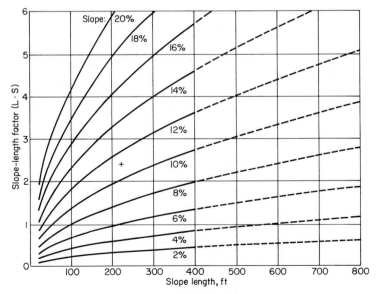

Figure 6.11   Slope—length factor $L \cdot S$. [SOURCE: Predicting Rainfall—Erosion Losses from Cropland East of the Rocky Mountains, *Agricultural Handbook No. 282, by Wischmeier and Smith (Ref. 53). Published by the Agricultural Research Service, U.S. Department of Agriculture.*]

Cropping Management Factor (C).   The cropping factor reflects the degree of erosion control inherent in growing various crops on a given plot. Wischmeier (Ref. 61) and Wischmeier and Smith (Ref. 53) present detailed tables of C factors for many agricultrual crops and cropping patterns. Table 6.8 presents a summary of typical cropping factors.

Erosion Control Practice Factor (P).   This factor incorporates the effects of various management techniques, such as contouring, into the analysis. Table 6.9 presents P factors for several management techniques.

TABLE 6.8   Typical Cropping Factors (C)

Crop	Notes	C	Reference
Bare ground		1.0	56
Grass and legume (hay)	All-year average	0.004–0.01	61
Clover	All-year average	0.015–0.025	61
Lespedeza	All-year average	0.01–0.02	61
Continuous corn	Rough fallow-residue removal	0.60–0.85	61
Continuous corn	Seedbed residue removal	0.70–0.90	61
Continuous corn	Growing crop residue removal	0.25–0.50	61
Continuous corn	Residue or stubble Residue removal	0.60–0.85	61
Continuous corn	Rough fallow-residue left	0.20–0.70	61
Continuous corn	Seedbed residue left	0.45–0.75	61
Continuous corn	Growing crop residue left	0.25–0.40	61
Continuous corn	Residue or stubble-residue left	0.15–0.65	61
Continuous cotton	Rough fallow	0.30–0.45	61
Continuous cotton	Seedbed	0.50–0.80	61
Continuous cotton	Growing crop	0.45–0.55	61
Continuous cotton	Residue or stubble	0.20–0.50	61
Grass cover	. . .	0.01	56
Land denuded by fire	. . .	1.00	56
Seed and fertilizer	18–20 month construction period	0.60	56
Seed, fertilizer, and straw mulch	18–20 month construction period	0.30	56

Example 4: Universal Soil Loss Equation.   Using the universal soil loss equation, estimate average annual erosion from an established meadow on a sandy loam soil having about 2 percent organic matter. The slope is 10 percent, the slope length is 150 ft and the rainfall factor R is 300. Contouring is practiced.

$$R = 300$$
$$K = 0.24 \text{ (Table 6.7)}$$
$$L \cdot S = 1.65 \text{ (Figure 6.11)}$$
$$C = 0.02 \text{ (Table 6.8 for clover)}$$

$$P = 0.60 \text{ (Table 6.9)}$$
$$E = R \cdot K \cdot L \cdot S \cdot C \cdot P$$
$$= (300)(0.24)(1.65)(0.02)(0.60)$$
$$= 1.4 \text{ ton/acre} \cdot \text{yr}$$

Example 5: Universal Soil Loss Equation. What would be the expected erosion rate if the area in the preceding example were bare earth?

$$R = 300$$
$$K = 0.24$$
$$L \cdot S = 1.65$$
$$C = 1.00 \text{ (Table 6.8)}$$
$$P = 1.00 \text{ (Table 6.9)}$$
$$E = (300)(0.24)(1.65)(1.00)(1.00)$$
$$= 119 \text{ ton/acre} \cdot \text{yr}$$

TABLE 6.9   Erosion Control Practice Factors (P)

Practice	Land slope, %	P
None	. . .	1.00
Contouring	1.1–2.0	0.60
Contouring	2.1–7.0	0.50
Contouring	7.1–12.0	0.60
Contouring	12.1–18.0	0.80
Contouring	18.1–24.0	0.90
Contour stripcropping	1.1–2.0	0.45
Contour stripcropping	2.1–7.0	0.40
Contour stripcropping	7.1–12.0	0.45
Contour stripcropping	12.1–18.0	0.60
Contour stripcropping	18.1–24.0	0.70
Terracing	1.1–2.0	0.45
Terracing	2.1–7.0	0.40
Terracing	7.1–12.0	0.45
Terracing	12.1–18.0	0.60
Terracing	18.1–24.0	0.70
Straight row farming	. . .	1.00

SOURCE: *Predicting Rainfall-Erosion Losses from Cropland East of the Rocky Mountains,* Agricultural Handbook No. 282 by W. H. Wischmeier and D. D. Smith (Ref. 53). Published by the Agricultural Research Service, U. S. Department of Agriculture.

The universal soil loss equation is also useful for estimating the difference in erosion rates in impact assessment. For a given soil and location, the R and K factors are constant and the $L \cdot S$ factors will normally be constant. As a result, the expected annual erosion rate with the project can be estimated by:

$$E_1 = E_o \frac{(C_1 \cdot P_1)}{(C_0 \cdot P_0)}$$

where E_0 = erosion rate before the project, ton/acre $\cdot$ yr
    E_1 = erosion rate after the project, ton/acre $\cdot$ yr
   C_0, C_1 = cropping factor before and after the project, respectively
   P_0, P_1 = erosion control factor before and after the project, respectively

Another source of sediment is erosion in gully systems. Normally, sediment loss from gully erosion is less than the loss associated with sheet erosion (Refs. 51,52), although gully erosion may be quite significant for high rainfalls and poor land management practices. No universal methods are available for predicting rates of sediment loss or rates of gully advance. However, several empirical models for estimating the rate of gully head advance are presented by Vanoni (Ref. 51).

Erosion can also take place within a stream channel itself. Erosion of stream bed and banks can be significant. Reliable quantitative models of channel erosion are not currently available. Lane (Ref. 62) developed the following qualitative relationship that can be used for channels in noncohesive materials:

$$QS = KG_s d_s$$

where Q = stream discharge
S = slope of the stream channel
K = constant of proportionality
G_s = bed sediment discharge
d_s = particle diameter of bed material

This expression can be useful in impact assessment if two of the variables can be taken as constants. Numerous sediment discharge formulas have been developed which may be of some use in estimating changes in scour or deposition when future changes in flow regime are anticipated. Sediment discharge equations are reviewed by Vanoni (Ref. 51) and Colby and Hembree (Ref. 63). The USEPA (Ref. 64) and Dawdy (Ref. 65) also present discussions of methods that can be used to estimate erosion and sedimentation.

Not all of the material eroded from the surface of a watershed is transported downstream. Much of this material is either deposited on the land's surface where slopes become more gentle, or on the stream bottom in slower moving areas of the stream. Sediment yield is the amount of sediment actually reaching a given point downstream. The sediment delivery ratio is given by (Ref. 51):

$$D = \frac{Y}{T} \times 100\%$$

where D = delivery ratio, %
Y = sediment yield, ton/acre $\cdot$ yr
T = total material eroded from the watershed, ton/acre $\cdot$ yr

Unfortunately, no universal models are available for predicting sediment yields or sediment delivery ratios. However, several empirical models are available for particular watersheds or regions. These models were reviewed by Vanoni (Ref. 51). An example of such a model is the equation developed by Maner (Ref. 66) for the Red Hills area in southern Kansas, western Oklahoma, and western Texas, given by:

$$D = 2.94259 - 0.82363 \log(R/L)$$

where D = sediment delivery ratio, %
R = watershed relief, difference between average elevation at the watershed divide and elevation at the point of concern, ft
L = watershed length, distance from point of concern to watershed divide along main stream channel, ft

If sediment yield information is available for similar nearby watersheds, this information can be used, with caution, to estimate sediment delivery ratios for the watershed of concern.

Reservoirs, lakes, and pools located on a stream represent significant sinks of sediment. The ability of a reservoir to retain sediment is reflected in the trap efficiency:

$$\text{Trap efficiency} = \frac{\text{Sediment deposited}}{\text{Sediment in inflow}}$$

Brune (Ref. 67) developed envelope curves that can be used to estimate the trap efficiency of reservoirs. Figure 6.12 presents these curves. Annual reservoir sedimentation can then be estimated by:

$$RS = T \cdot A \left(\frac{D}{100} \right) (\text{Trap Efficiency})$$

where RS = reservoir sedimentation, ton/yr
T = total watershed erosion, ton/acre $\cdot$ yr
A = watershed area, acre
D = sediment delivery ratio, %

Groundwater Models   Groundwater has held a historically important position as a source of potable water. For years, the major concern with groundwater has been the determination of permissible safe yields, i.e., the maximum sustained withdrawal rate from an aquifer without deleterious effects. Within the last several decades, the potential for adverse impacts on groundwater quantity and quality resulting from human activities has been observed. Excessive withdrawals result in severe drawdown of the water table and, in coastal areas, salt water intrusion. Deepwell disposal of wastewaters, septic tank systems, and agricultural practices have resulted in pollution of groundwaters.

Modeling groundwater hydrology and water quality is a difficult task. The chemistry of interactions between the groundwater and the aquifer are complex. A detailed discussion of groundwater modeling is beyond the scope of this chapter. However, if interested, the reader should consult such texts in groundwater hydrology as *Geohydrology* by De Wiest (Ref. 68). Summers and Spiegel (Ref. 69) present a bibliography on the subject of groundwater pollution, and Domenico (Ref. 70) presents descriptions of modeling techniques for groundwater hydrology and chemistry.

Survey of Existing Models   A large number of water quality and hydrologic models have been developed for numerous lakes, streams, estuaries, and aquifers. Many of these models have been designed to be site specific, that is, applicable to only one physical situation. Several generalized models have been developed that may have a wide range of application. Several of these so-called "canned programs" were created for, or by, govern-

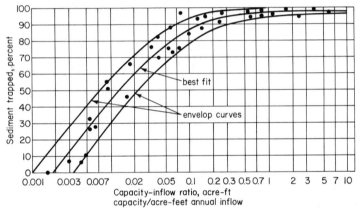

Figure 6.12   Reservoir trap efficiency. [SOURCE: *"Trap Efficiency of Reservoirs,"* by G. M. Brune, in Transactions, American Geophysical Union, *June 1953 (Ref. 67). Used with permission of the author.*]

mental agencies and are available to the public without restriction. Other canned programs are proprietary and require permission and possibly payment of royalties. The following are some of the models which have been developed and are generally available:

QUAL-II.   This water quality model is capable of simulating steady-state and dynamic response in flowing streams (Ref. 71). Chlorophyll *a*, ammonia, nitrite, nitrate, phosphorus, carbonaceous BOD, benthic oxygen demand, dissolved oxygen, coliforms, conservative substances, and temperature can be simulated.

Stanford Watershed Model.   This hydrologic model simulates streamflows, once historical precipitation data are supplied (Ref. 41). The major components of the hydrologic cycle are modeled, including: interception, surface detention, overland flow, interflow, groundwater, evapotranspiration, and routing of channel flows.

Hydrocomp Model.   The Hydrocomp Model is an expansion and refinement of the Stanford Watershed Model. The Hydrocomp Model can be used to simulate dynamic water quality conditions in streams and impoundments (Ref. 72). Temperature, total dissolved solids, dissolved oxygen, carbonaceous BOD, coliforms, algae-chlorophyll *a*, zooplankton, nitrate, nitrite, ammonia, total nitrogen, phosphate, and conservative substances can be simulated. Long-term meteorological and wastewater characterization data is used to simulate streamflows and stream water quality.

Stormwater Management Model (SWMM).   This model simulates runoff resulting from individual rainfall events (Ref. 40). Runoff is modeled from overland flow, through surface channels, and through the sewer network. Both combined and separate sewers can be modeled. Hydrographs and pollutographs are developed for specific points in the sewer system. The model also enables the user to simulate the water quality effects of stormwater or combined sewer discharges.

Storm.   This model, which was developed by the Hydrologic Engineering Center of the U. S. Army Corps of Engineers, simulates the dynamic behavior of stormwater runoff in urban and nonurban areas (Ref. 56). The model is usually run using a long record of hourly precipitation data. Hourly hydrographs and pollutographs are generated. The universal soil loss equation is utilized to predict soil erosion. The amounts of rainfall stored in storage facilities, treated, and directly discharged are displayed. However, neither the effect of treatment on the quality of the stormwater nor the effects of stormwater discharges on the receiving water are simulated.

Explore-I.   This model (Ref. 18) represents an expansion of the receiving water quality model contained within the Stormwater Management Model. This dynamic model simulates flows and stage, carbonaceuous BOD, nitrogenous BOD, benthic demands, total organic carbon, refractory organic carbon, phosphorus (sedimentary, soluble, organic), nitrogen (ammonia, nitrite, nitrate, organic), toxic substances, phytoplankton, zooplankton, and dissolved oxygen.

Battelle Reservoir Model.   This is a two-dimensional, multisegment model of water quality in a deep lake or reservoir (Ref. 31). The water body is divided into segments along the direction of flow, and each segment is divided into a number of horizontal layers. The model simulates temperature, dissolved oxygen, total and benthic BOD, phytoplankton, zooplankton, organic and inorganic nitrogen, organic and inorganic phosphorus, coliform bacteria, toxic substances, and hydro-dynamic conditions. The model was found to generate excellent simulation of temperature and good predictions of water quality parameters.

HEC-1.   This program was developed by the Hydrologic Engineering Center of the U.S. Army Corps of Engineers (Ref. 73). The program performs flood hydrograph computations for multi-subbasin, multichannel river basins. Computations are performed for a single storm event. The program can optimize certain parameters within the model to give the best fit to observed hydrographs.

HEC-2.   This model system computes and plots water surface profiles for stream channels (Ref. 49). Both subcritical and supercritical flow conditions can be accommodated. The program can consider the effects of structures, such as bridges, culverts, weirs, dams, levees, embankments, and flood walls, within the channel and floodplain. The model can also be used to plot flow profiles for various flood flows for natural and modified channels and floodways.

These models require significant computer resources. Most have been adapted for use on the equivalent of CDC 6600, IBM 360, and UNIVAC 1108 hardware. Keyes (Ref. 43) compares several canned water quality, runoff, and streamflow models. Eichert (Ref. 50) compares several canned models for predicting water surface profiles, while Heeps and Mein (Ref. 44) and Marsalek *et al.* (Ref. 45) compare urban runoff models.

Water Use and Wastewater Flows

A planned project may involve withdrawals of water and/or discharge of wastewater to a receiving stream. The estimation of water use and wastewater flows are actually part of the description of the project. However, these estimates are important in the assessment of water quality impacts. The following sections deal with methods for estimating water use and the quantity and characteristics of wastewater generated by a planned project.

Water Use   Estimates of water use and water consumption are important in impact assessments. Water use directly determines withdrawals required from surface waters or groundwaters. In addition, water use may have an impact on existing water supply systems, including water treatment plants, water distribution systems, and even the adequacy of water resources to meet expanded demand.

In most instances, existing water demand is known. Data concerning water use can usually be obtained from the local or regional water department. Existing average per capita consumption in gallons per day can be obtained by dividing the average flow of water (gal/day) by the population served by the water system. If historical data are available for water flow and service area population, the historical pattern of per capita water demand can be developed. The product of future population and future per capita consumption yields projections of future water demands.

If water demand and population data are not available, the water quality team must resort to formulating estimates. Data from nearby communities may be utilized, with caution, to generate per capita demand figures. Table 6.10 presents historical per capita demand figures for several cities across the United States that can provide some guidance in this task.

Water use figures for commercial, retail, and industrial facilities are often more difficult to obtain than figures for domestic use. Table 6.11 presents estimates of per capita water use by a wide variety of commercial, institutional, and retail activities. Industrial water use varies tremendously, even within a single industry, as a function of products manufactured, age and type of technology, and degree of water management practiced. Table 6.12 presents average water use on a per unit of product basis for several representative

TABLE 6.10   Average Per Capita Water Consumption

	Average consumption (gallons per capita per day)				
Location	1936	1946	1956	1966	1976
New York, N.Y.	134	146	138	153	162
Baltimore, Md.	131	150	159	174	188
Philadelphia, Pa.	...	163	171	180	188
Springfield, Mass.	100	127	175	230	278
Hartford, Conn.	80	100	111	122	134
Charlotte, N.C.	79	96	109	144	178
Lynchburg, Va.	83	95	114	134	156
Raleigh, N.C.	...	82	102	120	136
Baton Rouge, La.	88	89	85	93	102
Atlanta, Ga.	99	115	122	130	138
Buffalo, N.Y.	210	214	242	...	...
Toledo, Ohio	123	127	149	169	171
Akron, Ohio	91	120	138	150	162
Cedar Rapids, Iowa	...	91	136	170	200
Madison, Wis.	110	135	150	160	175
Des Moines, Iowa	100	100	114	117	125
Omaha, Neb.	140	156	181	190	207
Wichita, Kan.	93	117	146	148	160
Oklahoma City, Okla.	79	82	114	154	183
Dallas, Tex.	100	116	143	158	175
Austin, Tex.	99	122	140	140	138
Sacramento, Calif.	210	258	236	240	250
Oakland, Calif.	72	102	130	159	177
Portland, Oreg.	97	114	104	115	120
San Diego Co., Calif.	...	162	180	190	190
San Diego, Calif.	114	139	126	140	145
Salem, Oreg.	...	139	156	207	224

SOURCE: "Present and Future Estimates of Water Consumption," in *Public Works*, December 1956 (Ref. 74) Used with permission of Public Works Corporation.

industries. For more detailed information on a particular industry, the effluent guidelines development documents prepared by the USEPA are valuable sources of information.

Wastewater Flows and Loads   Wastewater flow and load information is of prime importance when assessing the impacts on water quality in the receiving water. When wastewater flow and wastewater characterization data are available from the local sewerage department, the basic procedures outlined for projecting water demands can be followed.

Domestic wastewater flows average about 100 gpcd (gallons per capita demand) (Ref. 77). However, this figure varies widely across the country. Normally, the peak-hour flow during the day is about 1.3 to 2.0 times the daily average flow (Refs. 35,36). In general, about 60 to 80 percent of water used is returned to the sewers as wastewater. This return percentage does not include an allowance for infiltration (the seepage of groundwater into

TABLE 6.11   Water Consumption for Various Types of Establishments

Type of establishment	Flow in gal/person or unit/day*
Dwelling units, residential:	
Private dwellings on individual wells or metered supply	50 to 75
Apartment houses on individual wells	75 to 100
Private dwellings on public water supply, unmetered	100 to 200
Apartment houses on public water supply, unmetered	100 to 200
Subdivision dwelling on individual well, or metered supply, per bedroom	150
Subdivision dwelling public water supply, unmetered, per bedroom	200
Dwelling units, transient:	
Hotels	50 to 100
Boarding houses	50
Lodging houses and tourist homes	40
Motels, without kitchens, per unit	100 to 150
Camps:	
Pioneer type	25
Children's central toilet and bath	40 to 50
Day, no meals	15
Luxury, private bath	75 to 100
Labor	35 to 50
Trailer with private toilet and bath, per unit (2½ persons)	125 to 150†
Restaurants (including toilet):	
Average	7 to 10
Kitchen wastes only	2½ to 3
Short order	4
Short order, paper service	1 to 2
Bars and cocktail lounges	2
Average type, per seat	35
Average type, 24-hour, per seat	50
Tavern, per seat	20
Service area, per counter seat (toll road)	350
Service area, per table seat (toll road)	150
Institutions:	
Average type	75 to 125
Hospitals	150 to 250
Schools:	
Day, with cafeteria or lunch room	10 to 15
Day, with cafeteria and showers	15 to 20
Boarding	75
Theaters:	
Indoor, per seat, two showings per day	3
Outdoor, including food stand, per car (3⅓ persons)	3 to 5
Automobile service stations:	
Per vehicle served	10
Per set of pumps	500
Stores:	
First 25-ft frontage	450
Each additional 25-ft frontage	400
Country clubs:	
Resident type	100
Transient type, serving meals	17 to 25
Offices	10 to 15
Factories, sanitary wastes, per shift	15 to 35
Self-service laundry, per machine	250 to 500
Bowling alleys, per alley	200
Swimming pools and beaches, toilet and shower	10 to 15
Picnic parks, with flush toilets	5 to 10
Fairgrounds (based on daily attendance)	1
Assembly halls, per seat	2
Airport, per passenger	2½

*Water under pressure, flush toilets and wash basins are assumed provided unless otherwise indicated. These figures are offered as a guide; they should not be used blindly. Add for any continuous flows and industrial usages. Figures are flows per capita per day, unless otherwise stated.

†Add 125 gallons per trailer space for lawn sprinkling, car washing, leakage, etc.

SOURCE "The Design of Small Water Systems," by J. A. Salvato, Jr., in *Public Works*, May 1960 (Ref. 75). Used with permission of Public Works Corporation.

TABLE 6.12 Industrial Water Use

Industry and product	Unit of product	Water required per unit, U.S. gallons
Chemical industries:		
Acetic acid	Ton of HAc	120,000–290,000
Alcohol	Gallon	52–138
Ammonia	Ton of NH$_3$	37,500
Ammonium sulfate	Ton	240,000
Calcium carbide	Ton	36,500
Calcium metaphosphate	Ton of Ca(Po$_3$)$_2$	2,800
Carbon dioxide	Ton	24,500
Caustic soda	Ton of NaOH (11%)	22,000–25,500
Cellulose nitrate	Ton	12,000
Charcoal and wood chemicals	Ton of CaAc$_2$	79,000
Corn refining	Ton of starch	333
Gasoline	Gallon	7–34
Gunpowder	Ton	200,000
Hydrochloric acid	Ton of 20 Bé°HCl	3,500
Hydrogen	Ton of H$_2$	800,000
Lactose	Ton	235,000
Oxygen	100 ft^3	65
Soap	Ton	300–600
Soda ash	Ton	18,000–22,000
Sulfuric acid	Ton of 100% H$_2$SO$_4$	800–6,000
Sulfur	Ton	3,000
Food and beverage industries:		
Beet sugar	Ton	20,000–25,000
Bread	Ton	600–1,200
Beans, green	Ton	20,000
Peaches and pears	Ton	5,300
Other fruits and vegetables	Ton	2,000–10,000
Gelatin	Ton	15,000–24,000
Meat packing	Ton live weight	5,000
Milk products	Ton	4,000–5,000
Oils, edible	Gallon	88
Sugar	Ton	1,200–2,600
Beer	Gallon	15
Whisky	Gallon	80
Pulp and Paper:		
Kraft pulp	Ton dry pulp	110,000
Sulfate pulp	Ton	82,000
Sulfite pulp	Ton	82,000–230,000
Soda pulp	Ton	101,000
Paper	Ton	47,000
Paperboard	Ton	17,500–103,000
Strawboard	Ton	31,500
Textile industries:		
Cotton	Ton	20,000–76,000
Cotton bleaching	Ton	72,000–96,000
Cotton dyeing	Ton	9,500–19,000
Linen	Ton	200,000
Rayon	Ton of yarn	105,000–240,000
Wool scouring	Ton	40,000–240,000
Metal and metal products:		
Rolled steel	Ton	96,000
Finished steel	Ton	79,000
Fabricated steel	Ton	52,000
Steel sheets	Ton	16,000
(Average all products)	Ton	20,000–35,000
Aluminum	Ton	360,000
Miscellaneous:		
Electric power	kWh	85–185
Coal washing	Ton	1,800–4,300
Leather tanning	Ton, raw hide	19,500
Synthetic rubber	Ton	24,000–800,000

SOURCE: *Engineering Management of Water Quality*, by P. H. McGauhey (Ref. 76). McGraw-Hill Book Company, New York, 1968. Used with permission of McGraw-Hill Book Company.

the sewers through cracked pipe and faulty joints), which normally ranges from 1000 to 40,000 gal per day/mi (Ref. 35). Wastewater flows from hotels, retail stores, and office space average about 60,000 gpd/acre; warehouse and wholesale establishments generate about 15,000 gal per day/acre; and light industrial areas produce about 14,000 gal per day/acre (Ref. 77). Wastewater generation rates for major industries are extremely variable. For this reason, the USEPA effluent guidelines development documents should be consulted for wastewater flow and load information on particular industries.

Table 6.13 presents estimates of typical domestic wastewater characteristics as published by Metcalf and Eddy, Inc. (Ref. 35). Values are shown for weak, average, and strong wastewater. Industrial wastewater characteristics vary widely even within a single indus-

TABLE 6.13   Characteristics of Typical Domestic Wastewater

	Concentration (mg/L)		
Constituent	Weak	Medium	Strong
Solids, total	350	700	1,200
Dissolved, total	250	500	850
Fixed	145	300	525
Volatile	105	200	325
Suspended, total	100	200	350
Fixed	30	50	75
Volatile	70	150	275
Settleable solids, (mL/liter)	5	10	20
Biochemical oxygen demand, 5-day, 20°C	100	200	300
Total organic carbon (TOC)	100	200	300
Chemical oxygen demand (COD)	250	500	1,000
Nitrogen, (total as N)	20	40	85
Organic	8	15	35
Free ammonia	12	25	50
Nitrites	0	0	0
Nitrates	0	0	0
Phosphorus, (total as P)	6	10	20
Organic	2	3	5
Inorganic	4	7	15
Chlorides*	30	50	100
Alkalinity, (as $CaCO_3$)*	50	100	200
Grease	50	100	150

*Values should be by amount in carriage water.
SOURCE: *Wastewater Engineering* by Metcalf and Eddy, Inc. (Ref. 35) McGraw-Hill Book Company, New York, 1972. Used with permission of McGraw-Hill Book Company.

try. For information regarding industrial wastes, the reader is referred to industrial waste textbooks such as *Theories and Practices of Industrial Waste Treatment* by Nemerow (Ref. 78) and the USEPA effluent guidelines documents.

Wastewater usually receives some degree of treatment before it is released to surface waters. Estimates of the effectiveness of various wastewater treatment processes for removing pollutants from wastewater are shown in Table 6.14.

WATER QUALITY IMPACTS BY PROJECT TYPE

Environmental impacts must be evaluated on a case-by-case basis. Impacts from identical projects in two different environmental settings may, and probably will, be considerably different. This final section of the chapter presents a discussion of the predominant types

TABLE 6.14 Efficiencies of Wastewater Treatment Processes

Unit processes	Percent removals[a]							
	BOD	Suspended solids	Metal	Phosphates	Total nitrogen	Bacteria	Viruses	Dissolved solids
Pre-treatment	5	5	0	0	0	0	0	0
Primary sedimentation	30	50	20	10	0	50	10	0
Chemical coagulation	75	90	50	90	0	60	90	0
Biological secondary treatment	85[b]	90[b]	50[b]	25	30	80	75	0
Secondary treatment in lagoons	80[b]	90[b]	10[b]	10	0	95	70	0
Microscreens (after secondary treatment)	93[b]	93[b]	0	0	0	90	90	0
Rapid sand filtration (after secondary treatment)	97[b]	98[b]	0	0	0	95	90	10[b]
Phosphate removal	0	0	0	95[b]	0	0	0	0
Nitrogen removal	0	0	0	0	85	0	0	0
Ion exchange	60	0	90	98	80	0	0	85
Adsorption	96[b]	96[b]	10	10	10	60	35	40
Disinfection	0	0	0	0	0	99.9[b]	99[b]	0
Storage	5	5	0	0	0	10	10	0
Irrigation and cropping	98[b]	97[b]	0	61	79[b]	99	99	0
Spreading-percolation	96[b]	94[b]	0	0	38[b]	0	0	0

NOTES: (a) Percent reduction by the particular unit process.
(b) Total percent reduction of this unit process and all other prior processes in the treatment system.
SOURCE: *Feasibility Study for Wastewater Management*, by Havens and Emerson, Ltd. (Ref. 79).
Used with permission of Havens and Emerson, Ltd.

of impacts on water resources/water quality environmental indicators for several major project types. These generalized impacts are summarized in Table 6.15. In each case, the emphasis is placed on primary impacts.

Construction

The major primary impacts associated with construction activities are usually temporary, lasting only about as long as the construction phase. The predominant impact is increased erosion from the construction site and increased sedimentation downstream from the site. Potential sedimentation problems are heightened if the site is adjacent to a waterway or is crossed by waterways. Wolman and Schick (Ref. 80) report that normal sediment yields for undisturbed watersheds in the Baltimore, Maryland–Washington, D.C. area were on the order of 200 to 500 ton/mi² · yr, and wooded areas had yields as low as 15 ton/mi² · yr. They report maximum yields from construction sites as high as 140,000 ton/mi² · yr. They also report that sediment concentrations in streams downstream of construction sites were on the order of 3000 to 150,000 mg/L, while streams in undisturbed areas had sediment loads of less than 2000 mg/L.

Erosion from construction sites can be minimized by implementing viable erosion control measures. Dallaire (Ref. 81) recommends the following:
1. Diversion of upstream runoff away from the site
2. Limitation of the area denuded of vegetation at any one time
3. Limitation of the time of exposure in a denuded state
4. Protection of bare soil from the full energy of rainfall
5. Slow the velocity of runoff from the site
6. Trap sediment in runoff from the site in retention basins

The SCS (Ref. 82) presents a detailed discussion of measures for the control of erosion from construction sites. Mitigative measures detailed include: temporary vegetative measures during construction, mulching, permanent vegetative cover after construction, sediment basins, runoff diversions, jute thatching, and grading.

A lesser impact of construction activities is the leaching of oil and chemicals used during construction into surface and groundwaters (Ref. 83). Their impact will be more significant for very small streams in the vicinity of the construction site.

The USEPA (Ref. 64) presents a detailed discussion of the impacts of construction activities.

Highways

Perhaps the most significant impacts of highways occur during the construction phase. These construction impacts closely parallel impacts discussed previously for construction activities. Erosion and sedimentation impacts are particularly acute since steep slopes are often evidenced along highway embankments, large areas may be denuded during construction, and highways will normally cross many drainage patterns. Burton, et al. (Ref. 84) reported a sediment yield of 82 metric tons/ha · yr or 750 metric tons/km · yr from the construction of a highway in northern Florida. Scheidt (Ref. 83) reported sediment yields from highway construction of about 3000 ton/mi · yr in Maryland. Scheidt also noted that highway construction resulted in increased turbidity and increased concentrations of total phosphorus and silica in nearby streams.

Vice, et al. (Ref. 85) observed sediment yields of 151 ton/acre · yr from highway construction in Fairfax Country, Virginia. Of this amount, 76 ton/acre · yr was transported from the watershed. The Highway Research Board (Ref. 86) states that highway construction may result in the loss of as much as 15,000 tons of soil per mile if no sediment control is practiced. This publication also presents measures for mitigating erosion impacts. Diseker, et al. (Ref. 87) observed that unprotected highway banks could los up to 300 ton/acre · yr. Diseker and Richardson (Ref. 88) noted annual erosion losses of 41 to 359 tons per acre from unprotected road cuts.

Parizek (Ref. 89) also discusses the possible impacts of highway construction on groundwater and surface water flow. A potential problem is the beheading of shallow aquifers in hilly and mountainous areas.

Once the highway is operational, other adverse impacts have been observed (Refs. 80, 83, and 89). Materials in automotive exhausts are released into the atmosphere, and small amounts of material may find their way into local watercourses. Spills from trucks carrying petroleum products and other chemicals pose potential problems. In northern climates,

TABLE 6.15 Potential Impacts from Major Projects

Project or Activity

Environmental indicator	Construction	Highways	Urban/suburban growth	Industrial expansion	Power plants	Dams and reservoirs	Dredging	Channelization	Mining	Agriculture/irrigation	Forest management
Surface stream discharge		●	●			●		●		●	
Surface water quality											
Temperature	●			●	●	●	●	●		●	●
BOD			●	●			●			●	
Dissolved oxygen			●	●	●	●				●	
Suspended solids	●	●	●	●			●	●	●	●	●
Turbidity	●	●	●	●	●		●	●	●	●	●
Total dissolved solids		●	●	●					●		
pH				●					●		
Bacteria and viruses			●	●		●				●	
Nitrogen		●	●	●	●	●	●	●		●	●
Phosphorus			●	●		●	●	●	●	●	
Hardness				●						●	
Iron and manganese				●		●	●				
Chlorides		●	●	●			●			●	
Heavy metals				●	●		●		●		
Radioactivity				●	●				●		
Pesticides				●		●	●			●	●
Toxic substances				●							
Stratification					●	●					
Flooding			●					●			
Groundwater											
Quantity		●	●							●	
Quality			●							●	
Erosion	●	●	●					●	●	●	●
Sedimentation	●	●	●	●		●	●	●	●	●	●
Water demand			●						●	●	
Wastewater system				●	●						

salt used for highway deicing may significantly increase chloride and total dissolved solids concentrations in streams. During flood periods, highway embankments may serve to increase the magnitude of flooding in areas upstream of the highway due to constriction of flow by culverts and bridges.

Urban/Suburban Growth

Urban/surburban growth and related development can have major impacts on water resources and water quality. This development may be the primary project being evaluated, or the urban/suburban expansion may represent a secondary impact resulting from the project of primary concern. Industrial expansion, highway construction, airports, and other types of development tend to spur additional urban/surburban expansion.

Urban/suburban growth results in increased water demand. This places additional stress on existing surface or groundwater sources, water treatment systems, and water distribution systems. Methods for estimating this additional water demand were presented in the impacts section of this chapter.

An impact of urban/suburban growth is the increase in wastewater flows and loads. Estimating wastewater flows and loads was the subject of a portion of the impacts section of this chapter. Increased wastewater flows may overload existing sewers and wastewater treatment facilities. Increased flows and loads may result in significant decreases in the water quality of receiving waters. A positive aspect of urban/suburban growth is the possible elimination of septic tanks and small, poorly-operated wastewater treatment plants in outlying areas, which are swallowed up by the expansion. Elimination of septic tanks may improve water quality in surface waters as well as groundwater.

Urban/suburban growth also influences runoff from the area. The proportion of impervious land area is increased, resulting in increased quantities of runoff and reduced flow time. Storm sewers may drastically alter natural drainage patterns. Stormwater from urban and suburban areas may add a significant pollution load to the receiving stream. Dust and chemicals deposited on the land surface will be transported by stormwater runoff. Increased urbanization and associated industrialization results in more rapid accumulation of various pollutants on the earth's surface. Sartor, et al. (Refs. 90, 91) found about 1400 lb of total solids, 95 lb of COD, 1.1 lb of phosphate, and 2.2 lb of Kjeldahl nitrogen per curb mile on the surface of city streets. Also present in material on urban streets were heavy metals, pesticides, and coliform organisms. Vitale and Sprey (Ref. 92) and Whipple, et al. (Ref. 93) discuss the impacts of urban runoff.

Urban/suburban development may also result in increased erosion, particularly during the development stage. Keller (Ref. 94) observed that the erosion of lands undergoing transition from rural to urban/suburban use in the vicinity of Washington, D.C. caused increases of about 17 ton/mi² · yr in suspended sediment in local streams. Scheidt (Ref. 83) reports increases of 1000 to 121,000 tons/mi² · yr in erosion from lands undergoing urban development. Dallaire (Ref. 81) reports erosion rates of 69 ton/acre · yr for disturbed urban lands and 0.5 ton/acre · yr for well established urban areas. Furthermore, Guy and Ferguson (Ref. 95) note that urban construction produced about 39 tons of sediment per acre.

Urban/suburban growth may also result in additional secondary impacts on water quality, since urban/suburban development normally spurs additional industrial, commercial, and institutional growth, all of which may result in increased pollution flows and loads. Leopold (Ref. 96) discussed the impacts of urban development.

Industrial Expansion

Each year, roughly 1000 new chemicals are produced by U.S. industry (Ref. 97). By the year 2000, production by the chemical and allied products industries is projected to increase by as much as six times the 1970 level. In the iron and steel and the petroleum products industries, production may reach three and one-half times 1970 levels (Ref. 98).

Future wastewater flows and loads resulting from the growth in production and advancing technology have been projected. In most industries, advancing technology leads to lower flows and waste loads. For example, in the petroleum industry, plants using new technology generate about one-fifth as much wastewater, about one-third the BOD and sulfides, and about one-fourth the phenols as other plants of similar capacity (Ref. 98).

Anticipated future advances in production technology will further reduce wastewater flows and loads per production unit, including the steel industry (Ref. 98). In addition, advances in wastewater treatment technology are expected to provide for increases in pollutant removal, internal recycle, and beneficial use of by-products. This will reduce the waste loads ultimately reaching the nation's waterways. A new "industrial revolution" is already emerging which takes what has previously been considered "wastes" and reprocesses them, recycles them, finds secondary uses for them, and reuses them (Ref. 99), all to the benefit of water quality.

The range of contaminants contained in raw industrial wastewaters is extensive and includes physical, chemical, biological, and radiological parameters. These wastes are derived from impurities contained in raw materials, undesirable byproducts from the processes, and unrecovered products. The number of important chemicals grows continuously. More than 12,000 chemicals are already on the government's toxic substances list, 1500 are suspected of causing tumors, and 30 compounds currently used in industry are known to cause cancer (Ref. 97).

Parameters identified by the USEPA in the effluent limitations guidelines documents for each industry should be included in the water quality assessment.

Power Plants

Electric power generation in the United States has doubled every 10 years since 1945, an annual growth rate of approximately 7 percent (Refs. 100, 101). Based on the shortage of some basic fuels, this is conservatively projected to increase at 4.5 percent annually through the year 2000 (Ref. 102).

In 1970, approximately 1.6 trillion kWh of electrical power were generated, 16.3 percent by hydroelectric stations, 81.8 percent by fossil-fueled steam electric stations, and 1.9 percent by nuclear-powered stations (Ref. 101). By 1990, 3.7 trillion kWh of power will be generated (Ref. 102), 12.1 percent by hydroelectric, 50.2 percent by fossil fuel, and 37.7 percent by nuclear stations (Ref. 101).

Both fossil fuel and conventional nuclear power plants generate electricity by expanding superheated steam through turbines which drive generators. Completion of the thermal cycle requires condensing the steam to the liquid phase. Modern fossil fuel plants operate at a thermal efficiency of approximately 40 percent, and 10 percent of the waste heat is discharged through the stacks. As a result, more than 54.5 percent of the energy entering those plants is released through the condenser cooling water. Since nuclear plants operate at a thermal efficiency of only about 33 percent and nearly all thermal wastes are released through the cooling waters, nuclear plants release nearly 50 percent more waste heat into the water than comparably sized fossil-fueled plants (Ref. 101).

Cooling water from the once-through cooling process is generally 15 to 20°F higher than the ambient temperature of the source when it is released. A variety of discharge configurations have been developed to minimize the effects of the heated discharge on the receiving water, including multiple-discharge ports, diffusers, vertical placement of intake, and discharge structures which take advantage of stratified conditions in lakes and impoundments.

In order to adequately assess the impacts of heated discharges on receiving waters, one should know the resulting temperature variations longitudinally, laterally, and vertically. Longitudinal temperature effects influence the extent of the thermal plume in the direction of flow. The existence of "zones of passage" (which provide for passage of fish around the zone of increased temperature) can be determined through lateral variations, and the extent of stratification induced by the heated discharge can be determined through vertical variations.

Although temperature is the primary water quality parameter used for assessing impacts from power plants, several other water quality parameters are affected by the change in temperature. Dissolved oxygen is probably the most important of these parameters. Oxygen solubility is lowered by higher temperatures, while the bacterial respiration rate is increased. In the presence of an organic load, oxygen resources can be reduced in the receiving waters. Other chemical reactions are also affected by temperature. Attention should be given to the ionic strength, dissociation, conductivity and solubility of chemicals entering the receiving water within the zone affected by the heated discharge (Ref. 100).

In 1965, more than 22.5 percent of the power plants in the United States with a capacity greater than 100 MW (megawatts) used cooling ponds or cooling towers, rather than once-through cooling systems to reduce or eliminate thermal pollution. Because wet-cooling towers release waste heat into the atmosphere through evaporation of water, makeup water must be continuously added. This evaporation causes natural salts in the makeup water to become so concentrated that chemicals may become deposited and inhibit the system's ability to limit the concentration of chemicals. A portion of the cooling water is removed from the system as blowdown, which amounts to 0.5 to 3 percent of the recirculating water flow. Blowdown contains concentrated amounts of salts from the makeup water, plus chemicals which are added to makeup water to prevent corrosion and/ or biological fouling. Those chemicals include acids, zinc, phosphates, nitrogen-based compounds, chlorine, and copper sulfate (Ref. 103). The assessment of water quality impacts from wet-cooling-tower blowdown should be similar to the assessment for an industrial effluent containing similar chemicals.

Hydroelectric power plants are not free of water quality impacts. If water is drawn from the hypolimnion of impoundments during summer, the temperature of the released water may be substantially lower than that of the ambient downstream waters, and the concentrations of some dissolved gases, particularly nitrogen, substantially higher. Since either of those conditions can apply stress to aquatic biota (Ref. 104), these parameters are of concern in water quality assessment.

Impoundments

The placement of a dam on a free-flowing stream creates an environment within the resulting impoundment which is significantly different from that of the natural stream. This, consequently, affects the environment of the downstream segment receiving water released from the impoundment. The direct physical effects of impoundments include increased water depth, increased detention time, and the potential for thermal stratification. These changes, in turn, affect a broad spectrum of water quality parameters within the impoundment, and the release of that water affects water quality downstream.

Impoundments produce both beneficial and adverse water quality effects. The adverse effects include: (1) decreased dispersion of waste discharges along the shoreline, resulting from reduced velocities, (2) reduced reaeration, caused by increased depths and reduced velocities; (3) tastes and odors resulting from increased algal activity, encouraged by reduced velocities; (4) accumulations of sludges, caused by the absence of bottom scour due to increased depth and reduced velocities; and (5) a variety of impacts resulting from thermal stratification.

The cold, denser water in the hypolimnion is effectively sealed off from the atmosphere by the epilimnion. As a result, the oxygen demand of the hypolimnion and bottom sediments removes dissolved oxygen from the hypolimnion. When anaerobic conditions prevail, iron and manganese are reduced and dissolve, sulfates are reduced, hydrogen sulfides are formed, and the pH of the water is lowered by the excess carbon dioxide that is created (Ref. 105).

Beneficial impacts of impoundments include: (1) reduced turbidity, resulting from long detention times and low velocities; (2) reductions in the hardness of the water from carbon dioxides produced by algae and precipitation of calcium carbonate; (3) reduced BOD, resulting from long detention times, which permit biodegradation; and (4) reduced coliform density, due to natural die-off resulting from long detention times (Ref. 105). Principal water quality impacts that take place downstream and are produced by reservoirs are caused by the release of hypolimnetic water. These impacts include reductions in nutrients and suspended solids concentrations (Ref. 106), increases in iron and manganese, low dissolved oxygen concentrations and temperatures, and the presence of hydrogen sulfide. These impacts are aggravated at peaking-power hydroelectric plants by subjecting downstream aquatic biota to wide variations in those parameters (Ref. 107).

Dredging

The primary purposes of dredging are to maintain, improve, or extend navigable waterways and to provide construction materials such as sand, gravel, or shell. As the agency responsible for the development of U.S. waterways, the Corps of Engineers dredges about 300 million yd³ of sediment annually for waterway maintenance and about 80 million yd³ in new work. Historically, the selection of dredging equipment and location of disposal sites have primarily been based on economic considerations. Open water disposal has been the dominant disposal method, with about two-thirds of the material dredged for waterway maintenance being disposed of in this manner. Some dredged materials have become polluted due to the concentration of industrial and population centers around the nation's navigable waterways (Ref. 108).

The potential for adverse impacts on water quality from dredging activities is influenced by many factors, including: the predominant soil types within the watershed; the magnitude of agricultural, urban, and industrial activity in the watershed; the quantity of material handled; the nature of the disposal site; and the equipment selected for removal, transportation, and disposal of dredged material (Ref. 109).

During the excavation process, the water quality impacts are principally short-term and include high turbidity, resuspension of contaminated materials into the water column, dissolved oxygen depletion, the release of nutrients and other materials from the sediments, and the creation of scum. When the dredged materials are transported, careless operation or poorly maintained equipment can produce water quality impacts which are similar to those produced by excavation (Ref. 109).

Unconfined disposal of dredged material on land minimizes the effects on water quality, especially if stormwater control is provided. Disposal of dredged material in confined disposal areas usually limits the sources of potential water quality impacts to seepage through confining dikes, which allows the water to percolate into groundwaters (Ref. 109) and returns the supernatant to the waterway. If adequate settling time is provided prior to release of the supernatant, the concentrations of constituents released to the waterway can be minimized (Ref. 110).

Open water disposal of dredged material may release ammonia and manganese to the water column. Under oxidizing conditions, orthophosphates may be removed from the water column by the hydrous ferric oxides and other mineral particulate matter. Under anoxic conditions, large amounts of orthophosphates may be released to the water column (Ref. 111).

While substantial amounts of manganese may be released to the water column from discharged sediment, the concentration of other heavy metals may be reduced in the water column. This phenomenon is caused by the scavenging of many heavy metals by ferric hydroxide, which is formed when anoxic sediments are disposed. Pesticides may be released from sediments having a low oil and grease content, but may be absorbed from the water column by discharged sediment containing high oil and grease concentrations (Ref. 98).

Channelization

Stream channelization is intended to increase the rate of flow through the channel, thereby decreasing bank overflow and reducing flood damages. The channel configuration usually determines whether pools and riffles will be replaced by uniformly shallow water depths and steady velocities. During construction, sediment is resuspended and downstream turbidity may be increased. Placement of silt curtains or isolation of the segments under construction are effective measures for reducing sediment transport.

The long-term effects of channelization on water quality are complex and subtle, and instantaneous measurements or short-term studies do not always reveal any impacts (Refs. 112,113). If oxygen-demanding materials are present, the steady velocities will cause steady concentrations of dissolved oxygen in channelized streams. In contrast, dissolved oxygen is depleted in pool areas of natural streams. Water temperatures in channelized streams may exceed those in natural channels since channelization clears away vegetation which shades the channel (Ref. 114).

Other water quality impacts result from the changes in land use which usually accompany channelization. Because of the reduced threat of flooding, bottomland forests often are cleared for agricultural use, which results in the removal of the stream's biological buffer zone. As a result, overland flow reaching the stream is subjected to heating from radiant energy. This flow will probably carry an increased sediment load, and will contain higher concentrations of nutrients (Ref. 114).

In natural floodplains, flooding often deposits rich sediments and reduces the need for additional fertilization. When the flood frequency is reduced by channelization, increased amounts of fertilizer are required for comparable crop yields and the runoff from these lands contains increased quantities of nutrients. In Illinois, the nitrate concentrations in some channelized streams have reached levels which interfere with the use of downstream water for public water supplies (Ref. 114).

Mining

About 200 of the 1600 known mineral species are considered economic minerals (Ref. 115). These economic minerals are mined by either underground or surface mining.

Prior to 1965, approximately 3.2 million acres of land had been disturbed by surface mining activities and an additional 10 million acres are expected to be disturbed before the year 2000 (Ref. 98). The principal surface mining methods include open pit, strip, quarry, hydraulic, dredging, and solution mining (Ref. 115).

Mining generates two major forms of water pollution—physical and chemical. Siltation constitutes the major physical impact on water quality from both surface and underground mines. The spoil banks, rock dumps, and tailings piles resulting from those operations are particularly vulnerable to erosion because of their inability to support vegetation and their fine-grained nature (Ref. 116).

Most chemical pollution results from the oxidation of sulfides which occurs with most minerals mined. Mining puts large quantities of sulfides in direct contact with oxygen forming sulfates which, in contact with runoff, form sulfuric acid. The acid then dissolves metals which are carried to waterways (Ref. 98, 116).

Water pollution is also caused by the exposure of water-leachable minerals to water. Leaching is principally associated with solution mining.

Unusual forms of chemical pollution include fluorides and radioactivity produced by some phosphate mining, radioactivity from uranium mining, and asbestos from taconite mining operations.

Agriculture and Irrigation

Agricultural activities can have significant impacts on water quality, including increases in stream sedimentation from erosion, and increases in nutrients, pesticides, and salt concentrations in runoff (Refs. 59, 76). The major areas of concern are irrigation practices and feedlot operations.

Erosion from agricultural lands was discussed in some detail in the impacts sections of this chapter. Dallaire (Ref. 81) reports the average erosion of farmlands was 4 ton/ acre·yr. Wolman and Schick (Ref. 80) suggest that agricultural lands in the Baltimore/ Washington, D.C. area may lose as much as 1000 tons/acre · yr due to erosion. As discussed in the impacts section, there are land management practices that can significantly reduce the rate of erosion from agricultural lands.

Irrigation is, of course, extremely important in the agricultural production of the United States. In 1971, about 48 million acres of land were irrigated in the United States (Ref. 117). Of this total, 43 million acres were located in 17 western states. The importance of irrigation is realized when one notes that about 10 percent of the total cropland in the United States is irrigated, yet irrigated lands account for about 25 percent of the total value of American crops.

Irrigation return flows can present a significant adverse water quality impact. Many of the problems with irrigation stem from the fact that only about 20 to 60 percent of water diverted for irrigation is returned to the stream (Ref. 76). The average return water flow in the western states is about 33 percent. Return flows contain significantly higher concentrations of salts, hardness-causing ions, nutrients, and pesticides (Refs. 76, 118). In addition, the temperature of the return flow may be elevated above the normal stream temperature (Ref. 76). The return flow may have salinity values 5 to 11 times those of the diverted flow. Concentrations of calcium and magnesium may increase by 3.5 to 5.5 and 4 to 14 times the concentrations in the diverted flow.

In 1968, an average of 135 lb of plant nutrients were applied to each acre of farmland in the north central states (Ref. 118). This represents a fourfold increase over application rates used in 1945. About 32 percent of this amount was applied during the Fall. In 1974, an annual average of 103 lb/acre of nitrogen and 27 lb/acre of phosphorous were applied to corn crops, while 46 lb/acre of nitrogen and 17 lb/acre of phosphorus were applied to wheat crops (Ref. 59). It is not surprising that some of these nutrients find their way into the groundwater and surface streams. Increased concentrations of nitrogen and phosphorus have been observed (Refs. 59, 118, 119). Goldberg (Ref. 119) reports total nitrogen concentrations of 1 to 60 mg/L in agricultural runoff.

Pesticides have also been reported in agricultural runoff (Refs. 59, 120). In fact, agricultural runoff represents the major source of pesticides in surface waters (Ref. 120). The concentration of pesticides in runoff is relatively low—normally, less than 5 percent of pesticides applied to croplands are found in runoff. However, herbicides, insecticides, and fungicides may be very toxic to fish and other aquatic animals, even in small concentrations.

Animal wastes may also substantially contribute to degradation of water quality. Feedlots containing high densities of animals are of particular concern. In 1973, there were approximately 101 million head of beef cattle in the United States (Ref. 121). Feedlots contained about 14 million head of beef cattle. One beef cow produces an average of 10 lb/day solids, 1.0 lb/day BOD_5, 0.3 lb/day nitrogen, and 0.1 lb/day of P_2O_5 (Ref. 122). Runoff from a dirt beef cattle feedlot may average 1500 mg/L BOD_5, 150 mg/L total nitrogen, and 80 mg/L total phosphorus (Ref. 121). Concentrations in runoff from paved feedlots average 3200 mg/L BOD_5, 1100 mg/L total nitrogen, and 110 mg/L total phosphorus. Feedlots for dairy cattle, hogs, and poultry also result in significant pollution loads. Animal wastes also contain relatively large amounts of pathogenic organisms that may harm other animals or, in some cases, humans (Refs. 122, 123). Feedlot wastes offer the advantages of being relatively easy to collect and treat prior to discharge to surface waters.

The USEPA (Ref. 64) presents a detailed discussion of the impacts of agricultural activities.

Forest Management

Undisturbed forested watersheds are an excellent source of high-quality water. Sediment yields from forest lands are on the order of 0.03 ton/acre · yr (Ref. 85). Human forest-management activities, however, can exert disruptive forces on the forest ecosystem in the form of erosion. Erosion problems are often caused by logging activity, particularly clearcutting. However, major erosion problems normally result from logging roads (Ref. 124). Sopper (Ref. 124) refers to an experiment where a clearcutting operation resulted in average stream turbidities of 490 Jackson Turbidity Units (JTU) during the logging operation, decreasing to 38 JTU after one year, and to 2 JTU after two years. A second clearcut employing a well-planned and well-maintained system of logging roads resulted in average turbidities of only 6 JTU during logging activities.

Logging activities may also cause increased water temperature because the removal of overstorey shade permits direct solar radiation to reach the stream. Mean monthly maximum temperature increases of 14°F have been observed by Brown and Krygier (Ref. 125). Swift and Messer (Ref. 126) noted temperature increases of as much as 12°F. These impacts can be minimized by leaving a strip of trees and brush along stream channels in the area.

Logging operations may also have an impact on nutrient concentrations in streams. Likens, et al. (Ref. 127) report increases in nitrate concentrations from 2 mg/L to 80 mg/L after clearcutting. It should be noted that herbicides were used to prevent regrowths for three years after the clearcut. Pierce, et al. (Ref. 128) report less significant nutrient increases from clearcutting. Stripcutting was shown to be an effective means of reducing nutrient loss (Ref. 124).

Area fertilization may significantly increase the nitrogen concentrations in streams within a 48-hour period after fertilization (Ref. 124). Aubertin, et al. (Ref. 129) observed significant increases in ammonia and nitrate concentrations in streams following a 500 lb/acre application of fertilizer to an area. The application of herbicides to forest lands may also result in short-term increases in the concentration of herbicides in a stream.

In general, the impacts of logging operations are short-term. Erosion and nutrient loss tend to peak during the logging operation itself. Within two years after the timber harvest, background levels are approached as revegetation occurs. The USEPA (Ref. 64) presents a detailed discussion of the water quality impacts of forestry practices.

SUMMARY

This chapter has dealt with procedures and methodologies for assessing the impacts on indicators of water quality. Methods for describing the existing environmental setting and for predicting the consequences of the proposed projects and alternatives have been presented. A significant amount of emphasis has been placed on field surveys and mathematical modeling. Finally, the potential impacts of several major projects were presented. These potential impacts were presented to provide guidance on the types of impact areas that should be considered for typical projects.

This chapter was designed as a guide for the individual who is relatively new to the field of water quality impact assessment. It is hoped that the wise veteran of many impact assessments may gain some fresh ideas or new methods for future work. The tables and figures presented should prove to be a ready reference in water quality analysis. In addition, the literature cited represents a stepping-off point to more detailed information.

REFERENCES

1. Fair, G. M. and J. C. Geyer, *Water Supply and Wastewater Disposal*, J. Wiley & Sons, Inc., New York, 1963.
2. American Public Works Association, *History of Public Works in the United States 1776–1976*, Ed. E. L. Armstrong, Chicago, 1976.
3. Sedgwick, W. T., *Principles of Sanitary Science and the Public Health*, Macmillan, New York, 1902.
4. American Water Works Association, *Water Quality and Treatment*, 2nd ed., New York, 1951.
5. U.S. Environmental Protection Agency, *Quality Criteria for Water*, Washington, D.C., 1976.
6. California State Water Pollution Control Board, *Water Quality Criteria*, Sacramento, Calif., 1952.
7. McKee, J. E. and H. W. Wolf, *Water Quality Criteria*, State Water Quality Control Board, Sacramento, Calif., Pub. 3-A, 1963.
8. *Water Quality Criteria*, Report of the National Technical Advisory Committee to the Secretary of the Interior, Federal Water Pollution Control Administration, Washington, D.C., April 1, 1968.
9. "EPA Regulations on Water Quality Standards approved by the Federal Government," *Federal Register*, 41:25000, June 22, 1976.
10. *Water and Sewage Works–Water Wastewater Handbook, Reference '76*, Scranton Gillette, Chicago, April 30, 1976.
11. "Water Programs, National Interim Primary Drinking Water Regulations," *Federal Register*, 40:248, December 24, 1975.
12. Kittrell, F. W., *A Practical Guide to Water Quality Studies of Streams*, USGPO, Washington, D.C., 1969.
13. Linsley, R. K., M. A. Kohler, and J. L. H. Paulhus, *Hydrology for Engineers*, McGraw-Hill, New York, 1958.
14. Kulin, G. and P. R. Compton, *A Guide to Methods and Standards for the Measurement of Water Flow*, National Bureau of Standards Special Publication 421, U.S. Government Printing Office, Washington, D.C., May, 1975.
15. *Standard Methods for the Examination of Water and Wastewater*, 14th ed., Washington, D.C., APHA, AWWA, WPCF, 1975.
16. *Methods for Chemical Analysis of Water and Wastes*, EPA-625-16-74-003a, Washington, D.C., U.S. Environmental Protection Agency, 1976.
17. Rabosky, J. G. and D. L. Koraido, "Gaging and Sampling Industrial Wastewaters," *Chemical Engineering*, 80(1):111–120, January, 1973.
18. Baca, R. G., W. W. Waddel, C. R. Cole, A. Brandstetter, and D. B. Cearlock, *Explore–I: A River Basin Water Quality Model*, Battelle Northwest Laboratories, Richland, Wash., 1973.
19. Warner, M. L., J. L. Moore, S. Chatterjee, D. C. Cooper, C. Ifeadi, W. T. Lawhon, and R. S. Reimers, *An Assessment Methodology for the Environmental Impact of Water Resource Projects*, EPA-600/5-74-016, USGPO, Washington, D.C., July 1974.
20. O'Connor, D. J. and W. E. Dobbins, "Mechanism of Reaeration in Natural Streams," *Transactions of ASCE*, 123:641–666, Paper No. 2934, 1958.
21. Churchill, M. A., H. L. Elmore, and R. A. Buckingham, "Prediction of Stream Reaeration Rates," *Transactions of ASCE*, 129:24–26, 1964.
22. Langbien, W. B. and W. H. Durum, *The Aeration Capacity of Streams*, U.S. Geological Survey Circular 542, 1967.
23. Thackston, E. L. and P. A. Krenkel, "Reaeration Prediction in Natural Streams," *Journal of Sanitary Engineering Division, ASCE*, 95(SA1):65–94, Paper 6407, February 1969.
24. Owens, M., R. W. Edwards, and J. W. Gibbs, "Some Reaeration Studies in Streams," *International Journal of Air and Water Pollution*, 8:469–486, 1964.
25. Fair, G. M., "The Dissolved Oxygen Sag-An Analysis," *Sewage Works Journal*, 11(3):445–461, May 1939.
26. Streeter, H. W. and E. B. Phelps, *A Study of the Pollution and Natural Purification of the Ohio River*, Public Health Service Bulletin 146, (reprinted 1958).
27. Edinger, J. E., D. K. Brady, and J. C. Geyer, *Heat Exchange and Transport in the Environment*, EPRI Publication No. 74-049-00-3, Electric Power Research Institute, Palo Alto, Calif., 1974.
28. Velz, C. J., *Applied Stream Sanitation*, Wiley-Interscience, New York, 1970.
29. Mosby, H., "Verdunstung und Strahlung auf dem Merre," *Annalen der Hydrographic und Maritimen Meteorologic*, 64:281, 1936.
30. Anderson, E. R., "Energy-Budget Studies," *Water-Loss Investigations: Volume 1–Lake Hefner Studies Technical Report*, Geological Survey Circular 299, U.S. Geological Survey, Washington, D.C., 1952.

31. Baca, R. G., M. W. Lorenzen, R. D. Mudd, and L. V. Kimmel, *A Generalized Water Quality Model for Eutrophic Lakes and Reservoirs,* Report to USEPA, Battelle Northwest Laboratories, Richland, Wash., 1974.
32. Kennedy, R. E., "Computation of Daily Insolation Energy", *Bulletin American Metorological Society,* **30**(6):208–213, June 1949.
33. Duffie, J. A. and W. A. Beckman, *Solar Energy Thermal Processes,* John Wiley and Sons, New York, 1974.
34. Mathur, S. P. "Thermal Discharges," *Handbook of Water Resources and Pollution Control,* Ed. H. W. Gehm and J. I. Bregman, Van Nostrand Reinhold, New York, pp. 719–779, 1976.
35. Metcalf and Eddy, Inc., *Wastewater Engineering,* McGraw-Hill, New York, 1972.
36. Joint Committee of ASCE and WPCF, *Design and Construction of Sanitary and Storm Sewers,* ASCE MOP No. 37, ASCE, New York, 1969.
37. Schaake, J. C., Jr., J. C. Geyer, and J. W. Knapp, "Experimental Examination of the Rational Method," *Journal of Hydraulics Division, ASCE,* **93**(HY6): 353–370. Nov. 1967.
38. Hershfield, D. M., *Rainfall Frequency Atlas of the United States for Durations from 30 Minutes to 24 Hours and Return Periods from 1 to 100 Years,* Technical Paper No. 40, Weather Bureau, U.S. Department of Commerce, USGPO, Washington, D.C., 1961.
39. U.S. Soil Conservation Service, *SCS National Engineering Handbook, Section 4, Hydrology,* USGPO, Washington, D.C., 1971.
40. Metcalf & Eddy, Inc., University of Florida, and Water Resources Engineers, Inc., *Storm Water Management Model,* Vol I, EPA 11024DCO7/71, USGPO, Washington, D.C., 1971.
41. Crawford, N. H. and R. K. Linsley, *Digital Simulation in Hydrology: Stanford Watershed Model IV,* Technical Report No. 39, Stanford University, Stanford, Calif., 1966.
42. Tholin, A. L. and C. J. Keifer, "Hydrology of Urban Runoff," *Transactions of ASCE,* **125**:1308–1379, 1960.
43. Keyes, D. L., *Land Development and the Natural Environment: Estimating Impacts,* The Urban Institute, Washington, D.C., 1976.
44. Heeps, D. P. and R. G. Mein, "Independent Comparison of Three Urban Runoff Models," *Journal of Hydraulics Division, ASCE,* **100**(HY7):995–1009, July 1974.
45. Marsalek, J., T. M. Dick, P. E. Wisner, and W. G. Clarke, "Comparative Evaluation of Three Urban Runoff Models," *Water Resources Bulletin,* **11**(2):306–328, April 1973.
46. Thomas, W. A., *Water Surface Profiles,* Vol. 6 of *Hydrologic Engineering Methods for Water Resources Development,* Hydrologic Engineering Center, U.S. Army Corps of Engineers, Davis, Calif., 1975.
47. U.S. Army Corps of Engineers, *Backwater Curves in River Channels,* EM 1110-2-1409, U.S. Army Corps of Engineers, 1959.
48. Chow, V. T., *Open-Channel Hydraulics,* McGraw-Hill, New York, 1959.
49. Hydrologic Engineering Center, U.S. Army Corps of Engineers, *HEC-2 Water Surface Profiles,* Computer Program 723-X6-L202, HEC, U.S. Army Corps of Engineers, Davis, Calif., 1973.
50. Eichert, B. S., "Survey of Programs for Water-Surface Profiles," *Journal of Hydraulics Division, ASCE,* **96**(HY2):547–563, Feb. 1970.
51. Vanoni, V., Ed., *Sedimentation Engineering,* MOP No. 54, ASCE, New York, 1975.
52. Johnson, H. P. and W. C. Moldenhauer, "Pollution by Sediment: Sources and the Detachment and Transport Processes," *Agricultural Practices and Water Quality,* T. L. Willrich and G. E. Smith, Eds., Iowa State University, Ames, Iowa, 1970.
53. Wischmeier, W. H. and D. D. Smith, *Predicting Rainfall–Erosion Losses from Cropland East of the Rocky Mountains,* Argicultural Handbook No. 282, Agricultural Research Service, U.S. Department of Agriculture, USGPO, Washington, D.C., 1965.
54. Olson, T. C. and W. H. Wischmeier, "Soil-Erodibility Evaluations for Soils on the Runoff and Erosion Stations," *Proceedings Soil Science Society of America,* **27**(5):590–592, September-October, 1963.
55. Wischmeier, W. H. and D. D. Smith, "Rainfall Energy and Its Relationship to Soil Loss," *Transactions of American Geophysical Union,* **39**(2):285–291, April 1958.
56. Hydrologic Engineering Center, U.S. Army Corps of Engineers, *Urban Storm Water Runoff–"STORM,"* Computer Program 723-58-L2520, HEC, U.S. Army Corps of Engineers, Davis, Calif., 1974.
57. Wischmeier, W. H., C. B. Johnson, and B. V. Cross, "A Soil Erodibility Nomograph for Farmland and Construction Sites," *Journal of Soil and Water Conservation,* **26**(5):189–192, September-October 1971.
58. Wischmeier, W. H. and J. V. Mannering, "Relation of Soil Properties to Its Erodibility," *Proceedings of Soil Science Society of America,* **33**(1):131–137, January–February 1969.
59. *Control of Water Pollution from Cropland,* Report No. ARS-H-5-1, Agricultural Research Service, U.S. Department of Agriculture and Office of Research and Development, USEPA, 1975.
60. Smith, D. D. and W. H. Wischmeier, "Factors Affecting Sheet and Rill Erosion," *Transactions of American Geophysical Union,* **38**(6):889–896, December, 1957.
61. Wischmeier, W. H., "Cropping-Management Factor Evaluations for a Universal Soil-Loss Equation," *Proceedings Soil Science Society of America,* **24**(4):322–326, July–August, 1960.
62. Lane, E. W., "Design of Stable Channels," *Transactions of ASCE,* **120**:1234–1260, 1955.

63. Colby, B. R. and C. H. Hembree, *Computations of Total Sediment Discharge Niobrara River Near Cody, Nebraska*, Water Supply Paper No. 1357, U.S. Geological Survey, Washington, D.C., 1955.

64. U.S. Environmental Protection Agency, *Methods for Identifying and Evaluating the Nature and Extent of Nonpoint Sources of Pollutants*, EPA-430/9-73-014, Washington, D.C., USGPO, Oct. 1973.

65. Dawdy, D. R., "Knowledge of Sedimentation in Urban Environments," *Journal of Hydraulics Division, ASCE*, 93(HY6):235–245, Nov. 1967.

66. Maner, S. B., "Factors Affecting Sediment Delivery Rates in the Red Hills Physiographic Area," *Transactions, American Geophysical Union*, 39(4):669–675, August 1958.

67. Brune, G. M., "Trap Efficiency of Reservoirs," *Transactions, American Geophysical Union*, 39(3):407–418, June 1953.

68. DeWiest, R. J. M., *Geohydrology*, John Wiley and Sons, New York, 1965.

69. Summers, W. K. and Z. Spiegel, *Ground Water Pollution: A Bibliography*, Ann Arbor Science, Ann Arbor, Mich., 1974.

70. Domenico, P. A., *Concepts and Models in Groundwater Hydrology*, McGraw-Hill, New York, 1972.

71. Roesner, L. A., J. R. Monser, and D. E. Evenson, *Computer Program Documentation for the Stream Qaulity Model QUAL–II*, Prepared for the USEPA, Water Resources Engineers, Inc., Walnut Creek, Calif., 1973.

72. Hydrocomp, Inc., *Hydrocomp Simulation Programming Mathematical Model of Water Quality Indices in Rivers and Impoundments*, Hydrocomp, Inc., Palo Alto, Calif.

73. Hydrologic Engineering Center, U.S. Army Corps of Engineers, *HEC-1 Flood Hydrograph Package*, Computer Program 723-X6-L2010, HEC, U.S. Army Corps of Engineers, Davis, Calif., 1973.

74. "Present and Future Estimates of Water Consumption," *Public Works*, 87(12):73–77, 152–156, December 1956.

75. Salvato, J. A., Jr., "The Design of Small Water Systems," *Public Works*, 91(5):109–133, May 1960.

76. McGauhey, P. H., *Engineering Management of Water Quality*, McGraw-Hill, New York, 1968.

77. Babbitt, H. E. and E. R. Baumann, *Sewerage and Sewage Treatment*, 8th ed., John Wiley and Sons, New York, 1958.

78. Nemerow, N. L., *Theories and Practices of Industrial Waste Treatment*, Addison-Wesley, Reading, Mass., 1963.

79. Havens and Emerson, Ltd., *Feasibility Study for Wastewater Management Program*, A Report to the Buffalo District, U.S. Army Corps of Engineers, 1971.

80. Wolman, M. G. and A. P. Schick, "Effects of Construction on Fluvial Sediment, Urban and Suburban Areas of Maryland," *Water Resources Research*, 3(2):451–464, 1967.

81. Dallaire, G., "Controlling Erosion and Sedimentation at Construction Sites," *Civil Engineering*, 47(10):73–77, October 1976.

82. Soil Conservation Service, *Guide for Sediment Control on Construction Sites in North Carolina*, U.S. Department of Agriculture, Raleigh, N.C., 1973.

83. Scheidt, M. E., "Environmental Effects of Highways," *Journal of Sanitary Engineering Division, ASCE*, 93(SA5):17–25, October 1967.

84. Burton, T. M., R. R. Turner, and R. C. Harris, "The Impact of Highway Construction on a North Florida Watershed," *Water Resources Bulletin*, 12(3):529–538, June 1976.

85. Vice, R. B., H. P. Guy, and G. E. Ferguson, *Sediment Movement in an Area of Suburban Highway Construction, Scott Run Basin, Fairfax County, Virginia, 1961–1964*, Geological Survey Water Supply Paper 1591-E, USGPO, Washington, D.C., 1969.

86. Highway Research Board, *Erosion Control on Highway Construction*, Report No. 18, National Academy of Sciences, Washington, D.C., 1973.

87. Diseker, E. G., E. C. Richardson, and B. H. Hendrickson, *Roadbank Erosion and Its Control in the Piedmont Upland of Georgia*, Agricultural Research Service, U.S. Department of Agriculture, ARS 41–73, 1963.

88. Diseker, E. G. and E. C. Richardson, "Erosion Rates and Control Methods on Highway Cuts," *Transactions, American Society of Agricultural Engineers*, 5:153–155, 1962.

89. Parizek, R. R., "Impact of Highways on the Hydrogeologic Environment," *Environmental Geomorphology*, D. R. Coats, ed., SUNY, Binghamton, NY, 1971.

90. Sartor, J. D., G. B. Boyd, and F. J. Agardy, "Water Pollution Aspects of Street Surface Contaminants," *Journal Water Pollution Control Federation*, 46(3):458–467, March 1974.

91. Sartor, J. D. and G. B. Boyd, *Water Pollution Aspects of Street Surface Contaminants*, EPA-R2-72-081, A Report to the U.S. Environmental Protection Agency, USGPO, Washington, D.C., Nov. 1972.

92. Vitale, A. M. and P. M. Sprey, *Total Urban Water Pollution Loads: The Impact of Storm Water*, PB-231-730, A Report prepared for the Council on Environmental Quality, Enviro Control, Inc., 1974.

93. Whipple, W. J., B. B. Berger, C. D. Gates, R. M. Ragan, and C. W. Randall, *Characterization of Urban Runoff*, Water Resources Research Institute, Rutgers University, New Brunswick, N.J., Sept. 1976.

94. Keller, F. J., Jr., "Effect of Urban Growth on Sediment Discharge, Northwest Branch Anacostia River Basin, Maryland," *Short Papers in Geology and Hydrology, Articles 60–119*, U.S. Geological Survey Professional Paper No. 450-C, 1962.

95. Guy, H. P. and G. E. Ferguson, "Sediment in Small Reservoirs Due to Urbanization," *Journal of Hydraulics Division*, ASCE, **88**(HY2):27–37, March 1962.

96. Leopold, L. B., *Hydrology for Urban Land Planning–A Guidebook on Hydrologic Effects of Urban Land Use*, Geological Survey Circular 554, U.S. Geological Survey, Washington, D.C., 1968.

97. *U.S. News and World Report*, February 7, 1977.

98. Van Tassel, A. J., Ed., *Environmental Side Effects of Rising Industrial Output*, D. C. Heath and Company, 1970.

99. *A Look at Business in 1990*, "Summary of the White House Conference on the Industrial World Ahead," Washington, D.C., February 7–9, 1972, U.S. Government Printing Office, November 1972.

100. Parker, F. L. and P. A. Krenkel, *Thermal Pollution: Status of the Art Report*, National Center for Research and Training in the Hydrologic and Hydraulic Aspects of Water Pollution Control, Report No. 3, December 1969.

101. Scott, D. L., *Pollution in the Electric Power Industry*, D. C. Heath and Company, 1973.

102. Tuve, G. L., *Energy, Environment, Populations, and Food, Our Four Interdependent Crises*, John Wiley & Sons, 1976.

103. Stratton, C. L. and G. F. Lee, "Cooling Towers and Water Quality," *Journal Water Pollution Control Federation*, **47**(7):1901–1912, July 1975.

104. Speakman, J. N. and P. A. Krenkel, *Quantification of the Effects of Rate of Temperature Change on Aquatic Biota*, National Center for Research and Training in the Hydrologic and Hydraulic Aspects of Water Pollution Control, Report No. 6, May 1971.

105. Symons, J. M., "Paper No. 17, The Project Report," *Water Quality Behavior in Reservoirs*, U.S. Department of Health, Education, and Welfare, Public Health Service, U.S. Government Printing Office, 1969.

106. Battelle Columbus Laboratories, *An Assessment Methodology for the Environmental Impact of Water Resource Projects*, EPA-600/5-74-016, July 1974.

107. Symons, J. M., S. K. Weibel, and G. G. Robeck, *Influence of Impoundments on Water Quality*, U.S. Department of Health, Education, and Welfare, Public Health Service, October 1964.

108. Boyd, M. B., et al., *Disposal of Dredge Spoil, Problem Identification and Assessment and Research Program Development*, U.S. Army Engineer Waterways Experiment Station, Technical Report H-72-8, November 1972.

109. *Report of the International Working Group on the Abatement and Control of Pollution from Dredging Activities*, May 1975.

110. Ritchie, G. A. and J. N. Speakman, "Effects of Settling Time on Quality of Supernatant from Upland Dredge Disposal Facilities," Proceedings, 16th Conference, Great Lakes Research, 1973.

111. Lee, G. F., et al., *Research Study for the Development of Dredged Material Disposal Criteria*, U.S. Army Engineer Waterways Experiment Station, Contract Report D-75-4, November 1975.

112. Duvel, W. A., et al., "Environmental Impact of Stream Channelization," *Water Resources Bulletin*, **12**(4), August 1976.

113. Arner, D. H., et al., *Effects of Channelization of the Luxapalila River on Fish, Aquatic Invertebrates, Water Quality and Furbearers*, prepared for Office of Biological Services, Fish and Wildlife Service, June 1976.

114. Minutes from the hearing "State of Alabama, ex rel. William J. Baxley, Attorney General, Plaintiff, vs. Corps of Engineers of the United States Army: Martin R. Hoffman, *et al.*, Defendants," Civil Action No. 75-H-2343-J, U.S. District Court for the Northern District of Alabama, Jasper Division.

115. Deju, R. A., et al., *Environment and its Resources*, Gordon and Breach, New York, 1972.

116. U.S. Environmental Protection Agency, *Processes, Procedures, and Methods to Control Pollution from Mining Activities*, EPA-430/9-73-011, October 1973.

117. Skogerboe, G. V. and J. P. Law, Jr., *Research Needs for Irrigation Return Flow Quality Control*, Project No. 13030, Office of Research and Monitoring, 1971.

118. Martin, W. P. et al., "Fertilizer Management for Pollution Control," *Agricultural Practices and Water Quality*, T. L. Willrich and B. E. Smith, eds., Iowa State University Press, Ames, Iowa, 1970.

119. Goldberg, M. C., "Sources of Nitrogen in Water Supplies," *Agricultural Practices and Water Quality*, T. L. Willrich and G. E. Smith, eds., Iowa State University Press, Ames, Iowa, 1970.

120. Nicholson, H. P., "The Pesticide Burden in Water and its Significance," *Agricultural Practices and Water Quality*, T. L. Willrich and G. E. Smith, eds., Iowa State University Press, Ames, Iowa, 1970.

121. U.S. Environmental Protection Agency, Effluent Guidelines Division, Office of Air and Water Programs, *Development Document for Effluent Limitations Guidelines and New Source Performance Standards–Feedlot Point Source Category*, 1973.

122. Miner, J. R. and T. L. Willrich, "Livestock Operations and Field-Spread Manure as Sources of Pollutants," *Agricultural Practices and Water Quailty*, T. L. Willrich and G. E. Smith, eds., Iowa State University Press, Ames, Iowa, 1970.

123. Diesch, S. L., "Disease Transmission of Water-borne Organisms of Animal Origin," *Agricultural Practices and Water Quality*, T. L. Willrich and G. E. Smith, eds., Iowa State University Press, Ames, Iowa, 1970.

124. Sopper, W. E., "Effects of Timber Harvesting and Related Management Practices on Water Quality in Forested Watersheds," *Journal of Environmental Quality*, 4(1):24–29, January–March 1975.

125. Brown, G. W. and J. T. Krygier, "Effects of Clearcutting on Stream Temperature," *Water Resources Research*, 6(4):1133–1139, August 1970.

126. Swift, L. W., Jr. and J. B. Messer, "Forest Cuttings Raise Temperatures of Small Streams in the Southern Appalachians," *Journal of Soil and Water Conservation*, **26**(3):111–116, May–June 1971.

127. Likens, G. E., et al., "Effects of Forest Cutting and Herbicide Treatment on Nutrient Budgets in the Hubbard Brook Watershed Ecosystem," *Ecology Monograph* No. 40, 1970.

128. Pierce, R. S., et al., "Effects of Elimination of Vegetation on Stream Water Quantity and Quality," *International Symposium Results Representing Expansion Basins*, International Association of Scientific Hydrology, Wellington, New Zealand, 1970.

129. Aubertin, G. M., D. W. Smith, and J. H. Patrick, "Quantity and Quality of Streamflow after Urea Fertilization on a Forested Watershed: First-Year Results," *Proceedings of Northeastern Forest Soils Conference*, U.S. Forest Service, General Technical Report NE-3, 1973.

Chapter 7

Vegetation and Wildlife Impact Analysis

TED HANES

In recent years we have "rediscovered" our environment. A series of wars, the atomic and space ages, pollution, and the demands of an ever-increasing human population have jolted us into the realization that our environment is precious, finite, and deteriorating. This realization came upon us at the very time great numbers of people were experiencing tremendous benefits from our burgeoning technologies—benefits, however, which often resulted in significant losses to the natural environment.

Basic to human survival is a favorable environment composed of nonliving factors such as air, temperature, water, and soils rich in mineral nutrients; living factors of food supply from plants and animals; and a total environmental setting suitable to the human psyche. Yet the human environment is not one of separate, compartmentalized factors. Rather, it is composed of a network of interrelationships that are easily disturbed by people.

The way people view and use the land is a major part of our environmental problem. If viewed as an unlimited and expendable resource, all forms of nature, both physical and biological, give way to "development" by people. If viewed as a limited and precious resource, natural features are retained and "no development" by people results. Between these two extremes lies an ethic which allows for rational development in harmony with nature.

GENERAL BACKGROUND

Vegetation and wildlife are important features of the environment, and they present special problems in environmental assessment. Living things are adapted to their setting. They are organized into natural groupings (communities) with mutual dependencies among their members, and they show various responses and sensitivities to outside influences. Plants and animals have countless life-cycle modes, forms, and activities that may be important considerations in their assessment. Most organisms are native (indigenous) to the area in which they are found, but some may be alien and perhaps troublesome. Retention or removal of natural communities and their replacement with domestic forms have numerous implications that must be considered both ecologically and economically.

Assessment of the biological portion of the environment must include what is present, its value, and its response to impacts. Various methods are available to describe the natural community and its components. The assessment should provide a description of community uniqueness, the dominant species, and an evaluation of rare or endangered species. Further, the assessment should consider the vulnerability to and the outcome of various human impacts. Finally, the biotic assessment should predict the recovery potential of the natural community from disturbance.

Biological Concepts and Terms

The recent emergence of environmental science as a recognized discipline has drawn heavily upon biological concepts. These "ways of nature" are basic to an understanding of environmental assessment studies. Special terms and concepts needed to assist in this understanding are as follows.

Environment   "Environment" is the sum of all factors that influence organisms, and includes more than that. For example, the nonliving (abiotic) and the living (biotic) factors that exist in a place create a total environment that equals more than the sum of its component parts. Environment is not haphazard but is interacting and functional. Each factor in an environment is considered an integral part, whether living or nonliving.

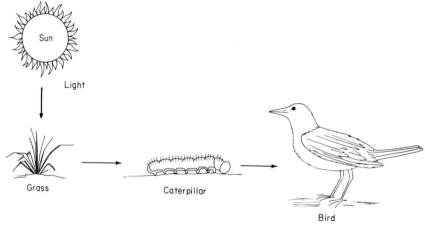

Figure 7.1   Typical food chain.

Human beings add certain factors to the environment not present in a natural environment. To the abiotic or physical environment, they may add factors of energy and chemicals, and to the biotic environment, they may add species of organisms not previously present. They further add social, political, and economic factors to the environment. Environment is therefore a multivariate concept that needs to be defined carefully when used in environmental assessment studies.

Physical Factors   The framework of nature is set by the nonliving physical factors of the environment. Sun, shade, heat, cold, rain, drought, soils, altitude, topography and many more factors provide the conditions and commodities, as well as the limits for life.

The physical factors of an environment are generally not delicately balanced and are only upset by some overt act of nature or human beings. Flooding, severe freezing, tornadoes diverting rivers, and air and water pollution are examples of changes in physical factors that cause upset to the living components of the environment.

Biological Factors   Living things everywhere interact with their nonliving surroundings, and the nature of the inanimate world largely determines which organisms live where; that is, organisms are adapted to function in their physical setting. Any alteration in physical factors will bring about a concomitant effect on living creatures. In turn, living things have subtle effects on the physical components of their environment. A tree intercepts the rays of the sun, shade is produced, and a new microenvironment results. Organisms, therefore, create countless highly specific conditions for other organisms by modifying the general physical factors of an area.

There is thus developed an intimate relationship between organisms and their immediate environment. Green plants draw their energy from the sun and raw materials from soil, air, and water, and assemble them into their bodies. Some animals (herbivores) depend upon plants for their food and shelter. Other animals (carnivores) feed on other animals. Simple organisms (decomposers) reduce dead plants and animals to simple substances which are returned to the nonliving environment. Such relationships can be pictured as simple "food chains" where one kind of organism depends on another kind of organism for its food and energy (Figure 7.1). Yet, in nature the relationships and interdependencies are usually more complex and can be pictured as "food webs" (Figure 7.2).

The degree of intimacy between organisms and their environmental factors (both living and nonliving) varies with the species involved and from place to place. For instance, some plant-eating animals thrive on only one or two species of plants, whereas other animals can consume literally hundreds of different kinds of plants.

Each kind of organism lives in and is adapted to a particular set of environmental factors collectively called its "habitat." Within its habitat, each organism carries out its own

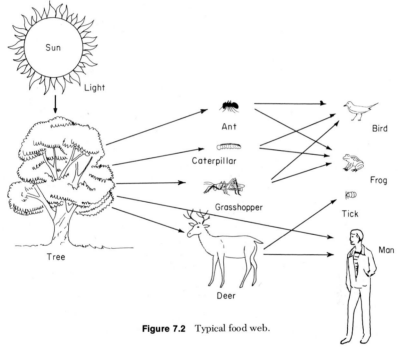

Figure 7.2 Typical food web.

activities (functional niche). In a broader sense there are many living places and activities in an area. A forest is composed of many habitats for simple and complex plants, insects, birds, and snakes. When speaking of small creatures, such as termites and centipedes, having small spatial requirements, it is appropriate to use the term "microhabitat." A few organisms, such as flies, mice, and men, are successful in a wide variety of climatic, geographic, and other environmental conditions. Such organisms are referred to as having a "wide range of tolerance." Most organisms, however, have a "narrow range of tolerance" for the various components of their habitat. As a consequence, most organisms cannot be uprooted from their habitat and be expected to survive in another kind of habitat. Knowing the environmental factors of an area and the ecology of the organisms involved allows ecologists to predict with a fairly high degree of certainty the outcome of altering a habitat or its inhabitants.

Within a habitat or area, there are different kinds of plants and animals. All of the individuals of one kind of organism make up a "population." For example, all the wild roses of one kind in an area make up a population, as do garter snakes, deer, or katydids.

The population size, age classes, and vitality are important considerations in assessing natural populations in an area.

"Species" is the technical word used to describe all of the members of a population that can reproduce freely with each other, but which usually cannot breed successfully with members of another species. There are exceptions to this general definition, however, and the offspring is called a "hybrid." A species is made up of one or more populations and may be few in number and distribution (as with rare or endangered species) or may be even worldwide in distribution.

The plant and animal populations in an area form recognizable associations called "natural communities." These are characterized by a few species called "dominants." For example, a spruce-fir forest is a community dominated by spruce and fir, but has in it numerous other plant and animal species.

Natural communities have structure based on the life forms (e.g., grass) of the species that make them up. A hardwood forest has a given structure by virtue of the trees and shrubs that compose it. The "species composition" refers to the kinds of species making up the community. The variety of species and their relative numbers are referred to as "species diversity." A community composed of few species is called "simple" or one of "low diversity." A community composed of many species is called "complex" or one of "high diversity." The greater the biotic diversity, the greater the number and kind of habitats for the inhabitants of the community. Conversely, a reduction in structural or species diversity results in a loss of habitats, with a further loss in species.

The meeting ground of two different natural communities is called an "ecotone." Such boundaries may be very sharp (narrow ecotones), reflecting some abrupt change in a physical factor such as soil moisture. Poorly defined boundaries (broad ecotones) indicate a gradual change in some physical factor such as elevation.

Ecosystem   An "ecosystem" is composed of plant and animal populations, and it differs from a natural community designation in that it involves the total nutrient and energy economies of the system as well as the organisms involved. Ecosystems are self-contained and self-maintaining. Natural ecosystems are invariably richer in species and more stable than those artificially developed, due to their many interdependencies and interrelationships. The city is an ecosystem contrived by people. All of the facilities and arrangement of parts is for the exclusive use of people. The material and energy flow come from far beyond the city limits. The city can be pictured as a huge mechanical octopus composed of a complex body with long arms reaching out and drawing in life-supporting materials from great distances. But being artificial, interdependencies are missing and it is not self-sustaining. Energy and materials are not recycled efficiently, and constant maintenance is required or the city ecosystem will deteriorate.

Succession   Natural communities are not static but pass through a series of recognizable changes called "ecological succession." The organisms living in them modify their environment through time in such a way that more complex communities arise composed of more specialized organisms. "Primary succession" occurs in an area where life has not existed before, such as on bare rocks, tallus slopes, lava flows, cinder cones, sand bars, and sand dunes. "Secondary succession" occurs on bare sites previously vegetated. Because of the natural forces of change, such as floods, fire, and hurricanes, and the disruptive practices of people, secondary succession is far and away the most common and widespread type of community change.

The first organisms to occupy a bare area are "pioneer species." These are generalists, having few, if any, specific requirements and making minimal demands on the site for resources. Pioneer plant species range from lichens and mosses to annual flowering plants. Pioneer animal species forage on a wide variety of foods and have few specific habitat requirements. Pioneer species can endure harsh conditions and perpetuate their kind over years, yet their very presence leads to gradual site changes that favor the entrance of other kinds of plants and animals. This replacement sequence may repeat itself several times. Each (seral) stage is composed of larger plants and animals and of an ever-increasing number of kinds. Eventually the natural community comes in harmony with its environment and a stable community results, called the "climax community." It is characterized by having species perpetuating their kind indefinitely, without giving way to other species. Arriving at the climax community stage in succession may take decades or centuries depending upon the climate, geography, and biological resources of the area. Any disturbance or disruption of a natural community at any stage of succession will, in effect, set the clock back to some earlier stage of succession, and the process of secondary succession will proceed from there.

Natural Biotic Associations   Such physical factors as moisture, temperature, and sunlight set the broad environmental limits in which natural biotic associations develop. The broadest biotic assemblages are aquatic (mainly freshwater and marine) and terrestrial (land) "realms" (Table 7.1). Terrestrial realms contain about 80 percent of all the species of living organisms that inhabit the earth,[1] and here we find the greatest variety of organisms and ecosystems on earth. Although marine habitats occupy over 70 percent of the earth's surface, the vast majority of marine organisms are confined to shallow water.

The water and land biotic realms are subdivided into regional groupings called "biomes," which are products primarily of regional climates and are characterized by

TABLE 7.1   Natural Biotic Associations

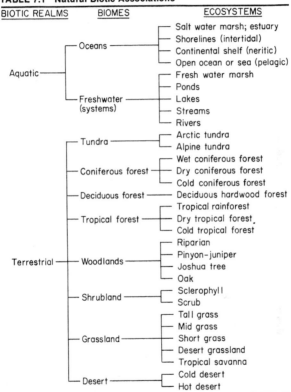

BIOTIC REALMS	BIOMES	ECOSYSTEMS
Aquatic	Oceans	Salt water marsh; estuary
		Shorelines (intertidal)
		Continental shelf (neritic)
		Open ocean or sea (pelagic)
	Freshwater (systems)	Fresh water marsh
		Ponds
		Lakes
		Streams
		Rivers
Terrestrial	Tundra	Arctic tundra
		Alpine tundra
	Coniferous forest	Wet coniferous forest
		Dry coniferous forest
		Cold coniferous forest
	Deciduous forest	Deciduous hardwood forest
	Tropical forest	Tropical rainforest
		Dry tropical forest
		Cold tropical forest
	Woodlands	Riparian
		Pinyon-juniper
		Joshua tree
		Oak
	Shrubland	Sclerophyll
		Scrub
	Grassland	Tall grass
		Mid grass
		Short grass
		Desert grassland
		Tropical savanna
	Desert	Cold desert
		Hot desert

dominant life forms. For example, a grassland biome is dominated by grasses and large grazing animals. Biomes in turn are composed of smaller but recognizable natural communities (discussed earlier).

Environmental Impacts

Environmental impacts vary in their directness, intensity, and duration depending upon both the nature of the action and of the biotic community. Every environment has some recovery potential, given time, and the degree depends upon life forces, location, and the type and degree of impact.

The complexity and variability of ecosystems and their recovery capabilities make precise quantitative predictions impossible. Nevertheless, the repetition of environmental impacts has escalated in recent years so that our predictions are becoming more reliable.

Projects and activities usually produce adverse biological consequences of two types, direct or indirect, and of varying duration, short-term or long-term (Fig. 7.3). "Direct impacts" are those that destroy, displace, or in some way adversely affect plants and animals. Examples are logging, clearing for agriculture, land grading, channelization, and hunting. "Indirect impacts" are those that destroy or disrupt habitats, ecosystems, or other

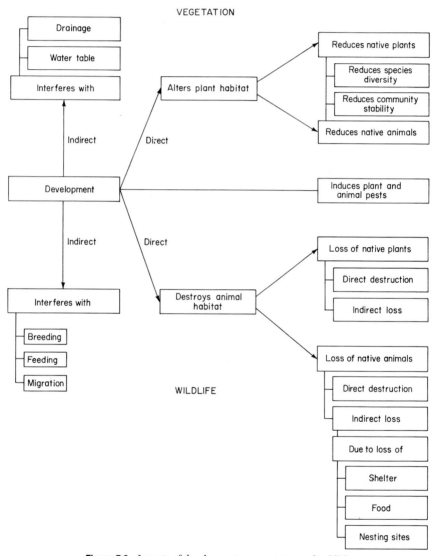

Figure 7.3 Impacts of development on vegetation and wildlife.

physical or biological factors upon which a species depends. Examples are livestock grazing, altering water table or drainage, eliminating nesting or resting sites, breaking food chains or webs upon which an organism depends, using biocides, introducing species, dumping pollutants into the air or waterways, causing noise, and blocking animal movement. "Short-term impacts" on nature relate to the immediate and direct environmental changes that occur at the inception of a project or action, but end or are corrected soon after the completion of the project or termination of the action. "Long-term impacts" on nature result from either major, direct environmental change or chronic perturbations resulting from the operational phase of a completed project.

The full consequence of many impacts on organisms may not be obvious, but the long-term impact may be more profound in the life of the organism than is evident. As an illustration, Table 7.2 presents typical human-induced environmental impacts (both adverse and beneficial) on biota. Each of these human practices is undertaken with the notion that the results will be of benefit to humanity. Unfortunately, we are finding that some of our most complex environmental problems are the result of environmental and ecological backlash.[20] As a general rule we find that artificial projects and technological additions lead to the simplification of natural systems. This reductionism results in losses in biological efficiency, diversity, balance, and self-sufficiency of the biological community, and concomitant increase in pest species of plants and animals as escapees and weeds.

Alteration or removal of natural vegetation has been the primary cause of habitat destruction, reduction in native plants and animals, and species extinctions. Any proposed project that will alter or remove native vegetation must consider the impacts listed in Table 7.2. The indirect impacts caused by the replacement of native vegetation with impervious roofs, roads, highways, and parking lots should also be considered, such as:

1. Reduction in the amount of food and oxygen production (through plant photosynthesis) per unit area covered, as well as the dissipation of heat and gases back into the atmosphere from the decomposition of dead vegetation.

2. Increase in rates of surface water runoff from rainfall and melting snow, which in turn increases the problem of handling wastewater disposal and flood control and limits the recharging of underground water reserves and wetlands.

3. Stopping of normal nutrient recycling between vegetation and the soil.

Projects which involve altering water resources, such as building dams and reservoirs, diverting rivers and streams, dredging harbors, draining and filling wetlands, and using water for the discharge of solid, chemical, and heat wastes, have generated tremendous positive (beneficial) and negative (adverse) impacts (Table 7.2). Most of these projects are massive engineering projects and are often referred to as "public works." Since aquatic realms are complex and life-giving to not only the forms that dwell there but to the land biota (including human beings), their alteration has profound effects on the total environment. Some of the negative impacts that must be considered when dealing with proposed water projects are:

1. Direct habitat destruction, both terrestrial and aquatic.

2. Alteration of the physical properties of the aquatic habitat resulting from impoundment of water.

3. Clogging of stream and river channels and reservoirs with silt and debris. This problem is compounded by the removal of vegetation and timber upstream, and other types of watershed and drainage system management.

4. Irrigation in semiarid and arid lands leads to the accumulation of salts in surface soils. As the salinity increases, crop production declines until the salt content becomes so high that only salt tolerant, noncrop plants survive.

5. Water sources high in total dissolved salts (TDS) may exceed public health limits for human consumption.

6. Discharge of certain urban waste substances into rivers, streams, and lakes. For example: (a) Certain inorganic and organic chemicals are toxic to aquatic life. Others may cause an overfertilization of the water resulting in massive algal growth (blooms) and other water-contamination problems. The rapid aging of aquatic ecosystems is known as "cultural eutrophication." (b) Excessive amounts of organic waste, such as sewage, reduce the oxygen content (biological oxygen demand goes up) in the water which may lead to a dying-off of all those organisms in the aquatic habitat that require oxygen for life.

Countless human activities, especially in industrialized societies, result in air pollution. The discharge of gaseous, liquid, and particulate wastes into the atmosphere result in a variety of subtle negative environmental impacts such as:

1. Reducing solar energy (sunlight) absorption by plants and hence reducing photosynthesis and oxygen production in the polluted area.

2. Adversely affecting certain plants (and perhaps animals) susceptible to various air pollutants. Some researchers believe that most plants are adversely affected to some degree.

3. Significantly altering the climate of urban centers from that of surrounding rural

TABLE 7.2 Typical Impacts of Activities on Biota

Activity	Types of biota impacts	Adverse	Beneficial
Clearing	Creates new environment.	X	X
	Creates conditions suitable for rodent outbreaks.	X	
	Habitat destruction.	X	
	Loss of shelter and food.	X	
	Loss of native plants and animals.	X	
	Reduced species diversity.	X	
Timber harvest	Increased "edge effect."		X
	Habitat destruction.	X	
	Loss of climax species.	X	
Agriculture	Encourages a few species.	X	X
	Habitat destruction.	X	
	Loss of native plants and animals.	X	
	Increase in weedy species.	X	
Grazing livestock	Habitat destruction.	X	
	Loss of native plants and animals.	X	
	Increase in weedy species.	X	
Dams and reservoirs	Creates shoreline ecosystem.		X
	Potential increase in species types.	X	X
	Habitat destruction.	X	
	Loss of native plants and animals.	X	
Diversion of streams and rivers	Habitat destruction.	X	
	Change in migratory patterns.	X	X
Dredging of harbors	Benthic habitat destruction.	X	
Draining and filling	Habitat destruction.	X	
	Loss of native plants and animals.	X	
	Reduced species diversity.	X	
Power plant construction and operation	Alteration of breeding and feeding activities by noise pollution.	X	
	Change in form of aquatic life due to heating of adjacent waters.	X	X
	Potential loss of wildlife due to radiation effects.	X	
Discharge of pollutants into water bodies	Disturbance of wetland habitat.	X	
	Loss of native plants and animals.	X	
	Reduced species diversity.	X	
	Potential for species extinction.	X	
	Loss of fisheries.	X	
Air pollution	Potential crop damage.	X	
	Loss of timber and natural foliage.	X	
Pumping and removal of ground water	Enhancement of plant growth and productivity.	X	
	Loss of deep-rooted trees and shrubs.	X	
Fishing, hunting, and trapping	Keeps population size below carrying capacity of the site.		X
	Eliminates old and infirmed individuals.		X
	Creates nature imbalance.	X	

Activity	Types of biota impacts	Adverse	Beneficial
Wildlife management	Favors selected species.	X	X
	Enhances habitat.		X
	Hinders nonmanaged species.	X	
Resource exploration and development	Habitat destruction.	X	
	Wildlife disturbance.	X	
Roads, highways, railroads, and airports	Increased "edge effect."		X
	Habitat destruction.	X	
	Interference with migration routes.	X	
	Loss of native plants and animals.	X	
	Creation of hedgerows, windbreaks, roadsides, and embankments.	X	X
Communications and utility towers	Creates new roosting sites for birds.		X
	Interferes with migratory birds.	X	
Gas and oil pipelines	Interferes with daily and seasonal animal migration.	X	
Industrial, commercial, and residential development	Enhances site for weedy species of plants and animals.	X	
	Habitat destruction.	X	
	Loss of native plants and animals.	X	
	Increased "edge effect" in the form of hedgerows, windbreaks, roadsides, and embankments.		X
Offshore drilling	Habitat disturbance.	X	
	Potential hazard to intertidal organisms and birds from oil spills and leaks.	X	
Landfills	Favors scavenger species.	X	X
	Habitat disturbance and destruction.	X	
	Loss of native plants and animals.	X	
Conservation and restoration	Increases numbers of native plants and animals.		X
	Increases species diversity.		X
	Restores balanced ecosystems.		X

areas. Notable changes are increased frequency of light rainfall; altered relative humidity of air; acidified rainwater, snow, and soils; and elevated levels of heavy metals and pesticides that may reach toxic levels in plants and animals.

Both industrial and urban developments have the worst potential impacts on natural systems and communities. Table 7.2 lists some of these, but there are other subtle impacts such as:

1. Accumulation of solid wastes that results from the inflow of food and material without a concomitant recycling within the urban center.

2. Direct impact on plants in urban areas by human activities—overpruning, restricting tops and confining root systems by buildings and pavement, trampling and eroding open space, arranging uniform street tree and ornamental plantings (uniform both in age and kind), eliminating mature trees and leaf litter, overwatering or underwatering, putting plants in environments (habitats) unsuited to their requirements, lowering water table, and exposing susceptible plants to gaseous and particulate forms of air pollution.

3. Animals in the urban environment are adversely affected by human activities, through deprivation of natural food supplies; poisoning of nontarget organisms; illegal shooting (poaching, vandalism, target practice); damage and destruction of nests and young; restriction of movement by highways, fences, power lines, poles, and antennae; killing by land and air vehicles; noise; and other activities.

4. Urban developments usually produce direct negative impacts on the native animals and plant life of an area. Established and necessary habitats are destroyed, and their destruction is the major cause of declining animal populations and local extinctions. Many plants have specialized environmental requirements, and when the environment is altered these plants cannot survive.

5. The removal of native vegetation has profound effects not only upon the natural community structure but also upon the physical environment and animal life. The immediate result is an intensification of physical factors, such as increased sunlight, drying effects, and wind and water erosion. A concomitant effect is the paucity of food, shelter, and nesting materials and sites for certain animals. This in turn leads to their overexposure, disease, starvation, and increased hazards as animals attempt to migrate out of the development site.

By altering the character of the environment, human beings bring about changes in the behavioral patterns within and between species so that most species are unsuccessful. However, the few that are successful take over, sometimes in explosive fashion. If the urban development leaves some of the natural community structure, it is usually composed of pioneer-type animals that tolerate changes in food types, shelter, and have only limited relationships with other organisms.

Most domestic livestock, agricultural crops, and pets have resulted from the introduction of foreign or exotic species by people. Some introductions, however, have been accidental or mere escapees from domestication and have become major pest species. Some have become crop destroyers, others compete with native species, and others may be either disease carriers (vectors) or poisonous. The more natural areas that are perturbed by people, the more successful will be the alien species. For example, if a 100-acre parcel is cleared for development, this may lead to weedy and pest species moving in from surrounding areas, creating maintenance problems.

Few deny the benefits to human well-being resulting from the varied human activities that produce impacts on the natural environment. Some of these benefits are: (1) a stable and varied food supply; (2) abundant and varied natural resources; (3) a stable but diminishing energy supply; (4) reduced environmental hazards such as freezing, flooding, and fire; (5) a stable water supply for domestic, industrial, and recreational purposes; (6) increased ease and speed of movement, both short and long distances; and (7) efficient and rapid means of communication.

These benefits are offset to varying degrees by such adverse effects as: (1) human health hazards from some pesticides, heavy metals, asbestos, air and water pollutants, pest species, disease organisms, and noise; (2) property damage and loss by intruding into certain natural systems such as floodplains, coastal strands, and fire-prone vegetation types; and (3) loss of contact with natural processes, cycles, and beauty.

Altering Natural Communities Humanity must pay a price for altering natural communities. This may be illustrated by a sequence of events as illustrated in Figure 7.4.

1. Natural communities are stable in structure and composition, in harmony with their environment, and self-sustaining for long periods of time.

2. Any alteration of this natural state by people requires energy and money, as attested to by the expense of clearing a forest or filling a swamp. During the development phase of a new urban development, the natural community is disrupted or destroyed. Often the net results of such activities are almost indistinguishable from the aftermath of war.

3. Yet, in both cases, if left to natural healing processes, the disturbed area first protects itself with a mantle of pest or weedy species having short life cycles and minimal requirements. These assemblages of plants create semistable communities with limited structure and interdependencies.

4. If left untended for many years, these weedy communities will gradually be succeeded by native plants which finally rebuild the natural plant community to its original state.

5. Or, by spending money and energy, the restoration process may be accelerated or

a natural community may be "constructed" in a desired location by the planting of appropriate native species and by the addition of soil amendments, irrigation, and maintenance practices.

6. Often, people choose to remove and replace the native vegetation completely in order to create a landscape design with a heavy emphasis on introduced or exotic species of turf, groundcover, shrubs, and trees. Such a transformation requires the expenditure of money, time, and energy.

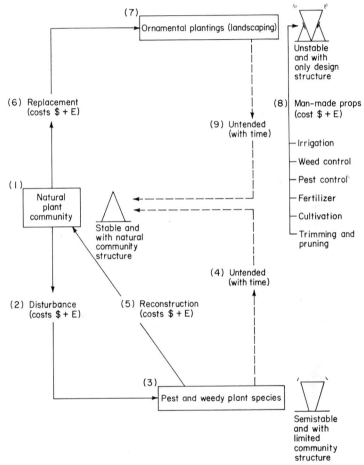

Figure 7.4 Consequences of disturbing or replacing natural plant communities. $ = money expended, E = energy expended (human and fossil fuel).

7. The resulting ornamental landscaping is a trademark of most urban areas in the U.S. and many other countries. Such artificial assemblages of plants are characteristically unstable, having only design structure. Further, they are distinguished by high energy inputs and turnover rates, low species diversity, and a high degree of orderliness. Human-contrived plantings are often ecological "misfits," the plants being out of harmony with each other and with their environment.

8. In order to maintain them, people must provide various artificial props which cost money and energy. This is necessary since the normal environmental factors and checks and balances that operate on certain insects and diseases may be absent with respect to

exotics. As a consequence, the area may fall prey to heavy infestations of pests that must then be controlled by biocides. Weeds are introduced species which have to be controlled by hand or with equipment or herbicides. In many localities, irrigation is required and fertilizers have to be applied to maintain the landscaping. The high cost of maintaining municipal grounds and the landscaping associated with freeways (especially in the semiarid Southwest) is a major concern among various governmental officials. In the final analysis, once people try to mold nature to their liking, it acquires characteristics of human products. These imitations of nature are replete with repetition and incompleteness.

9. If maintenance costs get too high and landscaping is abandoned, depreciating forces set in. Weeds invade and successional processes of nature reclaim the site.

The standard by which all contrived plantings should be measured is self-maintenance and renewal. If plants are able to propagate themselves, if they can complete their life cycle without the artificial props of irrigation, fertilizer, and pest control, then the plantings can survive without the constant care of people and the constant infusion of money and fossil-fuel energy.

Special Cases of Biotic Impact   In land use planning and biological assessment studies at all levels of government, there are special cases of biotic impact that need to be considered. Illustrations of these are as follows.

Migratory Game Birds.   Migratory game birds are a natural resource found at some time of year in nearly all major habitat types in North America, including forests and offshore waters. Timing of fall and spring migration is species specific, but migration success is closely linked with appropriate food, shelter, nesting sites and material, as well as many other environmental factors.

The 45 species of waterfowl native to the continental U.S. and Canada are composed of ducks, geese, and swans. Waterfowl abundance and distribution are closely correlated with habitat. An estimated 127 million acres of wetland once existed in the contiguous U.S. By 1953, 82 million acres remained intact, of which less than ¼th was judged to be of significant value to waterfowl. Wetland drainage in the glaciated prairie pothole region of western Canada has not been as widespread as in the U.S., but localized drainage has reduced such production. Recent hydroelectric projects, oil exploration, and other mineral developments pose serious threats to a number of the most productive areas of the far north.

Estuaries and coastal wetlands are wintering grounds for a major segment of North American waterfowl. During the period from 1922 to 1954, at least 25 percent of shallow, coastal water and marsh have been destroyed. Further, 73 percent of the nation's estuaries have been moderately or severely degraded. Besides dredging and filling for housing and industrial developments, coastal wetlands have been lost because of ditching and drainage for conversion to agricultural land and cattle ranges, and to control mosquitoes. In addition, these wetlands are used as garbage dumps and are subject to pollution from various sources. Recent estimates place the nationwide rate of loss of coastal wetlands at 0.5 to 1.0 percent annually.

Watershed protection, flood control projects, expanding agriculture, urban and industrial use, and irrigation and water diversion projects have resulted in the clearing of millions of acres of bottomland hardwood forest and the channelization of thousands of miles of streams in the United States, primarily in the Southeastern and North-central sections.

Some 20 million acres of wooded swamps and overflow bottomland remained along the Mississippi River and its tributaries during the 1950s. During the 1960s, an average of 200,000 acres of bottomland hardwoods were cleared annually so that about two-thirds of the lower Mississippi waterfowl habitat has been converted to cotton, rice, and soybean production.

In contrast, people have created or improved waterfowl habitats by developing livestock ponds and irrigating formerly arid lands. Large reservoirs are important feeding and sanctuary habitat for migratory waterfowl. Federal, state, and privately funded habitat acquisition and management programs provide additional food and sanctuary in key locations.

The widespread planting of pine plantations in the Southeast has created an extensive homogeneous type inferior in quality for many bird species to the former complex of cropland and woodland. In the Midwest, an increasing threat to the mourning dove population is the removal of shelterbelts. These are the shrub-tree zones of vegetation along roadways, creeks, and other land margins. The clearing of salt cedar and other woody plants in Texas and other Southwestern states has destroyed thousands of acres of extremely valuable nesting habitat for both the mourning and white-winged doves.

Native Vegetation Types.   A mixture of vegetative types and biotic habitats presents special problems when land use changes are anticipated. The biotic impacts may include: (1) loss of open space, (2) changes in landforms, (3) direct removal or loss of native plant and animal life, (4) major ecological changes which may adversely affect the remaining plant and animal life (such as loss of raptors and other large carnivores, and change of drainage affecting riparian species), (5) increased erosion and siltation, (6) increased numbers of a few species of plants and animals (such as weeds, deer, house finch, sparrows, starlings, ground squirrels, gophers, and snakes), (7) reduction in shelter and feeding areas used by avifauna by removal of vegetation, (8) loss of quiet areas upon which many wildlife species depend, and (9) loss of natural biotic resources for both consumptive and nonconsumptive use.

Certain native vegetation may pose additional impacts. For instance, grassland types, dense shrublands that grow in areas of seasonal drought (like California chaparral), and dry coniferous forests may produce massive and potentially destructive fires. Their modification or removal reduces or eliminates this natural threat to people and their developments.

Aquatic Ecosystems.   Since aquatic ecosystems are among the most productive and diverse habitats, it is not surprising that they are easily upset by a variety of perturbations caused by human beings. For example, a trout stream is a complex of several habitats and physical factors that must be present and must interact if the trout are to thrive in the stream. The stream is a great deal more than just water running in a channel.

The stream itself has a variety of bottom characteristics, temperatures, velocities, bank features, riparian vegetation, and invertebrates that serve as fish food. Spawning must occur in the stream's incubators (gravel beds) where eggs are well oxygenated and are hidden from predators. The newly hatched fish must escape from the gravel beds and seek out the stream nurseries. These are the shallow impoundments of quiet water in which there are warmer temperatures, protective vegetation cover, and abundant invertebrates for food. Here the fish grow rapidly, and many serve as food for a variety of waterfowl and other aquatic vertebrates, such as frogs, turtles, and larger fish. Those fish that develop beyond the fingerling stage eventually leave the nurseries to fend for themselves in the swift reaches of the stream. The deep pools, shaded by the overhanging riparian shrubs and trees, are vital resting sites for the trout.

Such a complex relationship between physical and biological factors can be easily upset. Disturbance of the bottom gravels interferes with spawning; filling or polluting the impoundments eliminates the juvenile development stage; and slowing the water or removing the shading effect of the riparian vegetation leads to a reduction in oxygen and interferes with the normal activities of the adult fish. Sewage effluent, fertilizer runoff from farms, silt, pesticides, and many industrial chemicals greatly reduce or eliminate invertebrate life in the stream upon which fish and other vertebrates depend directly or indirectly. Ideally, no direct losses to fisheries should occur if normally accepted construction and operation practices are followed to prevent stream habitat destruction and pollution.

When public agencies such as the U.S. Army Corps of Engineers, the U.S. Soil Conservation Service, and the Bureau of Reclamation have engaged in stream channelization in the name of channel "improvement," stream disasters have often been the result for the aquatic habitat and associated plant and animal life. Fisheries are destroyed, wildlife habitat along the stream is eliminated, water velocity is accelerated, and bottom features for various organisms are removed.

In addition to the more obvious direct destruction of the plants and animals, there is the containment of the dynamic actions of the stream. The normal mature stream develops a lateral (meanders) pattern. This pattern is continually changing as the fast side of each meanders loop undercuts its bank and carries away the resulting sediment. These sediments are redeposited on the slow side of meanders downstream. These newly deposited sediment bars are quickly stabilized by pioneer aquatic plants and in turn are replaced by a series of emergent plants, herbs, shrubs, and trees. This dynamic reworking of sediments and the creation of new habitats by the meandering stream is stifled by channelization or similar channel modification.

A further impact of stream channelization involves the groundwater table. During spring runoff, extra amounts of water will move out of the stream channel and into older portions of the drainage system or floodplain. This water moves slowly, allowing maximum percolation into the groundwater table. Also of vital importance is the introduction of new nutrient supplies into these flooded portions of the drainage system from upstream. Channelization, stream realignment by dredging, and bank stabilization with riprap or concrete terminate these two life-sustaining processes for the streamside ecosystems. If the water table is lowered as a consequence of channelization, remaining riparian and nonriparian trees and deep-rooted shrubs may die off for lack of suitable water supplies.

Reservoir and Dam Projects.   Table 7.3 provides an illustration of the general ecological impacts which might occur due to reservoir creation. Particular attention should be given to the geographical area before construction of proposed reservoirs in: (1) reaches of watersheds used for spawning by migratory species, (2) reaches of watersheds which must

TABLE 7.3   Biotic Impacts Resulting from Reservoir and Dam Projects*

INFLUENCING FACTORS	IMPACT FACTORS	BIOTIC IMPACT
Upstream from dam		
A. Purpose of reservoir → Inundation ——————→		Destroys terrestrial plants and displaces wildlife
B. Reservoir design		
C. Topography of region ——→ Seasonal runoff ——————→ and flooding		Removes microorganisms with the topsoil
D. Impounded water ◄———— Temperature increases → Oxygen concentrations ⟋ decrease		Changes species composition, reduces species diversity
	Sediments and nutrients entrapped	
E. Changes in water level ◄——► Creates drawdown ————→ (denuded) zone		Induces weedy species
	Creates "edge" effect	
Downstream from dam		
A. Stream characteristics ——→ Altered ecology ————→ (temperature, oxygen, nutrients, organic content)		Changes species composition, reduces species diversity
B. Flow regimen ◄———— Temperature, oxygen, and nutrients fluctuate abruptly		

*Summary of information from *An Assessment Methodology for the Environmental Impact of Water Resources Projects,* Environmental Protection Agency, 1974.

be traversed by migratory species to upstream/downstream spawning areas, (3) reaches of watersheds immediately upstream from coastal plain estuaries and/or marshes, (4) reaches of watersheds used by migratory and/or resident waterfowl, and (5) reaches of watersheds which are used by rare or endangered species.

Legal Aspects

Laws protecting natural biota in this country, especially wildlife, have accumulated into a significant body of legal environmental armament in the past several years (see Chapter 1). This body of law is varied and interesting, and represents a highly challenging arm of environmental law.

Historically, environmental awareness as it relates to wildlife legislation dates to the latter half of the nineteenth century. Rapid industrial expansion and unrestricted hunting in the United States brought the passenger pigeon and heath hen to extinction and threatened the existence of our national animal, the American bison (buffalo). The U.S. Congress and the U.S. government have enacted laws and international treaties for the protection of wildlife. This body of law dealing with wildlife protection has no unified structure or coherent principles since it came about so sporadically. Even though it represents legislative reaction to single problems of wildlife conservation, in total it represents a significant expression of legislative concern for distinctive components of the American biota. Land use planners need to be aware of these laws in order to protect biota of national value and avoid federal violations of biotic statutes.

The National Environmental Policy Act lists among its purposes "to promote efforts which will prevent or eliminate damage to the environment and biosphere" of which wildlife is surely an integral part, and "to enrich the understanding of the ecological systems and natural resources important to the Nation." In the declaration of national environmental policy, NEPA pledges "to use all practicable means and measures . . . to create and maintain conditions under which man and nature can exist in productive harmony . . . to the end that the Nation may . . . fulfill the responsibilities of each generation as trustee of the environment for succeeding generations." Ostensibly these purposes and policies embrace wildlife protection.

NEPA is the most comprehensive and all-encompassing legal restriction on the destruction of wildlife habitats in the course of federal projects. Other federal legislation deals either with the specific protection of wildlife types (e.g., Endangered Species Conservation Act, the Migratory Bird Treaty Act, and the Fish and Wildlife Coordination Act) or the protection of wildlife habitat areas (e.g., the Wild and Scenic Rivers Act, the Anadromous Fish Conservation Act, the Migratory Bird Conservation Act, the Federal Aid in Wildlife Restoration Act, and the Recreational Use of Conservation Areas Act). Table 1.1 in Chapter 1 provides a brief overview of the specific features of these laws.

In addition to federal laws, federal lands provide significant protection to wildlife habitats. Nearly a third of the land area of the United States is in the hands of the U. S. government.[10] When viewed from the perspective of wildlife habitat, wildlife resources, and wilderness, the proportion of federal government land holdings is much greater and significant. Generally speaking, the federal government leaves game conservation and management to the states, even on federally owned lands such as national forests and Bureau of Land Management (BLM) lands.

The passage of the Endangered Species Act of 1973 was the most significant wildlife event during 1973.[6] In addition to considering all animal life and recognizing the importance of wild plants, it provides for the protection of "threatened" as well as "endangered" species. This permits preventive action before a critical stage is reached and thereby enhances the likelihood of successful recovery. It requires coordination among all federal agencies whose activities may impact threatened or endangered species or their habitats, and it authorizes a grant program to assist state endangered species programs.

Two lists are particularly important in designating rare and endangered species. The International Union for the Conservation of Nature and Natural Resources (IUCN) publishes a *Red Data Book*—a late warning system of rare and endangered species around the world.[16] This publication (in constant revision) is the most comprehensive international guide to endangered species. The U.S. Fish and Wildlife Service (USFWS), Department of Interior publishes categories and lists of endangered species for the U.S.[29] A comparison of these two listings is shown in Table 7.4.

The reader should recognize that state and local governments have also established statutes for the protection of natural areas with their wildlife and plant life. In many instances, parallel or supportive legislation has been written to the federal acts. In others, the local statutes have preceded federal enactments and may be more restrictive or in some cases more permissive than their federal counterparts. It is not practical to review this plethora of ordinances, restrictions, regulations, statutes, and laws pertaining to the protection of animals, plant life, and wildlife habitats in the states, counties, and municipalities of the country. The interested reader should become familiar with these at the local level.

ENVIRONMENTAL ASSESSMENT TOPICS

The scope of an environmental impact assessment study varies with the size of the site, complexity of the habitats, diversity and rarity of the plant and animal life, type of proposed project, alternatives presented, mitigating measures described, concerns and agencies involved, public interests, and other considerations. The investigator or team needs to apply judgment in establishing the depth of the study, or the level of detail suited to the study scope and task assignments. For example, the assessment of a 100-acre site for a nature park should entail a detailed study of the species present, their abundance, any unique habitats, and interrelationships. An assessment of the same site for residential development should identify natural areas worthy of retention as pocket parks or as part of community greenbelts, and should also identify rare or endangered species worthy of retention or preservation.

In most environmental studies, the report is based on a combination of existing information and on-site visits and investigations. An inventory of existing data and studies

TABLE 7.4   Classification of Status of Rare and Endangered Animal Species*

Red data book	Rare and endangered fish and wildlife of the United States
E *Endangered.* In immediate danger of extinction: continued survival unlikely without the implementation of special protective measures.	E *Endangered.* An endangered species or subspecies is one whose prospects for survival and reproduction are in immediate jeopardy. Its peril may result from one or many causes—loss of habitat or change in habitat, overexploitation, predation, competition, disease. An endangered species must have help or extinction will probably follow.
R *Rare.* Not under immediate threat of extinction, but occurring in such numbers and/ or in such restricted or specialized habitat that it could quickly disappear. Requires careful watching.	R *Rare.* A rare species or subspecies is one that, although not presently threatened with extinction, is in such small numbers throughout its range that it may be endangered if its environment worsens. Close watch of its status is necessary.
D *Depleted.* Although still occurring in numbers adequate for survival, the species has been heavily depleted and continues to decline at a rate which gives cause for serious concern.	P *Peripheral.* A peripheral species or subspecies is one whose occurrence in the U.S. is at the edge of its natural range and which is rare or endangered within the U.S. although not in its range as a whole. Special attention is necessary to assure retention in our nation's fauna.
I *Intermediate.* Apparently in danger, but insufficient data currently available on which to base a reliable assessment of status. Needs further study.	U *Status Undetermined.* A status-undetermined species or subspecies is one that has been suggested as possibly rare or endangered, but about which there is not enough information to determine its status. More information is needed.

*Adapted from Nobile and Deedy, 1972.
NOTE: The Office of Endangered Species of the Department of Interior (USDI) issues periodic lists of "threatened" or "endangered" species of plants and animals. This practice represents a practical step toward implementing the Endangered Species Act of 1973 (Federal Register, 1975).

of the area is a good starting point for an environmental assessment study. This establishes a baseline of the physical, biological, social, and economic environments. It should provide indications of local concerns and pertinent community interests.

Environmental studies large in scope may justify a visual or aerial photographic survey in order to identify characteristics of the study area on a broad scale, including land forms, land use, and status of land development. Considerable time may be required to conduct walking surveys to prepare an accurate description of the present environment. One criterion against which proposed changes are measured is an accurate description of the existing environment.

Proposed actions in or near the proposed development site should be assessed for probable environmental impact by measurement against environmental quality objectives derived from goals and standards expressed by governmental agencies and citizen groups. Each potential impact should be evaluated as to its nature and relative importance. The nature of an impact can be expressed as beneficial, adverse, or neutral (i.e., resulting in no net change).

Of necessity, the depth of a study is usually limited to examining site resources, the activities affected in the area of the proposed development, and existing literature. Detailed analysis of every environmental parameter, every element of the proposed project, and all possible uses of the study area is usually not feasible within the time and budgetary limitations of most environmental studies. For this reason, it is imperative to utilize all available sources of information.

Biotic assessment information and data may be from primary and/or secondary sources. "Primary" information and data are obtained by the biologist on the project site or similar adjacent areas. Direct observations and identifications, sample plots to determine composition and population characteristics of plants and animals, and professional judgment provide most of the primary information. "Secondary" information and data are from published and unpublished sources, and from interviews with authorities or other people recognized to be knowledgeable of the local biota.

Botanical and zoological articles, guides, books, and monographs have been published on local and regional plant and animal life. However, most of these publications demand a considerable degree of familiarity with specific organisms before they are of full value. Local reference collections of preserved plants and animals may be valuable resources and are usually found in museums, field stations, educational institutions, and the like.

Also, there is a scarcity of urban naturalists and, as a consequence, the natural history of most city environs is unknown to most residents. The best sources of information on urban plant and animal life are the biology departments of local schools and colleges, natural history museums, or nature clubs. Individuals who are local authorities are important resources and should not be overlooked.

To a large extent there has always been a gulf between the professionally trained biologist and the layperson. This gulf became abundantly clear when environmental assessment studies were mandated in the 1970s. The jargon and intimate knowledge about organisms and their environment freely shared among biologists were unfamiliar to others. Biologists had to come up with descriptions of the plant and animal life of an area that nonbiologists could understand and act upon. Such descriptions should present the information and observations of biotic assessments in a factual, concise, understandable manner. Furthermore, all opinions, whether those of the biologist/investigator or others, should be identified as opinions in the assessment report. Care should be taken to index the source of the data used, such as primary or secondary.

Existing Environment

Assessing the plant and animal life of an area involves defining the total environment as a necessary first step. This may not be easy, but it is an essential part of an environmental assessment study. As generally used, "environment" means the physical and biological factors or conditions existing in an area which will be affected by a proposed project or activity. These include land, air, water, minerals, plant and animal life (both native and introduced, wild and domestic), ambient noise, and objects of historic and aesthetic significance. A current trend is to include people and their institutions in "environment," taking into account cultural factors (both natural and artificial) which affect the health, senses, and intellects of area residents.

Difficulties arise in trying to define the natural environment. It has no legal or physical boundaries—birds fly to-and-fro freely, rodents move over and under the ground, both surface and ground water move in and out of a given site, and various gases are exchanged between plants and animals and the mobile atmosphere above them. Where does the assessor draw the line when considering a particular study site? Should only the large, obvious plants and animals be noted, or should lesser forms, such as lichens, annual flowering plants, and transient birds, be evaluated?

A second difficulty in defining the environment is that the factors must be analyzed separately, but yet be synthesized into a clear statement that adequately describes and evaluates the total environment. Furthermore, as a consequence of the ambiguity of the biotic environment, the environmental assessment report must state the parameters used in the study.

The biologist/investigator should obtain answers to a series of questions before the actual biotic assessment study begins (Table 7.5). This information will insure an understanding of (1) the project site, (2) the scope of the project, (3) possible project impacts, (4) how involved the biotic assessment will be in time and extent, (5) the investigator's involvement, (6) whether the investigator is capable of accomplishing the biotic assess-

TABLE 7.5   Questions the Biologist Investigator Should Ask Before the Biotic Assessment Begins

1. What is the geographic size of the proposed project site?
2. What part of the site is involved in the proposed project?
3. What is the type of project (e.g., residential, commercial, resource exploration, etc.)?
4. Is the project to be short-term final or long-term incremental?
5. If long-term incremental, is the biotic assessment to be done in increments or at one time?
6. What is the biotic character of the part of the site involved?
7. How will the project influence the plants?
8. How will the project influence the animals?
9. How will the project influence the ecology of the various habitats?
10. What use is to be made of the biotic assessment (decision document, information document, proforma)?
11. What legal or jurisdictional limitations apply?
12. What information exists on the biota of the area?

ment alone or whether assistance will be needed, (7) the level of assistance (from technician to expert), and (8) availability of information.

The site being assessed should be given a name; often this is a geographic or biotic name based on the dominating feature. For example, names such as dune, alkali sink, bog, swamp, sagebrush scrub, and pine barrens, convey a necessary word picture of the site. Often the name assigned to the assessment site is the name of the project or action— "North Slope Pipeline" or "Fire Ant Eradication Project," for example.

Assessing the Physical Environment   Biotic assessments should include a description of the total physical setting. Table 7.6 summarizes the physical features commonly used in describing the existing physical environment. The physical features on the site should be presented in a regional context. This insures that their uniqueness and integrity (or the lack thereof) to the region is properly conveyed. Vegetation and physical setting create the habitat for wildlife and should be described as such. Finally, the influence of the physical setting upon the biota must be interpreted, whether it be moisture, light, or food supply. It is imperative that the biologist provides this insight for those who will review the report and for those who will use it as a basis for decision making.

Topographic maps published by the U.S. Geologic Survey (USGS) are the best single source of geographic information. These maps are available commonly in 7½ and 15 minute sizes. The study site can be put into a regional context readily by using these maps. Water drainages, topography, and geographic features from these maps can be used to prepare a plot map of the study site as shown in Fig. 7.5. Aerial photographs also may be useful in preparing plot maps.

The geologic history and soils of an area can have direct and indirect effects on the plants and animals, hence they must be described. The occurrence of certain types of plants and animals in some locations may best be explained by geologic events such as glaciers, volcanoes, mud and landslides, seismic activity, and mountain building. Rock outcrops and promontories often provide suitable habitats for specific kinds of reptiles, mammals, and raptorial birds. Some soils sustain only particular species of plants and their color-matched animals. Among the key soil factors that may need assessing are texture (rock, sand, silt, clay), acid-alkaline reaction (pH), nutrients, and ability to hold water. Burrowing invertebrates, reptiles, birds, and mammals may be restricted by soil that is too crumbly, rocky, or hard, or by mineral deficiency in the plants they feed upon.

Geologic and soils information can be obtained from the U.S.G.S., U.S. Soil Conservation Service, State Department of Geology, State University Extension Service, and

---

TABLE 7.6   Summary of Physical Features Used in Describing the Existing Environment as they Relate to Plant and Animal Life

A. Geography—Include present and past use of the land
1. Elevation off- and on-site—show detail of the site on a topographic map
2. Drainage—water and air (including pollution entering study site from outside)
3. Presence of aquatic systems off- and on-site such as ocean, lake, reservoir, river, stream, aqueduct, pond, estuary, swamp, marsh, or lagoon

B. Geology and soils
1. Geologic formation in a regional context
    a. Include interesting, unique geologic, and biologically important features
    b. Dominant rock types
    c. Activity—faults, uplifts, folding, subsidence
    d. Soil analysis—profile, depth, age, stability, mineral content, pH, angle of repose
2. Soil types as they relate to plant and animal distribution

C. Climate
1. Precipitation—average annual rainfall and snowfall, storm characteristics
2. Temperature—average annual, seasonal changes, extremes
3. Wind—prevailing (direction, speed, biotic influence), windstorms, tornadoes

D. Other
1. Utilities—power poles and lines
2. Noise—highways, aircraft landings and take-offs, factories, construction phase
3. Sewage and solid waste disposal
4. Flood-control structures—dams, channels, percolation basins, dikes

---

private soils laboratories. Some of this information may be available in state, county, and municipal departments and offices. The information may indicate how much abuse of the land or how much pedestrian or other traffic an area can assimilate before root systems, animal burrows, etc., are destroyed. If available, this information should be included in the report. It is as important as species lists.

The climate of the area should also be included in the description of the environment. The principal components of climate are temperature and moisture and their yearly distribution. Daily weather conditions have little meaning in the existence of natural ecosystems; rather, climate sets the broad limits for life in the ecosystems. The amount of precipitation and its seasonal distribution, the frequency and intensity of storms, the form of the precipitation (i.e., rain, snow, hail, ice), rate of penetration into the soil, runoff, accumulation in surface impounds, and influence on the water table all have important effects on the plant and animal life of an area.

Temperature also influences seasonal activity patterns of plants and animals as well as the rates of snow melt and moisture evaporation, relative humidity, air movement, and temperature inversions. These climatic factors in turn are influenced by regional and local geographic features, such as mountains, lakes, rivers, streams, and oceans. Climatic data and information are available from local sources, such as individuals, cities, newspapers, and weather stations maintained by the U.S. Department of Commerce. This agency publishes monthly weather summaries of all their weather stations within each state.

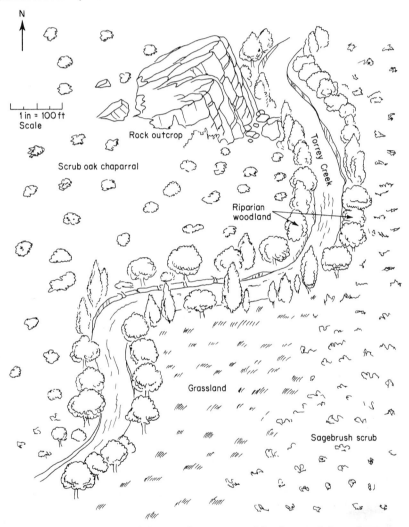

Figure 7.5 Sample sketch of plot map of project site with key biotic and abiotic features.

Although most projects do not influence the regional climate, it should be appreciated that organisms live in their own habitats or microhabitats and are adapted to these specific conditions. For example, wind velocities decrease within a tree canopy and toward the ground. If an area has some important species, climatic data must be supplemented with microclimatic data or other information before it can be properly determined if the project will change the local physical parameters of the environment to which the species is adapted.

A variety of other factors and features present in the physical setting need to be described, depending on the types of setting and project. These may include utility structures and rights-of-way, sewage and waste disposal facilities, flood control structures, and noise levels. Each potential and real biological hazard or barrier should be described and assessed, as well as the benefits. Information on some of these is usually lacking, and more research is needed to clarify their importance and impact on biota.

Assessing the Biotic Environment   One approach to assessing the biotic environment is a sequence of questions arranged in flowchart fashion, starting from a regional perspective and proceeding to specific site factors. Such a flowchart serves as a guide in identifying biotic problem areas, insures that no factors are overlooked, assists the biologist in selecting appropriate assessment methodologies, and provides for the assessment of possible biotic impacts resulting from the proposed project or activity. See Tables 7.7 and 7.8 for an illustration of a sample field form used in a preliminary assessment.

General questions to ask in assessing the existing biotic environment are as follows:

1. *What is the regional setting for the site or action?*

The regional setting should be evaluated in order to properly assess the character of the natural environment of the site in context. Further questions to be asked include:

   a. Is the site unique regionally?

   b. How is the site related to other regional features?

     (1) Physically

     (2) Biologically

2. *What is the existing vegetation?*

This question should lead to the characterization of what type(s) of vegetation exist(s) in the general region and the study site. General types of vegetation are:

   a. Weeds

   b. Ornamentals (both native and introduced)

   c. Grasslands

   d. Shrublands

   e. Forest

   f. Other vegetations

3. *Are there unique vegetative features?*

This question should call attention to any intrinsic or extrinsic vegetation features such as:

   a. Rare or endangered species

   b. Species of high visual, historic, and/or aesthetic appeal

   c. Plants associated with particular habitat features (for example, a stream, substrate type, acid soil, shade)

   d. Threats posed by individual plant species or vegetation (for example, poisonous, fire potential, diseased)

4. *Should any plants or communities be retained or preserved?*

Often answers to questions 1 and 2 will lead to the answer of this question or to further questions:

   a. Should specimens be saved?

   b. Should a representative plant community be saved?

   c. Should a particular natural habitat be saved?

In each case the reason for retention or preservation should be given. It may be purely visual or scenic, or may involve historic value. There may be some unique ecological relationship between plants and site factors or between plant and animal species.

5. *What wildlife is present or uses the area seasonally?*

This question correlates closely with the existing vegetation in the region and on the study site. By knowing the vegetative cover present and the food and nesting requirements of various animal species, the biologist/investigator can make a tentative assessment of the animal life. General animal groups are:

   a. Birds (both game and nongame species)

   b. Mammals (both game and nongame species)

   c. Reptiles and amphibians

   d. Fish

   e. Insects and other arthropods

   f. Pests

   g. Other animals

TABLE 7.7   Sample Field Form Used in Biotic Assessments

Name of project_____Date_____

Type of project_____

Prepared for _____

Investigator_____

I. Reconnaissance: Record key features on plot map or sketch to show location.

   A. Physical features

      1. Geography_____

      2. Geology and soils_____

      3. Climate_____

      4. Water_____

      5. Other_____

   B. Vegetation

      1. Plant communities — dominant plants

         _____

         _____

         _____

         _____

      2. Unique vegetative features_____

   C. Fauna

      1. Animal populations — dominant animals

        Mammals_____

        Birds_____

        Reptiles_____

        Amphibians_____

        Fish_____

        Insects_____

        Other invertebrates _____

      2. Unique wildlife_____

   D. Unique habitats for:

	Important plant species	Important animal species	Important wildlife productivity	Visual, historic, and/or aesthetic values
_____				
_____				
_____				

   E. Condition of unique habitats:

	Disturbed to pristine	Fragile to durable	Recovery potential and speed
_____			
_____			
_____			

F. Special relationships

   1. Vegetation to substrate _____

   2. Animal to substrate _____

   3. Plant to animal _____

   4. Animal to plant _____

   5. Animal to animal _____

   6. Other _____

G. Aquatic habitat

   1. Present condition _____

   2. Visual, historic, and/or aesthetic features _____

   3. Sport, commercial, recreational, or educational values _____

   4. Rare or endangered species _____

6. *Are there any unique wildlife features?*

This question should call attention to any intrinsic or extrinsic wildlife features such as:

   a. Rare or endangered species

   b. Species of high visual, historic, or aesthetic appeal

   c. Threats posed by animal species (for example, poisonous, large carnivores, disease carriers)

7. *What natural habitats are present?*

This question focuses attention on natural assemblages of plants and animals in relation to specific physical factors. It leads to further questions such as:

   a. Are any of the habitats suitable for special, important, or rare or endangered species? (For example, rocky cliffs for nesting, gravel beds for spawning, marshy area for shelter, prairie grassland for breeding)

   b. Are any of the habitats important in wildlife productivity? (For example, sport fish and wildlife, spectators, commercial, educational)

8. *What is the disturbance level?*

This question is basic in assessing not only the existing environment but also its response potential to various perturbations. The well-trained biologist uses such clues as the presence of weedy species, successional species, lack of young climax species, and the response of indicator species to various kinds of impacts to assess both the durability of a natural community and the level of disturbance. Further questions may be asked:

   a. Is the area or site pristine or relatively undisturbed by people or natural forces?

   b. Is the community or habitat fragile or durable?

   c. What is the recovery potential?

      (1) Areas of low vegetative recovery potential may be due to such site factors as soil compaction, erosion, low soil fertility, low rainfall, lack of seed, and severe climate.

      (2) Areas of high vegetative recovery potential possess deep and rich soil, good seed source, and moderate climate.

9. *Are there any special relationships?*

This question gets to the heart of the ecology of the area; that is, it goes beyond naming the biotic components of the area by pointing out the interdependencies that exist between organisms and between organisms and other environmental factors. Examples are the interdependencies between:

   a. Vegetation and substrate—tallus slope, dune, chemical composition

   b. Animal and substrate—burrowing animals

   c. Plant and animal—pollination, shelter, food

   d. Animal and plant—herbivore

   e. Animal and animal—parasite

10. *Are sonic conditions a problem to the wildlife?*

This question addresses the problem of people and their machines producing noises sufficient to disturb or displace certain wildlife species. This factor needs to be assessed

TABLE 7.8   Completed Field Form Used in Biotic Assessments

Name of project _Blue Creek_ _____ Date _21 July_ __

Type of project _Low density housing_ _____

Prepared for _Happy Homes, Inc._ _____

Investigator _I. C. Livermore_ _____

I. Reconnaissance: Record key features on plot map or sketch to show location.

   A. Physical features

      1. Geography _gentle hill country & creek drainage_

      2. Geology and soils _sandstone & limestone – sandy loam_

      3. Climate _semi-arid; Mediterranean_

      4. Water _flowing creek_

      5. Other _sandstone rock outcrop in north-central portion_

   B. Vegetation

      1. Plant communities —dominant plants

         Scrub Oak Chaparral – scrub oak, ceanothus, mt. mahogany

         Riparian Woodland – sycamore, alder, live oak

         Sagebrush Scrub – sagebrush, buckwheat, white sage

         Grassland – bromegrass, wild oats

      2. Unique vegetative features _Beautiful stand of live oaks associated with the creek below rock outcrop_

   C. Fauna

      1. Animal populations —dominant animals

        Mammals _mule deer, wood rats, coyote_

        Birds _red-tailed hawk, thrasher, common towhee, roadrunner_

        Reptiles _gopher snake, garter snake, blue-tailed skink_

        Amphibians _western newt, tree frog, western toad_

        Fish _none_

        Insects _numerous aquatic & flying insects_

        Other invertebrates _aquatic crustaceans_

      2. Unique wildlife _burrowing owl associated with rock outcrop area_

   D. Unique habitats for:

	Important plant species	Important animal species	Important wildlife productivity	Visual, historic, and/or aesthetic values
Rock Outcrop		burrowing owl		
Riparian Woodland				especially the live oaks
Scrub Oak Chaparral			deer bedding area	
Sagebrush Scrub		dove & quail		

   E. Condition of unique habitats:

	Disturbed to pristine	Fragile to durable	Recovery potential and speed
Riparian Woodland	intact	durable	fair, slow
Scrub Oak Chaparral	intact	durable	good, medium
Grassland	intact but weedy	fragile	good, fast
Sagebrush Scrub	moderately disturbed	semi-fragile	good, fast

F. Special relationships

   1. Vegetation to substrate _*none*_

   2. Animal to substrate _*pocket gopher in grassland soil*_

   3. Plant to animal _*none*_

   4. Animal to plant _*flicker in sycamore*_

   5. Animal to animal _*none*_

   6. Other _*red-tailed hawk searches over grassland & sagebush scrub*_

G. Aquatic habitat

   1. Present condition _*poor – contaminated and misused*_

   2. Visual, historic, and/or aesthetic features _*good*_

   3. Sport, commercial, recreational, or educational values _*poor*_

   4. Rare or endangered species _*none*_

in both the construction and operational phases of the project or activity and must be related to the life history and behavior of affected species. Further questions will delineate sonic problems:

   a. What is the source of the sound? Is it stationary or moving (e.g., rock crusher, motorcycles)?

   b. Is the area pristine and serene or already subject to noises?

   c. Will disturbing noises be generated only during the construction phase, or will they be an ongoing operational perturbation to the area?

   d. Which species of wildlife will be sensitive to the noise generated by the project or activity?

11. *What are the visual conditions that exist?*
This question directs the investigator to look both at the study site from the outside and at the off-site conditions through the eyes of the wildlife.

   a. Is the visual image of the site an important consideration? If so, in what respects?

   b. Will any of the wildlife within view of the proposed project be adversely impacted visually?

12. *Are any plants susceptible to project-created air pollution?*
This question will cause the investigator to consider the existing plants in relation to their possible susceptibility to air pollutants as well as the amount and types of pollutants that will be emitted by the proposed project.

13. *Is there an aquatic habitat present?*
This question will cause the investigator to focus on the biota associated with a key terrestrial physical factor, water. The presence of a pond, creek, stream, lake, bog, marsh, lagoon, or estuary on or near a project site inevitably increases the species diversity of the area as well as its biotic productivity. A detailed assessment of such a habitat and its relation to the adjacent terrestrial habitats is essential to any such environmental analysis. Further questions are:

   a. What is the condition of the aquatic system? Both biotic and abiotic parameters should be assessed.

   b. Does the aquatic habitat have high visual, historic, and/or aesthetic features?

   c. Does the aquatic habitat have sport, commercial, recreational, or educational value?

   d. Are there rare or endangered species associated with the aquatic habitat?

In the body of the assessment report, it is incumbent upon the biologist to provide an evaluation of the key plant and animal species, to give an ecological perspective of important species present, and to evaluate the biota in a regional context. This information comes from direct observation and study on the site, from experience, and from secondary sources. Special attention should be given to the relationships between plants and animals, such as food source and habitat. Important features of animals that need to be discussed are: (1) the role of key species (mostly carnivores), (2) numbers as they relate to commonness or rarity, (3) seasonal changes (such as migration, hibernation, or breeding) in activity or locality, (4) sensitivity to various impacts, and (5) aesthetic value. Without such evaluations by the biologist, the assessment of plant and animal life is incomplete and of questionable value.

Assessing Environmental Impact

In the ideal world, it would be desirable to know ecological relationships so exactly that environmental impacts could be quantitatively determined by inserting them into a formula. At present this is not attainable. Even though ecologists have developed considerable data on certain species of organisms and their habitats, they lack complete knowledge about most ecosystems. As a consequence, ecological assessments use empirical estimates based on measured parameters.[12]

It is possible to anticipate or predict ecological changes that will result from a project or action with some degree of accuracy. This necessitates a survey of the site and its natural resources before any alterations occur on or near the site. In the hands of a qualified investigator, this information can lead to useful predictions of biotic impact as well as to the establishment of baseline conditions and population numbers. Monitoring programs can be instituted during both the construction and operation phases to properly assess biotic changes as they occur and to take corrective actions as necessary. This standard of impact assessment should be used whenever possible. Haste and expediency, however, often result in impact assessments far below this standard.

At the present time, ecological impacts are predicted by using either of two information sources: (1) professional knowledge of plant and animal life and their habitat requirements, professional judgment of the biotic community's ability to withstand or respond to disturbance, professional experience with the impending changes and impacts, and results from similar studies, and common sense (a biologist who simply lists the names of organisms observed on the site—without an interpretation of key life histories, ecological interrelationships, and habitat requirements—misses the primary intent of environmental impact reports); (2) computer simulation models based upon a few variables, assumptions, and minimal on-site information of the biota and habitat being emulated.[12] In the latter case, such models allow the testing of variables and their possible impact or mitigating effects. For example, a simulation model could be used to test alternate sitings and the impact of a new highway on animal migration, habitat destruction, and rare or endangered species.

The investigator needs to consider possible impacts as to their temporary or short-term aspects as well as their permanent or long-term impacts. Further, in many projects attention should be given to the project or activity phases in relation to their biotic impact. Such phases are site preparation, construction, operation, phasing-out, removal, and site restoration. The degree of project impact depends upon several variables that need to be considered, such as site location, nature and scale of the project, and the type of biotic communities involved.

Several deficiencies in biotic impact assessment reporting should be avoided, including: (1) evasion of possible impacts and lack of their assessment, (2) omission of pertinent information necessary for unbiased evaluation of impacts, (3) inadequate description of adverse impacts, and (4) a plethora of biotic data or information without interpretation or correlation with possible impacts.

Site features that relate to the biota and that may be impacted by the project or action should be noted. This may be accomplished quickly by the aid of a site features checklist such as that presented in Table 7.9.

As an illustration, suppose that in the siting of a large dam several sites could be considered. Checklists similar to Table 7.9 could then be prepared for each site. By comparing each checklist for potential impact features, one would then select the site with the fewest critical site features as the best candidate. A possible summary of the checklist showing critical site features by site is given in Table 7.10. In this case, site R is the best site according to our criterion.

The biologist should then assess the biotic community in relation to the proposed project. Such variables as size of the natural community, its present condition, its durability or fragility, and its unique features need to be equated against the size of the development, type of development, short-term and long-term changes, and postdevelopment uses. This phase of the study is important in that it evaluates the existing biological resource in relation to the proposed changes in land use.

In order to provide a suitable assessment discussion of possible significant environmen-

TABLE 7.9 Site Features That May Be Impacted by a Project or Action

Will the proposed project or action have negative impacts on the following site features?

Landforms (shapes in and on earth's crust):	Yes	No
Sandy beach		
Rocky beach		
Coastal shore (rock sculptures, wave cut benches, sea grottos)		
Coastal dunes		
Valley		
Canyon		
Rock outcrop		
Cave, cavern		
Butte		
Mesa		
Natural bridge		
Pinnacles, towers		
Buttresses		
Alluvial cones and fans		
Alluvial terrace		
Desert (bajada, playa, dune, pavement)		
Plains		
Plateau		
Glacial deposits (moraine, esker, kame, kettle, drumlin, till plain, erratic)		
Volcanic cones (parasitic, spatter, pit crater, plug, dome, spine, caldera)		
Erupted material (volcanic blocks, volcanic bombs, scoria, cinders, ash, tuff, breccia)		
Lava flows (pahoehoe, pillow lava, caverns, tubes)		
Minor volcanic features (geyser, fumarole, hot spring, mudpot, rim, dome, terrace)		
Land:		
Slopes (stability)		
Soils (subsidence, landslide, erosion)		
Geology (faults, seismic activity)		
Water:		
Floodplain (meanders, cutoffs, oxbow lakes and ponds, meander scars, natural levees, deltas)		
Spring		
Stream		
River		
Drainage channel		
Groundwater		
Waterfall (plungepool, rapids, cascades)		
Lake		
Glacier		
Natural features and resources (removal for commercial purposes):		
Rock		
Gravel		
Sand		
Oil		
Coal		
Minerals		
Trees		
Biota:		
Rare or endangered species		
Mature trees or shrubs		
Habitats		

TABLE 7.9   Site Features That May Be Impacted by a Project or Action   (*Continued*)

	Yes	*No*
Food source	_____	_____
Water source	_____	_____
Breeding ground	_____	_____
Wildlife refuge or preserve	_____	_____
Migration route (daily or seasonal movement)	_____	_____
Mammals	_____	_____
Birds	_____	_____
Reptiles	_____	_____
Amphibians	_____	_____
Fish	_____	_____
Insects	_____	_____
Other invertebrates	_____	_____
Plants	_____	_____
Archaeology:		
Petroglyphs	_____	_____
Middens	_____	_____

tal effects or "negative impacts," the biologist should ask and provide complete answers to the following questions:

1. *What are all possible negative impacts that might adversely affect the biota of the area?*
2. *Which biotic habitat(s) will be impacted?*
   a. What is the degree of impact?
   b. Of what significance will the individual impacts be to the biotic habitat(s)?
   c. Will there be any "pinball effect" resulting from an impact?
   d. What mitigations are needed?
3. *Within each affected habitat, which species have biological significance, if any?*
4. *Are individual specimens involved and, if so, how will they be impacted?*
   a. Do they have historic value?
   b. Do they have aesthetic value?
   c. Do they have scientific or educational value?
   d. Do they have sport or commercial value?
   e. What mitigations are needed?
5. *Will there be short-term biotic impacts?*
   a. Which species will be displaced from the site?
      (1) Will they pose a problem to surrounding areas?
      (2) Will their displacement create ecological problems?

TABLE 7.10.   Illustrative Critical Site Features by Site

	Site *A*	Site *N*	Site *R*	Site *S*
Canyon	X	X	X	X
Rock outcrop		X	X	
Cave				X
Butte	X			
Natural bridge				X
Pinnacles				X
Slopes			X	X
Fault		X		X
Spring		X		X
Waterfall		X		X
Endangered species (bird)	X			X
Migration route	X			X
Fish	X			X
Petroglyphs				X

    b. Which species will be destroyed on site?
      (1) Will their destruction have adverse ecological consequences?
    c. Which new species will invade the disturbed site?
      (1) Do these have ecological, economic, aesthetic, or other consequences?
    d. What would be the short-term consequences of the project or action without mitigating measures? (Examples are noise, water pollution, and human activity.)
    e. What mitigations are needed?
  6. *Will there be long-term biotic impacts?*
    a. What effects will a loss of native plants and animals have on the site and surrounding area or region?
    b. What problems may result from the encroachment of weedy species?
    c. In what ways will diurnal or seasonal animal movement be disrupted?
    d. In what ways will animals' food and water supplies or sources be affected? What is the biological significance of these changes?
    e. To what extent will there be a loss of animal shelter and nesting sites? What is the biological significance of these changes?
    f. What will be the ecological consequences of loss of breeding grounds?
    g. What problems will result from loss of fisheries?
    h. What will be the consequences resulting from the loss of endemic, unique, rare, endangered, or otherwise special species from the site or area?
    i. What will be the biological consequences of noise from fixed or moving sources on bird rookeries and large mammals? Will the noise drive wildlife from the area, mask out mating and other signals, disrupt inter- and intra-population spacing, impact different species in different ways?
    j. What mitigations are needed?

To illustrate the use of such a series of questions in determining negative impacts, suppose a project called for the construction of a deep water tanker port facility and oil pipeline. The biologist should assess the impact of a major oil spill on marine birds, mammal life, fish yield, shellfish and crustaceans, and recreational use. The recovery rate and potential of the biota and their habitat should also be assessed.

The biotic assessment should also consider "positive impacts" that may result from the project or program, or that could be accomplished as alternatives or mitigating measures. A series of questions and examples will illustrate this important aspect of any biotic assessment.

    1. *What are all possible positive impacts that might occur as a result of the project or program?* (Examples are retention of natural areas, reduction or elimination of noxious plants and animals, and reduced fire hazard.)

    2. *Will beneficial species be enhanced?* (Examples are (*a*) mourning doves eat weed seeds and insects, (*b*) humming birds, honey bees, and a large variety of insects pollinate many fruit trees and other flowering plants, thus fruit and seed set is greatly enhanced.)

    3. *Can wildlife management programs result from the project or program?* Numerous species of wildlife are a significant natural resource for food, sport, pleasure, and study.

    4. *Can nature preserves be established?* Certain developments, sites, and programs lend themselves to the retention, enhancement, or creation of nature preserves. These may be wildlife refuges or preserves, wilderness parks, or sanctuaries.

    These islands of wilderness may be left untended, provided with various forms of protection, or may be maintained through various natural and wildlife management programs. Besides providing protection for wildlife, they furnish educational, scientific, recreational, aesthetic, and open space values to humans. Such long-term natural area commitments involve questions of land value, size limits, maintenance costs, human hazards and liability, and nature interpretation and/or wildlife management programs. In some cases the establishment of nature preserves may be part of the mitigating measures proposed in a project or may be the entire project or program itself.

Environmental impacts affect organisms directly or set in motion environmental changes that indirectly affect organisms. The evaluator must assess not only both of these causes of biotic change but also the kind and extent of the biotic change. Table 7.11 summarizes these assessment considerations for selected terrestrial and aquatic projects that create unique impacts needing special attention. Not all variables will occur in every study, but neither should any be overlooked.

TABLE 7.11 Impact Assessment Considerations in Selected Terrestrial and Aquatic Projects

Project	Project activities	Potential impacts to be assessed
Expansion and construction of new airports	Construction phase	1. Vegetation destroyed or disturbed. 2. Wildlife habitat reduced and broken up. 3. Wildlife destroyed or displaced. 4. Migration routes for wildlife disrupted or destroyed. 5. Nesting, mating, and other wildlife behavior patterns disrupted or destroyed.
	Operation phase	1. Bird and other wildlife migration disrupted. 2. Wildlife displaced. 3. Animal behavior disrupted by noise and activities.
Construction of new highways	Construction phase	1. Vegetation destroyed or disturbed. 2. Wildlife habitat reduced or broken up 3. Wildlife displaced. 4. Migration routes for wildlife disrupted or destroyed.
	Operation phase	1. Bird and other wildlife migration disrupted. 2. Wildlife displaced. 3. Roadkills. 4. Animal behavior disrupted by noise and traffic activity.
Construction of nuclear power plants	Construction phase	1. Normal project impacts to vegetation and wildlife due to construction.
	Operation phase (potential problems such as radiation leakage, air pollution, water pollution, contaminated forage)	1. Radiation damage to vegetation and wildlife resulting from direct radiation exposure or radioactive materials.
Construction of new oil pipelines	Construction phase	1. Vegetation destroyed. 2. Wildlife killed or displaced. 3. Migration routes disrupted or destroyed if pipeline is large and above ground.
	Operation phase	1. Destruction or disruption of plants and animals from maintenance activities. 2. Destruction of vegetation and wildlife from pipeline rupture and leakage.
Transfer and transport of crude oil	Loading and unloading Tankers and barges Drilling platforms Tanker mishaps	1. Killing of intertidal and shore biota. 2. Killing of some open-water biota. 3. Killing and damage of marine and shore waterfowl.
Strip mining of coal and minerals	Operation phase	1. Vegetation destroyed. 2. Small animals destroyed. 3. Wildlife displaced. 4. Wildlife habitats destroyed. 5. Migration routes destroyed or disrupted. 6. Breeding grounds destroyed. 7. Wildlife ranges divided. 8. Isolation of animals with small home ranges.

Project	Project activities	Potential impacts to be assessed
	Reclamation phase	1. Soil erosion. 2. Regeneration potential of vegetation. 3. Loss of native forage. 4. Effects of air and water pollution on plants and animals. 5. Human impact on plant and animal life.
Construction of dams and reservoirs	Preparing dam site and clearing for reservoir	*Terrestrial habitat* 1. Vegetation destroyed. 2. Small animals destroyed. 3. Wildlife displaced. 4. Wildlife habitats destroyed. 5. Migration routes destroyed or disrupted. 6. Breeding grounds destroyed. 7. Wildlife ranges divided. 8. Isolation of animals with small home ranges. 9. Suitability of remaining habitats.
	Impoundment of water in reservoir	*Aquatic habitat* 1. Physical properties of impounded water. 2. Changes in aquatic species composition and number. 3. Changes in bottom versus sediment dwellers. 4. Fish stocking practices.
	Changes in impounded water level	*"Edge" Habitat and Drawdown Zone* 1. (Invasion of new plant species and plant communities. 2. Increased habitat diversity. 3. Invasion of new animal species. 4. Human impact on biota.
	Downstream water discharge	*Aquatic Habitat* 1. Physical properties of water. 2. Changes in aquatic species composition and number. 3. Alteration of stream and riparian habitats. 4. Migration and spawning of fish disrupted. 5. Reduction in fish sizes.
Channelization of streams and rivers	Straighten channel and line banks and bottoms	*Aquatic Habitat* 1. Physical properties of the water. 2. Changes in aquatic species composition and number. 3. Changes in bottom dwellers. 4. Fisheries destroyed. 5. Fish sizes decrease. 6. Aquatic animals without suitable habitat or food supply.
		Terrestrial Habitat 1. Vegetation destroyed or disturbed. 2. Wildlife habitat reduced or broken up. 3. Wildlife destroyed or displaced. 4. Migration routes for wildlife disrupted or destroyed. 5. Breeding and feeding grounds destroyed.

TABLE 7.11   Impact Assessment Considerations in Selected Terrestrial and Aquatic Projects (Continued)

Project	Project activities	Potential impacts to be assessed
Dredging harbors and coastal marshes	Opening harbors and channels for boat and barge traffic	1. Physical properties of the water. 2. Disruption or destruction of bottom sediment organisms. 3. Wildlife habitat disruption or destruction. 4. Wildlife breeding and feeding grounds disturbed or destroyed. 5. Broken life cycles and food chains.

To illustrate the kind of assessments that may be helpful in evaluating impacts, let us look at fish population data from streams in eastern North Carolina before and after channelization (Table 7.12). Fish censusing showed that the average poundage of game fish per surface acre was over 400 percent greater in natural streams than in channelized streams, the number of harvestable game fish was reduced by more than 75 percent by channelization, natural streams produced larger fish than did channelized streams, invertebrates were reduced by 78.8 percent in volume after channelization, and the overall quality of streams (based on species diversity) was reduced by 27.5 percent after channelization. In this assessment, the number of each fish species and size before and after channelization were key assessment factors.

It is the biologist's responsibility to state clearly all possible impacts of the proposed project or action on the plants, animals, biotic habitats, and ecosystems on and near the project site. It is through the biologist's eyes that the layperson sees the biological and

TABLE 7.12   Fish and Other Impact Assessment Data from Natural and Channelized Streams in Eastern North Carolina*

Stream name	Total lb fish/acre	Game fish, lb/acre	Harvestable game, fish/acre	Nongame fish (no/lb)	Volume of fish food, organisms/ft^2	Species diversity index
Ahoshi creek						
Natural	54.43	32.89	38	62	8.45	2.5364
Channelized	4.07	1.64	22	315	0.00	2.5440
Griddle creek						
Natural	41.09	17.79	40	99	3.60	2.9217
Channelized	105.36	38.96	70	65	0.05	2.7418
Mosely						
Natural	160.78	77.40	126	42	0.00	3.9172
Channelized	42.99	11.02	0	160	0.10	2.9949

*Extracted from Committee on Government Operations (1971). Streams selected from study only to illustrate data collected.

ecological significance of impacts and possible synergistic and cascading effects in the habitats and ecosystems directly and indirectly involved with the project or action. Disruption of cycles in nature, both biotic and abiotic, must be discussed in the light of short-term and long-term stability of natural ecosystems. The assessment of possible biotic impacts must explain how and why they are significant biologically.

Unavoidable Adverse Effects

Most land use changes result in various adverse impacts on the biota. Many of these can be mitigated to some extent. However, a few impacts produce unavoidable adverse effects on biota, and these should be discussed or summarized in the environmental assessment report. Typical unavoidable effects on the biotic environment are:

1. Decrease in the amount of open space
2. Changes in landforms, including water drainages
3. Loss of natural habitats and "wilderness" aspects
4. Reduction in the number of native animals in the area—with loss of quail, dove, owls, hawks, and large carnivorous mammals particularly noticeable

5. Reduction in the area occupied by native plants and the loss of plant species

6. Increased water runoff, erosion, and siltation which become chronic impacts to the biota of the area, especially riparian and other aquatic forms

7. Reduction in nonconsumptive use of wildlife

As an example, the U. S. Fish and Wildlife Service establishes annual regulations permitting the sport hunting of migratory birds. The Service recognizes that permitting hunting results in such unavoidable adverse environmental effects as: (1) reduction of the migratory bird population by several million birds annually; (2) disruption of migratory bird and other wildlife habitats by the hunter's presence and the direct impact on vegetation, littering, and pollution; (3) negative impacts of hunters on each other, especially where birds occur in high concentrations; (4) migratory birds becoming more difficult for the nonconsumptive user to locate because of hunting activities, with possible increase in travel required to pursue personal interest in birds; (5) private landowners having to contend with trespass, vandalism, and property damage in addition to loss of privacy, solitude, and wildlife; (6) states experiencing negative impacts as a result of federal hunting regulations; and (7) yearly consumption of an estimated 175 million gallons of fuel in the administration and enforcement of the annual hunting regulations of the Service, and in the harvest of migratory birds.

Irreversible and Irretrievable Commitments of Resources

Some environmental impacts commit natural resources irreversibly and irretrievably. For example, air is a simple natural system and will return to a relatively pure and stable state once pollutants are stopped at their source. On the other hand, streams and other aquatic ecosystems are infinitely complex and will not return to a natural state in a lifetime even when pollutants are no longer dumped into them.

Further, stream channelization, removal of forests over large areas, filling wetlands, dredging bays and salt water marshes, and draining swamps are other examples of committing natural resources to a use from which there is no repeal. However, the elimination of a species (population) in an area due to a project may be mitigated by reintroducing members of the species into the impacted area. The success of such an introduction will be assured only if the impacted area is restored in such a way that it will support the species.

Mitigating Measures

Depending on the kind and degree of anticipated negative impact on the biota, the biologist should provide a complete listing and description of mitigating measures to the proposed project. The intent of these measures is to decrease direct and indirect impacts on the plant and animal life of the project site or vicinity. It is hoped that the lessened impact will allow more of the existing ecosystem, with its multiple interrelationships, to continue. Reduced impact means retention of naturalness to a degree.

A basic concept that must be dealt with in alleviating possible impacts is the stability of the existing ecosystem. Natural systems of organisms are stable in that they are in harmony with their environment and with each other. Possible mitigations must therefore be suggested that will insure the integrity of the system both structurally and functionally. Saving one species of plant or animal would be a mitigation, but saving a large enough area in which all of the species can function and survive harmoniously would be a better mitigation.

Disturbed, altered, or fabricated ecosystems lack many structural and functional interrelationships and are less stable than natural ones if not artificially maintained. The organisms are not in harmony with their surroundings or their neighbors. Consequently, the net effect of a given impact will be less than that experienced by a natural ecosystem. In other words, if you start with less in the biological bank account, the loss (impact) will not be as great as if you start with a more adequate account. Mitigating measures for altered ecosystems need not be as closely prescribed as for natural ecosystems.

The suggested mitigations should range from those that should or must be implemented to those that could be implemented in an ideal world. It is up to the biologist to provide a spectrum of choices so that the developer or user of the land may become aware of biological parameters in formulating an approach to the project. Also, land use planners and the various governmental agencies that review environmental studies are sensitized to biological concerns by seeing a variety of mitigating measures. Local and higher levels of government are made aware of environmental and biological concerns and are empowered to counter the threat of environmental degradation.

Various actions and project operations create ongoing perturbations to the plant and animal life of a site. The amount of impact and the choice and implementation of appropriate mitigating programs are often dependent upon a well-designed and well-implemented monitoring program. Such a program detects population declines of important plant and animal species before they reach a critically low level. Unmonitored operations may result in a population decline to a critical level before it is generally realized. Then, it may be too late to save the species. Monitored operations may be the basis for instituting habitat restoration, preservation, and maintenance programs.

Open Space and Conservation Areas "Open space" to a naturalist means natural areas with few if any artificial features. Open space may mean the same thing to land planners, but usually includes all unbuilt land (as well as flood control channels), parking lots, utility rights-of-ways, cemeteries, and agricultural areas with less than one dwelling per acre. Golf courses, race tracks, and school grounds, are usually included in open space elements of municipal and county plans.

In the mind of a naturalist, open space lands should provide essential features for wildlife, such as feeding and nesting grounds for birds and nocturnal mammals. If an open area does not provide such amenities to wildlife, it is then simply an airshed or viewshed for people and should not be considered as a conservation component or element in a local general plan.

If open space areas are isolated one from another, their wildlife diversity and future intregrity is limited. Great benefit is derived from linking open spaces to one another contiguously or through some sort of open corridor. Open space units that are well-stocked with plant and animal species can serve as replacement reservoirs for the less populated or perennially impacted open space units within the system.

Open space should be designed as a network to enhance wildlife potential. Water courses, some roadways, railroads, utilities, and other rights-of-ways provide wildlife corridors between open space units scattered throughout metropolitan areas. Suburban bikeways and equestrian trails provide additional links in the urban open space network. These corridors also may be habitats in themselves and provide suitable homes for a variety of plants and animals.

Gill and Bonnett[15] show London as a city with an integrated suburban wildlife habitat. Many cities are naturally endowed with rivers, streams, lakes, hills, and other geographic features that create suitable habitats for a variety of plant and animal species. The urban area of London extends to a 15-mile radius, beyond which a "greenbelt" preserves significant amounts of open countryside.

Small animals, especially invertebrates, may do well wherever a suitable habitat is retained or recreated within the urban setting. Butterflies, perching birds, rodents, toads, and frogs, both resident and migrant species, are common urban animals.

Often the plant and animal components that attract people to an area are the first things to be removed in a development. Instead of blending the project design with the natural topography and biotic components, the tendency of most land developers has been to remove everything and start from ground zero.

Land once farmed but later abandoned gradually returns to a simulation of its original vegetation and wildlife. The character of this "returned" community depends on such variables as the size of the parcel, its proximity to similar natural areas, time since abandonment, climate, soils, and geography.

Can urbanized areas be naturalized? In urban renewal projects, the removal of old buildings, parking lots, and streets opens the land to nature's colonizers. Some plants and animals will quickly move into a vacant area that is left undeveloped. These are the "weedy species," equipped by nature to tolerate harsh, unnatural conditions. Land occupied by buildings for thousands of years in Europe proved productive and able to provide wildlife habitat within a few months after being exposed by World War II activities.[15] The resilience of plant populations to revegetate a denuded area is well known to anyone who farms or gardens. Weedy species are endowed with the ability to germinate and grow rapidly in disturbed conditions. If the area is not disturbed again, there is a normal progression of plants to a shrub or tree community over many years.

Wildlife Considerations Wildlife habitats of a project area, as well as the biota itself, should receive special attention as regards mitigations. Retention of significant areas of wildlife habitat will in turn aid in the retention of wildlife species. In some cases wildlife habitat can be re-established or restored. The addition of water to an area usually enhances its biotic potentials considerably. In other situations, new wildlife habitats can be created by such means as the construction of small islands or rock spits in bays and large reservoirs. The retention of trees is highly desirable for certain wildlife species.

Islands of vegetation provide shelter, escape cover, and forage. If made a part of an interconnected system, the vegetation lanes that tie such islands together provide a means of escape and migration as well as shelter and forage. The abundance of vegetative edges is an asset to many wildlife species such as deer.

Often wildlife species can be transplanted to save a local population from destruction or to increase the wildlife potential of an area. Valuable bird and mammal species may be transplanted from an area to a suitable habitat prior to initiating a project or activity. Similarly, an area suitable to a valuable wildlife species, but lacking that species or where numbers have been depleted, may be restocked by transplanting.

Vegetative Considerations Existing vegetations in some urban and rural development sites have special value. Often ornamental trees and shrubs in urban situations may have attained specimen size, or may be rare or exotic, and are worthy of saving. Upon completion of the project, the retention or re-establishment of these "old" specimens helps provide an element of familiarity and original charm.

The retention of intact and healthy specimen trees and shrubs may be achieved if grading of the site is not so excessive as to cut away a considerable amount of soil or cover the trunk with several feet of fill dirt. Some species lend themselves to being moved from the site during the site preparation and construction phases, and then returned to the site for replanting. Various techniques in moving large trees and shrubs include bare-rooting, balling, and boxing. The species of plant must be considered since there is considerable difference in tolerance of transplanting. The season of the year is another significant factor in determining suitability and method of transplanting. Frozen ground, actively growing plants, and summer hot or dry spells are among the conditions that make moving difficult if not impossible. The type of root system is also an important consideration. Fibrous-rooted trees, such as palms, move easily, whereas deep tap-rooted trees, such as walnuts and oaks, move poorly.

The environmental impact on both the aesthetic appearance of an area and its wildlife habitat potentials are modified to varying degrees by the planting of appropriate vegetative types to achieve stated mitigating purposes. These practices are particularly effective if tied into existing greenbelts, conservation zones, shelterbelts, and open space elements of local and regional plans.

Other considerations in mitigating with specimen plants are:

1. Incorporation of specimen plants into small groups, accessways, greenbelts, corridors, or shelterbelts to provide continuity and wildlife habitat.

2. Whether the plant is native to the area or whether it is an introduced or exotic species.

3. The number of plants to be retained or removed. This must be determined by a botanist, horticulturalist, or nursery expert knowledgeable of the species involved, their condition (presence or absence of disease, stunted, or vigorous), ecological requirements (water, light, nutrient, pest control), and the possible impacts of the project or action on the plants.

4. Specimen value, both aesthetic and biotic. For example, an investigator must properly assess the loss of a young seedling tree versus a 350-year old specimen of the same species.

Special note should be made of the oxygen production by green plants. Considerable misunderstanding has centered around the claim that in industrialized areas, the lack of green plants results in an unhealthful lowering of the oxygen level in the air and that the world will soon run out of oxygen. All green plants (one-celled algae, native trees and shrubs, as well as lawns and ornamental plants) produce oxygen. Hence, the removal of native vegetation and its replacement with ornamental plants poses no threat to the world's oxygen supply. There is such a mixing of air by atmospheric movement that the oxygen supply is relatively constant at given elevations.

The aesthetic value of green plants is uncontested. Appropriate landscaping provides visual screening and sound buffers, is pleasing to human residents, and modifies physical factors of the environment to make conditions more suitable for other kinds of living things. Anyone who has lived on an elm-lined street in the East and has experienced the visual, auditory, and environmental changes resulting from their loss to the Dutch Elm Disease can attest to their aesthetics, biological importance, and desirable moderating effects.

One practical application of plant community analysis (as discussed in the section on Assessment Methodologies) is that of synthesizing a community type. This involves planting container plants or compounding seed mixes of native species in the right proportions. Knowing the species present in a natural community type, their relative numbers, structural relationships (i.e., stratum relationships), and microhabitat requirements enables one to synthesize that community type in an appropriate setting. The whole concept of synthesizing a community type has been used (unknowingly perhaps) by the hydro-seed industry in recent years.

In the future, synthesizing plant communities may well become an important practice of plant landscape designers in large developments, new towns, and urban renewal projects. These projects are usually large in scope and provide ideal opportunities to plan mitigations with the whole ecology (both natural and human) of an area. It is imperative that all those concerned with land use be aware of the interdependencies and interactions (both positive and negative) of all the environmental components. All elements in the environment must be evaluated, not evaded. Changes must be purposeful, not random or capricious. A serious attempt should be made to plan for people, not as units or numbers, but as whole persons with physical, social, and psychological needs. Appropriate use of plants and provisions for certain birds and other wildlife enhances the human environment.

The plan should provide for appropriate landscape requirements to serve as necessary buffers and screening to create a sense of privacy and to reduce or eliminate undesirable urban noises and air pollution. Landscaping should fit the various land categories. These include slope of land, present vegetation, soil conditions, and aesthetic considerations. Developers should use their best efforts to assure the most creative site planning solutions for all buildings and other improvements necessary to preserve the wooded character of the site, to preserve the natural contour of the land, to promote good site drainage, to assure good access from and to the street system, and to permit good views to and from the site, buildings, and other improvements.

Alternatives

In assessing the biotic environment, the biologist must consider viable alternatives to the proposed project that would reduce or eliminate negative impact on the biota. Alternatives are derived in part, therefore, from the anticipated environmental impacts described earlier. The validity of the alternatives is incumbent on the quality of the biotic impact analysis. Generally, the more severe the anticipated impacts on the biota, the greater is the necessity for viable alternatives to the project. Negative impacts might be accepted as is or mitigated without the necessity of project alternatives.

Most environmental assessment reports must present alternatives. One of these is the "no action" alternative. This alternative is applicable in situations where the biotic habitat is of special value or uniqueness, or where rare or endangered species are involved. Nonstructural alternatives (such as floodplain zoning) favor natural areas with their plant and animal components over structural developments (such as dams and channelization).

In residential developments, larger lots and cluster developments favor the biotic environment and open space, whereas smaller lots and homogeneous lot developments tend to destroy the biotic environment.

Alternatives may be fundamentally or incrementally different. If differences are "fundamental" (e.g., preventing flood damage by levee construction versus floodplain zoning), impact significance can better be measured by the optimum method for each type of alternative rather than by direct comparison of alternatives. This is because environmental impacts will differ in kind and size. If differences are "incremental" (e.g., preventing flood damage by constructing a large dam versus constructing a series of small dams over a number of years), impact significance can better be measured by one method that generates ample quantitative data on the many facets of the project impact. A series of small dams will keep generating new alternatives that must be assessed in terms of fisheries, recreation, and wildlife habitat. A single large dam would eliminate future incremental alternatives by committing a reach of stream to one condition irreversibly.

Growth-inducing Considerations

Land planners and land managers must assume full responsibility for the environment. This fact necessitates their familiarity with the entire existing environment over which they have jurisdiction. Assuming such foreknowledge and responsibility, land planners and managers should then be able to evaluate proposed changes in land use. Environmental assessment necessitates that biotic and ecological growth-inducing considerations be appraised by the biologist investigator. Murphy[22] points out that many people pursue "an unfortunate insistence that every portion of nature be judged by its comparative ability to produce a cash flow, and a reluctance to realize that the changes produced have been profound, rarely foreseen, often unfortunate, and perhaps irreversible."

A change in land use often results in a change in the composition of the animal community and a shift in its numbers. Fruit orchards induce English sparrows, and new lawns support lawn moths. Often there is a population explosion of a few species (such as weeds, mice, ground squirrels, pocket gophers, and starlings) that are judged by people to be pests. In other situations some of the remaining species may pose a direct threat to domestic animals. For example, residential areas in rural settings often lose pets to coyotes. Rabbits and deer can ruin landscape plantings, young orchards, and gardens. Poison ivy, fleas, ticks, and mosquitoes are examples of biological backlash when humans intrude into nature.

ASSESSMENT METHODOLOGIES

Environmental assessment and impact analysis are presently more art than science. There are no universally acceptable procedures for conducting such studies. This is especially true for biological impact assessments. Few biological methodologies have been developed specifically for impact statement preparation. However, there are various biological methodologies that lend themselves to the environmental assessment process. The most commonly used ones will be discussed here.

Environmental impact statements should list the dominant plant and animal species found on the project site, as well as possible rare or endangered species. These should be discussed in a regional context as to their uniqueness or commonness. Sensitive or fragile habitats and their associated species should be studied and described in greater detail in order to establish baseline data which can be used to detect unforeseen project impacts.

Regardless of the methodology used, certain field deficiencies will be present. Examples of these are: (1) annual plants are present only during the growing season, (2) some animal species change location during the year, (3) other animals are seen only at night, and (4) most animals and simple organisms in the soil are difficult to determine and assess environmentally. Professional judgment must be used in coping with these deficiencies while, at the same time, the proper methodologies to overcome these field deficiencies are selected, if possible.

Studies of plant and animals and their environment may be described subjectively, based purely upon observation, judgment, and intuition. The value of this qualitative method is limited and varies with the experience and expertise of the one doing the study. Highly qualified field biologists can, in the course of a few hours, summarize the ecology of a limited area. In this regard, it is important that the qualifications of the individuals involved in the study and the assessment methodologies used be listed in the assessment report.

Some biotic areas are sufficiently researched so that secondary sources can be used rather than detailed on-site studies. Although such secondary sources may have a cost advantage, they should be used with caution since biological communities change continuously as does the status of certain species. However, some ecologists claim that even the most elementary study demands a quantitative approach. If summary statements or conclusions have to be verified (i.e., statements of fact or data to prove them), a more inclusive, quantitative study is necessary.

Quantitative studies generally necessitate sampling of one sort or another, and this suggests that the use of statistics is involved. Smith[25] cautions that in the hands of those who know little about them, statistics can be dangerous. They are often abused rather than used and are misapplied to the problems at hand. As a general rule, only the most elementary statistics need to be used in analyzing the data generated in biotic assessment studies. An exception would be in regional or long-term studies where the biological data becomes a part of the data base used to generate statistical models.

Other problems may arise in quantifying ecological assessments. How large should samples be? Where should they be taken? What are the samples to determine? To be statistically valid, samples must be random and should represent the important aspects of the entire population or community being studied.

This section attempts to show the reader some practical methodologies for assessing the biotic environment of an area. This discussion is not intended to teach the user how to carry out the specific methodologies—rarely would they be directly applicable. Instead, the key ideas and procedures are described in sufficient detail to help the reader determine their suitability to a given situation or to provide reviewers with tools and guidelines for evaluating completed studies.

Environmental Impact Assessment

Assessment of possible environmental impacts often entails the selection of a reference point from which biotic and abiotic changes can be measured. The reference point is usually an intact ecosystem which is compared with its altered or stressed counterpart. For example, by assessing an intact hardwood deciduous forest and a similar stand nearby that has been exposed to heavy foot traffic, the investigator gains some measure of that community's response to physical impact. If the project site is to be monitored for possible impact during the construction and/or operation phase, a suitable method of ongoing data acquisition must be selected and implemented in the project impact area.

In the selection of an assessment methodology, several impact assessment parameters should be considered, namely: (1) type and amount of input data needed, (2) assumptions and limitations of the technique(s), and (3) conditions under which a given technique would or would not be appropriate.

There is no single "best" methodology for environmental assessment or impact analysis. The characteristics of a given methodology appropriate for one type of impact assessment might be quite inappropriate for another type of assessment. Evaluating the merits of removing vegetation by means of a bulldozer or by using herbicides would require different assessment methodologies. Only the investigator can determine which tools may best fit a specific case. In selecting the most appropriate tools, the following considerations may be useful[12]:

1. *Use.* Is the assessment report primarily a "decision" or an "information" document? A "decision document" is vital to determining the best course of action and is based on an analysis that generally requires greater emphasis on the identification of key issues and on direct comparison of alternatives. An example would be the determination of whether a power plant on a site would cause major or minor upset to the biota. An "information document" reveals primarily the implications of an already agreed upon course of action and requires a more comprehensive analysis dealing with the significance of all possible impacts, such as in constructing a major highway or airport.

2. *Resources.* How much time, skill, money, data, equipment, and computer facilities are available? Generally, the more extensive the project or action is, the larger the study site. Further, the more complex the biota, the more quantitative analysis is required and the more specific details are needed in the impact assessment report. A transcontinental pipeline would demand a far more comprehensive study than a local pipeline.

3. *Familiarity.* How familiar are the investigators with both the type of action proposed and the environmental resources of the area? Greater familiarity generally improves the substance and validity of both subjective and objective analysis of impact significance.

4. *Significance.* The issue of "significant" environmental impact is site specific and also is dependent upon the way people perceive the environment involved in the proposed project or activity. All other things being equal, the bigger the issue, the greater the need for explicitness, quantification, and identification of key issues.

5. *Ecosystem(s) involved.* What is the nature of each ecosystem, i.e., is it common, unique, durable, or fragile? Are there critical habitat features or are rare or endangered species involved? The impact assessment methodology must be site specific depending on the ecosystem characteristics. Thus, commonness implies simple methods of assessment, and uniqueness or rarity demands detailed methods of assessment. A salt marsh

would require an extensive data base, whereas a vacant lot would require minimal data. The types of methodologies which can be employed in performing biotic (as well as other) assessments can be generally categorized as follows:

1. *Ad hoc.* These methodologies provide minimal guidance to impact assessment beyond suggesting broad areas of possible impacts (e.g., impacts on plants, animals, lakes, forests), rather than defining specific parameters to be investigated. An example would be the designation of major natural community types and their response to various kinds of urban development.

2. *Overlays.* These methodologies rely on a set of maps of environmental characteristics (physical, social, ecological, aesthetic) made for a project area. These maps are overlaid to produce a composite characterization of the regional environment. Impacts are identified by noting the areas of overlap and then analyzing all possible impacts in these areas.

3. *Checklists.* These methodologies present a specific list of environmental parameters to be investigated for possible impacts but do not require the establishment of direct cause-effect links to project activities. They may or may not include guidelines on how parameter data are to be measured and interpreted. (See Table 7.9.)

4. *Matrices.* These methodologies incorporate a list of project activities and possible impacts, and a list of potentially impacted environmental features. These two lists are cross-related in a matrix which identifies cause-effect relationships between specific activities and environmental impacts. See Table 7.13 for an example of a simple matrix showing various environmental factors and project actions. Matrices may *(a)* specify which actions impact which environmental components or *(b)* simply list the range of possible actions versus environmental components in an open matrix to be used as a guide by the biologist investigator.

5. *Networks.* These methodologies work from a list of project activities to establish cause-condition-effect networks. They are an attempt to recognize that a series of impacts may be triggered by a single project action. These network approaches generally define a set of possible networks and allow the analyst to identify impacts by selecting and tracing out the appropriate project action(s). In a sense, networks are an extension of matrices for long-term projects.

Physical Environment Assessment

The physical environment of an area can best be described by assembling geographic, geologic, soils, and climatological information and data. These can be obtained from U.S. Geologic Survey topographic maps, state and local geographic and geologic maps, U.S. Soil Conservation and university extension offices, U.S. Department of Commerce climatic reports, and other public and private organizations. This material should be digested into appropriate summaries that have a direct bearing on and pertinence to the biotic and environmental impact assessment. Some of the geographic information can be of assistance in the preparation of a plot map of the study site (see Fig. 7.5).

Flora Assessment

The plant life (vegetation or flora) of an area shows certain affinities with its environmental setting and between plant species. Normally vegetation analysis is directed toward the natural plant communities as they relate to the various site factors. Disturbed areas often have fragmentary plant associations having no community structure or integrity. But in either case, the analysis should consider the dominant species that characterize the site, the sensitivity of the vegetation to impact, and the presence of rare or endangered species.

A preliminary examination or "reconnaissance" is made of the study area to get a general picture of the landscape and its vegetation. Regional and local topographic maps, and in some cases aerial photographs, assist in determining access routes, topographic obstacles, and on-site features to be studied. In reconnaissance, the land is traversed rapidly and one searches for the most obvious and important features. These features should be indicated on a site map and may include:

1. Major vegetative patterns and plant communities, including their growth form (physiognomy) and dominant species.

2. Correlations between plant communities and features such as topography, geology, soils, and water.

3. Past and present human influence on the vegetation

Although reconnaisance is done quickly, it should not be done haphazardly. Preselec-

TABLE 7.13 Examples of an Environmental Factors/Project Action Matrix

	Project actions						
	Release of		Runoff	Removal of vegetation	Grading the site	Construction phase	Operation phase	Possible impacts
Environmental Factors	Noise	smoke						
Physical:								
Air	X					X	X	Noise pollution
Water			X					Water pollution
Soil				X				Erosion
Landform			X					Contour, alteration
Biological:								
Vegetation				X				Loss of native vegetation
Key Plants				X				Loss of key plants
Rare or endangered plants								None present
Animals				X	X			Loss of habitat
Key animals				X	X			Loss of key animals
Rare or endangered animals								None present

tion of the best route of travel is important to insure a complete examination of the salient features. Effective reconnaissance is greatly facilitated by previous knowledge of the flora, especially the bulk of common plants.[5] Identification of unknown plants may be done in the field, or plant materials may be collected and taken from the site for later determinations.

Plant Survey   Plants may be assessed in several descriptive ways. A "species list" includes both common and scientific names of the plants found or suspected to occur in the study area. This list is developed by site visits and by consulting published literature. Secondary sources are particularly important for out-of-season annual species or for possible rare or endangered plant species. Such sources must be referenced. A species list has limited value in the final environmental report because it does not provide any measure of relative plant numbers and does not convey ecological relationships between the plants and their environment. For this reason the species list should usually be appended to the report rather than included in the text. In small reports the list may be included in the text, but should be interpreted there too.

The "number of plants" provides information on composition, vegetation patterns, and relative abundance of each species in the area. Such information also allows comparisons to be made between stands of the same type or with other vegetation stands of different types in order to establish commonness or rarity of species of vegetation types on or near the study sites.

The number of plants can be expressed without actually counting each plant. A common method employed in limited assessment studies is to devise an "abundance scale" and then assign each species to an "abundance class." Abundance scales range from low to high abundance and may be a simple two or three point scale, or a complex one composed of numerous values of relative abundance. Each species is estimated as belonging to an abundance class. The ground area covered by the aerial portion of a plant is called its "cover" and is used as a measure of the plant's importance in the vegetation. Plant cover values are sometimes combined with abundance classes to give a composite measure of dominance. For example, an investigator wanted to determine what plant species were most important in the vegetation on a site. After selecting several observation points within the vegetation, estimating the percent cover of each species at each point, and placing them in abundance classes, the following values were obtained:

Abundance class	Abundance scale	Species
5	Species cover, 76–100%	Chamise
4	Species cover, 51–75%	None
3	Species cover, 26–50%	None
2	Species cover, 6–25%	Ceanothus and buckwheat
1	Species cover, up to 5%	All other species

The investigator's report described the vegetation as composed of 23 species and dominated by chamise, with buckwheat and ceanothus as important subdominants. This method allowed the investigator to develop a good description of the vegetation with minimum time expenditure.

Some studies may call for a more intensive quantitative survey of the vegetation, but do not require counting each plant in the entire vegetation. The number of plants by species may be estimated, based upon samples within a stand or between stands. Quantitative data may be obtained by using sample plots (to be discussed later). The data may deal with number, size, distribution, condition and health, age, rarity, and place of each species in the community structure. This data may be presented in tabular form, but is preferable in graphic form (Fig. 7.6). The data presented must be interpreted to provide ecological understanding, economic evaluation, aesthetic consideration, and conservation perspective. For example, a site that involves a grassland might necessitate the identification of grasses and other herbaceous species without counting each one. On the other hand, if the study calls for an evaluation of each species in the grassland, sample plots would provide counts of each species and their relative abundance. Key species should be discussed as to commonness or rarity, ecological role, and sensitivity to disturbance or change. Such studies provide baseline data for detection of environmental changes due to a project. A control plot or series of plots should be established to insure that nonproject variables are not influencing the data.

If the "absolute number of plants" is an important consideration, there are several techniques by which it can be derived. Specimen plants can be counted on a simple "walk-through" of the site, or plants may be counted within each community type. The exact location of each plant may be plotted on the site map and, as a result, such a map can serve as a voucher for the plants to be retained in a project.

Vegetation may be analyzed on a "structural" basis without considering the species involved (Fig. 7.7). Such a structural analysis relies on designating either the vertical layers of vegetation (such as herb, shrub, small tree, and canopy tree) or the horizontal distribution of vegetation or community types (such as grassland, herbland, shrubland, woodland, or forest). See Ref. 24 for details. These diagrams have value in designating key species, wildlife habitats, successional trends, and in describing community changes resulting from environmental impacts.

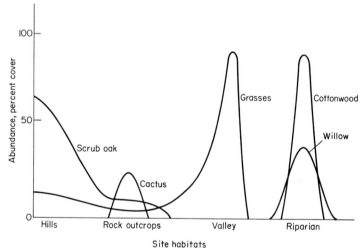

Figure 7.6   Abundance of key species by site habitats.

In more detailed studies of vegetation structure, the assessment might include the life form of dominant plants (e.g., erect woody, climbing, or decumbent, epiphytes, herbs, bryoids), plant coverage (e.g., barren or very sparse, interrupted or discontinuous, in patches, tufts, or clumps, continuous or dense), or leaf characteristics (e.g., leafless, deciduous or evergreen, succulent, needle, spine, scale, or broadleaf).

The plant survey technique just presented gives an easily understood visual interpretation of an area. The adage that one look (at a site) is worth 10 reports is met by these techniques, which give the reader a synopsis of an area. Techniques of this type are generally used in forested habitats since our present aesthetic views favor the visual impact of these vertical plant associations.

Quantitative Plant Assessment   Since plants occur in recognizable associations and interact with one another, special quantitative methods may be used in assessing these vegetative parameters. Beyond considering the species composition of a vegetation, these methods provide information on the role and relative importance of individual plant species (such as dominance or rarity), relationships between plants, species distribution and spacing, number of individuals of each species, and structure of the plant community.

The methods of analyzing vegetation quantitatively are too numerous to describe. Those presented here are relatively simple to use and are suited to most environmental assessment studies. They involve sampling the vegetation rather than measuring every plant in a stand—something that can be done, but usually is not practical or necessary. Samples may be distributed at random by using a grid or random-numbers table. Even though randomization permits statistical treatment of the data, it may be too laborious and time consuming for most impact studies. As alternatives, typical or representative sample

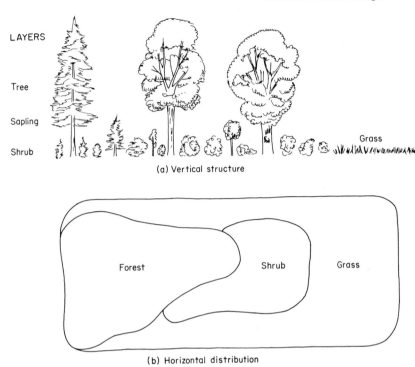

LAYERS

Tree

Sapling

Shrub

Grass

(a) Vertical structure

Forest

Shrub

Grass

(b) Horizontal distribution

Figure 7.7 Vegetation structure. (*a*) Profile diagram; (*b*) distribution diagram.

areas may be chosen subjectively by a trained biologist. Samples may be taken in an even, zigzag, or other prearranged fashion, or samples may be established along a predetermined line or series of lines. For complex environments and regional studies, it may be advisable to use more sophisticated methods of sampling and vegetation analysis. Examples of such methods are presented in Refs. 5, 9, 21, and 24.

   Sample Plots.   The most widely used technique of sampling vegetation is that of sample plots or quadrats. Their popularity rests on their ease of establishment and data acquisition. Care must be taken to insure that the sample is adequate to represent the entire stand. Choice of size, shape, number, and distribution of the sample plots or quadrats will influence confidence in the adequacy of the sample. The quadrat sampling technique is applicable in all types of plant communities and for the study of submerged, sessile (attached at the base), or sedentary plants and animals. The commonly accepted plot sizes are 0.1 m² for mosses, lichens, and other small mat-like plants; 1 m² for herbaceous vegetation, including grasses; 10–20 m² for shrubs and saplings up to about 3 m tall; and 100 m² for tree communities. In studies requiring an analysis of simple to complex plant forms, such as in a wet forest, nested quadrats may be useful (Fig. 7.8). The smaller quadrats are used for the simple plants, and progressively larger quadrats are used for the larger forms.

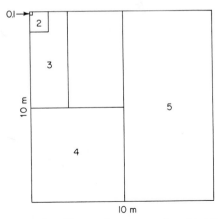

0.1

2

3

10 m

5

4

10 m

Figure 7.8 Nest quadrants ranging from 0.1–100 square meters.

The question of how many sample plots will adequately sample the vegetation is a persistent one. There should at least be enough to reveal the majority of the species present in the stand. Often, the number of plots is chosen more or less arbitrarily and is based on previous experience, time and money constraints, and, in some cases, prejudice.

The question of what plants are to be included in the sample must be considered. Are *all* plants going to be assessed—including mosses, mushrooms, and annuals—or will assessment involve only the woody shrubs and trees? The common practice is to place major importance on the common species and count only individuals that are rooted within the plot. In some cases, it is difficult to decide what constitutes an individual. For example, plants that reproduce vegetatively by means of runners, stolons, or rhizomes must be assessed by some arbitrary means. With all the above considerations resolved, the plants to be sampled within the plot are recorded by number of individuals of each species and by their individual size measurements.

Several parameters, such as height and cover may be measured on each plant within the plot. Plant "height" is usually measured directly by means of a meter tape, ranging pole, or meter stick. However, large trees are usually measured with the aid of a sighting instrument. In addition to knowing the plant's height, we need its "cover" or horizontal spread as a measure of size or dominance. Cover is derived by considering the area of ground surface covered or shaded by the leafy portion or crown of the plant and is used to describe both the crown of an individual and the horizontal dimension or canopy formed by many plants. The diameter at breast height (4.5 ft or about 135 cm) of tree trunks is used as an expression of cover or dominance. Knowing the trunk volume is an important measurement when assessing forests which contain commercially valuable timber species.

When actual counts are made of a species in sample plots of known size, the data can be expressed as "density," i.e., the average number of individuals by species per unit area sampled. The species within each sample plot are counted. Then, the total number of each species is divided by the total sample area.

Another descriptive parameter of the community is "dominance," i.e., the relative degree to which a species predominates a community by its sheer numbers, size, bulk, or biomass. Species that exert the greatest control or influence in the community are called "dominants" (key species). Dominant species generally indicate the ability to capitalize on the resources in their environment most efficiently. As mentioned earlier, plant cover is described as the aerial spread of the crown or trunk (basal area) of a species and it has become the main measure of dominance. It can represent percent ground cover per unit area. Data needed for these calculations include fixed sample plots of a determined size. In addition, it is generally necessary to have field personnel trained in analytical methodologies and taxonomy. Data output is exceptionally good and can be used in the preparation of vegetation maps and, as a result, will give some indication of community structure and the dominant plants.

Plant dispersion over an area or within a community is another vegetative parameter and can be derived by simple data analysis. Plants may be either widely dispersed or clumped. In a series of plots, a measure of species "frequency" is obtained which expresses the proportion of equal-size sample plots in which at least one plant of that species occurs relative to the number of plots taken.

In summarizing plant data, it is desirable to use as many values as possible. Knowing the density of one species gives an idea of the number of plants in a stand but tells nothing of how they are distributed (even vs. clumped) or their size or dominance. Combining relative density, dominance, and frequency values for a species into a single "importance value," and converting the sums to a 100 point scale, generates a composite relative value for each species in a stand or series of stands. The importance value allows quantitative comparison of each species in a stand with the other species in the stand, or allows comparison of the species in one stand with the species in other stands. This comparison capability is an important one in situations where stands must be assessed or ranked for their uniqueness, commonness, or degree of disturbance.

The approach to computing importance values is as follows:

$$\text{Density} = \frac{\text{number of species } A^*}{\text{area sampled}} \tag{1}$$

*Species A represents any species being considered.

$$\text{Relative density} = \left(\frac{\text{density of species } A}{\text{total density of all species}}\right) \times 100 \tag{2}$$

$$\text{Dominance} = \frac{\text{total cover or basal area of species } A}{\text{area sampled}} \tag{3}$$

$$\text{Relative dominance} = \left(\frac{\text{dominance for species } A}{\text{total dominance for all species}}\right) \times 100 \tag{4}$$

$$\text{Frequency} = \frac{\text{number of plots in which species } A \text{ occurs}}{\text{total number of plots sampled}} \tag{5}$$

$$\text{Relative frequency} = \left(\frac{\text{frequency value for species } A}{\text{total of frequency values for all species}}\right) \times 100 \tag{6}$$

$$\text{Importance value} = (\text{relative density} + \text{relative dominance} + \text{relative frequency}) \div 3 \tag{7}$$

Transects and Line-intercepts.   Long, narrow sample plots, such as transects and line-intercepts, are popular for sampling shrub and certain other vegetative types. They allow for rapid assessment of vegetation transition zones, require minimum time or equipment to establish, and lend themselves to sampling two or more vegetation strata simultaneously. "Transects" may be of any desired dimension, but are usually 10 to 100 times as long as wide. One meter is a convenient width since it allows for accurate accounting of each plant as it is encountered in the plot. Also, by subdividing the transect into subplots, a series of plots can be established along a meter tape and the lateral dimension can be obtained with the aid of a meter stick. "Line-intercepts" are linear plots usually 10 to 100 meters in length. Cover is recorded as the amount of line intercepted by a vertical projection of the plant crown onto the line; the portions of the plant not intercepted by the line are not considered. The total amount of line intercepted by each species divided by the total length of the line intercept gives its cover. For example, in a 10-m (1000-cm) intercept, three individuals of species A are intercepted: 120 cm, 55 cm, and 10 cm. The total cover for species A is therefore 185 cm, or the percent cover is 185 cm/1000 cm, or 18.5 percent. By using 1 to 5 m intervals along the line as subplots, frequency can be obtained. The interval size is determined by the plant sizes and must be as large as the largest plant being sampled.

Absolute density cannot be determined by the line-intercept method since the plot is one dimensional. However, relative density is commonly calculated by:

$$\text{Relative density} = \left(\frac{\text{total individuals of species } A}{\text{total individuals of all species}}\right) \times 100$$

Since this value is a percentage, it can be applied to an area of any size.

Plotless Sampling.   Several sampling techniques have been developed which do away with the necessity of laying out areas to be sampled. Vegetation measurements are determined from points rather than being determined in an area with boundaries. These so-called "plotless methods" are used mostly in grasslands, and open shrub and tree communities. All plotless methods allow more rapid and extensive sampling of a stand than the plot methods. See Refs. 5, 9, 18, 21, and 24.

The "point-quarter" plotless method is commonly used in woods and forests. A series of points randomly determined or at random intervals along a line of march is established. Each point is divided into four 90° quadrants. The tree in each quadrant is identified, its distance from the point is measured, and its basal area is determined. A minimum of 50 such point tallies are generally taken.

Calculations used in the point-quarter method are as follows:

$$\text{Mean point-plant distance} = \left(\frac{\text{sum of all point-plant distances}}{\text{number of point-plant distances}}\right) \times 100 \tag{1}$$

$$\text{Mean area per plant} = (\text{mean distance})^2 \tag{2}$$

$$\text{Total density of all species} = \frac{1000 \text{ m}^2 \text{ (hectare), or } 43{,}500 \text{ ft}^2 \text{ (acre)}}{\text{mean area per plant}} \tag{3}$$

$$\text{Relative density} = \left(\frac{\text{number of individuals of a species}}{\text{total number of individuals of all species}}\right) \times 100 \tag{4}$$

$$\text{Density} = \left(\frac{\text{relative density of a species}}{100}\right) \times \text{total density of all species} \tag{5}$$

$$\text{Relative dominance} = \left(\frac{\text{basal area for a species}}{\text{total basal area for all species}}\right) \times 100 \tag{6}$$

$$\text{Frequency} = \frac{\text{number of points at which a species occurs}}{\text{total number of points sampled}} \tag{7}$$

$$\text{Relative frequency} = \left(\frac{\text{frequency value of a species}}{\text{total number of points sampled}}\right) \times 100 \tag{8}$$

$$\text{Importance value (\%)} = (\text{relative density} + \text{relative dominance} + \text{relative frequency}) \div 3 \tag{9}$$

Vegetative Succession. Methods used in assessing vegetative succession vary with the vegetation types involved.

The techniques used in intertidal areas, herb and grasslands, or early stages of succession where the plants are small employ sample plots 1 m² or smaller. Small grids and photographic treatments are useful in determining relative numbers of species per plot, their age, size, growth per unit time, proportion dead, and proportion of pioneer, seral (intermediate stage), or climax species. This data can also be helpful in determining the age or stage of community succession. For example, in assessing the impact of sewage effluent on marine algae establishment growth, a biologist used a small grid mounted on a camera. Resulting photographs, taken over a period of two years at various sites from the sewer outfall, provided data necessary to establish successional trends and potentials.

Larger quadrats, transects, line intercepts, and plotless methods are used to evaluate perennial shrub or tree community types in successional studies. Plots are established in undisturbed vegetation and are compared with data collected from similar disturbed vegetation of known age. The relative frequency and age of each species are used to establish growth potentials and compositional changes of key species composing the vegetation. For example, a biologist sampled various pipeline routes of different ages to establish the growth potential of vegetation on a proposed new pipeline route. Age of trees and large shrubs may be determined from increment cores taken from trunks, the number of annual rings of cut stems, fire scars from known fires, and other established criteria.

Site Index. "Site index" is defined as the height reached by a tree species at a given age. It involves measuring the height and age of representative trees within a given area. The index is an indication of the "growth potential" of the site, i.e., the culmination of the abiotic factors influencing growth. For example, a site index in an area with both timber production and ski resort potentials was determined. Those areas with high site indices were used for buildings and ski run establishments. Site index is arbitrary, at best, but does allow a relatively quick evaluation of the potential for a site to produce timber.

Net Productivity. "Net productivity" is defined as the amount of growth or biomass produced per unit time. It is a valuable biological measurement in studies where growth potentials must be known in order to determine potential economic and ecological loss from removal of vegetation. This measurement can also be used as a baseline for monitoring environmental impacts during the operation phase of a project. In aquatic systems, net productivity is a valuable measure of the degree of pollution in both natural and artificial ponds and lakes.

In situations where one species is involved—or in plant systems where herbivore consumption is negligible—plants are harvested and used as a measure of net productivity. To calculate net productivity, biomass per unit area is collected, dried, weighed, and expressed as dry weight per unit area. This method is used in calculating the cropping capacity of agricultural lands, in natural grasslands and other vegetation used as forage, and for forest trees being cut for commercial purposes. Net productivity studies based on the harvest method have been used in comparing forage values on sites before and after strip mining.

The carbon dioxide method equates CO_2 uptake to photosynthetic rates minus respiration. As with the harvest method, productivity can be calculated for a single plant, species, or community. In general, the method necessitates physically enclosing a plant or part of the plant community in a CO_2-impermeable chamber. Air is pumped in through a CO_2 monitor, and the airstream is analyzed for its CO_2 content as it leaves the chamber. The relationship between the CO_2 intake (photosynthesis) and CO_2 production in the dark (respiration) is a measure of net productivity.

The carbon dioxide method is limited by the cost of electronic and other equipment and the "artificial" environment of the chamber. However, this method allows the investigator to assess the impact of air pollutants on the productivity of plants. If the air effluent from a proposed new factory can be shown greatly to reduce net productivity of the vegetation in the area, siting of the factory or mitigations of the effluent must be considered.

Species Diversity. The number of species and the number of individuals in each species comprise a measure of community richness called "species diversity." Even though the values derived are not absolute, they do provide a measure of community complexity and "health." Disturbed, polluted, or degraded natural communities—such as vacant lots and polluted or channelized streams—usually have low species diversity indices, whereas mature and intact communities such as natural woods, forests, grasslands, and natural streams often have high diversity indices. An index in common use is Simpson's index, calculated by the formula

$$D = \frac{N(N - 1)}{\Sigma n(n - 1)}$$

where D = diversity index
N = total number of individuals of all species
n = number of individuals of a species

Since the index values are influenced by sample size, comparisons between two or more communities must be based on equal simple sizes. Species diversity indices are applicable to both plant and animal communities. By knowing the species diversity index of two proposed highway routes, a biologist could better assess which route would result in the smallest biotic impact.

Fauna Assessment

The animal life (fauna) of an area is dependent upon the vegetation, and there are countless relationships between the species composing an animal community. Assessment techniques generally provide information on the dominant species involved, rare or endangered species, species sensitive to impacts, and important ecological interrelationships. Fauna assessment involves more problems than flora assessment by virtue of the greater variety of animal types, their mobility, and behavior. The techniques presented here are the most commonly used in environmental assessment studies. They start with the determination of the species present in the general area and are followed by techniques appropriate to the site, the size of the area, the number and kind of species, the proposed development, and time and monetary constraints. For a more detailed description of faunal sampling techniques, see Ref. 26.

Faunal assessment provides a basis for determining relative abundance and evaluating commonness or rarity of each species encountered or reputed to be present from the literature. Certain population parameters can be obtained by direct and indirect observation, trapping, and censusing. Age structure of game animals should be determined if the project or action will have long-term negative impact on their population. Other than a simple walk-through assessment, all faunal studies are time consuming, requiring days, weeks, or months to accomplish.

Animal Survey   There are several ways in which animals may be assessed. The choice of which method is used varies with the extent and purpose of the study and with the composition of the animal community involved, such as insects, fish, birds, and mammals. Table 7.14 lists faunal assessment information needed for various projects.

Species Lists. Animal communities can be assessed by observing the fauna directly. As with plant communities, a species list may be prepared by walking through the area, listing each species as encountered. Tabulation of number of species and their location is a valuable product of direct observation. In addition to seeing the animals themselves, a trained biologist can determine the presence of certain animal species by recognizing animal signs, such as tracks, scats, and bones in owl pellets. Published lists and descriptions of animal species should be consulted in preparing a species list. Knowing what species might be present aids the biologist both in preparing the list and in field observations.

Animal species lists, like plant species lists, have limited value if they are the extent of the faunal assessment. Such lists present common and scientific names of the species involved so that the faunal resources of the area are catalogued. Extensive species lists are usually placed in an appendix of the environmental impact assessment report.

Direct Contact. Sampling animal populations involves collection, study, and release. In some cases, it may be necessary to kill the animal in order to study its stomach contents or other study requirements, but this should be a rare necessity.

Ground organisms such as large arthropods, small reptiles, amphibians, and rodents may be trapped by using pitfalls—cans sunken into the ground with the open rim even with the ground surface. Soil and ground litter organisms, especially the small arthropods, are most easily collected with the aid of a Berlesee funnel. Larger organisms can be caught by hand or by means of hand sieves, cans, or pitfall traps. Certain objects, like rocks or boards, will attract certain spiders, centipedes, lizards, and other small animal forms. Manure, dead meat, and sugary baits will attract certain scavengers. These techniques are useful for long-term studies.

TABLE 7.14   Faunal Assessment Information Needed for Various Projects

Degree of assessment	Needed assessment information
Minimum (short-term)	Species list Environmental interactions Community type(s) Rare and endangered status (both species and communities)
Desired	Key species related to site map Species diversity and relative abundance Food chains and webs Vegetation cover as related to fauna Successional status
Specialized (long-term)	Absolute population estimates Population size, age structure Seasonal changes Migration routes Habitat structure Net productivity Data base

Insects and other arthropods require aerial and sweep nets for collecting flying insects during the day. Employing these nets allows rapid and extensive sampling of the insect fauna. An estimate of the population size of each species may be obtained by standardizing the number of strokes over a measured area. At night a lighted sheet provides an efficient means of attracting and sampling the insect fauna of an area. Unless voucher specimens are required, the collected insects may be released after the collecting period is over. See Ref. 19 for estimating insect population density by using sweep nets.

Birds are best studied by direct observation. Binoculars are valuable aids in the field identification of birds. Bird calls are meaningful only to a well-trained biologist or bird-watcher. If collecting birds is necessitated by the study, mist nets may be used on small birds. These nets must be attended at all times while in place so that each bird can be freed as it becomes entangled. Large birds need not be collected. State or federal laws restrict bird capture, and the field biologist conducting the avifauna survey must possess permits from the appropriate agencies. A good secondary source of information is provided by the National Audubon Society, which conducts and publishes yearly winter bird and breeding bird censuses.

The survey of birds can be systematized by preselecting routes to be walked through the study site. These routes of a given distance, time, direction, and in representative habitats, make it possible to quantify somewhat the abundance and range of each species. Dust baths, tracks, flight patterns, roosts and perches, food habitats, locations of sightings, owl-pellet analysis, nests and eggs, and flocking behavior augment the analysis of bird communities.

Mammals and other vertebrates such as reptiles and amphibians can be studied directly by observation or collection, or indirectly through their tracks, homes, or sound. Their numbers, frequency, and range are difficult to determine because of their elusiveness. In most environmental assessment studies, direct and indirect observations made on one or more visits to the site provide the field biologist with sufficient information to describe and evaluate the mammalian community adequately. In large projects, it may be necessary to sample the community in order to quantify the assessment.

Preselected walk-through routes or observation points should be established similar to or simultaneous with those used for bird study. The direct observation of mammals in the field must be carried out in a discreet manner and is time consuming. Binoculars are helpful aids in observing large animals at a distance or small animals nearby.

Direct observations of mammals may be augmented by indirect observations such as: tracks, droppings (scat), hair and fur, remains, claw marks and gnawings, cropped vegetation, harvested plants and stashes, holes, burrows, dens, nests and shelters, wallows, dust baths, yards, trails and runways, noises and calls.

To illustrate these techniques, consider the example of the biotic impacts caused by the construction of a new expressway. Suppose it has been determined that it will intersect a major water drainage in which a deer population resides. The biologist wants to determine the deer population size, its use of the drainage area, and diurnal and seasonal movements in and out of the drainage area. To establish these population characteristics, direct observations are made of the deer population for several mornings and evenings. By observing their tracks and droppings, the biologist is able to discover their daytime resting area. Further, tracks and browsing signs locate their feeding area. This information establishes that the expressway would sever the deer feeding area from their resting area and greatly impact the herd. A mitigation is suggested and adopted: that a dual-purpose culvert be constructed to allow free passage of both water and deer under the expressway.

In certain studies, capturing animals is necessary in order to identify the species or obtain an accurate sample of the population. Heavy-duty nets, nooses, and sticks are helpful in collecting various amphibians and reptiles. Small rodents may be trapped (dead) by means of snap traps, or (alive) by means of live traps made of wood, wire, or metal. Drift fences and funnel traps made of ¼-in mesh hardware cloth are sometimes employed in census studies. "Live trapping" is a great deal of work, requires close attention, and is time consuming. Traps are set and then checked on a 24-hour basis in order to avoid unnecessary escapes and deaths. All animals captured are tagged or marked and released daily. A record is kept of the animals caught each day, the number recaptured, and the number of animals newly tagged. "Marking" individual animals may be necessary in certain studies (such as above) when population dynamics is important. Paints, dye, metal and plastic tags or bands, toe and fin clipping, removal of fur or scales, notched ears, banding, telemetry devices, and radioactive tracers are available for marking various kinds of animals. The use of "baits" and other attractants often provides valuable information on the fauna of an area without the necessity of trapping the animals. These include sticky and sugary substances, grain, peanut butter, meat, peelings, and other scraps.

Animal Population Size Determining the number of individuals in a population of animals is difficult. In most environmental impact studies, only an estimate of the population size of a given species is needed. Such an estimate provides data for a species that can be either compared with data from other sites or used to picture the relative abundance between species on the site.

Animal biologists have developed several techniques and methods for determining the number of animals in an area. These techniques include the use of census data, sample plots, count indices, removal methods, and mark-recapture methods.

Census Data. Census data can be a complete count (true census) or a sample count of an animal population over a specified area at a given period of time. From these data, estimates are made of the total population being assessed or manipulated. The methods employed in collecting the data and the form of the calculating formulae will depend on the species, the season of the year, the habitat, the purpose of the study, and on any other feature of the problem that might influence the observations and the validity of the method. Herding animals and flocking birds lend themselves to this method. Direct counts are taken in areas of concentration such as feeding grounds, resting areas, wintering grounds, rutting and breeding grounds, rookeries, and roosts. It is assumed that the evaluator is familiar with the natural history, behavior, and ecology of the species involved as well as the general environmental features of the area and specific habitat requirements of each species.

In general, the goal of population estimation should be twofold: (1) to obtain the best possible estimates of the extant population commensurate with the objectives of the study and the time, resources, and personnel available; (2) to make the estimation within established limits of accuracy.

Sample Plots. Sample plots allow estimation of population size and are best obtained on sedentary or attached animals such as barnacles, mollusks, and sea stars. Sample plots are also useful in estimating the number of emergence holes, burrows, and other animal signs. The data can be converted to density or frequency values for a given habitat or study site much as plant data are handled.

Count Indices. Count indices provide estimates of animal populations and are obtained from counts of animal signs, calls, or trailside or roadside counts. These estimates, though they do not provide absolute population numbers, provide an index of the various species in an area. Such indices allow comparisons through the seasons or between sites or habitats. For limited environmental studies, count indices are often all the information needed.

Aerial counts are made from fixed or rotary winged aircraft and are used to estimate large mammal populations for occurrence, size, and distribution. The observer is limited by (1) type of aircraft available, (2) season and time of count, (3) altitude, (4) weather, (5) observer efficiency, and (6) pilot efficiency. A complete or random search design may be used depending upon the species involved and the accuracy needed. As individuals are sighted, a mark corresponding to location is placed on a site map. In this way, distribution, occurrence, and number of individuals by species are recorded. This technique has one major disadvantage: estimates of population densities are consistently low.

Roadside counts (sometimes called horseback or foot counts) of animals are made by traversing a given distance in which designated sampling areas occur along the way. This method is useful along roads and highways and also along existing or proposed powerlines, piplines, railroads, and other right-of-ways. For example, a count would be made at ½-mile intervals for a total distance of 5 miles. The data are expressed as a census index defined by:

$$\text{Census index} = \frac{N}{M}$$

where N = number of individuals of each species seen
M = number of miles traveled

For example, consider the following count data in Table 7.15. In this case, the census index is computed as:

$$\text{Census index:} \quad \frac{\text{Species } A}{3/5 = 0.6} \quad \frac{\text{Species } B}{11/5 = 2.2} \quad \frac{\text{Species } C}{2/5 = 0.4}$$

In this example, species B (a rabbit) was the dominant animal, whereas species A and C (a bird and a deer) were less common.

Factors affecting this technique are: (1) animal species and abundance; (2) activity of the animals as influenced by the time of day, food availability, weather, and season; (3) sex of the animal; (4) condition of the vegetation and terrain; (5) selection of representative sample sites.

Drive counts are conducted by driving the desired organisms past trained observers. Techniques vary according to the size of the area being studied and the number of personnel involved; motor vehicles and aircraft have been used in such studies. The minimum requirement is that two observer teams be present to verify the counts. The observer teams must be positioned along natural foraging or migratory routes. The drive is made in such a way as to move the largest possible number of organisms past the observers. Observers count or photograph the animals as they pass.

This technique is limited by the number of individuals and the species that can be herded and moved past the observers. Further, the numbers obtained are only relative, not absolute; that is, the actual size of the population is not determined. The technique works well in confined areas for large game animals but is useless for small and/or solitary animals and in large areas.

Temporal counts are analogous to the drive method. Here, the spatial dimension is a point and the count is made of all animals passing the point during some stated interval of time. Counts are usually recorded by visual observation, or perhaps by photographing large game animals. This technique is particularly useful for migrating animals that use a definite route.

Call counts use the breeding, feeding, or nesting calls of animals (particularly birds and some large ungulates) to determine presence. The investigator must be experienced and competent in recognizing species by the sound they produce. Sufficient time in the field must be devoted to such studies, since auditory activity is influenced by such variables as: (1) time of day, (2) weather conditions, (3) season, and (4) disturbance factors. The field specialist needs to design the sampling program around these restraints and variables. Such studies cannot be carried out on a moment's notice. When properly conducted, however, this method is excellent for estimating occurrence and population densities.

Scat boards provide an alternative to trapping or direct counts of animals. Small 4″ x 4″ (10 × 10 cm) plywood boards are distributed in a line or in a grid pattern through the sample area. Small animals use these boards as defecation sites. The scats are characteristic of each species, so the trained biologist can identify the species that deposited them.

TABLE 7.15 Illustrative Species Count Data

Miles from starting point	Count of species A	Count of species B	Count of species C
0.5	. . .	2	
1.0	1	1	
1.5	2	. . .	1
2.0	. . .	1	1
2.5	. . .	. . .	
3.0	. . .	1	
3.5	. . .	2	
4.0	. . .	4	
4.5	. . .	. . .	
5.0	. . .	. . .	
Total	3	11	2

The frequency of occurrence of scats is an indication of population size. At least 100 boards should be used, visited each 24-hour period, and left in place for at least three consecutive days. The use of colored baits in different parts of the study area allows some measure of the movements of mammals around the colored bait station, since the droppings of the animal will carry the colored bait.

Sand transects are strips of sandy soil which are cleaned and smoothed daily. Animal tracks are observed daily, providing an assessment of relative abundance by species.

Removal Methods.   Removal methods are used to obtain population estimates of small mammals, such as rodents, through baited snap traps. Running a trap line or grid several nights in a row makes it possible to estimate the exposed population size. See Ref. 25 for several variations of the removal method and data handling. The removal method is useful in assessment studies where an estimation of population size of rodents is needed but where retention of the species is not an issue.

Mark-recapture Methods.   Mark-recapture techniques are used in cases where absolute estimates of population size are precluded by species mobility or secretiveness. These involve capturing a portion of the population, marking each individual with its own distinguishing mark, releasing the individuals, and at some later date sampling the ratio of marked to total animals caught in the population. There are many variations of this technique; they vary in frequency and number of recaptures; all of them require much time and attention. An estimate of the total population is computed using the formula:

$$N = \frac{T}{t/n} \quad \text{or} \quad \frac{nT}{t}$$

where T = number originally marked
   t = number of marked animals recaptured
   n = total number of animals captured during recapture census
   N = population size estimate.

NOTE: Population refers to all the members of one species in an area.

Example: In a study, 24 ground squirrels are originally marked. The next trap-day, 12 marked ground squirrels are recaptured along with 6 unmarked. The population size estimate is:

$$N = \frac{24}{12/18} = 36$$

By knowing the size of the squirrel population and the food requirements of a hawk species in the study area, the biologist was able to predict that reduction of the squirrel population by a third would eliminate the hawk in that area.

Aquatic Assessment   The biota of aquatic ecosystems (including freshwater, brackish, and marine) differ in many ways from terrestrial organisms and terrestrial communities. Hence, they present special problems in sampling techniques and assessment. In certain environmental assessment studies of aquatic systems, it is desirable to analyze both the attached organisms (periphyton) and the suspended or floating plants (phytoplankton). Since algae are the dominant food producers of the aquatic realm, they must be assessed in situations where changes in any of the following are anticipated: water quality, temperature, level and/or depth, flow rate, organic content, source, pollutants, or fisheries potential. Generally such studies are time consuming and must be carried on for many months in order to assess seasonal changes.

The nonalgal plant species such as pondweeds may have aesthetic, habitat, or other desirable characteristics, or they may be considered noxious or in some way undesirable. Generally, the periphyton does not receive as much attention from aquatic biologists (limnologists) as the phytoplankton because it is not the primary food producer of the aquatic system nor is it as sensitive an indicator of water quality as is the phytoplankton. Exceptions would be salt marsh and estuarine plants, pondweeds, and water hyacinth. These sorts of plants are usually sampled by means of some modified sample-plot method. Algae growing on rock or aquatic animals may be scraped off for study. Collecting bottles, bags, and other gear must be used. If algal cell counts are desired for quantifying a species or as a measure of productivity or contamination, glass slides are placed at differing depths or in differing positions. These slides are allowed to remain in place and be colonized over a period of days or weeks; they are then removed and examined under a compound-light microscope. Grids on the slides or in the microscope facilitate the counting process. If biomass values or productivity measurements are desired, these properties of the periphyton may be obtained by such procedures as dry weight, loss through ignition in a muffle furnace to determine organic content, pigment extraction, and dark and light bottles with or without periphyton.

Plankton (phytoplankton and zooplankton) are sampled by taking water samples at various locations and depths or by using plankton nets. Cell counts of algae present in each sample can be made by using a counting chamber under the microscope. By tallying the numbers by species, an assessment of the community structure, composition, and biomass at each station is obtained. Procedures for analyzing plankton can be found in Ref. 25. One valuable measure in assessing the status of phytoplankton in ponds or lakes is the percentage frequency, calculated by:

$$\text{Frequency (\%)} = \left(\frac{\text{total number of occurrences of a species}}{\text{total number of samples taken}} \right) \times 100$$

By knowing the species frequencies, the limnologist can determine the water status, whether low in oxygen, high in organics, etc.

Composition.   The collection and identification of aquatic organisms within a project area are necessary in order to establish which organisms are present. Collection may be accomplished by a variety of techniques, such as dip nets, bottom sampler screens, and checking the underside of submerged rocks. Identification of the organisms collected may be accomplished by the use of appropriate taxonomic keys. Standard references for detailed sampling and identification methodologies should be consulted.[3,31] From this information alone, the freshwater biologist can make a preliminary assessment of the water quality.

Abundance. Determining the abundance of each species in relation to all other species of aquatic organisms within the study area requires the use of quantitative techniques. These techniques give such values as number (abundance/density) and weight (biomass) of individuals of a specific type per unit area or volume. There are many techniques to derive such values. Basically, they all are designed to register directly or indirectly the size (in terms of area or volume) of the proportion of the environment which is sampled and the organisms present. The efficiency of sampling processes must also be considered. The applicability of a given technique to the acquisition of particular information is a matter of professional judgment, with due consideration given to accuracy, assumptions involved, and sampling intensity which is required.[12]

The plethora of techniques (or their modifications) used for estimating abundance precludes an itemization or description, particularly since new ones are developed frequently. The most complete and up-to-date descriptions of newer techniques in use are presented in such journals as *Limnology and Oceanography* and *Transactions of the American Fisheries Society*. The investigator of aquatic impact assessment should always provide the reviewer with the following information[12]:

1. Documentation (from the investigator) of the specific technique and the basis for its selection, including identification of literature pertaining to the specific technique utilized and other techniques which could have been used.

2. Consultation with expert professionals as to the reliability of the information supplied in the documentation stage.

Temporal Change. Temporal change is obtained by spot sampling, but measurements taken at any one time are not necessarily applicable at a later time. Seasonal sampling or monitoring should be emphasized, and continuous monitoring programs of key organisms should extend well into the future (especially in new reservoir projects) if a true assessment of impacts is to be gained.

Short-term temporal factors also play an important role in the significance of composition and relative abundance of species within the aquatic realm. An organism's relative abundance is not necessarily an indication of its importance in the community's food web. Productivity of a given population also must be considered in determining an organism's significance as a material and energy sink and/or source. There are numerous techniques for determining rates-of-change of a population and/or food-web component or assemblages of numerous populations. Specific techniques are applicable to specific situations and circumstances. As in the case of techniques for determining standing crop, some latitude of choice is generally available; the final choice is usually a matter of personal or professional judgment. Reference 30 provides a variety of productivity techniques and background information on algal productivity.

Aquatic Animals. Aquatic animals range from one-celled protozoan animals, through sponges, coelenterates, worms, mollusks, crustaceans, insects, fish, reptiles, birds, and mammals. Many of these forms can be observed directly, collected barehanded or with the aid of plankton, bottom, dip, seine, or mist nets. Bottom (benthic) forms in lakes or the ocean may be collected by using various dredges lowered from a boat. Stream-bottom organisms may be colleccted by hand or with the aid of a bottom sampler.

The zooplankton, by virtue of their size, must be collected by taking representative samples of water or using plankton net trawls of given length and depth, and then assessing the number and species with the aid of a compound-light microscope. Standardized counting slides facilitate obtaining estimates of species numbers. Reference 4 is a standard reference for the study of freshwater organisms.

A recent publication assesses the entire freshwater wetland ecosystem by means of computer models.[17] One of the submodels was developed to treat wildlife based on numerous biological and physical data and information. The use of these models makes possible rapid and objective evaluation of complex ecosystems such as wetlands.

Radiation Impact on Biota

The impact of radioactive substances on biota has been determined by the Atomic Energy Commission.[2] A series of models includes the radioactive impact (dose) on vegetation from fallout, ingestion by animals, uptake by rootsystems of plants, secondary organisms feeding on contaminated vegetation, and whole body dose to aquatic biota from exposure to contaminated water and ingested aquatic foods. These models and equations are too complex to present here, but the methods are available to make useful predictions of radioactive impact on the flora and fauna of terrestrial and aquatic ecosystems.

ILLUSTRATIVE EXAMPLE OF BIOTIC ASSESSMENT

In order to understand the way a biologist carries out a biotic assessment and the methods that can be used, consider the example given by the Wonder Hills project. This project was an urban development proposed by the U.S. Department of Housing and Urban Development (HUD). The site was a 5000-acre area, formerly a grazing cattle ranch adjacent to a rapidly expanding metropolitan area. The topography included steep-sided hills and many small canyons and gulleys covered by grass, shrub, and tree areas. The Wonder Hills development was proposed as a "total community" project. It was to feature

TABLE 7.16   Questions the Biologist Investigator Should Ask Before the Biotic Assessment Begins

1. What is the geographic size of the proposed project site? *5000 acres*
2. What part of the site is involved in the proposed project? *All*
3. What is the type of project (e.g., residential, commercial, resource exploration)? *"Total community" development*
4. Is the project to be short-term final or long-term incremental? *L-t incremental*
5. If long-term incremental, is the biotic assessment to be done in increments or at one time? *At one time*
6. Of the part of the site involved, what is its biotic character? *Natural*
7. How will the project influence the plants? *Most will be destroyed, except in canyons.*
8. How will the project influence the animals? *Destroy or displace most by removing habitat and food supply.*
9. How will the project influence the ecology of the various habitats? *Most habitats will be destroyed. Canyon habitats will retain natural ecology.*
10. What use is to be made of the biotic assessment? (decision document, information document, proforma) *Information document*
11. What legal or jurisdictional limitations apply? *HUD wants attention paid to conservation of flora and fauna.*
12. What information exists on the biota of the area? *None specifically; only general field guides of the region.*
13. *Climatological information as well as geologic data and flood control information on the area to be supplied by the consulting firm.*

a complete residential community consisting of residential, recreational, and shopping facilities. It was also to include locations for schools, churches, and municipal facilities.

A field biologist at the local university was contacted by an environmental consulting firm to provide consulting services by carrying out the biotic assessment on the proposed Wonder Hills development site.

Before agreeing to perform the biotic assessment, the biologist obtained answers from the environmental consulting firm to several questions. (See Table 7.16 for the questions asked by the biologist—and the answers received—before entering into a consulting agreement.) A site visit by the biologist and an estimate of the time needed to carry out the assessment led to the drafting of a study proposal of: the estimated time required to conduct the preliminary study, the field investigations and specimen identifications, data summary, and report preparation. The time of the principal investigator and a graduate student assistant, their rate of pay, and the cost of supplies and other expenses were included in the study proposal. Upon acceptance of the proposal by the environmental consulting firm, the biological team carried out the biotic assessment over a three-week period in the following way:

Step 1

A thorough reconnaissance of the area was conducted. Using topographic maps and aerial photographs of the area, the biologist preselected two routes for the reconnaissance—one through the grass-shrub areas and the other along the stream and over the rocky cliffs.

A rough sketch map of the study area was made in the field; it showed vegetation patterns and types, as well as major topographic features (Fig. 7.9). Species lists of plants and animals were drafted, and relationships of important species to key features of the site such as topography, water drainage, and rocky cliffs were noted on a field form (See Table 7.7 and questions in the section of this chapter on Assessing the Biotic Environment for examples).

Finally, human presence, both past and present, was noted. It was found that little direct human impact was evident. The major impact had been indirect and was caused by years of cattle and sheep grazing. This impact had mainly affected the grass and shrub areas.

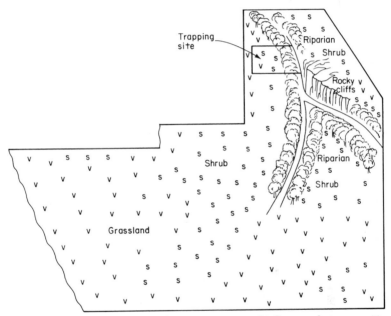

Figure 7.9 Wonder Hills study site map, showing main topographic features, plant communities, and mammal trapping site.

Step 2

Actual species lists were compiled over the three-week period of the study in conjunction with the vegetation and wildlife analysis phases. Specimens that were not known by the investigators were identified by use of field guides and reference to preserved specimens at the university.

Step 3

The investigator wanted to describe the plant communities quantitatively so that their species composition could be related to wildlife species and their habitat requirements. The line-transect method was used in the shrub areas, quadrats were used in the grass areas, and the point-quarter method (a plotless method) was used to analyze the riparian woodland. Results of the riparian study are given in Table 7.17. A total of 40 line transects, 1 x 10 m, were established in the shrub areas. Grass areas were sampled using a square meter frame subdivided into 100 square decimeters. Samples were taken every 100 m along predetermined lines of march that were 300 m apart. A total of 56 quadrats were

TABLE 7.17 Summary of the Riparian Woodland Vegetation Analysis by the Point-quarter Method.
(Abbreviated sample to show summary categories)

Species	Number of individuals	Density	Relative density	Dominance	Relative dominance	Frequency	Relative frequency	Importance value
Black walnut	10	34.67	62.5	168.98	89.6	0.625	62.4	214.5
Live oak	1	3.44	6.3	5.51	3.0	0.063	6.3	15.6
Laurel sumac	3	10.29	18.8	10.79	5.9	0.188	18.8	43.5
Lemonade berry	1	3.44	6.3	1.72	0.9	0.063	6.3	13.5
Scrub oak	1	3.44	6.3	1.03	0.6	0.063	6.3	13.2
Total		54.67	100.2	183.04	100.0	1.002	100.1	300.2

taken. The data from the line transects were summarized in a similar way to that in Table 7.17 to yield importance values for each plant species and each community sample.

Cover values were the only quadrat data collected in the grassland areas and are presented in Table 7.18.

TABLE 7.18   Cover Values of Grassland Species on the Wonder Hills Site.

Species	Cover (%)
Black mustard	37
Wild oats	20
Foxtail	17
Ripgut	12
Herons bill	4
Others	10

With this data the biologist established the composition and structure of the grassland community and the key species upon which birds and wildlife were dependent for food and shelter.

Step 4

A 100-acre site was selected for small mammal population study. It was selected because it was judged to be representative of the grass-shrub-woodland areas. A total of 3900 trap days (one trap set for a 24-hour period equals 1 trap day), using live traps, were employed. Traps were checked at daybreak and evening to insure the capture of both nocturnal and diurnal animals. Owl pellets were collected under roosting trees and analyzed for rodent species being utilized as food. Animal tracks, scat mapping, and visual sightings were used to augment the trapping data. Species lists of mammals, birds, reptiles, and amphibians were developed from these animal studies. Data from the trapping grid gave an estimate of the rodent population and are summarized in Table 7.19.

TABLE 7.19   Estimates of Rodent Populations from Live Trap Data

Species	Number caught	Estimate per 8000 m²	Estimate per acre
Brush mouse	1	2	1.5
California mouse	8	19	14.0
California pocket mouse	8	19	14.0
California mole	1	2	1.5
Deer mouse	6	14	10.0
Dusky footed woodrat	9	21	15.0
Western harvest mouse	2	5	3.5
TOTAL	35	82	59.0

Field investigations revealed an active red-tailed hawk nest in the rocky cliff area, and several riparian trees were being used by barn owls as roosting sites. Owl pellets revealed that the grassland and shrub areas were the main areas being used for the owls' rodent supply. Retention of the riparian woodland, rocky cliffs, and some grassland-shrub areas would be necessary to insure the continuance of the hawks and owls in the area. No rare or endangered plant or animal species were encountered on the study site.

SUMMARY

Since 1970, we have witnessed a renaissance in awareness of our enviroment. Once again, we have begun to realize the importance of open space and naturalness. Our awareness has been supported by environmental legislation mandating that the environment, including native plants and animals, must be properly considered in any land use change or action. Not only must the existing biota be considered, but possible adverse impacts upon it must also be assessed. Finally, these laws require that mitigating measures be presented in order to minimize adverse biotic effects of the proposed project or activity.

This turn of events has necessitated a rash of biotic surveys across the land unequalled in our history. These have ranged from listing the plants and animals on a parcel of land to be subdivided to major biological studies involving teams of highly trained biologists working for one or two years to complete the biotic assessment. Many assessment studies have been nothing more than short statements that dismiss the plants and animals as if they did not exist or had no merit. This has been due, in part, to inadequate evaluations of the biota by the person doing the study and, in part, to reviewers who lack a biological perspective. In some cases, the reverse has been true, with biotic studies far too detailed to serve the needs of the planner or decision maker.

Biological assessments should include a description of the existing environment. Habitats should be described and evaluated. Organisms that compose the biotic community on the site and those using it should be identified and described in a regional context. Interrelationships between organisms and site factors as well as other organisms should be presented. Those species that are dominant or that control the biotic community should be evaluated. The sensitivity and rarity of certain species should also be discussed.

An evaluation of possible impacts on the biota of the site or those involved in the proposed action should be included in environmental assessment reports. Such an evaluation is drawn from a study of the biotic components involved and a knowledge of the response of similar natural communities to various perturbations. Habitat destruction is the primary cause of negative biological impacts. It is possible, however, to create positive biotic impacts by such practices as the establishment of nature preserves, habitat enhancement, and wildlife management programs.

The biotic assessment must include measures that will reduce project impacts on the biota. These mitigations are based on a knowledge of the natural communities and habitats involved. Species and ecosystems vary not only in their structural nature but also in their abilities to undergo alterations and impacts. For this reason, mitigating measures need to be site specific and realistic. Various developments and activities will and must occur. It is the purpose of mitigating measures to insure that the resulting impacts on nature will be kept to a minimum.

Numerous methodologies are available for conducting biotic assessments. Both plant and animal assessment methodologies range from descriptive, nonquantitative to sophisticated, quantitative methods. The selection of the method to be used in a particular study depends upon the size of the study site, the scope of the project or program, the complexity of the biota, time and resource considerations, the purpose of the study, and personnel available. All studies should start with a reconnaissance of the study site that pinpoints the key physical features, vegetation types, and associated animals. This enables the investigator to select the appropriate assessment methodology for the study.

Furthermore, all assessments of the biotic environment should include a species list of observed and potential organisms on the site. By selecting from a variety of methods, the relative or absolute number of organisms should be determined. Several sampling techniques were described that will provide data on species composition and population size. Techniques were also given that allow assessment of productivity of both terrestrial and aquatic ecosystems. Such techniques are suitable in baseline and monitoring studies. The information gained from the biotic assessment should be interpolated into a meaningful statement of the extant biotic community. Based on this biotic assessment, the investigator should describe all possible impacts to the natural community which may result if the project or program is implemented.

In all of the assessment processes, it is imperative that the investigator be familiar with the biota involved and its response to various forms of impact.

REFERENCES

1. Alexander, T. R., and G. S. Fichter. 1973. *Ecology*. Golden Press, New York.
2. A. E. C. 1975. *Draft Environmental Statement*. Tokamak Fusion Test Reactor Facilities. Princeton, New Jersey. Wash–1544.
3. A. P. H. A. 1973. *Standard Methods for the Examination of Water and Wastewater*. American Public Health Association, New York.

4. Andrews, W. A. 1972*a*. *A Guide to the Study of Freshwater Ecology*. Prentice-Hall, Englewood Cliffs, N.J. 1972*b*. *A Guide to the Study of Soil Ecology*. Prentice-Hall, Englewood Cliffs, N.J.

5. Cain, S. A. and G. M. de Oliveira Castro. 1959. *Manual of Vegetation Analysis*. Harper & Row, New York.

6. C. E. Q. 1974. *Environmental Quality*. Fifth Annual Report. U.S. Printing Office, Washington, D.C.

7. C. E. Q. 1976. *Environmental Quality*. 1976: The seventh annual report of the Council on Environmental Quality. U.S. Printing Office, Washington, D.C.

8. Committee on Government Operations. 1971. Stream Channelization (Part 1). Hearings Before a Subcommittee of the Committee on Government Operations. House of Representatives, 92nd Congress–First Session, May 3–4, 1971.

9. Cox, G. W. 1976. *Laboratory Manual of General Ecology*. Wm. C. Brown Co., Dubuque, Iowa. 3rd ed.

10. Dolgin, E. L., and T. G. P. Guilbert, eds. 1974. *Federal Environmental Law*. West Publishing Co., St. Paul, Minn.

11. Ehrenfeld, D. W. 1976, "The Conservation of Non-Resources." *American Scientist*, **64**:648–656.

12. E. P. A. 1974. *An Assessment Methodology for the Environmental Impact of Water Resources Projects*. Report #EPA-600/5-74-016, July.

13. F. H. A. 1975. *Highway-Wildlife Relationships*. Federal Highway Administration Office of Research and Development, Washington, D.C. 20590. 2 vol.

14. Federal Register. 1975. "Threatened or endangered fauna and flora." Vol. 40, No. 127, Part V. July.

15. Gill, Don, and Penelope Bonnett. 1973. *Nature in the Urban Landscape: A Study of City Ecosystems*. York Press, Baltimore, Md.

16. IUCN. 1966. *Red Data Book*. Volumes 1–4. Survival Service Commission, International Union for the Conservation of Nature and Natural Resources. Morges, Switzerland.

17. Larson, J. S., ed. 1976. *Models for Assessment of Freshwater Wetlands*. Water Resources Research Center, University of Massachusetts at Amherst. Pub. No. 32.

18. Levy, E. B., and E. A. Madden. 1933. "The Point Method of Pasture Analysis." *New Zealand Journal of Agriculture* 46:267–279.

19. Menhinick, E. F. 1963. "Estimation of Insect Population Density in Herbaceous Vegetation with Emphasis on Removal Sweeping." *Ecology*. 44:617–621.

20. Miller, G. Tyler, Jr. 1975. *Living in the Environment: Concepts, Problems, and Alternatives*. Wadsworth Publishing Co., Inc. Belmont, Calif.

21. Mueller-Dobois, D. and H. Ellenberg. 1974. *Aims and Methods of Vegetation Ecology*. John Wiley and Sons, N.Y.

22. Murphy, E. F. 1971. *Man and His Environment: Law*. Harper & Row, New York.

23. Odum, E. P. 1973. "The Economic Value of the Tidal Marsh Estuary, as an Example of Extending Economic Evaluation to Include the Work of Nature," from a statement prepared by J. G. Gosselik, R. M. Pope, and E. P. Odum for *Georgia Conservancy Magazine*, July.

24. Phillips, E. W. 1959. *Methods of Vegetation Study*. Henry Holt and Co., Inc., New York.

25. Smith, Robert L. 1974. *Ecology and Field Biology*. Harper & Row, San Francisco, California. 2nd ed.

26. Southwood, T. R. 1966. *Ecological Methods: With Particular Reference to the Study of Insect Populations*. Barnes and Noble, N.Y.

27. Turk, Amos, J. Turk, J. T. Wittes, and R. Wittes. 1974. *Environmental Science*. W. B. Saunders Co., Philadelphia, Pa.

28. U. S. D. T. 1974. *Procedures for Considering Environmental Impacts*. U.S. Department of Transportation, Federal Register, Vol. 39, No. 190, Part II, Sept. 30, 1974.

29. U.S. Fish and Wildlife Service. 1968. *Rare and Endangered Fish and Wildlife of the United States*. Committee on Rare and Endangered Wildlife Species. U.S. Department of Interior. Revised Ed., Washington, D.C.

30. Vollenwider, R. A. 1969. *A Manual on Methods for Measuring Primary Production in Aquatic Environments*. International Biological Program (IBP) Handbook No. 12, Blackwell, Oxford.

31. Welch, P. S. 1964. *Limnological Methods*. McGraw-Hill, New York.

Chapter 8

Summarization of Environmental Impact

JOHN G. RAU

Federal and state requirements for environmental impact statements have stimulated the development of a number of techniques and methods for impact assessment, each displaying variety in conceptual framework, data format, data requirements, and technical sophistication. The preceding chapters have addressed techniques and methods for impact assessment in specific impact areas. However, the next question concerns how to assess collectively the results of these specific impact assessments in terms of an overall or summary evaluation. Because of the complexity of environmental systems and the specialized functions of the various public agencies involved in the environmental impact assessment process, it is unlikely that one universal method will ever be developed or would even be appropriate in all cases.

OVERVIEW OF IMPACT ASSESSMENT METHODOLOGIES

The process of environmental impact assessment involves the major elements of identification, measurement, interpretation, and communication of impacts. However, measurement techniques vary, interpretations vary from impacts which are adverse to those which are beneficial, and decision makers are faced with balancing these project pros and cons to reach an "equitable" or "compromise" decision. Therefore, a number of techniques have been developed for presentation of these impact results to decision makers and the general public. These techniques include ad hoc methods, map overlays, impact checklists, impact matrices, and cause-condition-effect networks.

"Ad hoc methods" provide minimal guidance for total impact assessment while suggesting the broad areas of possible impacts and the general nature of these possible impacts. For example, impacts on plant and animal life might be stated as minimal but adverse, whereas the impacts on the regional economy might be stated as significant and extremely beneficial. These statements are qualitative and could be based on subjective or intuitive assessments, or could be qualitative interpretations of quantitative results. The simplest approach to evaluating the total impact of a project by this method would be to consider each environmental area and identify the nature of the impact upon it, such as no effect, problematic, short- or long-term, and reversible or irreversible. An illustrative example of this approach is presented in Table 8.1.

"Overlay methods" generally rely on a set of maps of a project area's environmental characteristics (physical, social, ecological, aesthetic, etc.). These maps are overlaid to produce a composite characterization of the area's environment. Impacts are then identified by noting the impacted environmental characteristics within the project area boundaries. This presents a graphical display of the types of impacts, the impacted areas, and

TABLE 8.1   Illustrative Ad Hoc Approach to Environmental Impact Versus Environmental Area

Environmental Area \ Environmental Impact	No Effect	Positive Effect	Negative Effect	Beneficial	Adverse	Problematic	Short-term	Long-term	Reversible	Irreversible
Wildlife			X			X	X			
Endangered Species	X									
Natural Vegetation			X			X			X	
Exotic Vegetation	X									
Grading			X			X		X		X
Soil Characteristics	X									
Natural Drainage	X									
Groundwater		X		X						
Noise			X				X			
Surface Paving						X				
Recreation	X									
Air Quality			X	X				X		X
Visual Disruption	X									
Open Space			X	X				X		X
Health and Safety	X									
Economic Values		X		X				X		
Public Facilities (includes schools)						X	X	X		
Public Services	X									
Conformity to Regional Plans		X		X				X		

their relative geographical location. This method is sometimes referred to as the McHarg method (Ref. 18).

The use of "impact checklists" is a method of combining a list of potential impact areas that need to be considered in the environmental impact assessment process with an assessment of the individual impacts. This approach has been adopted by a number of public agencies since it insures that a prescribed list of areas is considered in the assessment process. Unfortunately, this type of method does not provide for the establishment of direct cause-effect links to the various project activities and, generally, does not include an overall interpretation of the collective environmental impacts. A further discussion of this type of method is presented in the section on the Checklist Method later in this chapter.

"Matrix methods" basically incorporate a list of project activities or actions with a checklist of environmental conditions or characteristics that might be affected. Combining these lists as horizontal and vertical axes for a matrix allows the identification of cause-effect relationships between specific activities and impacts. The entries in the cell of the matrix can be either qualitative estimates or quantitative estimates of these cause-effect relationships. The latter are in many cases combined into a weighting scheme leading to a

TABLE 8.2   Illustrative Matrix Approach to Comparing Environmental Impact of Actions on Existing Characteristics and Conditions of the Environment

Existing Environmental Conditions \ Proposed Actions	Modification of Habitat	Alteration of Hydrology and Drainage	Surface Paving	Noise and Vibration	Urbanization	Cut and Fill (Land Fill)	Erosion Control	Landscaping	Traffic Circulation
Land Form	B	C	B	A	B	C	C	D	B
Water Recharge	A	B	B			B	A	D	
Climate	A				A				
Floods - Stability	C	C	B			B	A	D	
Stress - Strain (Earthquake)	B	C			A	B	A		
Open Space	D		D	B	C			D	B
Residential	D				D				
Health and Safety	D	B	B		B	B	A		C
Population Density	B			A	B				
Structures	B	B	B		B	B	A		B
Transportation	B		C		B				C
TOTAL COMPUTATIONS	B	C	B	A	B	B	A	D	B,C

LEGEND: A - Insignificant low impact not injurious to land and environment.

B - Measurable impact, but with proper planning and building is not injurious to land.

C - High impact on environment, but can be curbed by taking proper precautionary measures.

D - Impact on environment, but considered good.

E - Impact that will be detrimental to environment.

total "impact score." Table 8.2 provides an illustrative example of the former approach, whereas the latter approach is discussed further in the section on the Matrix Method later in this chapter.

"Network methods" start with a list of project activities or actions and then generate cause-condition-effect networks (i.e., chains of events). This type of method is basically an attempt to recognize that a series of impacts may be triggered by a project action. Hence, this method provides a "roadmap" type of approach to the identification of second- and third-order effects. The idea is to start with a project activity and identify the types of impacts which would initially occur. The next step is to select each impact and identify the impacts which may be induced as a result. This process is repeated until all possible impacts have been identified. Sketching this in network form results in what is commonly referred to as an "impact tree." One advantage of this type of approach is that it allows the user to identify impacts by selecting and tracing out the events as they might be expected to occur. A major problem in constructing cause-condition-effect networks is achieving the degree of detail necessary for informed decision making. On the other hand, if the environmental condition changes are described in detail and all possible interrelationships are included, the resulting impact networks could be too extensive and complex to really be useful. An example of this approach is presented in Figure 8.1. This general method is discussed in detail in the section on the Network Method later in this chapter.

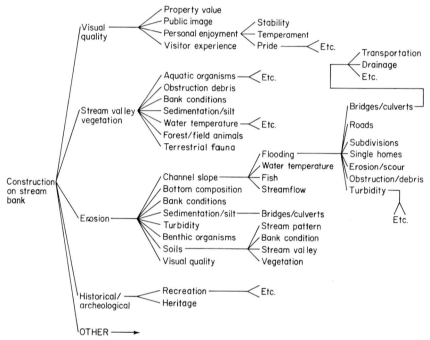

Figure 8.1   Impact tree for a hypothetical bank stabilization project. (NOTE: (1) The line in this illustration is to be read as "has an effect on." (2) It is emphasized that the cause-and-effect chain presented here should be viewed as only a small part of a larger overall impact tree, which would address the full range of economic, social, and environmental categories of human interest and concern.) (SOURCE: *Reference 24.*)

CHECKLIST METHOD

The checklist method is basically a variant of the ad hoc method for summarizing environmental impacts in the sense that it starts with a list of potential impact areas. The next step is to assess the character or nature of the impact. This is usually accomplished through the use of such descriptive terms as adverse or beneficial, short-term or long-term, no effect or significant effect. For example (Ref. 5), Table 8.3 provides an illustration of a typical checklist form which could be used to insure that all important aspects of an environment impact evaluation are considered. In the actual implementation of this form one would place a check mark or "X" opposite each item to indicate whether the proposed project will have an adverse effect, no effect, or a beneficial effect on the item in question.

The compilation of one general, all-inclusive list of impact areas with applicability to all projects, actions, and environmental conditions is likely to be very large, cumbersome to use, and may contain information too generalized to adequately describe the nature of the

TABLE 8.3   Typical Project Checklist by Impact Area

POTENTIAL IMPACT AREA	CONSTRUCTION PHASE			OPERATING PHASE		
	Adverse effect	No effect	Beneficial effect	Adverse effect	No effect	Beneficial effect
A. LAND TRANSFORMATION AND CONSTRUCTION						
a. Compaction and settling						
b. Erosion						
c. Ground cover						
d. Deposition (sedimentation, precipitation)						
e. Stability (slides)						
f. Stress-strain (earthquake)						
g. Floods						
h. Waste control						
i. Drilling and blasting						
j. Operational failure						
B. LAND USE						
a. Open space						
b. Recreational						
c. Agricultural						
d. Residential						
e. Commercial						
f. Industrial						
C. WATER RESOURCES						
a. Quality						
b. Irrigation						
c. Drainage						
d. Ground water						
D. AIR QUALITY						
a. Oxides (sulfur, carbon, nitrogen)						
b. Particulate matter						
c. Chemicals						
d. Odors						
e. Gases						
E. SERVICE SYSTEM						
a. Schools						
b. Police						
c. Fire protection						
d. Water and power systems						
e. Sewerage systems						
f. Refuse disposal						
F. BIOLOGICAL CONDITIONS						
a. Wildlife						
b. Trees, shrubs						
c. Grass						
G. TRANSPORTATION SYSTEMS						
a. Automobile						
b. Trucking						
c. Safety						
d. Movement						
H. NOISE AND VIBRATION						
a. On-site						
b. Off-site						
I. AESTHETICS						
a. Scenery						
b. Structures						
J. COMMUNITY STRUCTURE						
a. Relocation						
b. Mobility						
c. Services						
d. Recreation						
e. Employment						
f. Housing quality						
K. OTHER (List as appropriate)						

impacts. For these reasons, many federal and state agencies have prepared specific impact area lists that are applicable to the kinds of actions and activities within their jurisdiction. Examples would include checklists applicable to housing projects, highways, sewage treatment facilities, nuclear power plants, and airports. Typical impact areas relative to these types of projects were previously identified and discussed in Chapter 1.

For example (Ref. 25), the Department of Housing and Urban Development (HUD) has established a three-level environmental clearance process, including normal environmental clearance, special environmental clearance, and environmental impact statement clearance. Normal environmental clearance is essentially a check for consistency with HUD environmental policies and standards. Special environmental clearance requires an environmental evaluation of greater detail and depth, whereas an environmental impact statement clearance is a complete and fully comprehensive environmental evaluation. Table 8.4 presents a HUD checklist used for the determination of normal and special environmental clearance for subdivision and multifamily projects. In the use of this checklist, four ratings are assigned to component environmental factors associated with the project. A rating of "A" indicates that the component is acceptable—there are no special problems associated with this item, adverse impacts are negligible, and other effects are neutral or beneficial. "B" indicates that the component is questionable—problems associated with the item call for discretion in granting environmental approval to the project, and ameliorative measures should be pursued and may be mandated by specific environmental or program policies. "C" indicates that the component is undesirable or unacceptable—problems associated with this item are serious enough that rejection may be mandated by specific environmental or program policies, ameliorative measures should be vigorously pursued, and approval is allowed only when justified by a careful examination and comparison of alternatives. "NA" indicates that the environmental factor is not applicable to this project—for example, access to local schools is not applicable to elderly housing projects, coastal zone environmental policies do not apply to inland projects. For normal environmental clearance, if there are no "B" or "C" ratings on any item, the proposal is generally acceptable on environmental grounds. Marginal or "B" ratings could lead to project rejection or the preparation of an environmental impact statement. Unacceptable or "C" ratings could lead to rejection, modification of project, further study, or the preparation of an environmental impact statement.

An elaborate example (Ref. 26) of the application of the checklist method is presented in Table 8.5 in which the environmental impact of the Pauahi, Hawaii Neighborhood Development Project is assessed. In this example, the elements of the environment are listed on the left-hand side of the matrix and the impacting actions are listed across the top of the matrix. The entries in the matrix cells are based upon defining a "positive impact" as favorably improving the environment, including the reduction or elimination of blighting conditions. A "negative impact" is defined as disrupting or otherwise adversely affecting the existing environment or services. To use the matrix, one would begin at the left-hand side and, reading from left to right, determine the nature of the impacts of the project actions. For example, new residential buildings, parks and open space, and historical preservation would have a major positive impact on neighborhood viability, whereas business relocation, demolition, grading, and construction would have a major negative impact.

MATRIX METHOD

The environmental impacts of projects or actions generally encompass a broad range of impacts from air and noise pollution to effects on employment and neighborhood social structure. All of these impacts vary in magnitude as well as in their beneficial or adverse classification. As a result, a natural question arises as to what is the "collective" or "overall" environmental impact of the project or action taken. Is the project beneficial or is it adverse? To answer such a question requires a comparison of these impacts and, to some extent, a subjective evaluation of which impacts are more important than others. Generally, evaluations of this type are performed using numerical techniques.

The simplest technique which could be used to assess the overall impact would be to use a ranking method. For example, in the comparison of alternate highway improvement projects, one could rank each alternative with respect to its ability to satisfy the social, environmental, and economic factors under consideration. As shown in Table 8.6, if the

TABLE 8.4 Department of Housing and Urban Development Normal and Special Environmental Clearance Form for Subdivision and Multifamily Projects

A. PROJECT IDENTIFICATION:

Applicant's Name: _____ Street Address: _____

City or County: _____ State: _____ Zip Code: _____

Phone: _____ Project Name: _____ FHA File No. _____

Project/Subdivision Location: _____

Number of Lots or Units Proposed: _____ Size of Tract (acres/sq. ft.): _____

Demand for housing in this area: Adequate □ Reject □ If reject, go to Section 1.

For Subdivision Only:

Has work started? Yes □ No □ If work has started: Grading is ____ % Completed:

Street improvements are ____ % Completed. Number of homes under construction: ____

Number of homes completed: ____

ENVIRONMENTAL ANALYSIS

Evaluate project and assign a rating: A, B, C. or NA. (See Instructions.)

B. COMPLIANCE WITH STANDARDS:

1. Have A-95 review requirements been met? Yes □ No □ In process

2. Is the project in compliance with the local and regional comprehensive plans? Yes □ No □

3. Is the project in compliance with local zoning ordinances? Yes □ No □

4. Compliance with applicable standards:

	Rating	Source/ Documentation		Rating	Source/ Documentation
a. Historic Properties	____	____	e. Wetlands	____	____
b. Noise	____	____	f. Air Quality	____	____
c. Flood Plain	____	____	g. Other (specify)	____	____
d. Coastal Zone	____	____			

Is the project in violation of applicable standards? Yes □ No □

Should the project be rejected? Yes □ No □ If reject, go to Section I. If not, continue the environmental assessment (Section C).

TABLE 8.4 Department of Housing and Urban Development Normal and Special Environmental Clearance Form for Subdivision and Multifamily Projects (Continued)

C. SITE SUITABILITY ANALYSIS:

	Rating	Source/Documentation		Rating	Source/Documentation
1. Slope stability	___	___	6. Natural hazards	___	___
2. Foundation conditions	___	___	7. Man-made hazards	___	___
3. Terrain	___	___	8. Nuisances	___	___
4. Soil permeability	___	___	9. Compatibility in use and scale with environment	___	___
5. Ground water	___	___	10. Neighborhood character	___	___

Services and Facilities	Rating (Access)	Rating (Adequacy)	Source/Documentation
11. Elementary School	___	___	___
12. Junior and senior high school	___	___	___
13. Employment	___	___	___
14. Shopping	___	___	___
15. Park, playground and open space	___	___	___
16. Police and fire	___	___	___
17. Health care/ social services	___	___	___
18. Transportation	___	___	___
19. Other services:	___	___	___

Utilities	Rating	Source/Documentation		Rating	Source/Documentation
20. Water supply system	___	___	23. Solid waste disposal	___	___
21. Sanitary sewer system	___	___	24. Other utilities	___	___
22. Storm sewer system	___	___	25. Paved access to site	___	___

D. Does project size exceed special clearance size thresholds? Yes ☐ No ☐ If yes, continue review (Section E). If not, go to Section F. (See Chapter 8, Handbook 4010.1)

E. IMPACTS ON THE ENVIRONMENT (SPECIAL CLEARANCE):

	Rating	Source/Documentation
1. Impact on unique geological features or resources		_____
2. Impact on rock and soil stability		_____
3. Impact on soil erodability		_____
4. Impact on ground water (level, flow and quality)		_____
5. Impact on open streams and lakes		_____
6. Impact on plant and animal life		_____
7. Impact on energy resources		_____
8. Impact on social fabric and community structures		_____
9. Displacement of persons or families		_____
10. Impact on aesthetics and urban design		_____
11. Impact on existing or programmed community facilities:		

	Rating	Source/Documentation		Rating	Source/Documentation
a. Schools		_____	e. Transportation		_____
b. Parks, playgrounds & open space		_____	f. Water supply system		_____
			g. Sanitary sewer system		_____
c. Health care and social services		_____	h. Storm sewer system		_____
d. Community services		_____	i. Solid waste disposal system		_____

F. Will the project have notable impacts on the environment? Yes ☐ No ☐ If yes, is further analysis necessary? Yes ☐ No ☐ Are there alternative site designs that can be considered? Yes ☐ No ☐

COMMENT: _____

_____

TABLE 8.4 Department of Housing and Urban Development Normal and Special Environmental Clearance Form for Subdivision and Multifamily Projects (Continued)

G. Assess the following conditions: (a) Does the project form part of a larger development pattern? Yes □ No □: (b) Is the project likely to stimulate additional development? Yes □ No □: (c) Are there other developments planned which are or will be impacted by the project? Yes □ No □

If any of the above area is answered "Yes" indicate how the cumulative environmental impact of the larger development will be addressed. EIS _____ Special Environmental Clearance _____ 701 planning funds _____ other _____. Should this project be delayed until the cumulative impacts are accounted for? Yes □ No □

COMMENT:

H. LOCATION AND MARKET:

1. Marketability is: Acceptable □ Reject □ If reject, go to Section I.
2. Most marketable price or rental range is $ _____ to $ _____
3. Most marketable units 0–2 BR _____
   3 BR _____
   4 or more _____

4. For Subdivisions:
   Estimated market price of typical lot $ _____ to $ _____
   Typical lot size _____ ft. x _____ ft.
   Local Authorities:
1. Local authorities have □ have not □ approved tentative map.
2. Local officials contacted:
   Name: _____ Title: _____ Phone: _____
   Name: _____ Title: _____ Phone: _____
3. Information obtained and date obtained:

I. ENVIRONMENTAL FINDINGS: (Check applicable items)

☐ Reject

☐ EIS Required

☐ No EIS Required. Project is consistent with HUD environmental policies and requirements and is not a major Federal action significantly affecting the quality of the human environment.

☐ Further environmental review is required
Backup material is appended. Yes ☐ No ☐

For Subdivisions Only

☐ Issue Interim Form ASP-5.
Special problems involve:

Sanitary engineering ☐
Site engineering ☐
Site planning ☐
Architecture ☐

☐ Issue ASP-6.
VA has been contacted. Yes ☐ No ☐

COMMENT:

Field Inspection and Assessment made by: _____

Name _____ Title _____ Date _____
Name _____ Title _____ Date _____
Name _____ Title _____ Date _____

J. REVIEW AND COMMENT OF ENVIRONMENTAL CLEARANCE OFFICER:

_____ _____
Environmental Clearance Officer   Date

K. INSTRUCTIONS BY CHIEF UNDERWRITER:

_____ _____
                                     Date

SOURCE: HUD Transmittal No. 4, 4010.1 CHG, Nov. 1974, Appendix B.

TABLE 8.5   Illustrative Example of Checklist Approach to Neighborhood Development Project

		IMPACTING ACTIONS										
		ACTION PERIOD				EFFECTS OF COMPLETED ACTIONS						
	ELEMENTS	Residential Relocation	Business Relocation	Demolition, Grading, Construction	Interim Period (Temporary Uses)	New Utilities In Place	New Residential Buildings	New Commercial Buildings	Parking Structures	Parks and Open Space	Historical Preservation	Modifications to Street System
PHYSICAL	Soil & Geology	✻	✻	✻	✻	✻	✻	✻	✻	●	✻	✻
	Sanitary Sewer Systems	✻	✻	o	o	•	•	•	✻	✻	✻	•
	Water Systems	✻	✻	o	o	●	•	•	✻	✻	✻	•
	Vegetation	✻	✻	o	o	✻	●	•	✻	●	✻	✻
	Animal Life	✻	✻	✻	✻	✻	✻	✻	✻	o	✻	✻
	Air Quality	✻	✻	o	✻	✻	O	O	O	•	•	✻
	Adjacent Land Use	✻	✻	O	O	✻	●	✻	✻	●	●	X
	Storm Drainage	✻	✻	o	o	●	•	•	✻	•	✻	•
	Transportation System — Streets	✻	o	O	o	•	•	•	●	✻	✻	•
	Transportation System — Public Transportation	✻	✻	o	o	✻	X	X	X	✻	X	X
	Transportation System — Pedestrian	o	o	O	o	✻	●	●	•	●	X	X
	Open Space	✻	✻	✻	✻	✻	●	o	o	●	X	X
SOCIOECONOMIC	Demand for Ancillary Services	•	•	•	o	✻	•	•	✻	✻	•	•
	Tax Base	✻	✻	✻	o	•	●	●	•	✻	X	✻
	Health & Safety	✻	✻	o	o	●	•	•	✻	•	•	•
	Neighborhood Viability	o	O	O	O	✻	●	•	•	●	●	X
	Residents	o	O	O	O	•	●	•	•	●	•	X
	Public Schools	✻	✻	o	o	✻	•	✻	✻	•	•	X
	Police Services	o	o	O	o	•	•	•	•	X	✻	X
	Fire Services	o	o	O	o	•	•	•	•	X	•	X
AESTHETIC	View	✻	✻	o	O	✻	•	●	o	●	o	✻
	Historic Structures	✻	✻	o	o	•	✻	✻	X	•	●	✻
	Amenity	o	o	o	o	•	●	●	•	●	•	X
	Neighborhood Character	o	o	O	O	•	●	•	o	●	•	X

LEGEND	
o	indicates a minor negative impact.
O	indicates a major negative impact.
•	indicates a minor positive impact.
●	indicates a major positive impact.
X	indicates an undetermined impact.
✻	indicates no appreciable impact.

impact area of concern deals with the number of dwelling units destroyed, a rank of 1 is assigned to the alternative with the least impact and a rank of n (where n equals the number of alternatives) is assigned to the alternative that is least desirable.

When there is more than one type of environmental impact, ranking according to environmental impact area yields the best ordering of alternatives relative to each impact area, but does not enable one to distinguish incremental differences among alternatives or

TABLE 8.6   Sample Ranking of Highway Improvement Project Alternatives

Alternative	Number of dwelling units destroyed	Rank
V	0	1
W	2	2
X	20	3
Y	24	4

to recognize that the factors under consideration may not all be of equal importance. Table 8.7 illustrates this situation for the case of five alternative highway improvement projects versus seven environmental impact areas.

As can be seen, even though alternative No. 3 ranks first in three areas, it ranks last or next to last in two other areas. Alternative No. 2 ranks consistently at the middle level for all impact areas. Because no one alternative ranks first in all seven impact areas, there is no clear-cut choice as to which alternative is the best.

The next step in level of sophistication would be to recognize that the impact areas are not necessarily of the same importance to the community, which should be taken into consideration in deciding which alternative is best. As an example, suppose that a poll was conducted by interviewing residents and local business people in the impact area to determine the relative importance of each of these seven impact areas. Further, suppose

TABLE 8.7   Ranking Example for Five Highway Project Alternatives and Seven Impact Areas

Impact area	Alternatives				
	No. 1	No. 2	No. 3	No. 4	No. 5
Market access					
Rank	5	3	1	2	4
Average time to Civic Center (min)	20	16	12	15	19
Level of service					
Rank	1	3	4	4	2
Average travel speed (mi/h)	45	40	36	36	42
Provision of public service					
Rank	4	3	1	2	5
Police response time (min)	10	9	6	8	12
Disruption of homes					
Rank	2	3	5	4	1
Number of homes taken	12	14	40	20	4
User costs					
Rank	2	2	1	3	2
Annual dollars (millions)	1.0	1.0	0.8	1.6	1.0
Noise pollution					
Rank	5	3	4	1	2
Decibel level at 100 ft	75	65	70	50	60
Disruption of businesses					
Rank	1	3	5	2	4
Number of businesses lost	2	6	10	4	8

that on a scale of 1 to 10, where 10 represents the highest importance, the following results were obtained:

Market access 4
Level of service 5
Provision of public service 7
Disruption of homes 10
User costs 1
Noise pollution 6
Disruption of businesses 10

Applying these importance factors to the ranking results in Table 8.7 and multiplying the rank by the community importance of the impact area, one obtains the results in Table 8.8. This multiplication would result in a "rating" for each alternative relative to each impact

TABLE 8.8   Rating Example for Five Highway Project Alternatives and Seven Impact Areas

Impact areas	Alternatives				
	No. 1	No. 2	No. 3	No. 4	No. 5
Market access	20	12	4	8	16
Level of service	5	15	20	20	10
Provision of public service	28	21	7	14	35
Disruption of homes	20	30	50	40	10
User costs	2	2	1	3	2
Noise pollution	30	18	24	6	12
Disruption of businesses	10	30	50	20	40
Total score	115	128	156	111	125

area. A technique at this step for selecting which alternative is best would be to add the ratings to obtain a "total score." In this case, alternative No. 4 has the lowest total score and thus is most desirable from the point of view of its environmental impact.

The obvious weakness in the preceding level of analysis is the failure to recognize the incremental differences among rankings, that is, the inherent nonlinearity of the rating scale. Specifically, refering to Table 8.7, alternative No. 5 causes 4 homes to be removed and alternative No. 1 causes 12 homes to be removed. Hence, they are ranked in the order one and two, respectively. If alternative No. 1 had only caused 6 homes to be removed, it still would have been ranked second. However, it would have been more comparable in impact to that of alternative No. 5. In other words, a simple ranking fails to recognize the magnitude of the relative differences between alternatives. One way to remedy this would be to establish the ranking scale on the basis of relative differences such as, for example, a rank of 1 for the best score (smallest average time to civic center, largest average travel speed, etc.), a rank of 2 for any score within 20 percent of the best, a rank of 3 for any score greater than 20 percent but within 40 percent of the best, a rank of 4 for any score greater than 40 percent but within 70 percent of the best, and a rank of 5 for any score greater than 70 percent of the best. Using this *illustrative* method one would obtain the new rankings given by Table 8.9.

TABLE 8.9   Incremental Ranking Example for Five Highway Project Alternatives and Seven Impact Areas

Impact area	Alternatives				
	No. 1	No. 2	No. 3	No. 4	No. 5
Market access	4	3	1	3	4
Level of service	1	2	2	2	2
Provision of public service	4	4	1	3	5
Disruption of homes	5	5	5	5	1
User costs	3	3	1	5	3
Noise pollution	4	3	3	1	2
Disruption of businesses	1	5	5	5	5

Clearly, there is still no definitive choice as to which alternative is the best. If, however, we now apply the impact area ratings of importance as used in deriving Table 8.8, we obtain the ratings in Table 8.10. In this case, we observe that there is a tie between alternative No. 1 and No. 5 and, relative to Table 8.8, alternative No. 4 is no longer the most desirable.

The preceding illustrative discussion forms the basis for what is known as the "weighting scheme approach" in evaluating the environmental impacts of alternative projects or actions. In this context, the weighting scheme approach is based on the desire to assess quantitatively the impact and weight of that value by its "significance" or "importance." The idea is to require environmental impact analyses to define two aspects of each action which may have an impact on the environment. The first aspect is "magnitude" of the impact upon specific environmental factors. The term "magnitude" is used in the sense of degree, extensiveness, or scale. For example, highway development will alter or affect the existing drainage pattern and may thus have a large "magnitude" of impact on the drainage. The second is a weighting of the degree of "importance" (i.e., significance) of the particular action on the environmental factor in the specific instance under analysis. Thus, the overall "importance" of the impact of a highway on a particular drainage pattern may be small because the highway is very short or because it will not interfere significantly with the drainage. An arbitrary scale, say, from 1 to 10 could be used where 10 represents the greatest magnitude of impact and 1 the least, and, similarly, 10 represents

TABLE 8.10 Rating Example Based on Incremental Rankings for Five Highway Project Alternatives and Seven Impact Areas

	Alternatives				
Impact areas	No. 1	No. 2	No. 3	No. 4	No. 5
Market access	16	12	4	12	16
Level of service	5	10	10	10	10
Provision of public service	28	28	7	21	35
Disruption of homes	50	50	50	50	10
User costs	3	3	1	5	3
Noise pollution	24	18	18	6	12
Disruption of businesses	10	50	50	50	50
Total score	136	171	140	154	136

the greatest importance and 1 the least. An added degree of sophistication would be to place "+" in front of the magnitude number if the impact is beneficial and "−" if the impact is adverse. Unfortunately, such a scheme allows the possibility of introducing subjectivity into (1) the choice of a scale number for magnitude and importance and (2) assessment of whether the impact is beneficial or adverse. One must be aware of this shortcoming in using such a scheme. The value of this approach, however, is that it provides a way of quantitatively comparing alternatives merely by choosing as the total impact score of a project alternative the total weighted sum of the impact magnitudes. Mathematically, letting

$$m_{ij} = (+ \text{ or } -) \text{ magnitude of the } j^{th} \text{ action on the } i^{th} \text{ environmental factor}$$
$$w_{ij} = \text{importance weighting of the } j^{th} \text{ action on the } i^{th} \text{ environmental factor,}$$

we have

Total impact on the ith environmental factor from all actions $= \sum_j m_{ij} w_{ij}$

Total impact of the jth action on all environmental factors $= \sum_i m_{ij} w_{ij}$

Total project impact $= \sum_i \sum_j m_{ij} w_{ij}$

The preceding measure of total project impact is in essence a quality-of-life indicator (Ref. 8) in the sense that m_{ij} represents the magnitude of impact of the jth action on the ith

quality-of-life factor and w_{ij} represents the weighting of importance as viewed by members of society.

To illustrate the concept of the weighting approach using a numerical example, Table 8.11 corresponds to Table 8.2. However, the entries are of the form "$x(y)$" where "x" denotes the magnitude of the impact and "y" the importance, and "$+$" or "$-$" is used to denote beneficial or adverse impact, respectively. The convention used is based on $A = 1$, $B = 3$, $C = 7$ and $D = E = 10$. This choice is for illustrative purposes only and, in this example, w_{ij} is chosen independent of j for convenience. From Table 8.11, we observe that:

1. The total weighted impact of the project, 314, is positive, which means that it is beneficial to the environment.

TABLE 8.11   Illustrative Example of Weighted Impact on Actions on Existing Characteristics and Conditions of the Environment

Existing Environmental Conditions \ Proposed Actions	Modification of Habitat	Alteration of Hydrology and Drainage	Surface Paving	Noise and Vibration	Urbanization	Cut and Fill (Land Fill)	Erosion Control	Landscaping	Traffic Circulation	Total Factor Impact
Land Form	8(3)	−2(7)	3(3)	1(1)	9(3)	−8(7)	−3(7)	−3(10)	1(3)	3
Water Recharge	1(1)	1(3)	4(3)			5(3)	6(1)	1(10)		47
Climate	1(1)				1(1)					2
Floods-Stability	−3(7)	−5(7)	4(3)			7(3)	8(1)	2(10)		5
Stress-Strain (Earthquake)	2(3)	−1(7)			1(1)	8(3)	2(1)			26
Open Space	8(10)		6(10)	2(3)	−10(7)			1(10)	1(3)	89
Residential	6(10)				9(10)					150
Health and Safety	2(10)	1(3)	3(3)		1(3)	5(3)	2(1)		−1(7)	45
Population Density	1(3)			4(1)	4(3)					22
Structures	1(3)	1(3)	1(3)		3(3)	4(3)	1(1)		1(3)	34
Transportation	1(3)		−9(7)		7(3)				−10(7)	−109
TOTAL ACTION IMPACT	180	−47	42	11	97	31	−2	70	−68	314

2. Alternation of hydrology and drainage, erosion control, and traffic circulation have an adverse effect.

3. Transportation is adversely affected by the project.

Table 8.11 provides an illustration of the basic structure of the matrix method approach, namely, a matrix in which each proposed action (or its separate components) is identified as a column of the matrix and the environmental conditions or impacted areas are identified as the rows of the matrix. The entries in the matrix represent not only an indication of the areas impacted by each action but also a measure of the impact's extent. This method, attributed to Luna Leopold (Ref. 16) is basically an extension of the checklist approach in the sense that it combines the checklist of project elements with the checklist of impacts.

DETERMINATION OF ENVIRONMENTAL IMPACT IMPORTANCE

The matrix method approach discussed in the previous section and the need for tables presenting comparisons of alternatives both require a statement of the impact on the particular environmental area, given a specific action. This calls for some kind of measurement in the most general sense. First, one must measure the impact itself, that is the magnitude, and then one must evaluate that level of impact in terms of its relative value to the appropriate constituency. In the first case, one is looking for data about changes in the environment and must rely on scientific knowledge. In the second case, one is looking for the relative values of the society or segments of society concerned in the evaluation of a project. This latter situation is inherently "value judgment" and is not necessarily based on scientific knowledge. Furthermore, these values are generally based on a survey of constituencies in an attempt to determine the preferences of the affected groups.

These observations are of fundamental importance in comparing alternatives and in the selection of the best project from a given set of alternatives. Based on consideration of the total environmental impact, there is really no way to avoid transforming the magnitudes of impacts into their importance relative to values held, either explicitly or implicitly, by some constituency. When a choice is made from among alternatives, the relative values of each environmental impact factor are implicitly determined to a degree at least sufficient to have led to that decision.

To illustrate the implicit assignment of relative values to impact areas, consider the simple situation in which one must select from among five alternate highway corridors. The two impact areas of concern deal with displacement of dwelling units and removal of farm land. The potential impacts are as follows:

Impact area measures	Highway corridor alternatives				
	A	B	C	D	E
Number of dwelling units displaced	16	5	12	4	6
Number of units of farm land displaced	20	8	6	10	7

The choice from among the alternatives requires some statement of the relative values of the two types of environmental impacts. If alternative B is chosen, a unit of farm land is implied to be worth less than one dwelling unit, since the decision maker preferred to forego alternative E where an additional displaced dwelling unit could have been accepted in order to reduce farm land consumed by one unit. A unit of farm land is worth more than ½ dwelling unit since, otherwise, alternative D would be preferred to B. Hence, the unit of farm land is worth between ½ and 1 dwelling unit. Similar implied valuations can be derived for other choices of highway corridors.

The weighting scheme approach suggests the transformation of the degree of impact (as measured by the magnitude) into a value scale and the transformation of the value scale for each type of environmental impact into a composite value score. The latter is in effect what is implied when one adds the quantities given by $m_{ij}w_{ij}$. For example, relative to the preceding highway corridor alternative selection, Figure 8.2 shows two possible

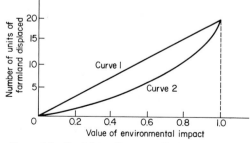

Figure 8.2 Candidate value curves for farmland impact.

transformations, or value curves, for the impact on farm land. For any number of units of farm land displaced, one can find the relative value by drawing a horizontal line from the vertical axis at the given number of units displaced to the appropriate value curve and then, at the point of intersection, drawing a vertical line down to the horizontal axis. The intersection point on this axis would represent the value (analogous to the product $m_{ij}w_{ij}$ in the weighting scheme) of the impact on farm land. Curve 1 illustrates a linear relationship in which value increases as the number of units of farm land displaced increases. Curve 2 illustrates a nonlinear value relationship where small displacements have little effect. However, as more and more farm land is displaced, the remainder becomes more valuable and, therefore, its displacement causes a greater increase in value. Of course, in this example high value is regarded as bad in the sense of large adverse impact.

One can make similar transformations for the other impact area given by displacement of dwelling units and then add the corresponding values for each highway corridor alternative. The one with the smallest total would be the proper selection.

The point of this discussion is that to compare different project alternatives or actions meaningfully, one must be able to recognize the relative value of each type of environmental impact. The use of weightings such as the w_{ij}'s or the use of value transformations such as previously discussed are techniques for accomplishing this. One must not lose sight of the inherent subjectivity in this process and the fact that in some cases inherent biases on the part of decision makers could influence the outcome of the overall selection process.

In determining the appropriate weights and the form of the value function for each impact measure, it is necessary to follow a procedure that will produce reliable results. Because these weights are essentially based on the judgmental values or attitudes of those surveyed, the selected procedure must be systematic and must be able to reduce all possible variation. The group of persons ultimately selected for the weighting should include a cross section of society such as individuals from governmental agencies, politicians and decision makers, experts in the field of environmental evaluation, representatives from special interest groups, and members of society in general. Groups of individuals representing this cross section must be sampled a number of times to obtain consistent estimates of the weights.

The procedure selected for determining the relative importance of each environmental impact area consists of ranking and pair-wise comparisons. Each individual is required to rank the impact areas and to compare in pair-wise fashion the degree of importance of highest rank with the one immediately following. If this procedure is followed in a systematic way, a weight will be developed for each area. The procedure is repeated a number of times for different groups in order to get the desired cross-sectional population representation and the reliability needed for an importance weighting.

Steps for Determining Weightings of Importance

The basic steps to be followed in determining the weightings of importance can be described as follows:

Step 1 Select a group of individuals for evaluation and explain to them in detail the weighting concept and the use of rankings and weightings.

Step 2 Prepare a table with columns corresponding to the range of values which can be assigned as a "score of importance" to each impact area—for example, if five values are possible, there would be five columns. The rows in the table would correspond to the impact areas being ranked as to importance.

Step 3 Give a copy of the table developed in Step 2 to each individual evaluator and repeat Steps 4–9 until no further changes in the table entries are desired.

Step 4 Ask each individual to place an "X," or other signifying mark, in each column for each impact area. Thus, a value of importance is assigned to each impact area.

Step 5 Ask all individuals to compare the marked columns on a pair-wise basis to insure that the impact areas are ordered on the proper relative basis in their opinion. If not, they should reassign their scores so as to have the desired relative ordering of impact areas. (For example, on a scale from 1 to 10, if a value of 10 has been assigned to impact area A and it appears that A is twice as important as B, impact area B should be assigned a value of 5.)

Step 6 Ask each individual to add the value (or importance score) selected for each of the impact areas to obtain a total.

Step 7 The individual should then divide the value selected for each impact area by the total obtained in Step 6 to determine the desired weighting for each area.

Step 8 Collect the tables from each individual evaluator and average the weightings determined for each impact area to obtain a "group or composite average."

Step 9 Present the averages obtained to the individual evaluators and ask them to compare the group weightings with those derived by each of them individually in Step 7.

Step 10 If any one or more individuals desires to change the assignment of scores based on what the group decided, go to Step 4 and repeat the entire process. If none desire to change their scores, stop the experiment, because the impact area relative weightings of importance will have been derived.

As an example, Table 8.12 illustrates a table of the type described in Step 2, in which there are thirteen impact areas of interest and five possible importance scores. By adding

TABLE 8.12 Illustrative Example of the Development of Impact Area Importance Weightings

| | Low importance ↓ | | Average importance ↓ | | High importance ↓ | | |
Impact area	1	2	3	4	5	Total	Weighting
Park requirements	X					2	$\frac{2}{43}$
School age students generated			X			3	$\frac{3}{43}$
Trips generated	X					2	$\frac{2}{43}$
Police protection				X		4	$\frac{4}{43}$
Fire protection				X		4	$\frac{4}{43}$
Public service costs					X	5	$\frac{5}{43}$
Total revenues					X	5	$\frac{5}{43}$
Employment (long-term jobs)				X		4	$\frac{4}{43}$
Electricity consumption			X			3	$\frac{3}{43}$
Natural gas consumption			X			3	$\frac{3}{43}$
Solid waste generated		X				2	$\frac{2}{43}$
Sewage discharge			X			3	$\frac{3}{43}$
Water consumption			X			3	$\frac{3}{43}$
						43	1.0

the scores corresponding to each "X," one obtains a total of 43 points. Dividing each score by 43, we obtain the relative importance weightings given by the last column in the table. These fractions would be used as values of the w_{ij}'s as presented in the section of this chapter on the Matrix Method for the total project impact score formulation.

Steps for Development of Value Functions

Scientific information should form the basis for the value function evaluation. This information would specify the form of the function and the points of inflection or change. In cases where this information is not available, estimation procedures are necessary. The suggested procedure for this estimation divides the environmental quality range (say 0–1) into an equal number of intervals. For each interval, an estimate of the functional relationship between the interval and the impact measure value is determined. Repeating this procedure a number of times makes it possible to define a representative value function.

In estimating the value function for each impact measure, five steps have to be followed:

Step 1 Obtain scientific information when available on the relationship between the measure or parameter and the quality of the environment.

Step 2 Order the impact measure scale so that the lowest value of the parameter is zero and it increases in the positive direction—no negative values.

Step 3 Divide the quality scale (0–1) into equal intervals and express the relationship between this interval and the parameter. Continue this procedure until a curve is constructed.

Step 4 Average these values as expressed in curves over all persons in the experiment. (For parameters based solely on judgment, value functions should be determined by a representative population cross section.)

Step 5 Replicate this experiment with the same group or another group of persons to increase the reliability of the functions.

TABLE 8.13   Illustrative Example of Dissolved Oxygen Level Versus Environmental Quality

Level of dissolved oxygen (mg/L)	Relative environmental quality value at each level
0	0
1	0.05
2	0.10
3	0.15
4	0.25
5	0.50
6	0.75
7	1.0
8	1.0
9	1.0
10	1.0

One impact measure or parameter that provides a relatively simple example for developing value functions is the level of dissolved oxygen, expressed in milligrams per liter. Suppose that a group of water quality specialists agreed on the relationships shown in Table 8.13 between dissolved oxygen and the overall value of environmental quality that dissolved oxygen levels represent. (Note: For dissolved oxygen, value would relate primarily to the support of aquatic life.)

In other words, using this example, dissolved oxygen at 4 mg/L is only valued at 25 percent of its maximum quality, whereas dissolved oxygen at 7 mg/L or up provides 100 percent quality. Based on these estimates, the value function for dissolved oxygen is shown in Figure 8.3.

EXAMPLE OF TOTAL IMPACT EVALUATION

To illustrate the techniques discussed in the preceding sections consider the following situation. A 100-acre area within a city presently consists of a mix of low-income family units, abandoned houses, grocery and liquor stores, and numerous light-manufacturing plants. This area, because of its present condition, is being considered by the local redevelopment agency for redevelopment. There are four candidate redevelopment configurations, which are as follows:

Configuration A: 100 acres of single-family detached homes at 5 DU/acre

Configuration B: 60 acres of single-family detached homes at 5 DU/acre; 20 acres of apartments at 20 DU/acre; 20 acres of townhouses at 7 DU/acre

Configuration C: 40 acre shopping center with 1,000,000 ft² gross leasable area (GLA); 160,000 ft² GLA of office space covering 5 acres; 30 acres of apartments at 20 DU/acre; 25 acres of townhouses at 7 DU/acre

Configuration D: 40 acre shopping center with 1,000,000 ft² GLA; 60 acres of apartments at 20 DU/acre

The selection of a redevelopment configuration will be based on a weighted evaluation of the impacts in thirteen potential impact areas, namely:

Park requirements	Employment (long-term jobs)
School age students generated	Electricity consumption
Trips generated	Natural gas consumption
Police protection	Water consumption
Fire protection	Solid waste generated
Public service costs	Sewage discharged
Total revenues	

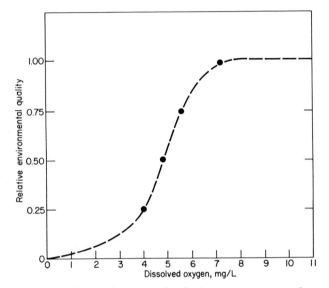

Figure 8.3   Example of a value function for dissolved oxygen as a measure of water quality.

To perform this weighted evaluation, a number of factors must be employed. These are presented in Tables 8.14 and 8.15. Additional assumptions are as follows for the present land use:

Natural gas consumption	$= 5,000,000$ ft³/month
Water consumption	$= 250,000$ gal/day
Electricity peak demand	$= 5,000$ kilowatts (kW)
Sewage discharge	$= 175,000$ gal/day
Solid waste generated	$= 20,000$ lb/day
Vehicle trips generated	$= 10,000$ trips/day
Public service cost	$= \$200,000$
Total revenues	$= \$250,000$
Students generated	$= 500$
Park requirements	$= 5$ acres
Policemen needed	$= 2$
Firemen needed	$= 1$
Employment (i.e., permanent jobs)	$= 500$

In addition, it is assumed that the residental land uses under consideration generate no long-term employment in the area, only short-term jobs due to construction.

Table 8.12 provides a framework which will be used to provide the weightings of importance for each environmental impact area. These weightings are to be multiplied by the value of each configuration's impact on each of the thirteen impact areas. The values will be determined according to the scheme presented in Table 8.16 in which "high" value is considered bad. In two cases, the signs must be reversed, namely in the case of

TABLE 8.14 Planning Factors for Land Use Configurations

	Single-family detached	Apartments	Townhouses	Shopping center	Office building
Natural gas consumption	9,000 ft³ per month per DU	4,750 ft³ per month per DU	6,250 ft³ per month per DU	20 ft³ per month per ft² GLA	3.5 ft³ per month per ft² GLA
Water consumption	125 gal per day per person	90 gal per day per person	100 gal per day per person	30 gal per day per employee	25 gal per day per employee
Electricity consumption (peak demand)	5 kW per DU	4 kW per DU	4.5 kW per DU	8 W per ft² GLA	7 W per ft² GLA
Sewage discharge	100 gal per day per person	72 gal per day per person	80 gal per day per person	24 gal per day per employee	16 gal per day per employee
Solid waste	5 lb per day per person	5 lb per day per person	5 lb per day per person	2 lb per day per 100 ft² GLA	1 lb per day per 100 ft² GLA
Vehicle trips	10 trips per day per DU	6 trips per day per DU	7.5 trips per day per DU	43 trips per 1000 ft² GLA	38 trips per 1000 ft² GLA
Public serivce cost	$1600 per acre	$2600 per acre	$1400 per acre	$2700 per acre	$1800 per acre
Nonproperty tax revenue	$800 per acre	$2200 per acre	$900 per acre	$7400 per acre	$300 per acre

employment impact and total revenue impact since in these cases an increase is generally regarded as "good." To illustrate the use of this table, suppose a configuration has an implied requirement of 6 acres of parks. This would represent a 20 percent increase over the present land use of 5 acres, and so a score of +1 would be assigned to the specific configuration's impact on park requirements.

Table 8.17 provides the results obtained using the planning factors of Tables 8.14 and 8.15. The percent change quantities are the result of comparing the magnitude of impact

TABLE 8.15. Additional Planning Factors

Number of policemen per 1000 population	=	1.8
Number of firemen per 1000 population	=	0.9
Number of employees per 500 ft² of shopping center GLA	=	1.0
Number of employees per 150 ft² of office GLA	=	1.0
Number of students per DU	=	$\begin{cases} 1.4 \text{ single-family detached} \\ 0.8 \text{ apartments} \\ 1.0 \text{ townhouses} \end{cases}$
Acres of local parks per 1000 population	=	4.0
Property tax rate	=	$1.50 per $100 assessed valuation
Tax assessment factor	=	25%
Population per DU	=	$\begin{cases} 4.0 \text{ single-family detached} \\ 2.2 \text{ apartments} \\ 3.0 \text{ townhouses} \end{cases}$
Market value per DU	=	$\begin{cases} \$60,000 \text{ single-family detached} \\ \$18,000 \text{ apartments} \\ \$45,000 \text{ townhouses} \end{cases}$
Market value per ft² GLA	=	$\begin{cases} \$30 \text{ shopping center} \\ \$25 \text{ office building} \end{cases}$

with the present level of impact and the value quantities are then derived from application of Table 8.16. For example, in the case of water consumption for Configuration C, this is computed as follows:

$$\left(\begin{array}{c}1{,}000{,}000\ ft^2 \\ of\ shopping \\ center\end{array}\right)\left(\begin{array}{c}1\ employee \\ per\ 500\ ft^2\end{array}\right)\left(\begin{array}{c}30\ gal \\ per\ day\ per \\ employee\end{array}\right) = \begin{array}{c}60{,}000\ gal \\ per\ day\end{array}$$

$$\left(\begin{array}{c}160{,}000\ ft^2 \\ of\ office \\ space\end{array}\right)\left(\begin{array}{c}1\ employee \\ per\ 150\ ft^2\end{array}\right)\left(\begin{array}{c}25\ gal \\ per\ day\ per \\ employee\end{array}\right) = \begin{array}{c}26{,}700\ gal \\ per\ day\end{array}$$

$$\left(\begin{array}{c}30\ acres \\ of\ apart- \\ ments\end{array}\right)\left(\begin{array}{c}20\ dwelling \\ units\ per \\ acre\end{array}\right)\left(\begin{array}{c}2.2\ people \\ per\ apart- \\ ment\end{array}\right)\left(\begin{array}{c}90\ gal \\ per\ day \\ per\ person\end{array}\right) = \begin{array}{c}118{,}800\ gal \\ per\ day\end{array}$$

$$\left(\begin{array}{c}25\ acres \\ of\ town- \\ houses\end{array}\right)\left(\begin{array}{c}7\ dwelling \\ units\ per \\ acre\end{array}\right)\left(\begin{array}{c}3.0\ people \\ per\ town- \\ house\end{array}\right)\left(\begin{array}{c}110\ gal \\ per\ day \\ per\ person\end{array}\right) = \begin{array}{c}52{,}500\ gal \\ per\ day\end{array}$$

$$TOTAL = \begin{array}{c}258{,}000\ gal \\ per\ day\end{array}$$

Comparing this total with the present usage of 250,000 gal implies a $\dfrac{258{,}000 - 250{,}000}{250{,}000} =$ 3.2% increase which results in a value of +1.

TABLE 8.16   Determination of Impact Values

Impact on present condition	Value
≥100% increase	+7
50–99.9% increase	+5
25–49.9% increase	+3
0–24.9% increase	+1
No change	0
0–24.9% decrease	−1
25–49.9% decrease	−3
50–99.9% decrease	−5
≥100% decrease	−7

NOTE: Reverse the signs for employment and total revenues.

Multiplying the weighting factors of Table 8.12 times the corresponding impact area value of Table 8.17 and then adding over all impact areas yields a total impact score for each redevelopment configuration. This results in a total score for Configuration A given by:

Park requirements	$(2/43) \times (+5) =$	0.235
School age students	$(3/43) \times (+3) =$	0.210
Trips generated	$(2/43) \times (-5) =$	−0.235
Police protection	$(4/43) \times (+5) =$	0.465
Fire protection	$(4/43) \times (+5) =$	0.465
Public service costs	$(5/43) \times (-1) =$	−0.116
Total revenues	$(5/43) \times (+1) =$	0.116
Employment	$(4/43) \times (+7) =$	0.651
Electricity consumption	$(3/43) \times (-5) =$	−0.350
Natural gas consumption	$(3/43) \times (-1) =$	−0.070
Solid waste generated	$(2/43) \times (-5) =$	−0.235
Sewage discharge	$(3/43) \times (+1) =$	0.070
Water consumption	$(3/43) \times (0)$ $=$	0
	TOTAL =	1.206

TABLE 8.17 Summary of Configuration Environmental Impacts

Impact area		Configuration			
		A	B	C	D
Park requirements	No. acres	8	10	7.38	10.56
	Percent change	+60	+100	+48	+111
	Value	+5	+7	+3	+7
School age students	No. generated	700	880	655	960
	Percent change	+40	+76	+31	+92
	Value	+3	+5	+3	+5
Trips generated	No. trips/day	5,000	6,450	53,993	50,200
	Percent change	−50	−36	>100	>100
	Value	−5	−3	+7	+7
Police protection	No. policemen	3.6	4.5	3.3	4.75
	Percent change	+80	>100	+65	>100
	Value	+5	+7	+5	+7
Fire protection	No. firemen	1.8	2.25	1.65	2.375
	Percent change	+80	>100	+65	>100
	Value	+5	+7	+5	+7
Public service costs	Amount ($)	160,000	176,000	230,000	264,000
	Percent change	−20	−12	+15	+32
	Value	−1	−1	+1	+3
Total revenues	Amount ($)	192,500	228,125	583,531	621,500
	Percent change	−23	−9	>100	>100
	Value	+1	+1	−7	−7
Employment	No. jobs	0	0	3,067	2,000
	Percent change	−100	−100	>100	>100
	Value	+7	+7	−7	−7
Electricity consumption	Peak demand in kW	2,500	3,730	12,308	12,800
	Percent change	−50	−25	>100	>100
	Value	−5	−3	+7	+7
Natural gas consumption	Million ft³/month	4.5	5.475	24.504	25.7
	Percent change	−10	+10	>100	>100
	Value	−1	+1	+7	+7
Solid waste generated	lb/day	10,000	12,500	30,825	33,200
	Percent change	−50	−38	+54	+66
	Value	−5	−3	+5	+5
Sewage discharge	Gal/day	200,000	217,000	202,000	238,080
	Percent change	+14	+24	+15	+36
	Value	+1	+1	+1	+3
Water consumption	Gal/day	250,000	271,200	258,000	297,600
	Percent change	0	+8	+3	+19
	Value	0	+1	+1	+1

In a similar way, one can compute the totals for configuration B, C, and D which are given by 2.350, 1.618, and 2.690. Hence, Configuration A is the best by virtue of having the lowest total.

NETWORK METHOD

Network approaches attempt to expand upon the matrix theme by introducing a cause-condition-effect network which allows identification of cumulative or indirect effects. The network is actually shown in the form of a tree, also called a relevance or impact tree, and is used to relate and record secondary, tertiary, and higher order effects. Figure 8.4 shows a conceptual framework for such a tree due to J. Sorensen (Refs. 11 and 22). To develop a network of this type basically requires answering a series of questions relative to each of the project activities such as what are the primary impact areas, what are the primary impacts within these areas, what are the secondary impact areas, what are the secondary impacts within these areas, what tertiary impacts flow from these, etc. This is the approach which must be followed. Figure 8.5 provides an illustrative example of this approach for the case of new freeway construction in an established downtown business district upon consideration of two of the many primary impacts given by the removal of homes and the removal of businesses.

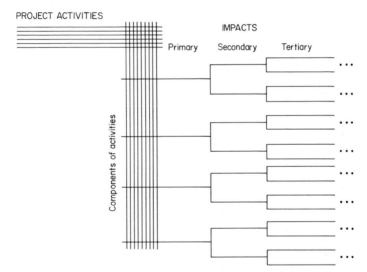

Figure 8.4   Conceptual framework of impact networks.

A network approach is appealing because the environment operates as such a complex system. An action causes one or more environmental condition changes which in turn will produce one or more subsequent condition changes that will ultimately result in one or more terminal effects. For example, highway cuts or fills could cause erosion of soil off slopes into a stream course. The added soil material could increase stream turbidity, shoal the channels, alter stream channel regime, and these, in turn, could increase flood potential, block passage of aquatic biota, or degrade stream habitat for aquatic biota.

Unfortunately, in the construction of impact networks, it may happen that cycles of effects may repeat in the expansion of the tree of impacts. This is to be expected when there exist complex interactions between effects and corresponding chain reactions. Other considerations in the use of this type of method deal with the probability that an identified condition change will produce a further condition change and whether or not the additional condition change that might be produced, regardless of low or high probability of occurrence, is significant enough to include in the impact network.

For example (due to Sorensen), a wastewater treatment plant may release a highly nutrified effluent (project action) into an estuary. The increase in nutrient concentration (initial condition change) will stimulate phytoplankton blooms in the estuary. Conceivably, a potential impact of the phytoplankton blooms could be increased sedimentation of the estuary from the accumulation of dead organisms. Sedimentation of the estuary could then be traced to decreased water depth. Decreased water depth, in turn, could produce a myriad of impacts (increased penetration of sunlight, increase of bottom plant growth, increased temperature of estuary, decreased flushing of the estuary—to list but a few). The key question is whether blooms of phytoplankton have been known to increase the sedimentation rate of an estuary to the extent that there will be a significant decrease in the water depth. If the effect of sedimentation from dead plankton is an imperceptible decrease in water depth over a period of a few years, the impact should not be included in the network.

An impact network does provide in a summary form an overview of the impacts caused and/or induced by the project and its related activities. For this reason it is a useful tool. However, this is only a qualitative summary that can be used to generate an overall impact score as was done with the use of impact matrices. The method of accomplishing this requires (1) estimation of the occurrence probabilities of the individual chain of events in a branch of the tree and (2) adding for each possible branch the product of the probability that the events on the branch occur and the total impact score using a measure of the type suggested in the section on the Matrix Method.

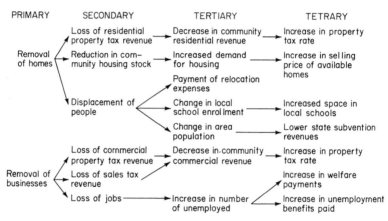

Figure 8.5   Example of impact tree for new freeway construction in established downtown business district.

To illustrate this technique, consider the impact tree given in Fig. 8.6, in which there are two basic project activities, say A and B. In Fig. 8.6a activity A has two primary impacts, three secondary impacts, and two tertiary impacts. Activity B has two primary impacts, four secondary impacts, and four tertiary impacts. There are ten branches of this tree given by the chains of events shown in Fig. 8.56b.

Now let

$$p_i = \text{probability that the events on branch } i \text{ occur}$$

for $i = 1, 2, \ldots, 10$. Also, for each impact X, define

$$M(X) = (+ \text{ or } -) \text{ magnitude of impact } X$$

and

$$I(X) = \text{importance weighting of impact } X,$$

where both $M(X)$ and $I(X)$ have values ranging over some arbitrary scale (for example, from 1 to 10). Then we define the impact score for a given branch of the impact tree to be

$$\sum M(X)I(X),$$

where the summation is over all impacts (events) X on the branch.

For example, the impact score for branch 1 would be given by

$$M(A_1)I(A_1) + M(A_{1,1})I(A_{1,1}) + M(A_{1,1,1})I(A_{1,1,1}).$$

In a similar way, one could compute the impact score of the other nine branches. Now, since there is some uncertainty as to whether or not the identified primary, secondary, and tertiary impacts will actually occur, one might weight these branch impact scores by their probability of occurrence. Adding these weighted scores over all branches (i.e., all combinations of events which could occur) leads to an "expected environmental impact score" given by

$$\text{Expected environmental impact} = \sum_{i=1}^{10} p_i \begin{pmatrix} \text{Impact score} \\ \text{for branch } i \end{pmatrix}$$

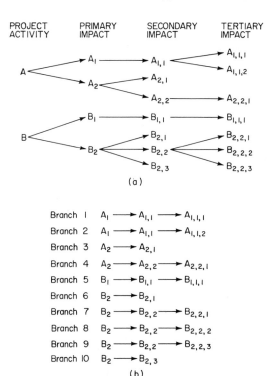

Figure 8.6   Illustrative impact tree (a) and corresponding branches (b).

To illustrate this technique, consider the example presented in Figure 8.5 for the case of typical impacts caused by new freeway construction in a downtown area. Suppose it has been determined that the magnitude and importance scores for these impacts are as shown in Table 8.18.

There are nine branches in the impact tree depicted in Figure 8.5. The probability of occurrence of the impacts on the branch given by

$$\begin{array}{c}\text{Removal}\\\text{of homes}\end{array} \rightarrow \begin{array}{c}\text{Loss of residential}\\\text{property tax}\\\text{revenue}\end{array} \rightarrow \begin{array}{c}\text{Decrease in}\\\text{community}\\\text{residential}\\\text{revenue}\end{array} \rightarrow \begin{array}{c}\text{Increase in}\\\text{property tax}\\\text{rate}\end{array}$$

is equal to $(1.0)(1.0)(1.0)(0.3) = 0.3$, and the total branch impact score is equal to $(-2)(4) + (-1.5)(5) + (-0.5)(10) + (-1)(3) = -23.5$. The weighted impact score would be

$(0.3)(-23.5) = -7.05$. Similarly, the probability of occurrence of the impacts on the branch given by

$$\begin{array}{c}\text{Removal}\\\text{of businesses}\end{array} \xrightarrow{} \begin{array}{c}\text{Loss of}\\\text{jobs}\end{array} \xrightarrow{} \begin{array}{c}\text{Increase in}\\\text{number of}\\\text{unemployed}\end{array} \xrightarrow{} \begin{array}{c}\text{Increase in}\\\text{unemployment}\\\text{benefits paid}\end{array}$$

is equal to $(1.0)(0.9)(0.9)(0.2) = 0.162$, and the total branch impact score is equal to $(-4)(5) + (-3)(6) + (-0.5)(7) + (-0.1)(0.2) = -41.52$. Hence, the weighted impact score is $(0.162)(-41.52) = -6.73$. Repeating these calculations for the other seven branches and adding the weighted impact scores for the nine branches, one obtains

$$\text{Expected environmental impact} = -54.93,$$

which implies a significant adverse impact.

TABLE 8.18   Illustrative Impact Frequency, Magnitude, and Importance Values for New Freeway Construction in a Downtown Area

Impact	Probability of occurrence	Magnitude	Importance
Removal of homes	1.0	−2	4
Loss of residential property tax revenue	1.0	−1.5	5
Decrease in community residential revenue	1.0	−0.5	10
Increase in property tax rate	0.3	−1	3
Reduction in community housing stock	1.0	−0.25	2
Increased demand for housing	0.4	+3	3
Increase in selling price of available homes	0.2	−1.2	1
Displacement of people	1.0	−1	7.5
Payment of relocation expenses	1.0	−0.7	0.5
Change in local school enrollment	0.8	+2.2	1
Increased space in local schools	0.8	+1.5	3.5
Change in area population	0.95	+0.2	1.5
Lower state subvention revenues	0.5	−1.1	9
Removal of businesses	1.0	−4	5
Loss of commercial property tax revenue	1.0	−4.8	6
Decrease in community commercial revenue	0.2	−1.5	10
Loss of sales tax revenue	0.2	−2.5	10
Loss of jobs	0.9	−3	6
Increase in number of unemployed	0.9	−0.5	7
Increase in welfare payments	0.1	−0.8	0.7
Increase in unemployment benefits paid	0.2	−0.1	0.2

NOTE: The convention employed is that "−" denotes an adverse impact to the community as a whole and "+" denotes a beneficial impact to the community as a whole.

Several important observations can be made regarding this attempt to obtain an overall quantitative score. First, one must be able to estimate meaningfully the chances of occurrence of individual impacts, as measured by the probability of occurrence. Second, the total score computed has no real value on an absolute basis—it is a relative score which can be used to compare various project alternatives or the results of implementing possible mitigation measures. Third, in order for the measure given by the expected environmental impact to have meaning, the underlying impact network must address all possible and significant cause-condition-effect sequences or chains of events—if any are left out, then clearly the score is incomplete.

SUMMARY

Since the enactment of the National Environmental Policy Act of 1969, a number of systematic checklists, matrix methods, and network-type procedures have been proposed and utilized as guides in environmental impact assessment. These devices play a fundamental role in the four basic aspects of environmental impact analysis given by (1) identification of impacts, (2) measurement of impacts, (3) interpretation of impacts, and (4) communication of the results. Furthermore, each method differs from the others vis-à-vis these four areas. For example, a checklist is primarily designed to aid in impact identification and, as a result, provides, because of its structure, communication of the results. The matrix method provides both identification and communication, but, coupled with some type of impact measure based on magnitude and importance of impact, can also provide information regarding measurement and interpretation. Similarly, the network or impact tree method provides both identification and communication, but, using an expected value type measure of environmental impact, can also provide information regarding measurement and interpretation.

It must be remembered that a neat structure for recording impacts does not eliminate the difficulties of actually determining what they are and then meaningfully assessing their extent. Even if one develops some type of checklist, matrix, or network as a guide in conducting the assessment, one must not lose sight of the fact that these impacts depend upon the particular type of project activity being undertaken as it relates to the place where it is occurring. In addition, many of these impacts are temporal in nature. In the latter case, the use of numerical techniques with matrices or impact trees to derive an impact score is not easily modified or amenable to time differences between impacts such as short-term versus long-term.

The point to be made here is that the tools and techniques presented and discussed in this chapter are intended to be used as an aid in conducting environmental impact analysis. They are appealing because they provide assistance in trying to grasp the overall effect of the project in the sense of assessing the collective impact of the "good" and the "bad" of the project. However, this overall assessment or summarization of the environmental impact should only serve as information for the general public and the decision makers involved. There are other considerations such as public opinion and local politics which will influence whether or not the project will be undertaken and, if so, how its activities might be altered and adverse impacts mitigated. A total impact score is, in itself, nothing more than a measure of what the overall environmental impact is on some specified scale. The significance of the score and how it is used is left to be determined by those individuals and agencies with jurisdiction over the project and its activities.

REFERENCES

1. Bisselle, C. A., S. H. Lubore, and R. P. Pikul, *National Environmental Indices: Air Quality and Outdoor Recreation,* Report No. MTR-6159, The Mitre Corp., McLean, Va., April 1972, PB 210 668.
2. Burnham, J. B. et al, *A Technique for Environmental Decision Making Using Quantified Social and Aesthetic Values,* BNWL-1787, Battelle Pacific Northwest Labs., Wash., Feb. 1974.
3. Canter, Larry W., *Environmental Impact Assessment,* McGraw-Hill, New York, 1977.
4. Carter, E. C., J. W. Hall, and L. E. Haefner, "Incorporating Environmental Impacts in the Transportation System Evaluation Process," Highway Research Record No. 467, 1973.
5. Carter, Steve, Murray Frost, Claire Rubin, and Lyle Sumek, *Environmental Management and Local Government,* Office of Research and Development, U.S. Environmental Protection Agency, Report No. EPA-600/5-73-016, Feb. 1974.
6. Denver Regional Council of Governments, *Guide to Preparation of Environmental Impact Statements,* Report No. DRCOG-73-004, May 1973, PB 221 262.
7. Dickert, Thomas G. and Katherine R. Domeny, *Environmental Impact Assessment: Guidelines and Commentary,* University Extension, University of California, Berkeley, 1974.
8. Environmental Protection Agency, *Quality of Life Indicators: A Review of State-of-the-Art and Guidelines Derived to Assist in Developing Environmental Indicators,* Dec. 1972, PB 225 034.
9. Hellstrom, David I., *A Methodology for Preparing Environmental Statements,* Arthur D. Little, Inc., Cambridge, Mass., Aug. 1975, AD AO30265.

10. Hill, Morris and Rachel Alterman, "Power Plant Site Evaluation: The Case of the Sharon Plant in Israel," *Journal of Environmental Management*, Vol. 2 (1974), pp 179–196.
11. Hopkins, Lewis D. et al, *Environmental Impact Statements: A Handbook for Writers and Reviewers*, Report No. IIEQ 73-8, Illinois Institute for Environmental Quality, Chicago, Ill., Aug. 1973, PB 226 276.
12. Hornback, Kenneth E., Joel Guttman, Harold L. Himmelstein, Ann Rappaport, and Roy Reyna, *Studies in Environment*, Volume II, *Quality of Life*, Report No. EPA-600/5-73-012b, Environmental Protection Agency, Feb. 1974.
13. Hyde, Luther W., *Environmental Impact Assessment by Use of Matrix Diagram*, Alabama Development Office, State of Alabama, June 1974, PB 235 221.
14. Jain, R. K. and L. V. Urban, *A Review and Analysis of Environmental Impact Assessment Methodologies*, Tech. Report E-69, Construction Engineering Research Laboratory, Champaign, Ill., June 1975, AD A013 359.
15. Jones & Stokes Associates, Inc., *Development Guidelines for Areas of Statewide Critical Concern*, Vol. I, *"Development Guidelines,"* Report No. OPR-74-10-V-1, Sacramento, Calif., July 1974, PB 237 319.
16. Leopold, Luna B., Frank E. Clarke, Bruce B. Hansaw, and James R. Balsely, *A Procedure for Evaluating Environmental Impact*, Geological Survey Circular No. 645, U. S. Department of Interior, 1971.
17. Malcolm, D. G. et al, *Environmental Indices for the Los Angeles Data Base*, California State University, Los Angeles, March 1975, PB 245 281.
18. McHarg, I., *Design with Nature*, Natural History Press, Garden City, N.Y., 1969.
19. Odum, Eugene P. et al, "Totality Indexes for Evaluating Environmental Impacts of Highway Alternatives," Transportation Research Record 561, *Transportation Energy Conservation and Demand*, pp. 57–67.
20. Schaenman, Phillip S., *Using an Impact Measurement System to Evaluate Land Development*, U.I. 203-214-6, The Urban Institute, Washington, D.C., Sept. 1976.
21. Schlesinger, B. and D. Daetz, "A Conceptual Framework for Applying Environmental Assessment Matrix Techniques," *Journal of Environmental Sciences*, July/August 1973, pp. 11–16.
22. Sorensen, Jens C. and Mitchell L. Moss, *Procedures and Programs to Assist in the Impact Statement Process*, University of California, Berkeley, April 1973, COM-73-11033.
23. The Futures Group, Glastonbury, Conn., *A Technology Assessment of Geothermal Energy Resource Development*, 15 April 1975.
24. U.S. Department of Defense, Corps of Engineers, "Environmental Considerations: Proposed Policies and Procedures," *Federal Register*, Vol. 42 No. 36, 23 Feb. 1977.
25. U.S. Department of Housing and Urban Development, "Procedures for Protection and Enhancement of Environmental Quality," *Federal Register* Vol 38, 18 July 1973 and as amended in *Federal Register* Vol. 39, 4 Nov. 1974.
26. ——, *Environmental Impact Statement for Pauahi Urban Renewal Project, Hawaii R-15*, Eis-HI-73-0851-F.
27. U.S. Department of the Interior, Bureau of Land Management, *Environmental Protection and Enhancement*, BLM Manual 1790, 13 June 1974.
28. Warner, L., Environmental Impact Analysis: An Examination of Three Methodologies, Department of Agricultural Economics, University of Wisconsin, 1973, PB 231 763.
29. Whitman, Ira L., Norbert Dee, John T. McGinnis, David C. Fahringer, and Janet K. Baker, *Design of an Environmental Evaluation System*, Battelle Columbus Laboratories, Columbus, Ohio, 30 June 1971, PB 201 743.
30. Yurman, Dan, "Focused Investments in the City," *Practicing Planner*, Feb. 1976, pp. 16–23.

Index

Index

3